WITHDRAWN
0 2 MAR 2022

Handbook of Experimental Pharmacology

Volume 127

Springer
Berlin
Heidelberg
New York
Barcelona
Budapest
Hong Kong
London
Milan
Paris
Santa Clara
Singapore
Tokyo

Quinolone Antibacterials

Contributors

D. Beermann, T. Bergan, W. Christ, W.A. Craig,
S.E. Critchlow, A. Dalhoff, M.J. Everett, K. Grohe,
J. Kuhlmann, H. Lode, A. Maxwell, U. Petersen,
L.J.V. Piddock, E. Rubinstein, P. Schacht, H.-G. Schäfer,
T. Schenke, S. Segev, J.T. Smith (†), R. Stahlmann,
C. Thornsberry, E. von Keutz, H.-J. Zeiler

Editors

J. Kuhlmann, A. Dalhoff and H.-J. Zeiler

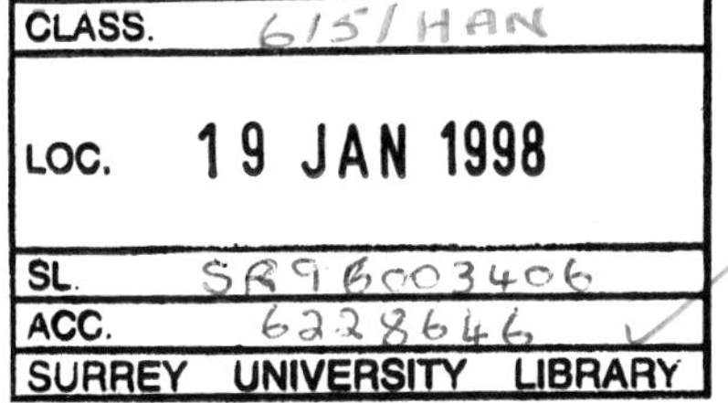

Springer

Professor Dr. J. Kuhlmann
Institut für Klinische Pharmakologie
Bayer AG
Geschäftsbereich Pharma
Pharma-Forschungszentrum,
Geb. 429
D-42096 Wuppertal
Germany

Dr. H.-J. Zeiler
Institut für Chemotherapie
Bayer AG
Geschäftsbereich Pharma
Pharma-Forschungszentrum,
Geb. 405
D-42096 Wuppertal
Germany

PD Dr. A. Dalhoff
Christian-Albrechts-Universität
Institut für Medizinische
Mikrobiologie und Virologie
Brunswiker Str. 4
D-24105 Kiel
Germany

With 115 Figures and 103 Tables

ISBN 3-540-62512-7 Springer-Verlag Berlin Heidelberg New York

Library of Congress Cataloging-in-Publication Data

Quinolone antibacterials / contributors, D. Beermann . . . [et al.]; editors, J. Kuhlmann, A. Dalhoff, and H.-J. Zeiler.
p. cm. — (Handbook of experimental pharmacology; v. 127)
Includes bibliographical references and index.
ISBN 3-540-62512-7 (hardcover)
1. Quinolone antibacterial agents. I. Beermann, D. II. Kuhlmann, J. (Jochen) III. Dalhoff, Axel. IV. Zeiler, H. J. (Hans-Joachim), 1947– . V. Series.
[DNLM: 1. Quinolones—pharmacology. 2. Quinolones—therapeutic use. 3. Anti-Infective Agents, Quinolone—therapeutic use. W1 HA51L v. 127 1997 / QV 250 Q69 1997]
QP905.H3 vol. 127
[RM666.Q55]
615′.1 s—dc21
[616.9′2061]
DNLM/DLC
for Library of Congress 97-455
CIP

Cover design: *design & production* GmbH, Heidelberg

Typesetting: Best-set Typesetter Ltd., Hong Kong

SPIN: 10492063 27/3020 – 5 4 3 2 1 0 – Printed on acid-free paper

Preface

It has been over 30 years since the first clinically important member of the quinolone class, nalidixic acid, was introduced into medical practice. The modification produced in the quinolone nucleus by introducing a fluorine at the 6-position led to the discovery of the newer fluoroquinolones with enhanced antibacterial activities as compared to nalidixic acid. By now a great deal of preclinical and clinical experience has been obtained with these agents.

The intense interest in this class of antibacterial agents by chemists, microbiologists, toxicologists, pharmacologists, clinical pharmacologists, and clinicians in various disciplines encouraged us to summarize the information on the history, chemistry, mode of action and in vitro properties, kinetics and efficacy in animals, mechanisms of resistance, toxicity, clinical pharmacology, clinical experience, and future prospects in one volume of the *Handbook of Experimental Pharmacology*. As this series deals predominantly with "experimental" characteristics of drugs, our volume is dedicated specifically to quinolones and emphasizes principally their preclinical and clinical pharmacological characteristics, despite the existence of several summaries on quinolones.

The chemistry of the quinolones is described in detail.

The chapter on the mode of action of quinolones reports the conclusive evidence that gyrase is the intracellular target of the quinolones; however, another enzyme, topoisomerase IV, may also be a target for quinolones, and the exact mechanisms by which quinolones act bactericidally are far from being understood.

The overview on the in vitro antibacterial activity of selected quinolones against various gram-negative and gram-positive species as well as anaerobes and some intracellular pathogens indicates that currently marketed quinolones are highly active against the gram-negative species; however, as gram-positive bacteria and anaerobes are less susceptible to these quinolones, new derivatives should have increased activity against gram-positive and anaerobic bacteria.

Evaluation of the pharmacokinetics and pharmacodynamics of these agents in animal models may bridge the gap between the in vitro and the clinical evaluation of a compound. However, species-specific differences in fluoroquinolone pharmacokinetics and differences in the pharmacokinetics between animals and humans are major drawbacks in the use of animal models and should be considered.

The pleiotropic interactions of fluoroquinolones with the host–parasite relationship may contribute to marked in vivo efficacy. Despite difficulties in defining the mechanisms of interactions with the host–parasite relationship and the relevance of experimental data to the clinical situation, fluoroquinolones may combine a direct antibacterial action with a direct or an indirect effect on host defense mechanisms.

The toxicological profile of the various quinolones has been well characterized in a large number of preclinical studies. Despite the impression that the overall pattern of potential toxicities is comparable for all quinolones there are marked differences in both the incidence and the type of reaction induced by certain compounds. The data derived from animal toxicology studies on the effects of quinolones at these target sites are comprehensively reviewed and their clinical implication discussed.

The section on clinical pharmacology of fluoroquinolone antibiotics focuses on pharmacokinetics of fluoroquinolones and interactions with other drugs. It also provides a general overview of the most common adverse effects of the newer fluoroquinolones, with an emphasis on some severe adverse events, which must be considered individually for the various members of the quinolone class.

The quinolones continue to make a significant contribution in the fight against community- and hospital-acquired infections and add a new dimension of antibacterial therapy to the well-established armamentarium of anti-infective drugs. Extensive clinical experience – especially with ciprofloxacin and also with norfloxacin and ofloxacin – together with a rich body of post-marketing data have confirmed the high tolerability and efficacy of the described quinolones. However, there are still open questions, for example, concerning their usefulness in mycobacterial infections, the development of resistance, and the appropriate dosing regimen for special conditions.

Increasing knowledge of the relationships of chemical structure and effects, both desirable and undesirable, will progressively heighten the ratio of benefit to risk with quinolones. The question is whether to seek more quinolones with broader antimicrobial activity or to develop new compounds designed for specific targets. In the 1980s many drug companies and researchers gave up trying to develop new approaches to antibiotics, believing that existing drugs or modified versions of them could keep pace with infectious bacteria. In the 1990s the number of bacterial strains that no longer succumb to antibiotics is rising, and it seems clear that antibiotics are losing their magic touch after decades of incautious prescription, improper use, and the inevitable spread of bacterial genes that confer drug resistance.

Regardless of whether the next antibiotic comes from the natural world or the artificial world of computer modeling, one thing that all researchers agree on is that something must be done soon.

We express our gratitude to the authors of this volume for their contributions and their cooperation. They have met the challenge of providing carefully written, up-to-date information on the science and practice of infectious diseases. Our hope is that readers will find this work a ready resource for new and helpful information, and that the efforts of our colleagues will therefore have been worthwhile.

We thank Mrs. D. WALKER from Springer-Verlag for her efficient management and editorial assistance through the entire project. We are also indebted to Mrs. G. LION for her secretarial and administrative help of very exceptional quality.

Wuppertal, Germany — J. KUHLMANN
Kiel, Germany — A. DALHOFF
Wuppertal, Germany — H.-J. ZEILER

List of Contributors

BEERMANN, D., Institut für Klinische Pharmakologie Bayer AG, Geschäftsbereich Pharma, Pharma-Forschungszentrum, Geb. 431, D-42096 Wuppertal, Germany

BERGAN, T., Rikshospitalet and University of Oslo, Institute of Medical Microbiology, N-0027 Oslo, Norway

CHRIST, W., Bundesinstitut für Arzneimittel und Medizinprodukte, Seestraße 10, D-13353 Berlin, Germany

CRAIG, W.A., William S. Middleton Memorial Veterans Hospital, Department of Veterans Affairs, 2500 Overlook Terrace, Madison, WI 53705, USA

CRITCHLOW, S.E., Department of Biochemistry, University of Leicester, Adrian Building, University Road, Leicester LE1 7RH, Great Britain

DALHOFF, A., Christian-Albrechts-Universität Kiel, Institut für Medizinische Mikrobiologie und Virologie, Brunswiker Str. 4, D-24105 Kiel, Germany

EVERETT, M.J., Antimicrobial Agents Research Group, Department of Infection, University of Birmingham, The Medical School, Edgebaston, Birmingham B15 2TT, Great Britain

GROHE, K, Bayer AG, ZF-Chemische Forschung, Werk Leverkusen, Geb. Q18-8. ET, D-51368 Leverkusen, Germany

KUHLMANN, J., Institut für klinische Pharmakologie, Bayer AG, Geschäftsbereich Pharma, Pharma-Forschungszentrum, Geb. 429, D-42096 Wuppertal, Germany

LODE, H., Städt. Krankenhaus Zehlendorf, Bereich Heckeshorn, Pneumologische Klinik 1, Zum Heckeshorn 33, D-14109 Berlin, Germany

MAXWELL, A., Department of Biochemistry, University of Leicester, Adrian Building, University Road, Leicester LE1 7RH, Great Britain

PETERSEN, U., Bayer AG, ZF-Chemische Forschung, Werk Leverkusen, Geb. Q18-6. ET, D-51368 Leverkusen, Germany

PIDDOCK, L.J.V., Antimicrobial Agents Research Group, Department of Infection, University of Birmingham, The Medical School, Edgebaston, Birmingham B15 2TT, Great Britain

RUBINSTEIN, E., Infectious Diseases Unit, The Chaim Sheba Medical Center, Tel Aviv University School of Medicine, Tel-Hashomer 52621, Israel

SCHACHT, P., Gellertweg 20, D-42115 Wuppertal, Germany

SCHAEFER, H.-G., Lilly Research Centre Limited, Erl Wood Manor, Windlesham, Surrey GV20 6PH, Great Britain

SCHENKE, T., Bayer AG, PH-R, Aprather Weg, Werk, Geb. 460, D-42096 Wuppertal, Germany

SEGEV, S., Infectious Diseases Unit, The Chaim Sheba Medical Center, Tel Aviv University School of Medicine, Tel-Hashomer 52621, Israel

SMITH, J.T., Department of Pharmaceutics, The School of Pharmacy, University of London, 29/39 Brunswick Square, London WC1N 1AX, Great Britain (†)

STAHLMANN, R., Institut für Klinische Pharmakologie und Toxikologie, Freie Universität Berlin, Universitätsklinikum Benjamin Franklin, Garystr. 5 , D-14195 Berlin, Germany

THORNSBERRY, C., MRL Pharmaceutical Services, 357 Riverside Drive, Franklin, TN 37064, USA

VON KEUTZ, E., Bayer AG, Institute of Toxicology, PO Box 10 17 09, Aprather Weg, D-42096 Wuppertal, Germany

ZEILER, H.-J., Bayer AG, PH-RESEARCH, Buildg. 402, Aprather Weg, D-42096 Wuppertal, Germany

Contents

CHAPTER 1

History and Introduction
J.T. SMITH and H.-J. ZEILER. With 1 Figure 1

A. Chemistry 1
B. Antibacterial Activity 4
C. Oxygen and 4-Quinolones 6
D. 4-Quinolone Kinetics and Distribution in Humans 6
E. Outlook 8
References 9

CHAPTER 2

The Chemistry of the Quinolones: Methods of Synthesizing the Quinolone Ring System
K. GROHE. With 42 Figures 13

A. Introduction 13
B. Methods of Synthesizing the Quinolone Ring System 16
I. Gould-Jacobs Reaction 16
II. Dieckmann Cyclization of Diesthers 21
III. Cycloaracylation Procedure 25
IV. Biere and Seelen Approach 35
V. Isatoic Anhydride Procedure 35
VI. Camps Quinolone Synthesis 39
VII. Meth-Cohn Quinolone Synthesis 41
VIII. Synthesis of Quinolone Analogues 42
IX. Synthesis of 4-Cinnolone-3-carboxylic Acids 45
X. Synthesis of 4-Pyridone-3-carboxylic Acids 49
References 52

CHAPTER 3

The Chemistry of the Quinolones: Chemistry in the Periphery of the Quinolones
U. PETERSEN and T. SCHENKE. With 35 Figures 63

A. Introduction 63
B. 1-Position 63
C. 2-Position 65
D. 3-Position 65
E. 4-Position 71
F. 5-Position 71
G. 6-Position 76
H. 7-Position 78
I. Synthesis of Specific Amines 78
1. Bicyclic Piperazine Derivatives 78
2. Aminopyrrolidine and Aminomethylpyrrolidine Derivatives 79
3. 3,4-Bridged Pyrrolidine Derivatives 83
4. Mixed Derivatives 88
II. C–N Linkage 89
III. C–S Linkage 94
IV. C–O Linkage 95
V. C–C Linkage 95
I. 8-Position 99
References 101

CHAPTER 4

Mode of Action

A. Maxwell and S.E. Critchlow. With 12 Figures 119

A. Introduction 119
B. Effects on Bacteria 124
C. Effects on DNA Gyrase 130
I. Reactions of Gyrase 130
II. Mechanistic Steps 131
1. DNA Binding 131
2. DNA Cleavage 133
3. ATPase 136
III. Illegitimate Recombination 137
IV. Summary 137
D. Mode of Binding 137
I. Binding of Quinolones to DNA 138
II. Effect of DNA Gyrase on Quinolone Binding 139
III. Cooperative Quinolone–DNA Binding Model 140
IV. DNA Cleavage is not an Absolute Requirement for Quinolone Binding to a Gyrase–DNA Complex 141
V. A Role for Magnesium Ions in the Binding of Quinolones to DNA 142
VI. Binding of Quinobenzoxazines to DNA 144

VII. Quinolone Binding to Quinolone-Resistant Mutants of DNA Gyrase 145
VIII. Mode of Binding of Topoisomerase II-Targeting Drugs ... 146
IX. Problems with Current Models 149
X. Conclusions 150
E. Mechanism of Cell Killing 150
I. Paradoxical Effects of Quinolones 150
II. Poison Hypothesis 153
III. Polymerase Blocking 153
F. Conclusions and Future Prospects 156
References 158

CHAPTER 5

The In Vitro Antibacterial Activity of Quinolones: A Review
C. THORNSBERRY 167

A. Introduction 167
B. In Vitro Activity 169
C. The Future 174
References 177

CHAPTER 6

Pharmacokinetics of Fluoroquinolones in Experimental Animals
A. DALHOFF and T. BERGAN. With 8 Figures 179

A. Introduction 179
B. Norfloxacin 180
C. Pefloxacin 181
D. Enoxacin 184
E. Ofloxacin 186
F. Ciprofloxacin 188
G. Temafloxacin 189
H. Tosufloxacin 191
I. Fleroxacin 191
J. Lomefloxacin 193
K. Sparfloxacin 195
L. Penetration of Quinolones at Sites of Infection 195
References 202

CHAPTER 7

Pharmacodynamics of Fluoroquinolones in Experimental Animals
W.A. CRAIG and A. DALHOFF. With 5 Figures 207

A. Introduction 207
B. Bacterial Killing In Vivo 208
C. In Vivo Postantibiotic Effects 208
D. Pharmacodynamic Parameters Determining Efficacy 208
E. Emergence of Resistance to Fluoroquinolones 217
I. *P. aeruginosa* Experimental Infections 220
II. Staphylococcal Infections in Experimental Animals 222
III. Miscellaneous Infection Models 223
IV. Factors Contributing to the Emergence of Resistance In Vivo 223
References 226

CHAPTER 8

Interaction of Quinolones with Host–Parasite Relationship
A. DALHOFF. With 3 Figures 233

A. Introduction 233
B. Effect on Adherence 235
C. Effect Against Slowly Growing Bacteria 241
D. Effect on Exoenzyme Production 245
I. *E. coli* 245
II. *P. aeruginosa* 245
E. Quinolone-Induced Endotoxin Release 248
F. Summary 252
References 253

CHAPTER 9

Mechanisms of Resistance to Fluoroquinolones
M.J. EVERETT and L.J. V. PIDDOCK. With 5 Figures 259

A. Introduction 259
B. Target Site Modification 260
I. Mutations in *gyrA* 260
II. Mutations in *gyrB* 267
III. Mutations in Other Topoisomerase Genes 268
C. Reduced Intracellular Accumulation 270
I. Decreased Uptake 273
II. Increased Efflux 275
III. The *mar* Operon 277
D. Reduced Killing 278
E. Prevalence of Fluoroquinolone Resistance 281
I. Community Acquired Pathogens 281
II. Nosocomial Pathogens 282

III. Impact of Fluoroquinolone Use in Agriculture 284
F. Distribution of Resistance Mechanisms 285
G. Summary 286
References 286

CHAPTER 10

Toxicology and Safety Pharmacology of Quinolones

E. von KEUTZ and W. CHRIST. With 1 Figure 297

A. Introduction 297
B. Arthropathy 297
C. Achilles Tendinitis and Rupture 302
D. Nephropathy 304
E. Effects on Central Nervous System 306
F. Ocular Toxicity 313
G. Impairment of Spermatogenesis 315
H. Cardiovascular Effects 316
I. Possible Mutagenic and Carcinogenic Effects 317
J. Phototoxicity 322
K. Photocarcinogenicity and Photomutagenicity 327
L. Drug Interactions 329
M. Metabolic and Nutritional Effects 330
N. Conclusion 330
References 331

CHAPTER 11

Clinical Pharmacology

J. KUHLMANN, H.-G. SCHAEFER, and D. BEERMANN 339

A. Introduction 339
B. Pharmacokinetics of Fluoroquinolone Antibiotics 340
I. Healthy Subjects 341
1. Absorption 343
2. Intravenous Administration 346
3. Bioavailability 347
4. Distribution 347
5. Disposition 349
II. The Elderly 350
III. Patients with Various Degrees of Renal Failure 353
1. Ciprofloxacin 353
2. Norfloxacin 354
3. Ofloxacin 354
4. Enoxacin 354

5. Fleroxacin ... 354
6. Pefloxacin ... 354
7. Lomefloxacin ... 355
8. Summary ... 355
IV. Patients with Hepatic Failure ... 355
V. Fluoroquinolones in Pediatric Patients ... 356
C. Interactions of Fluoroquinolone Antibiotics with Other Drugs ... 357
I. Interactions During the Absorption Process ... 357
1. Food and Dairy Products ... 358
2. Al^{3+}-, Ca^{2+}- and Mg^{2+}-Containing Antacids ... 359
3. Sucralfate ... 361
4. Didanosine ... 362
5. Other Metal Cations ... 362
6. Chemotherapy Treatment ... 363
7. Activated Charcoal ... 364
II. Interactions of Fluoroquinolones Due to Alterations in Metabolism ... 365
1. Theophylline, Caffeine and Structurally Closely Related Substances ... 365
2. Antipyrine ... 370
3. Phenytoin ... 370
4. H_2-Receptor Antagonists ... 371
5. K^+/Na^+-ATPase Inhibitors ... 372
6. Warfarin ... 372
7. Cyclosporine ... 372
8. Rifampin ... 373
9. Oral Contraceptive Steroids ... 374
10. Benzodiazepines (Diazepam, Temazepam) ... 374
III. Alterations in Renal Excretion ... 375
1. Probenecid ... 375
2. β-Lactam Antibiotics ... 375
IV. Pharmacodynamic Interactions ... 376
1. Nonsteroidal Anti-inflammatory Drugs ... 376
2. Metronidazole ... 377
V. Conclusions ... 377
D. Adverse Reactions of Fluoroquinolones ... 378
I. Gastrointestinal Tract ... 379
II. Central Nervous System ... 379
III. Skin and Allergic Reactions ... 381
1. Photosensitivity ... 381
2. Photoallergy ... 384
IV. Nephropathy and Crystalluria ... 384
V. Arthropathy and Musculoskeletal Disorders ... 385
VI. Body Systems ... 385
VII. Others ... 386
References ... 386

CHAPTER 12

Concentration–Effect Relationship of the Fluoroquinolones

R. STAHLMANN and H. LODE. With 3 Figures 407

A. Introduction 407
B. Pharmacodynamic Data of Antimicrobials as a Basis for Clinical Use 408
I. β-Lactams: Concentration-Independent Killing Rate 408
II. Aminoglycosides: Concentration-Dependent Killing Rate 409
III. Fluoroquinolones 410
1. In Vitro Models 411
2. Animal Models 411
3. Clinical Data 414
4. Overall Evaluation of Pharmacodynamics 416
C. Pharmacokinetic Aspects 416
I. Protein Binding 417
II. Tissue Concentrations and Volume of Distribution 417
D. Summary and Conclusion 419
References 419

CHAPTER 13

Clinical Use of Quinolones

P. SCHACHT 421

A. Introduction 421
B. Urinary Tract Infections and Prostatitis 421
I. Acute Uncomplicated Urinary Tract Infection 422
II. Complicated Urinary Tract Infection 422
III. Prostatitis 422
C. Gastrointestinal Infections and Traveller's Diarrhea 425
I. *Salmonella typhi* – Enteric Fever 425
II. *Salmonella typhi* Carriers 426
III. Salmonella Gastroenteritis Outbreaks 427
IV. Shigellosis 427
V. Cholera 428
VI. Traveller's Diarrhea – Prevention 429
VII. Traveller's Diarrhea – Therapy 429
D. Respiratory Tract Infections 429
I. Acute Exacerbation of Chronic Bronchitis 430
II. Pneumonia 432
III. Recurrent Respiratory Tract Infections in Patients with Cystic Fibrosis 435
IV. Sinusitis 435
V. Bacterial Otitis (Chronic Suppurative Otitis Media) 436
VI. Malignant External Otitis 436

E. Osteomyelitis 437
F. Skin and Skin Structure Infection 438
G. Sexually Transmitted Diseases 439
I. Gonorrhea 439
II. Chancroid 440
III. Nongonococcal Urethritis 440
H. Intra-abdominal Infections 440
I. Anaerobic Intra-abdominal Infections 440
II. Cholangitis 441
III. Peritonitis in Chronic Ambulatory Dialysis Patients 441
IV. Gynecological Infections 441
I. Bacteremia and Sepsis 442
J. Surgical Prophylaxis 443
I. Transurethral Prostatic Surgery 443
II. Endoscopic Retrograde Cholangiopancreatography 443
III. Abdominal Surgery 443
K. Infections in Neutropenic Patients 444
I. Empirical Treatment of Febrile Neutropenic Patients 444
II. Prophylaxis in Neutropenic Cancer Patients 444
L. Use of Quinolones in Pediatrics 445
M. Remarks 445
References 445

CHAPTER 14

Future Aspects
S. SEGEV and E. RUBINSTEIN 455

A. Introduction 455
B. Molecular Structure and Mechanism of Action 455
C. Antimicrobial Activity 458
I. Gram-Positive Bacteria 459
1. Staphylococci 459
2. Streptococci 460
3. Enterococci 462
II. Anaerobic Bacteria 463
III. *Mycobacterium tuberculosis* 464
D. Bacterial Resistance 465
E. Future Use of Quinolones in Pediatrics 466
F. Antitumor Potential 467
G. New Attitudes 468
H. Directions of Future Research on the Quinolones 469
References 469

Subject Index 477

CHAPTER 1

History and Introduction

J.T. Smith† and H.-J. Zeiler

A. Chemistry

Lesher et al. (1962) showed that 7-chloro-1,4-dihydro-1-ethyl-4-oxoquinoline-3-carboxylic acid, referred to as compound (a), which was obtained as an impurity during the manufacture of chloroquine, possessed antibacterial activity.

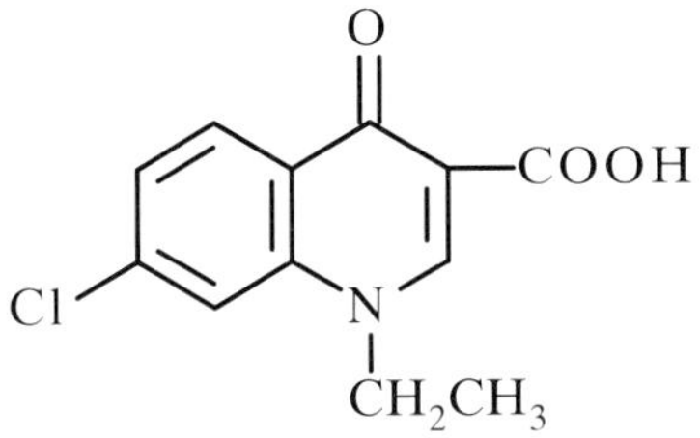

(a) 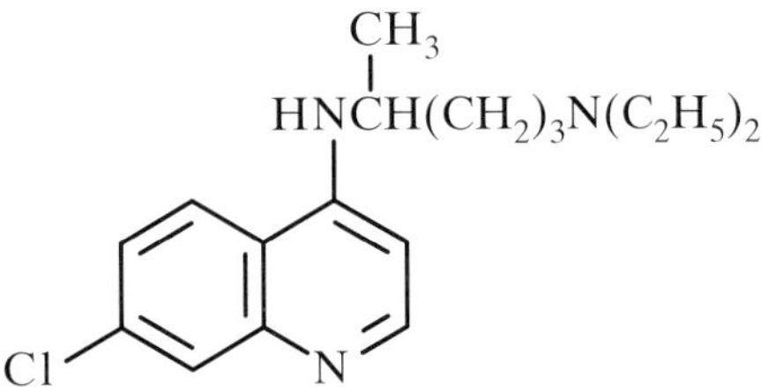 chloroquine

Consequently many derivatives of compound (a) were synthesized and evaluated for their antibacterial potency.

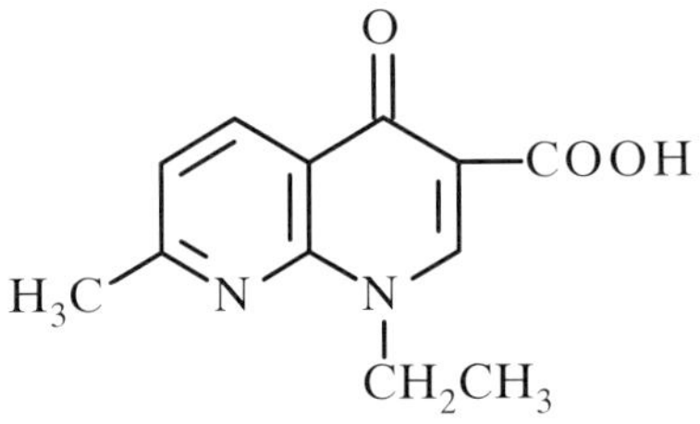

nalidixic acid 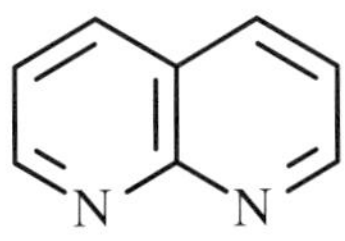 naphthyridine 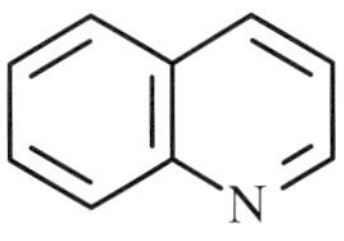 quinoline

Dedication. In memory of J.T. Smith, who died unexpectedly in 1996 during the publication of this book. From the beginnings of quinolones, he made many significant contributions to the understanding of the mechanism of this class of antibacterials. He was a wonderful person to work with and we are grateful to have known him for so many years. We will hold him and his work in high regard.

Ultimately nalidixic acid was chosen as the optimum compound to treat human bacterial infections. Nalidixic acid was patented in 1962 and launched in 1965. It is curious that a naphthyridine became the chosen congener of compound (a) because neither chloroquine nor compound (a) are naphthyridines but rather are quinolines. In later years quinolines proved to be the 4-quinolones of choice rather than naphthyridines.

It is also worth noting that compound (a) is halogenated at the 7-position, while nalidixic acid was not halogenated. The very successful modern 4-quinolone antibacterials are all halogenated at the 6-position and some are halogenated at position 8 as well. It is worth remembering that fluorine at C-6, which is now a common substituent in modern drugs, was first adopted when flumequine was synthesized by the 3M group (GERSTER 1973). This was done at their Riker laboratories, where this C-6 position was successfully fluorinated for the first time. It was rumoured that this occurred because 3M had assembled a team of fluorine chemists who so rapidly solved problems associated with the development of other products made by the group that they were asked to try to fluorinate 4-quinolone antibacterials. Flumequine was patented in 1973 and was launched as long ago as 1982. Its previous designation was R 802.

It is also interesting that a compound having another common substituent of modern 4-quinolones, the piperazine group at position 7, was first patented as early as 1974 in pipemidic acid (MATSUMOTO and MINAMI 1975); it was synthesized by the Dainippon laboratories and was used clinically for the first time in 1978.

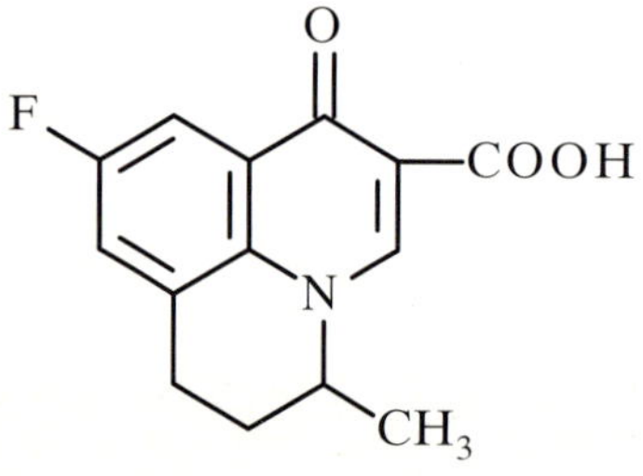

flumequine

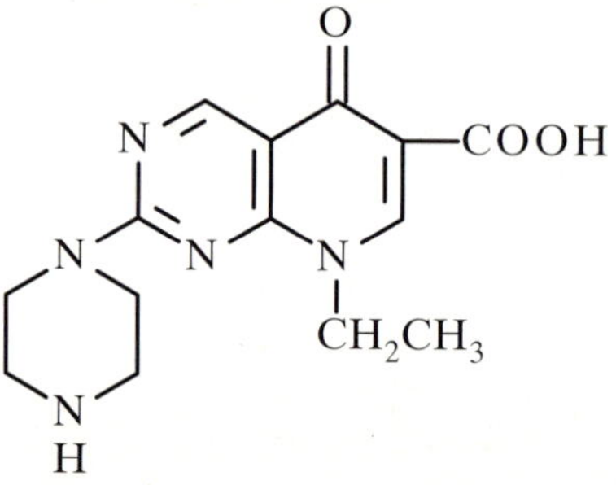

pipemidic acid

Another curious feature of the 4-quinolone antibacterials is that, as mentioned before, their origin was chloroquine, which was then and still is an effective treatment for malaria. SARMA (1989) accidentally found that norfloxacin cured malaria in patients being treated for *Pseudomonas aeruginosa* infections. It is also worth noting that ICI-56780 is an antimalarial drug about 50 times more potent than chloroquine (RYLEY and PETERS 1970), and this compound bears more ressemblance to the modern 4-quinolone antibacterials than to chloroquine.

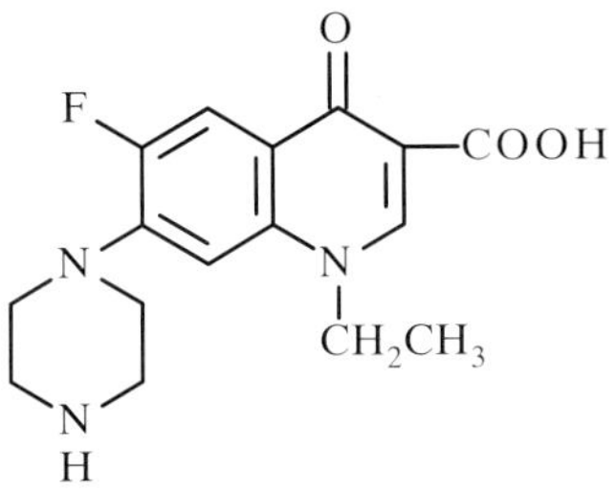

norfloxacin

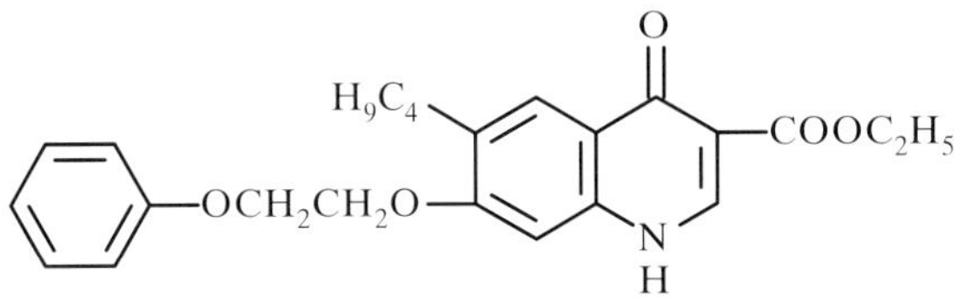

ICI-56780

Since the early days of the evaluation of the 4-quinolone antibacterials, several thousand derivatives have been studied and several structure – activity relationships have been reported (ALBRECHT 1977; MITSCHER et al. 1993; DOMAGALA 1994). However, it must be recognized that the structure activity relationship referred to in these reviews is that against bacteria and has been deduced from published data retrospectively. When it comes to relating 4-quinolone structure to the pharmacology and toxicity of such drugs in mammalian systems, correlations are virtually impossible to predict. Who could have forecasted from a structural standpoint that temafloxacin (HOOPER and WOLFSON 1993), which was patented in 1986, clinically launched in 1991 and then withdrawn in 1992, would suffer its fate, despite extensive safety testing in vitro, in vivo and even in human volunteers?

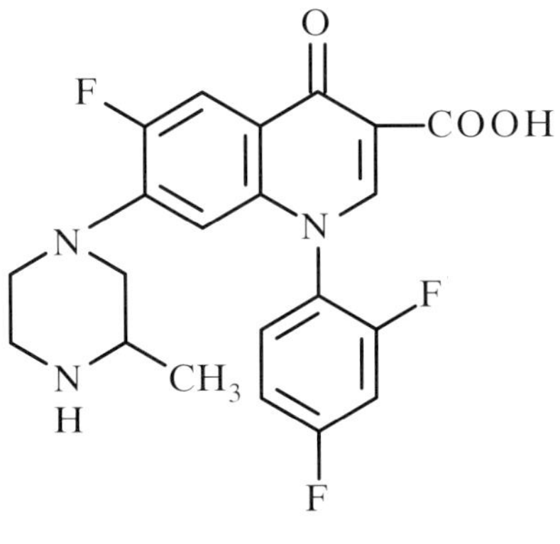

temafloxacin

This unpredictability of 4-quinolones in mammalian systems is perhaps best illustrated by examining acrosoxacin (LESHER and CARABATEAS 1973) which was patented in 1973 and launched in 1982. As this drug has since been in constant clinical use, it has therefore stood the test of time with respect to its safety for use in humans. Lomefloxacin (HOKURIKU PHARMACEUTICAL 1986), on the other hand, was patented in 1985 and launched in 1989 after the very rigorous safety assessments that applied at the time, which were much more rigorous than was the case at the time when the safety of acrosoxacin for human use was assessed.

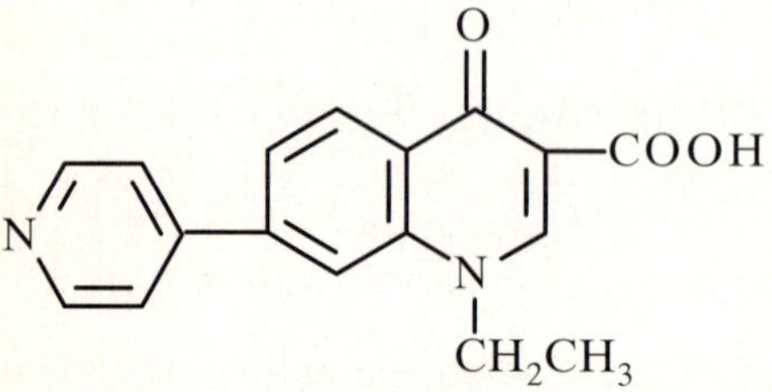

acrosoxacin

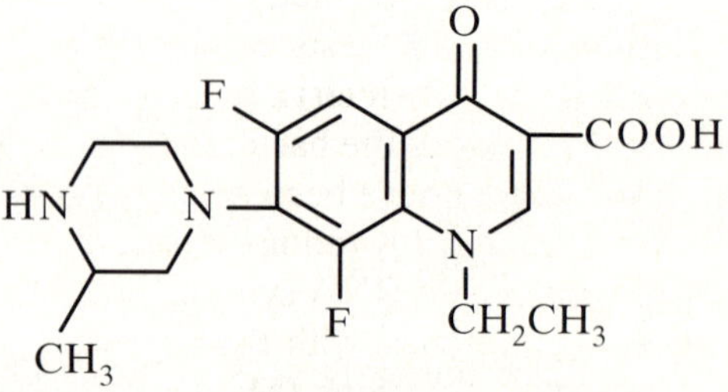

lomefloxacin

Yet despite the safety of these two drugs for human use, if we take the C-7 side chain of acrosoxacin and add it to the 4-quinolone nucleus of lomefloxacin we arrive at Pfizer compound CP 67015 (HOLDEN et al. 1989).

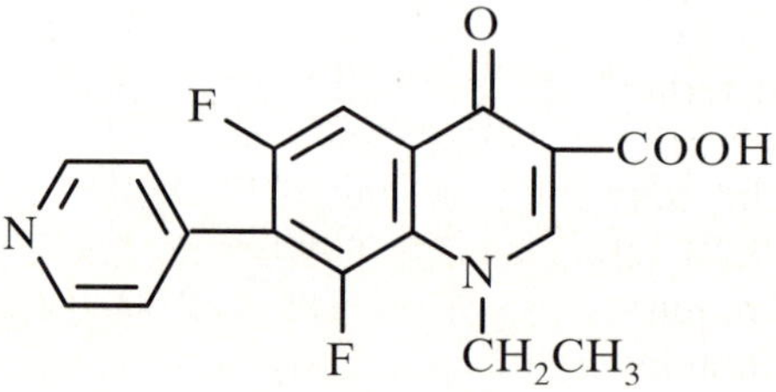

Pfizer CP 67015

There is no doubt that CP 67015 is extremely toxic to mammalian cells since in vitro and in vivo the drug causes devastating mutagenic and clastogenic effects at low concentrations (HOLDEN et al. 1989). This constraint means that to discover and to launch a new effective 4-quinolone which is safe to use in humans is a daunting technical and financial challenge.

B. Antibacterial Activity

At their minimum inhibitory concentrations (MICs) the 4-quinolones are merely bacteriostatic. At higher concentrations the drugs become increasingly bactericidal up to a point called the most bactericidal concentration (SMITH 1984), which was later renamed as the optimal bactericidal concentration (LEWIN and SMITH 1990). At still higher concentrations, progressively less kill is observed and this has been explained by the inhibition of RNA synthesis by high concentrations of 4-quinolones (CRUMPLIN and SMITH 1975); RNA synthesis is essential for their full bactericidal effect. The fact that 4-quinolones are merely bacteriostatic at their MIC values can cause confusion. For example, novobiocin and nalidixic acid are synergistic when studied by MIC tests (Chao 1978), but are very antagonistic to each other when bacterial kill is assessed (Lewin et al. 1991a). In addition several other phenomena, such as the effects of pH, metal ions and anaerobiasis, lead to different results when studied by MIC tests compared with bactericidal investigations (Lewin and Smith 1989; Ratcliffe and Smith 1983, 1985a; Smith and Lewin 1988).

As regards the bactericidal activity of 4-quinolones, four different bactericidal actions have been proposed, which are termed mechanisms A, B, B′ and C. Mechanism A requires bacteria which are capable of multiplication as well as protein and RNA synthesis for its activity. This mechanism is exhibited by all 4-quinolones, with the exception of 4-quinolones against *P. aeruginosa* at high concentrations (Morrissey and Smith 1994). Mechanism A is the sole bactericidal mechanism of older 4-quinolones such as nalidixic and oxolinic acids (Lewin et al. 1991b). Mechanism B is found as an additional bactericidal mechanism with some modern 4-quinolones such as ofloxacin, ciprofloxacin (Smith 1984), lomefloxacin (Lewin et al. 1989), levofloxacin (Lewin and Amyes 1989), fleroxacin, pefloxacin (Lewin and Amyes 1990a), PD131628 (Lewin 1992) and sparfloxacin (Lewin et al. 1992). Unlike mechanism A, mechanism B is active against nondividing *Escherichia coli* and most other gram-negative bacteria (Zeiler and Grohe 1984) and does not require active protein or RNA synthesis (Smith 1984; Lewin et al. 1991b). In addition, a related bactericidal mechanism, termed B′, has also been proposed for the 4-quinolone PD127391 (clinafloxacin; Lewin and Amyes 1990b). This mechanism, while not requiring active protein or RNA synthesis, is nevertheless inactive against nondividing *E. coli* and staphylococci (Lewin and Amyes 1990b). Mechanism C, which has so far been found in *E. coli* only with norfloxacin (Ratcliffe and Smith 1985b) and enoxacin (Lewin et al. 1989), does not require bacterial multiplication but does require active protein and RNA synthesis.

Are these bactericidal mechanisms of any practical significance? Perhaps they are since ciprofloxacin is much more active than ofloxacin against *E. coli* and both drugs exhibit mechanism B against this species. On the other hand, both drugs have a similar activity against staphylococci and this may occur because ofloxacin (but not ciprofloxacin) exhibits mechanism B against this genus (Lewin and Smith 1988). Other pointers are the findings that *Streptococcus faecalis* and *Streptococcus pneumoniae* are only very slowly killed by 4-quinolones and mechanism B is weak or absent against such species, at least as far as ciprofloxacin, ofloxacin (Lewin et al. 1991c; Morrissey and Smith 1993a) and even sparfloxacin (Wiedemann et al. 1994; Dalhoff 1994) are concerned.

Paradoxically, ciprofloxacin and ofloxacin exhibit mechanisms A and B against *P. aeruginosa* at low concentrations, but they only exhibit mechanism B at higher concentrations (Morrissey and Smith 1994). This species also presents mutational resistance problems (Scully et al. 1986; Smith 1990). The underlying principle in these examples is that, theroretically, mutational resistance is much less likely to occur against two mechanisms of action since two mutations would be required to occur simultaneously. A correlation has been found between the possession of extra bactericidal mechanisms and the binding of these drugs to their target site, DNA gyrase. It has been shown that the concentration of norfloxacin required to inhibit DNA supercoiling (0.2 μg/ml) is also the concentration at which cooperative norfloxacin binding to DNA gyrase occurs in vitro where four drug molecules associate with each DNA gyrase tetramer (SHEN et al. 1989). In addition, 0.2 μg/ml is precisely the same concentration at which norfloxacin starts to initiate its extra bactericidal mechanism C (HOWARD et al. 1993). With ofloxacin and ciprofloxacin, mechanism C suddenly initiates at the lower concentrations of 0.08 and 0.01 μg/ml, respectively (HOWARD et al. 1993). It would be interesting to know whether these concentrations are also those at which cooperative drug binding to DNA gyrase becomes apparent. Mechanism B of ofloxacin and ciprofloxacin suddenly initiates at 0.4 and 0.04 μg/ml, respectively (HOWARD et al. 1993); it is unknown what molecular events occur to DNA gyrase at these higher drug concentrations. There is no doubt that the target site of mechanism B resides within DNA gyrase (LEWIN et al. 1991d). We must await further in vitro tests to identify the chemical interactions which are required to trigger mechanism B.

C. Oxygen and 4-Quinolones

Oxygen is needed for the bactericidal activities of 4-quinolones but not for bacteriostasis (SMITH and LEWIN 1988). Even when cultures are exposed to air, high bacterial concentrations (about 10^{10} bacteria per ml) use up all available oxygen and hence prevent 4-quinolones from killing bacteria (MORRISSEY et al. 1990). It would seem that 4-quinolone bactericidal activity is prevented by anaerobiasis due to the reduced active drug uptake by bacteria which occurs in the absence of oxygen (MORRISSEY and SMITH 1993b); this phenomenon is similar to that already known to apply to the aminoglycosides (MATES et al. 1983). This has practical significance in that regions of poor blood supply (e.g. prostheses) may predispose to failure of 4-quinolone therapy. Indeed the development of resistance in such situations is noticeably higher (SMITH 1991) which is compatible with the reduced bactericidal activity of the drugs at such anaerobic infection sites.

D. 4-Quinolone Kinetics and Distribution in Humans

Table 1 shows that the serum concentrations of 4-quinolones are extremely low compared with those of the β-lactam and aminoglycoside antibiotics (BERGAN and WILLIAMS 1982; BERGAN et al. 1982; SCHENTAG 1982). Indeed the apparent volume of distribution for ciprofloxacin is about 2.4 times greater than the human body volume (see Fig. 1), whereas the apparent volumes of distribution for penicillins and aminoglycosides are very much lower (25%–

Table 1. 500 mg of drug administered to a 70-kg human

Drug	Peak serum level (μg/ml)	Peak tissue level (μg/ml = mg/l)		
Penicillins	20	Very low		
Aminoglycosides	29	Very low		
		C.S.F.	0.15–1	Below serum level
		Eye	0.1–0.6	Below serum level
		Fat	0.75–1.5	Below serum level
Ciprofloxacin	3	Sputum, pleural fluid, muscle → At serum level		
		Epithelial cells, gastric mucosa, prostate, macrophages	6–9	Above serum level
		Bile	15–30	Above serum level
		Lung	6–30	Above serum level
		Polymorphs	18–21	Above serum level

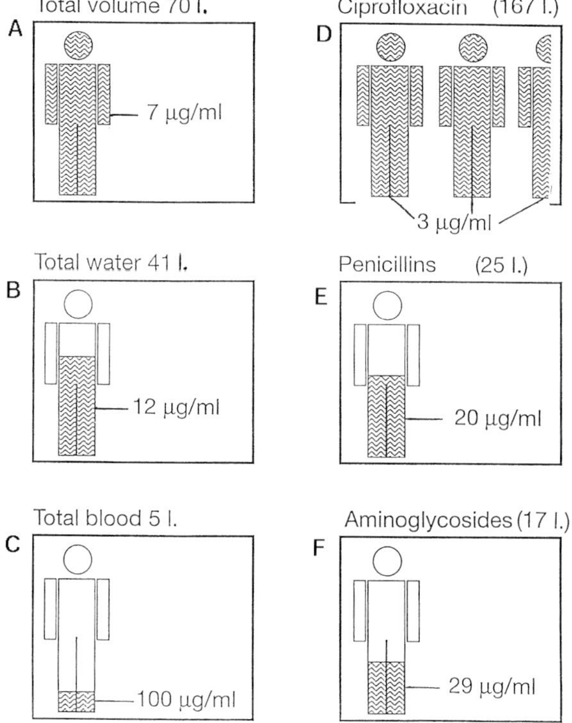

Fig. 1A–F. 500 mg of drug distributed in a 70-kg human. **A** If 500 mg of drug is distributed in the total volume (70 l) the concentration will be 7 μg/ml. **B** If 500 mg of drug is distributed in the total water (volume = 41 l) the concentration will be 12 μg/ml. **C** If 500 mg of drug is distributed in the total blood (volume = 5 l) the concentration will be 100 μg/ml. **D** With ciprofloxacin, a serum concentration of 3 μg/ml is reached after a 500-mg dose. It therefore seems that the drug is dissolved in an apparent volume of distribution of 167 l. **E** With penicillins, a serum concentration of 20 μg/ml would occur after a 500-mg dose. (Apparent volume of distribution is 25 l.) **F** With aminoglycosides, a serum concentration of 29 μg/ml would occur after a 500-mg dose. (Apparent volume of distribution is 17 l)

35%) than the body volume. So where do the 4-quinolones go in the body? It would seem that they are concentrated in certain cells of the body (see Table 1). This possibly explains why 4-quinolones are so effective in the treatment of bacterial infections, especially considering their concentration within macrophages and polymorphs, which are the phagocytes responsible for the first-line defences against bacterial invasion.

Yet another peculiar pharmacokinetic property of some of the 4-quinolones is their transmucosal transfer from the serum into the gut lumen (ROHWEDDER et al. 1990), which clearly occurs in the case of ciprofloxacin. This elimination route could well explain why these drugs are so effective in curing gastroenteritis. This is unexpected because the gut contents are highly anaerobic and, as mentioned earlier, the 4-quinolones are not bactericidal in the absence of oxygen. Could this paradox be explained by the process of transmucosal elimination whereby ciprofloxacin is presented on the surface of the gut mucosa together with oxygen (delivered by the arterioles of the villi), thus providing a bactericidal zone at the surface of the gut? It is well known that gut infections are surface infections; the fact that no kill could occur within the highly anaerobic gut lumen would be of little or no importance.

E. Outlook

From the foregoing analyses it is clear that the 4-quinolone antibacterials have unique properties which clearly distinguish them from other classes of front-line antimicrobial agents. In the last 5 years or so the success of the 4-quinolones has been dramatic. They have enjoyed stunning clinical success which has accelerated at unprecedented rates to encompass application in a widening variety of infectious diseases. Not surprisingly, this in turn has justifiably led to outstanding commercial success. The only worry with this explosion is that the drugs could be overused, thus leading to resistance as has already been seen with some species.

Why should the 4-quinolones be so successful? It would seem that the reasons for this are quite simple. Firstly, there is as yet no plasmid-mediated resistance to the 4-quinolones in clinical bacteria. Secondly, the side effects of the 4-quinolones are less frequent and less severe than those of other classes of antimicrobials. Thirdly, and perhaps most important of all, even for the most severe infections, the 4-quinolone antibacterials can be administered by the oral route. The cost of health care is currently an international preoccupation. In this respect it cannot be stressed too strongly that any drug given by the oral route will be cheaper than any parenteral treatment because the the latter entails the very high costs of administration. A further advantage of oral drugs is that they can be administered away from hospitals, which can be deadly sources of antibiotic-resistant pathogenic bacteria.

The present success of the 4-quinolone antibacterials is probably only a start; more potent derivatives are being developed which could reach bacteria either on the fringe of current 4-quinolone therapy or beyond the reach of present-day drugs. Indeed, as discussed in the chapter by Segev and Rubinstein, the 4-quinolones may ultimately find application as antiviral, antiparasitic and antifungal agents and even as anticancer chemotherapeutics.

References

Albrecht R (1977) Development of antibacterial agents of the nalidixic type. Prog Drug Res 21:9–104

Bergan T, Williams JD (1982) Dose-dependence of piperacillin pharmacokinetics. Chemotherapy 28:153–159

Bergan T, Thornsteinsson SB, Steingrimsson O (1982) Dose-dependent pharmacokinetics of azlocillin compared to mezlocillin. Chemotherapy 28:160–170

Chao L (1978) An unusual interaction between the target of nalidixic acid and novobiocin. Nature 271:385–386

Crumplin GC, Smith JT (1975) Nalidixic acid, an antibacterial paradox. Antimicrob Agents Chemother 8:251–261

Dalhoff A (1994) Activities of ciprofloxacin and sparfloxacin against strept. pneumoniae (Abstr 34). 5th International Congress on New Quinolones, Singapore, p 43

Domagala JM (1994) Structure-activity and structure side-effect relationships for the quinolone antibacterials. J Antimicrob Chemother 33:685–706

Gerster JF (1973) DOS 2 264 163. Substituted benzo[ij]quinazoline-2-carboxylic acids and their derivatives, Riker Laboratories, Northridge. Chem Abstr 79:92029y

Hokuriku Pharmaceutical (Japan) (1986) NY – 198. Drugs Future 11:578–579

Holden HE, Barrett JF, Huntington CM, Muehlbauer PA, Wahrenburg MG (1989) Genetic profile of nalidixic acid analog: a model for the mechanism of sister chromatid exchange induction. Environ Mol Mutag 13:238–252

Hooper DC, Wolfson JS (1993) Adverse effects. In: Hooper DC, Wolfson JS (eds) Quinolone antimicrobial agents, 2nd edn. American Society of Microbiology, Washington, pp 489–512

Howard BMA, Pinney RJ, Smith JT (1993) 4-quinolone bactericidal mechanisms. Arzneimittelforschung 43:1125–1129

Lesher GY, Carabateas PM (1972) DOS 2224090. 1-Alkyl-1,4-dihydro-4-oxo-3-quinolonecarboxylic acids and esters

Lesher GY, Carabateas PM (1973) 1-Alklyl-1,4-dihydro-4-oxo-3-quniolonecarboxylic acids and esters. Sterling Drug Inc. (C.A. 78, 84280 n (1973))

Lesher GY, Froelich EJ, Gruett MD, Bailey JH, Brundage RP (1962) 1,8-naphthyridine derivatives, a new class of chemotherapy agents. J Med Chem 5:1063–1065

Lewin CS (1992) Antibacterial activity of a 1,8-naphthyridine quinolone, PD 131628. J Med Microbiol 36:353–357

Lewin CS, Amyes SGB (1989) The bactericidal activity of DR-3355, an optically active isomer of ofloxacin. J Med Microbiol 30:227–231

Lewin CS, Amyes SGB (1990a) Conditions required for the bactericidal activity of fleroxacin and pefloxacin against Escherichia coli KL 16. J Med Microbiol 32:83–86

Lewin CS, Amyes SGB (1990b) Bactericidal action of PD 127391, an enhanced spectrum quinolone. J Med Microbiol 33:67–70

Lewin CS, Smith JT (1988) Bactericidal mechanisms of ofloxacin. J Antimicrob Chemother [Suppl C] 22:1–8

Lewin CS, Smith JT (1989) Interactions of 4-quinolones with other antibacterials. J Med Microbiol 29:221–227

Lewin CS, Smith JT (1990) Conditions requried for the bactericidal activity of 4-quinolones against Serratia marcescens. J Med Microbiol 32:211–214

Lewin CS, Amyes SGB, Smith JT (1989) Bactericidal activity of enoxacin and lomefloxacin against Escherischia coli K16. Eur J Clin Microbiol 8:731–733

Lewin CS, Howard BMA, Smith JT (1991a) 4-quinolone interactions with gyrase B subunit inhibitors. J Med Microbiol 35:358–362

Lewin CS, Morrissey I, Smith JT (1991b) The mode of action of quinolones: the paradox in activity of low and high concentrations and activity in the anaerobic environment. Eur J Clin Microbiol 10:240–248

Lewin CS, Morrissey I, Smith JT (1991c) The fluoroquinolones exert a reduced rate of kill against Enterococcus faecalis. J Pharm Pharmacol 43:492–494

Lewin CS, Howard BMA, Smith JT (1991d) Protein- and RNA-synthesis independent bactericidal activity of ciprofloxacin that involves the A subunit of DNA gyrase. J Med Microbiol 34:19–22

Lewin CS, Morrissey I, Smith JT (1992) The bactericidal activity of sparfloxacin. J Antimicrob Chemother 30:625–632

Mates SM, Patel L, Karback HR, Miller MH (1983) Membrane potential in anaerobically growing Staphylococcus aureus and its relationship with gentamicin uptake. Antimicrob Agents Chemother 23:526–530

Matsumoto J, Minami S (1975) Pyrido[2,3-d]pyrimidine antibacterial agents. 3. [1]8-alkyl- and 8-vinyl-5,8-dihydro-5-oxo-2-(1-piperazinyl)pyrido[2,3-d]pyrimidine-6-carboxylic acids and their derivatives. J Med Chem 18:74–79

Mitcher LA, Devasthale P, Zavod R (1993) Structure-activity relationships. In: Hooper DC, Wolfson JS (eds) Quinolone antimicrobial agents, 2nd edn. American Society of Microbiology, Washington, pp 3–51

Morrissey I, Smith JT (1993a) Activity of quinolone antibacterials against Streptococcus pneumoniae. Drugs 45[Suppl 3]:196–197

Morrissey I, Smith JT (1993b) Effect of inoculum size on 4-quinolone uptake by Escherischia coli KL 16 (Abstr 19). J Pharm Pharmacol 45[Suppl 2]:1106

Morrissey I, Smith JT (1994) The activity of 4-quinolones against Pseudomonas aeruginosa. Arzneimittelforschung 44:1157–1161

Morrissey I, Lewin CS, Smith JT (1990) The influence of oxygen upon bactericidal potency. In: Crumplin GC (ed) The 4-quinolone antibacterial agents in vitro. Springer, Berlin Heidelberg New York, pp 23–36

Ratcliffe NT, Smith JT (1983) Effects of magnesium on the activity of 4-quinolone antibacterial agents. J Pharm Pharmacol 35:61P

Ratcliffe NT, Smith JT (1985a) Ciprofloxacin's bactericidal and inhibitory actions in urine. Chemotherapia 4[Suppl 2]:385–386

Ratcliffe NT, Smith JT (1985b) Norfloxacin has a novel bactericidal mechanism unrelated to that of other 4-quinolones. J Pharm Pharmacol 37[Suppl]:92P

Rohwedder RW, Bergan T, Thorsteinsson SB, Scholl H (1990) Transintestinal elimination of ciprofloxacin. Diagn Microbiol Infect Dis 13:127–133

Ryley JF, Peters W (1970) The antimalarial activity of some quinolone esters. Ann Trop Med Parasitol 64:209–222

Sarma PS (1989) Norfloxacin a new drug in the treatment of falciparum malaria. Ann Intern Med 3:336–337

Schentag JJ (1982) Aminoglycosides. In: Evans WE, Schentag JJ, Jusko WJ (eds) Applied pharmacokinetics; principles of therapeutic drug monitoring. Applied Therapeutics, San Francisco, pp 174–209

Scully BE, Parry MF, Neu HC, Mandell W (1986) Oral ciprofloxacin therapy of infections due to Pseudomonas aeruginosa. Lancet 2:819–822

Shen LL, Baranowski J, Pernet AG (1989) Mechanism of inhibition of DNA gyrase by quinolone antibacterials: specificity and cooperativity of drug binding to DNA. Biochemistry 28:3879–3885

Smith JT (1984) Awakening the slumbering potential of the 4-quinolone antibacterials. Pharm J 233:299–305

Smith JT (1990) Mutation rates to 4-quinolones resistance. Arzneimittelforschung 40:65–68

Smith JT (1991) Ofloxacin, a bactericidal antibacterial. Chemotherapy 37[Suppl 1]:2–13

Smith JT, Lewin CS (1988) Chemistry and mechanisms of action of the quinolone antibacterials. In: Andriole VT (ed) The quinolones. Academic, London, pp 23–82

Wiedemann B, Rustige-Wiedeman C, Kratz B (1994) Comparison of the pharmacodynamic properties of quinolones (Abstr 67). 5th International Congress on New Quinolones, Singapore, p 84

Zeiler H-J, Grohe K (1984) The in vitro and in vivo activity of ciprofloxacin. Eur J Clin Microbial 3:339–343

CHAPTER 2

The Chemistry of the Quinolones: Methods of Synthesizing the Quinolone Ring System

K. GROHE

A. Introduction

The quinolones, similar to the sulfonamides and nitrofurans, are totally synthetic chemical compounds used to combat infections (ALBRECHT 1977). They are derived by substitution of 1,4-dihydro-4-oxo-quinoline-3-carboxylic acids (structural formula **1**) at the nitrogen atom of position 1.

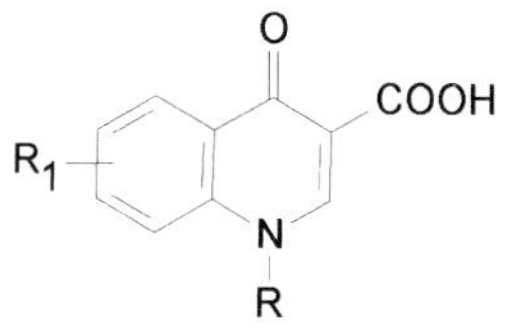

O COOH R_1 N R 1

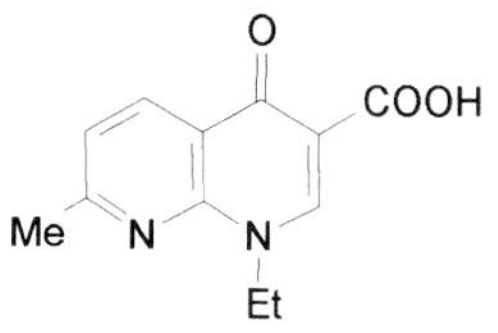

O COOH Me N N Et 2

The first commercial product, nalidixic acid (structural formula **2**; LESHER et al. 1962), a 1,8-naphthyridone or 8-azaquinolone carboxylic acid, and the structural variants which were systematically synthesized up to the end of the 1970s are shown in Table 1. Oxolinic acid **3**, rosoxacin **4**, piromidic acid **5**, flumequine **6**, and pipemidic acid **7** were obtained by means of the classical Gould-Jacobs reaction. For cinoxacin **8**, a cinnolone carboxylic acid, a special reaction sequence was developed (see Sect. B.IX).

These first-generation quinolones were followed at the beginning of the 1980s by the highly active broad-spectrum antibacterial fluoroquinolones. By replacing hydrogen by a fluorine atom at position 6 and substituting a diamine residue at position 7 and new residues at position 1 of the quinolone ring, the antibacterial efficacy was markedly enhanced. Important members of this new group of anti-infectives are norfloxacin **9**, enoxacin **10**, ofloxacin **11**, pefloxacin **12** and ciprofloxacin **13** (Table 1).

The fluoroquinolones **9–12** can be synthesized by the Gould-Jacobs reaction. Ciprofloxacin **13** was synthesized using the cycloaracylation procedure developed by GROHE et al. (GROHE et al. 1980; GROHE 1993).

The following gives an overview of the most important methods of synthesis based on the 4-quinolone and 4-pyridone ring systems. This is followed by a description of peripheral reactions at positions 1–8 of the quinolone ring.

Table 1. First- and second-generation quinolone and fluoroquinolone anti-infectives

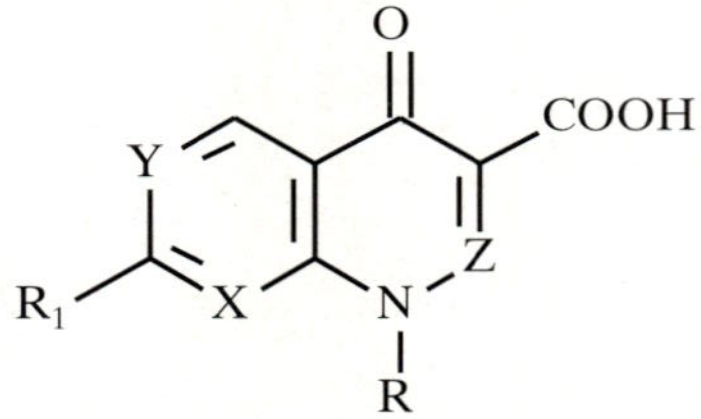

Compound	X	R	Y	R_1	Z	Reference
Nalidixic acid <u>2</u>	N	Et	CH	CH_3	CH	LESHER et al. (1962)
Oxolinic acid <u>3</u>	CH	Et	$C—OCH_2O—$		CH	KAMINSKY et al. (1968)
Rosoxacin <u>4</u>	CH	Et	CH	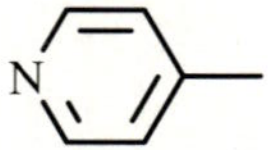	CH	LESHER et al. (1973)
Piromidic acid <u>5</u>	N	Et	N	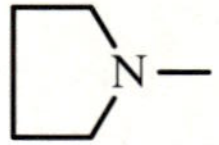	CH	MINAMI et al. (1971)
Flumequin <u>6</u>	$C—(CH_2)_2—CH$∼Me		CF	H	CH	GERSTER (1973); ROHLFING et al. (1976)

Pipemidic acid 7	N	Et	N	HN(piperazine)N—	CH	Matsumoto et al. (1975)
Cinoxacin 8	ĊH	Et	C—OCH_2O—		N	White (1970)
Norfloxacin 9	CH	Et	CF	HN(piperazine)N—	CH	Koga et al. (1980)
Enoxacin 10	N	Et	CF	HN(piperazine)N—	CH	Matsumoto et al. (1984a,b)
Ofloxacin 11	C—OCH_2—CH~Me		CF	MeN(piperazine)N—	CH	Sato et al. (1982); Hayakawa et al. (1984)
Pefloxacin 12	CH	Et	CF	MeN(piperazine)N—	CH	Gouefon et al. (1981)
Ciprofloxacin 13	CH	cyclopropyl	CF	HN(piperazine)N—	CH	Grohe et al. (1983); Zeiler and Grohe (1984)

B. Methods of Synthesizing the Quinolone Ring System

I. Gould-Jacobs Reaction

Until the end of the 1970s, the Gould-Jacobs reaction was the most commonly used procedure for synthesizing 1,4-dihydro-4-oxo-quinoline-3-carboxylic acid derivatives alkylated at the position 1 nitrogen.

The reaction sequence is illustrated by the synthesis of the fluoroquinolone Norfloxacin **9** (Koga et al. 1980; Fig. 1). 3-Chloro-4-fluoroaniline **14** is condensed at 80–120°C with ethoxymethylenemalonic ester **15** (EMME) to form arylaminomethylenemalonic ester **16**. This is followed by Gould-Jacobs ring closure of structure **16** to form 4-hydroxyquinoline carboxylic acid ester **17** by heating in boiling diphenyl ether or preferably Dowtherm A, a eutectic mixture of diphenyl ether and biphenyl (Price et al. 1946), at 250–270°C. The alkylation of **17** to **18** is carried out with ethyl iodide or ethyl bromide in the presence of potassium carbonate in dimethylformamide (DMF). The resulting quinolone carboxylic acid ester **18** is hydrolyzed by acid or alkali to the corresponding carboxylic acid **19**. Reaction of **19** with piperazine yields norfloxacin **9**.

Secondary aliphatic-aromatic amines (*N*-alkylaniline) can serve as starting compounds for the Gould-Jacobs procedure. Starting, for example, with the cyclic derivative **20**, one arrives via the intermediates **21**, **22**, and **23** at the racemic fluoroquinolone ofloxacin **11** (Sato et al. 1982; Hayakawa et al. 1984; Fig. 2).

The cyclocondensation of **21** to **22** must be carried out in this case according to Agui and colleagues (Agui et al. 1971; Harris 1971) in the presence of polyphosphoric acid. Other Friedel-Crafts catalysts, e.g., polyphosphate ester

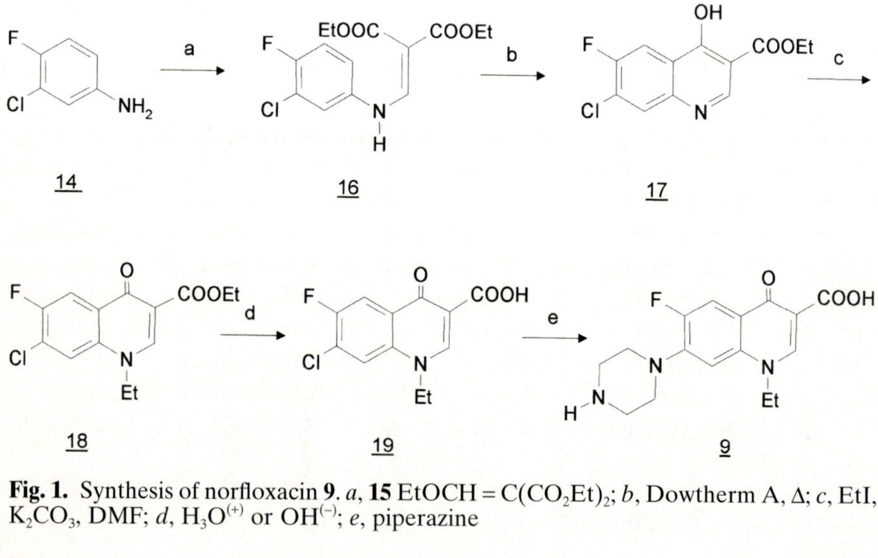

F Cl NH2 14 —a→ EtOOC COOEt F Cl N H 16 —b→ OH COOEt F Cl N 17 —c→

O COOEt F Cl N Et 18 —d→ O COOH F Cl N Et 19 —e→ O COOH F N N H N Et 9

Fig. 1. Synthesis of norfloxacin **9**. *a*, **15** $EtOCH=C(CO_2Et)_2$; *b*, Dowtherm A, Δ; *c*, EtI, K_2CO_3, DMF; *d*, $H_3O^{(+)}$ or $OH^{(-)}$; *e*, piperazine

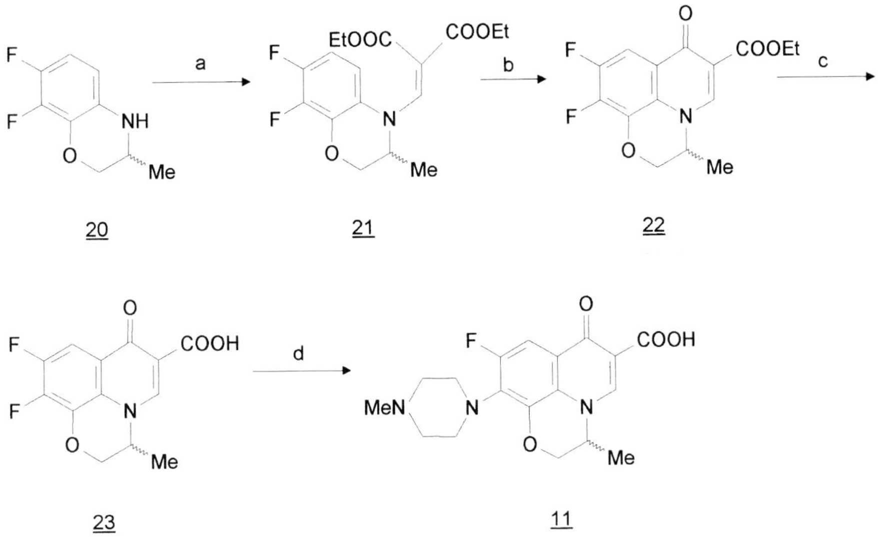

Fig. 2. Synthesis of ofloxacin **11**. *a*, Ethoxymethylenemalonic ester **15** (EMME), Δ; *b*, polyphosphoric acid; *c*, $H_3O^{(+)}$; *d*, 1-methylpiperazine

(Agui et al. 1975a; Dohmori et al. 1976) or phosphorus pentoxide (Markces et al. 1974), have also proved useful in such reactions.

Ofloxacin **11** has an asymmetric center at the C-3 atom of the oxazine ring and therefore consists of two optically active isomers. In order to synthesize the more active antibacterial S-enantiomer levofloxacin **26**, the compound **20** is replaced by the corresponding optically pure enantiomer **25** in the above reaction sequence (Fig. 3). A good yield of substance **25** can be obtained by asymmetric reduction of benzoxacine **24** with chiral triacyloxyborohydride (Atarashi et al. 1991).

An alternative method that can be used to produce **25** is chromatographic separation followed by saponification of the diastereomers of structure **27**, which are obtainable in a separable form by reaction of **20** with *N*-tosyl-L-prolinyl chloride (Atarashi et al. 1987). (For further possible synthesis routes for **26**, see Sect. B.III.)

Japanese researchers have recently reported an interesting modification of the Gould-Jacobs reaction. By reaction of the isothiocyanates **28** with the diethyl ester of malonic acid in the presence of sodium hydride, the sodium salts of aminomercaptomalonic esters **29** were obtained and could be alkylated to **30** using 4-methoxybenzyl chloride or chloromethyl methyl ether. After the usual cyclization of **30** to **31**, the acid-labile protective groups attached to the sulfur atom were cleaved off, thus obtaining the central intermediates **32** (Fig. 4).

The resulting cyclocondensation of **32** with 2-haloacetals, 1,2-dihaloalkanes, 1,1-dihaloalkanes, 1,3-dihaloketones and amines, or 2,3-

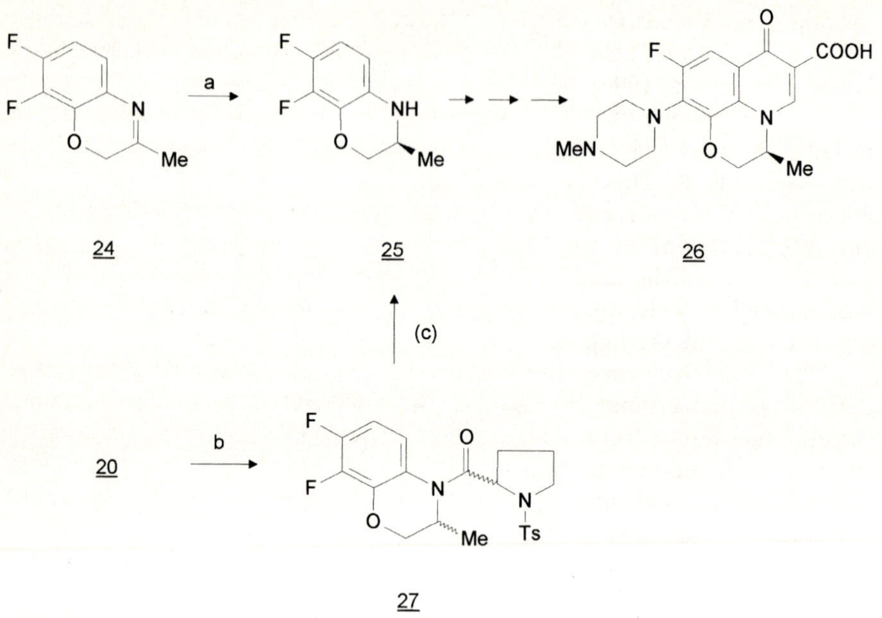

Fig. 3. Synthesis of levofloxacin **26**. *a*, Na-BH [N COO / COOCH$_2$CHMe$_2$]$_3$; *b*, *N*-tosyl-L-prolinyl chloride, pyridine; *c*, chromatography, OH$^-$

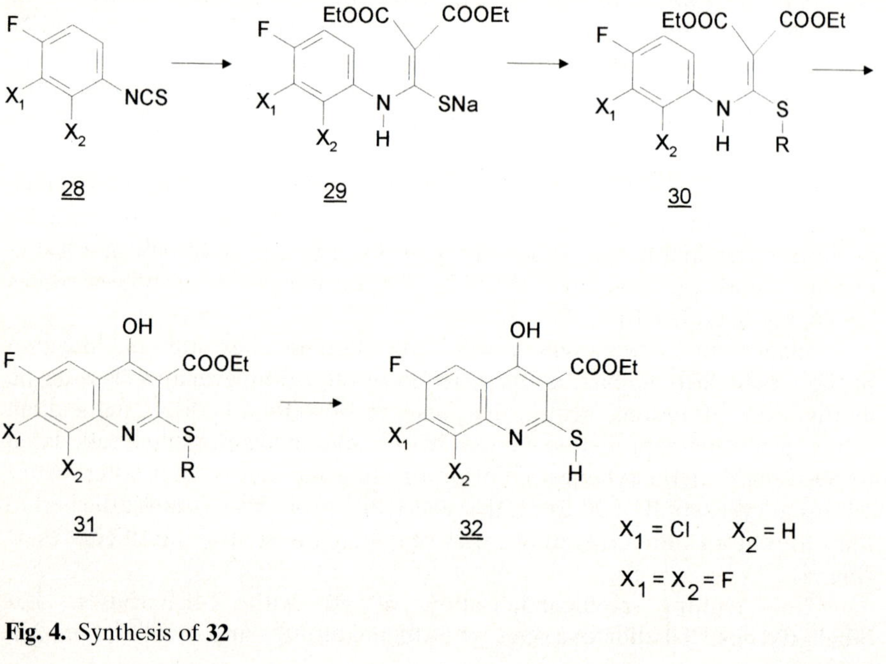

Fig. 4. Synthesis of **32**

dihalopropanol leads to the tri- or tetracyclic quinolone carboxylic acids **33** (SEGAWA et al. 1992a), **34** (SEGAWA et al. 1992a), **35** (SEGAWA et al. 1992b), **36** (JINBO et al. 1993a, 1994), and **37** (JINBO et al. 1993b; Fig. 5).

In the following outlined synthesis of the fluoroquinolone development product **46** (KB-5246; Fig. 6), TAGUCHI et al. (1992) start from 2,3,4-trifluoroaniline **38**. This is converted to the dithiocarbamate **39**. The tricyclic malonic ester derivative **43** is obtained from this via the intermediates **40**, **41**, and **42**. Treatment of **43** with polyphosphoric acid yields the tetracyclic quinolone carboxylic acid ester **44**. Hydrolysis of the ester **44** with fuming sulfuric acid to yield quinolone carboxylic acid **45** and its reaction with 1-methylpiperazine then gives the bioactive product **46**.

The Gould-Jacobs reaction sequence is a versatile method for annelating pyridone rings to aromatic systems. The advantage of the procedure is that one can frequently start from relatively easily available compounds and arrive at the desired quinolone carboxylic acid derivatives in a simple experimental procedure. The disadvantage of the thermal procedure is the high temperature required for cyclocondensation and the formation of ring-closure isomers substituted in position 3 of 2,6-unsubstituted arylaminomethylenemalonic

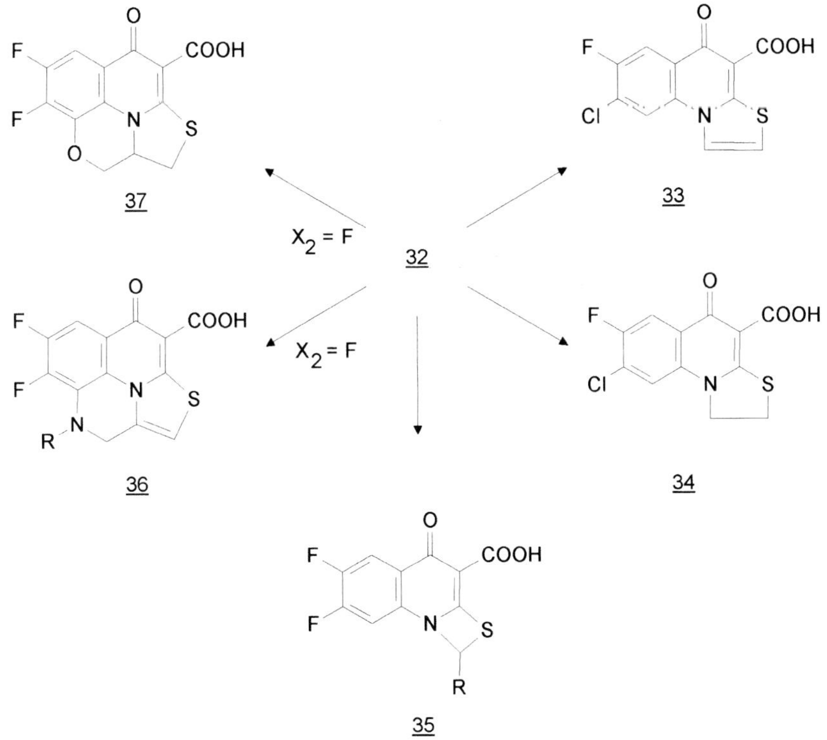

Fig. 5. Cyclocondensation of **32**

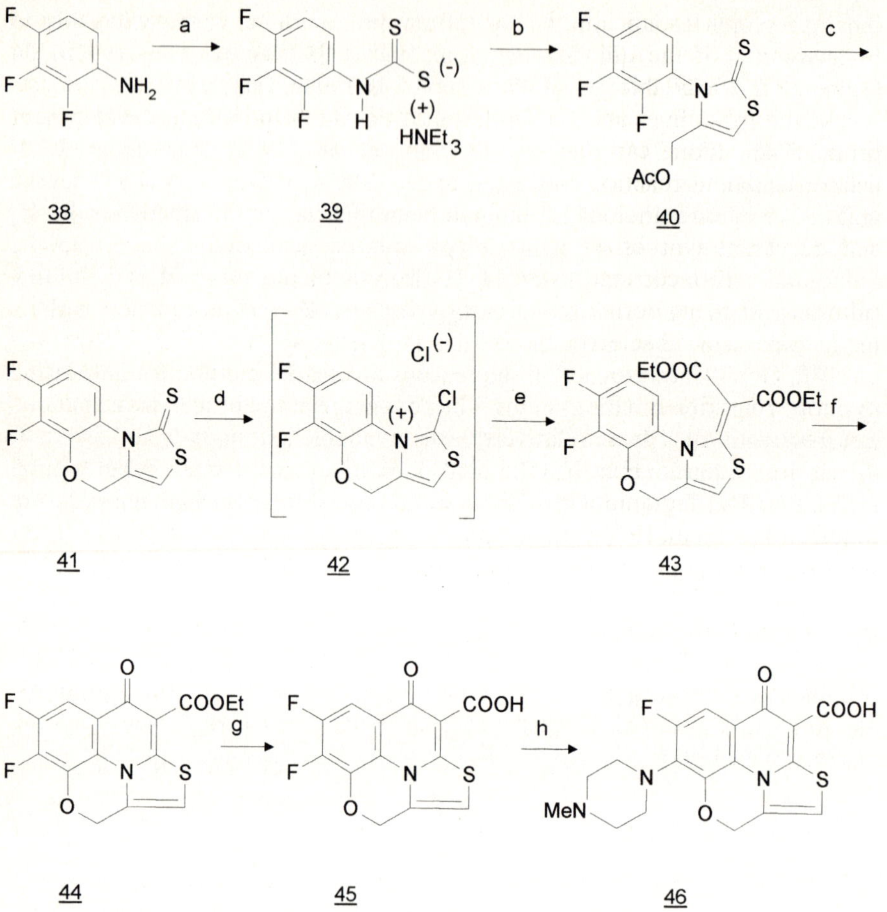

Fig. 6. Synthesis of **46** (KB-5246). *a*, CS_2, NEt_3; *b*, $AcOCH_2COCH_2Cl$/AcOEt-HCl/AcOEt; *c*, aqueous KOH/EtOH; *d*, $ClCO_2CC1_3$/toluene; *e*, $CH_2(CO_2Et)_2/NEt_3/CH_3CN$; *f*, polyphosphoric acid (PPA); *g*, fuming H_2SO_4; *h*, 1-methylpiperazine

esters (Agui et al. 1975; Hirose et al. 1982; Sauter et al. 1988a). The ester of 4-hydroxy-quinoline-3-carboxylic acids, e.g., **17**, cannot be N-alkylated or N-arylated with secondary and tertiary alkyl halides, cyclic alkyl halides, or the less reactive aryl halides (Grohe, unpublished results; Grohe 1993). Only in the case of 4-nitrofluorobenzene or 2,4-dinitrochlorobenzene does arylation at the position 1 nitrogen proceed, and even only a moderate yield is obtained (Radl and Zikan 1989).

In the case of the catalytic variant of the Gould-Jacobs reaction, apart from special cases (Ishizaki et al. 1985; Bridges et al. 1990), neither the cyclopropyl group nor the 2,4-difluorophenyl group can be introduced to position 1 of the quinolone ring structure. This is because no suitable methods are available yet to synthesize the necessary starting products *N*-cyclopropylanilines and *N*-(2,4-difluorophenyl) anilines. In the case of synthesis of quinolone carboxylic acids such as ofloxacin types, the competing cycloaracylation reaction should also be taken into consideration (see B.III). The fluroquinolone commercial and development products listed in Table 2 can be synthesized by the Gould-Jacobs reaction.

The Gould-Jacobs reaction has been used in recent years for the synthesis of many fused-ring derivatives of 4-pyridone-3-carboxylic acid. The following list includes a small selection: pyrido[1,2,3-de][1,4]benzothiazines (CECCHETTI et al. 1993), 7-(2-thiazolyl)-quinolones (ZHANG et al. 1991), 2-(4-pyridinyl)-thieno[2,3]pyridones (BACON and DAUM 1991), 1,7-naphthyridones (RADL and HRADIL 1991), pyrazolo[3,4-f]quinolones (SCHWAN et al. 1983), 1,6- and 1,8-naphthyridones (HIROSE et al. 1982), 1,5-naphthyridones (HEINDL et al. 1977), thieno[3,2-g]quinolones (SAUTER et al. 1988a), thiazolo[4,5-g]quinolones (SUZUKI et al. 1979), oxisoxazolo[5,4-b]pyridones (CHIARINO et al. 1988), pyrrolo[3,2-b]pyridones (BAYOMI et al. 1985; ABDALLA and SOWELL 1990), 7-[(1-imidazolyl)phenylmethyl]quinolones (FRIGOLA et al. 1987), pyrrolo[3,4-b]pyridones (TOJA et al. 1986a), thieno[2,3-f]quinolones (SAUTER et al. 1988b), pyrido[3,2,1-gh][1,7]phenanthrolones (SAUTER et al. 1989), pyrrolo[3,2,1-ij]quinolones (PARIKH et al. 1988; ISHIKAWA et al. 1990), benzo[ij]quinolizinones (ISHIKAWA et al. 1989), pyridoquinolones, and pyridophenanthrolones (JORDIS et al. 1988).

II. Dieckmann Cyclization of Diesters

A suitable procedure for the synthesis of aza- and diazaquinolone carboxylic acid derivatives was published by PESSON et al. (1974a). It is based on the base-catalyzed intramolecular cyclization of diesters. The process is described and exemplified by the alternative synthesis route for enoxacin **10** (MIYAMOTO et al. 1987a).

The ethyl ester of 2,6-dichloro-5-fluoronicotinic acid **47** reacts relatively regioselectively at position 6 with *N*-acetylpiperazine to form 6-piperazinyl compound **48**. The treatment of **48** with the ethyl ester of 3-ethylaminopropionic acid leads to the diester **49**, which in turn undergoes intramolecular ring closure with potassium tert-butylate to form the ethyl ester of tetrahydronaphthyridine carboxylic acid **50**. The dehydrogenation of **50** by means of chloranil yields the ethyl ester of naphthyridone carboxylic acid **51**, which can then be hydrolyzed to the fluoroquinolone enoxacin **10** (Fig. 7).

During the dehydrogenation of tetrahydroquinoline carboxylic acid esters, e.g., **50** to **51**, bromination in chloroform is frequently used, followed by dehydrohalogenation by means of triethylamine in ethanol (PESSON and

Table 2. Commercially available and experimental fluoroquinolones synthesized by the Gould-Jacobs reaction

O, F, COOH, R_1, X, N, R

Common name	X	R	R_1	Reference
Binfloxacin	CH	Et	N N—	Jefson and McGuirk (1988, 1989)
Benofloxacin	C—$(CH_2)_2$—CHMe		MeN N—	Otsubo et al. (1982)
Esafloxacin	N	Et	H_2N N—	Egawa et al. (1984)

Fleroxacin	CF	$F(CH_2)_2$—	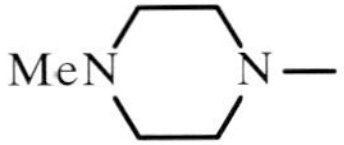 MeN N—	Irikura et al. (1981)
Irloxacin	CH	Et	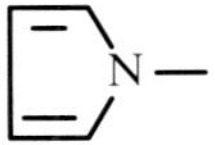 N—	Stefancich et al. (1985)
Lomefloxacin	CF	Et	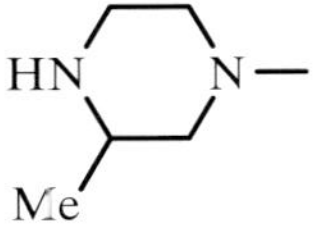 HN N— Me	Itoh et al. (1985)
Marbofloxacin	C—OCH_2—N—Me		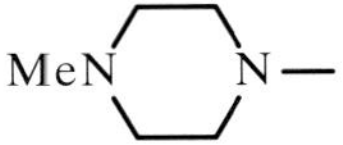 MeN N—	Aoki et al. (1988)
Nadifloxacin	C—$(CH_2)_2$—CHMe		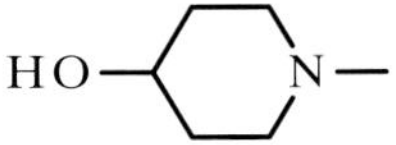 HO— N—	Ishikawa et al. (1989)
Rufloxacin	C—SCH_2CH_2—		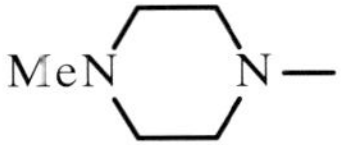 MeN N—	Cecchetti et al. (1987); Terni et al. (1988)

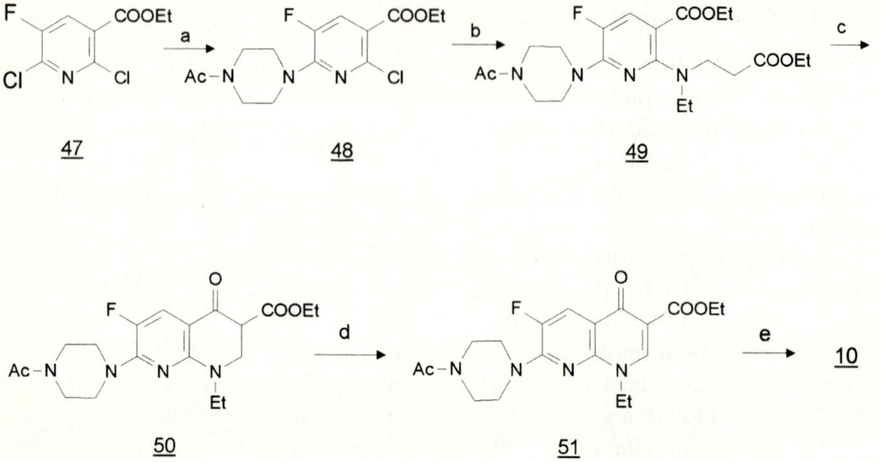

Fig. 7. Synthesis enoxacin **10**. *a*, *N*-acetylpiperazine, CH_3CN; *b*, $EtNH(CH_2)_2CO_2Et$, dimethylformamide (DMF), $NaHCO_3$, 120°C; *c*, $KOC(CH_3)_3$ or NaH, toluene; *d*, chloranil, pyridine; *e*, $OH^{(-)}$

Chabassier 1974; Pesson et al. 1974b). The double bond between carbons 2 and 3 can also be formed by heating with chloranil in ethanol or dioxane (Pesson et al. 1980) or by using thionyl chloride (Santilli et al. 1975).

We have not succeeded in applying the above reaction sequence, which is successful using **47** (K. Grohe, unpublished results), to ethyl 2,4-dichloro-5-fluorobenzoate or ethyl 2,4-dichloro-5-nitrobenzoate. However, back in 1979 we were able to synthesize for the first time 8-cyclopropyl-pyrido[2,3-d]pyrimidine-6-carboxylic acid according to the Pesson process (Grohe et al. 1981). In an analogous manner others (Nishimura and Matsumoto, 1987; Miyamoto et al. 1987b) later synthesized 1-cyclopropyl-1,8-naphthyridone carboxylic acid. However, the addition of the cyclopropyl residue to the 1 position of quinolone and azaquinolone carboxylic acids remains in the domain of the cycloaracylation procedure (see Sect. B.III).

Fused-ring 4-pyridone derivatives which have been produced in the last 20–25 years by means of the Dieckmann cyclization procedure include the following: pyrido[2,3-c]pyridazines (Boamah et al. 1990), benzo[h][1,6]-naphthyridines (Pesson et al. 1975), 1,6-naphthyridines (Strehlke 1977), pyrido[2,3-d]pyrimidines (Pesson et al. 1974c; Rufer and Schwarz 1977), pyrido[2,3-b]quinoxalines (Pesson et al. 1976), pyrido[2,3-e]asym. triazines (Pesson et al. 1980), pyrazo[5,4-h]-1,6-naphthyridines (Le Hao Dong et al. 1981), and pyrrolo[2,3-b]- and pyrrolo[3,4-b]pyridines (Toja et al. 1986a,b).

III. Cycloaracylation Procedure

The cycloaracylation procedure was developed by GROHE et al. (1980) in the mid-1970s and was published in a summarized form in GROHE (1993). Three variations of the cycloaracylation process are described below, exemplified by the synthesis of ciprofloxacin (Ciprobay), which was the first fluoroquinolone commercial product with a cyclopropyl group attached to the position 1 nitrogen of the ring.

2,4-Dichloro-5-fluorobenzoyl chloride **52** reacts with ethyl 3-cyclopropylaminoacrylate **53** in the presence of triethylamine to yield the acylated product **54**. Cyclization of **54** under the influence of potassium carbonate produces the ethyl ester of the quinolone carboxylic acid **55**, which is subsequently hydrolyzed to the carboxylic acid **56**. The reaction of **56** with piperazine gives rise to ciprofloxacin **13** with a good overall yield (GROHE and HEITZER 1987a; Fig. 8).

If 1-ethylpiperazine is introduced into **56** instead of piperazine, the veterinary antibacterial enrofloxacin (Baytril) is obtained (ALTREUTHER 1987; BAUDITZ 1987; SCHEER 1987).

The central intermediate **54** can also be obtained in an analogous manner by reacting **52** with ethyl 3-dimethylaminoacrylate **57**, thus yielding **58** and subsequent exchange of the dimethylamino group for the cyclopropylamino group (GROHE 1986, 1992). Another pathway for the synthesis of **54** is based on

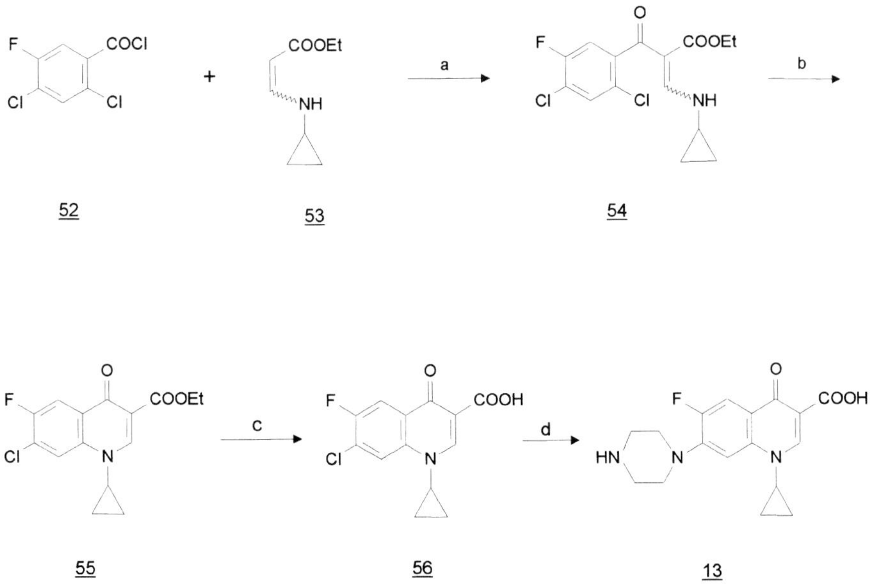

Fig. 8. Synthesis of ciprofloxacin **13**. *a*, NEt_3, toluene; *b*, K_2CO_3, dimethylformamide (DMF); *c*, KOH; *d*, piperazine, dimethyl sulfoxide (DMSO)

the use of ethyl 2,4-dichloro-5-fluorobenzoylacetate **59**. Reaction with ethyl orthoformate gives rise to the ethoxymethylene derivative **60**, which then reacts with cyclopropylamine to give a high yield of **54** (GROHE et al. 1983; Fig. 9).

New perspectives for the cycloaracylation procedure were opened by the reaction of the esters of 3-ethoxy-2-benzoylacrylic acid, which contain a highly halogenated aromatic ring, e.g. in **61**, with amino alcohols and diamines (Fig. 10). If, for example, **61** is treated with DL-2-amino-1-propanol, the aminoacryl ester derivative **62** is primarily obtained, which under the influence of potassium carbonate undergoes cyclocondensation and double ring closure to form the quinolone carboxylic acid ester **63** bridged between positions 1 and 8. Saponification of **63** leads to the tricyclic quinolone carboxylic acid **64**. If sodium fluoride is used for the cyclization of **62**, then the reaction proceeds only as far as the primary cyclization stage **65**. The second cyclization reaction, i.e., conversion of **65** to **63**, can then be mediated by potassium carbonate or sodium hydride, resulting in the bridging of positions 1 and 8.

The difluoroquinolone carboxylic acid **64** is converted to the fluoroquinolone ofloxacin **11** (Tarivid) by reaction with 1-methylpiperazine (GROHE and SCHRIEWER 1987; EGAWA et al. 1986). Ofloxacin **11** has an asymmetric center at position 3. Reaction of S-(+)- and R-(−)-2-amino-1-propanol instead of DL-2-amino-1-propanol with **61** yields the pure optical isomers of **11** in a relatively simple way (SCHRIEWER et al. 1987a; ATARASHI et al. 1987; MITSCHER et al. 1987). Microbiological tests have shown that the S-(−)-enantiomer of **11**, levofloxacin, is responsible for the antibacterial activity (HAYAKAWA et al. 1986).

The extended cycloaracylation process has been extensively used for the synthesis of polycyclic quinolone carboxylic acids, whereby the ethoxyacrylic

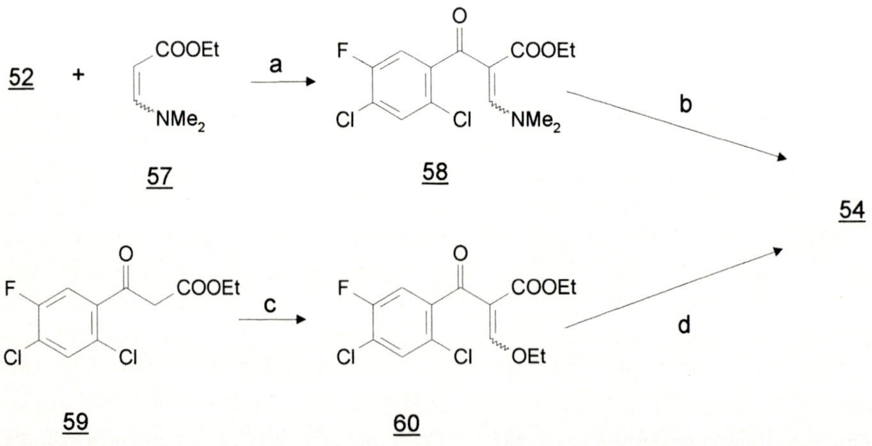

Fig. 9. Alternative methods of synthesis of the ciprofloxacin intermediate **54**. *a*, pyridine, toluene; *b*, c-$C_3H_5NH_2$, $CHCl_3$; *c*, $HC(OEt)_3$, Ac_2O; *d*, c-$C_3H_5NH_2$, EtOH

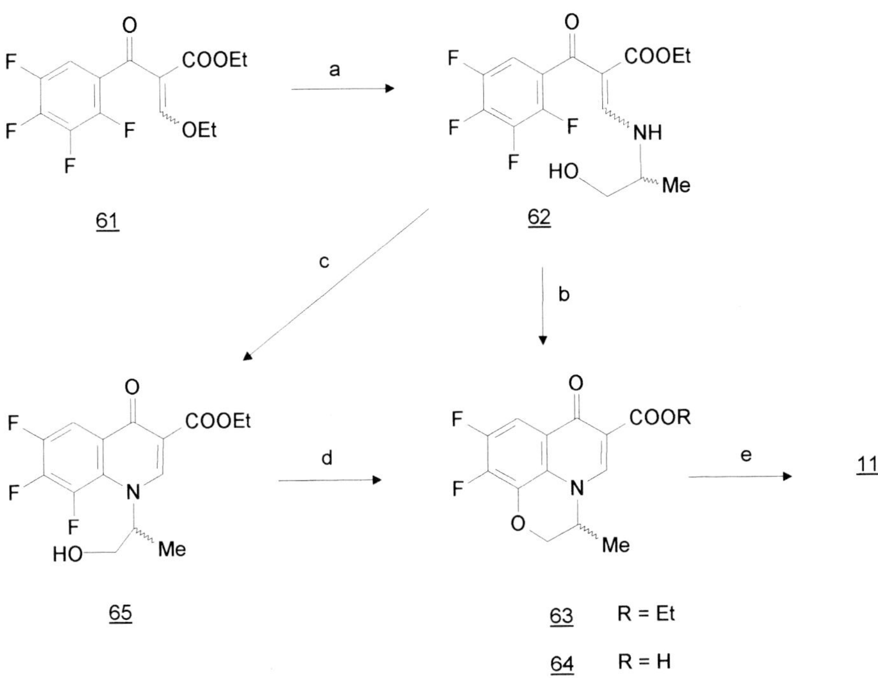

Fig. 10. Synthesis of ofloxacin **11**. *a*, H_2N-$CH(CH_3)CH_2OH$; *b*, K_2CO_3, dimethylformamide (DMF); *c*, NaF, DMF; *d*, K_2CO_3 or NaH, DMF; *e*, 1-methylpiperazine

acid ester **61**, which was first synthesized by GROHE et al. (1985) at the beginning of the 1980s from ethyl 2,3,4,5,-tetrafluorobenzoylacetate, has been frequently used as the starting material of choice.

On the basis of reaction types involving double or triple cyclization, **61**, for example, has been converted to the tricyclic or tetracyclic esters of quinolone carboxylic acids **66** (GROHE and SCHRIEWER 1987), **67** (GROHE and SCHRIEWER 1987), **68** (SCHRIEWER et al. 1987b), and **69** (GROHE and SCHRIEWER 1987) by reaction with 3-aminopropanol, *N*-methyl-ethylenediamine, *o*-aminophenol, or *N*-(2-hydroxyethyl)-ethylenediamine (Fig. 11).

Using 2-aminopropionaldehyde dimethyl acetal, **61** reacts in the usual manner to form the ester of quinolone carboxylic acid **70** by elimination of ethanol and subsequent primary cyclization with sodium fluoride. This is followed by acid hydrolysis to 1-(oxoalkyl)-quinolone carboxylic acid **71a** (SCHRIEWER et al. 1988). The corresponding ester of **71a**, designated **71b**, can also be synthesized in an alternative pathway by condensation of **61** with 3-amino-1-butene, ring closure to form 1-butenylquinolone carboxylic acid ester **72**, and subsequent ozonolysis (OKADA et al. 1991a). The second cyclization of **71a** or **71b** to pyridobenzoxazine carboxylic acid **73a** or to its corresponding

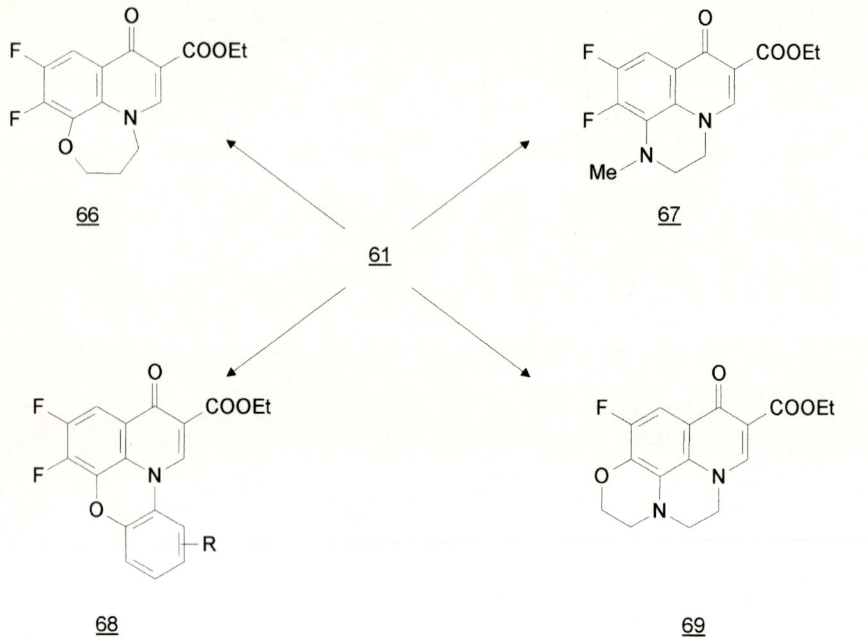

Fig. 11. Conversion of **61** to tricyclic or tetracyclic esters of quinolone carboxylic acids

ethyl ester **73b** can be carried out with alcoholic potassium hydroxide or sodium hydride. The pyridobenzoxazine carboxylic acid **73a**, obtained by acid hydrolysis of **73b** or directly from **71a**, subsequently reacts with 1-methylpiperazine, resulting in the exchange of the fluorine atom at position 10 for the piperazinyl group to form 2-3-dehydroofloxacin **74** (SCHRIEWER et al. 1988a; AUGERI et al. 1990; Fig. 12).

A further synthesis procedure for **71b** (Fig. 13) has recently been reported. The starting compound is the ester of 6,7,8-trifluoroquinolone carboxylic acid **75**, which is easily produced from compound **61** by the cycloaracylation process (SCHRIEWER and GROHE 1988) or from 2,3,4-trifluoroaniline by the Gould-Jacobs reaction (OKADA et al. 1991a). Compound **75** reacts with benzhydryl 2-bromopropionate to form the N-alkylated derivative **76**, which is hydolyzed to the carboxylic acid **77** by trifluoroacetic acid. The acid chloride **78**, which is obtained in the usual manner from the reaction of **77** with oxalyl chloride, is reduced to oxoalkylquinolone carboxylic acid ester **71b** with tributylstannic hydride. In addition to 2,3-dehydroofloxacin **74**, the sulfur analogue, with a sulfur atom replacing oxygen at position 1 of the ring system, was also produced by this process (OKADA et al. 1991a).

A structurally interesting fluoroquinolone is 9-fluoro-3-methylene-10-(4-methyl-1-piperazinyl)-7-oxo-2,3-dihydro-7H-pyrido[1,2,3-de][1,4]benzoxazine-6-carboxylic acid **79**. This compound, which is an isomer of 2,3-

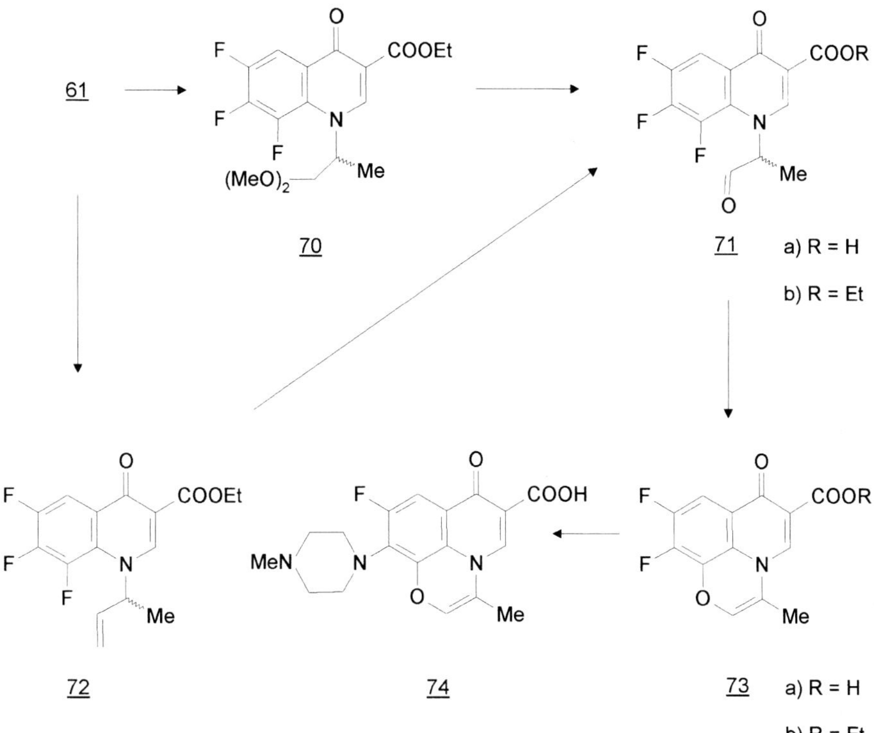

Fig. 12. Synthesis of 2–3-dehydroofloxacin **74**

Fig. 13. Synthesis of **71b**

dehydroofloxacin, has been described by several research groups. Synthesis can be carried out by the cycloaracylation procedure starting from the ester of ethoxyacrylic acid **61** (OKADA et al. 1991b; Fig. 14). Using 3-aminooxetane, the corresponding aminoacrylate ester is obtained, which undergoes cyclization to form 1-oxetanylquinolone carboxylic acid ester **80**. The oxetane ring is opened by hydrogen chloride in ethanol to form the chlorohydrin **81**, which in turn, at higher temperatures under the influence of potassium fluoride in dimethylformamide, is converted to 9,10-difluoro-3-methylene-pyridobenzoxazine carboxylic acid ester **82**. After acid hydrolysis of the ester to the acid **83**, the compound **79** is obtained by treatment with 1-methylpiperazine. The chlorohydrin **81** is also an important intermediate for the 1-sulfur analogue of **79** (OKADA et al. 1991b).

RADL et al. (1991) reported a simple synthesis of **82** by reaction of 8-hydroxyquinolone carboxylic acid ester **84** with 3-bromopropyne **85**. The cata-

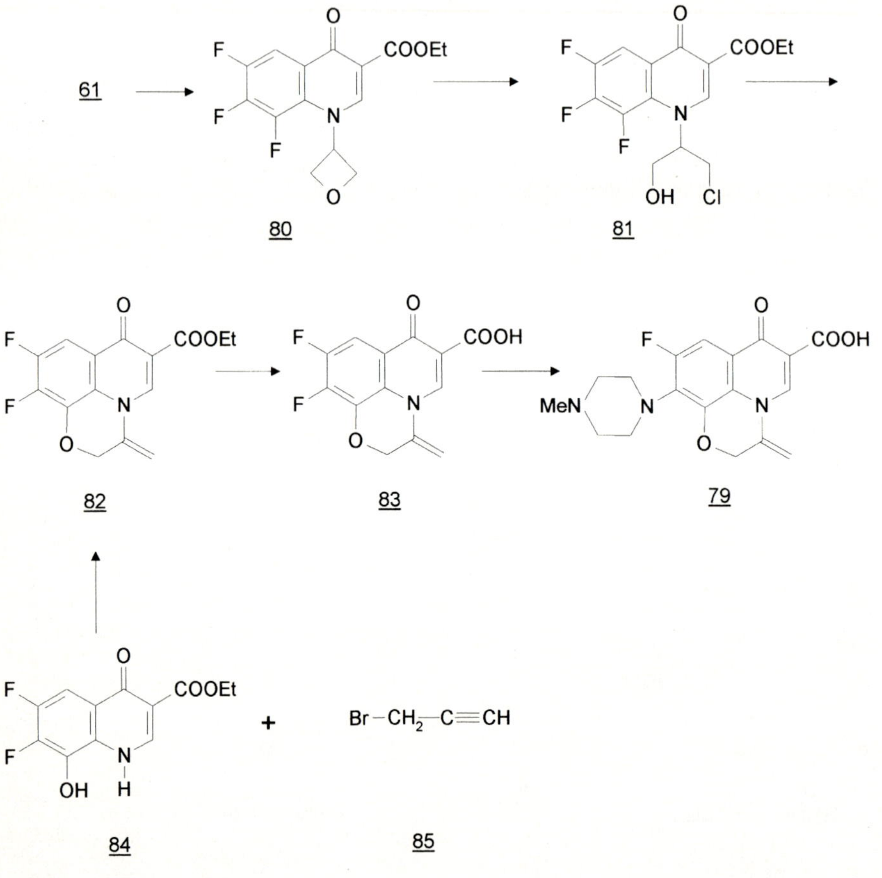

61 → 80 → 81 → 82 → 83 → 79

84 + 85 (Br–CH$_2$–C≡CH) → 82

Fig. 14. Synthesis of **79**

lytic hydrogenation of the exocyclic double bond of **82** leads directly to the ofloxacin precursor **63** (RADL et al. 1991).

The cycloaracylation procedure can also be used for the synthesis of quinolone carboxylic acids bridged between positions 1 and 2. The ester of benzo[c]quinolizine carboxylic acid **88** is obtained by the reaction of 2,4,5-trifluorobenzoyl chloride **86** with ethyl 2-pyridylacetate **87** (ZIEGLER et al. 1990; Fig. 15).

Ethyl 2,4-dichloro-5-fluorobenzoylacetate **59** is also a frequently used intermediate (GROHE et al. 1983). It reacts with the imidate ether **89a,b** in the presence of triethylamine to form the enamino esters **90a,b**, which undergo ring closure with sodium hydride, forming the esters of pyrrolo- or pyridoquinolone carboxylic acid **91a** and **91b** (CHU and CLAIBORNE 1987; Fig. 16).

Compound **59** reacts with 2-chlorobenzothiazole **92a** and 2-chlorobenzoxazole **92b** in the presence of 2 mol sodium hydride to form the esters of benzthiazolo- **93a** and benzoxazolo[3,2-a]quinolone carboxylic acid **93b**, respectively (CHU et al. 1986; CHU 1985a; Fig. 17).

Ethyl 3-(2,6-dichloro-5-fluoro-3-pyridyl)-3-oxopropionate **94** is first converted to the ketene dimethyl dithioacetal derivative **95** in a similar manner by carbon disulfide and sodium hydride. This is then reacted with 2-amino-1-ethanthiol, 2-amino-1-propanol, or *N*-methylethylenediamine **96a–c** to yield the intermediates **97a–c**, respectively, which are capable of undergoing cyclocondensation. When cyclocondensation is then carried out under the

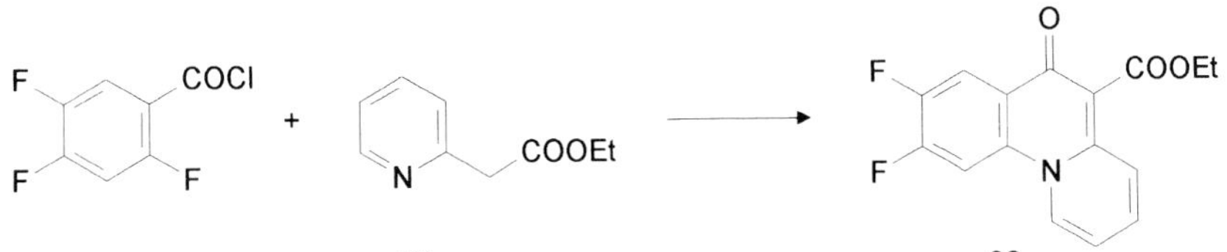

Fig. 15. Synthesis of the ester of benzo[c]quinolizine carboxylic acid **88**

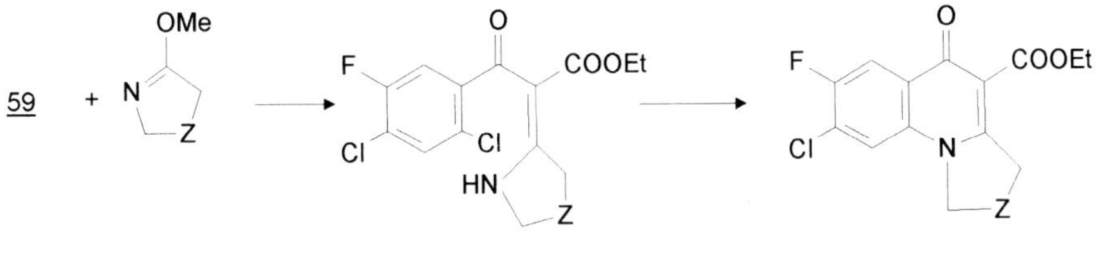

Fig. 16. Synthesis of esters of pyrrolo- and pyridoquinolone carboxylic acid **91a,b**

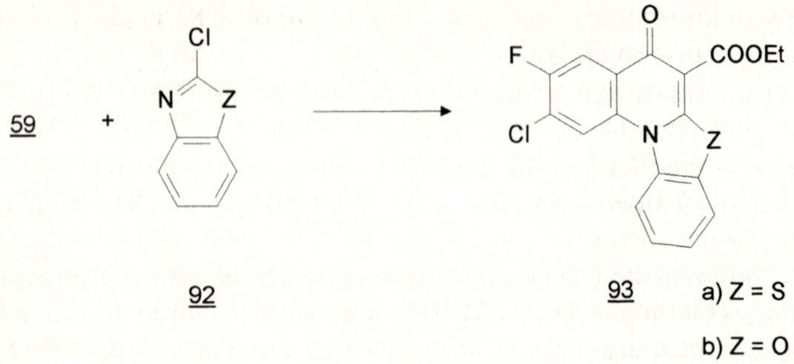

59 + 92 → 93 a) Z = S b) Z = O

Fig. 17. Synthesis of esters of benzthiazolo- **93a** and benzoxazolo-quinolone carboxylic acid **93b**

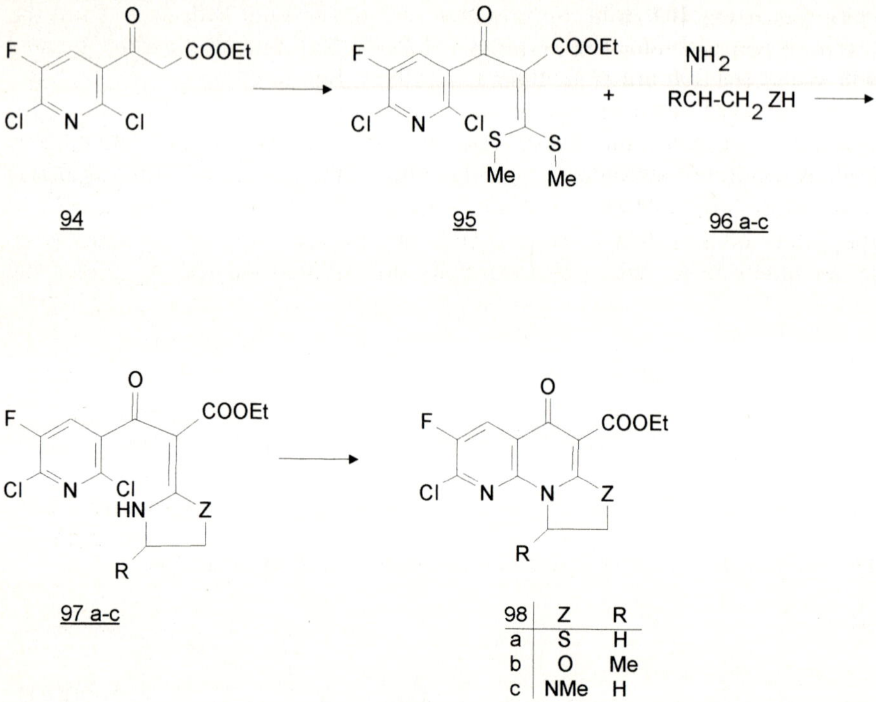

94 → 95 + 96 a-c →

97 a-c → 98

98	Z	R
a	S	H
b	O	Me
c	NMe	H

Fig. 18. Synthesis of ethyl esters of the thiazolo- **98a**, oxazolo- **98b**, and imidazolo[3,2-a][1,8]naphthyridine **98c** carboxylic acids

usual reaction conditions, the ethyl esters of the thiazolo-, oxazolo-, and imidazolo[3,2-a][1,8]naphthyridine carboxylic acids **98a–c** are produced (KONDO et al. 1990; Fig. 18).

Another possible way to bridge positions 1 and 2 is to convert the quinolone carboxylic acid ester **99**, which results from the reaction of **61** with

3-(1-aminocyclopropane)-propanol, to the iodide **100** and then allow it to undergo cyclization to the enol ester **101** under Grignard conditions. Reaction of this with sodium hydride and phenylselenium chloride and subsequent oxidation gives rise to the spirocyclic quinolone carboxylic acid ester **102** (SCHROEDER and KIELY 1988; Fig. 19).

The esters of quinolone carboxylic acids **88**, **91**, **93**, **98**, and **102** can be hydrolyzed to the corresponding carboxylic acids under acid or alkaline conditions. Exchange of the halogen atom attached to position 7 with cyclic amine residues leads to the antibacterial active agents.

The cycloaracylation procedure also forms the basis for the synthesis of quinolone carboxylic acid derivatives which are bridged between positions 2 and 3 of the ring structure. For example, by reacting the ethyl ester of 2,4,5-trifluorobenzoylacetate **103** (GROHE et al. 1985) with sodium hydride, cyclopropyl- or 4-fluorophenyl-isothiocyanate, and methyl iodide, the enamino esters **104**, which are suitable for the intramolecular nucleophilic cyclization, are obtained. The products **105** are obtained after cyclization, and the sulfoxides **106** are obtained by oxidation with 3-chloroperbenzoic acid. These are then converted to the mercapto derivatives **107** by reaction with sodium hydrogen sulfide. Compound **107** undergoes cyclization with hydroxylamine-*O*-sulfonic acid to the difluoro-isothiazolo[5,4-b]quinoline-3,4-diones **108** (CHU 1990; Fig. 20).

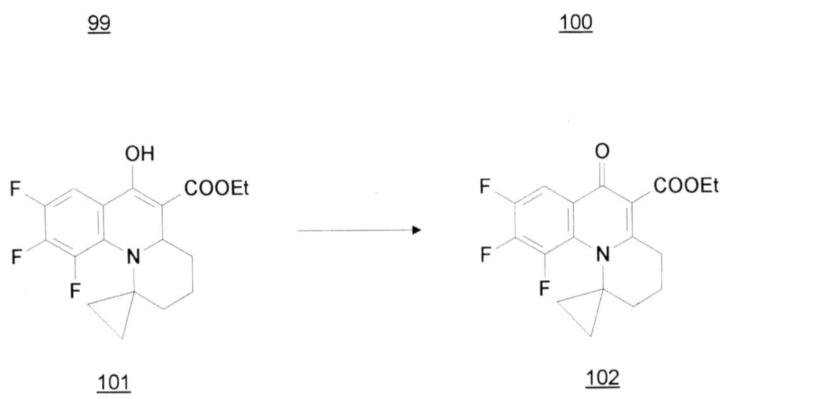

Fig. 19. Synthesis of spirocyclic quinolone carboxylic acid ester **102**

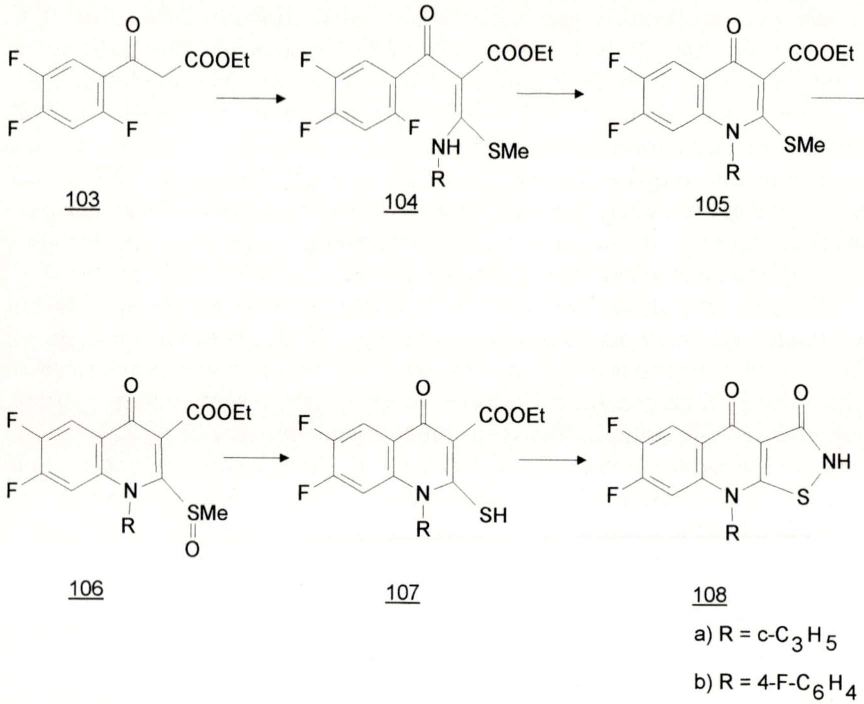

a) R = c-C_3H_5

b) R = 4-F-C_6H_4

Fig. 20. Synthesis of difluro-isothiazolo[5,4-b]quinoline-3,4-diones **108**

By exchanging the fluorine atom at position 7 in **108a** or **108b** with piperazine or pyrrolidine residues as appropriate, derivatives with excellent in vitro antibacterial activity are obtained. Aza analogues of **108** can be produced by a similar reaction pathway (CHU and CLAIBORNE 1990).

The cycloaracylation procedure permits a wide variety of substituents in the 1, 2, 3, 5–7, and 8 positions of the quinolone ring structure. For example, practically any desired residue can be introduced into position 1 of quinolone and azaquinolone carboxylic acids in the place of hydrogen (SCHRIEWER and GROHE 1988). Of special importance is the synthesis of 1-sec-alkyl-(DOMAGALIA et al. 1988; MORAN et al. 1989), 1-tert-alkyl-(GROHE et al. 1981; REMUZON et al. 1991; BOUZARD et al. 1992; DI CESARE et al. 1992), 1-cycloalkyl-(GROHE et al. 1981; SCHRIEWER et al. 1987c, 1988b; BOUZARD et al. 1989; ATARASHI et al. 1993), 1-aryl- (GROHE et al. 1981; CHU et al. 1985; XIAO et al. 1989; LIN and GUO 1992), and 1-amino-(aza)quinolone carboxylic acids (GROHE et al. 1981, 1986; GROHE and HEITZER 1987b; CHU 1985b), which cannot be synthesized by the Gould-Jacobs reaction. The cycloaracylation procedure is now the method of choice for such quinolone carboxylic acids substituted at position 1. This is especially true for the synthesis of highly active antibacterial 1-cyclopropyl- and 1-(2,4-difluorophenyl)quinolone carboxylic acids. Moreover, numerous quinolone carboxylic acids with new substituents in other positions are also being synthesized (MIYAMOTO et al. 1990; SANCHEZ et al. 1992; REMUZON et al. 1992; WENTLAND et al. 1993; LABORDE et al. 1993).

Methodological problems with the cycloaracylation procedure alone occur only when two different products can arise from the central cyclic condensation step. This may be the case, for example, when two leaving groups are present in positions 2 and 6 to the carbonyl group on the aromatic ring (GROHE 1993).

Furthermore, the extended cycloaracylation procedure is particularly suitable for the synthesis of 1,8-, 1,2-, and 2,3-bridged quinolone carboxylic acid derivatives. The method is often simpler, economically more advantageous, and preferable for the production of pure enantiomers from racemic mixtures of active quinolone compounds compared to alternative procedures, as demonstrated by the example of levofloxacin **26**. Only the 1,8-bridged quinolones of the types flumequine and benofloxacin cannot yet be obtained by the cycloaracylation procedure.

The summary in Table 3 of commercially available drugs and research products which have been produced by one of the described variant pathways demonstrates the importance of the cycloaracylation procedure.

IV. Biere and Seelen Approach

The Michael addition of esters of *o*-aminobenzoic acid **109** to esters of acetylene dicarboxylic acids **110** gives rise to the enamino esters **111**. These compounds undergo cyclization to the quinolone dicarboxylic acid esters **112** in the presence of a strong base such as sodium hydride or potassium tert-butoxide in dimethylformamide. The ester acids **113** or the dicarboxylic acids **114** are obtained after regioselective or complete alkaline hydrolysis of **112**. The carboxyl group at position 2 is selectively eliminated by thermal decarboxylation of **113** or **114**. The quinolone carboxylic acid esters **115** or the corresponding carboxylic acids **116** thus formed can then be N-alkylated (BIERE and SEELEN 1976; Fig. 21).

The Michael addition products **111** can also be obtained by reacting isatoic anhydrides with **110** (TAYLOR and HEINDEL 1967). According to the above reaction scheme, thieno[3,2-b]pyridone carboxylic acids would also be produced (BARKER et al. 1978).

The method of Biere and Seelen has not yet been used for the synthesis of fluoroquinolones. The disadvantages of this method are the expensive starting materials and the decarboxylation reaction, which only takes place at very high temperatures.

V. Isatoic Anhydride Procedure

STAIGER and MILLER (1959) discovered a further general synthesis of 4-quinolone-3-carboxylic acids. They reacted isatoic anhydride **117** with the

Table 3. Commercially available and experimental fluoroquinolones produced by the cycloaracylation procedure

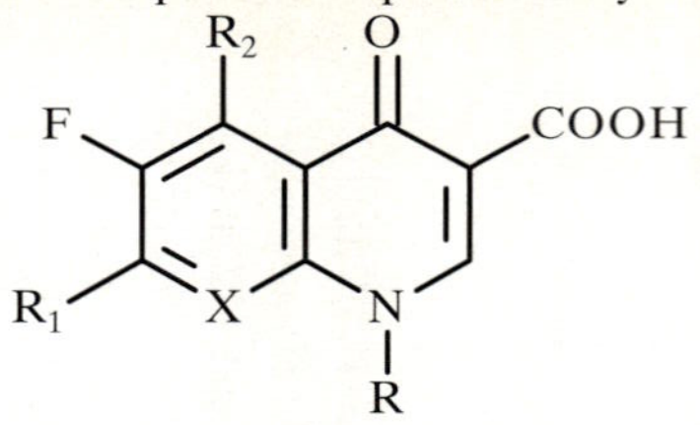

R₂ O F COOH R₁ X N R

Common name/no.	X	R	R_1	R_2	Reference
Ciprofloxacin	CH		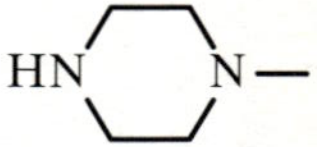 HN N—	H	GROHE et al. (1983)
Enrofloxacin	CH		 EtN N—	H	GROHE et al. (1983); GROHE (1986, 1992)
Tosufloxacin	N	2,4—$F_2C_6H_3$	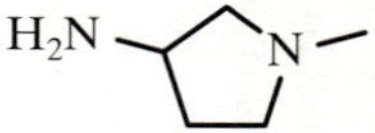 H_2N N—	H	NARITA et al. (1986d)
Danofloxacin	CH		 MeN N—	H	BRAISH and FOX (1990)
Sparfloxacin	CF		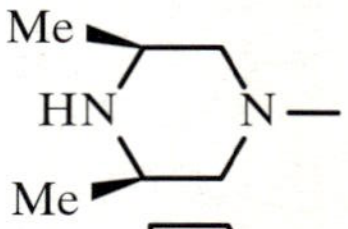 Me HN N— Me	NH_2	MIYAMOTO et al. (1990)
Clinafloxacin	CCl		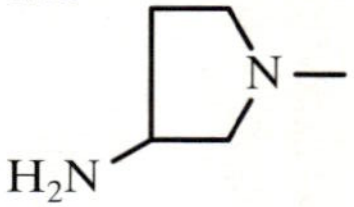 N— H_2N	H	PETERSEN et al. (1986, 1991); IRIKURA et al. (1987)
BAY 12-8039	C—OMe		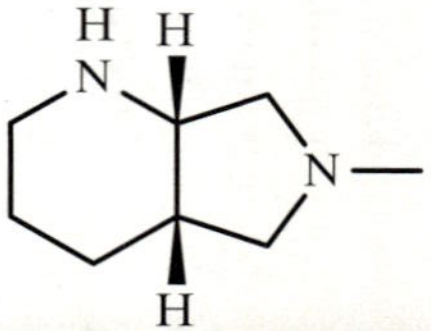 H N H N— H	H	PETERSEN et al. (1988, 1992); DALHOFF et al. (1996)

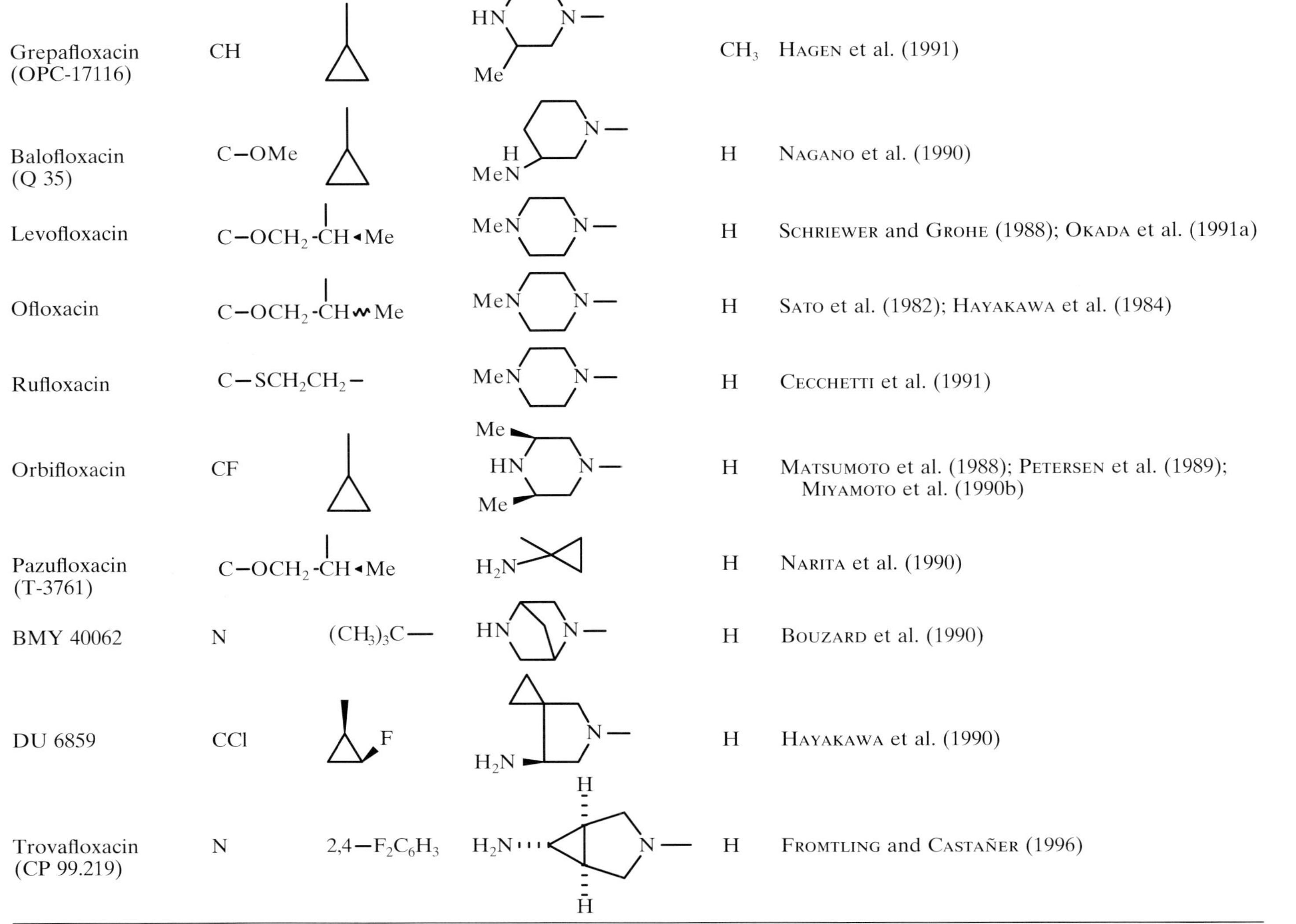

Grepafloxacin (OPC-17116)	CH			CH_3	HAGEN et al. (1991)
Balofloxacin (Q 35)	C—OMe			H	NAGANO et al. (1990)
Levofloxacin	C—OCH_2-CH◂Me			H	SCHRIEWER and GROHE (1988); OKADA et al. (1991a)
Ofloxacin	C—OCH_2-CH∿Me			H	SATO et al. (1982); HAYAKAWA et al. (1984)
Rufloxacin	C—SCH_2CH_2—			H	CECCHETTI et al. (1991)
Orbifloxacin	CF			H	MATSUMOTO et al. (1988); PETERSEN et al. (1989); MIYAMOTO et al. (1990b)
Pazufloxacin (T-3761)	C—OCH_2-CH◂Me			H	NARITA et al. (1990)
BMY 40062	N	$(CH_3)_3C$—		H	BOUZARD et al. (1990)
DU 6859	CCl			H	HAYAKAWA et al. (1990)
Trovafloxacin (CP 99.219)	N	2,4—$F_2C_6H_3$		H	FROMTLING and CASTAÑER (1996)

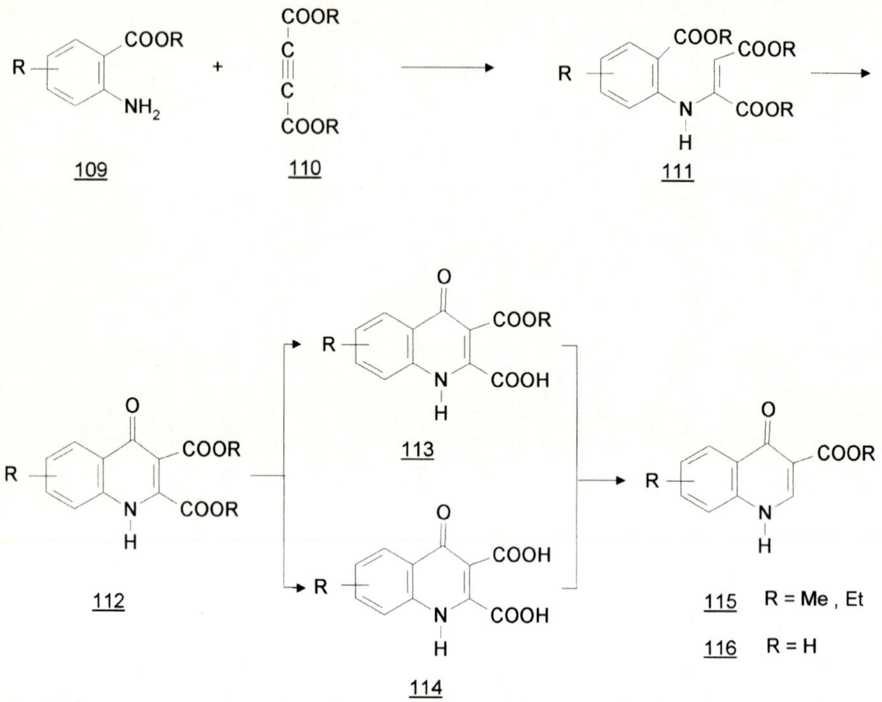

Fig. 21. Synthesis of quinolone carboxylic acid esters **115** and the corresponding carboxylic acids **116**

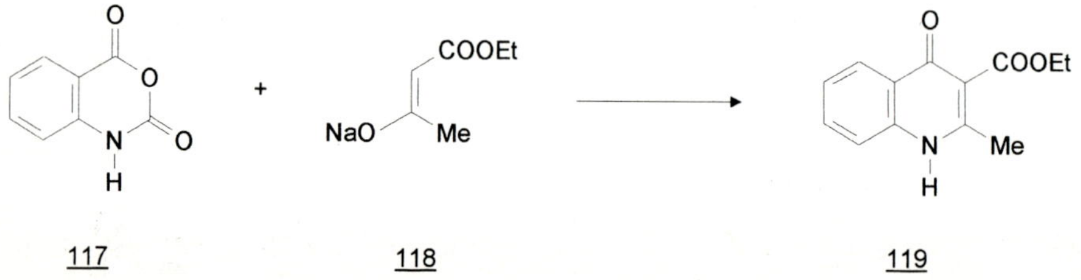

Fig. 22. Synthesis of the ethyl ester of 2-methylquinolone-3-carboxylic acid **119**

sodium salt of ethyl acetoacetate **118** and obtained a good yield of the ethyl ester of 2-methylquinolone-3-carboxylic acid **119** (Fig. 22).

The *N*-(4-methoxyphenyl)-isatoic anhydride reacts in a similar manner with the sodium salt of the benzoylacetic acid ester to form the ester of 1-(4-methoxyphenyl)-2-phenylquinolone carboxylic acid (BELL et al. 1970). Furthermore, derivatives of nalidixic acid (BRUNDAGE and LESHER 1976) and isomers of oxolinic acid (MITSCHER et al. 1979) were obtained when the corresponding isatoic anhydride analogues were reacted with the sodium salt of ethyl 3-hydroxyacrylate instead of with **118**.

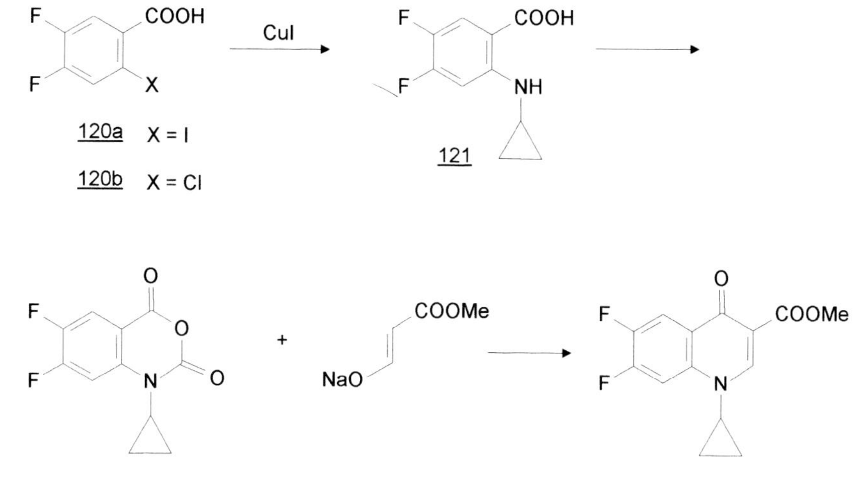

Fig. 23. Synthesis of the methyl ester of 1-cyclopropyl-6,7-difluoroquinolone-3-carboxylic acid **124**

Recently, the 2-cyclopropylamino-4,5-difluorobenzoic acid **121** was produced from the reaction of 2-iodo- or 2-chloro-4,5-difluorobenzoic acid **120a,b** with cyclopropylamine in the presence of Cu(I)iodide. This in turn gives rises to the substituted benzoxazine dione **122** by reaction with bis-(trichloromethyl)-carbonate. Compound **122** is then converted to the methyl ester of 1-cyclopropyl-6,7-difluoroquinolone-3-carboxylic acid **124** (Fig. 23) by treatment with the sodium salt of methyl 3-hydroxyacrylate **123** (O'NEILL 1990). Compound **124** is an important intermediate for fluoroquinolones of the ciprofloxacin type.

VI. Camps Quinolone Synthesis

The cyclocondensation of 2-formamido-acetophenone in alkaline medium to form 4-hydroxyquinoline was described by CAMPS in 1901. He obtained the ethyl ester of 4-quinolone-3-carboxylic acid **127a** and the corresponding free acid **127b** in a similar manner by heating the ethyl ester of 2-formamido-propynoic acid **125** with aqueous alcoholic sodium hydroxide, probably via the intermediary 2-formamido-benzoylacetic acid ester **126** (Fig. 24). The free acid **127b** was identical with kynurenic acid isolated by LIEBIG (1853, 1858) from the urine of dogs.

In the field of antibacterial quinolone research, the principle of the Camps reaction was first used at the end of the 1960s for synthesis in the area of the oxolinic acid compounds (OKUMURA et al. 1974a,b). In these experiments, the methine group was introduced between the amino and the active methylene group of the 2-aminoacetophenone derivative **128** by means of the triethyl ester of orthoformic acid **129** to form **130** (Fig. 25).

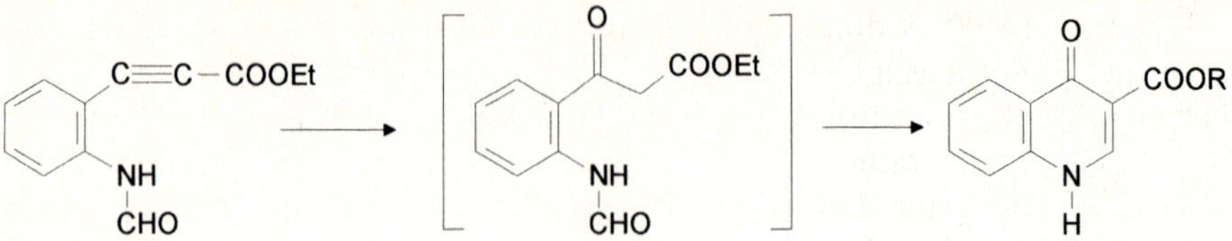

125 126 127a R = Et

127b R = H

Fig. 24. Synthesis of the ethyl ester of 4-quinolone-3-carboxylic acid **127a** and the corresponding free acid **127b**

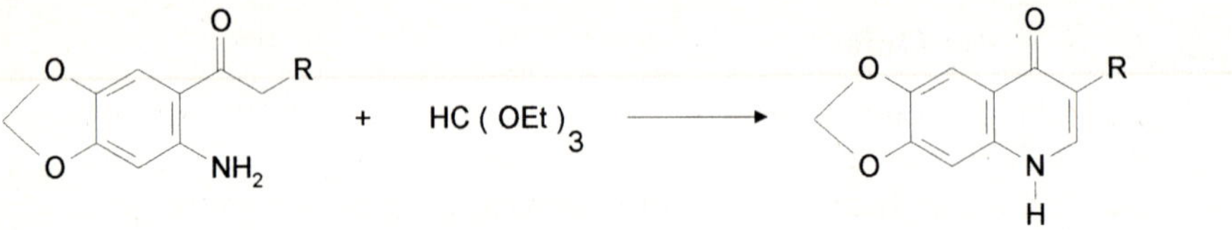

128 129 130 R = CN , CO_2CMe_3

Fig. 25. Synthesis of the quinolone carboxylic acid derivatives **130**

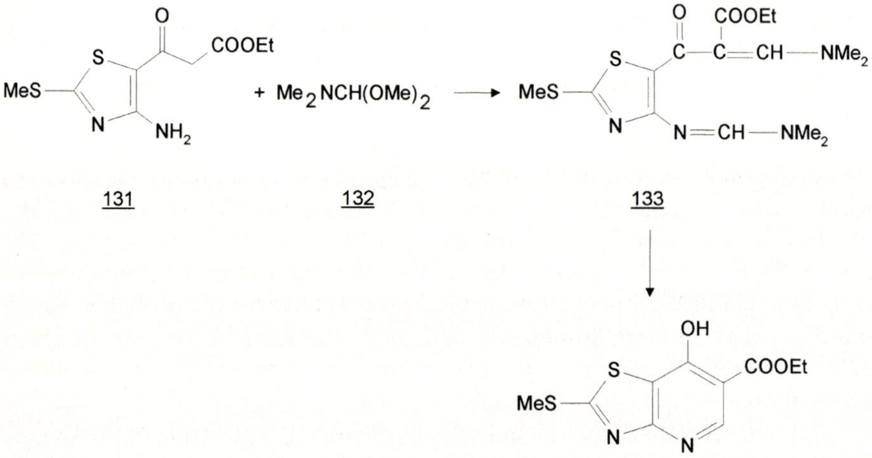

134

Fig. 26. Synthesis of thiazolo[4,5-b]pyridine carboxylic acid ester **134**

The reactive *N,N*-dimethylformamide dialkyl acetal was later found to be particularly useful as a C-1 building block during cyclization (STANOVNIK and TISLER 1974). Various research groups (HAYAKAWA and TANAKA 1984; LEYSEN et al. 1984) demonstrated examples of the formation of thiazolo[4,5-b]pyridine carboxylic acid ester **134** (Fig. 26) by the reaction of the ethyl ester of aminothiazole carboxylic acid **131** with *N,N*-dimethylformamide dimethyl acetal **132** via the intermediate **133**.

In the field of fluoroquinolone research, the modified Camps reaction has recently been used for the alternative synthesis of tosufloxacin (TODO et al. 1987). In the cyclization step of the reaction pathway, ethyl nicotinoylacetate **135** is reacted with **132** to form naphthyridone carboxylic acid ester **136a**. After hydrolysis of the ester **136a** to the corresponding acid **136b** and its subsequent substitution with 3-aminopyrrolidine at position 7, tosufloxacin **137** is obtained (Fig. 27; see also Table 3).

VII. Meth-Cohn Quinolone Synthesis

In Meth-Cohn quinolone synthesis (METH-COHN 1986), which was published in 1986, the Vilsmeier complex consisting of *N*-methylformanilide **138** and phosphoryl chloride is reacted with methylmalonyl chloride **139** to produce the quinolinium salt **140**, which is highly soluble in water and cannot be isolated. By treatment of **140** with ammonium hexafluorophosphate, the relatively insoluble 4-chloro-3-methoxycarbonyl-1-methylquinolinium hexafluorophosphate **141** is obtained. Compound **140** is converted to 1-methyl-4-quinolone-3-carboxylic acid **142** in almost quantitative yield by treatment with excess alkali followed by acidification (Fig. 28).

The limitations of this reaction, especially in terms of the synthesis of fluoroquinolones, have not yet been determined. The easy availability of appropriately substituted formanilides as starting reagents is required. Furthermore, it must be taken into consideration that ring-closure isomers may be

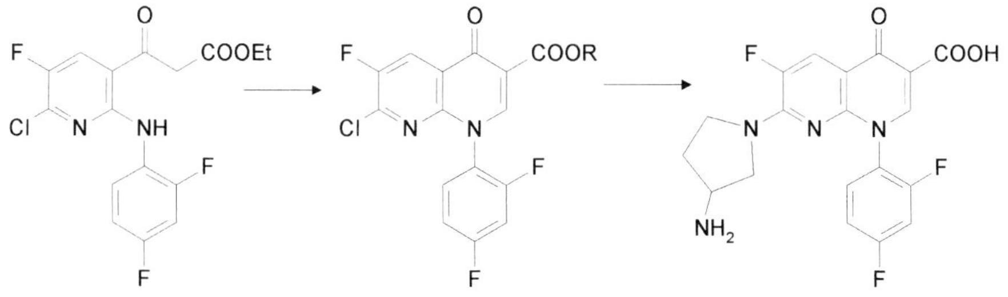

F O COOEt Cl N NH F F

135

F O COOR Cl N N F F

136a R = Et

136b R = H

F O COOH N N N F F NH$_2$

137

Fig. 27. Synthesis of tosufloxacin **137**

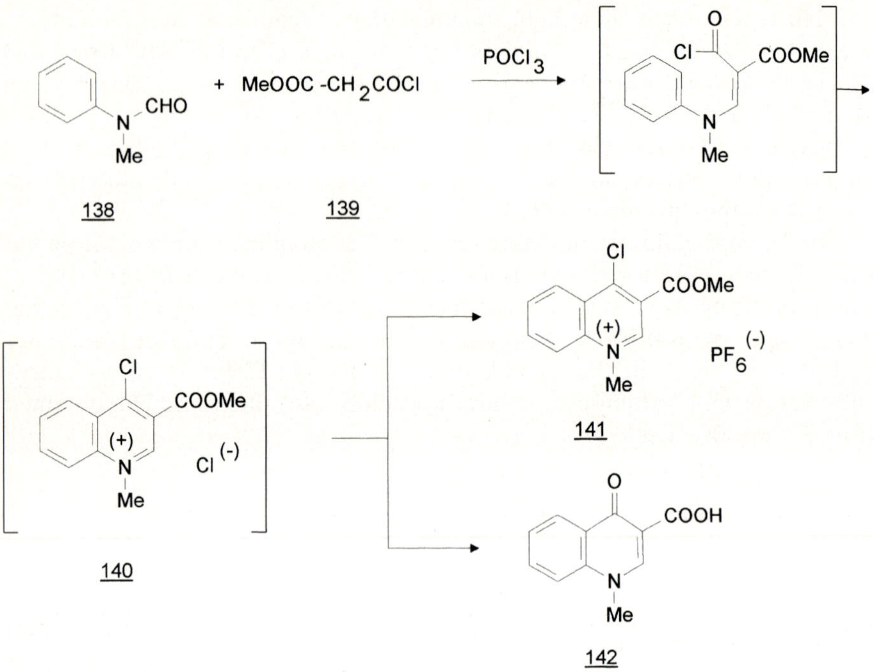

Fig. 28. Synthesis of 4-chloro-3-methoxycarbonyl-1-methylquinolinium hexafluorophosphate **141** and 1-methyl-4-quinolone-3-carboxylic acid **142**

formed arising from the cyclization of asymmetrically substituted aromatic formanilides.

VIII. Synthesis of Quinolone Analogues

The underlying reaction principle of the cycloaracylation procedure (see Sect. B.III) can also be applied to the synthesis of 1,4-benzothiazine derivatives (FENGLER and GROHE 1984; FENGLER et al. 1984a,b). If, for example, phenylsulfenyl chlorides **143a,b** are converted with methyl 3-ethylaminoacrylate **144** the vinylogenic sulfenamides **145a,b** are obtained, which can be oxidized with m-chloroperbenzoic acid to the sulfoxides **146a,b**. The cyclization of **146a,b** with n-butyllithium or potassium tert-butoxide leads to the 1-oxides of 1,4-benzothiazine-2-carboxylic acid methyl esters **147a,b** and, after their hydrolysis, to the corresponding carboxylic acids **148a,b** (Fig. 29).

At the beginning of the 1990s, 1-oxides and 1,1-dioxides of benzothiazine-2-carboxylic acids, containing a fluorine atom at position 7 and a 4-methylpiperazinyl residue at position 6, were also produced according to the above procedure in an analogous manner to the fluorquinolones (CULBERTSON 1991). However, these substances possess no antibacterial activity.

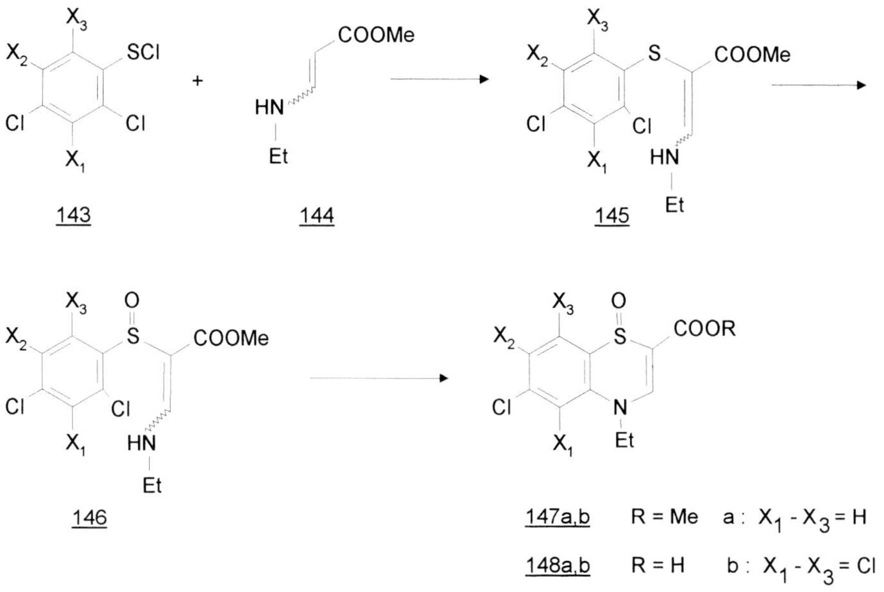

147a,b R = Me a: $X_1 - X_3 = H$

148a,b R = H b: $X_1 - X_3 = Cl$

Fig. 29. Synthesis of 1-oxides of 1,4-benzothiazine-2-carboxylic acid methyl esters **147a,b** and the corresponding carboxylic acids **148a,b**

In quinolizinones, the typical nitrogen atom in position 1 of quinolones is replaced by a carbon atom. The synthesis of this class of compounds is described in detail below, and an example is given (CHU et al. 1992). The starting material used, 2-benzylpyrimidine derivative **149**, is first converted with 1-methylpiperazine to the substituted product **150**. This is then converted to the addition product **151** by means of ethoxymethylene malonic ester in the presence of n-butyllithium. Heating **151** with DBU results in ring closure and the formation of quinolizinone carboxylic acid ester **152**. After trans-esterification of **152** with benzyl alcohol to **153**, the benzyl group is removed by catalytic hydrogenation. This results in 3-fluoro-9-(4-fluorophenyl)-2-(4-methylpiperazin-1-yl)-6H-6-oxo-pyrido[1,2-a]pyrimidine-7-carboxylic acid **154** (Fig. 30).

Although compounds of this class of agents show good antibacterial inhibitory effects in vitro, satisfactory in vivo effects could not be obtained.

In order to synthesize 1-carba-bioisosteric analogues of oxolinic acid (HÖGBERG et al. 1984), the Friedel-Crafts reaction was used as a first stage to convert veratrol **155** and 4,4-dimethylbutyrolactone **156** to 6,7-dimethoxy-4,4-dimethyl-1-naphthalenone **157** (SAWA et al. 1975). Methyl ether cleavage of **157** with boron tribromide followed by cyclization with methylene bromide leads to 6,7-methylenedioxynaphthalenone **158**, which is converted to 3,4-dihydronaphthalenone carboxylic acid ester **159** by diethyl carbonate and sodium hydride. Compound **159** is then dehydrogenated with 2,3-dichloro-5,6-

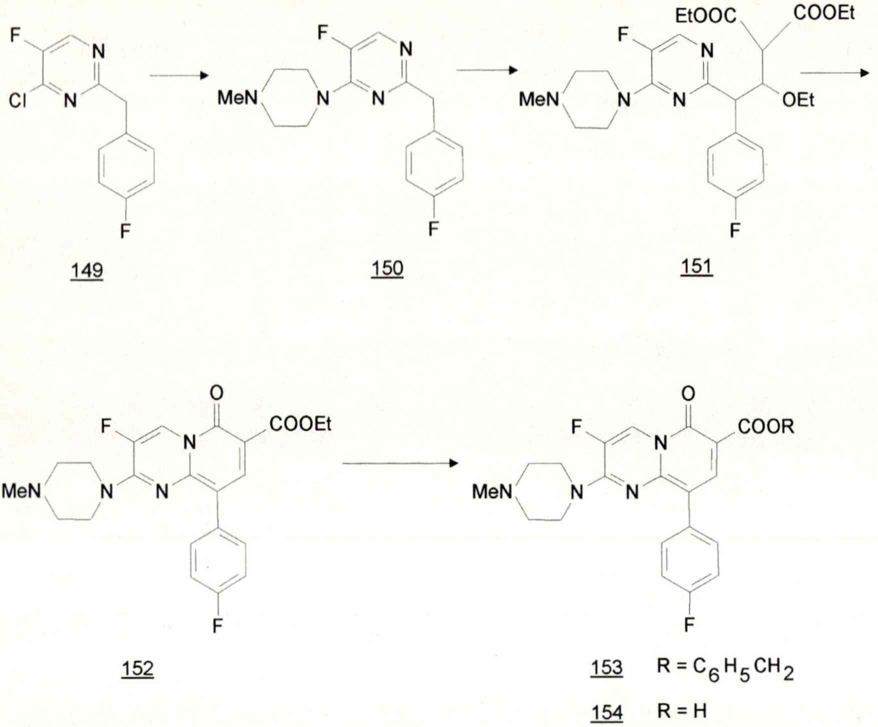

F Cl N N N F 149 → MeN N F N N F 150 → EtOOC COOEt F MeN N N N OEt F 151 →

O F N COOEt MeN N N F 152 → O F N COOR MeN N N F 153 $R = C_6H_5CH_2$ 154 R = H

Fig. 30. Synthesis of 3-fluoro-9-(4-fluorophenyl)-2-(4-methylpiperazin-1-yl)-6H-6-oxo-pyrido[1,2-a]pyrimidine-7-carboxylic acid **154**

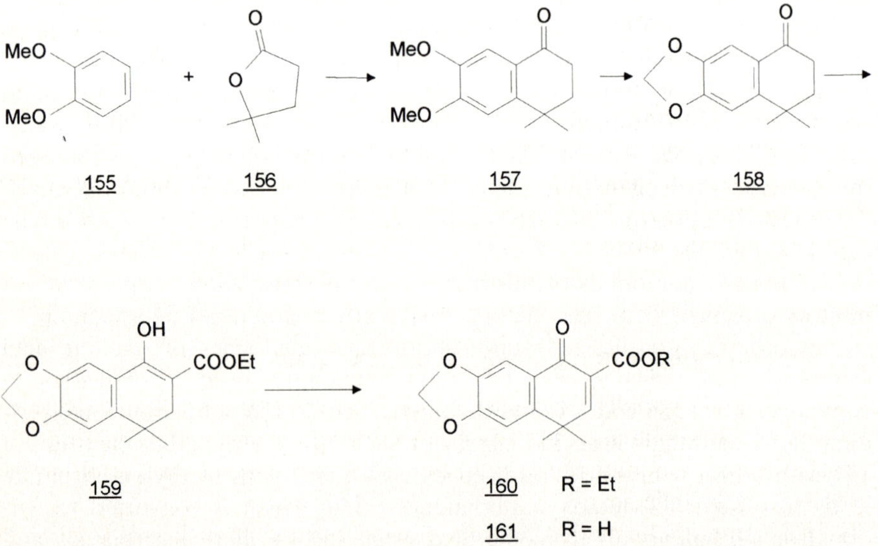

MeO MeO 155 + O O 156 → O MeO MeO 157 → O O O 158 →

OH O O COOEt 159 → O O O COOR 160 R = Et 161 R = H

Fig. 31. Synthesis of naphthalenone carboxylic acid ester **160** and the corresponding carboxylic acid **161**

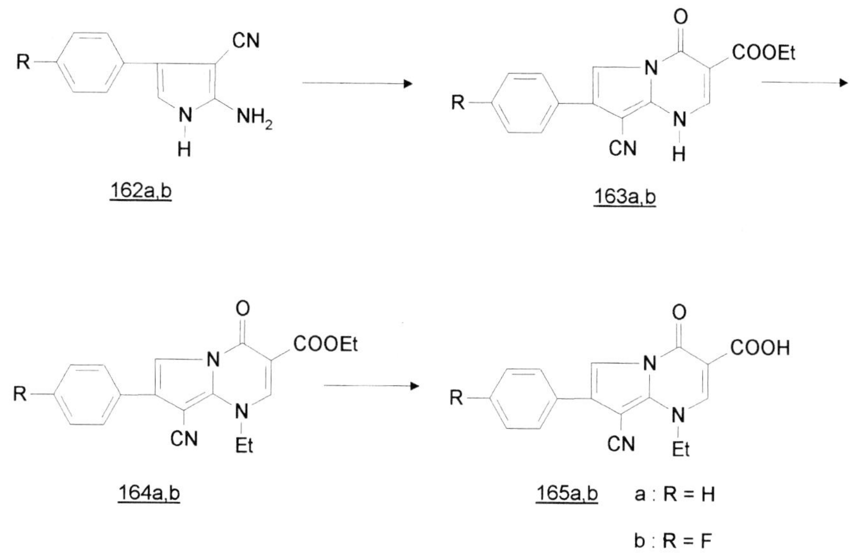

Fig. 32. Synthesis of 1-ethyl-8-cyano-1,4-dihydro-4-oxo-pyrrolo[1,2-a]pyrimidine-3-carboxylic acids **165a,b**

dicyanobenzoquinone to form naphthalenone carboxylic acid ester **160** (Fig. 31). Alkaline hydrolysis of **160** gives the desired 1,2-dihydro-4,4-dimethyl-1-oxo-6,7-(methylenedioxy)-2-naphthalene carboxylic acid **161**, which shows only very weak antibacterial activity.

4-Oxopyrrolo[1,2-a]pyrimidine carboxylic acids are synthesized by a modified Gould-Jacobs reaction (ABDALLA and SOWELL 1987). 2-Amino-3-cyanopyrroles **162a,b** are converted by means of EMME to the corresponding aminomethylene malonic esters, which then undergo cyclization by being heating with Dowtherm A or by reacting with potassium tert-butoxide in dimethylformamide at room temperature to produce pyrrolopyrimidine carboxylic acid esters **163a,b**. By selective ethylation of **163a,b**, the compounds **164a,b** are obtained, which in turn undergo alkaline hydrolysis to 1-ethyl-8-cyano-1,4-dihydro-4-oxo-pyrrolo[1,2-a]pyrimidine-3-carboxylic acids **165a,b** (Fig. 32).

IX. Synthesis of 4-Cinnolone-3-carboxylic Acids

As a result of the rapidly increasing research efforts in the 1970s and 1980s in the field of quinolone and fluoroquinolone anti-infectives, increasing interest was also focused on methods of synthesis of the analogous 2-azaquinolone or cinnolone carboxylic acids.

The first synthetic method for cinnolone-3-carboxylic acids was described by VON RICHTER in 1883. He observed that the 2-aminophenylpropynoic acids

166 were directly cyclized to 4-hydroxycinnoline-3-carboxylic acids **167** during diazotization. By alkylation of position 1 of **167**, the corresponding cinnolone carboxylic acids **168** were obtained (Fig. 33).

The von Richter reaction later found only limited use because of the difficulty of obtaining the starting materials (SCHOFIELD and SIMPSON 1945).

4-Hydroxycinnoline-3-carboxylic acids could be more easily obtained by Friedel-Crafts cyclization of mesoxalyl chloride phenylhydrazones. In this method, which was discovered by BARBER et al. (1961), the esters of mesoxalphenylhydrazones **169**, which are easily obtained by coupling of aryldiazonium salts with malonic esters, are hydrolyzed to the dicarboxylic acids **170** and then converted to the acid chlorides **171** with thionyl chloride. The cyclization of **171** to **167** is generally brought about by heating with titanium tetrachloride in nitrobenzene (COBURN and GALA 1982; Fig. 34).

The dicarboxylic acids **170** are also obtainable by treatment of arylhydrazines with mesoxalic acid. Furthermore, N-1-alkylated derivatives of **170** can be synthesized in this way and then undergo cyclization to the corresponding 1-alkylcinnolone carboxylic acids (AMES et al. 1964). One of the drawbacks of this method is that mesoxalyl chloride phenylhydrazones substituted in position 3 and unsubstituted in positions 2 and 6 undergoes cyclization, yielding a mixture of isomers (SHOUP and CASTLE 1965) in most cases.

The only cinnolone anti-infective which is commercially available at present is cinoxacin **8**, which is produced by the Borsche reaction (BORSCHE and HERBERT 1941). The 2-aminoacetophenone derivative **172**, which is used as the starting material, undergoes spontaneous cyclization to 4-hydroxycinnoline **173** during diazotization. Thereafter, **173** is brominated to **174**, and the bromine atom of **174** is subsequently exchanged for the cyano

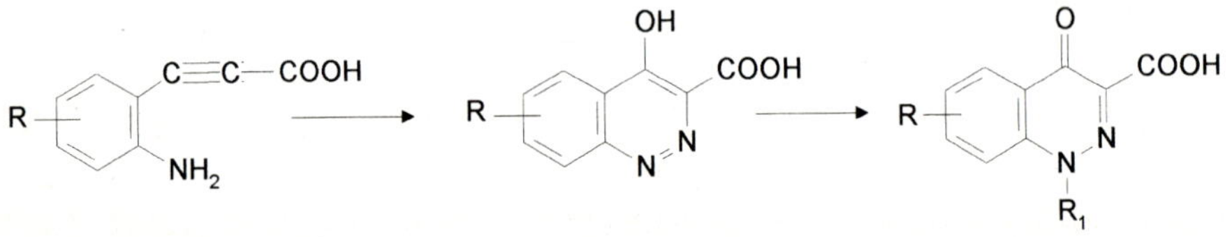

Fig. 33. Synthesis of cinnolone-3-carboxylic acids **168**

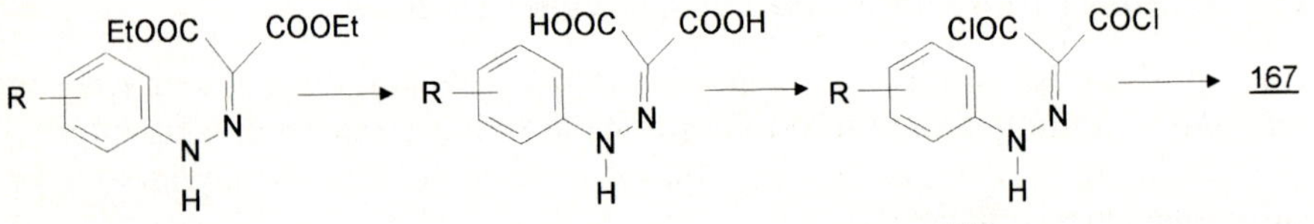

Fig. 34. Synthesis of **167**

group by means of Cu(I)cyanide in dimethylformamide to form **175**. Ethylation of **175** is then carried out, yielding the nitrile **176**, which is saponified under acidic conditions to cinoxacin **8** (WHITE 1970; Fig. 35).

Borsche synthesis is a versatile and frequently used method to produce derivatives substituted at the 3, 5, 6, 7, and 8 positions (OCKENDEN and SCHOFIELD 1953; DOWLATSHAHI 1985; SINGH 1991). Regiochemical complications are not to be expected using this procedure. However, as with the Gould-Jacobs reaction, 1-cycloalkyl-, 1-tert-alkyl-, and 1-aryl-cinnolone carboxylic acids cannot be synthesized by this method.

In a special synthesis procedure for 1-arylcinnolone carboxylic acids, 2-halobenzoylacetic acid esters **177** are coupled with aryldiazonium chlorides to form the hydrazones **178**, which then undergo base-induced ring closure to the esters of cinnolone carboxylic acids **179**. The corresponding acids **180** are obtained from the esters by alkaline or acid hydrolysis (Fig. 36).

This procedure, which was described by PRUDCHENKO et al. (1968; AMES et al. 1983) for the pentafluorobenzoylacetic acid esters, has also recently been used for the synthesis of 2-aza analogues of fluoroquinolone bioactive agents (RADL et al. 1990; MIYAMOTO and MATSUMOTO 1990a). The nitro group can replace the halogen as the leaving group during the cyclization step (SANDISON and TENNANT 1974).

In a new type of synthesis for 1-alkyl-(aza)cinnolone carboxylic acids, 2-diazo-3-[2-halo-(hetero)aryl]-3-oxopropionic acid esters **181** are used as starting compounds. These can be obtained either by acylation of diazoacetic acid esters with 2-halo-(hetero)aroyl halides (MIYAMOTO et al. 1978; SANCHEZ et al. 1987) or by reaction of 2-halo-(hetero)aroyl acetic acid esters with tosyl azide

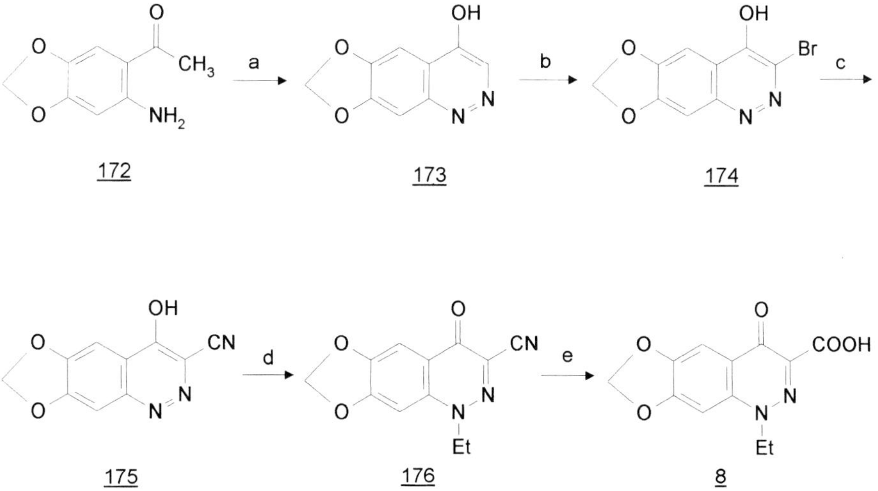

Fig. 35. Synthesis of cinoxacin **8**. *a*, $NaNO_2$, $H_3O^{(+)}$; *b*, Br_2; *c*, CuCN, dimethylformamide (DMF); *d*, EtI, NaH, DMF; *e*, $H_3O^{(+)}$

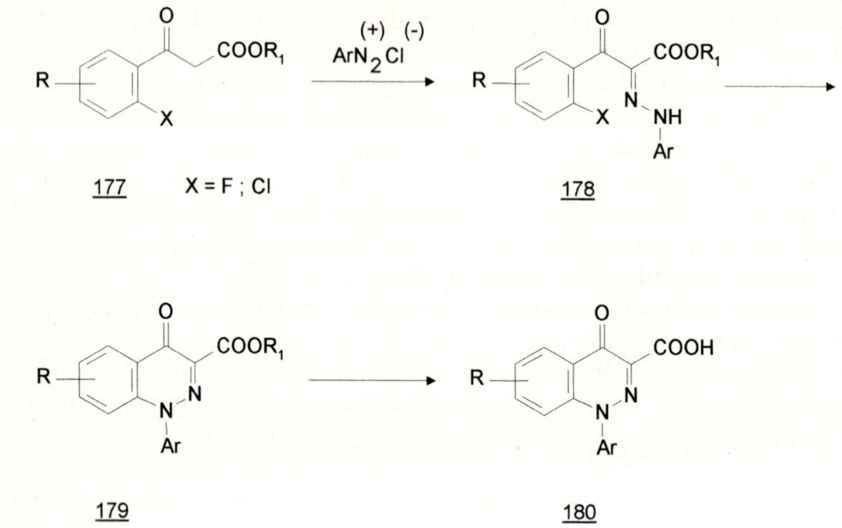

Fig. 36. Synthesis of 1-arylcinnolone carboxylic acids **180**

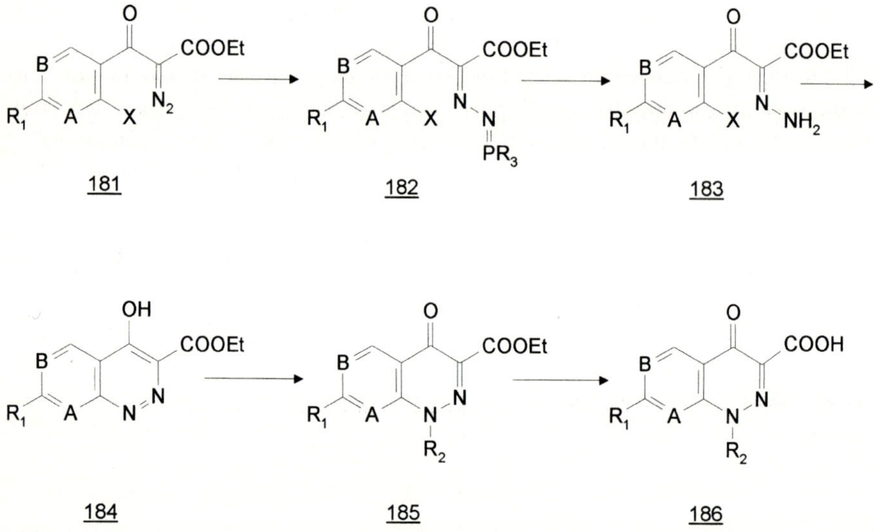

Fig. 37. Synthesis of 1-alkyl-(aza)cinnolone carboxylic acids **186**

(MIYAMOTO and MATSUMOTO 1988). By treatment with triphenylphosphine, the diazopropionic acid esters **181** are converted primarily to the phosphazines **182**, which are then easily hydrolyzed to the hydrazones **183**. The cyclization of **183** to the desired (aza)cinnolone carboxylic acid esters **184** can occur by heating when the leaving group X is very reactive, e.g., fluorine. Otherwise, potassium tert-butoxide in dioxane is used. The reaction of **184** with alkyl iodides in dimethylformamide in the presence of potassium carbonate yields the 1-alkyl-(aza)cinnolone carboxylic acid esters **185**, which are then hydrolyzed to the carboxylic acids **186** in the usual manner (Fig. 37).

If the strongly basic and nucleophilic tri-n-butylphosphine is used instead of triphenylphosphine for the conversion of **181**, the corresponding tributylphosphazine derivatives of **182** (R = n-butyl) can often be directly cyclized to **184** by heating in dioxane (MIYAMOTO and MATSUMOTO 1990b).

Another possibility is the reaction of aroyldiazopropionic acid esters **181** with hydrazine to give dihydro(aza)cinnolone carboxylic acid derivatives. Dehydrogenation to form (aza)cinnolones can then be carried out with benzoylperoxide (KIMURA et al. 1976).

X. Synthesis of 4-Pyridone-3-carboxylic Acids

A recognized aim of pharmaceutical research on active substances is to simplify the structures of active molecules. A common partial structure of all quinolone anti-infectives consists of 1-substituted 1,4-dihydro-4-oxo-pyridine-3-carboxylic acid. There has been no lack of effort to synthesize potent members of this class of compounds (GEORGOPAPADAKOU et al. 1987). For this purpose, a variety of synthetic methodologies, some of it new, has been developed.

By means of a variation of the Gould-Jacobs reaction, 5-mono- and 5,6-disubstituted 1,4-dihydro-4-oxo-pyridine-3-carboxylic acid esters **188** have been obtained by thermal ring closure of enaminomethylene malonic esters **187**. Their N-alkylation gives rise to the esters **189**, which then can be hydrolyzed to the desired pyridone carboxylic acids **190** (KEMETANI et al. 1977; Fig. 38).

The starting compounds **187** are produced in a relatively simple manner either by condensation of the respective primary enamines with ethoxymethylene malonic esters or condensation of the relevant ketones with aminomethylene malonic esters (AGUI et al. 1975b).

According to a method of synthesis attributed to HUFFMAN et al. 1962; BALOGH et al. 1980 (Fig. 39), acetoacetic acid derivatives **191** are converted

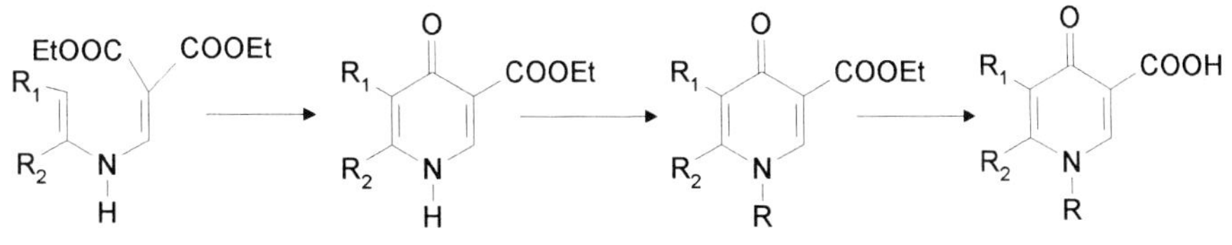

Fig. 38. Synthesis of 4-pyridone-3-carboxylic acids **190**

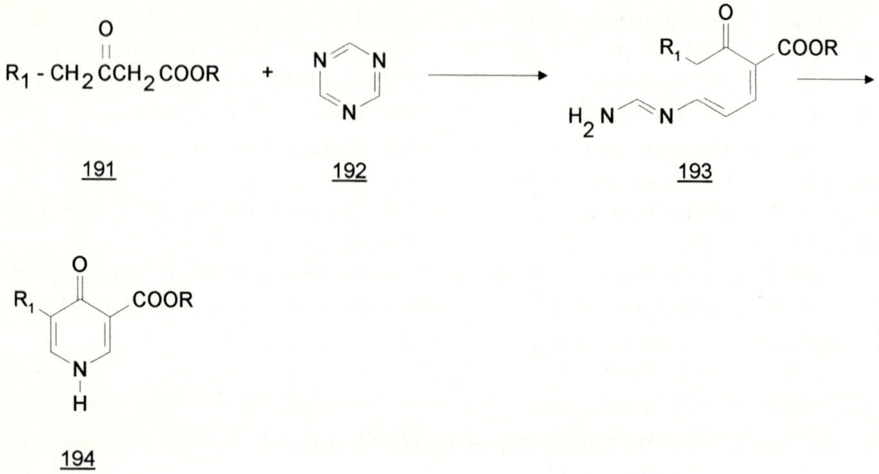

$R_1-CH_2\overset{O}{\overset{\|}{C}}CH_2COOR$ + 192 → 193 → 194

191 192 193

194

Fig. 39. Synthesis of 4-pyridone-3-carboxylic acid esters **194**

195 + 196 → 197 → 198 → 199, 200

199 R = Et

200 R = H

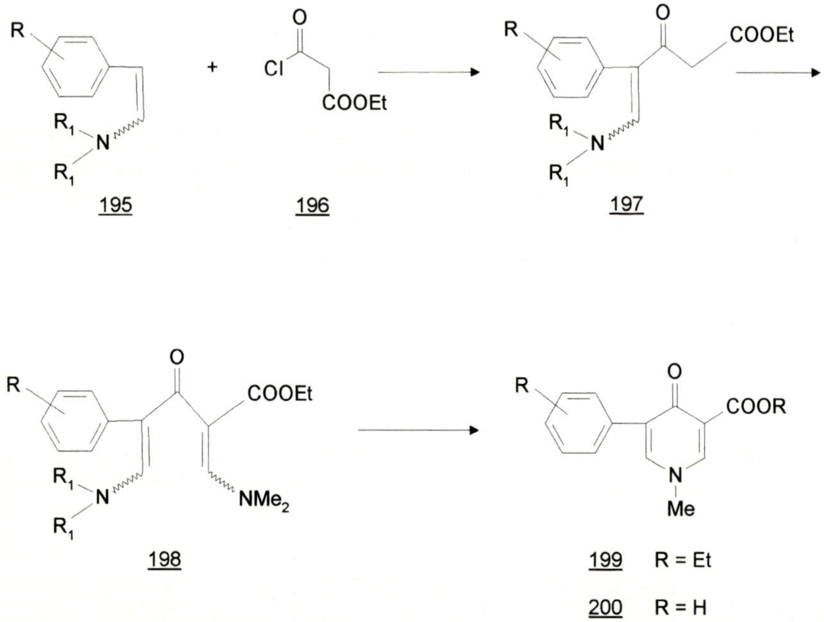

Fig. 40. Synthesis of 5-arylpyridone-3-carboxylic acid esters **199** and the corresponding carboxylic acids **200**

with s-triazine **192** via the intermediates **193** to the 4-pyridone-3-carboxylic acid esters **194** substituted in position 5.

Alkylation of **194** and saponification of the formed 1-alkylpyridone carboxylic acid esters occur as previously described.

In an interesting synthesis of 5-arylpyridone carboxylic acids, the enamines **195** are acylated with ethylmalonyl chloride **196** in position 3 to enaminones **197**, which are then condensed with dimethylformamide acetals to form the *bis*enaminones **198**. The latter undergo cyclization with the hydrochlorides of primary aliphatic amines, e.g., methylamine hydrochloride in boiling ethanol, to the pyridone carboxylic acid esters **199**, which are hydrolyzed in the usual way to the corresponding carboxylic acids **200** (ABDULLA et al. 1977a; Fig. 40).

An analogous reaction pathway leads from the starting compounds 3-dimethylaminoacrylonitrile and substituted phenylacetyl chlorides to 5-arylpyridone carboxylic acid nitriles (ABDULLA et al. 1977b).

After the discovery of the highly active antibacterial 1-arylfluoroquinolones, 1-arylpyridone carboxylic acids were increasingly produced.

4-Hydroxy-2-pyrones **201** can be used as starting materials. Their reaction with dimethylformamide acetals leads initially to the 3-dimethylamino-

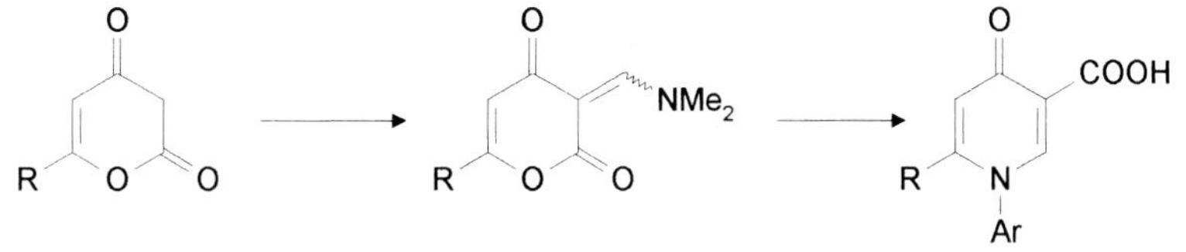

201 → 202 → 203 R = Alkyl

Fig. 41. Synthesis of 1-arylpyridone carboxylic acids **203**

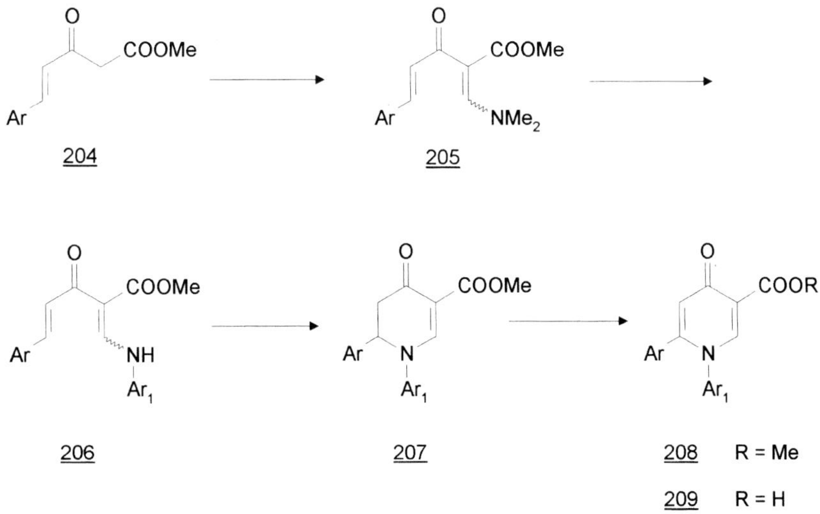

204 → 205 →

206 → 207 → 208 R = Me; 209 R = H

Fig. 42. Synthesis of 1,6-bisarylpyridone carboxylic acid esters **208** and the corresponding carboxylic acids **209**

methylenepyrones **202**, which undergo rearrangement with primary aromatic amines to the desired 1-arylpyridone carboxylic acids **203** (KILBOURN and SEIDEL 1972; RADL et al. 1988, 1989; BASSINI et al. 1993; Fig. 41).

In another versatile synthesis procedure for 1,6-*bis*arylpyridone carboxylic acids, Wittig reaction products **204** are used as starting compounds. Their reaction with dimethylformamide acetals and, thereafter, with arylamines leads via the intermediates **205** to the arylaminomethylene derivatives **206**. The subsequent ring closure of **206** to form **207** is then followed by its dehydrogenation to yield the pyridone esters **208** and the corresponding carboxylic acids **209** after hydrolysis (WICK 1979; NARITA et al. 1986a–c; Fig. 42).

References

Abdalla GM, Sowell JW Sr (1987) Synthesis of 8-cyano-1,4-dihydro-4-oxopyrrolo[1,2-a]pyrimidine-3-carboxylic acids as potential antimicrobial agents. J Heterocycl Chem 24:297

Abdalla GM, Sowell JW (1990) Synthesis of 3-benzoyl-4,7-dihydro-7-oxopyrrolo[3,2-b]pyridine-6-carboxylic acid derivatives as potential antimicrobial agents. J Heterocycl Chem 27:1201

Abdulla RF, Emmrick TL, Taylor HM (1977a) A new synthetic approach to 4(1h)-pyridone derivatives. I. 1-Alkyl-3,5-diaryl-4(1h)-pyridones. Synth Commun 7:305

Abdulla RF, Fuhr KH, Taylor HM (1977b) New synthetic approaches to 4(1h)-pyridones. II. Derivatives having nonaromatic substituents. Synth Commun 7:313

Agui H, Mitani T, Nakashita M, Nakagome T (1971) Studies on quinoline derivatives and related compounds. I. A new synthesis of 1-alkyl-1,4-dihydro-4-oxo-3-quinolinecarboxylic acids. J Heterocycl Chem 8:357

Agui H, Komatsu T, Nakagome T (1975a) Studies on quinoline derivatives and related compounds. II. Synthesis of 5-substituted 1-ethyl-1,4-dihydro-4-oxo-3-quinolinecarboxylic acid (1). J Heterocycl Chem 12:557

Agui H, Tobiki H, Nakagome T (1975b) Studies on quinoline derivatives and related compounds. III. A novel pyridine synthesis to give substituted 1,4-dihydro-4-oxonicotinic acids (1). J Heterocycl Chem 12:1245

Albrecht R (1977) Development of antibacterial agents of the nalidixic acid type. Prog Drug Res 21:9

Altreuther P (1987) Data on chemistry and toxicology of Baytril. Vet Med Rev 2:87

Ames DE, Chapman RF, Kucharska HZ (1964) Cinnolines. V.1. Methylation of some substituted cinnolines. J Chem Soc 5659

Ames DE, Leung OT, Singh AG (1983) Synthesis of 1-aryl-4-oxo-1H,4H-cinnoline-3-carboxylic acid esters. Synthesis 52

Aoki M, Kamata M, Otsuka T, Shimma N, Yokose K (1988) European patent 259 804 (12.09.1986), Hoffmann-La Roche. Chem Abstr 109:73478r

Atarashi S, Yokohama S, Yamazaki K, Sakano K, Imamura M, Hayakawa I (1987) Synthesis and antibacterial activities of optically active ofloxacin and its fluoromethyl derivative. Chem Pharm Bull (Tokyo) 35:1896

Atarashi S, Tsurumi H, Fujiwara T, Hayakawa I (1991) Asymmetric reduction of 7,8-difluoro-3-methyl-2H-1,4-benzoxazine. Synthesis of a key intermediate of (S)-(–)-Ofloxacin (DR-3355). J Heterocycl Chem 28:329

Atarashi S, Imamura M, Kimura Y, Yoshida A, Hayakawa I (1993) Fluorocyclopropyl quinolones. 1. Synthesis and structure-activity relationships of 1-(2-fluorocyclopropyl)-3-pyridonecarboxylic acid antibacterial agents. J Med Chem 36:3444

Augeri DJ, Fray AH, Kleinman EF (1990) Synthesis and antibacterial activity of 2,3-dehydroofloxacin. J Heterocycl Chem 27:1509

Bacon ER, Daum SJ (1991) Synthesis of 7-ethyl-4,7-dihydro-4-oxo-2-(4-pyridinyl)thieno[2,3-b]pyridine-5-carboxylic acid. J Heterocycl Chem 28:1953

Balogh M, Hermecz I, Meszaros Z, Simon K, Pusztay L, Horvath G, Dvortsak P (1980) Studies on chemotherapeutics. I. Synthesis of 5-substituted-4-oxo-1,4-dihydro-3-pyridinecarboxylic acid derivatives. J Heterocycl Chem 17:359

Barber HJ, Washbourn K, Wragg WR (1961) A new cinnoline synthesis. I. Cyclisation of mesoxalyl chloride phenylhydrazones to give substituted 4-hydroxycinnoline-3-carboxylic acids. J Chem Soc 2828

Barker JM, Huddleston PR, Jones AW (1978) Thienyl analogues of the alkaloids. 3. Analogues of echinorine and echinopsine; a convenient synthesis of thieno [3,2-b] pyridines. J Chem Res 393

Bassini C, Bismara C, Carlesso R, Feriano A, Gaviraghi G, Marchioro C, Perboni A, Shaw RE, Tamburini B, Tarzia G, Xerri L (1993) Synthesis and antimicrobial activity of DNA-gyrase inhibiting derivatives of 4-oxo-1,4-dihydro-3-pyridinecarboxylic acid. Farmaco [Sci] 48:159

Bauditz R (1987) Results of clinical studies with Baytril in calves and pigs. Vet Med Rev 2:122

Bayomi SM, Price KE, Sowell JW (1985) Synthesis of 7-oxopyrrolo[3,2-b] pyridine-6-carboxylic acid derivatives as potential antimicrobial agents. J Heterocycl Chem 22:83

Bell MR, Zalay AW, Oesterlin R, Schane P, Potts GO (1970) Basic ethers of 1-(p-hydroxyphenyl)-2-phenyl-1,2,3,4-tetrahydroquinoline and 1-(p-hydroxyphenyl)-2-phenylindole. Antifertility agents. J Med Chem 13:664

Biere H, Seelen W (1976) Verfahren zur Darstellung von 4-oxo-1,4-dihydropyridincarbonsäurederivaten. Liebigs Ann Chem 1972

Boamah PY, Haider N, Heinisch G (1990) Pyrido [2,3-c]pyridazines structurally related to nalidixic acid. Arch Pharm (Weinheim) 323:207

Borsche W, Herbert A (1941) Synthesen mit 5-Nitro-2-brom-acetophenon. Liebigs Ann Chem 546:293

Bouzard D, Di Cesare P, Essiz M, Jacquet JP, Remuzon P, Weber A, Oki T, Masuyoshi M (1989) Fluoronaphthyridines and quinolones as antibacterial agents. 1. Synthesis and structure-activity relationships of new 1-substituted derivatives. J Med Chem 32:537

Bouzard D, Di Cesare P, Essiz M, Jacquet JP, Kiechel JR, Remuzon P, Weber A, Oki T, Masuyoshi M, Kessler RE, Fung-Tomc J, Desiderio J (1990) Fluoronaphthyridines and quinolones as antibacterial agents. 2. Synthesis and structure-activity relationships of new 1-tert-butyl 7-substituted derivatives. J Med Chem 33:1344

Bouzard D, Di Cesare P, Essiz M, Jacquet JP, Ledonssal B, Remuzon P, Kessler RE, Fung-Tomc J (1992) Fluoronaphthyridines as antibacterial agents. 4. Synthesis and structure-activity relationships of 5-substituted-6-fluoro-7-(cycloalkylamino)-1,4-dihydro-4-oxo-1,8-naphthyridine-3-carboxylic acids. J Med Chem 35:518

Braish TF, Fox DE (1990) Synthesis of (S,S)- and (R,R)-2-alkyl-2,5-diazabicyclo[2.2.1]heptanes. J Org Chem 55:1684

Bridges AJ, Sanchez JP (1990) Synthesis of 7-amino-1,4-dihydro-4-oxo-6-(trifluoromethyl)-1,8-naphthyridines. The use of methylidenemalonate as an activating group and a sulfur assisted cyclization. J Heterocycl Chem 27:1527

Brundage RP, Lesher GY (1976) US patent 3 928 366 (10.03.1974), Sterling Drug. Chem Abstr 84:74253

Camps R (1901) Von der Amidophenylpropiolsäure zur Kynurensäure und deren Verwandten. Chem Ber 34:2703

Cecchetti V, Fravolini A, Fringuelli R, Mascellani G, Pagella PG, Palmioli M, Segre G, Terni P (1987) Synthesis of 7,12-dihydropyrido[3,4-b:5,4-b]diindoles. A novel class of rigid, planar benzodiazepine receptor ligands. J Med Chem 30:456

Cecchetti V, Fravolini A, Schiafella F (1991) One-pot synthesis of rufloxacin. Synth Commun 21:2301

Cecchetti V, Fravolini A, Pagella PG, Savino A, Tabarrini O (1993) Quinolinecarboxylic acids. 3. Synthesis and antibacterial evaluation of 2-substituted-7-oxo-2,3-dihydro-7H-pyrido[1,2,3-de] [1,4]benzothiazine-6-carboxylic acids related to rufloxacin. J Med Chem 36:3449

Chiarino D, Napoletano M, Sala A (1988) Synthesis of 4,7-dihydro-4-oxoisoxazolo[5,4-b]pyridine-5-carboxylic acid derivatives as potential antimicrobial agents. J Heterocycl Chem 25:231

Chu DTW (1985a) South African patent 852 802 (27.11.1985), Abbott Lab. Chem Abstr 106:213930x

Chu DTW (1985b) A regiospecific synthesis of 1-methylamino-6-fluoro-7-(4-methylpiperazin-1-yl)-1,4-dihydro-4-oxoquinoline-3-carboxylic acid. J Heterocycl Chem 22:1033

Chu DTW (1990) Syntheses of 6-fluoro-7-piperazin-1-yl-9-cyclopropyl-2,3,4,9-tetrahydroisothiazolo[5,4-b]quinoline-3,4-dione and 6-fluoro-7-piperazin-1-yl-9-p-fluorophenyl-2,3,4,9-tetrahydroisothiazolo[5,4-b] quinoline-3,4-dione. J Heterocycl Chem 27:839

Chu DTW, Claiborne AK (1987) Short syntheses of 1,2,3,5-tetrahydro-5-oxopyrrolo[1,2-a]quinoline-4-carboxylic acid and 1,2,3,4-tetrahydro-6H-6-oxopyrido[1,2-a]quinoline-5-carboxylic acid derivatives. J Heterocycl Chem 24:1537

Chu DTW, Claiborne AK (1990) Practical synthesis of iminochlorothioformates: application of iminochlorothioformates in the synthesis of novel 2,3,4,9-tetrahydroisothiazolo[5,4b][1,8]naphthyridine-3,4-diones and 2,3,4,9-tetrahydroisothiazolo[5,4-b]quinoline-3,4-dione derivatives. J Heterocycl Chem 27:1191

Chu DTW, Fernandes PB, Claiborne AK, Pihuleac E, Nordeen CW, Maleczka RE Jr, Pernet AG (1985) Synthesis and structure-activity relationships of novel arylfluoroquinolone antibacterial agents. J Med Chem 28:1558

Chu DTW, Fernandes PB, Pernet AG (1986) Synthesis and biological activity of benzothiazolo [3,2-a] quinolone antibacterial agents. J Med Chem 29:1531

Chu DTW, Lee CM, Li Q, Cooper CS, Plattner JJ (1992) WO 9 116 894 (02.05.1990), Abbott Lab. Chem Abstr 117:26583b

Coburn RA, Gala D (1982) Synthesis of 8-amino-6-methoxycinnoline. A precursor for 2-azaprimaquine. J Heterocycl Chem 19:757

Culbertson TP (1991) Synthesis of 4H-1,4-benzothiazine 1-oxide and 1,1-dioxide. Analogs of quinolone antibacterial agents. J Heterocycl Chem 28:1701

Dalhoff A, Petersen U, Endermann R (1996) In vitro activity of BAY 12-8039, a new methoxyquinolone. Chemotherapy (Basel) 42:410–425

Di Cesare P, Bouzard D, Essiz M, Jacquet JP, Ledonssal B, Kiechel JR, Remuzon P, Kessler RE, Fung-Tomc J, Desiderio J (1992) Fluoronaphthyridines and -quinolones as antibacterial agents. 5. Synthesis and antimicrobial activity of chiral 1-tert-butyl-6-fluoro-7-substituted-naphthyridones. J Med Chem 35:4205

Dohmori R, Kadoya S, Takamura I, Suzuki N (1976) Synthetic chemotherapeutic agents. I. Synthesis of 2-substituted thiazolo [5,4-f] quinoline derivatives. Chem Pharm Bull (Tokyo) 24:130

Domagala JM, Heifetz CL, Hutt MP, Mich TF, Nichols JB, Solomon M, Worth DF (1988) 1-substituted 7-[3-[(ethylamino)methyl]-1-pyrrolidinyl]-6,8-difluoro-1,4-dihydro-4-oxo-3-quinolinecarboxylic acids. New quantitative structure-activity relationships at N_1 for the quinolone antibacterials. J Med Chem 31:991

Dowlatshahi HA (1985) Synthesis of ethyl 7-methoxy-4-oxocinnoline-3-carboxylate. Synth Commun 15:1095

Egawa H, Miyamoto T, Minamida A, Nishimura Y, Okada H, Uno H, Matsumoto J (1984) Pyridonecarboxylic acids as antibacterial agents. 4. Synthesis and antibacterial activity of 7-(3-amino-1-pyrrolidinyl)-1-ethyl-6-fluoro-1,4-dihydro-4-oxo-1,8-naphthyridine-3-carboxylic acid and its analogues. J Med Chem 27:1543

Egawa H, Miyamoto T, Matsumoto J (1986) A new synthesis of 7H-pyrido[1,2,3-de][1,4]benzoxazine derivatives including an antibacterial agent, ofloxacin. Chem Pharm Bull (Tokyo) 34:4038
Fengler G, Grohe K (1984) DOS 3 229 126 (04.08.1982) Bayer AG. Chem Abstr 100:209385g
Fengler G, Arlt D, Grohe K (1984a) DOS 3 229 124 (04.08.1982) Bayer AG. Chem Abstr 101:90953u
Fengler G, Arlt D, Grohe K, Zeiler H-J, Metzger K (1984b) DOS 3 229 125 (04.02.1982) Bayer AG, Chem Abstr 101:7176z
Frigola J, Colombo A, Más J, Parés J (1987) Synthesis, structure elucidation and chemotherapeutic activity of 6-substituted 1-ethyl-1,4-dihydro-7-[(1-imidazoly)phenylmethyl]-4-oxo-3-quinolinecarboxylic acids. J Heterocycl Chem 24:399
Fromtling RA, Castañer J (1996) Trovafloxacin-Mesylate. Drugs Fut 21:496–505
Georgopapadakou NH, Dix BA, Angehrn P, Wick A, Olson GL (1987) Monocyclic and tricyclic analogs of quinolones: mechanism of action. Antimicrob Agents Chemother 31:614
Gerster JF (1973) DOS 2 264 163 (12.07.1973) Riker Laboratories Inc. Chem Abstr 79:92029y
Goueffon Y, Montay G, Roquet F, Pesson M (1981) New synthetic antimicrobial agent: 1,4-di-hydro-1-ethyl-6-fluoro-7-(4-methyl-1-piperazinyl)-4-oxo-quinoline-3-carboxylic acid (1589 RB). C R Acad Sci Biol (Paris) 292:37
Gould RG, Jacobs WA (1939) The synthesis of certain substituted quinolines and 5,6-benzoquinolines. J Am Chem Soc 61:2890
Grohe K (1986) DOS 3 502 935 (29.09.1984) Bayer AG. Chem Abstr 105:226051r
Grohe K (1992) DOS 4 015 299 (14.11.1991) Bayer AG. Chem Abstr 117:4811y
Grohe K (1993) The importance of the cycloaracylation process for the synthesis of modern fluoroquinolones. J Prakt Chem 335:397
Grohe K, Heitzer H (1987a) Synthese von 4-Chinolon-3-carbonsäuren. Liebigs Ann Chem 29
Grohe K, Heitzer H (1987b) Synthese von 1-Amino-4-chinolon-3-carbonsäuren. Liebigs Ann Chem 871
Grohe K, Schriewer M (1987) DOS 3 522 406 (22.06.1985) Bayer AG. Chem Abstr 107:23351g
Grohe K, Zeiler H-J, Metzger K (1980) DOS 2 808 070 (24.02.1978) Bayer AG. Chem Abstr 92:41916w
Grohe K, Zeiler H-J, Metzger K (1981) DOS 2 903 850 (01.02.1979) Bayer AG. Chem Abstr 95:43152e
Grohe K, Zeiler H-J, Metzger K (1983) DOS 3 142 854 (29.10.1981) Bayer AG. Chem Abstr 99:53790h
Grohe K, Petersen U, Zeiler H-J, Metzger K (1985) DOS 3 318 145 (18.05.1983) Bayer AG. Chem Abstr 102:78744q
Grohe K, Zeiler H-J, Metzger K (1986) DOS 3 409 922 (17.03.1984) Bayer AG. Chem Abstr 104:88589y
Hagen SE, Domagala JM, Heifetz CL, Johnson J (1991) Synthesis and biological activity of 5-alkyl-1,7,8-trisubstituted-6-fluoroquinoline-3-carboxylic acids. J Med Chem 34:1155
Harris ND (1991) Cyclization of diethyl 3,4-diisobutoxyanilino-methylenemalonate. Synthesis 256
Hayakawa I, Kimura Y (1990) European patent 341 493 (27.04.1988), Daiichi Seiyaku. Chem Abstr 113:40473f
Hayakawa I, Tanaka Y (1984) Facile synthesis of thiazolo[4,5-b]- and thieno[3,2-b]pyridine derivatives by a novel pyridine cyclization reaction via enamine intermediates. Heterocycles 22:1697
Hayakawa I, Hiramitsu T, Tanaka Y (1984) Synthesis and antibacterial activities of substituted 7-oxo-2,3-dihydro-7H-pyrido[1,2,3-de] [1,4]benzoxazine-6-carboxylic acids. Chem Pharm Bull (Tokyo) 32:4907

Hayakawa I, Atarashi S, Yokohama S, Imamura M, Sakano M, Furukawa M (1986) Synthesis and antibacterial activities of optically active ofloxacin. Antimicrob Agents Chemother 29:163

Heindl J, Kelm HW, Dogs E, Seeger A, Herrmann C (1977) Untersuchungen über die antibakterielle Aktivität von Chinoloncarbonsäuren. IX (1). Azaanaloga. Di- und trisubstituierte 1,4-Dihydro-4-oxo-1,5-naphthyridin-3-carbonsäuren und 1-Äthyl-4-pyridon-3-carbonsäure. Eur J Med Chem 12:549

Hirose T, Mishio S, Matsumoto J, Minami S (1982) Pyridone-carboxylic acids as antibacterial agents. I. Synthesis and antibacterial activity of 1-alkyl-1,4-dihydro-4-oxo-1,8- and 1,6-naphthyridine-3-carboxylic acids. Chem Pharm Bull (Tokyo) 30:2399

Högberg T, Khannal I, Drake SD, Mitscher LA, Shen LL (1984) Structure-activity relationships among DNA gyrase inhibitors. Synthesis and biological evaluation of 1,2-dihydro-4,4-dimethyl-1-oxo-2-naphthalenecarboxylic acids as 1-carba bioisosteres of oxolinic acid. J Med Chem 27:306

Huffman KR, Schaefer FC, Peters GA (1962) Reaction of s-triazine with acidic α-methylene compounds. J Org Chem 27:551

Irikura T, Koga H, Murayama S (1981) Belgian patent 887 574 (19.08.1980), Kyorin Pharm Co. Chem Abstr 95:187096n

Irikura T, Suzue S, Murayama S, Hirai K, Ishizaki T (1987) European patent 195 316 (08.03.1985), Kyorin Pharm. Chem Abstr 106:50068f

Ishikawa H, Tabusa F, Miyamoto H, Kano M, Ueda H, Tamaoka H, Nakagawa K (1989) Studies on antibacterial agents. I. Synthesis of substituted 6,7-dihydro-1-oxo-1H,5H-benzo[i,j]-quinolizine-2-carboxylic acids. Chem Pharm Bull (Tokyo) 37:2103

Ishikawa H, Uno T, Miyamoto H, Ueda H, Tamaoka H, Tominage M, Nakagawa K (1990) Studies on antibacterial agents. II. Synthesis and antibacterial activities of substituted 1,2-dihydro-6-oxo-6H-pyrrolo[3,2,1-ij]quinoline-5-carboxylic acids. Chem Pharm Bull (Tokyo) 38:2459

Ishizaki T, Suzue S, Irikura T (1985) New synthesis of fluoroquinolonecarboxylic acid. J Chem Soc Jpn 10:2054

Itoh Y, Kato H, Ogawa N, Koshinaka E, Suzuki T, Yagi N (1985) DOS 3 433 924 (19.09.1983), Horikuri Pharm Co. Chem Abstr 103:123517b

Jefson MR, McGuirk (1988) European patent 215 650 (18.05.1985), Pfizer Inc. Chem Abstr 108:94417q

Jefson MR, McGuirk PR (1989) US patent 4 775 668 (19.08.1986), Pfizer Inc. Chem Abstr 110:57534w

Jinbo Y, Kondo H, Inoue Y, Taguchi M, Tsujishita H, Kotera Y, Sakamoto F, Tsukamoto G (1993a) J Med Chem 36:2621

Itoh Y, Kato H, Ogawa N, Koshinaka E, Suzuki T, Yagi N (1985) DOS 3 433 924 (19.09.1983), Hokuriku Pharm Co. Chem Abstr 103:123517b

Jefson MR, McGuirk PR (1988) European patent 215 650 (18.05.1985), Pfizer Inc. Chem Abstr 108:94417q

Jefson MR, McGuirk PR (1989) US patent 4 775 668 (19.08.1986), Pfizer Inc. Chem Abstr 110:57534w

Jinbo Y, Kondo H, Inoue Y, Taguchi M, Tsujishita H, Kotera Y, Sakamoto F, Tsukamoto G (1993a) Synthesis and antibacterial activity of a new series of tetracyclic pyridone carboxylic acids. J Med Chem 36:2621

Jinbo Y, Taguchi M, Inoue Y, Kondo H, Miyasake T, Tsujishida H, Sakamoto F, Tsukamoto G (1993b) Synthesis and antibacterial activity of a new series of tetracyclic pyridone carboxylic acids. J Med Chem 36:3148

Jinbo Y, Kondo H, Inoue Y, Taguchi M, Tsujishita H, Kotera Y, Sakamoto F, Tsukamoto G (1994) Synthesis and antibacterial activity of thiazolopyrazine-incorporated tetracyclic quinolone antibacterials. J Med Chem 37:586

Jordis U, Sauter F, Rudolf M, Gan C (1988) Synthesen neuer Chinolon-Chemotherapeutika, 1. Mitt. Pyridochinoline und Pyridophenanthroline als "lin-benzo-Nalidixinsäure"-Derivate. Monatsschr Chem 119:761

Kametani T, Kigasawa K, Hiiragi M, Wakisaka K, Kusama O, Sugi H, Kawasaki K (1977) Synthetic studies on chemotherapeutics. II. (1) Synthesis of phenyl-substituted 1,4-dihydro-4-oxonicotinic acid derivatives. [Studies on the syntheses of heterocyclic compounds. Part 704 (2)]. J Heterocycl Chem 14:477

Kaminsky D, Meltzer RJ (1968) Quinolone antibacterial agents. Oxolinic acid and related compounds. J Med Chem 11:160

Kilbourn EE, Seidel MS (1972) Synthesis of n-alkyl-3-carboxy-4-pyridones. J Org Chem 37:1145

Kimura Y, Miyamoto T, Matsumoto T, Minami S (1976) Syntheses of pyrimido[4,5-c]pyridazine derivatives. I. A novel reaction of α-diazo-β-oxo-5-(4-chloropyrimidine)propionate with hydrazine leading to 1,2-dihydro-4-hydroxypyrimido[4,5-c]pyridazine-3-carboxamide. Chem Pharm Bull (Tokyo) 24:2637

Koga H, Itoh A, Murayama S, Suzue S, Irikura T (1980) Structure-activity relationships of antibacterial 6,7- and 7,8-disubstituted 1-alkyl-1,4-dihydro-4-oxoquinoline-3-carboxylic acids. J Med Chem 23:1358

Kondo H, Taguchi M, Inoue Y, Sakamoto F, Tsukamoto G (1990) Synthesis and antibacterial activity of thiazolo-, oxazolo-, and imidazolo[3,2-a][1,8]naphthyridinecarboxylic acids. J Med Chem 33:2012

Laborde E, Kiely JS, Culbertson TP, Lesheski LE (1993) Quinolone antibacterials: synthesis and biological activity of carbon isosteres of the 1-piperazinyl and 3-amino-1-pyrrolidinyl side chains. J Med Chem 36:1964

Le Hao Dong P, Coquelet C, Bastide JM, Lebecq JC (1981) Nouveaux agents antibactériens II. Dérivés pyrazolo-azaquinoléiniques. Eur J Med Chem Chim Ther 16:39

Lesher GY, Carabateas PM (1973) DOS 2 224 090 (30.11.1972), Sterling Drug Inc. Chem Abstr 78:84280n

Lesher GY, Froelich EJ, Gruett MD, Bailey JH, Brundage RP (1962) 1,8-naphthyridine derivatives. A new class of chemotherapeutic agents. J Med Pharm Chem 5:1063

Leysen DC, Haemers A, Bollaert W (1984) Thiazolopyridine analogs of nalidixic acid. 2. Thiazolo[4,5-b]pyridines. J Heterocycl Chem 21:1361

Liu J, Guo H (1992) Synthesis and antibacterial activity of 1-(substituted pyrrolyl)-7-(substituted amino)-6-fluoro-1,4-dihydro-4-oxo-3-quinolinecarboxylic acids. J Med Chem 35:3469

Markees DG, Schwab LS, Vegotsky A (1974) Synthesis and antibacterial activity of some substituted 4-quinolone-3-carboxylic acids. J Med Chem 17:137

Matsumoto J, Minami S (1975) Pyrido [2,3-d] pyrimidine antibacterial agents. 3.8-alkyl- and 8-vinyl-5,8-dihydro-5-oxo-2-(1-piperazinyl)pyrido[2,3-d]pyrimidine-6-carboxylic acids and their derivatives. J Med Chem 18:74

Matsumoto J, Miyamoto T, Minamida A, Nishimura Y, Egawa H, Nishimura H (1984a) Synthesis of fluorinated pyridines by the Balz-Schiemann Reaction. An alternative route to enoxacin, a new antibacterial pyridonecarboxylic acid. J Heterocycl Chem 21:673

Matsumoto J, Miyamoto T, Minamida A, Nishimura Y, Egawa H, Nishimura H (1984b) Pyridonecarboxylic acids as antibacterial agents. 2. Synthesis and structure-activity relationships of 1,6,7-trisubstituted 1,4-dihydro-4-oxo-1,8-naphthyridine-3-carboxylic acids, including enoxacin, a new antibacterial agent. J Heterocycl Chem 27:292

Matsumoto J, Miyamoto T, Egawa H, Nakamura S (1988) European patent 242 789 (25.04.1986), Dainippon Pharm. Chem Abstr 108:150325x

Meth-Cohn O (1986) The synthesis of quinolines from N-alkylformanilides and activated acetic acids. Synthesis 76

Minami S, Shono T, Matsumoto J (1971) Pyrido [2,3-d] pyrimidine antibacterial agents. II. Piromidic acid and related compounds. Chem Pharm Bull (Tokyo) 19:1426

Mitscher LA, Flynn DL, Gracey HE, Drake SD (1979) Quinolone antimicrobial agents. 2. Methylenedioxy positional isomers of oxolinic acid. J Med Chem 22:1354

Mitscher LA, Sharma PN, Chu DTW, Shen LL, Pernet AG (1987) Chiral DNA Gyrase inhibitors. 2. Asymmetric synthesis and biological activity of the enantiomers of 9-fluoro-3-methyl-10-(4-methyl-1-piperazinyl)-7-oxo-2,3-dihydro-7H-pyrido[1,2,3-de]-1,4-benzoxazine-6-carboxylic acid (ofloxacin). J Med Chem 30:2283

Miyamoto H, Ueda H, Otsuka T, Aki S, Tamaoka H, Tominaga M, Nakagawa K (1990) Studies on antibacterial agents. III. Synthesis and antibacterial activities of substituted 1,4-dihydro-8-methyl-4-oxoquinoline-3-carboxylic acids. Chem Pharm Bull (Tokyo) 38:2472

Miyamoto T, Matsumoto T (1988) A new cinnoline ring construction by the reaction of 2-diazo-3-(2-fluorophenyl)-3-oxopropionates with tri-n-butylphosphine. Chem Pharm Bull (Tokyo) 36:1321

Miyamoto T, Matsumoto J (1990a) Fluorinated pyrido[2,3-c]pyridazines. II. Synthesis and antibacterial activity of 1,7-disubstituted 6-fluoro-4-(1H)-oxopyrido[2,3-c]pyridazine-3-carboxylic acids. Chem Pharm Bull (Tokyo) 38:3359

Miyamoto T, Matsumoto T (1990b) Fluorinated pyrido[2,3-c]pyridazines. I. Reductive cyclization of ethyl 2-diazo-2-(5-fluoro-2-halonicotinoyl)acetate with trialkylphosphine. Chem Pharm Bull (Tokyo) 38:3211

Miyamoto T, Egawa H, Matsumoto J (1987a) Pyridonecarboxylic acids as antibacterial agents. VIII. An alternative synthesis of enoxacin via fluoronicotinic acid derivatives. Chem Pharm Bull (Tokyo) 35:2280

Miyamoto T, Egawa H, Shibamori K, Matsumoto J (1987b) Synthesis and reactions of 7-substituted 1-cyclopropyl-6-fluoro-1,4-dihydro-4-oxo-1,8-naphthyridine-3-carboxylic acids as an antibacterial agent. J Heterocycl Chem 24:1333

Miyamoto T, Matsumoto J, Chiba K, Egawa H, Shibamori K, Minamida A, Nishimura Y, Okada H, Kataoka M, Fujita M, Hirose T, Nakano J (1990) Synthesis and structure-activity relationships of 5-substituted 6,8-difluoroquinolones, including Sparfloxacin, a new quinolone antibacterial agent with improved potency. J Med Chem 33:1645

Miyamoto T, Kimura Y, Matsumoto J, Minami S (1978) Syntheses of pyrimido[4,5-c]pyridazine derivatives. II. A novel reaction of α-diazo-β-oxo-5-(4-chloropyrimidine)propionate with triphenyl phosphine leading to 1,4-dihydro-4-oxopyrimido-[4,5-c] pyridazine-3-carboxylate. Chem Pharm Bull (Tokyo) 26:14

Moran DB, Ziegler CB Jr, Dunne TS, Kuck NA, Lin Y (1989) Synthesis of novel 5-fluoro analogues of norfloxacin and ciprofloxacin. J Med Chem 32:1313

Nagano H, Yokota T, Katok Y (1990) European patent 342 675 (19.05.1988), Chugai Pharm. Chem Abstr 113:6175j

Narita H, Konishi Y, Nitta J, Nagaki H, Kitayama I, Watanabe Y, Saikawa I (1986a) Pyridonecarboxylic acids as antibacterial agents. I. Synthesis and structure-activity relationship of 1-aryl-6-(4-dimethylaminophenyl)-4-pyridone-3-carboxylic acids. Yakugaku Zasshi 106:775

Narita H, Konishi Y, Nitta J, Kobayashi Y, Watanabe Y, Minami S, Saikawa I (1986b) Pyridonecarboxylic acids as antibacterial agents. II. Synthesis and structure-activity relationship of 1-(4-hydroxyphenyl)-6-substituted-4-pyridone-3-carboxylic acids. Yakugaku Zasshi 106:782

Narita H, Konishi Y, Nitta J, Miyazima M, Watanabe Y, Yotsuji A, Saikawa I (1986c) Pyridonecarboxylic acids as antibacterial agents. III. Synthesis and structure-activity relationship of 1-(4-fluorophenyl)- and 1-(2,4-difluorophenyl)-6-substituted-4-pyridone-3-carboxylic acids. Yakugaku Zasshi 106:788

Narita H, Konishi Y, Nitta J, Kittayama I, Miyazima M, Watanabe Y, Yotsuji A (1986d) Pyridonecarboxylic acids as antibacterial agents. V. Synthesis and structure-activity relationship of 7-amino-6-fluoro-1-(fluorophenyl)-4-oxo-1,8-naphthyridine-3-carboxylic acids. Yakugaku Zasshi 106:802

Narita H, Todo J, Nitta J, Nakagi H, Lino F, Miyajima M, Fukuoka Y, Saikawa I (1990) DOS 3 913 245 (23.04.1988), Toyama Chem. Chem Abstr 113:78417n

Nishimura Y, Matsumoto J (1987) Pyridonecarboxylic acids as antibacterial agents. 9. Synthesis and antibacterial activity of 1-substituted 6-fluoro-1,4,-dihydro-4-oxo-7-(4-pyridyl)-1,8-naphthyridine-3-carboxylic acids. J Med Chem 30:1622

O'Neill BT (1990) European patent 370 686 (23.11.1988), Pfizer Inc. Chem Abstr 113:191182v

Ockenden DW, Schofield K (1953) Cinnolines. Part XXXIII. Some 3-aryl-4-hydroxycinnolines. J Chem Soc 3706

Okada T, Tsuji T, Tsushima T, Ezumi K, Yoshida T, Matsuura S (1991a) Synthesis and antibacterial activities of novel oxazine and thiazine ring-fused tricyclic quinolonecarboxylic acids: 10-(alicyclic amino)-9-fluoro-7-oxo-7H-pyrido[1,2,3-de][1,4]benzoxazine-6-carboxylic acids and the corresponding 1-Thia congeners. J Heterocycl Chem 28:1067

Okada T, Tsuji T, Tsushima T, Yoshida T, Matsuura S (1991b) Synthesis and antibacterial activities of novel dihydrooxazine and dihydrothiazine ring-fused tricyclic quinolonecarboxylic acids: 9-fluoro-3-methylene-10-(4-methylpiperazin-1-yl)-7-oxo-2,3-dihydro-7H-pyrido[1,2,3-de][1,4]benzoxazine-6-carboxylic acid and its 1-Thia congeners. J Heterocycl Chem 28:1061

Okumura K, Inoue K, Tomie M, Adachi T, Kondo K (1974a) Japanese patent 7 410 519 (19.02.1969), Tanabe Seiyaku. Chem Abstr 81:120503k

Okumura K, Inoue K, Tomie M, Adachi T, Kondo K (1974b) Japanese patent 7 410 520 (19.02.1969), Tanabe Seiyaku. Chem Abstr 81:120504m

Otsubo J, Manabe Y, Kazuyuki K (1982) Belgian patent 891 537 (18.12.1980), Otsuka Pharm Co. Chem Abstr 97:92321j

Parikh VD, Fray AH, Kleinman EF (1988) Synthesis of 8,9-difluoro-2-methyl-6-oxo-1,2-dihydropyrrolo[3,2,1-ij]quinoline-5-carboxylic acid. J Heterocycl Chem 25:1567

Pesson M, Chabassier S (1974) Chimie organique – nitriles et amide-oximes d'acides alkyl-8 oxo-5 dihydro-5.8 pyrido [2,3-d] pyrimidine-6 carboxyliques. C R Acad Sci [C] 279:413

Pesson M, Antoine M, Chabassier S, Geiger S, Girard P, Richer D, de Lajudie P, Horvath E, Leriche B, Patte S (1974a) Antibactériens dérivés des acides alkyl-8-oxo-5-dihydro-5,8 pyrido [2,3-d] pyrimidine-6 carboxyliques. I. Nouveau procédé de préparation. Eur J Med Chem Chim Ther 9:585

Pesson M, Antoine M, Chabassier S, Girard MP, Richer D (1974b) Chimie organique. Synthèse d'acides alkyl-8 oxo-5 dihydro-5.8 pyrido-[2.3-d] pyrimidine-6 carboxyliques. C R Acad Sci [C] 278:717

Pesson M, Antoine M, Chabassier S, Geiger S, Girard P, Richer D, de Lajudie P, Horvath E, Leriche B, Patte S (1974c) Antibactériens dérivés des acides alkyl-8 oxo-5 dihydro-5,8 pyrido [2,3-d] pyrimidine-6 carboxyliques. II. Dérivés pipérazinyl-2 et (alkyl-4 pipérazinyl)-2. Eur J Med Chem Chim Ther 9:591

Pesson M, de Lajudie P, Antoine M, Girard P, Chabassier S (1975) Synthèse d'acides alkyl-1 oxo-4 dihydro-1,4 benzo [h] naphtyridine-1.6 carboxyliques-3 à action antibactérienne. C R Acad Sci [C] 280:1385

Pesson M, de Lajudie P, Antoine M, Chabassier S, Girard P (1976) Acides alkyl-1 oxo-4 dihydro-1,4 pyrido [2.3-b]quinoxaline-3 carboxyliques à action antibactérienne. C R Acad Sci [C] 282:861

Pesson M, Antoine M, Benichon J-L, de Lajudie P, Horvath E, Leriche B, Patte S (1980) Antibactériens de synthèse – dérivés des acides pyrido (2,3-e) as.triazine-7 carboxyliques. Eur J Med Chem 15:269

Petersen U, Grohe K, Zeiler H-J, Metzger K (1986) European patent 167 763 (04.06.1984), Bayer AG. Chem Abstr 104:186447v

Petersen U, Grohe K, Schriewer M, Schenke T, Haller I, Metzger K, Endermann R, Zeiler H-J (1989) DOS 3 711 193 (02.04.1987), Bayer AG. Chem Abstr 110:114697c

Petersen U, Schenke T, Krebs A, Grohe K, Schriewer M, Haller I, Metzger KG, Endermann R, Zeiler HJ (1988) Bayer AG, EP 350 733

Petersen U, Krebs A, Schenke T, Grohe K, Schriewer M, Haller I, Metzger K, Endermann R, Zeiler H-J (1991) European patent 401 623 (07.06.1989), Bayer AG. Chem Abstr 114:207234x

Petersen U, Krebs A, Schenke T, Philipps T, Grohe K, Bremm KD, Endermann R, Metzger KG, Haller I (1992) Bayer AG, EP 550 903; Chem Abstr 120:8616x

Price CC, Roberts RM (1946) The synthesis of 4-hydroxyquinolines. I. Through ethoxymethylenemalonic ester. J Am Chem Soc 68:1204

Prudchenko AT, Shchegoleva GS, Barkhash VA, Vorozhtsov NN Jr (1967) Some reactions of pentafluorobenzoylacetic ester. Zh Obshch Khim 37:2366

Prudchenko AT, Shchegoleva GS, Barkhash VA, Vorozhtsov NN Jr (1968) Reactions of ethyl pentafluorobenzoylacetate. Chem Abstr 69:36059q

Radl S, Hradil P (1991) Synthesis of some 1-alkyl-1,4-dihydro-4-oxo-1,7-naphthyridine-3-carboxylic acids. Collect Czech Chem Commun 56:2420

Radl S, Zikan V (1989) Synthesis and antimicrobial activity of some 3-oxo-3H-pyrido[3,2,1-kl]phenoxazine-2-carboxylic acids. Collect Czech Chem Commun 54:506

Radl S, Zikan V (1989) Synthesis of some 1-aryl-1,4-dihydro-4-oxoquinoline-3-carboxylic acids and their antibacterial activity. Collect Czech Chem Commun 54:2181

Radl S, Houskova V, Zikan V (1988) Priprava nekterych nekondenzovanych derivatu 1,4-dihydro-4-oxo-3-pyridin-A 3-pyridazinkarboxylove kyseliny jako potencialnich antimikrobialne ucinnych latek. Cesk Farm 37:71

Radl S, Houskova V, Zikan V (1989) Priprava substituovanych 1,6-difenyl-1,4-dihydro-4-oxo-3-pyridazinkarboxylovych kyselin. Cesk Farm 38:114

Radl S, Moural J, Bendova R (1990) Synthesis and antibacterial activity of some 1-aryl-1,4-dihydro-4-oxocinnoline-3-carboxylic acids. Collect Czech Commun 55:1311

Radl S, Kovarova L, Moural J, Bendova R (1991) Structural modification and new methods for preparation of ofloxacin analogs. Collect Czech Commun 56:1937

Remuzon P, Bouzard D, Di Cesare P, Essiz M, Jacquet JP, Kiechel JR, Ledonssal B, Kessler RE, Fung-Tomc J (1991) Fluoronaphthyridines and -quinolones as antibacterial agents. 3. Synthesis and structure-activity relationships of new 1-(1,1-dimethyl-2-fluoroethyl), 1-[1-methyl-1-(fluoromethyl)-2-fluoroethyl], and 1-[1,1-(difluoromethyl)-2-fluoroethyl] substituted derivatives. J Med Chem 34:29

Remuzon P, Bouzard D, Di Cesare P, Dussy C, Jacquet J-P, Jaegly A (1992) Synthesis and antibacterial activity of new 5-substituted 1-cyclopropyl-6-fluoro-7-piperazinyl-1,4-dihydro-4-oxo-1,8-naphthyridine-3-carboxylic acids. J Heterocycl Chem 29:985

Rohlfing SR, Gerster JF, Kvam DC (1976) Bioevaluation of the antibacterial flumequine for urinary tract use. Antimicrob Agents Chemother 10:20

Rufer C, Schwarz K (1977) Untersuchungen über die antibakterielle Aktivität von Chinoloncarbonsäuren. VII (1). Azaanaloga. In 2-Position durch Pyrazole substituierte 8-Äthyl-5-oxo-5,8-dihydropyrido[2,3-d]pyrimidin-6-carbonsäuren. Eur J Med Chem Chim Ther 12:236

Sanchez JP, Ashok K, Trehan K, Nichols JB (1987) The synthesis of a pyrido[2,3-c]pyridazine: a cinnoline related to 6-fluoronalidixic acid. J Heterocycl Chem 24:55

Sanchez JP, Bridges AJ, Busch R, Domagala JM, Gogliotti RD, Hagen SE, Heifetz CL, Joannides ET, Sesnie JC, Shapiro MA, Szotek DL (1992) New 8-(trifluoromethyl)-substituted quinolones. The benefits of the 8-fluoro group with reduced phototoxic risk. J Med Chem 35:361

Sandison AA, Tennant G (1974) A new heterocyclisation reaction leading to cinnolin-4(1H)-one derivatives. J Chem Soc Chem Commun 752

Santilli AA, Wanser SV, Kim DH, Scotese AC (1975) Synthesis of 5,6,7,8-tetrahydro-5-oxopyrido[2,3-d]pyrimidine-6-carbonitriles and -6-carboxylic acid esters. J Heterocycl Chem 12:311

Sato K, Matsuura Y, Inoue M, Une T, Osada Y, Ogawa Y, Ogawa H, Mitsuhashi S (1982) In vitro and in vivo activity of DL-8280, a new oxazine derivative. Antimicrob Agents Chemother 22:548

Sauter F, Jordis U, Tanyolac S, Martinek P (1988a) Thieno[3,2-g]chinolin- und [1] Benzothieno-[5,6,7-ij]chinolizincarbonsäurederivate. Arch Pharm (Weinheim) 321:241
Sauter F, Jordis U, Tanyolac S (1988b) Synthesen neuer Chinolon-Chemotherapeutika, 3. Mitt. Sci Pharm 56:73
Sauter F, Jordis U, Martinek P (1989) Synthesis of novel quinolone-type drugs. Part 4. Pyrido[3,2,1-gh][1,7]phenanthroline- and benzo[i,j]quinolizine-carboxylic acids. Sci Pharm 57:7
Sawa Y, Kato T, Masuda T, Hori M, Fujimora H (1975) Studies on the syntheses of analgesics. IV. Syntheses of 1,2,3,4-tetrahydro-5H-benzazepine derivatives. Chem Pharm Bull (Tokyo) 23:1917
Scheer M (1987a) Studies on the antibacterial activity of Baytril. Vet Med Rev 2:90
Scheer M (1987b) Concentrations of active ingredient in the serum and in tissues after oral and parenteral administration of Baytril. Vet Med Rev 2:104
Schofield K, Simpson JCE (1945) Cinnolines. III. The Richter reaction. J Chem Soc 512
Schriewer M, Grohe K (1988) DOS 3 615 767 (10.05.1986), Bayer AG. Chem Abstr 108:94416p
Schriewer M, Grohe K, Zeiler H-J, Metzger K (1987a) DOS 3 543 513 (10.12.1985), Bayer AG. Chem Abstr 107:154342c
Schriewer M, Grohe K, Zeiler H-J, Metzger K (1987b) DOS 3 600 891 (15.01.1986), Bayer AG. Chem Abstr 107:198345k
Schriewer M, Grohe K, Zeiler H-J, Metzger K (1987c) DOS 3 509 546 (16.03.1985), Bayer AG. Chem Abstr 106:4900y
Schriewer M, Grohe K, Hagemann H, Zeiler H-J, Metzger K (1988a) DOS 3 623 757 (15.07.1986), Bayer AG. Chem Abstr 109:86326q
Schriewer M, Grohe K, Petersen U, Haller I, Metzger K, Endermann R, Zeiler H-J (1988b) DOS 3 702 393 (28.01.1987), Bayer AG. Chem Abstr 109:230824v
Schroeder MC, Kiely JS (1988) Synthesis of a novel tricyclic 4-quinolone. Incorporation of a spiro-cyclopropyl group at N1 by bridging to C2. J Heterocycl Chem 25:1796
Schwan TJ, Freedman R, Pollack JR (1983) Synthesis, structure elucidation and antibacterial activity of 6-ethyl-6,9-dihydro-9-oxopyrazolo[3,4-f]quinoline-8-carboxylic acid. J Heterocycl Chem 20:1351
Segawa J, Kitano M, Kazuno K, Tsuda M, Shirahase I, Ozaki M, Matsuda M, Kise M (1992a) Studies on pyridonecarboxylic acids [1]. 2. Synthesis and antibacterial activity of 8-substituted 7-fluoro-5-oxo-5H-thiazolo[3,2-a] quinoline-4-carboxylic acids. J Heterocycl Chem 29:1117
Segawa J, Kitano M, Kazuno K, Matsuoka M, Shirahase I, Ozaki M, Matsuda M, Tomii Y, Kise M (1992b) Studies on pyridonecarboxylic acids. 1. Synthesis and antibacterial evaluation of 7-substituted-6-halo-4-oxo-4H-[1,3] thiazeto[3,2-a] quinoline-3-carboxylic acids. J Med Chem 35:4727
Shoup RR, Castle RN (1965) Cinnoline chemistry. XI. The ultraviolet spectra of halogen substituted 4-hydroxy- and 4-mercaptocinnolines. J Heterocycl Chem 2:63
Singh B (1991) Synthesis of 2-aza analog of rosoxacin. J Heterocycl Chem 28:881
Staiger RP, Miller EB (1959) Isatoic anhydride. IV. Reactions with various nucleophiles. J Org Chem 24:1214
Stanovnik B, Tisler M (1974) Convenient cyclization of o-difunctional heterocycles with N,N-dimethylformamide dimethyl acetal. Synthesis 120
Stefancich G, Artico M, Corelli F, Massa S, Panico S, Simonetti N (1985) 1-ethyl-6-fluoro-1,4-dihydro-4-oxo-7-(1H-pyrrol-1-yl)-quinoline-3-carboxylic acid, a new fluorinated compound of oxacin family with high broad-spectrum antibacterial activities. Farmaco [Sci] 40:237
Strehlke P (1977) Untersuchungen über die antibakterielle Aktivität von Chinoloncarbonsäuren. VIII (1) Azaanaloga. Eine neue Synthese antibak-

teriell wirksamer 1,7-disubstituierter 1,4-Dihydro-4-oxo-1,6-naphthyridin-3-carbonsäuren. Eur J Med Chem Chim Ther 12:541

Suzuki N, Tanaka Y, Dohmori R (1979) Synthesis of antimicrobial agents. I. Synthesis and antimicrobial activities of thiazoloquinoline derivatives. Chem Pharm Bull (Tokyo) 27:1

Taguchi M, Kondo H, Inoue Y, Kawahata Y, Jinbo Y, Sakamoto F, Tsukamoto G (1992) Synthesis and antibacterial activity of new tetracyclic quinolone antibacterials. J Med Chem 35:94

Taylor EC, Heindel ND (1967) Cyclizations of anthranilate-acetylenedicarboxylate adducts. A facile route to 2,8-dicarboalkoxy-4(1H)-quinolinones. J Org Chem 32:3339

Terni P, Rugarli PL, Maiorana S, Pagella PG, Fusco R (1988) European patent 252, 352 (01.07.1986), Mediolanum Farm. Chem Abstr 109:129034n

Todo Y, Yamafuji T, Nagumo K, Kitayama I, Nagaki H, Miyajima M, Konishi Y, Narita H, Takano S, Seikawa I (1987) DOS 3 601 517 (20.01.1986), Toyama Chem. Chem Abstr 107:77777u

Toja E, Kettenring J, Goldstein B, Tarzia G (1986a) Pyrrolopyridine analogs of nalidixic acid. 2. Pyrrolo[3,4-b]pyridines. J Heterocycl Chem 23:1561

Toja E, Tarzia G, Ferrari P, Tuan G (1986b) Pyrrolopyridine analogs of nalidixic acid. 1. Pyrrolo [2,3-b] pyridines. J Heterocycl Chem 23:1555

Von Liebig J (1853) Ueber Kynurensäure. Ann Chem 86:125

Von Liebig J (1858) Ueber Kreatin und Kynurensäure im Hundeharn. Ann Chem 108:354

Von Richter V (1883) Ueber Cinnolinderivate. Chem Ber 16:677

Wentland MP, Lesher GY, Reumann M, Gruett MD, Singh B, Aldous SC, Dorff PH, Rake JB, Coughlin SA (1993) Mammalian topoisomerase II inhibitory activity of 1-cyclopropyl-6,8-difluoro-1,4-dihydro-7-(2,6-dimethyl-4-pyridinyl)-4-oxo-3-quinolinecarboxylic acid and related derivatives. J Med Chem 36:2801

White WA (1970) DOS 2 005 104 (06.08.1970), Eli Lilly & Co. Chem Abstr 73:77269j

Wick AE (1979) DOS 2 901 868 (18.01.1978), Hoffmann-La Roche. Chem Abstr 91:211273h

Xiao W, Krishnan R, Lin Y-I, Delos Santos EF, Kuck NA, Babine RE, Lang SA Jr (1989) Synthesis and in vitro antibacterial activity of some 1-(difluoromethoxyphenyl)quinolone-3-carboxylic acids. J Pharm Sci 78:585

Zeiler H-J, Grohe K (1984) The in vitro and in vivo activity of ciprofloxacin. Eur J Clin Microbiol 3:339

Zhang MQ, Haemers A, Vanden Berghe D, Pattyn SR, Bollaert W (1991) Quinolone antibacterials. 2. 6-substituted-7-(2-thiazolyl and thiazolidinyl)quinolones. J Heterocycl Chem 28:685

Ziegler CB Jr, Moran DB, Fenton TJ, Lin Y-I (1990) The Synthesis and biological activity of 8-fluoro-9-(4-methyl-1-piperazinyl)-6-oxo-6H-benzo[c]quinolizine-5-carboxylic acid. J Heterocycl Chem 27:587

CHAPTER 3

The Chemistry of the Quinolones: Chemistry in the Periphery of the Quinolones

U. PETERSEN and T. SCHENKE

A. Introduction

Numerous reviews have been published to date on the structure–activity relationships of quinolones (ANDRIOLE 1988; ASAHINA et al. 1992; DOMAGALA 1994; MITSCHER et al. 1990, 1993; RÁDL 1990; ROSEN 1990; SCHENTAG and DOMAGALA 1985; WENTLAND 1990); therefore these will be mentioned only briefly in connection with the discussion of chemical reactions. This chapter focuses on the chemistry of the quinolone parent substance. There have also been several reviews published on the chemistry of quinolones (ALBRECHT 1977; BOUZARD 1990; CHU 1993; CHU and FERNANDES 1991; LESHER 1978; LEYSEN et al. 1991a,b; RÁDL and BOUZARD 1992; MITCHER et al. 1988).

B. 1-Position

A prerequisite for potent antibacterial activity of quinolones is that the N atom in the 1-position is substituted. For this, radicals such as ethyl, 2-fluoroethyl, tert-butyl, vinyl, cyclopropyl, 1*R*,2*S*-2-fluorocyclopropyl, 4-fluorophenyl, 2,4-difluorophenyl and 5-fluoro-2-pyridyl are particularly suitable. Earlier calculations of structure–activity relationships, which suggest that substituents such as ethyl or groups of similar bulk such as vinyl, or fluoroethyl in the 1-position result in an optimum activity (KOGA et al. 1980), have not been confirmed. Significant increases in activity have been achieved by radicals which cause more steric hindrance, for example cyclopropyl, tert-butyl or 2,4-difluorophenyl. By introducing a fluorine atom into the cyclopropyl ring, the lipophilicity is reduced compared with non-fluorinated quinolones (ATARASHI et al. 1993; KIMURA et al. 1994b). It was shown that increasing lipophilicity at N-1 of 1-aryl-quinolones does not correlate with increased potency against mycobacteria (RENAU et al. 1995).

Replacing the nitrogen atom with oxygen (HÖGBERG et al. 1984a) or carbon (HÖGBERG et al. 1984b) leads to compounds which have no activity. On the other hand, some 1-sulphur analogues display an in vitro activity comparable to nalidixic acid, but which is much lower than that of the correspondingly substituted quinolones such as lomefloxacin (CECCHETTI et al. 1993a; cf. also WENTLAND et al. 1993b; TOYAMA 1994). In contrast, the "2-pyridones" (HECK and THORSETT 1987; CHU et al. 1993a,b; LI et al. 1996), such as A-

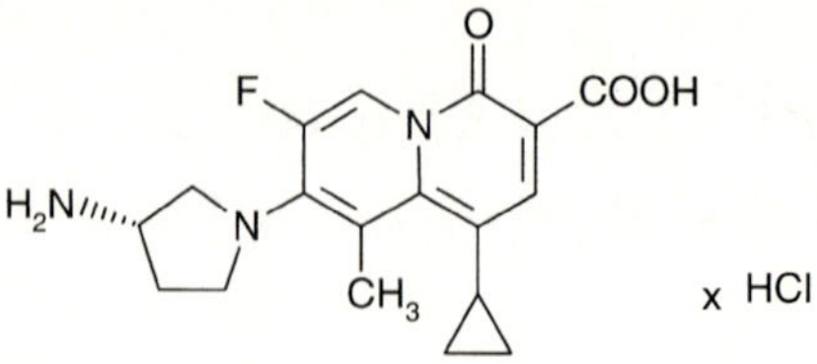

Fig. 1. Chemical structure of A-86719.1

86719.1 [8-(3*S*-aminopyrrolidin-1-yl)-1-cyclopropyl-7-fluoro-9-methyl-4-oxo-4*H*-quinolizine-3-carboxylic acid hydrochloride] (ABT-719; Fig. 1; CHU et al. 1994) and A-104954 [8-(3*R*-[1*S*-aminoethyl]-pyrrolidin-1-yl)-1-cyclopropyl-7-fluoro-9-methyl-4-oxo-4*H*-quinolizine-3-carboxylic acid], and the tricyclic pyridone A-84066 [(3S)-9-fluoro-3-methyl-10-(4-methyl-1-piperazinyl)-2*H*,3*H*,6*H*-6-oxo-pyrano[2,3,4-i,j]quinolizine-5-carboxylic acid] (LI et al. 1995a) have a high in vitro and in vivo activity. These 2-pyridones differ from the quinolones in that, in the former, the 1-substituent is on a carbon atom and the nitrogen atom has been shifted from the 1-position into the bridge-head position alongside the carbonyl group (see also Chap. 2BVIII).

As described in Chap. 2 BI, in the earlier processes the substituents were introduced at the 1-position of the quinolones by alkylation of the 4-hydroxyquinolinecarboxylic acid esters after the quinolone nucleus had been built up (for the tautomerism of 1-unsubstituted quinolones, see DE LA CRUZ et al. 1992). One of the first antibacterially active quinolones, 1,4-dihydro-1-methyl-6-nitro-4-oxo-3-quinolinecarboxylic acid, was prepared as early as 1954 via 1-methylation with dimethyl sulphate (ICI 1957). Direct alkylation is now usually carried out with alkyl halides, for example with ethyl iodide, in DMF in the presence of potassium carbonate (see, for example, NISHIMURA et al. 1988b) or sodium hydride (RADL 1994a), and with alkylsulphonic acid esters, for example with 2-fluoroethyl tosylate (see, for example, EGAWA et al. 1984). Other alkylating agents, such as triethyl phosphate (GAREMSZEGI et al. 1980; HERMECZ et al. 1987b), are also occasionally used. Unsaturated radicals, such as 1-vinyl and 1-isopropenyl, are introduced into the 1-position via dehydrohalogenation of 1-haloalkylquinolones (MATSUMOTO and MINAMI 1975; KOGA et al. 1980; EGAWA et al. 1984; MATSUMOTO et al. 1984a; NISHIMURA et al. 1988b; DOMAGALA et al. 1988b), by catalytic vinylation in the presence of Na_2 $(PdCl_4)$ (MCGUIRK et al. 1992), or via elimination reactions of quaternary ammonium iodides or by thermal decomposition of sulphoxides (BOUZARD et al. 1989). Substitution at the 1-position is also possible with activated haloaromatics, such as 2,4-dinitro-chlorobenzene (CHEIL PHARM. 1990), 2,4-dinitro-fluorobenzene (ZIKAN and RADL 1987) and 4-nitro-fluorobenzene (ZIKAN et al. 1987). The 1-methylamino group in amifloxacin has been introduced in three steps via an amination reaction with *O*-(2,4-dinitrophenyl)-hydroxylamine and subsequent methylation (WENTLAND et al. 1984). Here also, it is simpler to use the cycloaracylation method (GROHE et al. 1979, 1984a; GROHE and HEITZER 1987; CHU 1985). For a review of other methods see LEYSEN et al. (1991b).

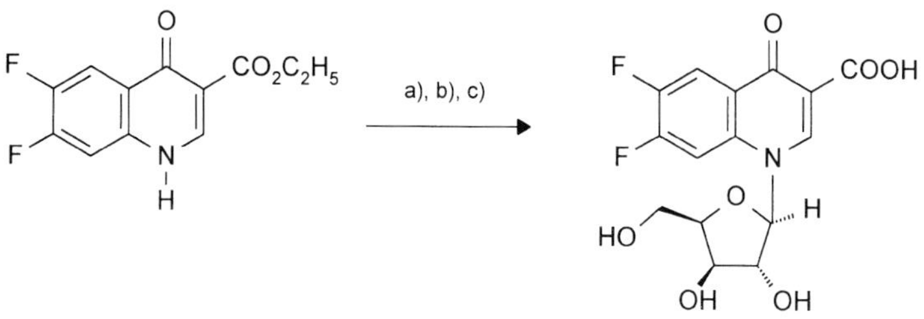

Fig. 2. Synthesis of a quinolone *N*-nucleoside. *a*, Tms-Cl, $NH(Tms)_2$, reflux; *b*, , $SnCl_4$, $Cl\text{-}CH_2CH_2\text{-}Cl$, 20°C; *c*, NaOH, MeOH, 20°

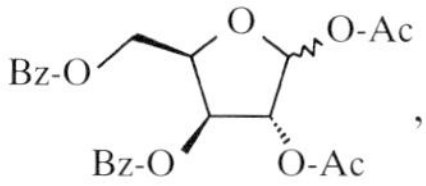

An *N*-nucleoside from a quinolone and a xylofuranose was prepared under known conditions (NIEDBALLA and VORBRÜGGEN 1974) by reaction of ethyl 6,7-difluoro-4-trimethylsilyloxy-3-quinolinecarboxylate with 1,2-di-*O*-acetyl-3,5-di-*O*-benzoyl-D-xylofuranose in dichloroethane in the presence of $SnCl_4$ (Fig. 2). Surprisingly, with penta-*O*-acetyl-α-D-glucopyranose, the corresponding C_8-glycoside is obtained under these conditions (TOLSTIKOV et al. 1993). On the other hand, an ethyl 6,7,8-trifluoro-1-(2,3,4,6-tetra-*O*-acetyl-β-D-glucopyranosyl)-3-quinolonecarboxylate was prepared via the cycloaracylation method starting from 2,3,4,6-tetra-*O*-acetyl-β-D-glucopyranosylamine (DE LA CRUZ et al. 1990).

C. 2-Position

Replacing the ring carbon atom at the 2-position of quinolones with a nitrogen atom leads to the cinnolones (see Chap. 2BIX), which generally have a weaker activity. In contrast to these and to some 1,2- or 2,3-bridged quinolones (see Chap. 2BI, 2BIII), the 2-substituted quinolones prepared to date have proven to have no activity. A general method has been described for introducing substituents (methyl, vinyl, phenyl) into the 2-position, for example by reacting a Grignard reagent with quinolone- or naphthyridonecarboxylic acid esters in the presence of copper(I)iodide in a Michael addition and reintroducing the 2,3-double bond by phenylselenation of the 3-position, oxidation with H_2O_2 and in situ syn-elimination (KIELY et al. 1989; WENTLAND et al. 1993b).

D. 3-Position

The carboxyl group at the 3-position of quinolones can form hydrogen bridges to the adjacent 4-carbonyl group, which are regarded as being necessary for bonding to DNA gyrase (SCHENTAG and DOMAGALA 1985). It was demonstrated earlier that modifications of the 3-carboxyl group or its replacement by other radicals causes a loss in the activity of quinolones (Albrecht 1977). This applies just as much to quinolones in which the 3-carboxyl group has been replaced by the 1*H*-tetrazol-5-yl radical (Gilis et al. 1980) as it does to the 3-nitro-quinolones prepared recently by the cycloaracylation method (Radl and Chan 1994) and to the 3-aminoquinolones obtained from these by catalytic hydrogenation (Radl 1994b).

On the other hand, prodrugs with 3-substituents which can be converted into a carboxyl group in vivo often show potent antibacterial activity together with modified pharmacokinetic properties. These prodrugs include esters such as the relatively unstable 5-methyl-2-oxo-1,3-dioxol-4-ylmethyl ester of norfloxacin (Sakamoto et al. 1985), the phenoxymethyl ester or alkoxy-polyethylene glycol esters of oxolinic acid (Loubinoux et al. 1991) and the 3-formyl-quinolones. For example, "3-formyl-norfloxacin", which has a weak antibacterial activity in vitro and in which the formyl group is metabolized rapidly in vivo to give the carboxyl group, reaches serum levels of norfloxacin in mice which are twice as high as those with norfloxacin itself (Kondo et al. 1988). "3-Formyl-tosufloxacin" has a better water solubility than tosufloxacin and reaches serum levels which are four times higher than those of tosufloxacin after oral administration in the dog (Chu et al. 1990). Reference may be made to the literature regarding 1-cyclopropyl-6-fluoro-1,4-dihydro-4-oxo-7-(1-piperazinyl)-quinoline-3-carbaldehyde ("3-formyl-ciprofloxacin"; Kondo et al. 1988), 1-cyclopropyl-6-fluoro-5-methyl-7-(3-methyl-1-piperazinyl)-1,4-dihydro-4-oxo-quinoline-3-carbaldehyde ("3-formyl-grepafloxacin"; Miyamoto et al. 1988/1989) and 1-cyclopropyl-6-fluoro-1,4-dihydro-7-[(1*S*,4*S*)-5-methyl-2,5-diazabicyclo[2.2.1]hept-2-yl]-4-oxo-quinoline-3-carbaldehyde ("3-formyl-danofloxacin"; Jefson 1992).

Simple alkyl esters are obtained not only by reacting a corresponding quinolone ester or naphthyridone ester precursor with amines in the 7-position, but also by direct esterification of the carboxylic acids with alcohols in the presence of strong acids such as hydrochloric acid (Grohe et al. 1985). Other ester radicals are introduced via the alkali metal salts (Na, K, Cs) of the quinolonecarboxylic acids by reaction with alkyl halides (e.g. Grohe et al. 1985; Ochi and Shimizu 1992). A titanate-catalysed transesterification (Seebach et al. 1982) with benzyl alcohol and subsequent cleavage of the resulting benzyl ester in methanol with formic acid/Pd-C can be carried out in those cases in which acid or alkaline ester hydrolysis presents problems, as in the case of the 6*H*-6-oxopyrido[1,2-a]pyrimidine-7-carboxylic acid esters (Chu et al. 1993a).

In specific types of the intensively investigated dual action compounds (cephalosporins, penems, carbapenems), the 3-carboxyl group of the quinolones is linked as an ester (Demuth and White 1988/1989b,c; Albrecht 1990; Okabe 1990; Demuth et al. 1991; Corraz et al. 1992; Keith et al. 1993; Okabe and Sun 1992; Perrone et al. 1992), thioester (Chan and Keith 1987/1988; Demuth and White 1988/1989a), amide or hydrazide (Demuth and White 1988/1989a) with a β-lactam unit. Two possibilities for the synthesis of quinolonecarboxylic acid esters linked with the 3-hydroxymethyl group of cephalosporins have been described:

1. The nucleophilic substitution reaction whereby a 3-halomethyl-cephalosporin, in which the carboxyl group must be protected, for example as the t-butyl or benzhydryl ester, is reacted with the alkali metal salt of a quinolone. In this reaction under alkaline conditions migration of the double bond in the 2,3-position of the cephalosporin may occur. The synthesis of Ro 23-5068 (Fig. 3) for example has been carried out in this way (ALBRECHT 1990; KEITH et al. 1993).
2. The direct acylation of the 3-hydroxymethyl group of a cephalosporin with an unprotected carboxyl group is possible if the quinolone is activated by conversion into a mixed anhydride (preferably with cyclohexyl chloroformate). As an example of this reaction, the preparation of Ro 23-9424 (as the dihydrochloride) from *N*-trityl-deacetylcefotaxime and a mixed anhydride of fleroxacin is shown in Fig. 4 (OKABE and SUN 1992; KEITH et al. 1993). In this reaction, which has been carried out on the kilogram scale, migration of the double bond in the cephalosporin presents no problem.

For the synthesis of 3-formyl-quinolones, quinolonecarboxylic acids are reduced with sodium borohydride in methanol to give 1,2,3,4-tetrahydroquinolone-3-carboxylic acids. These are decarboxylated under acid catalysis and then formylated in the 3-position with ethyl formate/sodium methylate. Subsequently, the 2,3-double bond is introduced again by oxidation with manganese dioxide; see the equation (Fig. 5) for the synthesis of 3-formyl-ciprofloxacin (KONDO et al. 1988). Another method for the synthesis of 3-formyl-quinolones is thermal decomposition of sulphonylhydrazides of quinolonecarboxylic acids in ethylene glycol in the presence of sodium carbonate (DAINIPPON 1990).

Other groups in the 3-position also have the ability to form hydrogen bridges. For example, fusion of an isothiazolone ring capable of tautomerism in the 2,3-position leads to highly active quinolones (CHU et al. 1988; CHU 1990, 1992; CHU and CLAIBORNE 1990). Quinolones which are capable of tautomerism and have an antibacterial activity are also obtained by condensation of the carboxyl group with nitromethane, diethyl malonate or other CH-acid compounds (KIM et al. 1992a–d, 1992/1993, 1993).

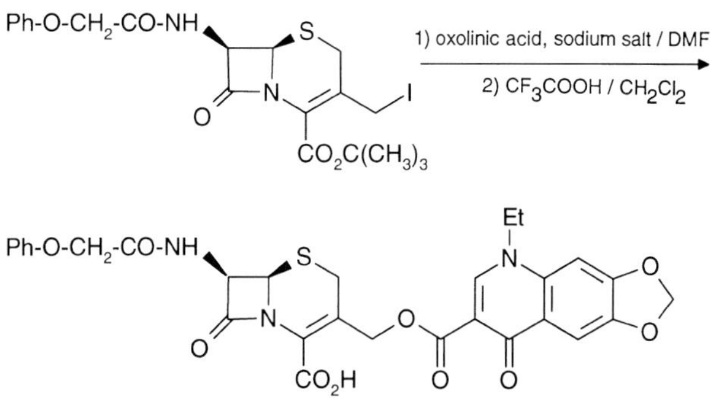

Ph-O-CH$_2$-CO-NH
S
N
O
I
CO$_2$C(CH$_3$)$_3$
1) oxolinic acid, sodium salt / DMF
2) CF$_3$COOH / CH$_2$Cl$_2$
Ph-O-CH$_2$-CO-NH
S
N
O
O
CO$_2$H
O
O
Et
N
O
O

Fig. 3. Synthesis of Ro 23-5068. *DMF*, dimethylformamide

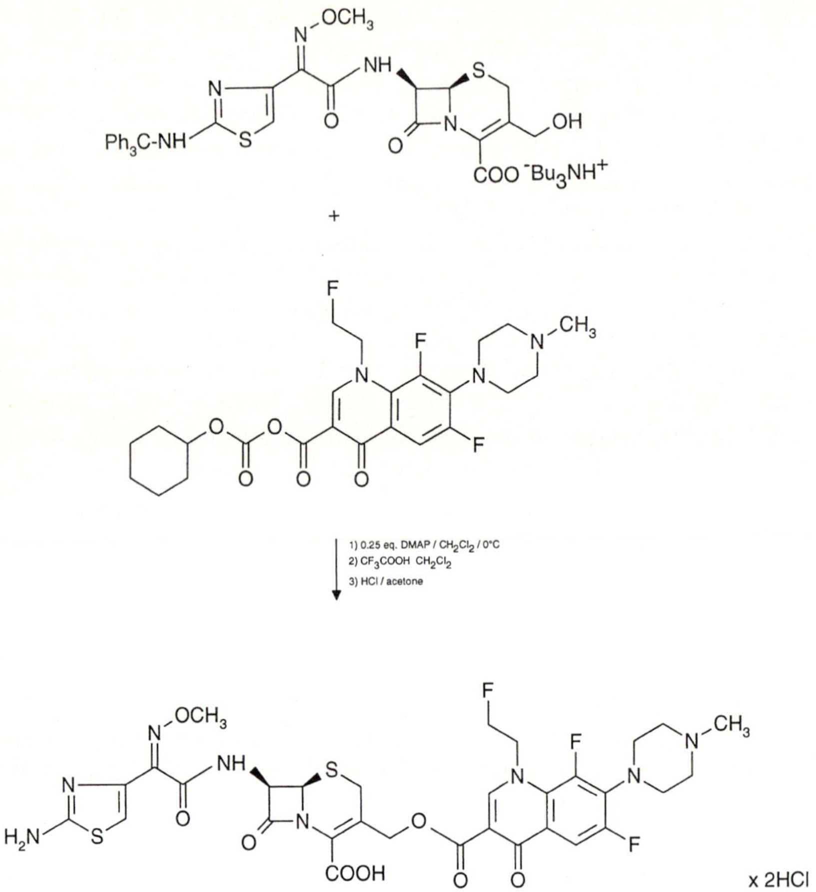

Fig. 4. Synthesis of Ro 23-9424 (dihydrochloride). *DMAP*, 4-dimethylamino-pyridine

The preparation of boronate complexes with the 3-carboxyl group of quinolones (Fig. 6) plays a role in the reaction of quinolones of low reactivity (8-methyl, 8-alkoxy etc.) with amines in the 7-position in particular (see Chap. 3HII). The chelates are synthesized by reaction of the quinolonecarboxylic acids or their esters with for example:

1. Acetic acid/boric acid/zinc chloride to give a (1-cyclopropyl-6-fluoro-7-chloro-1,4-dihydro-4-oxo-3-quinoline-O^3,O^4-carboxylic acid) bis(O-acetate)-borate (HERMECZ et al. 1987a)
2. Acetic anhydride/boric acid/zinc chloride to give a (1-cyclopropyl-6,7-difluoro-1,4-dihydro-8-methoxy-4-oxo-3-quinoline-O^3,O^4-carboxylic acid) bis(*O*-acetate)-borate (TAGAKI et al. 1990)
3. Acetic anhydride/boric acid to give a (1-cyclopropyl-6,7-difluoro-1,4-dihydro-8-methyl-4-oxo-3-quinoline-O^3,O^4-carboxylic acid) bis(O-acetate)-borate (H. MIYAMOTO et al. 1990)

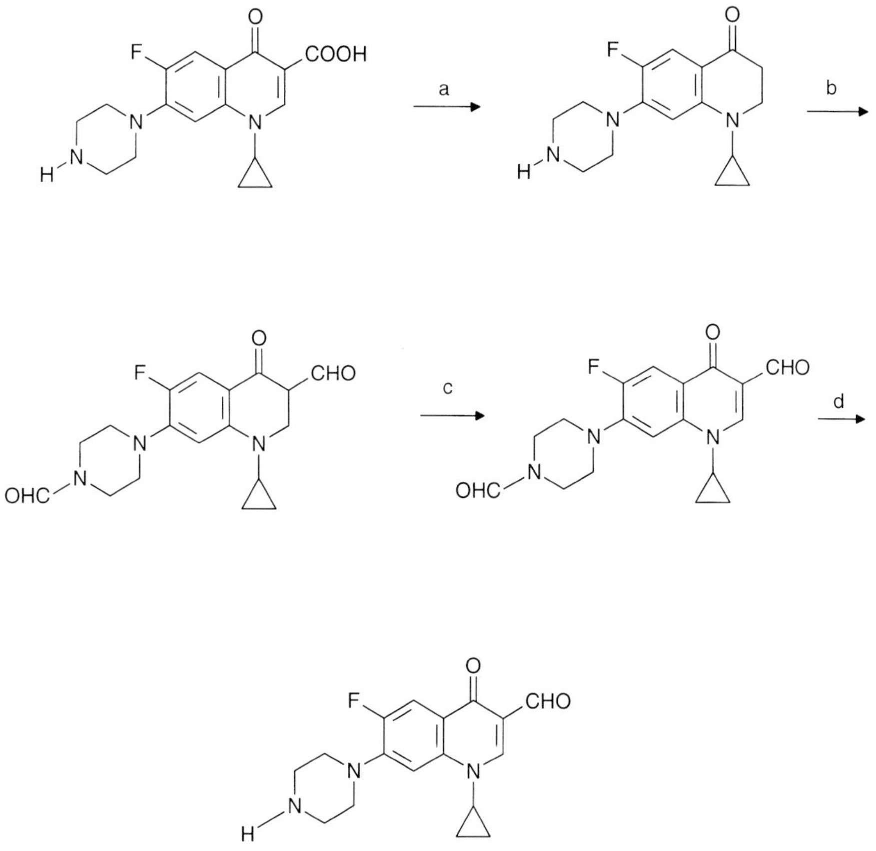

Fig. 5. Synthesis of 3-formyl-quinolones. *a*, $NaBH_4$/p-TsOH; *b*, $NaOCH_3/HCOOC_2H_5$; *c*, MnO_2/MeOH; *d*, 3n HCl/EtOH

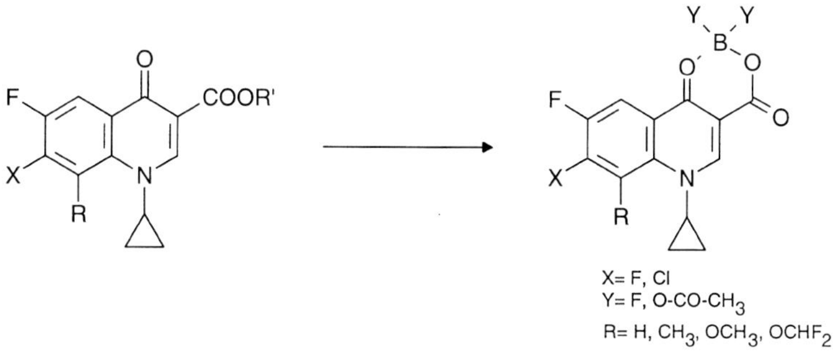

X= F, Cl
Y= F, O-CO-CH_3
R= H, CH_3, OCH_3, $OCHF_2$

Fig. 6. Synthesis of boranate complexes of quinolones

4. 42% Tetrafluoroboric acid to give 6,7-difluoro-1-(cis-2-fluoro-1-cyclopropyl)-8-methoxy-4-oxo-1,4-dihydro-3-quinolinecarboxylic acid BF_2-chelate (HAYAKAWA and KIMURA 1988a)
5. BF_3-etherate to give 1-cyclopropyl-8-difluoromethoxy-6,7-difluoro-1,4-dihydro-4-oxo-3-quinolinecarboxylic acid BF_2-chelate (IWATA et al. 1988), 1-cyclopropyl-6,7-difluoro-1,4-dihydro-8-methoxy-4-oxo-3-quinolinecar boxylic acid BF_2-chelate (SANCHEZ et al. 1995), or 5-amino-1-cyclopropyl-6,7-difluoro-1,4-dihydro-8-methyl-4-oxo-3-quinolinecarboxylic acid BF_2-chelate (ITO et al. 1993/1994; YOSHIDA et al. 1996a,b); an alternative route for the preparation of 1-cyclopropyl-6,7-difluoro-8-methyl-4-oxo-1,4-dihydro-3-quinolinecarboxylic acid BF_2-chelate via the Sandmeyer reaction should also be mentioned (CECCHETTI et al. 1996)

The following is a brief description of other reactions at the 3-position:

- Oxidative degradation of a 3-acetyl-naphthyridone with sodium hypochlorite to give 3-naphthyridone-carboxylic acid (DAINIPPON 1979).
- Oxidative degradation of 3-acetyl-quinolones with m-chloroperbenzoic acid to give 3-hydroxy-quinolones (RADL and JANICHOVA 1992).
- Acid hydrolysis of a 3-(oxazolin-2-yl) radical to give the 3-carboxyl group (DAINIPPON 1978).
- Preparation of 3-quinolinecarboxamides from the corresponding carboxylic acids with 1,1′-carbonyl-diimidazole as the coupling reagent and amines (CONRAD and WHITE 1981; WENTLAND et al. 1993a) or from the carboxylic acid esters with ethanolic ammonia in an autoclave (WENTLAND et al. 1993a).
- When quinolonecarboxylic acids are heated to high temperatures (about 300°C), decarboxylation occurs (MARKEES and SCHWAB 1972; LEYSEN et al. 1991b). Decarboxylation in the presence of cyanide is a milder method (WENTLAND et al. 1993b; REUMAN et al. 1994).
- Conversion of 3-quinolonecarboxylic acid alkyl esters with iodo-trimethylsilane into the 3-quinolonecarboxylic acid trimethylsilyl esters and in situ reaction of these with cyclic amines with simultaneous ester cleavage (DOMAGALA and SCHROEDER 1985; WARNER-LAMBERT 1986).
- The carboxylic group may be substituted by a benzyl residue via the following reaction sequence: (a) $NaBH_4$ reduction of the 2,3-double bond, (b) acid-catalyzed decarboxylation of the β-ketocarboxylic acid and (c) aldolcondensation with aldehydes; compared with the corresponding carboxylic acid 1-cyclopropyl-3-(2,6-dihydroxybenzyl)-7-(2,6-dimethyl-4-pyridyl)-6,8-difluoro-1,4-dihydro-4-oxo-quinoline shows an improvement of the topoisomerase II potency of approximately 80-fold with a corresponding increase in cytotoxicity in vitro (EISSENSTAT et al. 1995).
- ^{13}C-labelling of the 3-carboxyl group of some quinolones has been achieved by converting the quinolonecarboxylic acid on which they are based with lead tetraacetate/I_2 into 3-iodo-quinolone. This can be reacted with $K^{13}CN$/CuI to give the 3-[^{13}C]-cyanoquinolone, which can be hydrolysed under acidic conditions to yield the quinolone-3-[^{13}C]-carboxylic acid (CARR and SUTHERLAND 1994).

– Reference may be made to the literature and the references cited therein for a discussion of complexes of quinolones with metal ions (Ca^{2+}, Mg^{2+}, Zn^{2+}, Mn^{2+}, Cu^{2+}, Fe^{2+}, Fe^{3+}, Al^{3+}), in which the metal ion is located between the carboxyl group and the oxo group (Cole et al. 1984; Kara et al. 1991; Ross et al. 1992, 1993a,b; Ross and Riley 1994; Lecomte et al. 1994; Turel et al. 1994; Helena et al. 1995; Fan et al. 1995).

E. 4-Position

Reactions at the 4-position have been investigated to a lesser extent because the presence of the 4-oxo group is regarded as being necessary for inhibition of DNA gyrase. As stated in Chap. 2BVIII, the quinolone analogues benzothiazine 1-oxides and 1,1-dioxides have no antibacterial activity, i.e. a sulphoxide or sulphone group cannot be considered as a bioisosteric group for the 4-oxo group in quinolones (Culbertson 1991). Reduction of the 4-oxo group with $NaBH_4$–BF_3 leads to the CH_2 group (Miyamoto and Matsumoto 1990). The 4-oxo group has been converted into a 4-thioxo group with Lawesson's reagent (Wentland 1992) or P_2S_{10} (Wentland et al. 1995). Such 4-thioxo derivatives can be reacted with hydroxylamine or hydrazines to give 3,4-bridged isoxazolo[4,3-c]quinolin-3-ones or pyrazolo[4,3-c]quinolin-3-ones (Fig. 7), which inhibit topoisomerase II. After *S*-methylation, a 4-imino derivative can be obtained with glycine benzyl ester, and this product can be cyclized to pyrrolo[3,2-c]quinolin-3-ol (Wentland 1992; Wentland et al. 1995).

F. 5-Position

Although most commercial and development products are unsubstituted at the 5-position, there have been numerous attempts to influence the properties

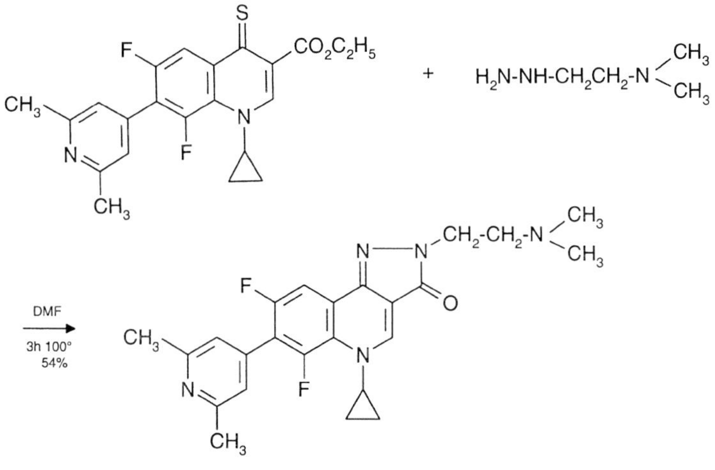

S
F
$CO_2C_2H_5$
+ H_2N-NH-CH_2CH_2-N CH_3 CH_3
CH_3
N
N
F
CH_3
CH_3
N—N CH_2-CH_2-N CH_3 CH_3
F
O
DMF
3h 100°
54%
CH_3
N
N
F
CH_3

Fig. 7. Synthesis of a pyrazolo[4,3-c]quinoline-3-one. *DMF*, dimethylformamide

of quinolones favourably by substitution at this position. It has been found that substituents such as fluorine, chlorine, alkyl or nitro lead to compounds of reduced antibacterial activity (JACK 1986; VERBIST 1986), while a 5-amino group in a certain substituent combination (1-cyclopropyl-6,8-difluoroquinolones) can, as in the case of sparfloxacin, lead to an increase in the in vitro activity against Gram-positive bacteria (DOMAGALA et al. 1988a, 1991). In contrast, 8-amino-ofloxacin (DAIICHI 1981) has a reduced activity compared with ofloxacin (DOMAGALA et al. 1986b), and the corresponding 4-amino-pyrido[2,3-d]pyrimidines have no activity (DOMAGALA et al. 1986b). 5-Methyl substitution also shows how decisive the combination of substituents in the quinolone molecule is: compared with the 5-H analogues, a 5-methyl group leads to an increase in the in vitro activity in 1-cyclopropylquinolones, but to a reduction in activity in 1-ethylquinolones (HAGEN et al. 1991).

Also in the naphthyridone series, the nature of the 1-substituent has a great influence on the activity if a 5-methyl group is introduced (BOUZARD et al. 1992a). A 5-methyl group in 1-tert-butyl- and 1-(fluoro-tert-butyl)-naphthyridones leads to a decrease in activity. In contrast, 1-cyclopropyl-5-methyl-naphthyridones exhibit a better in vitro activity than the 5-H analogues, although this is not achieved to the same extent in vivo. It was found in 5-substituted 1-cyclopropyl-6-fluoronaphthyridones that when radicals larger than methyl are introduced, the activity decreases drastically: Me > H >> Et > Ph. The influence of 5-methyl in 1-(2,4-difluorophenyl)-naphthyridones is to be regarded somewhat differently. 5-Methyl substitution also has a clear influence on physicochemical properties: with the introduction of 5-methyl the water solubility and lipophilicity increase in the 1-tert-butyl and 1-(2,4-difluorophenyl) series but decrease in the 1-cyclopropyl series (BOUZARD et al. 1992a).

Various methods are available for synthesizing 5-substituted quinolones and depend on the substitution pattern. Introduction of a 5-amino group is certainly the most important among these:

1. Starting from 2,3,4,5-tetrafluoro-6-nitro-benzoic acid, 6,7,8-trifluoro-5-nitro-quinolonecarboxylic acid is built up via the cycloaracylation method (see Chap. 2BIII) and is reduced catalytically to 5-amino-6,7,8-trifluoro-quionolonecarboxylic acid (DOMAGALA et al. 1988a, 1991). Substitution at the 7-position takes place in the last step.
2. 5,6,7,8-Tetrafluoro-quinolonecarboxylic acids can be reacted regio-selectively step by step in polar solvents, such as DMSO or ethanol, first in the 7-position with a cyclic amine and then in the 5-position with ammonia to give 5-aminoquinolones (Fig. 8; MATSUMOTO et al. 1985/1986; PETERSEN et al. 1987; T. MIYAMOTO et al. 1990; SHIBAMORI et al. 1990) or with hydrazine to give 5-hydrazino-quinolones (DEMUTH and WHITE 1992). Other nucleophiles (such as primary and secondary amines, alcoholate, thiolate) can also be used analogously. After 5-substitution with 4-methoxybenzylamine, 5-(4-methoxybenzylamino)-quinolones are obtained, and these can be deblocked in toluene with 36% hydrochloric acid

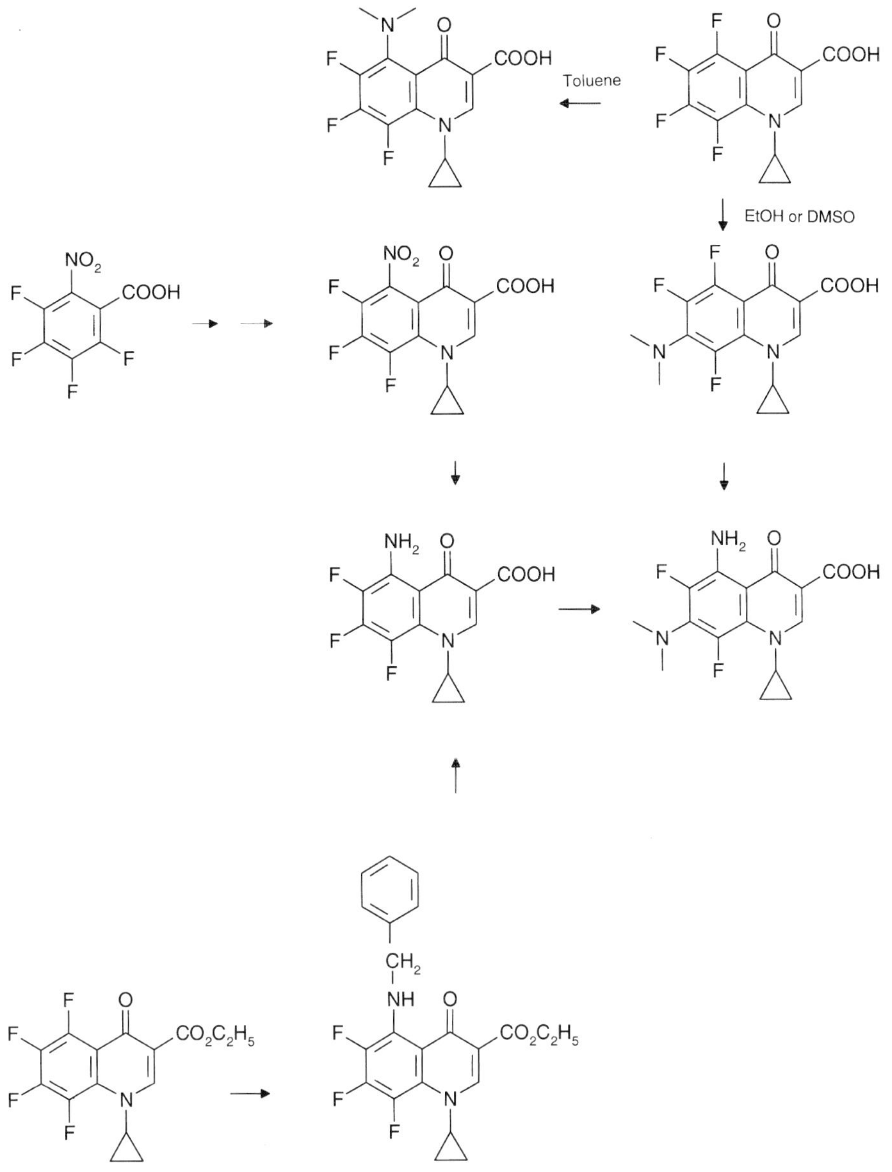

Fig. 8. Different routes for the synthesis of 5-amino-quinolones. *DMSO*, dimethylsulfoxide

to give 5-amino-quinolones (WEMPLE 1988/1989). In the naphthyridone series, the replacement of methylsulphonyl by benzylamine and subsequent hydrogenolytic debenzylation has been described (BOUZARD et al. 1992b). To replace the fluorine atom in the 5-position of ethyl 1-cyclopropyl-7-(2,6-dimethyl-4-pyridinyl)-5,6,8-trifluoro-1,4-dihydro-4-

oxo-3-quinolinecarboxylate by hydrogen, the compound was first reacted with benzylthiol or thiophenol in the presence of sodium hydride to give the 5-benzylthio- or 5-phenylthio-quinolonecarboxylic acid ester, and this was then desulphurized with Raney nickel in ethanol (LESHER et al. 1989a; WENTLAND et al. 1995).

3. 5,6,7,8-Tetrafluoro-quinolonecarboxylic acid can also be reacted regioselectively in the non-polar solvent toluene with amines in the 5-position (Fig. 8; SHIBAMORI et al. 1990). The 5,6,7,8-tetrafluoro-quino-lonecarboxylic acid ester, which cannot form a hydrogen bridge to the 4-carbonyl group, preferentially reacts with nucleophiles in the 5-position (MATSUMOTO et al. 1985/1986; T. MIYAMOTO et al. 1990; MORAN et al. 1989). 5-Amino-quinolonecarboxylic acid can also be prepared in this manner via reaction of a 5,6,7,8-tetrafluoro-quinolonecarboxylic acid ester with benzylamine and subsequent catalytic debenzylation and hydrolysis (T. MIYAMOTO et al. 1990; SHIBAMORI et al. 1990). The 5-hydroxy-quinolonecarboxylic acids may also be synthesized in a similar manner (T. MIYAMOTO et al. 1990). By acid ether cleavage of a 5-methoxy-quinolonecarboxylic acid, the corresponding 5-hydroxy derivative has been obtained (DAINIPPON 1987).
4. Direct nitration in the 5-position has been described for quinolones which are substituted by an electron donor, such as oxygen, in the 6- or 8-position or by carbon in the 8-position. For example, by reduction via the corresponding nitro-quinolones (Fig. 9) it is thus possible to prepare 5-amino-oxolinic acid (FRANK and RAKOCZY 1979), 8-amino-9,10-difluoro-3-methyl-7-oxo-2,3-dihydro-7*H*-pyrido-[1,2,3-de][1,4]benzoxacin-6-carboxylic acid (ofloxacin type; DAIICHI 1981), 5-amino-8-methoxy-quinolones (MASUZAWA et al. 1986a; SANCHEZ et al. 1995), 7-amino-8,9-difluoro-1,2-dihydro-2-methyl-6-oxo-6*H*-pyrrolo[3,2,1-ij]quinoline-5-carboxylic acid (ISHIKAWA et al. 1990), 8-amino-9,10-difluoro-3-methyl-7-oxo-2,3-dihydro-7*H*-pyrido[1,2,3-de][1,3,4]-benzoxadiazine-6-carboxylic acid (JAETSCH et al. 1994a,b), or (2S)-7-amino-8,9-difluoro-1,2-dihydro-2-methyl-6-oxo-6H-pyrrolol [3,2,1-ij]-quinoline-5-carboxylic acid (TSUJI et al. 1995). 6-Amino-naphthyridonecarboxylic acid esters can also be nitrated to the corresponding 5-nitro-naphthyridones after acetylation of the amino group, and subsequently reduced (CECCHETTI et al. 1991).
5. 5-Methyl- (HAGEN and DOMAGALA 1990; HAGEN et al. 1991; BOUZARD et al. 1992a; JACQUET et al. 1992; MATSUMOTO et al. 1987/1988; UEDA et al. 1987), 5-ethyl- (HAGEN et al. 1991; BOUZARD et al. 1992a), 5-phenyl- (BOUZARD et al. 1992a), 5-methoxy- (DOMAGALA et al. 1987), 5-dimethyl-amino- (DOMAGALA et al. 1987), 5-fluoro- (PETERSEN et al. 1987; MORAN et al. 1989), 5-chloro- (PETERSEN et al. 1991b) and 5-bromo-quinolonecarboxylic acids (PETERSEN et al. 1991b) or -naphthyridonecarboxylic acids are also prepared by the cycloaracylation method. For introduction of the CF_3 group into suitable nicotinic acids to build up the naphthyridone system, see REMUZON et al. (1993a).

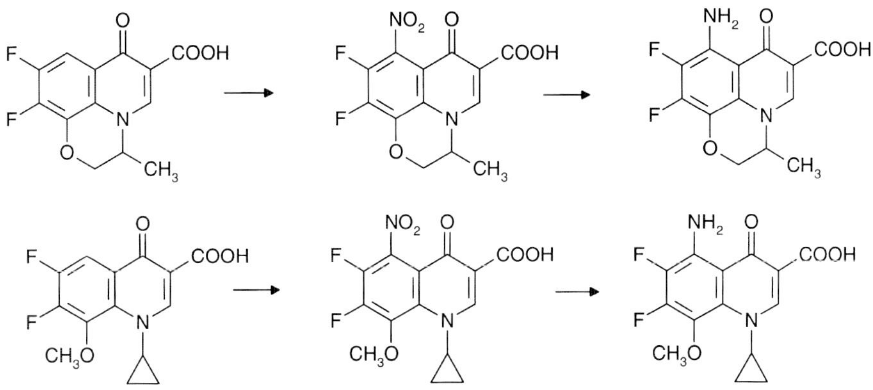

Fig. 9. Synthesis of 5-amino-quinolones by direct nitration and reduction

6. In the naphthyridone series, direct methylation of the 5-position is carried out starting from ethyl 7-chloro-1-cyclopropyl-6-fluoro-1,4-dihydro-4-oxo-1,8-naphthyridine-3-carboxylate (1; Fig. 10) by masking the C-2/C-3 double bond by reduction to (2), regioselective methylation of the 5-position by butyllithium/methyl iodide to give (3) and re-establishing the naphthyridone system (4; KIELY 1991).
7. 5-Chloro- and 5-bromo-quinolones have also been prepared via a Sandmeyer reaction from 5-aminoquinolones (T. MIYAMOTO et al. 1990).
8. 5-Formyl-naphthyridones are prepared via ethyl 5 (trimethylsilyl)-naphthyridone-carboxylates by ipso substitution with Vilsmeier reagent. The corresponding 5-hydroxymethyl-naphthyridones are prepared from these by sodium boronate reduction. The activity falls dramatically by the introduction of oxygen into the 5-methyl group (REMUZON et al. 1992c).
9. C–C linkage via palladium-catalysed reactions starting from 5-bromo-quinolonecarboxylic acid esters is suitable for the synthesis of 5-aryl- (HIMMLER 1991, unpublished results) and 5-vinyl-quinolones (HIMMLER et al. 1994; Fig. 11).
10. In the 7-(2,6-dimethyl-4-pyridinyl)-6,7-difluoro-quinolone series, synthesis of 5-amino-quinolones via the following reaction sequence is reported (WENTLAND et al. 1993b): reaction of a 5-fluoroquinolonecarboxylic acid ester with NaN_3 in DMF at 100°C leads to an isoxazoloquinolinecarboxylic acid ester in a yield of 40%, which is formed via an electrocyclic process from a 5-azido derivative which cannot be isolated. This isoxazole-bridged quinolone can easily be split by hydrogenolysis to give the 5-amino-quinolonecarboxylic acid ester.
11. The 5-amino group can be dimethylated with formic acid/formaldehyde following the Eschweiler-Clarke conditions and also monoalkylated after prior trifluoroacetylation (DOMAGALA et al. 1987, 1988a). It can also be converted into the 5-pyrrolyl derivative by reaction with 2,5-dimethoxytetrahydrofuran (T. MIYAMOTO et al. 1990a).

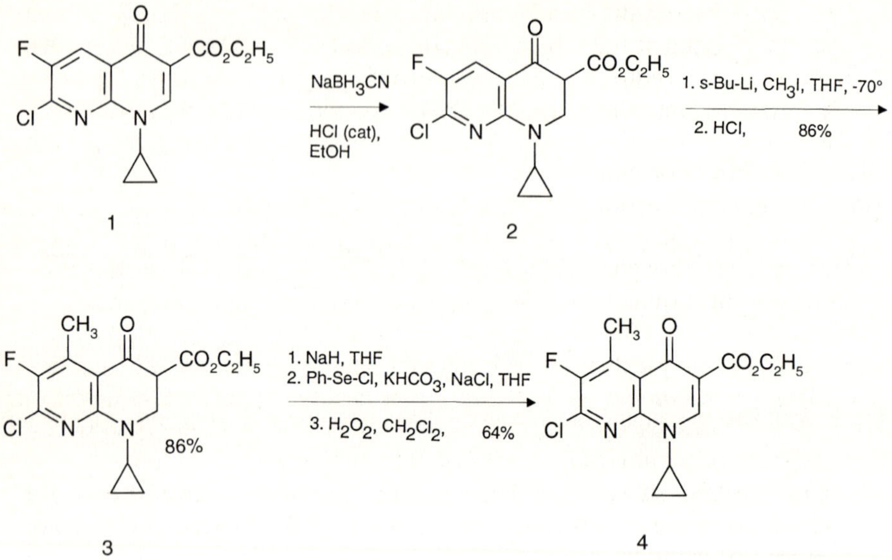

Fig. 10. Synthesis of ethyl 7-chloro-1-cyclopropyl-6-fluoro-1,4-dihydro-5-methyl-4-oxo-1,8-naphthyridine-3-carboxylate. *THF*, tetrahydrofuran

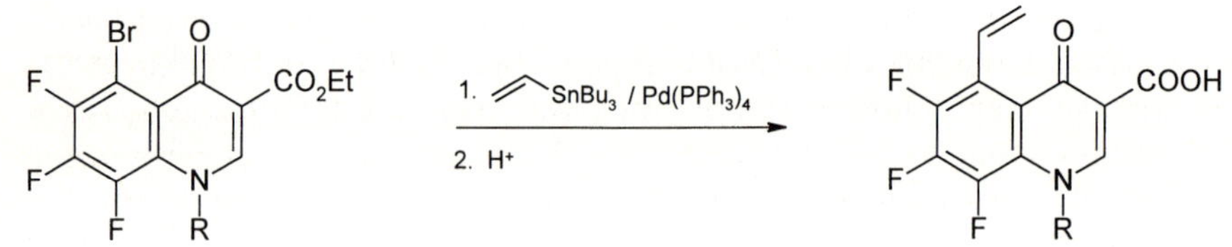

R = cyclopropyl, 2,4-difluorophenyl

Fig. 11. Synthesis of 5-vinylquinolones

12. 8-Substituted (Cl, CF_3) 7-oxo-2,3-dihydro-7*H*-pyrido-[1,2,3-de][1,4]benzothiazine-6-carboxylic acids are built up via the Gould-Jacobs reaction (CECCHETTI et al. 1987).

G. 6-Position

Modern quinolones have a fluorine atom at the 6-position ("fluoroquinolones"), which has proved to have advantages over other C-6 substituents (H, Cl, Br, CH_3, CN, SCH_3, NO_2, $COCH_3$; KOGA et al. 1980). The inhibition of DNA gyrase and above all the ability to penetrate through the cell membrane are improved dramatically by introducing a fluorine atom (DOMAGALA et al. 1986b).

The fluorine atom is usually already introduced before the quinolone system is built up, so that later reactions at the 6-position play a minor role. Nevertheless, some reactions may be of interest in individual cases:

Ethyl 1-ethyl-1,4-dihydro-7-methoxy-4-oxo-1,8-naphthyridine-3-carboxylate can be nitrated at the 6-position with a yield of 81% (HNO_3/H_2SO_4). The nitro group can be reduced to the amino group (Fe/glacial acetic acid), and this can be converted into the 6-cyano group or a 6-chlorine atom via a Sandmeyer reaction. In this reaction series, however, conversion of the corresponding diazonium tetrafluoroborate into the 6-fluorine compound in the context of a Baltz-Schiemann reaction fails (MATSUMOTO et al. 1984a).

The in vitro activity of various 6-amino-quinolonecarboxylic acids or -naphthyridonecarboxylic acids (Fig. 12), which are prepared via the corresponding 6-nitro-quinolone- or -naphthyridonecarboxylic acid esters by catalytic reduction (Ra-Ni/H_2), is approximately of the order of that of rufloxacin, i.e. they exhibit a high activity against Gram-negative bacteria, excluding *Pseudomonas aeruginosa*. Their activity against Gram-positive bacteria decreases, except in the case of compounds with thiomorpholine as a 7-substituent (CECCHETTI et al. 1991, 1995). In 6-amino-quinolones, the additional 8-methyl substitution leads to an increase, whereas mono- and dimethylation of the amino group results in a decrease in activity (CECCHETTI et al. 1996). Starting from 5,6-diamino-quinolones, 5,6-bridged imidazo- and triazolo-quinolones with good antibacterial activity were synthesized (FUJITA et al. 1995).

6-Methoxylation of clinafloxacin at a yield of 27% by an $F \rightarrow OCH_3$ exchange can be achieved under drastic conditions (autoclave, in methanol with sodium methylate, 140–150°C, 121 h; KYORIN 1986). Catalytic 6-dechlorination (Pd-C/H_2) to form 6-*H*-quinolones has also been reported (DAINIPPON 1987).

Direct bromination of 6-*H*-naphthyridonecarboxylic acid esters leads to 6-bromine derivatives, which can be converted into 6-methoxy-naphthyridones with sodium methylate (DAIICHI 1980). According to a recent publication, the 1-unsubstituted compound ethyl 7-fluoro-4-oxo-1,4-dihydroquinolinecarboxylate can indeed also be brominated at the 6-position with *N*-bromosaccharin with a yield of 68%, but the analogous 1-ethylquinolone can not (MOZEK and SKET 1994).

1,6-Naphthyridones, which are built up by the cycloaracylation method (GROHE et al. 1984b), also have an antibacterial activity. The 7-amino-1-cyclopropyl-8-fluoro-1,4-dihydro-4-oxo-1,6-naphthyridine-3-carboxylic acids are about four times less active against both Gram-positive and Gram-negative bacteria, however, than the isomeric 7-amino-1-cyclopropyl-6-

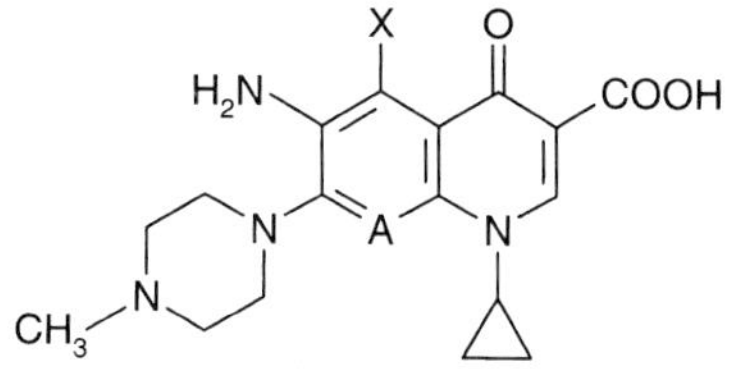

Fig. 12. Chemical structure of the 6-amino-quinolone or -naphthyridone type (*A* = CH, CF, CCH_3, N, *X* = H, NH_2)

fluoro-1,4-dihydro-4-oxo-1,8-naphthyridine-3-carboxylic acids (SANCHEZ and GOGLIOTTI 1993).

H. 7-Position

The 7-position is the position on the quinolone molecule which has by far been most varied and via which both the activity and the pharmacokinetic properties of the active compound can be influenced greatly. The substituent at the 7-position can be linked with the quinolone nucleus via N, C, S or O. While nalidixic acid carries a simple 7-methyl group and flumequine is unsubstituted at the 7-position, the modern quinolones usually have a cyclic diamine radical linked via nitrogen (see Table 1, Chap. 2). The most common substituents at the 7-position are cyclic amines, for example (substituted) piperazines, piperidines, morpholines, thiomorpholines, pyrrolidines, azetidines and azoles, with which further ring systems can be fused or linked spirocyclically. In recent years, however, some study groups have been showing greater interest again in 7-C–C-substituted quinolones.

I. Synthesis of Specific Amines

Over the last 10 years it has been found that the heterocyclic substituent of the 7-position can be varied widely in structure and significantly affects the activity/side effect ratio. The development and commercial products of the second generation quinolones were all substituted with piperazine derivatives at the 7-position. Simple piperazine derivatives such as 2-methylpiperazine or 2,6-dimethylpiperazine have not been modified further, apart from a few exceptions. These exceptions include *S*-2-methylpiperazine for the *S*-enantiomer of the former commercial product temafloxacin (CHU et al. 1991), which has been withdrawn from the market, 2-fluoromethylpiperazine, 2-hydroxymethylpiperazine and 2-aminomethylpiperazine (IWATA et al. 1988; T. MIYAMOTO et al. 1990; ZIEGLER et al. 1990b). The focal point of quinolone evolution has meanwhile shifted to pyrrolidine derivatives which exhibit an increased Gram-positive activity. To report on all the aspects of the extensive synthesis work in this field would go beyond the scope of this chapter. We will therefore limit ourselves to presenting the most important developments.

1. Bicyclic Piperazine Derivatives

Bicyclic piperazine and homopiperazine derivatives have acquired a certain importance for the development of quinolones (JEFSON and MCGUIRK 1985, 1986; PETERSEN et al. 1986; HUTT and KIELY 1987; FREIDMANN et al. 1988; FRAY et al. 1988; KIELY et al. 1991b). The 2,5-diazabicyclo[2.2.1]heptane system is of particular interest as the amine component of danofloxacin (MCGUIRK et al. 1992), an antibacterial for veterinary medicine. Synthesis of the bicyclic piperazine substituent (Fig. 13; BRAISH and FOX 1990) is based on the synthesis of

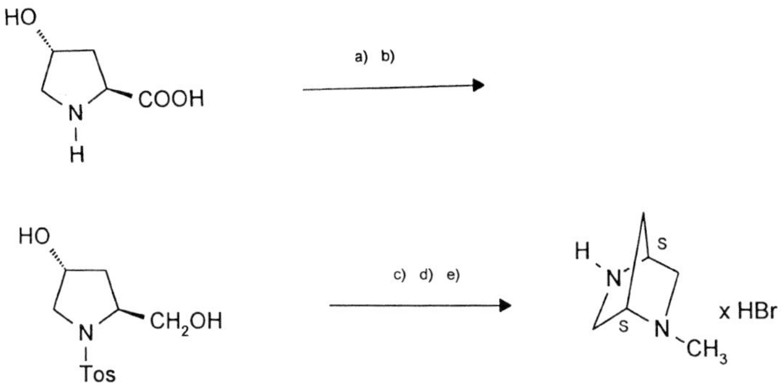

Fig. 13. Synthesis of the 2,5-diazabicyclo[2.2.1]heptane system. *a*, TosCl/OH^-; *b*, $NaBH_4/BF_3$; *c*, TosCl/pyridine; *d*, CH_3NH_2; *e*, HBr/AcOH

the parent compound starting from L-*trans*-4-hydroxyproline (PORTOGHESE and MIKHAIL 1966).

This method was used in modified form for the preparation of the enantiomers (JORDIS et al. 1990; BOUZARD et al. 1990) and methylated derivatives (REMUZON et al. 1992a,b,d, 1993b).

2. Aminopyrrolidine and Aminomethylpyrrolidine Derivatives

The chemistry of the quinolones has been greatly enriched by the introduction of aminopyrrolidine and aminomethylpyrrolidine derivatives as substituents of the 7-position. Modern methods based on intra- and intermolecular cycloaddition reactions allow a wide structural variability of the pyrrolidinyl radical. These studies were stimulated by the high antibacterial activities of many 7-pyrrolidinyl-substituted quinolones, especially against Gram-positive bacteria, but also by high rates of side effects, which needed to be reduced by further optimization. In contrast, pyrrolidine derivatives without a second basic group are mentioned only rarely as quinolone substituents in more recent literature, for example 3-pyrrolidinone ketals, oximes (COOPER et al. 1992a), 3-methylmercapto- and 3-methylmercaptomethylpyrrolidines (YOON et al. 1990). The combination of a methoximino group with a basic amino(methyl) radical exists in 4-amino-3-methoximino- and 4-aminomethyl-3-methoximino-pyrrolidine, of which the latter is the amine unit of LB 20304, a naphthyridone with a very good antibacterial activity, in particular against Gram-positive pathogens (Y.-K. KIM et al. 1995; OH et al. 1995). A pyrrolidinone ketal with an additional amino group is a unit of the 1-cyclopropyl-6,8-difluoroquinolone A-77143 (COOPER et al. 1992b).

Derivatives of 3-aminopyrrolidine were introduced earlier as a substituent (CHU 1983/1984; MATSUMOTO et al. 1983; CULBERTSON et al. 1984). In the racemic form, it is the amine component of tosufloxacin and of the development product clinafloxacin (CI 934/BAY V 3545/AM-1091; NEU et al. 1989; see Table 3, Chap. 2). Since the old synthesis method (IWANAMI et al. 1981) is not suitable for the production of larger quantities, alternative syntheses have been developed. The method of Tokyo Kasei (HOJO et al. 1985) starts from 1,2,4-trihydroxybutane, which, after conversion into the 1,2,4-trihalobutanes, is reacted with ammonia or benzylamine. In contrast, addition of primary amines to *N*-benzylmaleimide with subsequent $LiAlH_4$ reduction (KREBS and SCHENKE 1987) allows different alkyl radicals to be introduced on the amino group and on the pyrrolidine nitrogen.

In order to determine possible differences in antibacterial activity between enantiomerically pure 3-aminopyrrolidinyl-substituted quinolones, selective preparation of *R*- and *S*-3-aminopyrrolidine derivatives was required. Most of the published syntheses start from readily accessible enantiomerically pure 3-hydroxypyrrolidine derivatives, the different preparation methods of which have been summarized and reviewed (FLANAGAN and JOULLIE 1987). The amino function can then be introduced stereoselectively by the method of Mitsunobu (PETERSEN et al. 1989a) or by azide exchange (ROSEN et al. 1988b; CHU and ROSEN 1988a,b; SANCHEZ 1989; SANCHEZ et al. 1992b; DI CESARE et al. 1992), with inversion of the configuration. Other chiral starting materials are L-2,4-diaminobutyric acid (PETERSEN et al. 1989a) or L-aspartic acid (FERNANDES and CHU 1987; MADDALUNO et al. 1992; VAN LE et al. 1992). The *S*-3-aminopyrrolidinylquinolones have the highest antibacterial activity.

Alkyl radicals can be introduced at all positions of the pyrrolidine ring. Suitable starting materials are 3-oxopyrrolidine derivatives, which, on the one hand, can easily be prepared with 2-, 4- or 5-alkyl substituents and, for example, can be aminated by reduction via the oxime step (IWANAMI et al. 1981; IWATA et al. 1986). On the other hand, the keto function can be reacted with Grignard reagents, after which the tertiary alcohol obtained can be converted into the amino group via a Ritter reaction (MATSUMOTO et al. 1983).

Cis- and *trans*-4-hydroxyproline are suitable educts for enantiomerically pure 2-methyl-4-aminopyrrolidine derivatives. The carboxyl group can be reduced to the methyl group via the alcohol step, and the hydroxyl group can be converted stereoselectively into the amino group (ROSEN and CHU 1987; ROSEN et al. 1988c; DI CESARE et al. 1992). An alternative synthesis for (2*S*,4*S*)-4-acetamido-2-methyl-pyrrolidine proceeds from commercially available L-alaninol with a yield of 26% via eight steps (CHU et al. 1992). Two practical syntheses of (2S,4S)-4-tert-butoxycarbonylamino-2-methylpyrrolidine have been developed through the combination of diastereoselective and enantioselective reactions starting from ethyl crotonate and L-alanin, respectively (LI et al. 1995b).

1-Benzyl-3,4-epoxypyrrolidine can also be used as the starting material for introducing methyl groups into the 4-position of pyrrolidine by reacting it with methylmagnesium iodide in a Grignard reaction. After resolution of the racemate of the resulting alcohol, all four diastereomers of 3-amino-4-methylpyrrolidine can be obtained by azide exchange and reduction (DI CESARE et al. 1992). Another synthesis uses the chiral pool with aspartic acid and malic acid, to obtain the enantiomerically pure *cis*- and *trans*-3-amino-4-methylpyrrolidine derivatives after methylation of suitable intermediates

(Asahina et al. 1990). It is also possible for 2-pyrrolidone-4-carboxylic acid esters, which play an important role as intermediates for aminomethylpyrrolidines, to be methylated specifically in the 3-position and to be degraded to the amino compound (Masuzawa et al. 1985). 3-Amino-4,4-dimethylpyrrolidine can be synthesized in both enantiomeric forms from pantolactone (Di Cesare et al. 1992) or from ethyl acetoacetate (Hayakawa et al. 1990).

Pyrrolidines spiro-anellated at the 4-position are an important further development of 4,4-dialkylated aminopyrrolidine derivatives and have led to the development product DU 6859 (see Table 3, Chap. 2). To prepare these, ethyl acetoacetate is subjected to cyclizing alkylation with α,ω-dihalo-alkanes. After bromination of the methyl group, the pyrrolidine ring can be formed with primary amines which may be chiral. The amino group is then introduced via the oxime by reduction (Hayakawa and Kimura 1988; Hayakawa et al. 1988a,b, 1991; Fig. 14). 3-Amino-4-exomethylenepyrrolidines (Nishitani et al. 1988a,b; Okada and Tsushima 1990; Tsushima et al. 1990) and some 3-amino-4-arylpyrrolidine derivatives (Hagen et al. 1990; Bucsh et al. 1993), which can be prepared by azomethinylide addition to cinnamic acid derivatives, have also been described.

It is also possible to introduce other functional groups into the pyrrolidine skeleton. 3,4-Epoxypyrrolidine derivatives are particularly suitable for this, since epoxides can easily be opened by nucleophiles. Any substituted amino groups can be introduced by adding on azide, ammonia or primary or secondary amines, and the resulting hydroxyl group can be modified further, for example by alkylation (Iwata et al. 1986; Petersen et al. 1988b; Tsuji et al. 1988; Okada et al. 1993b) or fluorination (Bouzard et al. 1990). If the epoxide is first opened with an alcoholate, the hydroxyl group can be converted into the amino group by various methods, with reversal of the configuration (Petersen et al. 1988b; Tsuji et al. 1988).

H_3C COOC$_2$H$_5$ → a) b) c) → d) e) f) → NH$_2$ → g) h) → NH-Boc, N-H

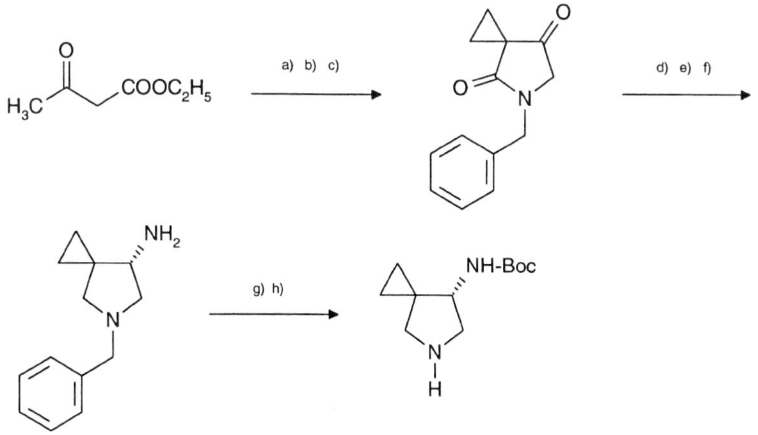

Fig. 14. Synthesis of 7-(S)-*tert*-butoxycarbonylamino-5-azaspiro[2,4]heptane. *a*, $BrCH_2CH_2Br/K_2CO_3$; *b*, Br_2; *c*, benzylamine; *d*, NH_2OH; *e*, $LiAlH_4$; *f*, L-tartaric acid; *g*, Boc_2O; *h*, H_2/Pd

Functional groups can also be introduced onto alkyl radicals. In the 4-position of the pyrrolidine, this is achieved by an intramolecular nitrone cycloaddition (PETERSEN et al. 1988b), or 1-benzyl-3-hydroxy-4-hydroxymethylpyrrolidine can be activated selectively on the primary OH group and converted into the methylthiomethyl derivative (KIM et al. 1991), while *trans*-4-hydroxyproline is the starting basis for aminopyrrolidine derivatives having a functionalized 2-methyl group (ROSEN et al. 1988a–c).

Derivatives of aminomethylpyrrolidine vary as much as the aminopyrrolidine derivatives. The simplest synthesis starts from itaconic acid or its esters, which react with primary amines by addition and cyclization to give 2-pyrrolidone-4-carboxylic acid derivatives. These can be reduced via the amides with $LiAlH_4$ to give the aminomethyl compounds with various radicals on the amino group (CULBERTSON 1982/1983; Fig. 15). The pure enantiomers can also be prepared easily by this route by reacting enantiomerically pure amines, such as phenethylamine with itaconic acid (CULBERTSON et al. 1987a).

This synthesis can be modified so that methyl groups are introduced on the alkyl chain (SCHROEDER et al. 1992; DOMAGALA et al. 1993; KIMURA et al. 1994a; HAGEN et al. 1994). Two further synthesis routes have been developed for 3-(1-amino-1-methylethyl)-pyrrolidine. This is obtained in the racemic form by addition of nitromethane onto a suitably substituted acrylic acid ester and reductive cyclization (HAYAKAWA and ATARASHI 1985). An enantioselective synthesis starts from enantiomerically pure 1-benzyl-3-hydroxypyrrolidine, which is converted into the 1-benzyl-3-cyanopyrrolidine derivative after activation. The key step is then a double addition of methylcerium dichloride onto the nitrile function (FEDIJ et al. 1994). Various syntheses have also been described for racemic and enantiomerically pure 3-(1-Boc-amino-cyclopropyl)-pyrrolidine, the unit for the "2-pyridone" A-101 211 (FUNG et al. 1995).

Alkyl or aryl radicals can best be introduced in the 3- or 4-position by azomethinylide addition reactions on acrylic acid derivatives and nitroolefins (DOMAGALA et al. 1986a; HAGEN et al. 1990; BUCSH et al. 1993). Some pyrrolidine derivatives fused spirocyclically in the 3-position, in which the amino group can be endocyclic (CULBERTSON et al. 1982/1983, 1990) or exocy-

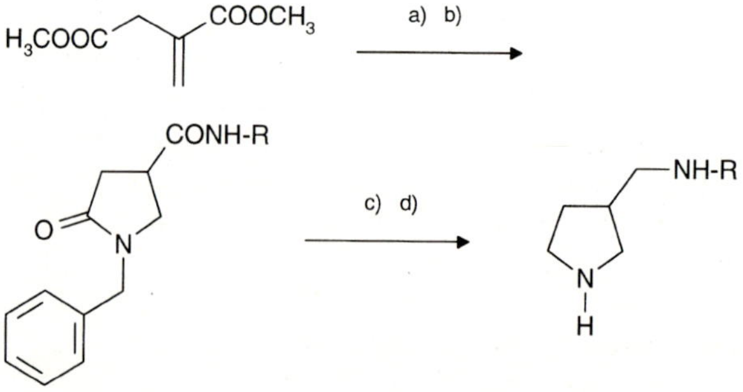

H3COOC COOCH3 a) b) CONH-R O N c) d) NH-R N H

Fig. 15. Synthesis of 3-aminomethyl-pyrrolidines ($R = H$, alkyl). *a*, Benzylamine; *b*, R-NH_2; *c*, $LiAlH_4$; *d*, H_2/Pd

clic (KIM et al. 1991d,f), should also be mentioned. Various processes for introducing halogen substituents or other functional groups have been described. 1-Benzyl-3-hydroxy-4-hydroxymethylpyrrolidine is often used as the starting substance here. The primary OH group can be converted selectively into an optionally substituted amino group, while the secondary OH group can be replaced, for example by fluorine or chlorine (MATSUMOTO et al. 1985; UEDA and MIYAMOTO 1985; DI CESARE et al. 1986; BOUZARD et al. 1990).

The Corey reaction of 3-pyrrolidone derivatives offers access to 3-aminomethylpyrrolidines functionalized in the 3-position. The epoxides obtained by this reaction can easily be reacted with any amines (PETERSEN et al. 1988a). (*S*)-3-Aminomethyl-3-fluoromethyl-pyrrolidine (SAITO et al. 1992/1993) and 3-aminomethyl-4-trifluoromethyl-pyrrolidine (NAKANO et al. 1993) are the amine units of the 1-cyclopropyl-8-methoxy-quinolones Y-688 (KITANI et al. 1995; YOKOYAMA et al. 1995) and S-32730 (MAEJIMA et al. 1995). Finally, 3-aminomethylpyrroline derivatives have also been described as quinolone substituents (KIM and LEE 1991).

3. 3,4-Bridged Pyrrolidine Derivatives

Syntheses in the field of 3,4-bridged pyrrolidine derivatives are an exceptionally widely researched area. The numerous different structures can be classified into diazabicycloalkanes, with a second nitrogen atom in the annelated ring, and azabicycloalkanes, which can carry an amino or aminomethyl group either on the annelated ring or on the bridgehead. The size of the annelated ring can be varied between three and seven ring atoms. The fused-on ring can also contain one or more double bonds in various positions, or heteroatoms, for example oxygen. The situation is further complicated by the fact that the number of all the conceivable stereoisomers increases exponentially with the number of chiral carbon atoms, and their specific preparation often requires completely different synthesis routes. The structural diversity offered by the 3,4-bridged pyrrolidine derivatives becomes clear from these considerations.

a) Diazabicycloalkanes

Azetidinopyrrolidines have been used for quite some time for the synthesis of new quinolones (CHIBA et al. 1987). Linkage with the quinolone can be either via the pyrrolidine nitrogen or via the azetidine nitrogen (JACQUET et al. 1991). 2,7-Diazabicyclo[3.3.0]octanes can be prepared by intramolecular azomethinylide addition reactions, in which any position of the bicyclic ring can be substituted (SCHENKE and PETERSEN 1989). The easily synthesized *N*-allylaminoacetaldehyde derivatives are reacted with *N*-alkylamino acids for this purpose (Fig. 16).

These substituents can also be linked with the quinolone via either of the two nitrogen atoms (PETERSEN et al. 1990). The preparation of 2,7-diazabicyclo[3.3.0]oct-4-enes has also been described. The second five-membered ring is fused onto the pyrrolidine skeleton via cyclizing ester con-

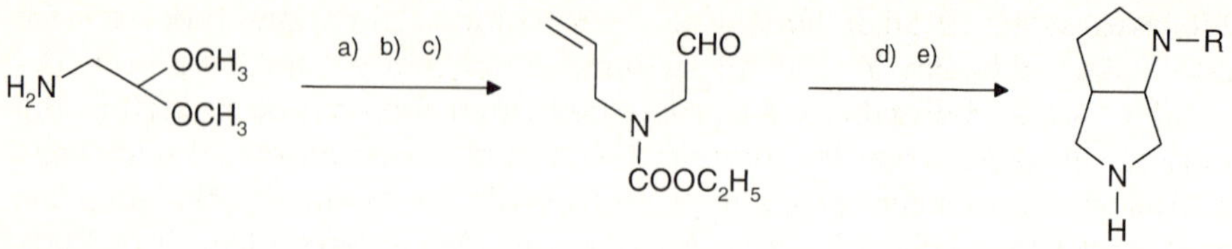

Fig. 16. Synthesis of 2,7-diazabicyclo[3.3.0]octanes (R = H, CH_3). *a*, $ClCOOC_2H_5$; *b*, allylbromide/OH^-; *c*, H^+; *d*, R-NH-CH_2COOH; *e*, H^+

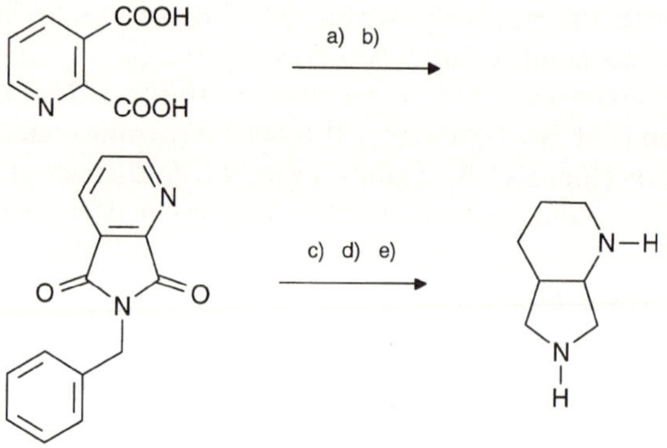

Fig. 17. Synthesis of *cis*-2,8-diazabicyclo[4.3.0]nonane. *a*, Acetanhydride; *b*, benzylamine/acetanhydride; *c*, H_2/Ru; *d*, $LiAlH_4$; *e*, H_2/Pd

densation, and the double bond is then introduced via an elimination reaction (KIM et al. 1992a,b). 3,7-Diazabicyclo[3.3.0]oct-1,5-enes have also been linked with quinolones (KIM et al. 1989a–c). Tetrakis(bromomethyl)-ethylene is used as the starting substance for the preparation of these bicyclic amines and is cyclized with toluenesulphonamide (KIM et al. 1989a–c; JENDRALLA and FISCHER 1995). The 7-(2,8-diazabicyclo[4.3.0]non-8-yl)-quinolones have an outstanding antibacterial activity. 2,8-Diazabicyclo[4.3.0]nonane can be prepared from pyridine-2,3-dicarboxylic acid, which is converted into the imide via its anhydride with benzylamine. The pyridine ring is then hydrogenated catalytically, and the imide is reduced to the pyrrolidine ring with $LiAlH_4$ (PETERSEN et al. 1988b). The *cis*-linked ring system is formed by this route (Fig. 17). The 3,8-diazabicyclo[4.3.0]nonane can be built up analogously from pyridine-3,4-dicarboxylic acid.

The enantiomers of cis-2,8-diazabicyclo[4.3.0]nonane are prepared by resolution of the racemate of a precursor with tartaric acid (PETERSEN et al. 1992b). The development product BAY Y 3118 (BREMM et al. 1992; PETERSEN et al. 1992d; PANKUCH et al. 1993; NORD et al. 1993; WISE et al. 1993; BAUERNFEIND 1993; FASS 1993; WEXLER et al. 1994; MOLINARI and SCHITO 1995; Fig. 26), which was prepared from the S,S-isomer and, to the knowledge of the authors, exceeds all commercial and development products in its antibacterial activity, had to be withdrawn from development because of phototoxicity findings. A product without phototoxicity is Bay 12-8039 (PETERSEN et al. 1988,

1992b; DALHOFF et al. 1996; ICAAC 1996; MARTEL et al. 1997; Fig. 18), the hydrochloride of Bay Y 6957 (PETERSEN et al. 1992d).

The *trans*-2,8-diazabicyclo[4.3.0]nonane has also been prepared (KIM et al. 1993), and a double bond has been introduced into the ring system (KIM et al. 1992a,b). It has been possible to synthesize the two isomeric dihydro-pyrrolopyridines (Fig. 19) via an intramoleular hetero-Diels-Alder reaction (PETERSEN et al. 1992a). On the basis of this, a further route to the 3,8-diazabicyclo[4.3.0]nonane system based on partial hydrogenation of a quaternary pyridinium salt to give the tetrahydropyridine derivative (Fig. 20) was opened up (PETERSEN et al. 1992a). It has also been possible to build up this structure starting from tetrabromo-2,3-dimethylbut-2-ene (KIM et al. 1989a–c).

Some oxa analogues to the structures mentioned so far have also been synthesized: it was possible to prepare the isomeric hexahydro-pyrroloisoxazoles (Fig. 21) by inter- and intramolecular nitrone additions (PETERSEN et al. 1988b; ZIEGLER et al. 1988). 1-Cyclopropyl-6,8-difluoro-quinolones with a 2-methyl- or 2-methylamino-4,6-dihydro-1*H*-pyrrolo[3,4-d]thiazol-5-yl radical in the 7-position show a very potent in vitro activity against Gram-positive bacteria, while the activity against Gram-negative bacteria is substantially lower than that against Gram-positive bacteria. The synthesis of the thiazolopyrrolidines proceeds in four or five steps starting from 1-tosyl-3-pyrrolidinone, with the tosyl radical as a protective group (W.J. KIM et al. 1995).

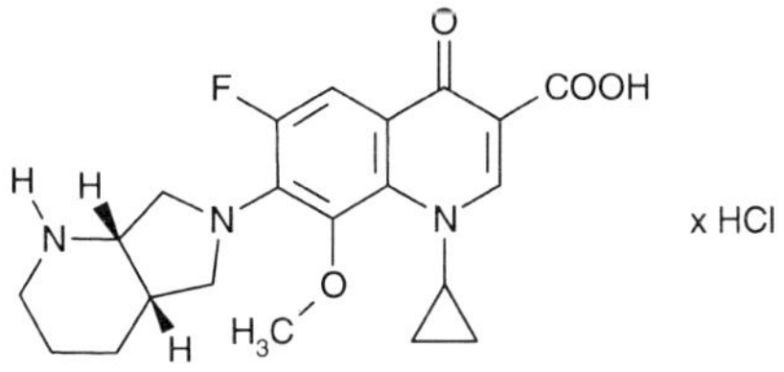

Fig. 18. Chemical structure of BAY 12-8039

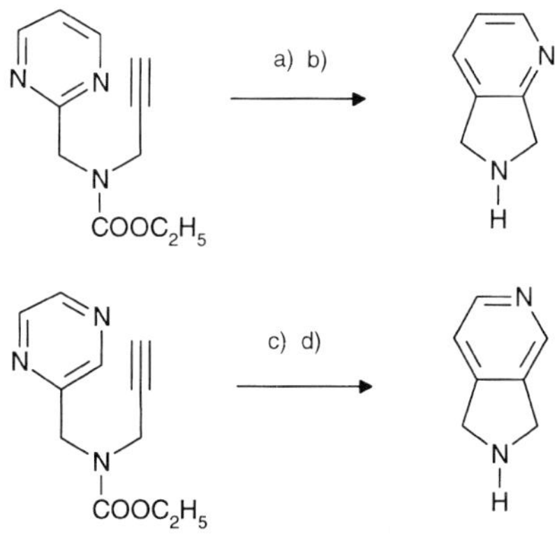

Fig. 19. Synthesis of the isomeric dihydro-pyrrolopyridines. *a*, Dimethylformamide/reflux; *b*, $Ba(OH)_2$; *c*, trifluoroacetic acid/reflux; *d*, $Ba(OH)_2$

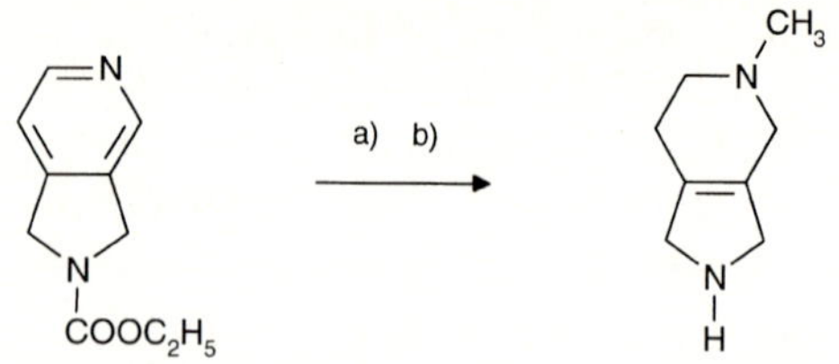

Fig. 20. Synthesis of 5-methyl-2,3,4,5,6,7-hexahydro-1H-pyrrolo[3,4-c]pyridine. *a*, $CH_3J/NaBH_4$; *b*, $Ba(OH)_2$

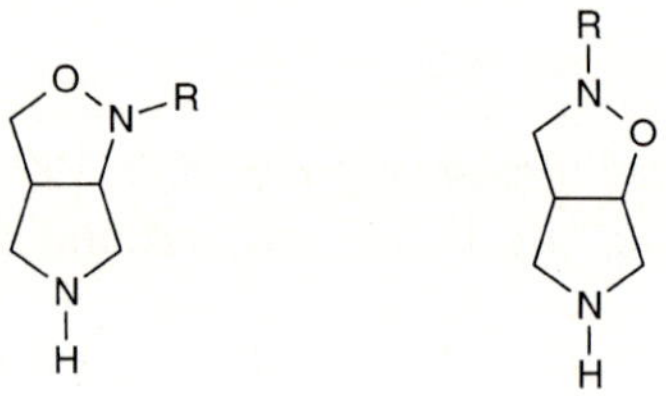

Fig. 21. Chemical structure of isomeric hexahydro-pyrroloisoxazoles

Finally, several different synthetic routes enable the 2-oxa-5,8-diazabicyclo[4.3.0]nonane skeleton (Fig. 22) to be prepared as a stereochemically uniform system. The *cis*-stereoisomers may also be prepared in an enantiomerically pure form by resolution of a racemic intermediate with tartaric acid and by an enantio-selective synthesis starting from tartaric acid (PETERSEN et al. 1988b; NAKAGAWA et al. 1989; SCHENKE et al. 1992). 4-Oxa-2,8-diazabicyclo[4.3.0]nonanes have also been described (PETERSEN et al. 1988b).

b) 1-Amino- or 1-Aminomethylazabicycloalkanes

The simplest representatives of this class of substance, 3-azabicyclo[3.1.0] hexane derivatives, have been described elsewhere in connection with antibacterial quinolonecarboxylic acids (CHIBA et al. 1987; BRIGHTY 1989). The bicyclic system is built up by an intramolecular alkylation reaction after addition of β-benzylaminopropionitrile to glycidol. All other compounds from this series have been prepared by different variants of azomethinylide addition reactions on cycloalkenylcarboxylic acid derivatives (OGATA et al. 1988a,b, 1991; PETERSEN et al. 1992c). As an example of such a reaction, the azomethinylide addition onto 4,5-dihydrofuran-2-carbonitrile is illustrated in Fig. 23.

c) Azabicycloalkanes with Amino Groups on the Annelated Ring

Bridged pyrrolidine derivatives with amine groups on the fused-on ring which can comprise three to seven atoms and have double bonds in various positions, are a class of quinolone substituents which have been the subject of particularly intensive research. The various synthesis methods are correspondingly diverse. Pyrrolidine derivatives bridged by three-membered rings may be obtained from benzylmaleimide by addition of bromonitromethane and subse-

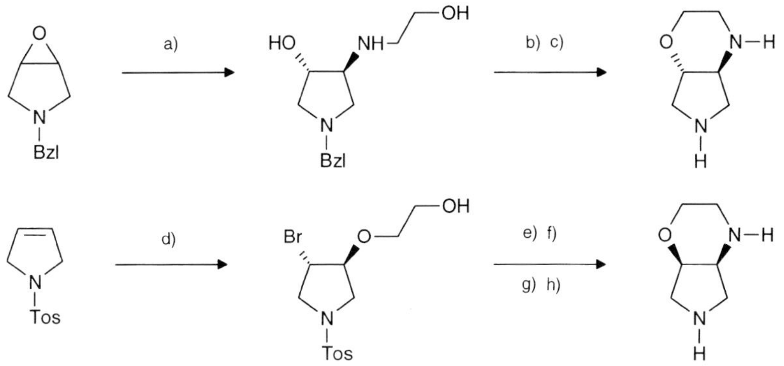

Fig. 22. Synthesis of *trans*- and *cis*-2-oxa-5,8-diazabicyclo[4.3.0]nonane. *a*, Ethanolamine; *b*, H_2SO_4; *c*, H_2/Pd; *d*, NBS/glycol; *e*, TosCl/NEt_3; *f*, benzylamine; *g*, HBr; *h*, H_2/Pd

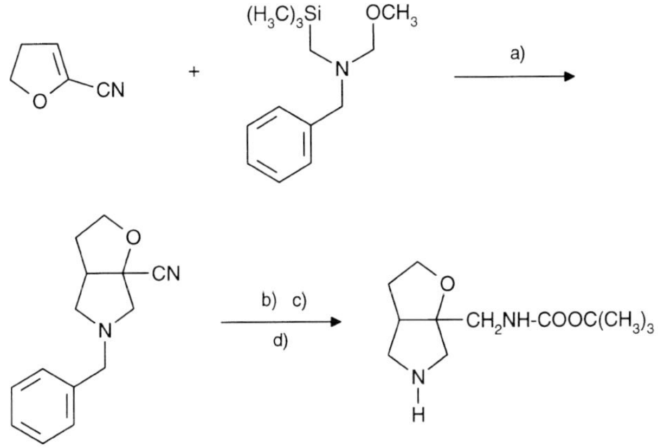

Fig. 23. Synthesis of 1-(*tert*-butoxycarbonylaminomethyl)-2-oxa-7-azabicyclo[3.3.0]-octane via an azomethinylide addition. *a*, Trifluoroacetic acid; *b*, $LiAlH_4$; *c*, Boc_2O; *d*, H_2/Pd

quent reduction (Braish 1992; Braish et al. 1996). Alternatively, they may be obtained by addition of diazoacetate with elimination of nitrogen and subsequent degradation of the acid group to the amine function (Brighty 1989; Brighty and Castaldi 1996). (1α,5α,6α)-6-Amino-3-azabicyclo[3.1.0]hexane is the substituent of the development product trovafloxacin (CP 99219; Fromtling and Castañer 1996).

A bicyclic ketone (Gobeaux and Ghosez 1989), which serves as the educt for optically uniform 6-amino-1-methyl-3-azabicyclo[3.2.0]heptane, which is the amine unit of the naphthyridone CFC-222 (Lee et al. 1995), can be built up by an intramolecular [2+2]cycloaddition. Azomethinylide addition onto cyclopentenone, cyclohexenone and cycloheptenone yields bicyclic ketones, which can be converted into the corresponding amino compounds by various methods (Ogata et al. 1988a,b, 1991). Pyrrolidines bridged by five- and six-membered rings which have a double bond between the bridge-head atoms are obtained from 1-cycloalkenyl-1,2-dicarboxylic acid derivatives. The amino group can be introduced into the allyl position after bromination, after which the two carboxyl groups are cyclized via a diol to give the pyrrolidine (KIM et al. 1991a–f; KIM and LEE 1991).

So far, two double-bond isomers of the pyrrolidines bridged by a six-membered ring, which can be prepared by inter- or intramolecular Diels-Alder reactions, have been described. Protected pentadienylamines are accessible by Curtius degradation of pentadienecarboxylic acids and can be cyclized with maleimides by a Diels-Alder reaction. Subsequent to the optional cleavage of the amine protective group, aminohexahydroisoindole derivatives (Fig. 24) are obtained by reduction with $LiAlH_4$ (OKURA et al. 1990; PETERSEN et al. 1991b).

Substituents of this type lead to quinolones with the highest antibacterial activity, but which very often also have a high genotoxic potential. The aim of a wide-ranging programme was therefore to identify among the double-bond isomers unknown to date those structures which have a lower toxicity coupled with the same activity. Pentadienylamine derivatives react with maleic or fumaric acid derivatives, for example maleic anhydride, in an intramolecular Diels-Alder reaction to give hexahydroisoindolonecarboxylic acid derivatives. The latter can be degraded to give the amino compounds and reduced to the bridged pyrrolidines (Fig. 25). All four possible diastereomers can be obtained by choosing the appropriate starting materials (PHILIPPS et al. 1992; BARTEL et al. 1993; JAETSCH et al. 1993).

4. Mixed Derivatives

Although the development of new 7-substituents for quinolines focused on the pyrrolidine derivatives, some other heterocyclic compounds have also been described for this application. 3-Methylaminopiperidine is the amine component of the development product balofloxacin (Q 35; see Table 3, Chap. 2; NAGANO et al. 1988). Bicyclic or spirocyclically fused aminopiperidines have also been described (HUTT et al. 1984; BRIGHTY et al. 1987; KLEINMAN 1988; JORDIS et al. 1991; YANG 1991). Aminoazetidines often lead to more active quinolones than piperidines. They can be synthesized by conventional synthesis (GAERTNER 1967; ANDERSON and LOK 1972) from substituted epihalohydrins

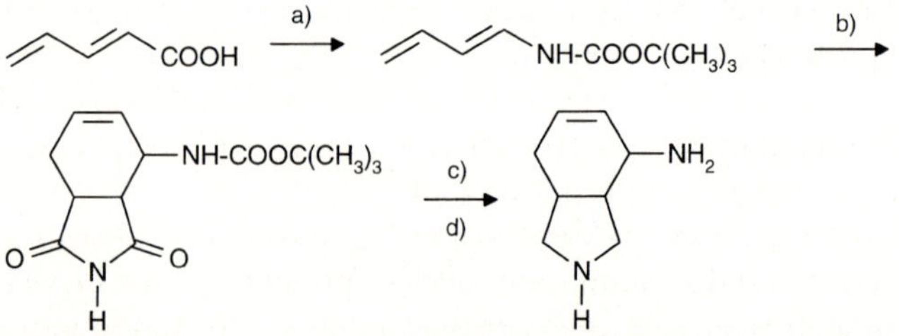

Fig. 24. Synthesis of 4-amino-1,3,3a,4,7,7a-hexahydroisoindole. *a*, $ClCOOC_2H_5/NaN_3$/*tert*-butanol; *b*, maleimide; *c*, trifluoroacetic acid; *d*, $LiAlH_4$

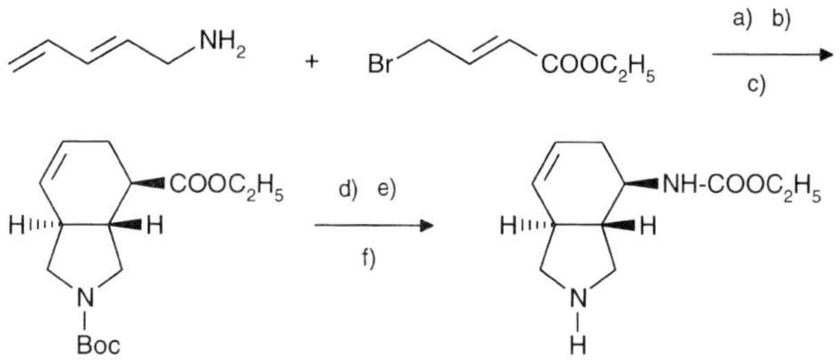

Fig. 25. Synthesis of (1SR,2RS,6SR)-2-ethoxycarbonylamino-8-azabicyclo[4.3.0]non-4-en. *a*, NEt_3; *b*, Boc_2O; *c*, 100°C; *d*, OH^-; *e*, $(PhO)_2PON_3$/Ethanol; *f*, HCl

and primary amines (PARES COROMINAS et al. 1989; OKADA et al. 1993a; FRIGOLA et al. 1993, 1994, 1995). 2-Aminomethyl-morpholine derivatives (ARAKI et al. 1987, 1993), which are prepared by addition of epichlorohydrin to *N*-benzylethanolamine, cyclization with sulphuric acid and conversion of the chloromethyl group into the aminomethyl group, are precursors of quinolones with high Gram-positive activity. 4-Amino-1-methylpyrazolidine (KIM and RYAN 1990) and 4-amino- and 5-aminomethylisoxazolidine derivatives have also been described as quinolone substituents (KIM 1990; ZIEGLER et al. 1990a).

II. C–N Linkage

The 7-amino-substituted quinolones are usually obtained with a high regioselectivity by reaction of 7-halo-quinolonecarboxylic acids or -carboxylic acid derivatives with cyclic amines in the presence of an auxiliary base. A fluorine atom on the quinolone nucleus is a better leaving group than a chlorine atom. For synthesis of 7-substituted naphthyridonecarboxylic acids or their aza analogues, a sulphonyl or sulphonyloxy group (MATSUMOTO and MINAMI 1975; EGAWA et al. 1984; MATSUMOTO et al. 1984b; MIYAMOTO and MATSUMOTO 1990) or a methyoxy group (MATSUMOTO et al. 1984a) can also serve as the leaving group. In reactions of ethyl 7-ethylsulphonyl-naphthyridone-carboxylate with morpholine or piperidine, substitition of the 6-fluorine atom has also been observed as a side reaction (NISHIMURA et al. 1988a).

If the cyclic amines contain further amino groups which could lead to side reactions, these must be blocked before the reaction by suitable protective groups, such as formyl, acetyl, *tert*-butoxycarbonyl or trichloroacetyl. The cyclic diamines which can be reacted regioselectively without the introduction of a protective group include 2-methylpiperazine (in temafloxacin, lomefloxacin), 2,6-dimethylpiperazine (in sparfloxacin) and *S,S*-2,8-diazabicyclo[4.3.0]nonane (in Bay Y 3118, Bay 12-8039; Figs. 18 and 26; PETERSEN et al. 1992d).

In contrast, for example before reaction with 3-amino-pyrrolidine (in tosufloxacin, clinafloxacin), it is advisable to protect the free amino group and to eliminate the protective group again after the linking with the quinolone. The use of the *tert*-butoxycarbonyl protective group (SANCHEZ et al. 1988) and an easily prepared azomethine protective group with 3-nitro-benzaldehyde, which can be removed again under mild conditions with dilute hydrochloric acid at room temperature (PETERSEN et al. 1989b), is illustrated for example in the synthesis of clinafloxacin (AM 1091, Bay V 3545, PD 127 391) in Fig. 27.

The reaction with the amines may be carried out in polar solvents, such as pyridine, DMSO, DMF, acetonitrile, HMPA, *N*-methyl-2-pyrrolidone, ethanol, 3-methoxy-1-butanol, water or mixtures of such solvents. Alternatively, it may be performed entirely without an added solvent in a melt of excess amine, such as piperazine or imidazole. Widely varying organic and inorganic bases have been described as auxiliary bases for this reaction. It is usual to use, for example, triethylamine, 1,4-diazabicyclo[2.2.2]octane (DABCO), 1,5-diazabicyclo[4.3.0]non-5-ene (DBN), 1,8-diazabicyclo[5.4.0]undec-7-ene (DBU), sodium carbonate, sodium bicarbonate, potassium carbonate or sodium hydride.

The reactivity of the 7-position depends very greatly on its steric and electronic environment: electron-withdrawing groups in the 8-position promote the reaction, while electron-rich substituents such as oxygen or sulphur, or groups which cause more steric hindrance, such as methyl, vinyl or phenyl, render it more difficult or impossible. Nevertheless, the desired substitution can in many cases be obtained by alternative experimental routes:

1. Boron complexes: while 7-fluoro-8-methyl-quinolonecarboxlic acids react hardly at all with cyclic amines, such a reaction proceeds with moderate yields after activation of the 7-position via suitable boron complexes (H. MIYAMOTO et al. 1990). Boron complexes are also very often employed for the synthesis of 8-methoxy-quinolones or pyrido[1,2,3-de][1,4]benzoxacin derivatives (see Chap. 3D). They are not only used as purified compounds, but can but can also be formed in situ by addition of trialkyl borates to the corresponding quinolonecarboxylic acid (UBE 1992; OCHI and SHIMIZU 1992).
2. Quinolones of the pyrido[1,2,3-de][1,4]benzothiazine type (rufloxacin type) can be built up by intermediate oxidation of the sulphide sulphur to sulphoxide, since the 10-position is activated sufficiently in this way. In the last step, the sulphoxide must be reduced to sulphide again with PBr_3 (Fig. 28; CECCHETTI et al. 1987, 1993b).

7-Azolyl-substituted quinolones are either prepared by reaction of 7-halo-quinolonecarboxylic acids with azoles, or, in certain cases, can be built up directly on the quinolone base skelecton. For example, the pyrrolyl radical is introduced by condensation of 7-amino-quinolones with 2,5-dimethoxy-

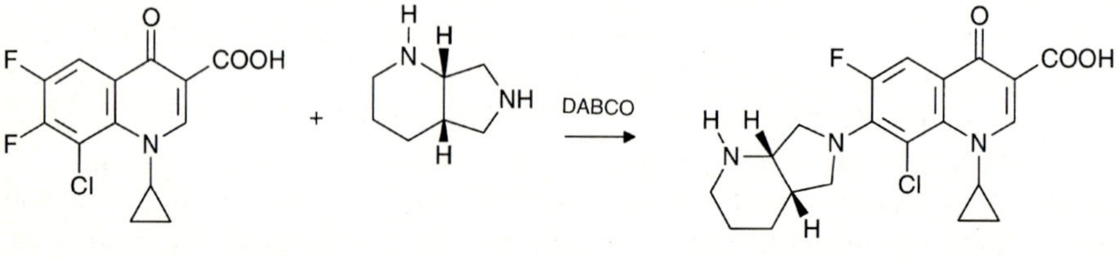

Fig. 26. Chemical structure of BAY Y 3118

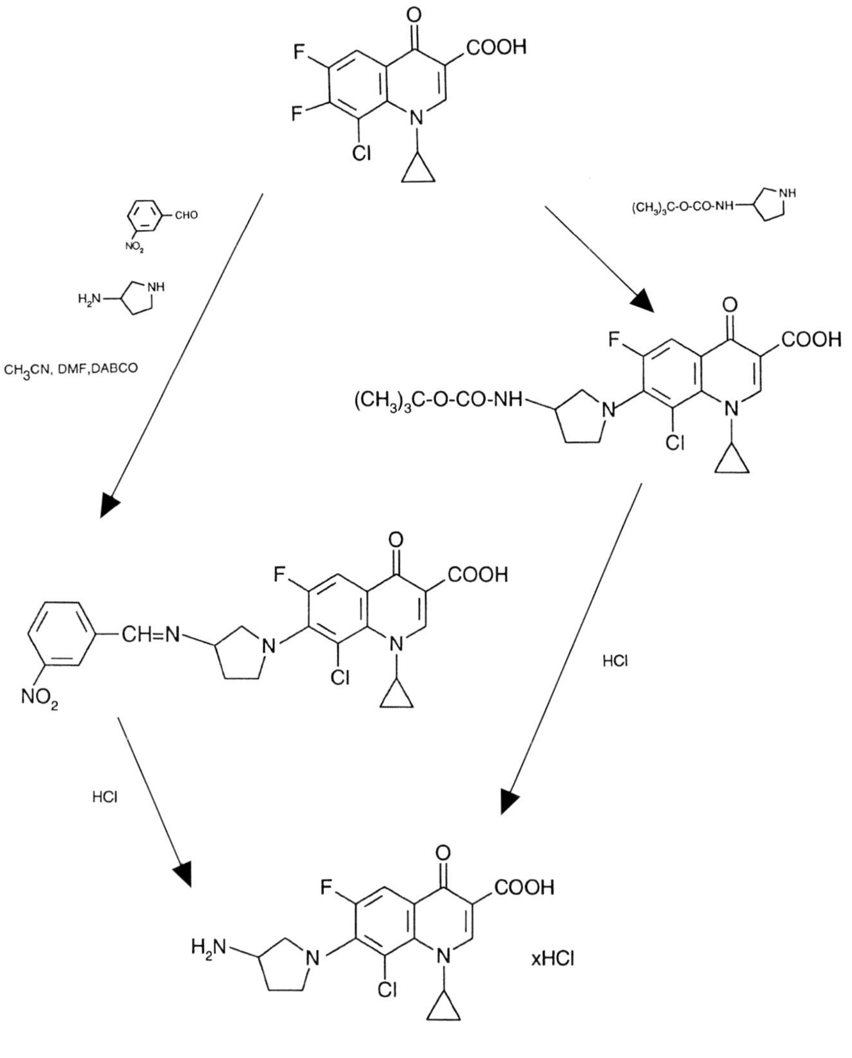

Clinafloxacin

Fig. 27. Examples of synthetic routes for clinafloxacin. *DMF*, dimethylformamide

tetrahydrofuran (UNO et al. 1987; ESTEVE SOLER 1983/1984; PETERSEN et al. 1985), the 1,2,4-triazol-4-yl radical is introduced by reaction with *N,N*-dimethylformamidazine (UNO et al. 1987) and the pyrazolyl radical is introduced by condensation of 7-hydrazinoquinolones with 1,1,3,3-tetramethoxypropane (PETERSEN et al. 1985). Starting from a 7-hydroxylamino-quinolone, cycloaddition reactions can be carried out with olefins via an intermediate 7-nitroso- or 7-methylene-nitrone derivative (ZIEGLER et al. 1988).

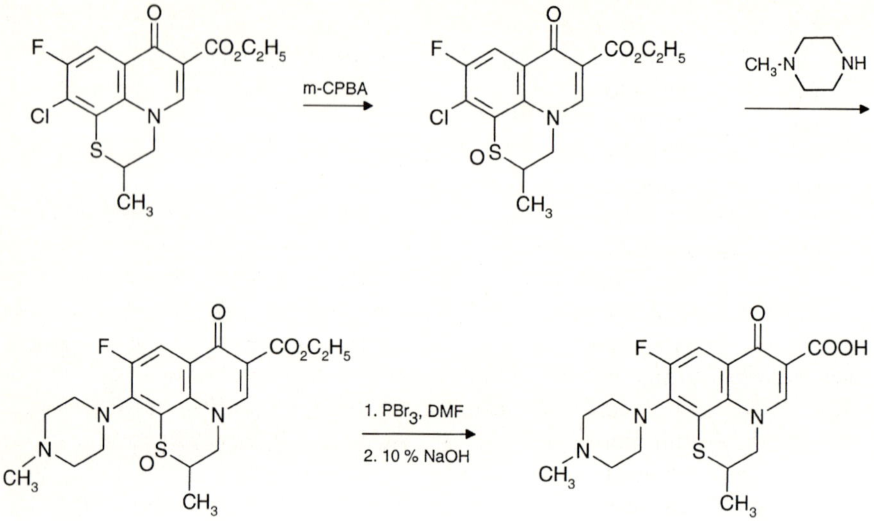

Fig. 28. Activation of the 10-position of the pyrido[1,2,3-de][1,4]benzothiazine-type by oxidation. *DMF*, dimethylformamide

After an amine has been linked with the quinolone parent substance, secondary reactions are often carried out on the amino function, some examples of which are:

1. Alkylation. Compounds which are prodrugs in some cases are obtained by *N*-alkylation of 7-piperazinyl-quinolones with 4-bromomethyl-5-methyl-2-oxo-1,3-dioxole (SAKAMOTO et al. 1985; NISHINO et al. 1989), halo-ketones, halocarboxylic acid derivatives, α,β-unsaturated ketones or α,β-unsaturated carboxylic acid derivatives (PETERSEN et al. 1982, 1983; KONDO et al. 1986). The *N*-[(5-methyl-2-oxo-1,3-dioxol-4-yl)methyl]-norfloxacin accessible by this route can be converted by oxidation with m-chloroperbenzoic acid into *N*-[(4-methyl-5-methylene-2-oxo-1,3-dioxolan-4-yl)oxy]norfloxacin. The latter compound also has prodrug properties and has a higher in vivo activity than norfloxacin after oral administration because of the higher plasma level which is attained (KONDO et al. 1989). Substituted *N*-vinyl groups have been introduced by reaction with ethoxymethylenemalonic ester derivatives (PETERSEN et al. 1983), propiolic acid esters or acetylene dicarboxylic acid esters (PETERSEN et al. 1992b; GAMMILL et al. 1993). *N*-alkylation with 3-chloromethyl- or 3-iodomethylcephalosporin derivatives (also prepared in situ) leads to dual action cephalosporins (DACs) in which the β-lactam is linked with the quinolone via the bond of a tertiary amine (Fig. 29; ALBRECHT et al. 1994) or of a quaternary ammonium compound (ALBRECHT et al. 1991a). Such DACs are chemically more stable than the compounds linked via an ester bond (see Chap. 3D).
2. Acylation. Coupling products with various dipeptides which are obtained via acylation with activated dipeptide derivatives can, as prodrugs with improved water solubility, be hydrolysed again in vivo to give the active

compound (Fig. 30; CHU and HALLAS 1988/1989). In the DACs of the carbamate type, the amine part of the quinolone and the 3-hydroxymethyl group of the cephalosporin are linked to one another via a carbamic acid ester bond (ALBRECHT et al. 1991b).

3. Other *N*-functionalizations. *N*-hydroxylated norfloxacin and ciprofloxacin have been obtained by oxidation of the corresponding *N*-(3-oxobutyl) derivatives with m-chloro-perbenzoic acid and subsequent thermal rearrangement of the intermediate *N*-oxide. These compounds are distinguished by the fact that as a result of the hydroxylation, the in vitro activity decreases approximately by a factor of 2–8, the acute toxicity remaining unchanged, while the in vivo activity following oral administration increases by a factor of 1.5–3. Splitting back into the quinolone on which they are based is discussed as the reason (UNO et al. 1990, 1993). Other 7-(4-hydroxy-1-piperazinyl)-quinolones have been prepared by direct reaction with 1-hydroxypiperazine (UNO et al. 1989). Norfloxacin and pipemidic acid have been converted into the *N*-chlorine derivatives by chlorination with tert-butyl hypochlorite and into the *N*-nitroso derivatives by nitrosation with sodium nitrite/acetic acid. *N*-Amino-norfloxacin is formed from *N*-nitroso-norfloxacin by reduction with zinc and can be converted into the hydrazones (DESIDERI et al. 1985). Similarly, 7-alkylamino-naphyridones have been reacted to 7-(1-alkylhydrazino)-naphthyridones via the *N*-nitroso derivatives, and converted into hydrazones (NISHIGAKI et al. 1985).

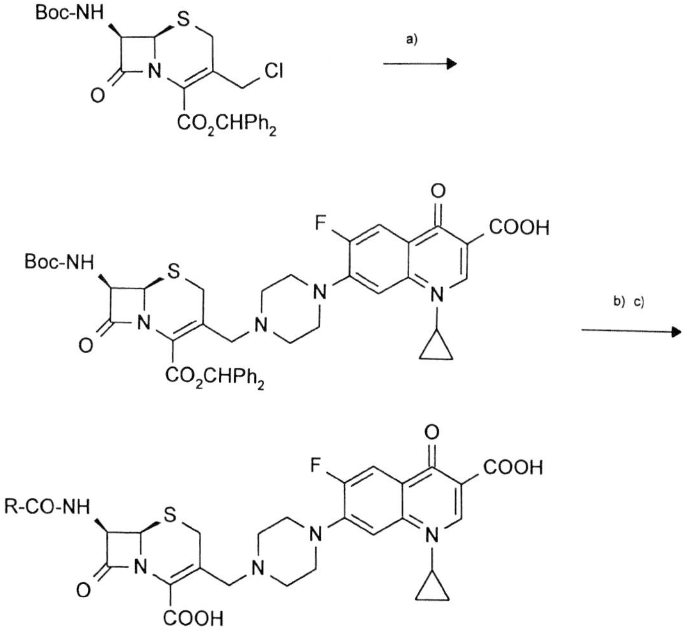

Fig. 29. Synthesis of dual-action cephalosporins (tertiary amine type). *a*, NaJ, ciprofloxacin, $NaHCO_3$, dimethylformamide; *b*, CF_3COOH, anisole, CH_2Cl_2; *c*, acylation

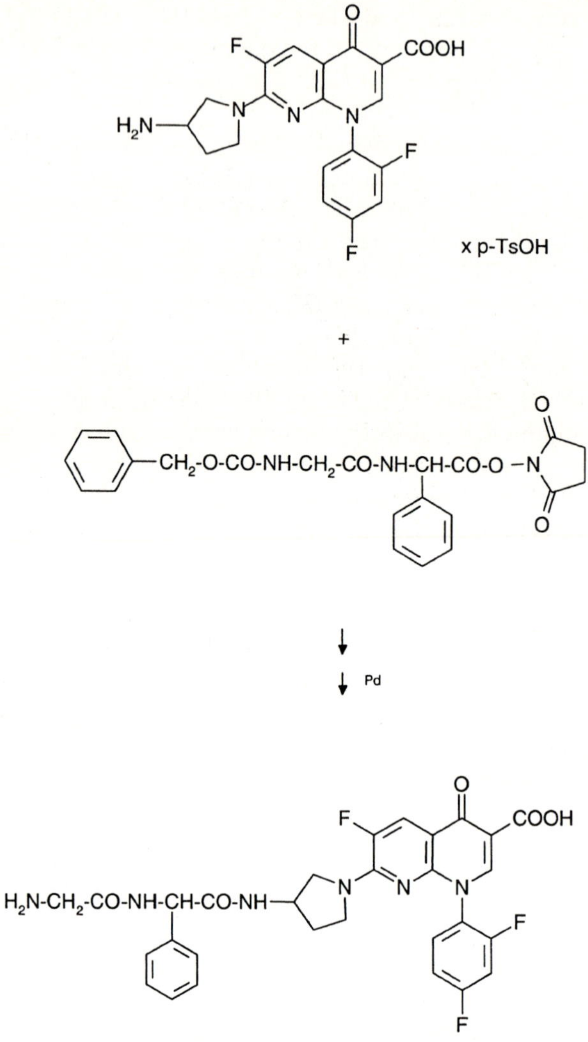

Fig. 30. Synthesis of a dipeptide derivative of tosufloxacin

III. C–S Linkage

For 7-substitution with thiols, these are linked directly with 7-halo-quinolones via the thiolate anion formed with bases (ZIEGLER et al. 1989; NISHIMURA et al. 1990). Alkylation of a 7-mercapto-quinolone represents an alternative for thiols of difficult accessibility (NISHIMURA et al. 1990). Of the various 7-alkylthio-, 7-arylthio-, heteroarylthio- and 7-(cyclic amino)thio-quinolones, 5-amino-7-(2-aminoethylthio)-1-cyclopropyl-6,8-difluoro-1,4-dihydro-4-oxo-3-quinolinecarboxylic acid has exhibited the highest in vitro activity (NISHIMURA et al. 1990).

IV. C–O Linkage

C–O substitution products have also occasionally been described. The reaction of 7-fluoro-quinolonecarboxylic acids with aqueous sodium hydroxide solution gives the 7-hydroxy-quinolones (ZIEGLER et al. 1989). These have only an extremely weak activity and are precursors for the C–C linkage after trifluoromethylsulphonylation (KIELY et al. 1991a). Better in vitro activity against Gram-positive bacteria is displayed by some 7-alkoxy-quinolones, which are obtained analogously by reaction with alcohols in the presence of bases (ZIEGLER et al. 1989). The reaction of hydroxylated cyclic amines proceeds via the oxygen if the nitrogen is substituted (YATSUNAMI et al. 1989); see also the reaction of amino alcohols having a free amino group in DMF in the presence of sodium hydride (YOSHIDA et al. 1990, 1991a,b).

V. C–C Linkage

Various processes are suitable for building up quinolones which have C–C linked radicals in the 7-position. 7-Aryl- and 7-heteroaryl-quinolonecarboxylic acids are made accessible by a palladium-catalysed coupling reaction of 7-bromo-quinolonecarboxylic acid derivatives with tributyltin aromatics or heteroaromatics (Fig. 31; Stille reaction). However, this C–C linkage reaction can also already take place at the step of the aroylacetate, which is then reacted by the cycloaracylation method to give the quinolone (WENTLAND et al. 1993b).

Rosoxacin, which is accessible via the Gould-Jacobs synthesis (CARABATEAS et al. 1984), is an early commercial product from this series (WENTLAND and CORNETT 1985). One of the most active derivatives of this type is 1-cyclopropyl-6,8-difluoro-1,4-dihydro-7-(2,6-dimethyl-4-pyridinyl)-4-oxo-3-quinolinecarboxylic acid, which has an excellent in vivo and in vitro activity against Gram-positive pathogens, including methicillin-resistant *Staphylococcus aureus* and anaerobic organisms (REUMAN et al. 1989, 1995; LESHER et al. 1987, 1989a,b). However, this compound also shows a pronounced interaction with mammalian topoisomerase II (WENTLAND et al. 1993b).

Direct palladium-catalysed coupling of acyclic and cyclic tributl-vinyl-tin derivatives with naphthyridonecarboxylic acids (LABORDE et al. 1991, 1993) leads to 7-ethenyl-, 7-(3-aminocycloalkenyl)- and 7-(1,2,3,6-tetrahydro-4-pyridinyl)-1,8-naphthyridonecarboxylic acid derivatives. These derivatives have an antibacterial in vitro activity which corresponds to that of the C–N linked analogues. In the quinolonecarboxylic acid series, the readily accessible 7-fluorine derivatives are usually used for C–N linkages. However, since these are unsuitable for the palladium-catalysed Stille reaction, in contrast to aryl iodides, bromides and chlorides and in spite of the electron-withdrawing substituents, they have been converted via the 7-hydroxy-quinolones into the 7-quinolyl-triflates. These have been used for C–C linkages with suitable substituted vinylstannanes (KIELY et al. 1991a).

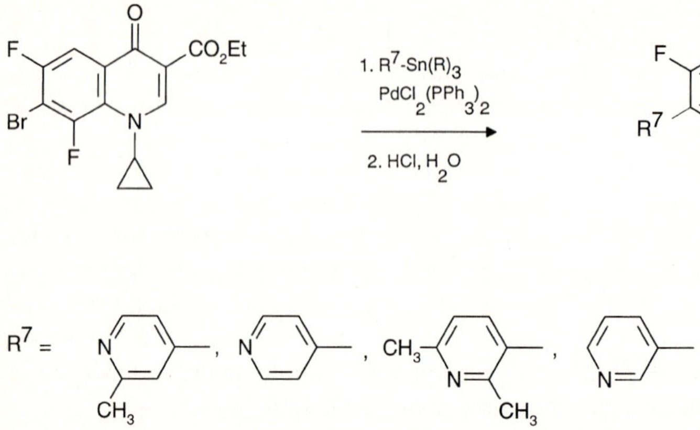

Fig. 31. Synthesis of 7-pyridyl-quinolonecarboxylic acids

In the context of a Kumada reaction, for example, chlorobenzene can also be reacted with ethyl 7-chloro-1-cyclopropyl-6-fluoro-1,4-dihydro-4-oxo-3-quinolinecarboxylate in the presence of nickel(II) chloride, metallic zinc and triphenylphosphine to give ethyl 1-cyclopropyl-6-fluoro-1,4-dihydro-4-oxo-7-phenyl-3-quinolinecarboxylate (HIMMLER et al. 1988). Subsequent reactions have been described on the 4-pyridyl radical (N-oxidation, chlorination, amination, methoxylation; CARABATEAS et al. 1984; WENTLAND et al. 1995), and also on the phenyl radical of 7-arylquinolones (nitration, reduction; HIMMLER et al. 1988). 7-(2-Thiazolyl)- and 7-(4-thiazolyl)-quinolones, some of which have a high antibacterial activity, are prepared by reaction of corresponding α-halo-ketones with thioamides (Fig. 32; CULBERTSON et al. 1987b).

Ethyl 1-ethyl-6-fluoro-7-vinyl-1,4-dihydro-4-oxo-3-quinolinecarboxylate is also obtained via a palladium-catalysed reaction of ethyl 7-bromo-1-ethyl-6-fluoro-1,4-dihydro-4-oxo-3-quinolinecarboxylate with vinyl-magnesium bromide. The vinyl group can be converted into a cyclopropyl group in a secondary reaction with diazomethane in the presence of palladium(II) acetate (Fig. 33; MCGUIRK 1988).

The quinolone ester (2 in Fig. 34), accessible via C–C linkage of ethyl 1-cyclopropyl-6,7-difluoro-quinolonecarboxylate (1 in Fig. 34) with malonic acid tert-butyl ester, can be converted into the intermediate (3 in Fig. 34), which can serve as a precursor for 7-alkyl-, 7-cycloalkyl- and 7-vinyl-quinolones (4 in Fig. 34). 7-Cyclopropyl- and 7-vinyl-quinolones have an in vitro activity against Gram-positive and Gram-negative bacteria – with the exception of *P. aeruginosa* – which corresponds to that of ciprofloxacin. Compared with ciprofloxacin, they have reduced side effects on the CNS, but a significantly lower recovery in the urine. This is probably caused by the lipophilic 7-substituent (TODO et al. 1994a). Starting from the same intermediate (3), it is possible to introduce hydrophilic hydroxyl and amino groups via suitable reaction steps and to sometimes improve the activity against Pseudomonas in

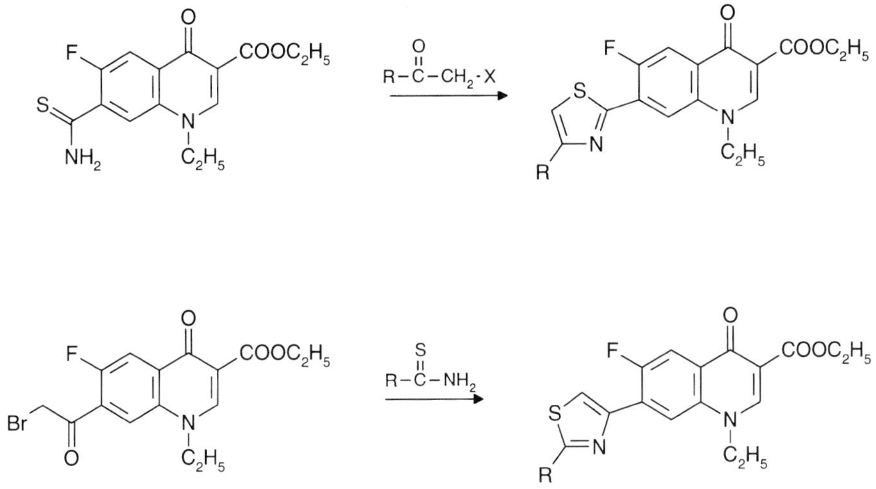

Fig. 32. Synthesis of 7-thiazolyl-quinolonecarboxylic acids

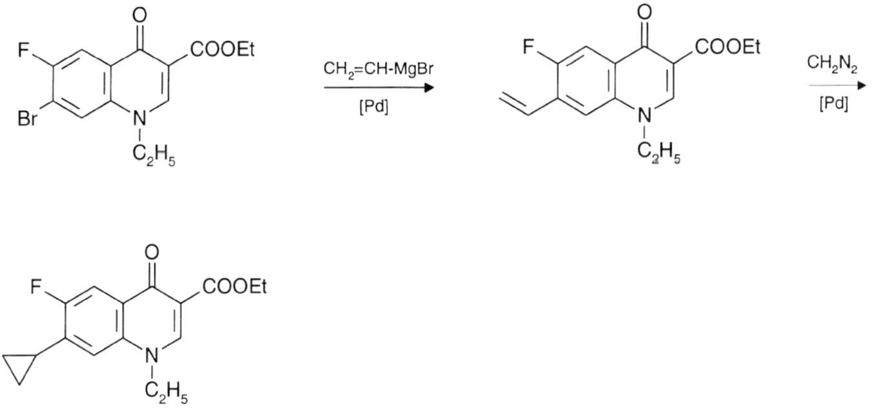

Fig. 33. Synthesis of 7-vinyl- and 7-cyclopropyl-quinolonecarboxylic acids

this way (TODO et al. 1994b). The quinolone with the best tolerability carries the 7-(1-aminocyclopropyl) substituent, which has also been linked with other quinolone parent substances (TODO et al. 1994c).

During synthesis of the enantiomerically pure development product T-3761 (see Table 3, Chap. 2), the (1-aminocyclopropyl) substituent is introduced in protected form at a precursor step (TODO et al. 1994d,e). The corresponding 7-(1-aminocyclopropyl)-1-cyclopropyl-6-fluoro-1,4-dihydro-8-methoxy-4-oxo-3-quinolinecarboxylic acid is obtained by building up the 7-substituent on the quinolone parent substance. This compound is more active than the homologues with a larger 7-(1-amino-cycloalkyl) radical. It has an activity of the order of ofloxacin, with a reduced cytotoxicity and phototoxicity

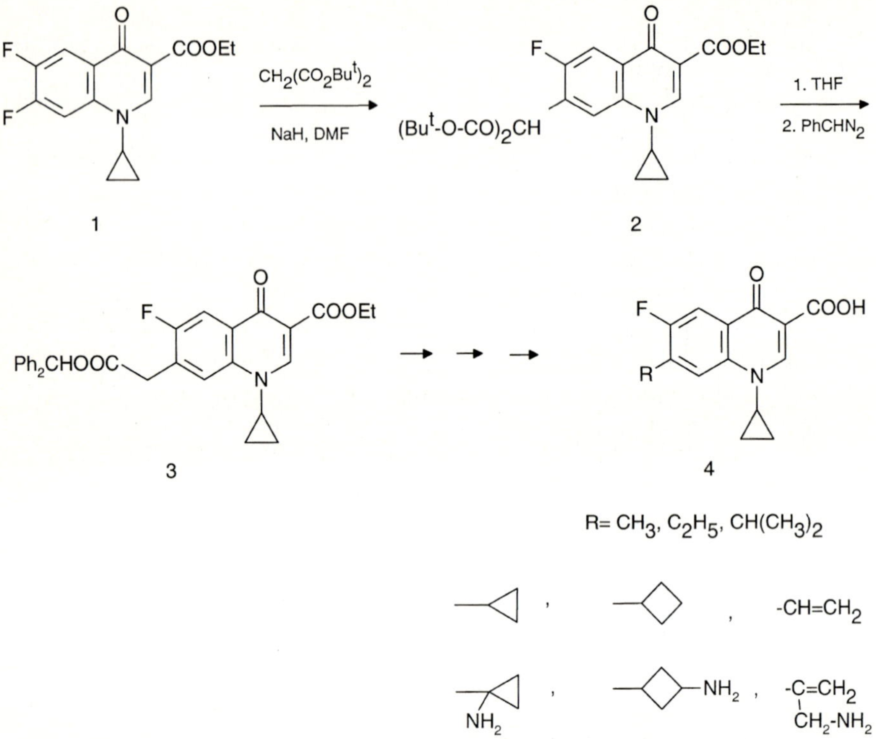

Fig. 34. Synthesis of 7-vinyl- and 7-cycloalkyl-quinolonecarboxylic acids via C–C linkage with malonic acid *tert*-butyl ester; *DMF*, dimethylformamide; *THF*, tetrahydrofuran

and fewer CNS interactions (YOKOTA et al. 1992; HARAMURA et al. 1994). Substituted 7-vinyl-naphthyridones of the nalidixic acid type are obtained by condensation of a 7-methyl group with aldehydes such as 5-nitro-2-furaldehyde (DOHMORI et al. 1969; LEYSEN et al. 1991b).

A few other reactions in the 7-position are mentioned briefly below:

- In the naphthyridonecarboxylic acid (MATSUMOTO et al. 1984a) and 6*H*-6-oxopyrido[1,2-a]pyrimidone-7-carboxylic acid (CHU et al. 1993a) series, a piperazinyl substituent can be split off by heating with alkali, the corresponding hydroxy derivative being obtained.
- Conversion of 7-nitro- into 7-fluoro-quinolonecarboxylic acid esters is achieved by reaction with KF in dimethyl sulphoxide at 128°–165°C (EGAWA et al. 1987).
- Conversion of 7-nitro- via 7-amino- into 7-chloro-quinolonecarboxylic acid esters via a Sandmeyer reaction has been described (EGAWA et al. 1987).
- The following exchange reactions in the 7-position have been carried out with ethyl 1-cyclopropyl-6,7,8-trifluoro-1,4-dihydro-4-oxo-3-quinolinecarboxylate with good yields: F → N_3 → NH_2 → Br (LESHER et al. 1989b).

I. 8-Position

The 8-position of quinolones with a good antibacterial activity preferably carries radicals such as H, F, Cl, CH_3, CF_3, OCH_3, $OCHF_2$, or an N atom in the naphthyridones. For information regarding the influence of the 8-substituent on the photostability and phototoxicity of quinolones see DOMAGALA (1994) and TIEFENBACHER et al. (1994).

The substituents at the 8-position are normally introduced via synthesis of the quinolone nucleus by suitable substitution patterns of the precursors. Examples of such substitutions are:

8-Bromo-1-cyclopropyl-6,7-difluoro-1,4-dihydro-4-oxo-3-quinolinecarboxylic acid (IRIKURA et al. 1984/1985; RENAU et al. 1996)

8-Chloro-1-cyclopropyl-6,7-difluoro-1,4-dihydro-4-oxo-3-quinolinecarboxylic acid (PETERSEN et al. 1984)

1-Cyclopropyl-6,7,8-trifluoro-1,4-dihydro-4-oxo-3-quinolinecarboxylic acid (GROHE et al. 1983)

Ethyl 7-chloro-1-cyclopropyl-6-fluoro-1,4-dihydro-8-nitro-4-oxo-3-quinolinecarboxylate (SANCHEZ et al. 1988)

1-Cyclopropyl-6,7-difluoro-1,4-dihydro-8-methyl-4-oxo-3-quinolinecarboxylic acid (H. MIYAMOTO et al. 1990)

5-Amino-1-cyclopropyl-6,7-difluoro-1,4-dihydro-8-methyl-4-oxo-3-quinolinecarboxylic acid (YOSHIDA et al. 1996a)

5-Chloro-1-cyclopropyl-6,7-difluoro-1,4-dihydro-8-methyl-4-oxo-3-quinolinecarboxylic acid (YOSHIDA et al. 1996a)

1-Cyclopropyl-6,7-difluoro-8-trifluoromethyl-1,4-dihydro-4-oxo-3-quinolinecarboxylic acid (SANCHEZ et al. 1992a)

1-Cyclopropyl-6,7-difluoro-1,4-dihydro-8-methoxy-4-oxo-3-quinolinecarboxylic acid (IWATA et al. 1986)

1-Cyclopropyl-8-difluoromethyoxy-6,7-difluoro-1,4-dihydro-4-oxo-3-quinolinecarboxylic acid (IWATA et al. 1988)

1-Cyclopropyl-6,7-difluoro-1,4-dihydro-8-methylthio-4-oxo-3-quinolinecarboxylic acid (YOSHIDA et al. 1996a)

7-Chloro-8-cyano-1-cyclopropyl-6-fluoro-1,4-dihydro-4-oxo-3-quinolinecarboxylic acid (SCHRIEWER et al. 1987)

8-Chloro-6,7-difluoro-1-(cis-2-fluoro-1-cyclopropyl)-1,4-dihydro-4-oxo-3-quinolinecarboxylic acid (HAYAKAWA and KIMURA 1988)

However, there are also numerous reactions at the 8-position which are worth noting for interesting derivatizations. Direct nucleophilic exchange of a fluorine atom at the 8-position by sodium methylate or sodium methylthiolate in 1-ethyl-6,8-difluoro-1,4-dihydro-7-(1-imidazolyl)-4-oxo-3-quinolinecarboxylic acid leads to the corresponding 8-methoxy- and 8-methylthio-quinolones, respectively (50%–55% yield; UNO et al. 1987). Such a substitution reaction has also been carried out with higher alcoholates such as sodium ethylate, propylates and butylates (MASUZAWA et al. 1986b). For the synthesis of 8-methoxy-quinolones by reaction with sodium methylate, see also SHIMIZU et al. (1992). 8-Methoxylation with methanol/potassium tert-butylate (CULBERTSON et al. 1982/1983) is also possible. A nucleophilic substitution reaction with benzyl alcohol/sodium hydride to give the 8-benzyloxy-quinolonecarboxylic acid ethyl ester and subsequent hydrogenolytic debenzylation thereof to give the 8-hydroxy-quinolone has been described in connection with marbofloxacin synthesis (ANONYMOUS 1988).

In contrast to the substitution reactions described for the 8-position, only the 6-dimethylamino derivative is isolated after reaction of a 6,8-difluoroquinolone with dimethylamine as the nucleophile (22% yield; UNO et al. 1987). 8-Dimethylamino substitution was achieved by reductively methylating an 8-aminoquinolone using formaldehyde, formic acid and sodium acetate (MASUZAWA et al. 1986b). In attempting to substitute an 8-fluorine atom by cyanide, decarboxylation readily occurs, especially in electron-deficient quinolones (REUMAN et al. 1994).

8-Bromo-quinolonecarboxylic acid ethyl esters are suitable intermediates for C–C linkage reactions catalysed by transition metals. For example, 8-vinyl-, 8-ethinyl- and 8-aryl-quinolones can be prepared from them (PETERSEN et al. 1991a). Some of these quinolones are also accessible via the corresponding benzoic acids (TURNER and SUTO 1993).

Electrophilic bromination of 1-unsubstituted derivatives, such as ethyl 6-fluoro-1,4-dihydro-4-oxo-quinoline-3-carboxylate and ethyl 7-methoxy-1,4-dihydro-4-oxo-quinoline-3-carboxylate, with *N*-bromosaccharin in the presence of pyridinium polyhydrofluoride takes place with a yield of 64% and 66%, respectively. However, the reaction does not succeed with the analogous 1-ethyl-quinolones (MOZEK and SKET 1994).

Direct 8-chlorination of 7-substituted quinolones is also possible, for example by reaction with sulphuryl chloride (HOKURIKU 1986, 1990a; KYORIN 1987) or elemental chlorine in chlorosulphonic acid in the presence of catalytic amounts of iodine (HOKURIKU 1988). Reductive dechlorination in the 8-position by reaction with zinc/ammonium chloride in dioxane/water has been described (HOKURIKU 1990b).

8-Nitro-quinolonecarboxylic acids can be obtained not only via synthesis of the suitable parent substance, but also via subsequent nitration. For example, 7-(4-acetyl-1-piperazinyl)-1-ethyl-6-fluoro-1,4-dihydro-8-nitro-4-oxo-3-quinolinecarboxylic acid is obtained from 1-ethyl-6-fluoro-1,4-dihydro-4-oxo-7-(1-piperazinyl)-3-quinolinecarboxylic acid by nitration with fuming nitric acid/acetic anhydride, after intermediate *N*-acetylation in acetic acid/acetic anhydride (MASUZAWA et al. 1986b). 8-Amino-quinolones are obtained by catalytic hydrogenation of 8-nitro-quinolones (SANCHEZ et al. 1988).

The “tandem cyclizations” described in Chap. 2BI, III, which lead to 1,8-bridged tri- or tetracyclic quinolones, are special examples for substitution reactions in the 8-position. A practical alternative synthesis for the pyrido-[3,2,1-ij]cinnoline ring system (YOKOMOTO et al. 1990/1991), from which compounds of high antibacterial activity are derived, proceeds via 1,8-cyclization of the Michael adduct of di-tert-butyl methylene-malonate on ethyl 6,7,8-trifluoro-1,4-dihydro-1-methylamino-4-oxo-3-quinolinecarboxylate with

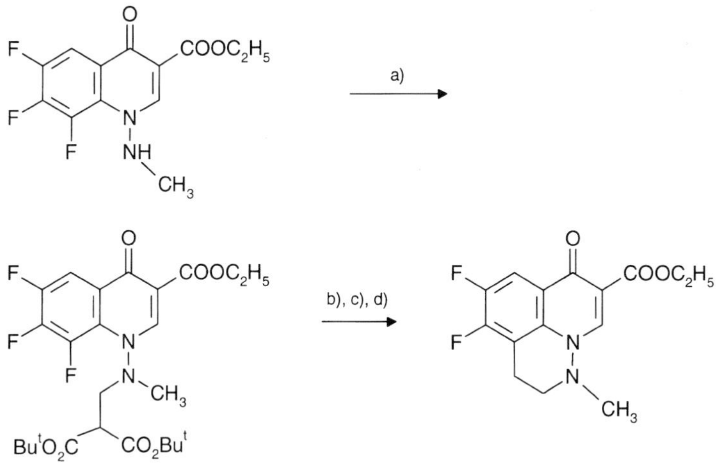

Fig. 35. Synthesis of the pyrido[3,2,1-ij]cinnoline nucleus. *a*, $=\!\!<$ CO_2Bu^t, O_2Bu^t, $TiCl_4/CH_2Cl_2$, methyloxiran; *b*, Cs_2CO_3/dimethylsulfoxide (DMSO); *c*, tetrahydrofuran; *d*, 150°C/DMSO (decarboxylation)

Cs_2CO_3/DMSO, with subsequent hydrolysis and decarboxylation (Barrett et al. 1995a; Fig. 35). The efficient synthesis of 2,7-dioxopyrido[3,2,1-ij]cinnolines by an intramolecular arylation reaction of *N*-1 tethered quinolone malonamides is described by a similar route (Barrett et al. 1995b).

References

Albrecht R (1977) Development of antibacterial agents of the nalidixic acid type. Prog Drug Res 21:9–104

Albrecht HA (1990) Cephalosporin 3′-quinolone esters with a dual mode of action. J Med Chem 33:77–86

Albrecht HA, Beskid G, Christenson JG, Durkin JW, Fallat V, Georgopapadakou NH, Keith DD, Konzelmann FM, Lipschitz ER, McGarry DH, Siebelist JA, Wei CC, Weigele M, Yang R (1991a) Dual-action cephalosporins: cephalosporin 3′-quaternary ammonium quinolones. J Med Chem 34:669–675

Albrecht HA, Beskid G, Christenson JG, Georgopapadakou NH, Keith DD, Konzelmann FM, Pruess DL, Rossman PL, Wei CC (1991b) Dual-action cephalosporins: cephalosporin 3′-quinolone carbamates. J Med Chem 34:2857–2864

Albrecht HA, Beskid G, Christenson JG, Deitcher KH, Georgopapadakou NH, Keith DD, Konzelmann FM, Pruess DL, Wei CC (1994) Dual-action cephalosporins incorporating a 3′-tertiary-amine-linked quinolone. J Med Chem 37:400–407

Anderson AG, Lok R (1972) The synthesis of azetidine-3-carboxylic acid. J Org Chem 37:3953–3955

Andriole VT (ed) (1988) The quinolones. Academic, London

Anonymous (1988) RD 291 097

Araki K, Kuroda T, Uemori S, Moriguchi A, Ikeda Y (1987) EP 311 948

Araki K, Kuroda T, Uemori S, Moriguchi A, Ikeda Y, Hirayama F, Yokoyama Y, Iwao E, Yakushiji T (1993) Quinolone antimicrobial agents substituted with morpholines at the 7-position. Syntheses and structure-activity relationships. J Med Chem 36:1356–1363

Asahina Y, Fukuda Y, Fukuda H, Kyorin Pharmaceutical (1990) EP 443 498

Asahina Y, Ishizaki T, Suzue S (1992) Recent advances in structure activity relationships in new quinolones. Prog Drug Res 38:57–106

Atarashi S, Imamura M, Kimura Y, Yoshida A, Hayakawa I (1993) Fluorocyclopropyl quinolones. I. Synthesis and structure-activity relationships of 1-(2-fluorocyclopropyl)-3-pyridonecarboxylic acid antibacterial agents. J Med Chem 36:3444–3448

Barrett D, Sasaki H, Tsutsumi H, Murata M, Terasawa T, Sakane K (1995a) A concise, practical synthesis of the pyrido[3,2,1-ij]cinnoline ring system of potent DNA gyrase inhibitors. J Org Chem 60:3928–3930

Barrett D, Tsutsumi H, Kinoshita T, Murata M, Sakane K (1995b) Synthesis of new 1,8-bridged tricyclic quinolones by a novel intramolecular arylation of N-1 tethered malonamides. Tetrahedron 51:11125–11140

Bartel S, Krebs A, Kunisch F, Petersen U, Schenke T, Grohe K, Schriewer M, Bremm KD, Endermann R, Metzger KG (1993) Bayer AG, DE 4 301 246

Bauernfeind A (1993) Comparative in-vitro activities of the new quinolone BAY Y 3118, and ciprofloxacin, sparfloxacin, tosufloxacin, CI-960 and CI-990. J Antimicrob Chemotherapy 31:505–522

Bouzard D (1990) Recent advances in the chemistry of quinolones. In: Lukacs G (ed) Recent progress in the chemical synthesis of antibiotics. Springer, Berlin Heidelberg New York, pp 249–283

Bouzard D, Di Cesare P, Essiz M, Jacquet JP, Remuzon P, Weber A, Oki T, Masuyoshi M (1989) Fluoronaphthyridines and quinolones as antibacterial agents. I. Synthesis and structure-activity relationships of new 1-substituted derivatives. J Med Chem 32:537–542

Bouzard D, Di Cesare P, Essiz M, Jacquet JP, Kiechel JR, Remuzon P, Weber A, Oki T, Masuyoshi M, Kessler RE, Fung-Tomc J, Desiderio J (1990) Fluoronaphthyridines and quinolones as antibacterial agents. II. Synthesis and structure-activity relationships of new 1-tert-butyl 7-substituted derivatives. J Med Chem 33:1344–1352

Bouzard D, Di Cesare P, Essiz M, Jacquet JP, Ledoussal B, Remuzon P, Kessler RE, Fung-Tomc J (1992a) Fluoronaphthyridines as antibacterial agents. IV. Synthesis and structure-activity relationships of 5-substituted-6-fluoro-7-(cycloalkylamino)-1,4-dihydro-4-oxo-1,8-naphthyridine-3-carboxylic acids. J Med Chem 35:518–525

Bouzard D, Di Cesare P, Hoffmann P, Fung-Tomc J, Kessler R (1992b) In vitro and in vivo evaluation of BMY 45243, a new 5-amino-naphthyridone derivative. Drugs Exp Clin Res 18:291–294

Braish TF (1992) Pfizer Inc., WO 93/18001

Braish TF, Fox DE (1990) Synthesis of (S,S)- and (R,R)-2-alkyl-2,5-diazabicyclo[2.2.1]heptanes. J Org Chem 55:1684–1687

Braish TF, Fox DE, Norris T, Rose PR (1994) Pfizer Inc., WO 95/19361

Braish TF, Castaldi M, Chan S, Fox DE, Keltonic T, McGarry J, Hawkins JM, Norris T, Rose PR, Sieser JE, Sitter BJ, Watson H Jr (1996) Construction of the (1α,5α,6α)-6-amino-3-azabicyclo[3.1.0]hexane ring system. Synlett 1100–1102

Bremm KD, Petersen U, Metzger KG, Endermann R (1992) In vitro evaluation of BAY Y 3118, a new full-spectrum fluoroquinolone. Chemotherapy (Basel) 38:376–387

Brighty KE (1989) Pfizer Inc., EP 413 455

Brighty KE, Lowe JA, McGuirk PR (1987) Pfizer Inc., EP 321 191

Brighty KE, Castaldi MJ (1996) Synthesis of (1α,5α,6α)-6-amino-3-azabicyclo-[3.1.0]hexane, a novel achiral diamine. Synlett 1097–1102

Bucsh RA, Domagala JM, Laborde E, Sesnie JC (1993) Synthesis and antimicrobial evaluation of a series of 7-[3-amino (or aminomethyl)-4-aryl (or cyclopropyl)-1-pyrrolidinyl]-4-quinolone- and 1,8-naphthyridone-3-carboxylic acids. J Med Chem 36:4139–4151

Carr RM, Sutherland DR (1994) A novel synthesis of carbon-labelled quinolone-3-carboxylic acid antibacterials. J Label Compound Radiopharm 34:961–971

Carabateas PM, Brundage RP, Gelotte KO, Gruett MD, Lorenz RR, Opalka CJ Jr, Singh B, Thielking WH, Williams GL, Lesher GY (1984) 1-Ethyl-1,4-dihydro-4-oxo-7-(pyridinyl)-3-quinolinecarboxylic acids. II. Synthesis. J Heterocycl Chem 21:1857–1863

Cecchetti V, Fravolini A, Fringuelli R, Mascellani G, Pagella P, Palmioli M, Segre G, Terni P (1987) Quinolonecarboxylic acids. II. Synthesis and antibacterial evaluation of 7-oxo-2,3-dihydro-7H-pyrido[1,2,3-de][1,4]benzothiazine-6-carboxylic acid. J Med Chem 30:465–473
Cecchetti V, Fravolini A, Terni P, Pagella PG, Tabarrini O. Mediolanum Farmaceutici (1991) EP 531 958
Cecchetti V, Fravolini A, Fringuelli R, Schiaffella F (1993a) 4H-1-benzothiopyran-4-one-3-carboxylic acids and 3,4-dihydro-2H-isothiazolo[5,4-b][1]benzothiopyran-3,4-diones as quinolone antibacterial analogs. J Heterocycl Chem 30:1143–1148
Cecchetti V, Fravolini A, Pagella PG, Savino A, Tabarrini O (1993b) Quinolinecarboxylic acids. III. Synthesis and antibacterial evaluation of 2-substituted-7-oxo-2,3-dihydro-7H-pyrido[1,2,3-de][1,4]benzothiazine-6-carboxylic acids related to rufloxacin. J Med Chem 36:3449–3454
Cecchetti V, Clementi S, Cruciani G, Fravolini A, Pagella PG, Savino A, Tabarrini O (1995) 6-Aminoquinolones: a new class of quinolone antibacterials? J Med Chem 38:973–982
Cecchetti V, Fravolini A, Lorenzini MC, Tabarrini O, Terni P, Xin T (1996) Studies on 6-aminoquinolones: synthesis and antibacterial evaluation of 6-amino-8-methylquinolones. J Med Chem 39:436–445
Chan K-K, Keith DD (1987/1988) F Hoffmann-La Roche & Co, EP 322 810
Cheil Pharm (1990) KR 9 206 780
Chiba K, Nishimura Y, Nakano J, Matsumoto J, Nakamura S, Dainippon Pharmaceutical (1987) JP 01 56.673 (Chem Abstr 111:153779w)
Chu DTW, Abbott Laboratories (1983/1984) EP 131 839
Chu DTW (1985) A regiospecific synthesis of 1-methylamino-6-fluoro-7-(4-methylpiperazin-1-yl)-1,4-dihydro-4-oxoquinoline-3-carboxylic acid. J Heterocycl Chem 22:1033–1034
Chu DTW (1987) Synthesis and structure-activity relationship of 1-aryl-6,8-difluoroquinolone antibacterial agents. J Med Chem 30:504–509
Chu DTW (1990) Synthesis of 6-fluoro-7-(piperazin-1-yl) 9 cyclopropyl-2,3,4,9-tetrahydroisothiazolo[5,4-b]quinoline-3,4-dione and 6-fluoro-7-(piperazin-1-yl)-9-(p-fluorophenyl)-2,3,4,9-tetrahydroisothiazolo[5,4-b]quinoline-3,4-dione. J Heterocycl Chem 27:839–843
Chu DTW (1992) Isothiazoloquinolones: antibacterial and antineoplastic agents. Drugs Fut 17:1101–1109
Chu DTW (1993) Fluoroquinolone carboxylic acids as antibacterial drugs. In: Filler R, Kobayashi Y, Yagupolskii LM (eds) Organofluorine compounds in medicinal chemistry and biomedical applications. Elsevier, Amsterdam, pp 165–207
Chu DTW, Claiborne AK (1990) Practical synthesis of iminochlorothioformates: application of iminochlorothioformates in the synthesis of novel 2,3,4,9-tetrahydroisothiazolo[5,4-b][1,8]naphthyridine-3,4-diones and 2,3,4,9-tetrahydroisothiazolo[5,4-b]quinoline-3,4-dione derivatives. J Heterocycl Chem 27:1191–1195
Chu DTW, Fernandes PB (1991) Recent development in the field of quinolone antibacterial agents. Adv Drug Res 21:39–144
Chu DTW, Hallas R, Abbott Laboratories (1988/1989) EP 360 258
Chu DTW, Rosen TJ, Abbott Laboratories (1988a) EP 331 960
Chu DTW, Rosen TJ, Abbott Laboratories (1988b) US 4 859 776
Chu DTW, Fernandes PB, Claiborne AK, Shen L, Pernet AG (1988) Structure-activity relationships in quinolone antibacterials: design, synthesis and biological activities of novel isothiazoloquinolones. Drugs Exp Clin Res 14:379–383
Chu DTW, Lico IM, Swanson RN, Marsh KC, Plattner JJ, Pernet AG (1990) Synthesis and biological properties of A-71497: a prodrug of tosufloxacin. Drugs Exp Clin Res 16:435–443
Chu DTW, Nordeen CW, Hardy DJ, Swanson RN, Giardina WJ, Pernet AG, Plattner JJ (1991) Synthesis, antibacterial activities, and pharmacological properties of enantiomers of temafloxacin hydrochloride. J Med Chem 34:168–174

Chu DTW, Li Q, Tanaka K, Alder J, Claiborne A, Lico I, Raye K, Rosen T, Plattner JJ (1992) Synthesis and biological properties of A-80556: a potent antibacterial fluoroquinolone (Abstr 652). 32nd Interscience Conference on Antimicrobial Agents and Chemotherapy, Anaheim

Chu DTW, Li Q, Lee CM, Raye K, Tanaka K, Alder J, Plattner JJ (1993a) Novel quinolone antibacterial agents: synthesis and biological activity of 6H-6-oxopyrido[1,2-a]pyrimidine-7-carboxylic acids. In: Bentley PH, Ponsford R (eds) Recent advances in the chemistry of anti-infective agents. Royal Society of Chemistry, Cambridge, pp 93–105 (Special publication 119)

Chu DTW, Li Q, Cooper CS, Fung AKL, Lee CM, Plattner JJ, Abbott Laboratories (1993b) WO 95/10519

Chu DTW, Li Q, Claiborne A, Raye-Passarelli K, Cooper C, Fung A, Lee C, Tanaka SK, Shen LL, Donner P, Armiger YL, Plattner JJ (1994) Synthesis and antibacterial activity of A-86719.1 and related 2-pyridones. A novel series of potent DNA gyrase inhibitors (Abstr F41). 34th Interscience Conference on Antimicrobial Agents and Chemotherapy, Orlando

Cole A, Goodfield J, Williams DR, Midgley JM (1984) The complexation of transition series metal ions by nalidixic acid. Inorg Chim Acta 92:91–97

Conrad RA, White WA, Eli Lilly (1981) US 4 379 929

Cooper CS, Klock PL, Chu DTW, Hardy DJ, Swanson RN, Plattner JJ (1992a) Preparation and in vitro and in vivo evaluation of quinolones with selective activity against gram-positive organisms. J Med Chem 35:1392–1398

Cooper CS, Donner PK, Chu DTW, Clement J, Alder J, Plattner JJ (1992b) Synthesis and biological activity of A-77143, a potent new quinolone antibacterial agent (Abstr 763). 32nd Interscience Conference on Antimicrobial Agents and Chemotherapy, Anaheim

Corraz AJ, Dax SL, Dunlap NK, Georgopapadakou NH, Keith DD, Pruess DL, Rossmann PL, Then R, Unowsky J, Wei C-C (1992) Dual-action penems and carbapenems. J Med Chem 35:1828–1839

Culbertson TP (1991) Synthesis of 4H-1,4-benzothiazine 1-oxide and 1,1-dioxide. Analogs of antibacterial agents. J Heterocycl Chem 28:1701–1703

Culbertson TP, Mich TF, Domagala JM, Nichols JB, Warner-Lambert (1982/1983) EP 106 489

Culbertson TP, Mich TF, Domagala JM, Nichols JB (1984) EP 153 163

Culbertson TP, Domagala JM, Nichols JB, Priebe S, Skeean RW (1987a) Enantiomers of 1-ethyl-7-[3-[(ethylamino)methyl]-1-pyrrolidinyl]-6,8-difluoro-1,4-dihydro-4-oxo-3-quinolinecarboxylic acid: preparation and biological activity. J Med Chem 30:1711–1715

Culbertson TP, Domagala JM, Peterson P, Bongers S, Nichols JB (1987b) New 7-substituted quinolone antibacterial agents. The synthesis of 1-ethyl-1,4-dihydro-4-oxo-7-(2-thiazolyl and 4-thiazolyl)-3-quinolinecarboxylic acids. J Heterocycl Chem 24:1509–1520

Culbertson TP, Sanchez JP, Gambino L, Sesnie JA (1990) Quinolone antibacterial agents substituted at the 7-position with spiroamines. Synthesis and structure-activity relationships. J Med Chem 33:2270–2275

Daiichi Seiyaku (1980) JP 56 118 081

Daiichi Seiyaku (1981) JP 57 149 286 (Chem Abstr 98:72117q)

Dainippon (1978) JP 55 036 436

Dainippon (1979) JP 56 045 473

Dainippon (1986) JP 62 226 962

Dainippon (1987) JP 1 016 767

Dainippon (1990) JP 4 049 765

Dalhoff A, Petersen U, Endermann R (1996) In vitro activity of BAY 12-8039, a new methoxyquinolone. Chemotherapy (Basel) 42:410–425

De la Cruz A, Elguero J, Goya P, Martinez A (1990) Synthesis of a valuable precursor for the preparation of novel quinolone glycosides. Synlett:753–754

De la Cruz A, Elguero J, Goya P, Martinez A, Pfleiderer W (1992) Tautomerism and acidity in 4-quinolone-3-carboxylic acid derivatives. Tetrahedron 48:6135–6150

Demuth TP Jr, White RE, Norwich Eaton Pharmaceuticals (1988/1989a) EP 366 193
Demuth TP Jr, White RE, Norwich Eaton Pharmaceuticals (1988/1989b) EP 366 640
Demuth TP Jr, White RE, Norwich Eaton Pharmaceuticals (1988/1989c) EP 366 641
Demuth TP, White RE, Procter & Gamble (1992) WO 94/10163
Demuth TP, White RE, Tietjen RA, Storrin RJ, Skuster JR, Andersen JA, McOsker CC, Freedman R, Rourke FJ (1991) Synthesis and antibacterial activity of new C-10 quinolonylcephem esters. J Antibiot 44:200–209
Desideri, Stradi R, Milanese A, Rorer Italiana (1985) EP 224 121
Di Cesare P, Jacquet J-P, Essiz M, Remuzon P, Bouzard D, Weber A (1986) EP 266 576
Di Cesare P, Bouzard D, Essiz M, Jacquet JP, Ledoussal B, Kiechel JR, Remuzon P, Kessler RE, Fung-Tomc J, Desiderio J (1992) Fluoronaphthyridines and -quinolones as antibacterial agents. V. Synthesis and antimicrobial activity of chiral 1-tert-butyl-6-fluoro-7-substituted-naphthyridones. J Med Chem 35:4205–4213
Dohmori R, Kadoya S, Tanaka Y, Takamura I, Yoshimura R, Naito T (1969) Synthesis of 1-substituted-1,4-dihydro-7-[2-(5-nitro-2-furyl)vinyl]-4-oxo-1,8-naphthyridine Derivatives. II. Chem Pharm Bull (Tokyo) 17:1832–1838
Domagala JM (1994) Rewiew: structure-activity and structure-side-effect relationships for the quinolone antibacterials. J Antimicrob Chemother 33:685–706
Domagala JM, Schroeder MC, Warner-Lambert (1985) EP 198 678
Domagala JM, Hagen SE, Sanchez JP, Warner-Lambert (1986a) EP 255 908
Domagala JM, Hanna LD, Heifetz CL, Hutt MP, Mich TF, Sanchez JP, Solomon M (1986b) New structure-activity relationships of the quinolone antibacterials using the target enzyme. The development and application of a DNA gyrase assay. J Med Chem 29:394–404
Domagala JM, Hagen SE, Sanchez JP, Warner-Lambert (1987) US 4 771 055
Domagala JM, Hagen SE, Heifetz CL, Huff MP, Mich TF, Sanchez JP, Trehan AK (1988a) 7-Substituted 5-amino-1-cyclopropyl-6,8-difluoro-1,4-dihydro-4-oxo-3-quinolinecarboxylic acids: synthesis and biological activity of a new class of quinolone antibacterials. J Med Chem 31:503–506
Domagala JM, Heifetz CL, Hutt MP, Mich TF, Nichols JB, Solomon M, Worth DF (1988b) 7-[3-[(Ethylamino)methyl]-1-pyrrolidinyl]-6,8-difluoro-1,4-dihydro-4-oxo-3-quinolinecarboxylic acids. New quantitative structure-activity relationships at N_1 for the quinolone antibacterials. J Med Chem 31:991–1001
Domagala JM, Suto MJ, Turner WR (1990) Warner Lambert, US 5 047 538
Domagala JM, Bridges AJ, Culbertson TP, Gambino L, Hagen SE, Karrick G, Porter K, Sanchez JP, Sesnie JA, Spense FG, Szotek D, Wemple J (1991) Synthesis and biological activity of 5-amino- and 5-hydroxyquinolones, and the overwhelming influence of the remote N_1-substituent in determining the structure-activity relationships. J Med Chem 34:1142–1154
Domagala JM, Hagen SE, Joannides T, Kiely JS, Laborde E, Schroeder MC, Sesnie JA, Shapiro MA, Suto MJ, Vanderroest S (1993) Quinolone antibacterials containing the new 7-[3-(1-aminoethyl)-1-pyrrolidinyl]side chain: the effects of the 1-aminoethyl moiety and its stereo-chemical configurations on potency and in vivo efficacy. J Med Chem 36:871–882
Egawa H, Miyamoto T, Minamida A, Nishimura Y, Okada H, Uno H, Matsumoto J (1984) Pyridonecarboxylic acids as antibacterial agents. IV. Synthesis and antibacterial activity of 7-(3-amino-1-pyrrolidinyl)-1-ethyl-6-fluoro-1,4-dihydro-4-oxo-1,8-naphthyridine-3-carboxylic acid and its analogues. J Med Chem 27:1543–1548
Egawa H, Kataoka M, Shibamori K, Miyamoto T, Nakano J, Matsumoto J (1987) A new synthetic route to 7-halo-1-cyclopropyl-6-fluoro-1,4-dihydro-4-oxoquinoline-3-carboxylic acid, an intermediate for the synthesis of quinolone antibacterial agents. J Heterocycl Chem 24:181–185
Eissenstat MA, Kuo G-H, Weawer III JD, Wentland MP, Robinson RG, Klingbeil KM, Danz DW, Corbett TH, Coughlin SA (1995) 3-Benzyl-quinolones: novel, potent inhibitors of mammalian topoisomerase II. Bioorg Med Chem Lett 5:1021–1026
Esteve Soler J (1983/1984) Provesan S. A., EP 134 165
Fan J-Y, Sun D, Yu H, Kerwin SM, Hurley LH (1995) Self-assembly of a quinobenzoxazine-Mg^{2+} complex on DNA: a new paradigm for the structure of a

drug-DNA complex and implications for the structure of the quinolone bacterial gyrase-DNA complex. J Med Chem 38:408–424

Fass RJ (1993) In vitro activity of BAY Y 3118, a new quinolone. Antimicrob Agents Chemother 37:2348–2357

Fedij V, Lenoir EA, Suto MJ, Zeller JR, Wemple J (1994) An efficient method for the synthesis of (R)-3-(1-amino-1-methylethyl)pyrrolidines for the antiinfective agent, PD 138 312. Tetrahedron Asymmetry 5:1131–1134

Fernandes P, Chu DTW, Abbott Laboratories (1987) EP 302 372

Flanagan DM, Joullie MM (1987) Synthetic strategies for the construction of 3-pyrrolidinol, a versatile nitrogen heterocycle. Heterocycles 26:2247–2265

Frank J, Rakoczy P (1979) Synthesis and antibacterial activity of some 5-substituted 6,7-methylenedioxy-4-quinolone-3-carboxylic acid derivatives. Eur J Med Chem Chim Ther 14:61–65

Fray AH, Augeri DJ, Kleinman EF (1988) A convenient synthesis of 3,6-disubstituted 3,6-diazabicyclo[3.2.2]nonanes and 3,6-diazabicyclo[3.2.1]octanes. J Org Chem 53:896–899

Freidmann RC, O'Neill BT, Lackey JW, Pfizer (1988) EP 366 301

Frigola J, Parés, Corbera J, Vañó D, Mercè R, Torrens A, Más J, Valentí E (1993) 7-Azetidinylquinolones as antibacterial agents. Synthesis and structure-activity relationships. J Med Chem 36:801–810

Frigola J, Torrens A, Castrillo JA, Mas J, Vaño D, Berrocal JM, Calvet C, Salgado L, Redondo J, García-Granda S, Valenti E, Quintana JR (1994) 7-Azetidinylquinolones as antibacterial agents. Synthesis and biological activity of 7-(2,3-disubstituted-1-azetidinyl)-4-oxoquinoline and 1,8-naphthyridine-3-carboxylic acids. Properties and structure-activity relationships of quinolones with an azetidine moiety. J Med Chem 37:4195–4210

Frigola J, Vaño D, Torrens A, Gómez-Gomar A, Ortega E, García-Granda S (1995) 7-Azetidinylquinolones as antibacterial agents. III. Synthesis, properties and structure-activity relationships of stereoisomers containing a 7-(3-amino-2-methyl-1-azetidinyl) moiety. J Med Chem 38:1203–1215

Fromtling RA, Castañer J (1996) Trovafloxacin mesylate. Drugs Fut 21:496–505

Fujii T, Nishida H, Abiru Y, Yamamoto M, Kise M (1995) Studies on synthesis of the antibacterial agent NM 441. II. Selection of a suitable base for alkylation of 1-substituted piperazine with 4-(bromomethyl)-5-methyl-1,3-dioxol-2-one. Chem Pharm Bull 43:1872–1877

Fujita M, Egawa H, Kataoka M, Miyamoto T, Nakano J, Matsumoto J (1995) Imidazo- and triazoloquinolones as antibacterial agents. Synthesis and structure-activity relationships. Chem Pharm Bull 43:2123–2132

Fung AKL, Chu DT, Armiger YL, Li Q, Tananka SK, Flamm RK, Shen L, Baranowski J, Marsh K, Crowell D, Plattner JJ (1995) Synthesis and structure-activity-relationships of 8-[3-(1-amino-alkyl)pyrrolidinyl]- and 8-[3-(1-aminocycloalkyl) pyrrolidinyl]-2-pyridopyridone antibacterials (Abstr F9) 35th Interscience Conference on Antimicrobial Agents and Chemotherapy, San Francisco

Gaertner VR (1967) Cyclization of 1-alkylamino-3-halo-2-alkanols to 1-alkyl-3-azetidinols. J Org Chem 32:2972–2976

Gammill RB, Bisaha SN, Timko JM, Barbachyn MR, Kim KS, Upjohn Company (1993) US 5 385 906

Garemszegi F, Lehoczk G, Somfai E, Ban K, Hernadi G (1980) Chinoin, HU T 30 014 (Chem Abstr 101:130 678s)

Gilis PM, Haemers A, Bollaert W (1980) 1H-tetrazol-5-yl derivatives of chemotherapeutic agents of the nalidixic acid type. Eur J Med Chem Chim Ther 15:499–502

Gobeaux B, Ghosez L (1989) Intramolecular [2+2]cycloadditions of keteniminium salts derived from α- and β-amino acids. A route to azabicyclic ketones. Heterocycles 28:29–32

Grohe K, Heitzer H (1987) Synthese von 1-Amino-4-chinolon-3-carbonsäuren. Liebigs Ann Chem 871–879

Grohe K, Zeiler H-J, Metzger KG, Bayer (1979) EP 14390

Grohe K, Petersen U, Zeiler H-J, Metzger K-G, Bayer (1983) EP 126 355
Grohe K, Zeiler H-J, Metzger K-G, Bayer (1984a) EP 155 587
Grohe K, Zeiler H-J, Metzger K-G, Bayer (1984b) US 4 670 444
Grohe K, Petersen U, Zeiler H-J, Metzger K-G, Bayer (1985) DE 3 525 108
Hagen SE, Domagala JM (1990) Synthesis of 5-methyl-4-oxo-quinolinecarboxylic acids. J Heterocycl Chem 27:1609
Hagen SE, Domagala JM, Heifetz CL, Sanchez JP, Solomon M (1990) New quinolone antibacterial agents. Synthesis and biological activity of 7-(3,3- or 3,4-disubstituted-1-pyrrolidinyl)quinoline-3-carboxylic acids. J Med Chem 33:849–854
Hagen SE, Domagala JM, Heifetz CL, Johnson J (1991) Synthesis and biological activity of 5-alkyl-1,7,8-trisubstituted-6-fluoroquinoline-3-carboxylic acids. J Med Chem 34:1155–1161
Hagen SE, Domagala JM, Gracheck SJ, Sesnie JA, Stier MA, Suto MJ (1994) Synthesis and antibacterial activity of new quinolones containing a 7-[3-(1-amino-1-methylethyl)-1-pyrrolidinyl] moiety. Gram-positive agents with excellent oral activity and low side-effect potential. J Med Chem 37:733–738
Haramura M, Okamachi A, Makino T, Kohda A, Katoh Y, Munemura K, Ito T, Matsumoto M, Kojima K, Yokota T (1994) Synthesis and biological activity of new 8-methoxyquinolones possessing 7-aminoalkyl group. 34th Interscience Conference on Antimicrobial Agents and Chemotherapy, Orlando
Hayakawa I, Atarashi S, Daiichi Seiyaku (1985) EP 207 420
Hayakawa I, Kimura Y, Daiichi Seiyaku (1988) EP 341 493
Hayakawa I, Atarashi S, Imamura M, Kimura Y, Daiichi Seiyaku (1988a) EP 357 047
Hayakawa I, Atarashi S, Imamura M, Kimura Y, Daiichi Seiyaku (1988b) EP 529 688
Hayakawa I, Atarashi S, Kimura Y, Kawakami K, Daiichi Seiyaku (1990) EP 488 227
Hayakawa I, Atarashi S, Kimura Y, Kawakami K, Saito T, Yafune T, Sato K, Une T, Sato M (1991) Design and structure-activity relationship of new N_1-cis-2-fluorocyclopropyl quinolones (Abstr 1504). 31st Interscience Conference on Antimicrobial Agents and Chemotherapy, Chicago
Heck JV, Thorsett ED, Merck (1987) EP 308 019
Helena M, Teixeira SF, Vilas-Boas LF, Gil VMS, Teixeira F (1995) Complexes of ciprofloxacin with metal ions contained in antacid drugs. J Chemother 7:126–132
Hermecz I, Kereszturi G, Vasári L, Horváth Á, Balogh M, Ritli P, Sipos J, Chinoin Gyogyszer (1987a) WO 88/07998
Hermecz I, Lehoczki G, Kereszturi G, Ritii P, Sipos J, Garamszegi F, Horvath A, Vasvari Debreczi L, Pajor A, Balogh A, Chinoin Gyogyszer (1987b) HU T 46 312 (Chem Abstr 111:97 087e)
Himmler T, Schriewer T, Petersen U, Grohe K, Haller I, Metzger KG, Endermann R, Zeiler H-J, Bayer (1988) EP 343 398
Himmler T, Petersen U, Bremm K-D, Endermann R, Stegemann M, Wetzstein H-G, Bayer (1994) EP 671 391
Högberg T, Vora M, Drake SD, Mitscher LA, Chu DTW (1984a) Structure-activity relationships among DNA-gyrase inhibitors. Synthesis and antimicrobial evaluation of chromones and coumarins related to oxolinic acid. Acta Chem Scand B38:359–366
Högberg T, Khanna I, Drake SD, Mitscher LA, Shen LL (1984b) Structure-activity relationships among DNA gyrase inhibitors. Synthesis and biological evaluation of 1,2-dihydro-4,4-dimethyl-1-oxo-2-naphthalenecarboxylic acids as 1-carba bioisosters of oxolinic acid. J Med Chem 27:306–310
Hojo K, Sakamoto A, Tsutsumi M, Yamada T, Nakazono K, Ishimori K (1985) EP 218 249
Hokuriku (1986) JP 62 178 586
Hokuriku (1988) JP 2 174 783, 2 196 786
Hokuriku (1990a) ZA 9 008 298
Hokuriku (1990b) JP 4 253 982
Hutt MP, Kiely JS, Warner-Lambert (1987) US 4 923 879

Hutt MP, Mich TF, Culbertson TP, Warner-Lambert (1984) EP 159 174
ICAAC (1996) Data on BAY 12-8039 were published on the 36th Interscience Conference on Antimicrobial Agents and Chemotherapy, New Orleans, Abstracts no B 45, F1–F26
ICI (1957) BE 564 863
Irikura T, Suzue S, Hirai K, Ishizaki T, Kyorin Pharmaceutical (1984/1985) EP 184 035
Ishikawa H, Uno T, Miyamoto H, Ueda H, Tamaoka H, Tominaga M, Nakagawa K (1990) Studies on antibacterial agents. II. Synthesis and antibacterial activities of substituted 1,2-dihydro-6-oxo-6H-pyrrolo[3,2,1-ij]quinoline-5-carboxylic acids. Chem Pharm Bull (Tokyo) 38:2459–2462
Ito Y, Kato H, Yasuda S, Kado N, Yoshida T, Yamamoto Y, Hokuriku Seiyaku (1993/1994) EP 641 793
Iwanami S, Takashima M, Hirata Y, Hasegawa O, Usuda S (1981) Synthesis and neuroleptic activity of benzamides. cis-N-(1-benzyl-2-methylpyrrolidin-3-yl)-5-chloro-2-methoxy-4-(methylamino)benzamide and related compounds. J Med Chem 24:1224–1230
Iwata M, Kimura T, Fujiwara Y, Katsube T, Ube Industries (1986) EP 241 206
Iwata M, Kimura T, Inoue T, Fujihara Y, Katsube T, Ube Industries/Sankyo (1988) EP 352 123
Jack DB (1986) Recent advances in pharmaceutical chemistry. The 4-quinolone antibiotics. J Clin Hosp Pharm 11:75–93
Jacquet J-P, Bouzard D, Kiechel J-R, Remuzon P (1991) Synthesis of a new bridged diamine, 3,6-diazabicyclo[3.2.0]heptane: applications to the synthesis of quinolone antibacterials. Tetrahedron Lett 32:1565–1568
Jacquet J-P, Bouzard D, Di Cesare P, Dolcnic N, Massoudi M, Remuzon P (1992) Synthesis and biological activity of novel 1-aryl-6-fluoro-5-methyl-1,8-naphthyridine-3-carboxylic acids. Heterocycles 34:2301–2311
Jaetsch T, Mielke B, Petersen U, Philipps T, Schenke T, Bremm KD, Endermann R, Scheer M, Stegemann M, Wetzstein H-G, Bayer (1993) DE 4 329 600
Jaetsch T, Mielke B, Petersen U, Schenke T, Bremm KD, Endermann R, Metzger K-G, Scheer M, Stegemann M, Wetzstein H-G, Bayer (1994a) EP 682 030
Jaetsch T, Petersen U, Bremm KD, Endermann R, Metzger K-G, Scheer M, Stegemann M, Wetzstein H-G, Bayer (1994b) EP 683 169
Jefson MR, Pfizer (1992) US 5 235 054 (1992)
Jefson MR, McGuirk PR, Pfizer (1985) EP 215 650
Jefson MR, McGuirk PR, Pfizer (1986) US 4 775 668
Jendralla H, Fischer G (1995) Synthesis of 1,2,3,4,5,6-hexahydropyrrolo[3,4-c]pyrrole dihydrobromide and 1,2,3,5-tetrahydro-2-[(4-methylphenyl)sulfonyl]pyrrolo[3,4-c]pyrrole. Heterocycles 41:1291–1298
Jordis U, Sauter F, Siddiqi SM, Küenburg B, Bhattacharya K (1990) Synthesis of (1R,4R)- and (1S,4S)-2,5-diazabicyclo[2.2.1]heptanes and their N-substituted derivatives. Synthesis 925–930
Jordis U, Sauter F, Siddiqi SM (1991) Synthesis of (1R,4S,5R)-endo-N,N-Dimethyl-2-azabicyclo[2.2.1]methanamine. J Heterocycl Chem 28:2045–2047
Kara M, Hasinoff BB, McKay DW, Campbell NRC (1991) Clinical and chemical interactions between iron preparations and ciprofloxacin. Br J Clin Pharmacol 31:257–261
Keith DD, Albrecht HA, Beskid G, Chan K-K, Christenson JG, Cleeland R, Deitcher K, Delaney M, Georgopapadakou NH, Konzelmann F, Okabe M, Pruess D, Rossman P, Specian A, Then R, Wei C-C, Weigele M (1993) Mechanism-based dual-action cephalosporins. In: Bentley PH, Ponsford R (eds) Recent advances in the chemistry of anti-infective agents. Royal Society of Chemistry, Cambridge, pp 79–92 (Special publication 119)
Kiely JS (1991) The preparation of ethyl 7-chloro-1-cyclopropyl-6-fluoro-1,4-dihydro-5-methyl-4-oxo-1,8-naphthyridine-3-carboxylate. J Heterocycl Chem 28:541–543
Kiely JS, Huang S, Leshesky LE (1989) A general method for the preparation of 2-substituted-4-oxo-3-quinolinecarboxylic acids. J Heterocycl Chem 26:1675–1681

Kiely JS, Laborde E, Lesheski LE, Bucsh RA (1991a) Synthesis of 7-(alkenyl, cycloalkenyl, and 1,2,3,6-tetrahydro-4-pyridinyl)quinolones. J Heterocycl Chem 28:1581–1585

Kiely JS, Hutt MP, Culbertson TP, Bucsh RA, Worth DF, Lesheski LE, Gogliotti RD, Sesnie JA, Solomon M, Mich TF (1991b) Quinolone antibacterials: preparation and activity of bridged bicyclic analogues of the C_7-piperazine. J Med Chem 34:656–663

Kim DY, Lee JW, Lee KS, Son HJ, Kang TC, Dae Woong Pharmaceutical (1991) WO 93/03026

Kim KS (1990) Synthesis of an aminoisoxazolidine substituted quinolone acid. Heterocycles 31:87–95

Kim KS, Ryan PC (1990) Synthesis of an aminopyrazolidine substituted quinolone acid. Heterocycles 31:79–86

Kim WJ, Lee TS (1991) KR 9 306 163

Kim WJ, Park MH, Oh JH, Jung MH, Kim BJ, Korea Research Institute of Chemical Technology (1989a) EP 424 850

Kim WJ, Park MH, Oh JH, Jung MH, Kim BJ, Korea Research Institute of Chemical Technology (1989b) EP 424 851

Kim WJ, Park MH, Oh JH, Korea Research Institute of Chemical Technology (1989c) EP 424 852

Kim WJ, Park MH, Ha JD, Baik KU, Hoechst (1991a) WO 93/08163

Kim WJ, Park MH, Ha JD, Baik KU, Korea Research Institute of Chemical Technology (1991b) EP 537 556

Kim WJ, Park MH, Ha JD, Baik KU, Lee TS, Park TH, Nam KS, Kim BJ, Korea Research Institute of Chemical Technology (1991c) EP 549 857

Kim WJ, Park MH, Ha JD, Baik KU, Korea Research Institute of Chemical Technology (1991d) EP 550 016

Kim WJ, Park MH, Ha JD, Baik KU, Lee TS, Park TH, Nam KS, Kim BJ, Korea Research Institute of Chemical Technology (1991e) EP 550 019

Kim WJ, Park MH, Ha JD, Baik KU, Korea Research Institute of Chemical Technology (1991f) EP 550 025

Kim WJ, Park TH, Kim MH, Korea Research Institute of Chemical Technology (1992a) WO 94/02479

Kim WJ, Park TH, Kim MH, Korea Research Institute of Chemical Technology (1992b) WO 94/02487

Kim WJ, Park MH, Ha JD, Kim BJ, Nam KS, Kong JY, Korea Research Institute of Chemical Technology (1992c) EP 574 231

Kim WJ, Lee TS, Park MH, Ha JD, Kim BJ, Nam KS, Kong JY, Korea Research Institute of Chemical Technology (1992d) WO 93/25545

Kim WJ, Park TH, Park JG, Kim MH, Ha JD, Pearson N, Korea Research Institute of Chemical Technology – SmithKline Beecham PLC (1992/1993) WO 94/14813

Kim WJ, Park TH, Kim BJ, Kim MH, Pearson N, Korea Research Institute of Chemical Technology – SmithKline Beecham PLC (1993) WO 94/15938

Kim WJ, Kim BJ, Lee TS, Nam KS, Kim KJ (1995) Synthesis and antibacterial activity of 1-cyclopropyl-6,8-difluoro-7-(2-substituted 4,6-dihydro-1H-pyrrolo[3,4-d]thiazol-5-yl)-1,4-dihydro-4-oxoquinoline-3-carboxylic acid. Heterocycles 41: 1389–1397

Kim Y-K, Choi H, Kim S-H, Chang J-H, Nam D-H, Kim Y-Z, Lee Y-H, Kwak J-H, Hong C-Y (1995) Synthesis and antibacterial activities of LB 20304: a new fluoronaphthyridone antibiotic containing novel oxime functionalized pyrrolidine (Abstr F204). 35th Interscience Conference on Antimicrobial Agents and Chemotherapy, San Francisco

Kimura Y, Atarashi S, Takahashi M, Hayakawa I (1994a) Synthesis and structure-activity relationships of 7-[3-(1-aminoalkyl)-pyrrolidinyl]- and 7-[3-(1-aminocycloalkyl)pyrrolidinyl]-quinolone antibacterials. Chem Pharm Bull (Tokyo) 42:1442–1454

Kimura Y, Atarashi S, Kawakami K, Hayakawa I (1994b) (Fluorocyclopropyl)quinolones. II. Synthesis and stereochemical structure-activity relationships of chiral 7-(7-amino-5-azaspiro[2.4]heptan-5-yl)-1-(2-fluorocyclopropyl) quinolone antibacterial agents. J Med Chem 37:3344–3352

Kitani H, Kuroda T, Moriguchi A, Hikida K, Ao H, Yokoyama Y, Hirayama F, Ikeda Y (1995) Novel 7-substituted-fluoroquinolones as potent antibacterial agents: synthesis and structure-activity relationships (Abstr F190). 35th Interscience Conference on Antimicrobial Agents and Chemotherapy, San Francisco

Kleinman EF, Pfizer (1988) EP 356 193

Koga H, Ito A, Murayama S, Suzue S, Irikura T (1980) Structure-activity relationships of antibacterial 6,7- and 7,8-disubstituted 1-alkyl-1,4-dihydro-4-oxo-quinoline-3-carboxylic acids. J Med Chem 23:1358–1363

Kondo H, Sakamoto F, Kodera Y, Tsukamoto G (1986) Studies on prodrugs. V. Synthesis and antimicrobial activity of N-(oxo-alkyl)norfloxacin derivatives. J Med Chem 29:2020–2024

Kondo H, Sakamoto F, Kawakami K, Tsukamoto G (1988) Studies on prodrugs. VII. Synthesis and antimicrobial activity of 3-formyl-quinolone derivatives. J Med Chem 31:221–225

Kondo H, Sakamoto F, Uno T, Kawahata Y, Tsukamoto G (1989) Studies on prodrugs. XI. Synthesis and antimicrobial activity of N-[(4-methyl-5-methylene-2-oxo-1,3-dioxolan-4-yl)oxy]norfloxacin. J Med Chem 32:671–674

Krebs A, Schenke T, Bayer (1987) EP 310 854

Kyorin (1986) JP 62 255 482

Kyorin (1987) JP 1090 183

Laborde E, Kiely JS, Lesheski LE, Schroeder NC (1991) Novel 7-substituted quinolone antibacterial agents. Synthesis of 7-alkenyl-, cycloalkenyl-, and 1,2,3,6-tetrahydro-4-pyridinyl-1,6-naphthyridines. J Heterocycl Chem 28:191–198

Laborde E, Kiely JS, Culbertson TP, Lesheski LE (1993) Quinolone antibacterials: synthesis and biological activity of carbon isosteres of the 1-piperazinyl and 3-amino-1-pyrrolidinyl side chains. J Med Chem 36:1964–1970

Lecomte S, Baron MH, Chenon MT, Coupry C, Moreau NJ (1994) Effect of magnesium complexation by fluoroquinolones on their antibacterial properties. Antimicrob Agents Chemother 38:2810–2816

Lee KH, Kim JW, Cho IW, Lee JM, Yoon YH, Lee KH, Song SB, Hwang HS (1995) Practical synthesis of CFC-222, a new fluoroquinolone (Abstr F198). 35th Interscience Conference on Antimicrobial Agents and Chemotherapy, San Francisco

Lesher GY (1978) Nalidixic acid and other quinolone carboxylic acids. In: Kirk-Othmer (ed) Encyclopedia of chemical technology, 3rd edn, vol 2. Wiley, New York, pp 782–789

Lesher GY, Singh B, Reuman M, Sterling Drug (1987) EP 309 789 (Chem Abstr 112: 7389y)

Lesher GY, Singh B, Reuman M, Daum SJ, Sterling Drug (1989a) US 5 075 319

Lesher GY, Singh B, Reuman M, Daum SJ, Sterling Drug (1989b) EP 417 669

Leysen DC, Zhang MQ, Haemers A, Bollaert W (1991a) Synthesis of antibacterial 4-quinolone-3-carboxylic acids and their derivatives. Part 1. Pharmazie 46:485–501

Leysen DC, Zhang MQ, Haemers A, Bollaert W (1991b) Synthesis of antibacterial 4-quinolone-3-carboxylic acids and their derivatives. Part 2. Pharmazie 46:557–572

Li Q, Claiborne A, Chu DTW, Lee CM, Raye KA, Sneller K, Ma Z, Wang W, Shen LS, Flamm R, Shortridge VD, Tanaka SK, Seif L, Cooper C, Fung A, Tufano M, Melcher LM, Henry R, Klein L, Plattner JJ (1995a) Synthesis and in vitro antibacterial activity of novel DNA gyrase inhibitors (Abstr F8). 35th Interscience Conference on Antimicrobial Agents and Chemotherapy, San Francisco

Li Q, Chu DTW, Raye K, Claiborne A, Seif L, Macri B, Plattner JJ (1995b) A practical stereoselective synthesis of (2S,4S)-4-tertbutoxycarbonylamino-2-methylpyrrolidine. Tetrahedron Lett 46:8391–8394

Li Q, Chu DTW, Claiborne A, Cooper CS, Lee CM, Raye K, Berst KB, Donner P, Wang W, Hasvold L, Fung A, Ma Z, Tufano M, Flamm R, Shen LL, Baranowski J, Nilius A, Alder J, Meulbroek J, Marsh K, Crowell DA, Hui Y, Seif L, Melcher

LM, Henry R, Spanton S, Faghih R, Klein LL, Tanaka SK, Plattner JJ (1996) Synthesis and structure-activity relationships of 2-pyridones: a novel series of potent DNA gyrase inhibitors as antibacterial agents. J Med Chem 39:3070–3088
Loubinoux B, Colin JL, Thomas V (1991) Synthesis and biological activities of oxolinic acid pro-drugs. Eur J Med Chem 26:461–467
Maddaluno J, Corruble A, Leroux V, Ple G, Duhamel P (1992) Handy access to chiral N,N′-disubstituted 3-aminopyrrolidines. Tetrahedron Asymmetry 3:1239–1242
Maejima T, Senda H, Iwatani W, Tatsumi Y, Arika T, Fukui H, Shibata T, Nakano J, Naito T (1995) Potent antibacterial activity of S-32730, a new fluoroquinolone, against Gram-positive bacteria including quinolone-resistant MRSA (Abstr F189). 35th Interscience Conference on Antimicrobial Agents and Chemotherapy, San Francisco
Markees DG, Schwab LS (1972) The synthesis and some reactions of N-Alkyl-4-quinolone-3-carboxylic acids. Helv Chim Acta 55(4):1319–1326
Martel AM, Leeson PA, Castañer J (1997) BAY 12-8039. Drugs Fut 22:109–113
Masuzawa K, Suzue S, Hirai K, Ishizaki T, Kyorin Pharmaceutical (1985) EP 208 210
Masuzawa K, Suzue S, Hirai K, Ishizaki T, Kyorin Pharmaceutical (1986a) EP 230 295
Masuzawa K, Suzue S, Hirai K, Ishizaki T, Kyorin Pharmaceutical (1986b) EP 235 762
Matsumoto J, Minami S (1975) Pyrido[2,3-d]pyrimidine antibacterial agents. III. 8-Alkyl- and 8-vinyl-5,8-dihydro-5-oxo-2-(1-piperazinyl)pyrido[2,3-d]pyrimidine-6-carboxylic acids and their derivatives. J Med Chem 18:74–79
Matsumoto J, Nakamura S, Miyamoto T, Uno H, Dainippon Pharmaceutical, (1983) EP 132 845
Matsumoto J, Miyamoto T, Minamida A, Nishimura Y, Egawa H, Nishimura H (1984a) Pyridonecarboxylic acids as antibacterial agents. II. Synthesis and structure-activity-relationships of 1,6,7-trisubstituted 1,4-dihydro-4-oxo-1,8-naphthyridine-3-carboxylic acids, including enoxacin, a new antibacterial agent. J Med Chem 27:292–301
Matsumoto J, Miyamoto T, Minamida A, Nishimura Y, Egawa H, Nishimura H (1984b) Synthesis of fluorinated pyridines by the Balz-Schiemann reaction. An alternative route to enoxacin, a new antibacterial pyridonecarboxylic acid. J Heterocycl Chem 21:673–679
Matsumoto J, Nakano J, Chiba K, Nakamura S, Dainippon Pharmaceutical (1985) EP 191 451
Matsumoto J, Miyamoto T, Egawa H, Nakamura T, Dainippon Phamaceutical (1985/1986) US 4 795 751
Matsumoto J, Minamida A, Hirose T, Nakano J, Nakamura S, Dainippon Pharmaceutical (1987/1988) EP 319 906
McGuirk PR, Pfizer (1988) EP 348 088
McGuirk PR, Jefson MR, Mann DD, Elliott NC, Chang P, Cisek EP, Cornell CP, Gootz TD, Haskell SL, Hindahl MS, LaFleur LJ, Rosenfeld MJ, Shryock TR, Silvia AM, Weber FH (1992) Synthesis and structure-activity relationships of 7-diazabicycloalkylquinolones, including danofloxacin, a new quinolone antibacterial agent for veterinary medicine. J Med Chem 35:611–620
Mitscher LA, Zavod RM, Sharma PN, Chu DTW, Shen LL, Pernet AG (1988) Recent advances on quinolone antimicrobial agents. In: Davis BD, Ichikawa T, Maeda K, Mitscher LA (eds) Horizons on antibiotic reserach. Japan Antibiotics Research Association, Tokyo, pp 166–193
Mitscher LA, Devasthale PV, Zavod RM (1990) Structure-activity relationships of fluoro-4-quinolones. In: Crumplin GC (ed) The 4-quinolones: antibacterial agents in vitro. Springer, Berlin Heidelberg New York, pp 115–146
Mitscher LA, Devasthale P, Zavod R (1993) Structure-activity relationships. In: Hooper DC, Wolfson JS (eds) Quinolone antimicrobial agents, 2nd edn. American Society for Microbiology, Washington, pp 3–51
Miyamoto H, Yamashita H, Tominaga M, Yabuuchi Y, Otsuka Pharmaceutical (1988/1989) EP 364 943

Miyamoto H, Ueda H, Otsuka T, Aki S, Tamaoka H, Tominaga M, Nakagawa K (1990) Studies on antibacterial agents. III. Synthesis and antibacterial activities of substituted 1,4-dihydro-8-methyl-4-oxoquinoline-3-carboxylic acids. Chem Pharm Bull (Tokyo) 38:2472–2475

Miyamoto T, Matsumoto J (1990) Fluorinated pyrido[2,3-c]pyridazines. II. Synthesis and antibacterial activity of 1,7-disubstituted 6-fluoro-4(1H)-oxopyrido[2,3-c]pyridazine-3-carboxylic acids. Chem Pharm Bull (Tokyo) 38:3359–3365

Miyamoto T, Matsumoto J, Chiba K, Egawa H, Shibamori K, Minamida A, Nishimura Y, Okada H, Kataoka M, Fujita M, Hirose T, Nakano J (1990) Synthesis and structure-activity relationship of 5-substituted 6,8-difluoroquinolones, including sparfloxacin, a new quinolone antibacterial agent with improved potency. J Med Chem 33:1645–1656

Molinari G, Schito GC (1995) Comparative in vitro activity of BAY Y 3118 with other fluoroquinolones. Drugs 49 [Suppl 2]:222–225

Moran DB, Ziegler CB Jr, Dunne TS, Kuck NA, Lin Y (1989) Synthesis of novel 5-fluoro analogues of norfloxacin and ciprofloxacin. J Med Chem 32:1313–1318

Mozek I, Sket B (1994) Direct monobromination of substituted 4-oxoquinoline-3-carboxylic acid derivatives. J Heterocycl Chem 31:1293–1295

Nagano H, Yokota T, Katoh Y (1988) EP 342 675

Nakagawa S, Mano E, Ushijima R (1989) JP 3 188 080

Nakano J, Fukui H, Shibata T, Senda H, Maejima T, Arika T, Kaken Pharmaceutical (1993) WO 95/11902

Neu HC, Novelli A, Chiu N-X (1989) Comparative in vitro activity of a new quinolone, AM-1091. Antimicrob Agents Chemother 33:1036–1041

Niedballa U, Vorbrüggen H (1974) A general synthesis of N-glycosides. IV. Synthesis of nucleosides of hydroxy and mercapto N-heterocycles. J Org Chem 39:3668–3671

Nishigaki S, Mizushima N, Kanazawa H, Ichiba M, Senga K (1985) Synthetic antibacterials. VIII. 7-(1′-Alkylhydrazino)-1,8-naphthyridines and related compounds. J Heterocycl Chem 22:1029–1032

Nishida H, Fujii T, Abiru Y, Yatsuki K, Yamamoto M, Shimizu N, Kakemi K, Mikawa M, Kise M (1994) Studies on synthesis of antibacterial agent (NM 441). I. Kinetics and mechanism of the reaction of 4-(bromomethyl)-5-metyl-1,3-dioxol-2-one with 1-substitued piperazine (NM 394). Bull Chem Soc Jpn 67:1419–1426

Nishimura Y, Minamida A, Matsumoto J (1988a) An intramolecular cyclisation of 7-substituted 6-fluoro-1,8-naphtyridine and –quinoline derivatives. J Heterocycl Chem 25:479–485

Nishimura Y, Minamida A, Matsumoto J (1988b) Pyridonecarboxylic acids as antibacterial agents. XII. Synthesis and antibacterial activity of enoxacin analogues with a variant at position 1. Chem Pharm Bull (Tokyo) 36:1223–1228

Nishimura Y, Hirose T, Okada H, Shibamori K, Nakano J, Matsumoto J (1990) Synthesis of 7-thio-substituted 4-oxoquinoline-3-carboxylic acids with antibacterial activity. Chem Pharm Bull (Tokyo) 38:2190–2196

Nishino T, Otsuki M, Ozaki M, Matsuda M, Kimura K (1989) Antimicrobial activities of NAD-441A, a new type of quinolone antibacterial (Abstr 1253) 29th International Conference on Antimicrobial Agents Chemotherapy, Houston

Nishitani Y, Nishino Y, Irie T (1988a) EP 362 759

Nishitani Y, Irie T, Nishino Y (1988b) JP 1 265 092

Nord CE, Lindmark A, Persson I (1993) In vitro activity of the new quinolone BAY Y 3118 against anaerobic bacteria. Eur J Clin Microbiol Infect Dis 12:640–642

Ochi K, Shimizu H, Chugai Seiyaku Kabushiki Kaisha (1992) EP 641 782

Ogata M, Matsumoto H, Shimizu S, Kida S (1988a) EP 343 524

Ogata M, Matsumoto H, Shimizu S, Kida S (1988b) EP 359 172

Ogata M, Matsumoto H, Shimizu S, Kida S, Nakai H, Motokawa K, Miwa H, Matsuura S, Yoshida T (1991) Synthesis and antibacterial activity of new 7-(aminoazabicycloalkanyl)-quinolonecarboxylic acids. Eur J Med Chem 26:889–906

Oh J-I, Ahn M-J, Paek K-S, Kim M-Y, Seo M-K, Lee Y-H, Nam D-H, Kim Y-Z, Kim I-C, Kwak J-H (1995). In vitro and in vivo antibacterial activities of LB 20304, a new fluoronaphthyridone (Abstr F206). 35th Interscience Conference on Antimicrobial Agents and Chemotherapy, San Francisco

Okabe M, F. Hoffmann–La Roche (1990) EP 453 952

Okabe M, Sun R-C (1992) Synthesis of a dual action cephalosporin: a novel approach to 3-acyloxymethyl-3-cephems. Synthesis 1160–1164

Okada T, Tsushima T, Shionogi Seiyaku Kabushiki Kaisha (1990) EP 487 030

Okada T, Ezumi K, Yamakawa M, Sato H, Tsuji T, Tsushima T, Motokawa K, Komatsu Y (1993a) Quantitative structure-activity relationships of antibacterial agents, 7-heterocyclic amine substituted 1-cyclopropyl-6,8-difluoro-4-oxoquinoline-3-carboxylic acids. Chem Pharm Bull (Tokyo) 41:126–131

Okada T, Sato H, Tsuji T, Tsushima T, Nakai H, Yoshida T, Matsuura S (1993b) Synthesis and structure-activity relationships of 7-(3′-amino-4′-methoxypyrrolidin-1′-yl)-1-cyclopropyl-6,8-difluoro-1,4-dihydro-4-oxoquinoline-3-carboxylic acids. Chem Pharm Bull (Tokyo) 41:132–138

Okura A, Yoshinari T, Arakawa H, Nakagawa S, Mano E, Ushijima R (1990) WO 92/12146

Pankuch GA, Jacobs MR, Appelbaum PC (1993) Susceptibility of 428 Gram-positive and negative anaerobic bacteria to BAY Y 3118 compared with their susceptibilities to ciprofloxacin, clindamycin, metronidazole, piperacillin, piperazillin-tazobactam, and cefoxitin. Antimicrob Agents Chemother 37:1649–1654

Pares Corominas J, Colombo Pinol A, Frigola Constansa J, Laboratorios del Dr. Esteve (1989) EP 388 298

Perrone E, Jabes D, Alpegiani M, Andreini BP, Della Bruna C, Del Nero S, Rossi R, Visentin G, Zarini F (1992) Dual-action penems. J Antibiot (Tokyo) 45:589–594

Petersen U, Grohe K, Zeiler H-J, Metzger KG, Bayer (1982) EP 113 092

Petersen U, Grohe K, Zeiler H-J, Metzger KG, Bayer (1983) EP 117 474

Petersen U, Grohe K, Zeiler H-J, Metzger KG, Bayer (1984) EP 167 763

Petersen U, Schriewer M, Grohe K, Zeiler H-J, Metzger KG, Bayer (1985) EP 203 488

Petersen U, Grohe K, Schenke T, Hagemann H, Zeiler HJ, Metzger KG, Bayer (1986) DE 3 601 567

Petersen U, Grohe K, Schriewer M, Schenke T, Haller I, Metzger K-G, Endermann R, Zeiler H-J, Bayer (1987) EP 284 935

Petersen U, Schenke T, Grohe K, Schriewer M, Haller I, Metzger KG, Endermann R, Zeiler HJ, Bayer (1988a) EP 326 916

Petersen U, Schenke T, Krebs A, Grohe K, Schriewer M, Haller I, Metzger KG, Endermann R, Zeiler HJ, Bayer (1988b) EP 350 733

Petersen U, Schenke T, Schriewer M, Krebs A, Grohe K, Haller I, Metzger KG, Endermann R, Zeiler HJ, Bayer (1989a) EP 391 169

Petersen U, Krebs A, Schenke T, Grohe K, Schriewer M, Haller I, Metzger KG, Endermann R, Zeiler H-J, Bayer (1989b) EP 401 623

Petersen U, Schenke T, Schriewer M, Grohe K, Krebs A, Haller I, Metzger KG, Bremm KD, Endermann R, Zeiler HJ, Bayer (1990) DE 4 032 560

Petersen U, Himmler T, Schenke T, Krebs A, Grohe K, Bremm K-D, Metzger KG, Endermann R, Zeiler H-J, Bayer (1991a) EP 523 512

Petersen U, Krebs A, Schenke T, Kunisch F, Philipps T, Grohe K, Bremm K-D, Endermann R, Metzger K-G, Haller I, Zeiler H-J, Bayer (1991b) DE 4 120 646

Petersen U, Krebs A, Schenke T, Grohe K, Bremm KD, Endermann R, Metzger KG, Zeiler HJ (1992a) EP 520 277

Petersen U, Krebs A, Schenke T, Philipps T, Grohe K, Bremm KD, Endermann R, Metzger KG, Haller I, Bayer (1992b) EP 550 903

Petersen U, Krebs A, Schenke T, Grohe K, Endermann R, Bremm KD, Metzger KG, Bayer (1992c) EP 589 318

Petersen U, Bremm KD, Krebs A, Metzger KG, Philpps T, Schenke T (1992d) Bay Y 3118, a novel 4-quinolone: synthesis and in vitro activity (Abstr 642). 32nd Interscience Conference on Antimicrobial Agents and Chemotherapy, Anaheim

Philipps T, Bartel S, Krebs A, Petersen U, Schenke T, Bremm KD, Endermann R, Metzger KG, Bayer (1992) DE 4 230 804

Portoghese PS, Mikhail AA (1966) Bicyclic bases. Synthesis of 2,5-diazabicyclo[2.2.1]heptanes. J Org Chem 31:1059–1062

Rádl S (1990) Structure-activity relationships in DNA gyrase inhibitors. Pharmacol Ther 48:1–17

Rádl S (1994a) Synthesis of ethyl 1-ethyl-6-fluoro-1,4-dihydro-8-hydroxy-4-oxoquinoline-3-carboxylate. Collect Czech Chem Commun 59:2119–2122

Rádl S (1994b) Synthesis of 3-amino- and 3-acyclamino-1-cyclopropylquinolin-4(1H)-ones. Collect Czech Chem Commun 59:2123–2126

Rádl S, Bouzard D (1992) Recent advances in the synthesis of antibacterial quinolones. Heterocycles 34:2143–2177

Rádl S, Janichová M (1992) Synthesis and antibacterial activity of some 3-hydroxyquinolones. Collect Czech Chem Commun 57:188–193

Rádl S, Chan K (1994) Synthesis of 1-substituted 3-nitroquinolin-4(1H)-ones. J Heterocycl Chem 31:437–440

Remuzon P, Bouzard D, Dussy C, Jacquet J-P, Massoudi M (1992a) Preparation of (6R)- and (6S)-(1R,4R)-6-Methyl-2-(p-toluenesulfonyl)-5-phenylmethyl-2,5-diazabicyclo[2.2.1]heptanes, intermediates in a synthesis of new quinolones. Heterocycles 34:241–245

Remuzon P, Massoudi M, Bouzard D, Jacquet J-P (1992b) Preparation of (1R,4R)-1-methyl-2-(p-toluenesulfonyl)-5-phenylmethyl-2,5-diazabicyclo[2.2.1]heptane, intermediate in a synthesis of new naphthyridones. Heterocycles 34:679–684

Remuzon P, Bouzard D, Di Cesare P, Dussy C, Jacquet J-P, Jaegly A (1992c) Synthesis and antibacterial activity of new 5-substituted 1-cyclopropyl-6-fluoro-7-piperazinyl-1,4-dihydro-4-oxo-1,8-naphthyridine-3-carboxylic acids. J Heterocycl Chem 29:985–989

Remuzon P, Bouzard D, Guiol C, Jacquet J-P (1992d) Fluoronaphthyridines as antibacterial agents. VI. Synthesis and structure-activity relationships of new chiral 7-(1-, 3-, 4-, and 6-methyl-2,5-diazabicyclo[2.2.1]heptan-2-yl)naphthyridine analogues of 7-[(1R,4R)-2,5-diazabicyclo[2.2.1]heptan-2-yl]-1-(1,1-dimethylethyl)-6-fluoro-1,4-dihydro-4-oxo-1,8-naphthyridine-3-carboxylic acid. Influence of the configuration on the blood pressure in dogs. A quinolone-class effect. J Med Chem 35:2898–2909

Remuzon P, Bouzard D, Jacquet J-P (1993a) Preparation of new 2-chloro-5-fluoro-6-(4-phenylmethylpiperazinyl)-4-trifluoromethyl-3-nicotinic acid. Heterocycles 36: 431–434

Remuzon P, Bouzard D, Clemencin C, Dussy C, Essiz M, Jacquet J-P, Saint-Germain J (1993b) Synthesis of (1R,4R,7S)- and (1S,4S,7S)-2-(4-tolylsulfonyl)-5-phenylmethyl-7-methyl-2,5-diazabicyclo[2.2.1]heptanes via regioselective opening of 3,4-epoxy-D-proline with lithium dimethyl cuprate. J Heterocycl Chem 30:517–523

Renau TE, Sanchez JP, Shapiro MA, Dever JA, Gracheck SJ, Domagala JM (1995) Effect of lipophilicity at N-1 on activity of fluoroquinolones against mycobacteria. J Med Chem 38:2974–2977

Renau TE, Sanchez JP, Domagala JM (1996) The synthesis of 3-bromo-2,4,5-trifluorobenzoic acid and its conversion to 8-bromoquinolonecarboxylic acids. J Heterocyclic Chem 33:1407–1411

Reuman M, Daum SJ, Singh B, Coughlin SA, Sedlock DM, Rake JB, Lesher GY (1989) Synthesis and Antibacterial Activity of some Novel 1-Substituted-7-pyridinyl-1,4-dihydro-4-oxoquinoline-3-carboxylic Acids (Abstr 1193) 29th Interscience Conference on Antimicrobial Agents and Chemotherapy, Houston

Reuman M, Eissenstat MA, Weaver JD III (1994) Cyanide mediated decarboxylation of 1-substituted-4-oxoquinoline and 4-oxo-1,8-naphthyridine-3-carboxylic acids. Tetrahedron Lett 35:8303–8306

Reuman M, Daum SJ, Singh B, Wentland MP, Perni RB, Pennock P, Carabateas PM, Gruett MD, Saindane MT, Dorff PH, Coughlin SA, Sedlock DM, Rake JB, Lesher GY (1995) Synthesis and antibacterial activity of some novel 1-substituted 1,4-dihydro-4-oxo-7-pyridinyl-3-quinolinecarboxylic acids. Potent antistaphylococcal agents. J Med Chem 38:2531–2540

Rosen TJ (1990) The fluoroquinolone antibacterial agents. Prog Med Chem 27:235–295

Rosen TJ, Chu DT, Abbott Laboratories (1987) EP 302 371 (Chem Abstr 111: 39348e)

Rosen TJ, Fesik SW, Chu DTW, Pernet AG (1988a) Asymmetric synthesis of 2-substituted (4S)-4-aminopyrrolidines. S_N2 displacement at the 4-position of the pyrrolidine moiety. Synthesis 40–44

Rosen TJ, Chu DTW, Lico IM, Fernandes PB, Shen L, Borodkin S, Pernet AG (1988b) Asymmetric synthesis and properties of the enantiomers of the antibacterial agent 7-(3-aminopyrrolidinyl)-1-(2,4-difluorophenyl)-1,4-dihydro-6-fluoro-4-oxo-1,8-naphthyridine-3-carboxylic acid hydrochloride. J Med Chem 31:1586–1590

Rosen TJ, Chu DTW, Lico IM, Fernandes PB, Marsh K, Shen L, Cepa VG, Pernet AG (1988c) Design, synthesis, and properties of (4S)-7-(4-amino-2-substituted-pyrrolidin-1-yl)quinolone-3-carboxylic acids. J Med Chem 31:1598–1611

Ross DL, Riley CM (1993) Physicochemical properties of the fluoroquinolone antimicrobials. V. Effect of fluoroquinolone structure and pH on the complexation of various fluoroquinolones with magnesium and calcium ions. Int J Pharm 93:121–129

Ross DL, Riley CM (1994) Dissociation and complexation of the fluoroquinolone antimicrobials – an update. J Pharm Biomed Anal 12:1325–1331

Ross DL, Elkinton SK, Riley CM (1992) Physicochemical properties of the fluoroquinolone antimicrobials. IV. 1-Octanol/water partition coefficients and their relationships to structure. Int J Pharm 88:379–389 (erratum 90:179)

Ross DL, Elkinton SK, Knaub SR, Riley CM (1993) Physicochemical properties of the fluoroquinolone antimicrobials. VI. Effect of metal-ion complexation on octan-1-ol-water partitioning. Int J Pharm 93:131–138

Saito A, Uesato S, Iwata H, Ao H, Kuroda T, Kawasaki K, Moriguchi A, Ikeda Y, Yoshitomi Pharmaceutical Industries (1992/1993) EP 677 522

Sakamoto F, Ikeda S, Kondo H, Tsukamoto G (1985) Studies on prodrugs. IV. Preparation and characterization of N-(5-substituted 2-oxo-1,3-dioxol-4-yl)methyl norfloxacin. Chem Pharm Bull (Tokyo) 33:4870–4877

Sanchez JP, Warner-Lambert (1989) US 4 916 141 (Chem Abstr 113:40658d)

Sanchez JP, Gogliotti RD (1993) The synthesis of a series of 7-amino-1-cyclopropyl-8-fluoro-1,4-dihydro-4-oxo-1,6-naphthyridine-3-carboxylic acids as potential antibacterial agents. J Heterocycl Chem 30:855–859

Sanchez JP, Domagala JM, Hagen SE, Heifetz CL, Hutt MP, Nichols JB, Trehan AK (1988) Quinolone antibacterial agents. Synthesis and structure-activity relationships of 8-substituted quinolone-3-carboxylic acids and 1,8-naphthyridine-3-carboxylic acids. J Med Chem 31:983–991

Sanchez JP, Bridges AJ, Bucsh R, Domagala JM, Gogliotti RD, Hagen SE, Heifetz CL, Joannides ET, Sesnie JC, Shapiro MA, Szotek DL (1992a) New 8-(trifluoromethyl)-substituted quinolones. The benefits of the 8-fluoro group with reduced phototoxic risk. J Med Chem 35:361–367

Sanchez JP, Domagala JM, Heifetz CL, Priebe SR, Sesnie JA, Trehan AK (1992b) Quinolone antibacterial agents. Synthesis and structure-activity relationships of a series of amino acid prodrugs of racemic and chiral 7-(3-amino-1-pyrrolidinyl)-quinolones. Highly soluble quinolone prodrugs with in vivo pseudomonas activity. J Med Chem 35:1764–1773

Sanchez JP, Gogliotti RD, Domagala JM, Gracheck SJ, Huband MD, Sesnie JA, Cohen MA, Shapiro MA (1995) The synthesis, structure-activity, and structure-side effect relationships of a series of 8-alkoxy- and 5-amino-8-alkoxyquinolone antibacterial agents. J Med Chem 38:4478–4487

Schenke T, Petersen U, Bayer (1989) EP 393 424 (Chem Abstr 114:122348n)

Schenke T, Krebs A, Petersen U, Bayer (1992) DE 4 200 415 (Chem Abstr 119:271179p)

Schentag JJ, Domagala JM (1985) Structure-activity relationships with the quinolone antibiotics. Res Clin Forums 7:9–13

Schriewer M, Grohe K, Petersen U, Haller I, Metzger KG, Endermann R, Zeiler HJ, Bayer (1987) DE 3 702 393 (Chem Abstr 109:230824v)

Schroeder MC, Kiely JS, Laborde E, Johnson DR, Szotek DL, Domagala JM, Stickney TM, Michel A, Kampf JW (1992) Synthesis of the four stereoisomers of several 3-(1-aminoethyl)pyrrolidines. Important intermediates in the preparation of quinolone antibacterials. J Heterocycl Chem 29:1481–1498

Seebach D, Hungerbühler E, Naef R, Schnurrenberger P, Weidmann B, Züger M (1982) Titanate-mediated transesterifications with functionalized substrates. Synthesis 138–141

Segawa J, Kitano M, Kazuno K, Matsuoka M, Shirahase I, Ozaki M, Matsuda M, Tomii Y, Kise M (1992) Studies on pyridonecarboxylic acids. 1. Synthesis and antibacterial evaluation of 7-substituted-6-halo-4-oxo-4H-[1,3]thiazeto[3,2-a]quinoline-3-carboxylic acids. J Med Chem 35:4727–4738

Segawa J, Kazuno K, Matsuoka M, Shirahase I, Ozaki M, Matsuda M, Tomii Y, Kitano M, Kise M (1995) Studies on pyridonecarboxylic acids. III. Synthesis and antibacterial activity evaluation of 1,8-disubstituted 6-fluoro-4-oxo-7-piperazinyl-4H-[1,3]thiazeto[3,2-a]quinoline-3-carboxylic acid derivatives. Chem Pharm Bull (Tokyo) 43:63–70

Shibamori K, Egawa H, Miyamoto T, Nishimura Y, Itokawa A, Matsumoto J (1990) Regioselective displacement reactions of 1-cyclopropyl-5,6,7,8-tetrafluoro-4(1H)-oxoquinoline-3-carboxylic acid with amine nucleophiles. Chem Pharm Bull (Tokyo) 38:2390–2396

Shimizu H, Miura Y, Fujimura Y, Chugai Seiyaku Kabushiki Kaisha (1992) WO 93/22 308

Tagaki N, Fubasami H, Matsukubo H, Kyorin Pharmaceutical (1990) EP 464 823

Takács K, Józan M, Hermecz I, Szász G (1992) Lipophilicity of antibacterial fluoroquinolones. Int J Pharm 79:89–96

Tiefenbacher E-V, Haen E, Przybilla B, Kurz H (1994) Photodegradation of some quinolones used as antimicrobial therapeutics. J Pharm Sci 83:463–467

Todo Y, Takagi H, Iino F, Fukuoka Y, Ikeda Y, Tanaka K, Saikawa I, Narita H (1994a) Pyridonecarboxylic acids as antibacterial agents. VI. Synthesis and structure-activity relationship of 7-(alkyl, cycloalkyl, and vinyl)-1-cyclopropyl-6-fluoro-4-quinolone-3-carboxylic acids. Chem Pharm Bull (Tokyo) 42:2049–2054

Todo Y, Nitta J, Miyajima M, Fukuoka Y, Ikeda Y, Yamashiro Y, Saikawa I, Narita H (1994b) Pyridonecarboxylic acids as antibacterial agents. VII. Synthesis and structure-activity relationship of amino- and hydroxyl-substituted 7-cycloalkyl and 7-vinyl derivatives of 1-cyclopropyl-6-fluoro-4-quinolone-3-carboxylic acid. Chem Pharm Bull (Tokyo) 42:2055–2062

Todo Y, Nitta J, Miyajima M, Fukuoka Y, Yamashiro Y, Nishida N, Saikawa I, Narita H (1994c) Pyridonecarboxylic acids as antibactrial agents. VIII. Synthesis and structure-activity relationship of 7-(1-aminocyclopropyl)-4-oxo-1,8-naphthyridine-3-carboxylic acids and 7-(1-aminocyclopropyl)-4-oxoquinoline-3-carboxylic acids. Chem Pharm Bull (Tokyo) 42:2063–2070

Todo Y, Takagi H, Iino F, Fukuoka Y, Takahata M, Okamoto S, Saikawa I, Narita H (1994d) Pyridonecarboxylic acids as antibacterial agents. IX. Synthesis and structure-activity relationship of 3-substituted 10-(1-aminocyclopropyl)-9-fluoro-7-oxo-2,3-dihydro-7H-pyrido[1,2,3-d,e]-1,4-benzoxacine-6-carboxylic acids and their 1-thio- and 1-aza analogues. Chem Pharm Bull (Tokyo) 42:2569–2574

Todo Y, Takagi H, Iino F, Hayashi K, Takata M, Kuroda H, Momonoi K, Narita H (1994e) Practical synthesis of T-3761, (S)-10-(1-aminocyclopropyl)-9-fluoro-3-methyl-7-oxo-2,3-dihydro-7H-pyrido[1,2,3-d,e]-1,4-benzoxacine-6-carboxylic acid. Chem Pharm Bull (Tokyo) 42:2629–2632

Tolstikov GA, Mustafin AG, Yghibaeva GK, Gataullin RR, Spirikhin LV, Sultanova VS, Abdrakhmanov IB (1993) N- and C-glycosylation of 6,7-difluoro-1,4-dihydro-4-oxo-3-quinoline carboxylic acid ethyl ester. Mendeleev Commun:194

Toyama Chemical (1994) JP 7 242 660

Tsuji T, Sato H, Okada T, Shionogi Seiyaku Kabushiki (1988) EP 347 851

Tsuji K, Tsubouchi H, Ishikawa H (1995) Synthesis and antibacterial activities of optically active substituted 1,2-dihydro-6-oxo-6H-pyrrolo[3,2,1-i,j]quinoline-5-carboxylic acids. Chem Pharm Bull (Tokyo) 43:1678–1682

Tsushima T, Okada T, Nishitani Y, Shionogi Seiyaku Kabushiki Kaisha (1990) EP 485 952

Turel I, Leban I, Bukovec N (1994) Synthesis, characterisation, and crystal structure of a copper (II) complex with quinolone family member (ciprofloxacin): bis(1-cyclopropyl-6-fluoro-1,4-dihydro-4-oxo-7-piperazin-1-yl-quinoline-3-carboxylate)copper(II) chloride hexahydrate. J Inorg Biochem 56:273–282

Turner WR, Suto MJ (1993) 3-Ethenyl-, 3-ethynyl, 3-aryl, and 3-cyclopropyl-2,4,5-trifluorobenzoic acids: useful intermediates in the synthesis of quinolone antibacterials. Tetrahedron Lett 34:281–284

Ube Industries (1992) JP 61 574 464

Ueda H, Miyamoto H (1985) Otsuka Pharmaceutical, EP 228 035

Ueda H, Miyamoto H, Yamashita H, Tone H, Otsuka Pharmaceutical (1987) EP 287 951

Uno T, Takamatsu M, Inoue Y, Kawahata Y, Iuchi K, Tsukamoto G (1987) Synthesis of antimicrobial agents. I. Syntheses and antibacterial activities of 7-(azole substituted)quinolones. J Med Chem 30:2163–2169

Uno T, Okuno T, Taguchi M, Iuchi K, Kawahata Y, Sotomura M, Tsukamoto G (1989) Synthesis of antimicrobial agents. IV. Synthesis of 1-hydroxypiperazine dihydrochloride and its applications to pyridone carboxylic acid antibacterial agents. J Heterocycl Chem 26:393–396

Uno T, Kondo H, Inoue Y, Kawahata Y, Sotomura M, Iuchi K, Tsukamoto G (1990) Synthesis of antimicrobial agents. III. Syntheses and antibacterial activities of 7-(4-hydroxypiperazin-1-yl)quinolones. J Med Chem 33:2929–2932

Uno T, Okuno T, Kawakami K, Sakamoto F, Tsukamoto G (1993) Synthesis of antimicrobial agents. V. In vivo metabolism of 7-(4-hydroxypiperazin-1-yl)quinolones. J Med Chem 36:2712–2715

Van Le T, Spence FG, Wemple JN (1992) US 5 177 217

Verbist L (1986) Quinolones in perspective. Quinolones: pharmacology. Pharm Weekbl 8:22–25

Warner-Lambert (1986) AU 8 666 870

Wemple JN, Warner-Lambert (1988/1989) EP 342 649

Wentland MP (1990) Structure-activity relationships of fluoroquinolones. In: Siporin C, Heifetz CL, Domagala JM (eds) The new generation of quinolones. Dekker, New York, pp 1–43

Wentland MP, Sterling Winthrop (1992) US 5 334 595

Wentland MP, Cornett JB (1985) Quinolone antibacterial agents. Annu Rep Med Chem 20:145–154

Wentland MP, Bailey DM, Cornett JB, Dobson RA, Powles RG, Wagner RB (1984) Novel amino-substituted 3-quinolinecarboxylic acid antibacterial agents: synthesis and structure-activity relationships. J Med Chem 27:1103–1108

Wentland MP, Perni RB, Dorff PH, Brundage RP, Castaldi MJ, Bailey TR, Carabateas PM, Bacon ER, Young DC, Woods MG, Rosi D, Drozd ML, Kullnig RK, Dutko FJ (1993a) 3-Quinolinecarboxamides. A series of novel orally-active antiherpetic agents. J Med Chem 36:1580–1596

Wentland MP, Lesher GY, Reuman M, Gruett MD, Singh B, Aldous SC, Dorff PH, Rake JB, Coughlin SA (1993b) Mammalian topoisomerase II inhibitory activity of 1-cyclopropyl-6,8-difluoro-1,4-dihydro-7-(2,6-dimethyl-4-pyridinyl)-4-oxo-3-quinolinecarboxylic acid and related derivatives. J Med Chem 36:2801–2809

Wentland MP, Carlson JA, Dorff PH, Aldous SC, Perni RB, Young DC, Woods MG, Kingsley SD, Ryan KA, Rosi D, Drozd ML, Dutko FJ (1995) Cyclic variations of 3-quinolinecarboxamides and effects on antiherpetic activity. J Med Chem 38:2541–2545

Wexler HM, Molitoris E, Finegold SM (1994) In vitro activity of BAY Y 3118 against anaerobic bacteria. Antimicrob Agents Chemother 37:2509–2513

Wise R, Andrews JM, Brenwald N (1993) The in vitro activity of BAY Y 3118, a new chlorofluoroquinolone. J Antimicrob Chemother 31:73–80

Yang BV (1991) WO 92/22550

Yatsunami T, Yamamoto H, Kuramoto Y, Hayashi N, Yazaki A, Inoue S, Noda S, Amano H, Wakunaga Seiyaku Kabushiki Kaisha (1989) EP 390 215

Yokomoto M, Yazaki A, Hayashi N, Hatono S, Inoue S, Kuramoto Y, Wakunaga Seiyaku Kabushiki Kaisha/Fujisawa Pharmaceutical (1990/1991) EP 470 578

Yokota T, Haramura M, Okamachi A, Makino T, Chugai Seiyaku Kabushiki Kaisha (1992) EP 664 288

Yokoyama Y, Morimoto M, Iwao E, Yamamoto K, Honjo K, Hirayama F, Ikeda Y (1995) Y-688, a new fluoroquinolone with high activity against quinolone resistant Gram-positive bacteria (Abstr F191) 35th Interscience Conference on Antimicrobial Agents and Chemotherapy, San Francisco

Yoon GJ, Kim DY, Lee JW, Park NJ, Lee KS, Kang TC (1990) WO 92/04342

Yoshida T, Yamamoto Y, Yagi N, Yasuda S, Katoh H, Itoh Y (1990) Studies on quinolone antibiotics. I. Synthesis and antibacterial activity of 7-(2-aminoethoxy)-, 7-(2-aminoethylthio)-, and 7-(2-aminoethylamino)-1-cyclopropyl-6-fluoro-1,4-dihydro-4-oxoquinoline-3-carboxylic acids and their derivatives. Yakugaku Zasshi 111:258–267

Yoshida T, Yamamoto Y, Yagi N, Takahashi Y, Yasuda S, Katoh H, Itoh Y (1991a) Studies on quinolone antibiotics. II. Synthesis and antibacterial activity of 7-aminoalkoxy-1-cyclopropyl-6-fluoro-1,4-dihydro-4-oxoquinoline-3-carboxylic acids and their derivatives. Yakugaku Zasshi 111:19–31

Yoshida T, Yamamoto Y, Yagi N, Takahashi Y, Yasuda S, Katoh H, Itoh Y (1991b) Studies on quinolone antibiotics. III. Synthesis and antibacterial activity of 5-amino-7-(2-aminoalkoxy)-1-cyclopropyl-6-fluoro-1,4-dihydro-4-oxoquinoline-3-carboxylic acids and their derivatives. Yakugaku Zasshi 111:386–392

Yoshida T, Yamamoto Y, Orita H, Kakiuchi M, Takahashi Y, Itakura M, Kado N, Mitani K, Yasuda S, Kato H, Itoh Y (1996a) Studies on quinolone antibacterials. IV. Structure-activity relationships of antibacterial activity and side effects for 5- or 8-substituted and 5,8-disubstituted-7-(3-amino-1-pyrrolidinyl)-1-cyclopropyl-1,4-dihydro-4-oxoquinoline-3-carboxylic acids. Chem Pharm Bull (Tokyo) 44:1074–1085

Yoshida T, Yamamoto Y, Orita H, Kakiuchi M, Takahashi Y, Itakura M, Kado N, Yasuda S, Kato H, Itoh Y (1996b) Studies on quinolone antibacterials. V. Synthesis and antibacterial activity of chiral 5-amino-7-(4-substituted-3-amino-1-pyrrolidinyl)-6-fluoro-1,4-dihydro-8-methyl-4-oxoquinoline-3-carboxylic acids and derivatives. Chem Pharm Bull (Tokyo) 44:1376–1386

Ziegler CB Jr, Bitha P, Lin Y (1988) Synthesis of some novel 7-substituted quinolonecarboxylic acids via nitroso and nitrone cycloadditions. J Heterocycl Chem 25:719–723

Ziegler CB Jr, Curran WV, Kuck NA, Harris SM, Lin Y (1989) Synthesis and antibacterial activity of some 7-substituted 1-ethyl-6-fluoro-1,4-dihydro-4-oxoquinoline-3-carboxylic acids: ethers, secondary amines and sulfides as C-7 substituents. J Heterocycl Chem 26:1141–1145

Ziegler CB, Kuck NA, Strohmeyer TW, Lin Y (1990a) Synthesis and in vitro biological activity of some 7-(5-aminomethyl-2-isoxazolidinyl)quinolone-3-carboxylic acids. J Heterocycl Chem 27:2077–2079

Ziegler CB, Bitha P, Kuck NA, Fenton TJ, Petersen PJ, Lin Y (1990b) Synthesis and structure-activity-relationships of new 7-[3-(fluoromethyl)piperazinyl]- and (fluorohomopiperazinyl)-quinolone antibacterials. J Med Chem 33:142–146

Zikan V, Radl S (1987) Czech. CS 261 328 (Chem Abstr 112:198398h)

Zikan V, Radl S, Hola V (1987) Czech. CS 262 587 (Chem Abstr 112:216721q)

CHAPTER 4

Mode of Action

A. MAXWELL and S.E. CRITCHLOW

A. Introduction

Over recent years there has been an enormous explosion of interest in the quinolone drugs in general and the fluoroquinolones in particular. This is manifested by the production of this volume and other works on this subject (e.g. FERNANDES 1989; CRUMPLIN 1990; SIPORIN et al. 1990; HOOPER and WOLFSON 1993a). Details of the structures of a wide range of quinolones will be found elsewhere in this book, but the structures of the principal quinolones mentioned in this chapter are shown in Fig. 1. In this chapter, we discuss the mode of action of quinolones, i.e. how they are thought to act on their intracellular target, DNA gyrase. Other recent reviews which cover this topic include those by DRLICA et al. (1990, 1995), HOOPER and WOLFSON (1991, 1993b), REECE and MAXWELL (1991a), MAXWELL (1992) and PALUMBO et al. (1993). It is worth noting at this point the existence of a second target, DNA topoisomerase IV, which is discussed below. Before describing the effects of quinolones on bacteria and on gyrase, we will discuss the basic properties of DNA gyrase.

DNA gyrase is a member of a family of enzymes called DNA topoisomerases which can catalyse linking number changes in DNA (WANG 1985; MAXWELL and GELLERT 1986; WIGLEY 1995a). There are two types, I and II, which are characterised by reactions involving single- or double-stranded breaks in DNA, respectively; DNA gyrase is a type II enzyme. Gyrase consists of two subunits, GyrA and GyrB, encoded by the *gyrA* and *gyrB* genes; the active enzyme is an A_2B_2 complex. The basic properties of the *Escherichia coli* enzyme are given in Tables 1 and 2, and a summary of the proposed domain organisation of the enzyme is given in Fig. 2. The reactions listed for gyrase are probably all manifestations of the same mechanistic steps (reviewed in MAXWELL and GELLERT 1986; REECE and MAXWELL 1991a; WIGLEY 1995b). Briefly, gyrase binds to DNA, and a segment of approximately 130 bp is wrapped around the protein. This wrapped DNA is cleaved in both strands, with a 4-base stagger between the break sites, which results in the formation of DNA–protein covalent bonds between the GyrA subunits and the 5′-phosphates. Another segment of DNA is passed through this double-stranded break which may then be resealed. Catalytic supercoiling requires the hydrolysis of ATP. As discussed below, the quinolone drugs interrupt this process at the DNA breakage–reunion step.

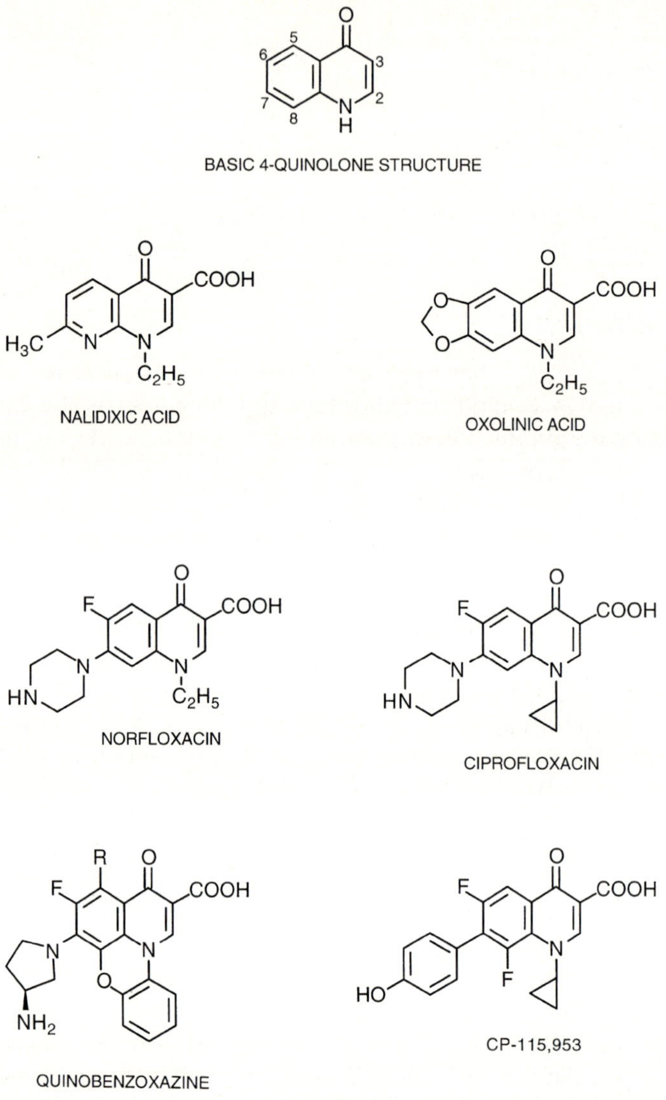

Fig. 1. Structures of quinolone drugs mentioned in this chapter. The quinobenzoxazines are structural analogues of quinolones where *R* is H in A-62176 and *R* is NH_2 in A-85226. The structure of CP-115,953, a quinolone active against eukaryotic topoisomerase II, is also shown

Gyrase is essential in bacteria as it is involved in a number of cellular processes, including DNA replication and transcription. However, it is apparently absent from eukaryotes. Therefore, gyrase is an ideal target for antibacterial agents. Evidence that gyrase is the target of the quinolones comes from various types of experiment. Early studies, described below, had shown

Table 1. Properties of *Escherichia coli* DNA gyrase subunits

Subunit	Gene	M_r	Major role	Drug interactions
GyrA	*gyrA* (2625 bp, formerly *nalA*)	96 887 (875 amino acids)	Breakage and reunion of DNA	Likely target of quinolone drugs (e.g. nalidixic acid, ciprofloxacin)
GyrB	*gyrB* (2412 bp, formerly *cou*)	89 893 (804 amino acids)	ATPase activity	Target of coumarin drugs (e.g. coumermycin A_1, novobiocin)

Table 2. Properties of *Escherichia coli* DNA gyrase

Subunit structure	M_r	Reactions
A_2B_2	373 560	Supercoiling (ATP-dependent) Relaxation of positive supercoils (ATP-dependent) Relaxation of negative supercoils (ATP-independent) Catenation/decatenation Unknotting

that nalidixic acid was an inhibitor of bacterial DNA replication. Prior to the discovery of DNA gyrase, a number of possible targets were found not to be inhibited by nalidixic acid, including DNA polymerase I, endonuclease I, exonucleases I, II and III from *E. coli*, DNA ligase and DNA methyltransferase from T4-infected *E. coli* and DNA polymerase from *Bacillus subtilis* (Pedrini et al. 1972). Following the isolation of gyrase from *E. coli* by Gellert et al. (1976a), it was shown that its supercoiling activity could be inhibited by oxolinic acid (Gellert et al. 1977; Sugino et al. 1977). Moreover, gyrase extracted from a nalidixic acid-resistant mutant was found to be resistant to oxolinic acid.

Since that time a large number of quinolone-resistance mutations have been mapped to the gyrase genes, principally to *gyrA* but also *gyrB* (see Maxwell 1992; Fisher et al. 1993; Hooper and Wolfson 1993b; Nakamura et al. 1993; and Everett and Piddock, Chap. 7, this volume for reviews). One example of such a mutation is the Ser83 to Trp mutation of GyrA found in *E. coli* (Yoshida et al. 1988; Cullen et al. 1989; Oram and Fisher 1991). Enzyme bearing this mutation is resistant to a range of quinolones in vivo and in vitro (Yoshida et al. 1988; Cullen et al. 1989; Oram and Fisher 1991; C.J.R. Willmott and A. Maxwell, unpublished data), and, when complexed with DNA, shows a greatly reduced ability to bind the drugs (Willmott and Maxwell 1993). It seems likely that Ser83 and other residues that confer drug resistance are involved in protein–drug interactions (see below). A more detailed discussion of quinolone-resistance mutations can be found elsewhere in this volume (Everett and Piddock, Chap. 7, this volume).

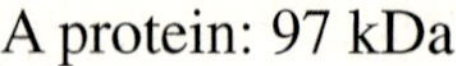

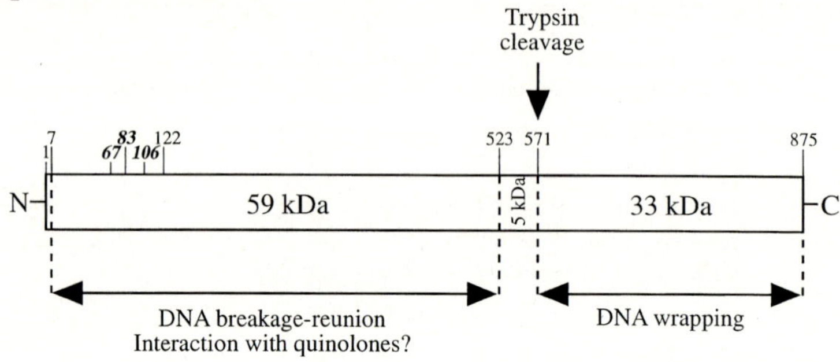

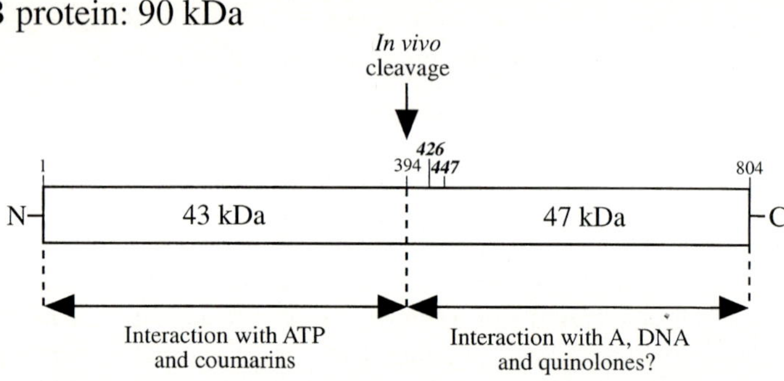

Fig. 2. Proposed domain structure of *E. coli* DNA gyrase. The gyrase proteins are represented as linear blocks with the positions of various amino acids indicated. Sites of some of the mutations leading to quinolone resistance (*67*, *83*, *106* of the *A protein*, and *426* and *447* of the *B protein*) are indicated in bold. The active site tyrosine residue is also shown at position *122* of the *A protein*. (Adapted from REECE and MAXWELL 1991a)

Although the evidence that gyrase is the intracellular target of the quinolones is compelling, the existence of a second target, DNA topoisomerase IV (topo IV), has recently been established. Like gyrase, topo IV is a bacterial type II DNA topoisomerase, but unlike gyrase it cannot supercoil DNA (KATO et al. 1990, 1992). Topo IV carries out the ATP-dependent relaxation of DNA and has been found to be a more potent decatenase than DNA gyrase (HOSHINO et al. 1994). Topo IV is composed of two subunits which in *E. coli* are encoded by the *parC* and *parE* genes. The *parC* gene encodes the ParC subunit (equivalent to GyrA) and the *parE* gene encodes the ParE subunit (equivalent of GyrB). Originally it was thought that *E. coli* DNA topo IV was not readily inhibited by the fluoroquinolones, with 30 times more drug required to inhibit topo IV-mediated relaxation than to inhibit relaxation by DNA gyrase (PENG and MARIANS 1993). However, HOSHINO et al. (1994) have recently shown that fluoroquinolones may have significant activity against topo IV-mediated decatenation of DNA.

Further evidence for a possible interaction between topo IV and the quinolone drugs has come from the analysis of fluoroquinolone-resistant isolates. FERRERO et al. (1994) have demonstrated that topo IV is a primary target of the fluoroquinolones in *Staphylococcus aureus*. Clinical isolates with high levels of fluoroquinolone resistance were found to have mutations in both *gyrA* and *grlA* (equivalent to *parC* in *E. coli*). Furthermore, clinical and laboratory isolates with low-levels of fluoroquinolone resistance were found to have mutations in only *grlA* (FERRERO et al. 1994; FERRERO et al. 1995). These data suggest that topo IV is also a target of the fluoroquinolones and that mutations in *grlA* may be a prerequisite for fluoroquinolone resistance before mutations in *gyrA* occur (FERRERO et al. 1994; FERRERO et al. 1995). Interestingly, all of the GrlA mutations in clinical isolates were found to map to Ser80, which corresponds to Ser83 of *E. coli* GyrA, indicating that fluoroquinolones may interact with GrlA in a similar manner as with GyrA (FERRERO et al. 1994). Further evidence in support of this proposal has come from the analysis of ciprofloxacin-resistant laboratory isolates which have GrlA mutations at the amino acid positions Ser80 and Glu84, which correspond to Ser83 and Asp87 of *E. coli* GyrA (FERRERO et al. 1995).

Fluoroquinolone-resistant isolates of *Neisseria gonorrhoeae* have also been shown to acquire mutations in both *gyrA* and *parC* (BELLAND et al. 1994). However, unlike the case for *S. aureus*, low levels of resistance were associated with mutations in *gyrA*, whereas strains with high-level resistance acquired analogous mutations in both *gyrA* and *parC*. Similar observations have been made by DEGUCHI et al. (1996) who found that clinical strains of *N. gonorrhoeae* which have alterations in both GyrA and ParC were significantly more resistant to fluoroquinolones than those with alterations in GyrA alone. BELLAND et al. (1994) reported that mutations occurred at Ser91 and Ser88 in *N. gonorrhoeae* GyrA and ParC, respectively. In both proteins these residues are the equivalent of Ser83 of *E. coli* GyrA. These results have led the authors to propose that *parC* can also be mutated in fluoroquinolone-resistant cells, but only when GyrA is already resistant to the drugs (BELLAND et al. 1994; DEGUCHI et al. 1996). This is different from the situation in *S. aureus* where it has been proposed that topo IV is the primary target of fluoroquinolones (FERRERO et al. 1994; FERRERO et al. 1995). This raises the possibility that the primary target of fluoroquinolones could be different in different bacterial species.

Analysis of quinolone resistance associated with mutations in GyrA and ParC in *E. coli* by HEISIG (1996) and VILA et al. (1996) have revealed that topo IV is likely to be a secondary target of the fluoroquinolones. Neither of these studies were able to detect quinolone resistance associated with ParC alone. However, it does appear that in highly quinolone-resistant *E. coli* strains, at least one *parC* mutation is required in addition to *gyrA* mutation(s) (HEISIG 1996; VILA et al. 1996). Quinolone-resistance mutations of *E. coli* ParC have been found to map to Ser80 and Glu84 which correspond to Ser83 and Asp87 of *E. coli* GyrA. Given that all of the quinolone-resistance mutations identified in *E. coli*, *S. aureus* and *N. gonorrhoeae* ParC (GrlA) are equivalent to quinolone-resistance mutations of GyrA, it is likely that fluoroquinolones interact in a similar way with gyrase and topo IV.

In general quinolones do not affect topoisomerase II (topo II; the counterpart of DNA gyrase in eukaryotes), except at very high concentrations (GOOTZ et al. 1990), although there are a few quinolones which do have activity against this enzyme, e.g. CP-115,953 (Fig. 1; ROBINSON et al. 1991). MOREAU et al. (1990) surveyed the effects of quinolones on DNA gyrase and topoisomerase I from *E. coli* and topoisomerases I and II from calf thymus and found significant effects of the drugs only on bacterial gyrase. There are a large number of anticancer drugs targeted against topo II (e.g. acridines and epipodophyllotoxins) whose mode of action appears to be similar to that of quinolones on gyrase (reviewed in LIU 1989; CORBETT and OSHEROFF 1994; CHEN and LIU 1994; FROELICH-AMMON and OSHEROFF 1995). Indeed it has recently been shown that a mutation in yeast topo II (Ser741 to Trp) which is homologous to the GyrA mutation, Ser83 to Trp, confers resistance to the quinolone-like drug CP-115,953 and hypersensitivity to the topo II inhibitor etoposide (HSIUNG et al. 1995). These results suggest that the binding site for the cleavable-complex-forming drugs is similar in gyrase and topo II.

The remainder of this chapter focuses largely on the interaction of quinolones with DNA gyrase as there has been a large amount of work done in this area. It must however be remembered that recent work on topo IV has identified this enzyme as a second target.

B. Effects on Bacteria

Before discussing the molecular basis of the action of quinolone drugs, it is important to establish the effects that the drugs have on bacteria. Other reviews on this aspect include those by SMITH (1984), SONSTEIN (1990), ROHATGI and COURTRIGHT (1990) and HOOPER and WOLFSON (1990, 1993b). The effects of a number of quinolone drugs on *E. coli* cells is shown in Fig. 3 (SMITH 1986). These survival curves show that bacterial killing increases up to a certain drug concentration (the optimum bactericidal concentration) and can then decrease at higher drug concentrations. (The way in which this behaviour can be rationalised is described below.) The exposure of bacteria to quinolones is accompanied by a range of physiological effects which are discussed in the following paragraphs:

- Filamentation and loss of septation
- Inhibition of nucleoid segregation
- Vacuole formation
- Inhibition of DNA synthesis
- Inhibition of conjugation

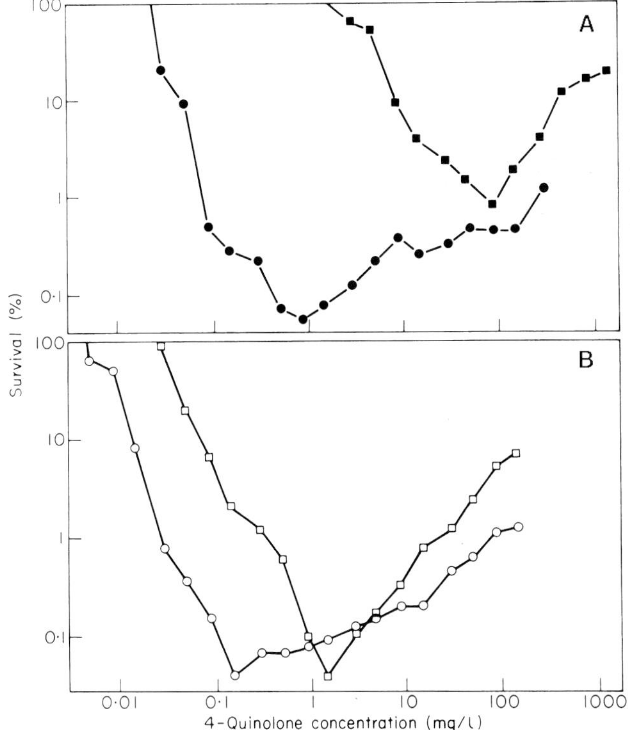

Fig. 3A,B. Effect of quinolone drugs on the growth of bacteria. *E. coli* KL16 cells grown overnight and diluted 1 in 50 into broth containing various concentration of: ■, nalidixic acid: •, ofloxacin; □, norfloxacin: ○, ciprofloxacin; and allowed to grow for a further 3 h. (From SMITH 1986)

- Inhibition of transcription and protein synthesis
- Loss of supercoiling
- DNA damage
- Induction of SOS response
- Induction of heat-shock response
- Increase in mutation rate
- Plasmid loss

One of the most striking consequences of exposing bacteria to quinolones is elongation of the cells (filamentation). This phenomenon was first noted by Goss et al. (1964) who found that *E. coli* cells treated with nalidixic acid became greatly elongated. Similar effects have been observed on the treatment of *E. coli* with norfloxacin and ciprofloxacin (Fig. 4; CRUMPLIN et al. 1984; DIVER and WISE 1986; ELLIOTT et al. 1987). These filaments have been shown to lack septae (WALKER and PARDEE 1968) and the elongation process may be a consequence of the induction of the SOS system (see below). The filaments have also been found to contain large nucleoids indicative of a defect in

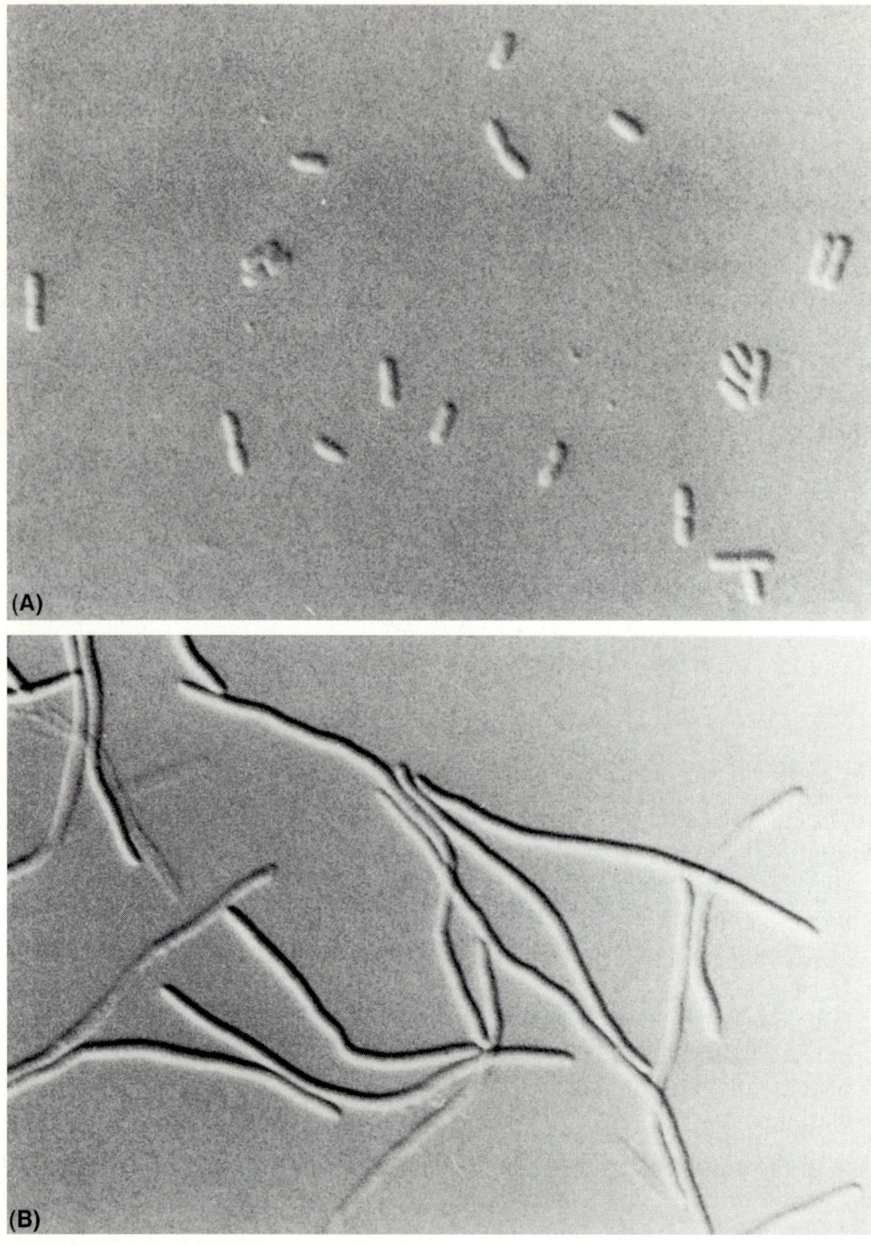

Fig. 4A,B. Morphological response of *E. coli* KL16 cells to ciprofloxacin. **A** Drug-free control. **B** 2-h incubation with ciprofloxacin at 60 μg/ml. (From DIVER and WISE 1986)

segregation (GEORGOPAPADAKOU and BERTASSO 1991). An additional morphological consequence of the exposure to quinolones is the appearance of intracellular vacuoles. These have been observed after treatment of *E. coli* with nalidixic acid (DOUGHERTY and SAUKKONEN 1985) and with norfloxacin and ciprofloxacin (ELLIOTT et al. 1987).

The most important intracellular consequence of the effects of quinolones is inhibition of DNA synthesis. This was first noted with nalidixic acid (GOSS et al. 1965; PEDRINI et al. 1972) but has also been observed with other drugs, e.g. oxolinic acid (ENGLE et al. 1982), norfloxacin and ciprofloxacin (PIDDOCK et al. 1990). Inhibition is rapid (DRLICA et al. 1980) and there is a good correlation between MICs and inhibition of DNA synthesis (CHOW et al. 1988). Progressive inhibition of DNA synthesis is thought to account for the increase in bacterial killing with increasing drug concentration (Fig. 3). With nalidixic acid, little inhibition of RNA and protein synthesis is found at low drug concentrations (GOSS et al. 1965; CRUMPLIN and SMITH 1975), but these processes are inhibited at higher drug concentrations. Similar results have also been found with a number of fluoroquinolones (PIDDOCK et al. 1990). It is thought that inhibition of RNA and protein synthesis may account for the second phase of quinolone activity (Fig. 3) in which increasing drug concentrations lead to less killing (CRUMPLIN and SMITH 1975). This implies that protein synthesis may be required for quinolone-mediated cell death. Indeed chloramphenicol (a protein synthesis inhibitor) and rifampin (an RNA synthesis inhibitor) reduce killing by nalidixic acid (DEITZ et al. 1966) and also fluoroquinolones such as norfloxacin (SMITH 1984). Recent work (CHEN et al. 1996) suggests that, at least with oxolinic acid, inhibition of DNA synthesis may not be itself be responsible for the bactericidal effects of the drugs.

A phenomenon related to the inhibition of DNA synthesis by quinolones is the inhibition of bacterial conjugation by nalidixic acid (HOLLOM and PRITCHARD 1965; BARBOUR 1967; BOUCK and ADELBERG 1970). This can be attributed to a secondary effect of the inhibition of DNA synthesis and indeed has been cited as evidence for the involvement of DNA synthesis in conjugal transfer (BARBOUR 1967).

A further consequence of quinolone exposure that has been noted is DNA damage. Treatment of *E. coli* cells with nalidixic acid has been found to result in DNA degradation (COOK et al. 1966a; CRUMPLIN and SMITH 1976). Such damage is distinct from drug-dependent cleavage revealed by the addition of SDS to bacteria exposed to quinolones (SNYDER and DRLICA 1979), which is discussed later. However, it is relevant to point out here that complexes between oxolinic acid, gyrase and DNA have been found to form very quickly (correlating with rapid inhibition of DNA synthesis, described above) and under conditions where no DNA relaxation is observed (discussed further below; SNYDER and DRLICA 1979). Other evidence suggests that the drug-induced chromosome damage occurs preferentially close to the replication fork (RAMAREDDY and REITER 1969), suggesting the existence of DNA lesions associated with DNA replication. Work by LEWIN and SMITH (1990) has shown that nalidixic-acid-induced DNA degradation occurs even under conditions where the drug is not bactericidal, suggesting that DNA breakdown may not contribute to the lethal action of the drug.

A related phenomenon is the induction of the SOS DNA repair system by quinolones. This is an inducible system in which a range of genes are expressed in response to DNA damage or lesions in DNA replication (WALKER 1984). The induction of SOS involves the derepression of genes controlled by the LexA repressor including *recA* and *lexA* itself. It has been demonstrated that nalidixic acid induces the SOS response in *E. coli* and that this induction requires the *recA*, *lexA*, and *recBC* functions (GUDAS and PARDEE 1975). Several other quinolones, including norfloxacin and ciprofloxacin, have also been found to induce SOS (PHILLIPS et al. 1987; PIDDOCK and WISE 1987). The level of drug found to maximally induce SOS corresponds to the optimum bactericidal concentration. It is likely that the DNA damage induced by quinolones activates RecA which then turns on the SOS genes via interaction with LexA. The action of quinolones on *E. coli* strains with mutations in SOS genes has been studied (PIDDOCK and WALTERS 1992); it was concluded that recombination and excision repair are involved in the repair of quinolone-induced DNA damage and that filamentation is a consequence of SOS induction. Mutations in SOS genes (e.g. *recA* and *lexA*) can influence the sensitivity of bacteria to the drugs, but the response depends upon the drug (MCDANIEL et al. 1978; LEWIN et al. 1989; HOWARD et al. 1993). This phenomenon may be connected with the proposal that there are three separate mechanisms of killing for quinolones (see below; SMITH 1984).

COOK et al. (1966b) suggested that nalidixic acid may be mutagenic and this effect has been linked to its ability to induce SOS (PHILLIPS et al. 1987). Although some doubts have been expressed about this idea (reviewed in PHILLIPS 1987), it has been concluded from more recent work using a range of newer quinolones, including norfloxacin and ciprofloxacin, that the drugs may be mutagenic (GOCKE 1991; FUNG-TOMC et al. 1993; MAMBER et al. 1993). The exact cause of the mutagenicity of quinolones is uncertain but it is likely to be a consequence of the induction of the error-prone SOS repair system. Evidence in support of this is the ability of quinolones to induce the *umuCD* genes in *E. coli*, which are essential for SOS-mediated mutagenesis (YSERN et al. 1990; POWER and PHILLIPS 1992, 1993a,b).

In addition to stimulating the SOS response, quinolones have also been shown to induce the heat-shock response (KRUEGER and WALKER 1984). Exposure of *E. coli* to nalidixic acid causes the induction of at least two genes, *groEL* and *dnaK*, which are members of the heat-shock regulon and whose expression is dependent on the *htpR* gene product. Treatment of cells bearing an *htpR* mutation with nalidixic acid fails to induce GroEL and DnaK (KRUEGER and WALKER 1984). The significance of heat-shock induction is not clear but it may reflect a role of the *htpR*-controlled gene products in the degradation of certain SOS proteins (WALKER 1984). It seems likely that induction of both SOS and heat shock are secondary effects of quinolone action which are part of the cell's survival response to the potentially lethal effects of the drugs.

Another effect of quinolones is to eliminate some plasmids from their host cells. Nalidixic acid has been reported to result in the loss of R-factor R1 from *Salmonella typhymurium* (HAHN and CIAK 1976), and both nalidixic acid and oxolinic acid were found to produce low levels of elimination of the wild-type plasmid pM6110 from *E. coli* (HOOPER et al. 1984). WEISSER and WIEDEMANN (1985, 1986) found that nalidixic acid and several other quinolones, including norfloxacin and ciprofloxacin, eliminated F'*lac* and various R plasmids from *E. coli*. As plasmid loss occurs at drug concentrations which still allow cells to grow, this suggests that the replication of certain plasmids is more sensitive to drugs than chromosome replication (WEISSER and WIEDEMANN 1986). It seems that plasmid loss is not a general phenomenon and is dependent on the plasmid, the host and the drug concentration used (PLATT and BLACK 1987; COURTRIGHT et al. 1988; HOOPER and WOLFSON 1993b). Coumarin drugs (the other group of gyrase-specific antibacterial agents) are more effective than quinolones in plasmid elimination (MCHUGH and SWARTZ 1977; TAYLOR and LEVINE 1979; DANILEVSKAYA and GRAGEROV 1980; CEJKA et al. 1982; Hooper et al. 1984).

One slightly surprising aspect of quinolones is that they have been reported not to produce significant relaxation of bacterial DNA in vivo. Even at relatively high concentrations oxolinic acid was not found to relax bacterial DNA (SNYDER and DRLICA 1979). However, very high concentrations of the drug do cause some relaxation of the bacterial chromosome (MANES et al. 1983). This is in marked contrast to the coumarins which can profoundly affect in vivo supercoiling (DRLICA and SNYDER 1978; MANES et al. 1983) and indicates different modes of action for the two groups of drugs. However, more recently, ALEIXANDRE et al. (1991) have demonstrated relaxation of plasmid DNA by treatment of cells with ciprofloxacin.

Although the morphological and physiological effects of quinolones on bacteria are seemingly quite diverse, they can largely be rationalised by considering the effects of the drugs on DNA gyrase. If, as described below, the gyrase–drug complex on DNA can form a barrier to the passage of polymerases, then it is clear how replication and transcription (and thence protein synthesis) could be affected. More specifically, it has been reasoned that rapid inhibition of DNA synthesis may be a consequence of gyrase being clustered around replication forks (DRLICA et al. 1980) while a slower onset of inhibition arises from the encounter of replication complexes with gyrase elsewhere on the chromosome (SNYDER and DRLICA 1979). Plasmid loss may also be a consequence of the inhibition of replication. Similarly, filamentation is thought to be a nonspecific response to interference with DNA synthesis (ELLIOTT et al. 1987) as other agents are known to induce this effect. More recently filamentation has been correlated with SOS induction (PIDDOCK and WALTERS 1992). If the interaction of the drugs with gyrase at the replication fork does lead to DNA damage, then induction of SOS and the heat-shock response may follow. Another response to DNA damage may be stimulation of the mutation rate.

However, we should be cautious about oversimplifying the explanations for the effects of quinolones on bacteria. Fluoroquinolones are thought to possess up to three mechanisms of cell killing (SMITH 1984; HOWARD et al. 1993): mechanism A (common to all quinolones) requires RNA and protein synthesis and is only effective against dividing bacteria; mechanism B does not require RNA and protein synthesis and can act on bacteria unable to multiply; mechanism C requires RNA and protein synthesis but does not require cell division. Recent work by CHEN et al. (1996) also supports the idea of three mechanisms of cell killing. From their data, mechanism A would be the blocking of replication by the gyrase–quinolone complex on DNA, mechanism B (chloramphenicol insensitive) can be correlated with dissociation of the gyrase subunits which constrain the ternary complex, and mechanism C may correlate with trapping of topo IV complexes on DNA.

Clearly there is a cascade of effects of quinolones on bacteria which are likely to stem from the action of the drugs on DNA gyrase and/or topo IV. We will first consider the action of the drugs on the reactions of gyrase before going on to examine the molecular events which lead to the toxic action of the drugs.

C. Effects on DNA Gyrase

I. Reactions of Gyrase

Prior to 1976 the target of the quinolone drugs was unknown although it had been reasoned that it might be a protein involved in DNA replication. For example, GOSS et al. (1965) proposed that the primary action of nalidixic acid was the inhibition of DNA biosynthesis and that the likely target was an enzyme involved in this process. In 1976, GELLERT et al. (1976a) published the discovery of DNA gyrase, an enzyme which could introduce negative supercoils into double-stranded closed-circular DNA. Gyrase was found to be the target of the coumarin drugs coumermycin A_1 and novobiocin (GELLERT et al. 1976b) and also the quinolone drugs nalidixic acid and oxolinic acid (GELLERT et al. 1977; SUGINO et al. 1977). Specifically, nalidixic acid and oxolinic acid were shown to be able to inhibit gyrase-catalysed supercoiling in vitro. DNA gyrase isolated from bacteria bearing a *nalA*^{r} mutation was found to be resistant to the drugs. In the absence of ATP, gyrase relaxes supercoiled DNA and this reaction has also been found to be inhibited by nalidixic and oxolinic acids. In addition, the drugs were shown to stimulate DNA cleavage by gyrase in a reaction that depended upon the addition of a protein denaturant (SDS) and proteinase K (GELLERT et al. 1977; SUGINO et al. 1977).

Taken together, these results suggested that the quinolones could interfere with the DNA breakage and reunion reaction carried out by gyrase in a manner that would inhibit both DNA supercoiling and relaxation and, under appropriate conditions, could induce double-stranded cleavage of DNA. Subsequently, the effects of the drugs on the reactions of gyrase were examined in more detail and with a wider range of agents.

Aside from supercoiling and relaxation of DNA, gyrase can also catalyse catenation and decatenation reactions as well as the unknotting of DNA (Kreuzer and Cozzarelli 1980; Liu et al. 1980). The catenation/decatenation reaction has been shown to be inhibited by oxolinic acid (Kreuzer and Cozzarelli 1980) and ciprofloxacin (Hallett and Maxwell 1991; A.P. Tingey and A. Maxwell, unpublished data) at concentrations similar to those required to inhibit supercoiling and relaxation. It therefore appears that the drugs will interrupt any reaction which requires breakage of DNA and strand passage.

The DNA relaxation reaction of topoisomerase II′ (consisting of the gyrase A subunit and the C-terminal 47-kDa fragment of the B subunit) was also found to be sensitive to oxolinic acid (Brown and Cozzarelli 1979; Gellert et al. 1979). DNA cleavage by this enzyme is also induced by drugs following SDS addition. Similarly, a deletion analysis of the gyrase A protein has shown that a 59-kDa N-terminal fragment is able to carry out quinolone-induced cleavage of DNA, in the presence of the B protein, as efficiently as the intact enzyme (Reece and Maxwell 1991b). These results suggest that the interaction with quinolone drugs involves the N-terminal domain of GyrA and the C-terminal domain of GyrB (Fig. 2). This contention is supported by the fact that point mutations leading to quinolone resistance in DNA gyrase occur exclusively in these domains (see Fig. 2 and Chap. 7, this volume).

II. Mechanistic Steps

The topoisomerisation reaction of gyrase can be broken down into discrete steps and the effect of quinolone drugs on these individual processes can be considered.

1. DNA Binding

The first step in the gyrase supercoiling reaction is the formation of a DNA–protein complex in which ~130 bp of DNA is wrapped around the gyrase tetramer (reviewed in Reece and Maxwell 1991a). A number of approaches have been used to study the formation of this complex and its interaction with quinolone drugs. The complex between gyrase and DNA can be trapped by binding to a nitrocellulose filter (Higgins and Cozzarelli 1982). The addition of oxolinic acid results in a complex which is more salt stable. Addition of SDS to the gyrase–DNA complex in the presence of oxolinic acid reveals cleavage of the DNA and covalent attachment of the enzyme to the broken DNA (discussed below). The 64-kDa N-terminal fragment of GyrA is able to form a complex with GyrB and DNA (Reece and Maxwell 1989). However, this complex is relatively unstable, as judged by gel retardation assays. The addition of a quinolone drug (ciprofloxacin) improves the stability of this complex (CRITCHLOW and MAXWELL 1996).

The gyrase–DNA complex has also been analysed by footprinting methods. DNase I footprinting (FISHER et al. 1981; KIRKEGAARD and WANG 1981; MORRISON and COZZARELLI 1981; RAU et al. 1987) showed that the enzyme protects about 100–155 bp of DNA with enhanced sensitivity within the protected region, characteristic of DNA which is wrapped on a surface. The addition of quinolones produced variable results in these experiments. FISHER et al. (1981) found marked changes in the degree and extent of protection in the presence of oxolinic acid, whereas MORRISON and COZZARELLI (1981) found no dramatic changes in the DNase I pattern but did see changes in the exonuclease III digestion pattern. These changes were interpreted as stabilisation of gyrase binding to a small region around the cleavage sit by the drug. In contrast, RAU et al. (1987) found no significant drug-induced changes in the DNase I footprints.

Recently the gyrase–DNA complex has been analysed by hydroxyl radical footprinting (ORPHANIDES and MAXWELL 1994). In these experiments, gyrase was found to protect 128 bp of DNA from the hydroxyl radical with the central 13 bp (adjacent to the gyrase cleavage site) being most strongly protected. Flanking this central region are arms showing periodic protection consistent with a helical repeat of 10.6 bp. The addition of ciprofloxacin caused striking changes in the footprint. The changes were observed in one arm only and consisted of extended protection and a shift in the phase of the cleavage maxima. In addition, decreased protection adjacent to the cleavage site was observed. Similar changes were also observed in the presence of ADPNP (5′-adenylyl β,γ-imidodiphosphate: ORPHANIDES and MAXWELL 1994). These changes can be interpreted as representing an altered conformation of the gyrase–DNA complex which is stabilised by drug binding. Specifically, it is proposed that one of the arms which comprise the DNA wrap is manipulated to facilitate strand passage, and the addition of drug (or nucleotide) traps such a conformational state.

A further technique which has been used to analyse the gyrase–DNA complex is transient electric dichroism (RAU et al. 1987). Data from these experiments support a model of the complex in which the DNA is wrapped around the protein core with two protruding tails. The change in dichroism found upon addition of ADPNP was interpreted as the two tails being wrapped upon the protein core suggestive of an intermediate in the strand-passage reaction. The addition of norfloxacin had only minor effects on the dichroism of the complex, which could be interpreted as skewing of the tails relative to the core DNA. However, addition of both norfloxacin and ADPNP prevented the nucleotide-induced change in dichroism (RAU et al. 1987), i.e. the drug appeared to inhibit a conformational change of the complex.

No clear consensus has emerged regarding the effect of quinolones on the complex between gyrase and DNA from the above studies. There are clearly unresolved differences in results obtained from different laboratories and by different methods. This probably reflects the fact that there has not been a systematic biophysical study of the gyrase–DNA–drug complex. It does seem, however, that quinolones tend to stabilise the gyrase–DNA complex and may favour an altered conformation of the complex.

2. DNA Cleavage

Following DNA binding, gyrase cleaves DNA in both strands. Although the occurrence of this reaction can be inferred from the ability of gyrase to catalyse catenation and decatenation (Kreuzer and Cozzarelli 1980; Liu et al. 1980), direct evidence has come from studies with quinolone drugs. Incubation of gyrase and DNA in the presence of quinolone drugs and addition of SDS results in double-strand DNA cleavage with a four-base stagger between the cut sites (Morrison and Cozzarelli 1979; Fig. 5). Following cleavage, the

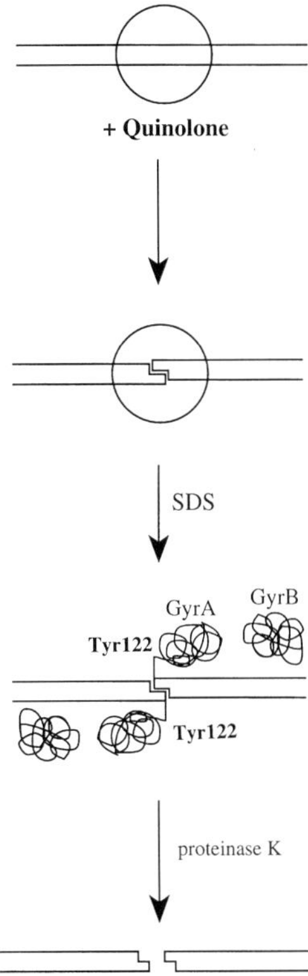

Fig. 5. Quinolone-induced cleavage by DNA gyrase. On addition of sodium dodecyl sulphate (SDS) and proteinase K to a gyrase–quinolone–DNA complex, double-stranded cleavage of DNA is revealed. At present it is unclear whether a cleaved complex is formed prior to SDS treatment (as shown here) or whether cleavage is revealed after SDS treatment

gyrase A proteins are covalently attached to the 5′-ends of the DNA and can be removed by proteinase treatment (Gellert et al. 1977; Sugino et al. 1977; Morrison and Cozzarelli 1979). Specifically, a covalent bond is formed between the 5′-phosphate on DNA and Tyr122 of the A subunit (Horowitz and Wang 1987). DNA cleavage by gyrase has also been shown to be induced in the absence of quinolone when Mg^{2+} is replaced by Ca^{2+} (L.M. Fisher, M.H. O'Dea and M. Gellert, unpublished data). Drug-dependent DNA cleavage by gyrase can be observed in vivo with fragmentation of the chromosome into ~100 kb pieces (Snyder and Drlica 1979). Quinolone-induced DNA cleavage by gyrase does not occur at prescribed sequences in DNA, but there are preferences, and a consensus sequence has been derived from analysis of oxolinic-acid-induced cleavage in vivo (Lockshon and Morris 1985). This comprises a 20-bp sequence containing many degeneracies and is consistent with sites found in vitro (Fisher et al. 1981; Kirkegaard and Wang 1981; Morrison and Cozzarelli 1981):

```
             ↓      T
5′-RNNRNNRTGRYC YNYNGNY-3′
             (G)    G     (T)
```

where R = purine, Y = pyrimidine, N = any nucleotide; T and G at the 13th position are equally preferred, and G and T in brackets are preferred secondarily to T and G, respectively. The arrow indicates the site of cleavage by DNA gyrase. The significance of this consensus sequence with respect to sites of gyrase action is discussed below.

Although the quinolone-induced cleavage of DNA by gyrase is well established, a number of fundamental issues concerning this reaction remain unresolved. One of these is whether cleavage of the DNA occurs prior to, or subsequent to, the addition of protein denaturant. This uncertainty is illustrated by experiments in which reactions containing gyrase, DNA and oxolinic acid are heated briefly to 80°C prior to the addition of SDS (Gellert et al. 1977). In this case no DNA cleavage occurs, i.e. protein can be removed without resulting in cleaved DNA. Thus it is formally possible that quinolones either stabilise the gyrase–DNA complex at a stage prior to cleavage, and this can lead to cleavage when SDS is added, or that they stabilise a cleaved complex in which the DNA–protein covalent bond is formed and in which religation can efficiently occur after heat treatment (Fig. 5). At present there are no experiments which distinguish these two possibilities.

A second unresolved issue is whether the sites of quinolone-induced cleavage are identical to sites at which cleavage would occur in the absence of drugs. A number of experiments have addressed this issue. Sites of drug-induced cleavage generally occur near the centre of the regions protected by gyrase from DNA-cleaving agents (Fisher et al. 1981; Kirkegaard and Wang 1981; Morrison and Cozzarelli 1981; Rau et al. 1987; Orphanides and Maxwell 1994). However, there is one instance of a gyrase binding site (detected by filter retention) which did not contain an oxolinic-acid-induced cleavage site (Kirkegaard and Wang 1981). In addition, the DNA cleavage patterns induced by different drugs (or Ca^{2+}) have been compared. One study (L.M. Fisher, M.H. O'Dea and M. Gellert, unpublished data) has found that gyrase cleaves linear pBR322 DNA at the same locations in the presence of either oxolinic acid or Ca^{2+}. In other work (C.J.R. Willmott, A.P. Tingey and A. Maxwell, unpublished data), comparison of the cleavage patterns in the presence of oxolinic acid, ciprofloxacin or Ca^{2+} has revealed significant differences. Moreover, a 147-bp fragment containing the preferred quinolone-induced gyrase cleavage site from plasmid pBR322 is cleaved inefficiently by gyrase in the presence of Ca^{2+} (Dobbs et al. 1992), and a 39-bp segment of DNA, which can be cleaved by gyrase in the presence of oxolinic acid or ciprofloxacin, is not cleaved in the presence of Ca^{2+} (A.P. Tingey and A. Maxwell, unpublished data).

The ideal experiment to test this issue would be one in which the cleavage sites in the absence and presence of drugs are compared. Such an approach has been feasible with eukaryotic topo II where low-level cleavage is seen in the absence of drug. For example, the cleavage sites in the absence of drugs and in the presence of compounds such as *m*AMSA, teniposide and doxorubicin have been compared (Capranico et al. 1990; Pommier et al. 1991a). These studies showed that the drugs can indeed influence the site of DNA cleavage. With DNA gyrase such an approach is not so readily applied as the level of drug-independent cleavage observed after the addition of SDS is very low. However, there are claims that drug-independent DNA cleavage by gyrase can occur (reviewed in Drlica and Franco 1988). For example, unpublished data (cited in Lockshon and Morris 1985) suggest that DNA fragments of comparable sizes are produced by gyrase in the absence and presence of oxolinic acid.

Taken together, the above observations suggest that quinlones can at least modify the DNA cleavage preference of gyrase. It is likely that the selection of a particular site is a consequence of a combination of several factors including the flexibility of the DNA surrounding the cleavage site (the arms) and the local DNA sequence at the cleavage site. The effect of point mutations around the major gyrase cleavage site in plasmid pBR322 (at 990 bp) has been studied (Fisher et al. 1986). Certain point mutations were found to have profound effects on oxolinic-acid-induced cleavage, supporting the idea that the local sequence around the cleavage site is important. By analogy with eukaryotic type II topoisomerases and antitumour drugs (discussed further below), it is likely that certain quinolone drugs will favour cleavage at particular sites based on local DNA sequence considerations. However, what actually constitutes a gyrase cleavage site still remains an unresolved issue.

Although gyrase has been shown to bind stably only to DNA fragments of >100 bp (Morrison et al. 1980; Maxwell and Gellert 1984), quinolone-induced DNA cleavage can occur with shorter fragments of DNA. Gmünder et al. (1995) have shown DNA cleavage by gyrase in the presence of fluoroquinolones with a fragment as short as 71 bp. Recent experiments (M.E.

Cove, A.P. Tingey and A. Maxwell, unpublished work) have shown that oxolinic acid and ciprofloxacin can induce DNA cleavage by gyrase within DNA fragments which are <40 bp, which, in principle, are too short to form a gyrase–DNA complex. The smallest DNA fragment with which DNA cleavage could be detected in these experiments was 20 bp. Such results are in accord with the DNA-binding experiments discussed above, where the drug can apparently stabilise the interaction of gyrase with DNA. In contrast, FISHER et al. (1986) found no cleavage of a 34-bp fragment in the presence of gyrase and oxolinic acid. Although the DNA fragments used in these two latter studies are based on the same sequence (the preferred gyrase cleavage site from pBR322) it may be that the differences are a consequence of different reaction conditions used in the two sets of experiments.

A further issue is what the drug is actually doing in terms of the chemistry of the DNA cleavage/religation process. The interaction of gyrase with DNA at the cleavage site can be viewed as a cleavage–religation equilibrium. Quinolone drugs can be considered to push the equilibrium towards cleavage. Arguably, this could be achieved by accelerating the rate of cleavage or inhibiting religation. It is not clear at present which of these is correct. Indeed it is still formally possible (see above) that quinolones block the gyrase reaction prior to DNA cleavage.

With regard to the ability of quinolones to induce DNA cleavage by gyrase, it is interesting to draw comparisons with the CcdB protein of the F plasmid. The proteins CcdB and CcdA, constitute a two-component system which is involved in maintaining the F plasmid in bacteria. In the absence of CcdA, expression of CcdB protein causes cell filamentation and cell death. This bactericidal action is thought to occur via a mechanism involving DNA gyrase. Evidence that gyrase is the target of the CcdB protein has come from genetic experiments which have mapped CcdB-resistant mutants to *gyrA* (BERNARD and COUTURIER 1992; MIKI et al. 1992). Furthermore, in vitro experiments have shown that the CcdB protein induces ATP-dependent cleavage of linear DNA by gyrase. This cleavage is only observed after treatment of a CcdB–gyrase–DNA complex with SDS and proteinase K (BERNARD et al. 1993). Unlike quinolone-induced cleavage, heating at 80°C does not prevent CcdB-induced cleavage. It appears that cleavage induced by CcdB is related but not identical to quinolone-induced cleavage.

3. ATPase

Catalytic supercoiling by gyrase requires the hydrolysis of ATP but the relaxation of negatively supercoiled DNA is ATP independent (GELLERT et al. 1977; SUGINO et al. 1977). This distinguishes DNA gyrase from other type II topoisomerases which cannot supercoil DNA and require ATP hydrolysis for DNA relaxation (WATT and HICKSON 1994). Quinolone drugs are not inhibitors of the ATPase reaction (MIZUUCHI et al. 1978; SUGINO and COZZARELLI 1980) but can be shown to modify the activity yielding altered kinetic parameters (S. Kampranis and A. Maxwell, unpublished data). This can be rationalised in terms of gyrase being a DNA-dependent ATPase and quinolones modifying the interaction of gyrase with DNA. Stimulation of the gyrase ATPase reaction normally requires gyrase to bind to fragments of >100 bp (Maxwell and Gellert 1984), but in the presence of norfloxacin shorter fragments (~50 bp) have been found to stimulate ATP hydrolysis (A. Maxwell and M. Gellert, unpublished data). Again this result suggests that quinolones can stabilise gyrase–DNA interaction enabling interaction with shorter DNAs.

III. Illegitimate Recombination

Experiments using a cell-free system from *E. coli* to study illegitimate recombination have suggested a role for DNA gyrase in this process (Ikeda et al. 1981, 1982). Oxolinic acid has been found to stimulate this process while coumermycin inhibits the reaction. One suggestion for the way in which gyrase facilitates recombination is via subunit exchange between two gyrase tetramers bound to different DNA molecules. Oxolinic acid may act to stabilise the intermediate with the DNA–protein covalent bond formed. However, the mechanistic details of this process remain to be resolved.

IV. Summary

It is very clear that the interaction of quinolones with the gyrase–DNA complex interrupts the DNA breakage–reunion step of the topoisomerisation reaction. This probably results in the stabilisation of an altered gyrase–DNA complex which may represent an intermediate along the strand-passage pathway. This intermediate may involve DNA–protein bonds and, in any case, would appear to be a more stable nucleoprotein complex than that without drug. A number of issues regarding the effects of quinolones on gyrase remain to be clarified. In part, this requires a further understanding of the mechanistic aspects of DNA gyrase, but also a better understanding of the nature of the complex of the drugs with gyrase and DNA. This latter issue is discussed in the next section.

D. Mode of Binding

The existence of quinolone-resistance mutations in both the *gyrA* and *gyrB* genes encoding DNA gyrase and the *parC* gene encoding the ParC subunit of topo IV, might suggest that quinolone drugs interact with one or both subunits of gyrase and the ParC protein of topo IV. However, numerous quinolone-binding studies have shown that the precise binding site of the quinolone drugs is not so easily defined. As it has only been recently recognised that topo IV is a target of the quinolones, little data is available at present concerning the mode of interaction of quinolones with this enzyme. Therefore, the following sections review some of the relevant studies of quinolone interaction with gyrase and DNA. However, given the sequence similarity between gyrase and topo IV, it is likely that the nature of the interaction between the drugs and the two enzymes will be very similar.

I. Binding of Quinolones to DNA

The first experiments investigating the possibility of quinolone binding to DNA were performed before the identification of DNA gyrase as a target of the quinolone drugs. Early studies involving absorption spectroscopy, measurement of DNA melting temperature and equilibrium dialysis found no evidence for the of binding of ^{3}H-labelled nalidixic acid to purified DNA (BOURGUIGNON et al. 1973). However, later studies suggested that nalidixic acid binds to guanine on single-stranded DNA in the presence of copper ions (CRUMPLIN et al. 1980; DREYFUSS and MIDGLEY 1983).

SHEN and PERNET (1985) investigated norfloxacin binding to gyrase and DNA using equilibrium dialysis and a membrane filtration technique. ^{3}H-labelled norfloxacin was found to bind to DNA and not to DNA gyrase or to its individual subunits. Binding of ^{3}H-labelled norfloxacin to ColE1 DNA and pBR322 occurred at a drug concentration near the apparent K_i of norfloxacin for the gyrase supercoiling reaction. ^{3}H-labelled norfloxacin was found to bind preferentially to single-stranded DNA, and binding to denatured DNA was found to be ten to 20 times stronger than that to relaxed or linear DNA. The preferential binding of norfloxacin to single-stranded DNA over double-stranded DNA may suggest that the binding is nonintercalative (SHEN and PERNET 1985).

Further characterisation of the binding of norfloxacin to DNA showed a preference for single-stranded, guanine-containing homopolymers, although no base preference was observed with double-stranded DNA. These results have been interpreted as an indication that quinolones interact with DNA by hydrogen bonding to the bases in a single-stranded region (SHEN et al. 1989a). Different topological forms of DNA also appear to have different binding properties. Binding of drug to relaxed double-stranded DNA is weak and noncooperative whereas binding of quinolones to supercoiled DNA occurs in a highly cooperative manner to a saturable site that may be a denatured bubble or cruciform produced as a direct consequence of negative supercoiling.

These data led SHEN and co-workers to suggest that quinolones are DNA-targeted drugs. This is supported to some degree by data from TORNALETTI and PEDRINI (1988a,b) who demonstrated that nalidixic acid and norfloxacin can cause unwinding of covalently-closed circular DNA as judged by electrophoretic assays. However, significant unwinding was found only at drug concentrations much greater than those required to inhibit gyrase. Data suggesting an absence of quinolone binding to DNA has also been published.

Using the techniques of fluorescence spectroscopy and equilibrium dialysis, no binding of norfloxacin to DNA was detected (PALÙ et al. 1988). However, the significance of these data has been contested (SHEN 1989) and subsequent experiments by the same group (PALÙ et al. 1992) using a variety of binding assays have detected binding of norfloxacin to DNA, but in a Mg^{2+}-dependent manner (see Sect. D.V).

Hence, there is evidence for the binding of quinolones to DNA. However, the evidence that DNA gyrase is the target of the quinolone drugs brings into question the physiological relevance of drug binding to DNA. The next section reviews the role of DNA gyrase in quinolone binding.

II. Effect of DNA Gyrase on Quinolone Binding

A number of independent studies have investigated the role of DNA gyrase in drug binding. It seems clear that quinolones do not show significant binding to DNA gyrase alone but do bind to a gyrase–DNA complex. However, there is some evidence for interaction between ciprofloxacin and the 64-kDa N-terminal fragment of GyrA from differential scanning calorimetry (BLANDAMER et al. 1994). Willmott and Maxwell have investigated the binding of ^{3}H-labelled norfloxacin to DNA gyrase and to a 147-bp DNA fragment containing the major gyrase cleavage site from pBR322 using rapid-gel filtration (WILLMOTT and MAXWELL 1993). Contrary to the binding studies performed by Shen and co-workers, no binding of ^{3}H-labelled norfloxacin was detected to DNA alone. This discrepancy may be the result of the different techniques used, as Shen et al. did not detect any binding of ^{3}H-labelled norfloxacin to linear plasmid DNA when using a gel filtration assay whereas a membrane filtration assay had detected binding (SHEN et al. 1989b). WILLMOTT and MAXWELL (1993) observed significant binding of ^{3}H-labelled norfloxacin to the complex of gyrase and DNA. Similar results have been obtained by YOSHIDA et al. (1993) who reported ^{3}H-labelled enoxacin binding to a complex between gyrase and linear pBR322 DNA, but not to gyrase or DNA alone.

A number of studies on the binding of quinolones to the gyrase–DNA complex have been performed by Shen and co-workers. Significant binding of ^{3}H-labelled norfloxacin to relaxed DNA was detected using a membrane filtration technique. Addition of DNA gyrase to relaxed DNA enhanced the amount of drug bound. Binding only occurred in the presence of a non-hydrolysable ATP analogue, ADPNP (SHEN et al. 1989b). This binding was found to be saturable and similar to the binding previously observed to supercoiled DNA in terms of the cooperativity and amount of binding. Experiments to determine the nucleotide requirement of cleavage of relaxed DNA by DNA gyrase revealed that quinolone-induced gyrase cleavage of relaxed DNA only occurred in the presence of ATP or ADPNP, thus confirming the nucleotide requirement of binding (SHEN et al. 1989b). However, it is known that quinolone-induced cleavage of relaxed DNA can occur in the absence of ATP (e.g. GELLERT et al. 1977; C.J. WILLMOTT and A. Maxwell, unpublished data).

Using a gel filtration assay no detectable binding of ^{3}H-labelled norfloxacin to linear DNA or to DNA gyrase was observed. Moreover, when complexed, gyrase and linear DNA were found to bind drug in an ATP-independent manner although it appeared that ATP caused binding to occur significantly faster (SHEN et al. 1989b). Binding of norfloxacin to this gyrase–DNA complex appears to be cooperative and similar to the binding observed to supercoiled DNA (SHEN et al. 1989a).

The similarities between quinolone binding to a saturable site on supercoiled DNA and the formation of a saturable quinolone binding site after binding of DNA gyrase to either relaxed or linear DNA has led Shen and his colleagues to suggest that these sites are closely related, and a model for the interaction of quinolone drugs with DNA and DNA gyrase was proposed (SHEN et al. 1989c).

III. Cooperative Quinolone–DNA Binding Model

A cooperative quinolone–DNA binding model has been proposed for the interaction of quinolones with gyrase and DNA (SHEN et al. 1989c). DNA gyrase binds to relaxed DNA and cleaves both strands with a 4-bp stagger in the presence of ATP. The drugs are proposed to bind to the exposed single-stranded regions at the active site of gyrase (Fig. 6). Quinolones are thought to bind to DNA via hydrogen bonding between donors from the single-stranded DNA and the carbonyl and carboxyl groups common to all quinolones (Fig. 1).

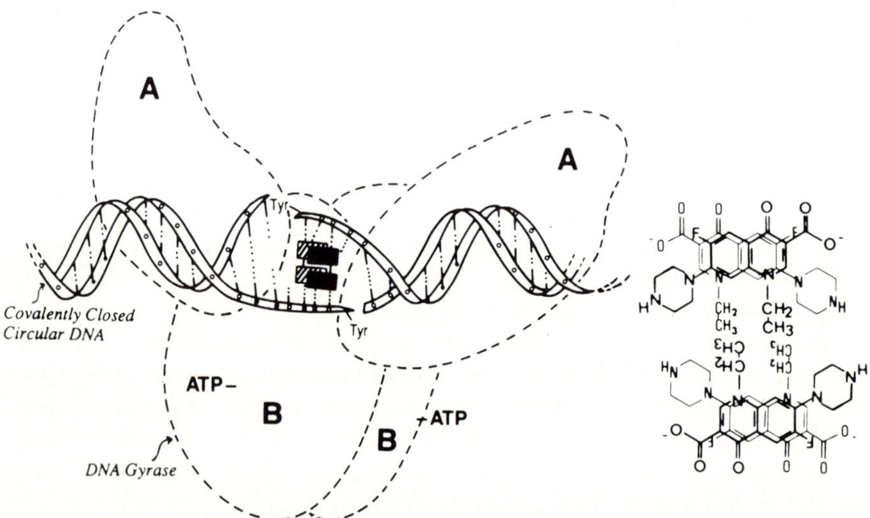

Fig. 6. Cooperative quinolone–DNA binding model proposed by SHEN et al. (1989c). *Filled* and *hatched boxes* represent the quinolone molecules inside a gyrase-induced single-stranded DNA binding site. Details of the self-association of quinolones are shown in the *right panel*. (From SHEN et al. 1989c)

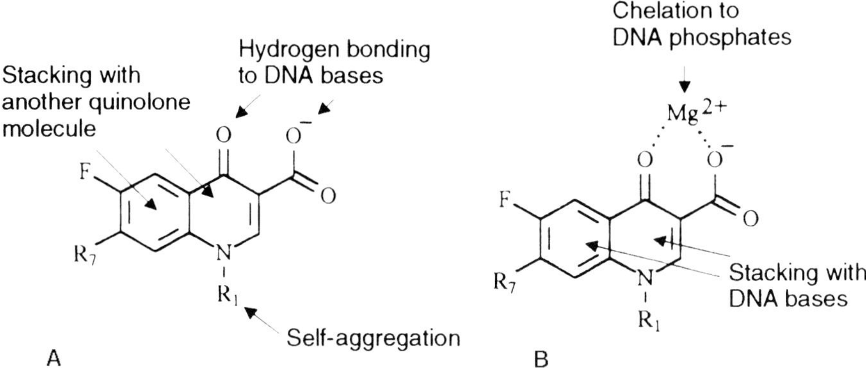

Fig. 7A,B. Models for quinolone–DNA interaction in the Shen model **A**, and in the Palumbo model **B**. In the Shen model the 3-carboxyl and 4-carbonyl groups hydrogen bond to DNA whereas the Palumbo model proposes that Mg^{2+} binds to these groups and forms a bridge between the quinolone drug and the phosphate groups of DNA. The planar ring system is proposed to be involved in stacking with other quinolone nuclei in **A** and with DNA bases in **B**. (From PALUMBO et al. 1993)

Four or more drug molecules are thought to bind DNA in the DNA pocket, forming strong intermolecular drug–drug interactions. Two types of drug interaction have been suggested based on interactions seen in the crystal structure of nalidixic acid: (1) ring stacking of two adjacent quinolone rings hydrogen bonded to the same DNA strand; (2) tail-to-tail hydrophobic interactions between the N-1 substituents of drugs hydrogen bonded to the two opposite DNA strands (Fig. 6). Furthermore, the model suggests that drug–gyrase interactions occur via the group at the C-7 position of the quinolone nucleus (SHEN et al. 1989c). Based on this model, functional domains of the quinolone drug have been assigned (Fig. 7a) and will be discussed in more detail later.

This model clearly provides a means by which quinolones can inhibit turnover by DNA gyrase and can also be applied to explain quinolone-induced cleavage of DNA following the addition of a protein denaturant. However, many of the detailed features of the model have a limited experimental basis and are therefore open to question. The next sections review other features of quinolone binding which have been investigated before a more detailed discussion of all of the available evidence on the mode of binding of quinolones (see Sect. D.IX).

IV. DNA Cleavage is not an Absolute Requirement for Quinolone Binding to a Gyrase–DNA Complex

According to the model of quinolone interaction proposed by SHEN et al. (1989c), drug molecules bind to the single-stranded regions revealed following

DNA cleavage by DNA gyrase. Recent experiments in our laboratory have sought to directly test this model by determining whether mutants which lack the active-site tyrosine, responsible for DNA cleavage, can bind drugs (CRITCHLOW and MAXWELL 1996).

Two mutations, Tyr122 to Ser and Phe, were generated in the 64-kDa N-terminal domain of GyrA; the Ser122 mutation was also made in full-length GyrA. In previous experiments the wild-type 64-kDa domain of GyrA, when complexed with GyrB, was shown to support quinolone-induced cleavage as efficiently as the intact GyrA protein (REECE and MAXWELL 1991b). These mutants were characterised for supercoiling, cleavage and DNA binding ability prior to use in quinolone-binding experiments. As expected both mutations abolished supercoiling and cleavage activities. DNA binding by the wild-type and mutant 64-kDa proteins was measured by gel retardation and filter binding and it was found that the complexes of the 64-kDa proteins with GyrB bound DNA weakly and that DNA binding by both the wild-type and mutant 64-kDa proteins is enhanced by the presence of ciprofloxacin. This result suggests that the Ser122 and the Phe122 proteins are able to interact with the quinolone drug.

In drug-binding experiments using ^{3}H-labelled ciprofloxacin it was found that mutant complexes between the 64-kDa fragments, GyrB and DNA bound less drug than the equivalent wild-type complex. However, in the presence of the 33-kDa GyrA domain, binding to the mutant proteins was increased to at least two-third of that of wild type. Analysis of drug binding to wild-type and the Ser122 mutant gyrase–DNA complexes involving full-length GyrA proteins revealed that both exhibit very similar levels of drug binding. The fact that quinolones can still bind to active-site mutants of DNA gyrase when the enzyme is complexed with DNA provides strong evidence that DNA cleavage is not a prerequisite for drug binding (CRITCHLOW and MAXWELL 1996). These data are in conflict with the Shen model for the interaction of quinolones with gyrase and DNA (SHEN et al. 1989c). The implications of these results will be discussed in Sect. IX.

V. A Role for Magnesium Ions in the Binding of Quinolones to DNA

The effect of magnesium ion concentration on norfloxacin binding to DNA has been investigated by fluorescence spectroscopy, electrophoretic DNA unwinding and affinity chromatography techniques (PALÙ et al. 1992). It was found that in the absence of Mg^{2+} no quinolone binding to plasmid DNA was observed. Similarly, in the presence of an excess of magnesium ions no binding was observed. However, within the range of 0.3 mM–5 mM Mg^{2+}, significant binding of quinolones to DNA was detected (PALÙ et al. 1992). Further experiments by the same group have shown that norfloxacin binding to single-stranded DNA is also Mg^{2+} dependent and that no binding of quinolone drug to linear DNA was observed at any Mg^{2+} concentration (PALUMBO et al. 1993). The latter result confirms data reported by WILLMOTT and MAXWELL (1993) and YOSHIDA et al. (1993). Furthermore, no evidence for cooperativity of quinolone binding to plasmid DNA was found (PALÙ et al. 1992).

These data have been interpreted in terms of the existence of three competing equilibria: Mg^{2+}–DNA, Mg^{2+}–quinolone and quinolone–Mg^{2+}–DNA adducts. The authors have suggested that Mg^{2+} acts as a bridge between the phosphate groups of the nucleic acid and the carbonyl and carboxyl moieties of norfloxacin (PALÙ et al. 1992). Hence, in the absence of Mg^{2+}, quinolones are unable to interact with DNA, and in the presence of excess Mg^{2+}, both the quinolone and the DNA are saturated with Mg^{2+} such that the Mg^{2+} is unable to act as a bridge between the drug and DNA. A model for quinolone binding to DNA has been proposed (Fig. 7b) whereby binding of drug to DNA occurs via a Mg^{2+} bridge rather than by direct hydrogen bonds. The drugs are thought to be stabilised through stacking interactions between a DNA base in a single-stranded region and the quinolone planar ring system (Fig. 8; PALUMBO et al. 1993). Unlike the Shen model, no cooperativity of drug binding is required for complex formation. The model suggests that quinolones may interact with DNA gyrase via the C-7 and N-1 substituents of the quinolone nucleus. Experiments which include gyrase would clarify this.

Studies determining the location of Mg^{2+} when bound to a quinolone lend support to the Palumbo model. ^{19}F-nmr experiments have located Mg^{2+} between the ketone and the carboxylate groups of the quinolone drug (LECOMTE et al. 1993). Studies using infrared spectroscopy and nmr have determined a 1:1 stoichiometry for Mg^{2+} and quinolone in agreement with the proposed model (LECOMTE et al. 1994). The Mg^{2+} dependence of the formation of a

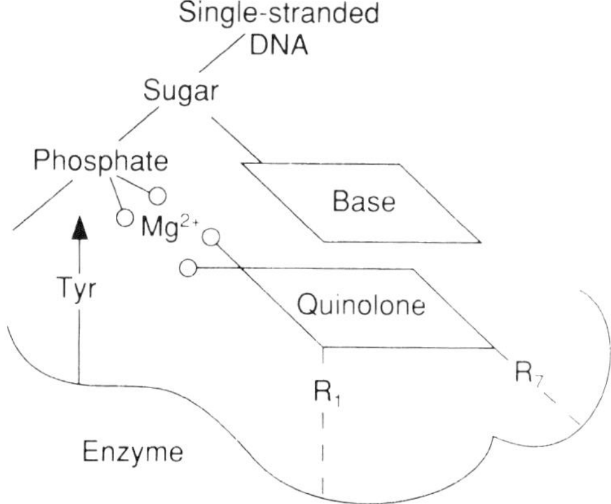

Fig. 8. Schematic representation of the Palumbo model. Quinolone drugs are proposed to bind to DNA via a Mg^{2+} bridge between the C-3 and C-4 groups and the phosphate groups of DNA. Interactions between gyrase and the drug are proposed to occur via the C-7 and N-1 substituents of the quinolone. (From PALUMBO et al. 1993)

gyrase–quinolone–DNA complex has been investigated (BAZILE-PHAM KHAC and MOREAU 1994). Using phase partitioning it was found that the binding of several fluoroquinolones to DNA was dependent on the ability of the drugs to bind Mg^{2+}. Using affinity chromatography with immobilised norfloxacin, it was found that gyrase was unable to bind fluoroquinolone in the absence of DNA, but that a DNA–quinolone–gyrase complex was formed in the presence of Mg^{2+}. The interpretation of these results can be questioned because norfloxacin is immobilised on epoxy-activated resin via the NH of the piperazine groups, which has been suggested to be important in the gyrase–quinolone interaction in the models which have been proposed.

To sum up, the two models discussed (SHEN et al. 1989c; PALUMBO et al. 1993) both propose that quinolones bind to a single-stranded DNA region produced after cleavage by DNA gyrase. However the details of the interactions of the drugs with DNA differ significantly between the two models and to date no experimental evidence can confirm either of these models. A further perspective on the possible mode of binding of quinolones has come from studies on a group of structurally related drugs, the quinobenzoxazines.

VI. Binding of Quinobenzoxazines to DNA

The quinobenzoxazines are a group of drugs that are structural analogues of norfloxacin (Fig. 1). These drugs are thought to be mammalian topo II inhibitors but do not exhibit antibacterial activity, although they are structurally related to quinolones. Clearly the structural differences must account for lack of interaction of quinobenzoxazines with DNA gyrase. An important difference between the action of the quinobenzoxazines and the quinolones is that the quinobenzoxazines do not induce enzyme-mediated DNA cleavage (PERMANA et al. 1994). This is in contrast to quinolones such as CP-115,953 (Fig. 1) which inhibits eukaryotic topo II and can also induce DNA cleavage by topo II (ROBINSON et al. 1991).

Like quinolones, quinobenzoxazines bind to DNA in a Mg^{2+}-dependent manner. A model for the interaction of the quinobenzoxazines with DNA has been proposed whereby a 4:4 quinobenzoxazine–Mg^{2+} complex is formed on DNA (FAN et al. 1995). It is proposed that a 2:2 drug–Mg^{2+} heterodimer complex is formed in which one of the drug molecules is intercalated into the DNA double helix and the second drug molecule is externally bound, held to the first by two Mg^{2+} bridges. The Mg^{2+} bridges are further stabilised by interacting with phosphate groups on the DNA (Fig. 9). Due to the cooperativity of binding observed, a 4:4 drug–Mg^{2+} complex is proposed in which the two externally bound molecules from different heterodimers interact via π–π interactions. This idea has been extrapolated to the mode of binding of quinolone drugs to DNA gyrase and DNA. FAN et al. (1995) have found that externally bound quinobenzoxazine can be replaced by norfloxacin to form mixed-structure dimers on DNA, which are proposed to bind to a supposed unwound region of DNA formed after binding of ATP to a gyrase–

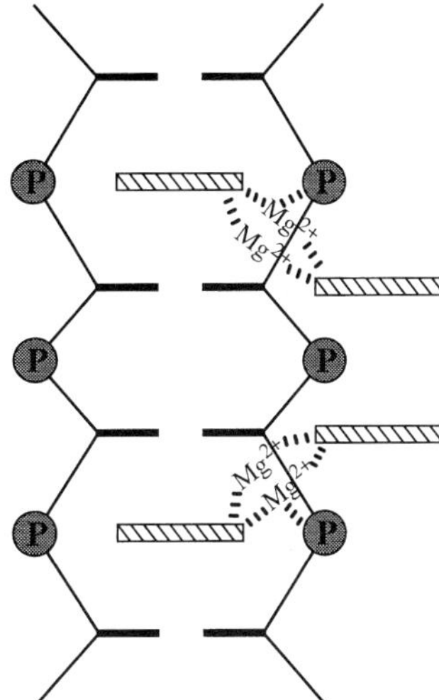

Fig. 9. Proposed model for the interaction quinobenzoxazines with DNA in the presence of Mg^{2+}. A 2:2 heterodimer is thought to form on DNA with one drug molecule intercalated into the double-helix and the second drug molecule being externally bound. A 4:4 drug:Mg^{2+} complex is proposed where the externally bound molecules from two different 2:2 dimers interact via π–π interactions (Redrawn from a figure provided by L.H. HURLEY)

DNA complex. This unwound region has been inferred from data obtained from hydroxyl radical footprinting of gyrase–DNA complexes in the presence of ADPNP and ciprofloxacin (see Sect. C; ORPHANIDES and MAXWELL 1994).

It is tempting to speculate that this type of binding is relevant to the mode of binding of quinolones with gyrase and DNA in the presence of magnesium ions. One of the key differences between this and the other models is that this model proposes that quinolones bind to unwound duplex DNA and not to single-stranded DNA. However, it has been claimed that quinolones do not intercalate into DNA (SHEN and PERNET 1985; TORNALETTI and PEDRINI 1988b). Further experiments are required to determine the relevance of this model to gyrase and quinolone drugs.

VII. Quinolone Binding to Quinolone-Resistant Mutants of DNA Gyrase

There is now considerable evidence that the quinolones bind to DNA and that their intracellular target is the gyrase–DNA complex. The issue of the interaction of the drugs with gyrase needs to be considered. A number of point mutations in the gyrase genes have been identified which give rise to amino acid substitutions conferring high levels of quinolone resistance, both in strains carrying the mutation and on isolated gyrase proteins (see Chap. 7). This suggests that these amino acid residues may be involved in quinolone binding. Recent studies have addressed the ability of quinolone-resistant mutants of DNA gyrase to bind quinolone drugs (Willmott and Maxwell 1993; Yoshida et al. 1993).

The most frequently isolated quinolone-resistant mutant of *E. coli* is a point mutation in the gyrase A protein which gives rise to an amino acid substitution from Ser83 to Trp. In a number of other organisms the equivalent amino acid is often found to be mutated in quinolone-resistant gyrases (see Chap. 7). This indicates that this residue may play a key role in quinolone binding. Willmott and Maxwell (1993) have investigated the ability of gyrase carrying a mutation of Ser83 to Trp to bind a quinolone drug in the presence of GyrB and DNA. The mutant protein–DNA complex showed greatly reduced norfloxacin binding. This result has led to the suggestion that Ser83 interacts directly with the quinolone drug (Maxwell 1992).

In other experiments Yoshida et al. (1993) have carried out a series of quinolone binding studies on DNA gyrases reconstituted from wild-type GyrA and GyrB and from GyrA and GyrB quinolone-resistant mutants. The results of these studies led to the proposal of a quinolone pocket model as the mechanism of action of the quinolones (Yoshida et al. 1993). In these experiments the GyrA mutant with the Ser83 to Leu substitution and the GyrB mutant with the Lys447 to Glu substitution bind approximately the same amount of enoxacin as the wild-type gyrase. In contrast, the GyrB mutant with the Asp426 to Asn substitution binds considerably less quinolone. These results have led the authors to suggest that the amount of quinolone bound to gyrase–DNA complexes does not necessarily correlate with the quinolone sensitivities of the gyrases. Furthermore, the loss of binding to the Asn426 mutant has led to the suggestion that the piperazinyl group of enoxacin is likely to be involved in an ionic interaction with the carboxyl group of Asp426. The result obtained for the Ser83 to Leu mutation is potentially in conflict with the result for the similar mutation Ser83 to Trp (Willmott and Maxwell 1993).

Yoshida et al. (1993) proposed a model in which the quinolones bind in a pocket of the gyrase–DNA complex that appears during the cleavage–reunion reaction. They propose that the quinolone-binding affinities are determined by both the GyrA and GyrB proteins in concert. The available data on binding of quinolones to quinolone-resistant mutants of DNA gyrase indicate that certain amino acids of both GyrA and GyrB may interact directly with quinolones. It is also possible that the mutations alter the conformation of a putative quinolone-binding pocket.

VIII. Mode of Binding of Topoisomerase II-Targeting Drugs

At this point it is useful to draw a comparison between the mode of binding of quinolone drugs to gyrase and DNA and the mode of binding of various

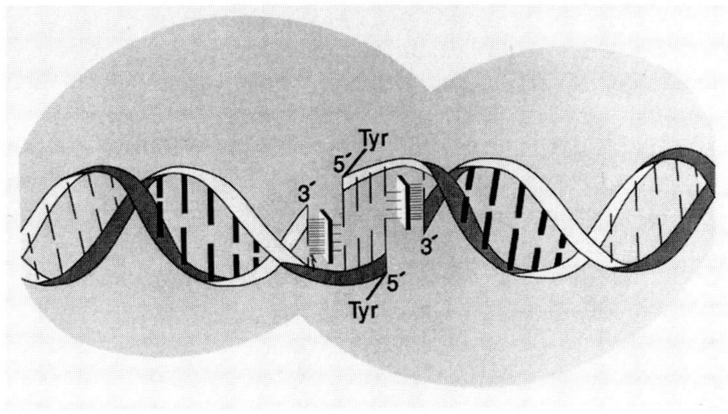

A

(a)

DNA gyrase-DNA complex

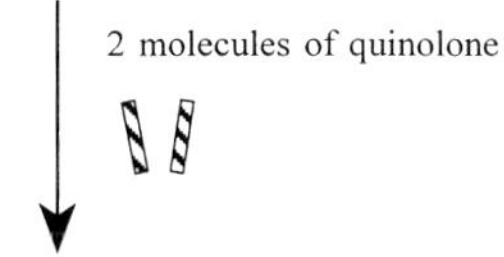

(b)

Gyrase-quinolone-DNA complex
in the absence of DNA cleavage

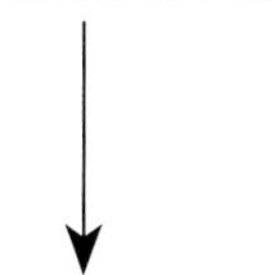

(c)

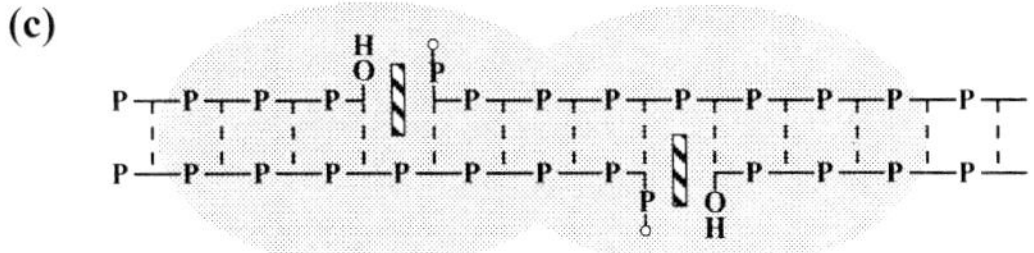

Gyrase-quinolone-DNA complex
in the presence of DNA cleavage

B

Fig. 10. A Model for the interaction of topoisomerase II inhibitors with DNA. The inhibitor (*rectangular box*) binds in the internucleotide space at the cleaved phosphodiester bonds. The *shaded area* represents the topoisomerase II dimer. Interaction take place between the inhibitor and the bases 5′ and 3′ to cleaved bond (from FREUDENREICH and KREUZER 1993). **B** A model for the interaction of quinolones with the gyrase–DNA complex. Quinolones (*hatched rectangles*) bind in the internucleotide spaces either before DNA cleavage in (*b*) or after cleavage and the formation of a phosphotyrosine linkage in (*c*). The *shaded area* represents the GyrA dimer and the active-site tyrosines (Tyr122) are represented by *white circles*

antitumour agents to DNA topo II and DNA. The eukaryotic and bacteriophage homologues of DNA gyrase are also inhibited by drugs which can reveal topoisomerase-mediated cleavage (Liu 1989; Corbett and Osheroff 1994). Mammalian type II topoisomerase is the target of several classes of antitumour agents that include anthracyclines, aminoacridines, ellipticines and epipodophyllotoxins. Evidence for binding of this type of drug to DNA at the cleavage site has been obtained from experiments which examined the DNA sequence preferences for cleavage by topo II in the presence of various inhibitors. The ability of different inhibitors to induce cleavage by eukaryotic topo II is dependent upon the identity of the base pairs at the cleavage site (Capranico et al. 1990, 1993; Pommier et al. 1991a,b). This has led to the proposal of a model in which a drug molecule is stacked between the bases immediately adjacent to the cleaved phosphodiester bond (Capranico et al. 1990).

Similar results have been obtained with bacteriophage T4 type II topoisomerase. In addition to specificity determined by the enzyme, drug-dependent specificity was observed with various inhibitors. Specifically, the bases at the 5′ side, and to a lesser extent at the 3′ side, of each cleaved phosphodiester bond were found to be critical in determining whether a given inhibitor would induce cleavage (Freudenreich and Kreuzer 1993). Hence, it appears that these drugs interact directly with DNA bases at the cleavage site (Fig. 10A). Freudenreich and Kreuzer (1994) have gone on to localize *m*AMSA (an aminoacridine) in a T4 topoisomerase–DNA complex. A photoactivatible analogue of *m*AMSA (3-azido-AMSA) has been used to precisely locate the binding site of this drug. It should be noted that this analogue can induce T4 topoisomerase cleavage of DNA as efficiently as the unmodified compound. Upon photoactivation, this drug becomes covalently attached to the DNA substrates but only in the presence of topoisomerase.

Primer extension analysis and piperidine cleavage were used to locate the site of attachment of the inhibitor. It was found that the 3-azido-AMSA became covalently attached to the two base pairs adjacent to both sites of phosphodiester bond cleavage by T4 topoisomerase II (Freudenreich and Kreuzer 1994). This represents the first direct evidence for the binding of these drugs precisely at the two sites of DNA cleavage in the presence of topo II. Freudenreich and Kreuzer (1994) have proposed that these kinds of inhibitors possibly intercalate in the internucleotide space next to the cleaved phophodiester bond and that this may represent a common mechanism for inhibition of topoisomerases. Given the similarities between the action of *m*AMSA and quinolones it is tempting to speculate that the interaction of quinolones with DNA may be analogous. Indeed, Freudenreich and Kreuzer (1994) used a mutant of T4 topoisomerase which is hypersensitive as regards DNA cleavage induced by oxolinic acid and found that the drug exerted base preferences on cleavage. This suggests that oxolinic acid acts in an analagous way to *m*AMSA.

IX. Problems with Current Models

A number of models for the interaction of quinolones with gyrase and DNA have now been proposed. So far, a relatively small amount of data concerning the role of DNA gyrase in drug binding has been obtained. This is reflected in the relative lack of information concerning which amino acids of DNA gyrase are directly or indirectly involved in quinolone binding and what substituents of the quinolone nucleus interact with DNA gyrase. Functional domains of the quinolone nucleus have been putatively assigned (Fig. 7). The C3 and C4 substituents are thought to be involved in drug–DNA interactions either by direct hydrogen bonding (SHEN et al. 1989c) or through a Mg^{2+} bridge between these groups and the phosphate backbone of DNA. The binding of Mg^{2+} between the carbonyl and ketone group of the quinolone nucleus tends to favour the latter hypothesis. Another possibility is that these substituents could participate directly in hydrogen bonding with hydrogen bond acceptors on the enzyme such as Ser83 of the GyrA protein (REECE and MAXWELL 1991a; MAXWELL 1992).

A potential problem with the Shen and Palumbo models is that they suggest that the C7 position of the quinolone nucleus interacts with DNA gyrase. From the structures of quinolones (Fig. 1) it can be seen that this position is highly variable such that this substituent seems an unlikely candidate for drug–gyrase interaction (MAXWELL 1992). However, evidence that this substituent is involved in gyrase interaction has been found by YOSHIDA et al. (1993) who have suggested that amino acid residues from GyrB may interact with the C7 piperazine group of the fluoroquinolones. Furthermore, studies on a quinolone which targets topo II, CP-115,953, fuel the argument that the C7 position interacts with DNA gyrase. This quinolone has a significant activity against eukaryotic topoisomerase II (ROBINSON et al. 1991). Figure 1 shows that the major difference between CP-115,953 and ciprofloxacin are around and include the C7 substituent. This indicates that this part of the quinolone nucleus is important in drug–enzyme interaction.

One of the most important features of the Shen and Palumbo models (SHEN et al. 1989c; PALUMBO et al. 1993) is that drug binding occurs to the single-stranded DNA sites revealed following DNA cleavage by gyrase. Recent drug binding experiments performed with active-site mutants of DNA gyrase (Tyr122 to Phe or Ser of the GyrA protein) have shown that proteins bearing these mutations are still able to bind drug, indicating that DNA cleavage is not required for drug binding (CRITCHLOW and MAXWELL 1996). Therefore, it is unlikely that quinolones bind in the manner suggested by these models as it is difficult to rationalise how quinolones bind in the absence of DNA cleavage. However, given that quinolones stabilise the formation of a cleavable complex between DNA gyrase and DNA, it is likely that quinolones are bound to the protein–DNA complex after DNA cleavage has occurred. Although it appears that DNA cleavage is not required for quinolone interaction with DNA gyrase and DNA, it is possible that a precleavage complex between gyrase, quinolone and DNA exists which is converted to a, perhaps more stable, complex following cleavage. Perhaps a way in which this could be achieved is if quinolones interact with a gyrase–DNA complex in a manner similar to that suggested for the interaction with topo II and DNA (Freudenreich and Kreuzer 1993; Fig. 10). The planar ring system of the quinolone drug could intercalate into the internucleotide space next to the cleaved phosphodiester bond and interact with amino acid residues of gyrase (Fig. 10B). This interaction would be very similar in the presence or absence of DNA cleavage, but the presence of the quinolones may push the cleavage equilibrium in the direction of cleavage.

In relation to the above, recent work using modelling techniques supports an intercalative mode of binding of quinolones to DNA in the gyrase–DNA complex (Llorente et al. 1996). In this work a set of 78 quinolone derivatives were used in a structure–activity study to identify structural features correlating with antibacterial activity. A three-dimensional model of a quinolone complexed to gyrase and DNA was built. In this model the drug intercalates into DNA and interacts with the enzyme via groups at positions 6 and 7 of the quinolone nucleus (Fig. 1). Mg^{2+} is bound at the 3-carboxy and 4-keto groups and also interacts with a purine base and phosphate of DNA.

X. Conclusions

The conclusion which can be drawn from the studies to date is that quinolones bind to a gyrase–DNA complex. The precise roles of gyrase and DNA in this interaction remain unclear. Furthermore, there is strong evidence for a role of Mg^{2+} ions which possibly results in the formation of a quaternary complex between quinolones, gyrase, DNA and Mg^{2+}. High-resolution structural data on such a complex or of its components would clearly provide an understanding of the molecular basis of these interactions.

E. Mechanism of Cell Killing

I. Paradoxical Effects of Quinolones

From the preceding sections it is clear that a primary intracellular target of the quinolone group of antibacterials is DNA gyrase. The fact that the drugs completely inhibit gyrase-catalysed supercoiling in vitro might suggest that this is how they act in vivo. However, observations made early on in the history of these drugs suggested that they were likely to act in some other way. In 1969, Hane and Wood (1969) showed that in partial diploids the *nalA*^{s} allele is dominant over *nalA*^{r} (conferring resistance to nalidixic acid; Hane and Wood 1969). The *nalA* gene was later shown to be the gene encoding the A subunit of DNA gyrase (*gyrA*; Gellert et al. 1977; Sugino et al. 1977;

KREUZER and COZZARELLI 1979). Mutations conferring resistance to nalidixic acid which map to *gyrB* have also been shown to be recessive to wild type (YAMAGISHI et al. 1986). Indeed transformation of a quinolone-resistant bacterial cell with a plasmid bearing the wild-type *gyrA* or *gyrB* genes and determining the quinolone susceptibility of the transformants is a method used to identify whether a quinolone-resistance mutation maps to *gyrA* or *gyrB* (NAKAMURA et al. 1989). If quinolones simply inhibited DNA supercoiling in vivo, this result would not be expected.

In other work it has been shown that the amount of quinolone drug required to inhibit supercoiling in vivo exceeds that required to inhibit bacterial growth. For example, GELLERT et al. (1977) showed that the concentration of oxolinic acid required to appreciably affect in vivo supercoiling was 50 times that required to block cell growth. Similarly, DNA supercoiling in vitro has also been found to be less sensitive to quinolones than bacterial growth. DOMAGALA et al. (1986) compared the concentrations of drug required to inhibit supercoiling in vitro by *E. coli* DNA gyrase with MICs (minimum inhibitory concentrations for cell growth) for a range of quinolones and found the latter to be lower than the former by factors ranging from ~40 to 300. WALTON and ELWELL (1988) showed that a 25- to 100-fold greater concentration of fluoroquinolones was required to inhibit DNA gyrase than was required to inhibit bacterial growth. In similar experiments carried out with the *Micrococcus luteus* enzyme and a range of drugs, MICs were found to be lower than the concentrations required to inhibit in vitro supercoiling by factors of ~2–100 (ZWEERINK and EDISON 1986). The wide variation in these numbers may reflect the differences in cell penetration of different quinolone drugs. In general, it appears that the supercoiling reaction of gyrase is less sensitive to the drugs than bacterial growth by one to two orders of magnitude. This is the opposite of what would be expected as regards in vitro supercoiling, in which case the isolated target (DNA gyrase) ought to be more sensitive.

There are a number of possible explanations which can be advanced to explain this latter phenomenon. One is that bacterial cells actively import quinolones. However, studies suggest that drug import is by passive diffusion (BEDARD et al. 1987; HOOPER et al. 1989) and drug efflux has been widely reported (PIDDOCK 1991). A study of quinolone accumulation indicated drug concentrations approximately four to five times higher inside the cell than outside the cell (HOOPER et al. 1989). Such levels would in general be insufficient to account for the above effect. Another possibility is that other reactions of gyrase, such as DNA relaxation and decatenation, are more sensitive to quinolones than supercoiling. However, in vitro decatenation and relaxation reactions of gyrase have been shown to be at least as sensitive to quinolones as DNA supercoiling (KREUZER and COZZARELLI 1980; HALLETT and MAXWELL 1991).

It is possible that these effects are due to quinolones inhibiting the reactions of topo IV in vivo. However, comparisons of inhibition of *E. coli* topo IV by quinolones, with DNA gyrase inhibition, has shown that approximately eight to 50 times more quinolone drug is required to inhibit topo-IV-mediated decatenation than is required to inhibit DNA-gyrase-catalysed DNA supercoiling in vitro (HOSHINO et al. 1994). Furthermore, KHORDURSKY et al. (1995) have shown that a number of different quinolones are approximately two-fold less effective at inhibiting the relaxation and decatenation reactions of topo IV than they are at inhibiting the supercoiling reaction of DNA gyrase in vitro. Recently, the effect of quinolones on topo IV activity was investigated in vivo (KHODURSKY et al. 1995). In a *gyrA'* (quinolone-resistant) strain, norfloxacin

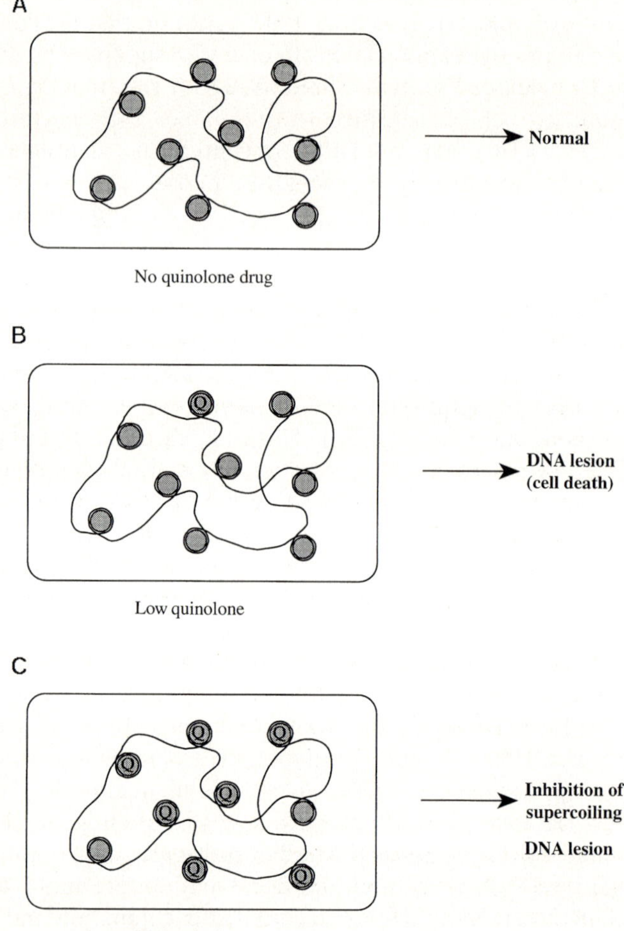

Fig. 11A–C. The poison hypothesis. A schematic representation of gyrase bound to DNA in the bacterial cell. In the absence of quinolone drug **A**, the bacteria have a normal phenotype. However, on addition of a low concentration of quinolone drug **B**, a DNA lesion is formed which initiates the events ultimately leading to cell death. Only at high quinolone concentrations **C** can inhibition of supercoiling by DNA gyrase be observed and at this concentration the lethal DNA lesion can still form

caused the accumulation of replication catenanes (consistent with topo IV inhibition) and cell death without a significant effect on DNA replication. However, in this case inhibition of topo IV only becomes apparent when gyrase is already quinolone resistant.

II. Poison Hypothesis

In summary, the two unexpected properties of the in vivo effects of quinolones are that a quinolone-sensitive allele of a DNA gyrase gene is dominant over its quinolone-resistant counterpart, and that the drugs inhibit bacterial growth at concentrations well below those at which an effect on supercoiling would be seen (the "MIC paradox"). Thus an explanation for the intracellular effect of quinolones on gyrase, other than simply inhibiting the enzymic reactions, is required.

Kreuzer and Cozzarelli (1979) reported a study on the effects of temperature-sensitive (ts) and nalidixic-acid-resistant mutations of the *gyrA* gene on the replication of bacteriophages øX174, T4 and T7. While øX174 and T4 growth was inhibited at the nonpermissive temperature in a *gyrA*(ts) strain, T7 growth was largely unaffected. This suggested a requirement for host gyrase in øX174 and T4 but not in T7. However, the growth of T7 is nalidixic acid sensitive and the drug markedly inhibits DNA synthesis (Baird et al. 1972). These observations prompted the suggestion that nalidixic acid converts DNA gyrase into a poison (the "poison hypothesis" see Fig. 11). The poison was thought to be a complex between gyrase and DNA, analogous to that induced by quinolones in vitro (Gellert et al. 1977; Sugino et al. 1977) and in vivo (Snyder and Drlica 1979). Indeed, Snyder and Drlica (1979) showed that inhibition of DNA synthesis correlated with the formation of oxolinic acid–gyrase complexes, leading to the idea that such complexes blocked replication fork movement. Sensitive alleles are dominant because the resistant gene product will not prevent the drug from binding the sensitive protein and forming a poison. The amount of drug–enzyme complex required to kill the cell could represent only a small fraction of the total gyrase in the cell and thus have very little impact on the overall supercoiling activity.

III. Polymerase Blocking

The question remains as to how this ternary complex (gyrase–drug–DNA) leads to cell death. There is much evidence that quinolones inhibit DNA replication and transcription. Early studies showed that nalidixic acid and oxolinic acid were inhibitors of DNA replication in vitro and in vivo (Goss et al. 1965; Bourguignon et al. 1973; Staudenbauer 1976; Gellert et al. 1977; Snyder and Drlica 1979). Moreover, studies of the antibacterial activity of nalidixic acid showed that it is bactericidal at low concentrations and bacteriostatic at higher concentrations (Crumplin and Smith 1975). These observations correlate with inhibition of DNA synthesis at low concentrations and inhibition of RNA and protein synthesis at high concentrations. Later studies with norfloxacin showed that this drug also exhibited a biphasic dose–response curve (CRUMPLIN et al. 1984). Work by ENGLE et al. (1982) showed that bacterial DNA synthesis is inhibited by both coumermycin A_1 and oxolinic acid and suggested that coumermycin inhibits DNA supercoiling but that oxolinic acid converts DNA gyrase into a barrier preventing replication fork movement. Thus, this work suggests that the "poison" may be a ternary complex which blocks the passage of polymerases.

In vitro studies with other topoisomerases have suggested that such a mechanism of killing may indeed operate. HSIANG et al. (1989) using an SV40 cell-free replication system found that addition of the topoisomerase-I-specific drug camptothecin inhibited replication only in the presence of topoisomerase I (topo I). Moreover, addition of excess purified calf-thymus topo I and camptothecin caused severe inhibition and the accumulation of linearised replication products. They proposed that collision between replication forks and topo I–camptothecin–DNA complexes caused fork arrest and possibly fork breakage. BENDIXEN et al. (1990) found that bacteriophage SP6 transcription in vitro could be inhibited by human topo I in the presence of camptothecin and attributed this effect to the physical blockage of RNA polymerase by the enzyme–drug complex on DNA. In the case of topo II, THOMSEN et al. (1990) found that the calf-thymus enzyme in the presence of an antitumour drug causes inhibition of transcription by SP6 RNA polymerase.

The idea of a drug-stabilised enzyme complex on DNA blocking polymerases has been tested in vitro with DNA gyrase and quinolone drugs (WILLMOTT et al. 1994). In these experiments the effects of gyrase and drugs on transcription by *E. coli* and T7 RNA polymerases were assessed. It was found that whereas neither the enzyme nor the drugs alone had an effect, the gyrase–quinolone complex on DNA could block transcription. This effect was found with a range of quinolone drugs, including nalidixic acid, oxolinic acid, norfloxacin and ciprofloxacin. Importantly, gyrase with a point mutation in the A subunit conferring quinolone resistance was found to cause blocking only at much higher drug concentrations. Other agents which block supercoiling by gyrase (novobiocin and ADPNP) had no effect on transcription (WILLMOTT et al. 1994). In more recent experiments (CRITCHLOW and MAXWELL 1996), the effect of a mutation of the active-site tyrosine (Tyr122 to Ser) in GyrA on transcription blocking was assessed. It was found that this mutant gyrase, although able to bind quinolones, is unable to block transcription. This result implies that the ability to cleave DNA (i.e. to form a cleavable complex) is required to block the passage of the polymerase.

This work also allowed a determination of the extent of "protection" of the DNA by gyrase from RNA polymerases (WILLMOTT et al. 1994). These experiments were carried out with transcription proceeding from either end of the DNA fragment such that the region made unavailable to the polymerase by gyrase could be determined. In the case of T7 RNA polymerase this was ~20 bp surrounding the cleavage site, and in the case of the *E. coli* enzyme,

~40 bp (WILLMOTT et al. 1994). These values are similar to those found with other topoisomerases in this type of experiment (BENDIXEN et al. 1990; THOMSEN et al. 1990) and with the highly protected region in DNase I and hydroxyl radical footprinting experiments with gyrase (FISHER et al. 1981; KIRKEGAARD and WANG 1981; MORRISON and COZZARELLI 1981; RAU et al. 1987; ORPHANIDES and MAXWELL 1994). In addition, the issue of whether the polymerase displaces a bound gyrase (without drug) was addressed. It appears that in these experiments the polymerase is likely to read through a bound gyrase without displacement (WILLMOTT et al. 1994), a phenomenon also found with nucleosomes (VAN HOLDE et al. 1992; LEWIN 1994).

Taken together, the above findings clearly supports the idea that a gyrase–quinolone complex on DNA can block the passage of RNA polymerase (Fig. 12), but the important issue is whether this is the lethal event in vivo. Given the observations discussed above that implicate inhibition of DNA synthesis as the primary response of quinolones, it seems that inhibition of DNA polymerase is more likely to be the lethal event. Unpublished data (reported in WILLMOTT et al. 1994) suggest that the gyrase–quinolone complex can indeed block the passage of DNA polymerases.

The possibility that quinolone binding to topo IV could also inhibit DNA synthesis and cause cell death should also be considered. However, the interaction of quinolones with topo IV leads only to a slow stop in DNA replication, with the effect being bacteriostatic in *E. coli* (KHODURSKY et al. 1995). This has led to the proposal that topo IV acts behind the replication fork (decatenating daughter chromosomes), forming a poison which is slow acting and may be repaired. For these reasons it is likely that topo IV is secondary to DNA gyrase as a quinolone target in vivo in *E. coli*. It is possible that in organisms other than *E. coli* the situation is different, but such experiments have not yet been reported.

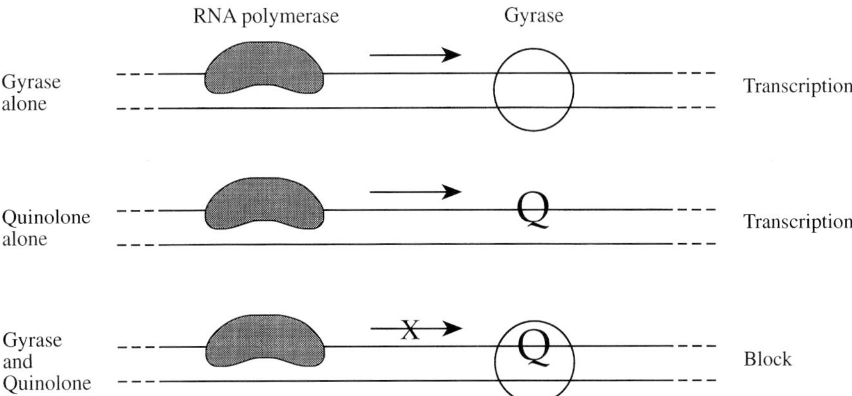

Fig. 12. Blocking of RNA polymerase by the gyrase–quinolone complex on DNA

Recently it was shown that a topoisomerase–DNA complex can be disrupted by a DNA helicase (Howard et al. 1994). In these experiments the effect of *E. coli* helicase II on a complex between T4 topoisomerase and *m*AMSA on DNA was investigated. It was found that the helicase could induce strand breakage in vitro by the topoisomerase–drug complex, i.e. the helicase can generate a potentially cytotoxic response to the inhibitor (Howard et al. 1994). Whether such a mechanism can operate in vivo remains to be determined.

Given the observations made with other topoisomerases on the effects of other "cleavable complex forming" drugs on replication and polymerases, it seems that the blocking of the replication complex is very likely to be a generally applicable mechanism by which such drugs initiate events leading to cell death. Not so clear is the precise mode of interaction of the drugs with the enzymes and DNA and the events subsequent to polymerase blocking which ultimately lead to cell death.

F. Conclusions and Future Prospects

We now understand some aspects of the mode of action of quinolone drugs on DNA gyrase and how this may lead to cell death. Upon entering a bacterial cell, it seems likely that the drugs bind to the gyrase–DNA complex and that Mg^{2+} plays an important role in this complex. There is evidence that the Mg^{2+} ion is bound at the 3-carboxy and 4-carbonyl positions of the drugs (Figs. 1 and 7b), but this is still a contentious issue. The precise interactions of the drug with the enzyme and the DNA remain to be determined. This gyrase–drug–DNA complex would appear to arrest the normal topoisomerisation cycle at the DNA breakage step and this trapped intermediate forms a lesion on DNA that can act as a barrier to polymerases. Given the well-documented effects of the drugs on DNA replication, it is likely that the blockage of the DNA polymerase III replication complex is the most important event. Subsequent to this, a range of responses are likely to occur including DNA damage and the induction of SOS (see Sect. B) which ultimately lead to cell death.

It is now possible to identify key areas where information is lacking and to suggest approaches to tackle these problems. It is an important priority to determine the structure of the gyrase–quinolone–DNA complex at high resolution. The method of choice here is X-ray crystallography, but given the size of the complex involved (molecular weight, approximately 500kDa), this is a daunting prospect. However, some success has already been achieved with individual domains of DNA gyrase. The 43-kDa N-terminal domain of GyrB (Fig. 2) has been crystallised with ADPNP and its structure solved to high resolution (Wigley et al. 1991). In addition, the structure of a 24-kDa N-terminal GyrB fragment complexed with novobiocin has been solved by X-ray crystallography (Lewis et al. 1996); this structure has given valuable insight into the mode of action of the coumarin drugs on gyrase. A 64-kDa N-terminal fragment of GyrA (Fig. 2) has been crystallised (REECE et al. 1990), but its structure has yet to be solved. However, recent work in this laboratory has focused on a 59-kDa N-terminal GyrA fragment (Fig. 2) which yields crystals which diffract beyond 2.5 Å; structure determination is currently underway. The structure of this fragment would give us the relative locations of the residues implicated in drug binding and allow the construction of a protein–drug complex by molecular modelling. One possibilty for determining the structure of the ternary complex would be to use the 59-kDa GyrA fragment complexed to the 47-kDa GyrB C-terminal fragment and DNA (Fig. 2). This would be a smaller complex (molecular weight, ~250 kDa) and should, in principle, contain all the necessary elements for quinolone interaction.

Recently the structure of a 92-kDa fragment of yeast topoisomerase II was solved by x-ray crystallography (BERGER et al. 1996). This fragment corresponds by sequence similarity to the 47-kDa C-terminal domain of GyrB fused to the 59-kDa N-terminal domain of GyrA (Fig. 2). As a consequence of the sequence similarity between gyrase and topo II it is possible to locate the sites of quinolone-resistance mutations in the crystal structure. The mutations in GyrA (which lie in the quinolone-resistance determining region of GyrA, amino acids 67–106) appear to lie close to the active-site tyrosine, such that if the quinolone drug binds at this site it would be also able to interact with the bound DNA. By contrast, the GyrB quinolone-resistance mutations lie at a different site, which may suggest that the effects of these muations are indirect, i.e. they are not involved in drug–protein contacts. However, these issues can only be properly addressed when the structural data on the corresponding regions of gyrase are available.

In lieu of high resolution structural data, other methods could be employed to probe the structure of the ternary complex. One example would be cross-linking of the drug to either the protein or the DNA which would provide useful information concerning drug–DNA and drug–protein contacts. This has already proved to be a successful approach with T4 topo II and *m*AMSA (FREUDENREICH and KREUZER 1994). Indeed, cross-linking of gyrase and DNA would also provide useful information about DNA–protein contacts which would help in solving the structure of the ternary complex.

More information is also required on the role of Mg^{2+} in the complex. This issue is complicated by the fact that the formation of the gyrase–DNA complex requires Mg^{2+} as does the binding of ATP to the enzyme. In addition, it is possible that Mg^{2+} plays a role in the chemistry of the DNA breakage–reunion reaction. Aside from high-resolution structural data, the clarification of the various roles played by Mg^{2+} in this process is required. For example, if the gyrase–DNA complex could be stabilised in some other way (i.e. without Mg^{2+}) then the Mg^{2+} requirement in the drug-binding process could be probed.

Another issue which needs to be resolved is the effects of quinolone-resistance mutations in gyrase on drug binding. Thus far YOSHIDA et al. (1993) have shown that the GyrA mutation Ser83 to Leu and the GyrB mutation Lys447 to Glu appear not to affect binding, whereas the GyrB mutation

Asp426 to Asn does. Willmott and Maxwell (1993) found that the GyrA mutation Ser83 to Trp greatly reduced drug binding and, in preliminary experiments, S.E. Critchlow and A. Maxwell (unpublished data) found that the GyrB mutation Asp426 to Asn did not appreciably affect drug binding. Clearly there are inconsistencies in these data which need to be resolved, and, in addition, mutations at other sites (e.g. Asp 87, Gln106) should be assessed in terms of their impact on drug binding. In combination with the emerging structural data such work will greatly assist in our understanding of the mode of action of these drugs.

A very important issue which has been recently raised is the role of topo IV as a target for quinolones. This has been convincingly demonstrated in several systems in vivo but what is lacking presently is the demonstration that the drugs can form a cleavable complex with topo IV which can lead to cell death in an analogous way to the gyrase–drug complex on DNA. Clearly this is a very important area and more work is required to assess the significance of this enzyme as a drug target.

Continued effort on the elucidation of the mode of action of the quinolones on gyrase is clearly required and is important in order to fully understand how these drugs work. This in turn may allow us to design, perhaps rationally, more potent antibacterial agents.

Acknowledgements. We would like to thank George Orphanides, Chris Willmott, Karl Drlica, Laura Piddock and J.T. Smith for helpful comments and suggestions. AM is a Lister Institute Jenner Fellow, SEC was funded by a Wellcome Trust Prize Studentship.

References

Aleixandre V, Herrera G, Urios A, Blanco M (1991) Effects of ciprofloxacin on plasmid DNA supercoiling of *Escherichia coli* topoisomerase I and gyrase mutants. Antimicrob Agents Chemother 35:20–23

Baird JP, Bourguignon GJ, Sternglanz R (1972) Effect of nalidixic acid on the growth of deoxyribonucleic acid bacteriophages. J Virol 9:17–21

Barbour SD (1967) Effect of nalidixic acid on conjugal transfer and expression of episomal *Lac* genes in *Escherichia coli* K12. J Mol Biol 28:373–376

Bazile-Pham Khac S, Moreau NJ (1994) Interactions between fluoroquinolones, Mg^{2+}, DNA and DNA gyrase, studied by phase partitioning in an aqueous two-phase system and by affinity chromatography. J Chromatog 668:241–247

Bedard J, Wong S, Bryan LE (1987) Accumulation of enoxacin by *Escherichia coli* and *Bacillus subtilis.* Antimicrob Agents Chemother 31:1348–1354

Belland RJ, Morrison SG, Ison C, Huang WM (1994) *Neisseria gonorrhoeae* acquires mutations in analogous regions of *gyrA* and *parC* in fluoroquinolone-resistant isolates. Mol Microbiol 14:371–380

Bendixen C, Thomsen B, Alsner J, Westergaard O (1990) Camptothecin-stabilized topoisomerase I–DNA adducts cause premature termination of transcription. Biochemistry 29:5613–5619

Berger JM, Gamblin SJ, Harrison SC, Wang JC (1996) Structure at 2.7 Å resolution of a 92K yeast DNA topoisomerase II fragment. Nature 379:225–232

Bernard P, Couturier M (1992) Cell killing by the F plasmid CcdB protein involves poisoning of the DNA–topoisomerase II complex. J Mol Biol 226:735–745

Bernard P, Kézdy KE, Van Melderen L, Steyaert J, Wyns L, Pato ML, Higgins NP, Couturier M (1993) The F plasmid CcdB protein induces efficient ATP-dependent DNA cleavage by gyrase. J Mol Biol 234:534–541

Blandamer MJ, Briggs B, Cullis PM, Jackson AP, Maxwell A, Reece RJ (1994) The domain structure of *Escherichia coli* DNA gyrase as revealed by differential scanning calorimetry. Biochemistry 33:7510–7516

Bouck N, Adelberg EA (1970) Mechanism of action of nalidixic acid on conjugating bacteria. J Bacteriol 102:688–701

Bourguignon GJ, Levitt M, Sternglanz R (1973) Studies on the mechanism of action of nalidixic acid. Antimicrob Agents Chemother 4:479–486

Brown PO, Cozzarelli NR (1979) A sign inversion mechanism for enzymatic supercoiling of DNA. Science 206:1081–1083

Capranico G, Kohn KW, Pommier Y (1990) Local sequence requirements for DNA cleavage by mammalian topoisomerase II in the presence of doxorubicin. Nucleic Acids Res 18:6611–6619

Capranico G, Tinelli S, Zunino F, Kohn KW, Pommier Y (1993) Effects of base mutations on topoisomerase II DNA cleavage stimulated by *m*AMSA in short oligomers. Biochemistry 32:145–152

Cejka K, Holubová I, Hubacek J (1982) Curing effect of chlorobiocin on *Escherichia coli* plasmids. Mol Gen Genet 186:153–155

Chen AY, Liu LF (1994) DNA topoisomerases: essential enzymes and lethal targets. Annu Rev Pharmacol Toxicol 34:191–218

Chen C-R, Malik M, Snyder M, Drlica M (1996) DNA gyrase and topoisomerase IV on the bacterial chromosome: quinolone-induced DNA cleavage. J Mol Biol 258:627–637

Chow RT, Dougherty TJ, Fraimow HS, Bellin EY, Miller MH (1988) Association between early inhibition of DNA synthesis and the MICs and MBCs of carboxyquinolone antimicrobial agents for wild-type and mutant [*gyrA nfxB(ompF) acrA*] *Escherichia coli* K-12. Antimicrob Agents Chemother 32:1113–1118

Cook TM, Deitz WH, Goss WA (1966a) Mechanism of action of nalidixic acid on *Escherichia coli*. IV. Effects on the stability of cellular constituents. J Bacteriol 91:774–779

Cook TM, Goss WA, Deitz WH (1966b) Mechanism of action of nalidixic acid on *Escherichia coli*. V. Possible mutagenic effect. J Bacteriol 91:780–783

Corbett AH, Osheroff N (1994) When good enzymes go bad: conversion of topoisomerase II to a cellular toxin by antineoplastic drugs. Chem Res Toxicol 6:585–597

Courtright JB, Turowski DA, Sonstein SA (1988) Alteration of bacterial DNA structure, gene expression, and plasmid encoded antibiotic resistance following exposure to enoxacin. J Antimicrob Chemother 21(Suppl B):1–18

Critchlow SE, Maxwell A (1996) DNA cleavage is not required for the binding of quinolone drugs to the DNA gyrase–DNA complex. Biochemistry 35:7387–7393

Crumplin GC (ed) (1990) The 4-quinolones. Antibacterial agents in vitro. Springer, London

Crumplin GC, Smith JT (1975) Nalidixic acid: an antibacterial paradox. Antimicrob Agents Chemother 8:251–261

Crumplin GC, Smith JT (1976) Nalidixic acid and bacterial chromosome replication. Nature 260:643–645

Crumplin GC, Midgley JM, Smith JT (1980) Mechanisms of action of nalidixic acid and its congeners. In: Sammes PG (ed) Topics in antibiotic chemistry vol 3. Harwood, Chichester, p 11

Crumplin GC, Kenwright M, Hirst T (1984) Investigations into the mechanism of action of the antibacterial agent norfloxacin. J Antimicrob Chemother 13(Suppl B):9–23

Cullen ME, Wyke AW, Kuroda R, Fisher LM (1989) Cloning and characterisation of a DNA gyrase A gene from *Escherichia coli* that confers clinical resistance to 4-quinolones. Antimicrob Agents Chemother 33:886–894

Danilevskaya ON, Gragerov AI (1980) Curing of *Escherichia coli* K12 plasmids by coumermycin. Mol Gen Genet 178:233–235

Deguchi T, Yasuda M, Nakano M, Ozeki S, Ezaki T, Saito I, Kawada Y (1996) Quinolone-resistant *Neisseria gonorrhoeae*: correlation of alterations in the GyrA subunit of DNA gyrase and the ParC subunit of topoisomerase IV with antimicrobial susceptibility profiles. Antimicrob Agents Chemother 40:1020–1023

Deitz WH, Cook TM, Goss WA (1966) Mechanism of action of nalidixic acid on *Escherichia coli*. III. Conditions required for lethality. J Bacteriol 91:768–773

Diver JM, Wise R (1986) Morphological and biochemical changes in *Escherichia coli* after exposure to ciprofloxacin. J Antimicrob Chemother 18(Suppl D):31–41

Dobbs ST, Cullis PM, Maxwell A (1992) The cleavage of DNA at phosphorothioate internucleotidic linkages by DNA gyrase. Nucleic Acids Res 20:3567–3573

Domagala JM, Hanna LD, Heifetz CL, Hutt MP, Mich TF, Sanchez JP, Solomon M (1986) New structure–activity relationships of the quinolone antibacterials using the target enzyme. The development and application of a DNA gyrase assay. J Med Chem 29:394–404

Dougherty TJ, Saukkonen JJ (1985) Membrane permeability changes associated with DNA gyrase inhibitors in *Escherichia coli*. Antimicrob Agents Chemother 28:200–206

Dreyfuss ME, Midgley JM (1983) Structure–activity relationships in aromatic ring fused 4-pyridones substituted at position 3. J Pharm Pharmacol 35:75P

Drlica K, Franco RJ (1988) Inhibitors of DNA topoisomerases. Biochemistry 27:2253–2259

Drlica K, Snyder M (1978) Superhelical *Escherichia coli* DNA: relaxation by coumermycin. J Mol Biol 120:145–154

Drlica K, Engle E, Manes S (1980) DNA gyrase on the bacterial chromosome: possibility of two levels of action. Proc Natl Acad Sci USA 77:6879–6883

Drlica K, Coughlin S, Gennaro ML (1990) Mode of action of quinolones: biochemical aspects. In: Siporin C, Heifetz CL, Domagala JM (eds) The new generation of quinolones. Dekker, New York, p 45

Drlica K, Malik M, Wang J-Y, Levitz R, Burger RM (1995) The fluoroquinolones as antituberculosis agents. In: Rom W, Garay S (eds) Tuberculosis. Little, Brown, New York, pp 817–827

Elliott TSJ, Shelton A, Greenwood D (1987) The response of *Escherichia coli* to ciprofloxacin and norfloxacin. J Med Microbiol 23:83–88

Engle EC, Manes SH, Drlica K (1982) Differential effects of antibiotics inhibiting gyrase. J Bacteriol 149:92–98

Fan J-Y, Sun D, Yu H, Kerwin SM, Hurley LH (1995) The self assembly of a 4:4 quinobenzoxazine–Mg^{2+} complex on DNA: implications for the structure of the quinolone bacterial gyrase–DNA complex. J Med Chem 38:408–424

Fernandes PB (1989) International telesymposium on quinolones. Prous Science, Barcelona

Ferrero L, Cameron B, Manse B, Lagneaux D, Crouzet J, Famechon A, Blanche F (1994) Cloning and primary structure of *Staphylococcus aureus* DNA topoisomerase IV: a primary target of fluoroquinolones. Mol Microbiol 13:641–653

Ferrero L, Cameron B, Crouzet J (1995) Analysis of *gyrA* and *grlA* mutations in stepwise-selected ciprofloxacin-resistant mutants of *Staphylococcus aureus*. Antimicrob Agents Chemother 39:1554–1558

Fisher LM, Mizuuchi K, O'Dea MH, Ohmori H, Gellert M (1981) Site-specific interaction of DNA gyrase with DNA. Proc Natl Acad Sci USA 78:4165–4169

Fisher LM, Barot HA, Cullen ME (1986) DNA gyrase complex with DNA: determinants for site-specific DNA breakage. EMBO J 5:1411–1418

Fisher LM, Oram M, Sreedharan S (1993) DNA gyrase: mechanism and resistance to 4-quinolone antibacterial agents. In: Andoh T, Ikeda H, Ogura M (eds) Molecular biology of topoisomerases and its application to chemotherapy. CRC, Boca Raton, p 145

Freudenreich CH, Kreuzer KN (1993) Mutational analysis of a type II topoisomerase cleavage site: distinct requirements for enzyme and inhibitors. EMBO J 12:2085–2097

Freudenreich CH, Kreuzer KN (1994) Localization of an aminoacridine antitumor agent in a type II topoisomerase–DNA complex. Proc Natl Acad Sci USA 91:11007-11011

Froelich-Ammon SJ, Osheroff N (1995) Topoisomerase poisons: harnessing the dark side of enzyme mechanism. J Biol Chem 270:21429–21432

Fung-Tomc J, Kolek B, Bonner DP (1993) Ciprofloxacin-induced, low-level resistance to structurally unrelated antibiotics in *Pseudomonas aeruginosa* and methicillin-resistant *Staphylococcus aureus*. Antimicrob Agents Chemother 37:1289–1296

Gellert M, Mizuuchi K, O'Dea MH, Nash HA (1976a) DNA gyrase: an enzyme that introduces superhelical turns into DNA. Proc Natl Acad Sci USA 73:3872–3876

Gellert M, O'Dea MH, Itoh T, Tomizawa J (1976b) Novobiocin and coumermycin inhibit DNA supercoiling catalyzed by DNA gyrase. Proc Natl Acad Sci USA 73:4474–4478

Gellert M, Mizuuchi K, O'Dea MH, Itoh T, Tomizawa J (1977) Nalidixic acid resistance: a second genetic character involved in DNA gyrase activity. Proc Natl Acad Sci USA 74:4772–4776

Gellert M, Fisher LM, O'Dea MH (1979) DNA gyrase: purification and catalytic properties of a fragment of gyrase B protein. Proc Natl Acad Sci USA 76:6289–6293

Georgopapadakou NH, Bertasso A (1991) Effects of quinolones on nucleoid segregation in *Escherichia coli*. Antimicrob Agents Chemother 35:2645–2648

Gmünder H, Kuratli K, Keck W (1995) Effect of pyrimido[1,6-α]benzimidazoles, quinolones, and Ca^{2+} on the DNA gyrase-mediated cleavage reaction. Antimicrob Agents Chemother 39:163–169

Gocke E (1991) Mechanism of quinolone mutagenicity in bacteria. Mutation Res 248:135–143

Gootz TD, Barrett JF, Sutcliffe JA (1990) Inhibitory effects of quinolone antibacterial agents on eucaryotic topoisomerases and related test systems. Antimicrob Agents Chemother 34:8–12

Goss WA, Deitz WH, Cook TM (1964) Mechanism of action of nalidixic acid on *Escherichia coli*. J Bacteriol 88:1112–1118

Goss WA, Deitz WH, Cook TM (1965) Mechanism of action of nalidixic acid on *Escherichia coli*. II. Inhibition of deoxyribonucleic acid synthesis. J Bacteriol 89:1068–1074

Gudas LJ, Pardee AB (1975) Model for regulation of *Escherichia coli* DNA repair functions. Proc Natl Acad Sci USA 72:2330–2334

Hahn FE, Ciak J (1976) Elimination of resistance determinants from R-factor R1 by intercalative compounds. Antimicrob Agents Chemother 9:77–80

Hallett P, Maxwell A (1991) Novel quinolone resistance mutations of the *Escherichia coli* DNA gyrase A protein: enzymatic analysis of the mutant proteins. Antimicrob Agents Chemother 35:335–340

Hane MW, Wood TH (1969) *Escherichia coli* K-12 mutants resistant to nalidixic acid: genetic mapping and dominance studies. J Bacteriol 99:238–241

Heisig P (1996) Genetic evidence for a role of *parC* mutations in development of high-level fluoroquinolone resistance in *Escherichia coli*. Antimicrob Agents Chemother 40:879–885

Higgins NP, Cozzarelli NR (1982) The binding of gyrase to DNA: analysis by retention by nitrocellulose filters. Nucleic Acids Res 10:6833–6847

Hollom S, Pritchard RH (1965) Effect of inhibition of DNA synthesis on mating in *Escherichia coli* K12. Genet Res 6:479–483

Hooper DC, Wolfson JS (1990) Mechanisms of killing of bacteria by 4-quinolones. In: Crumplin GC (ed) The 4-quinolones. Antibacterial agents in vitro. Springer, London, p 69

Hooper DC, Wolfson JS (1991) Mode of action of the new quinolones: new data. Eur J Clin Microbiol Infect Dis 10:223–231

Hooper DC, Wolfson JS (1993a) Quinolone antibacterial agents, 2nd edn. ASM, Washington DC

Hooper DC, Wolfson JS (1993b) Mechanisms of quinolone action and bacterial killing. In: Hooper DC, Wolfson JS (eds) Quinolone antibacterial agents, 2nd edn. ASM, Washington DC, pp 53–75

Hooper DC, Wolfson JS, McHugh GL, Swartz MD, Tung C, Swartz MN (1984) Elimination of plasmid pMMG110 from *Escherichia coli* by novobiocin and other inhibitors of DNA gyrase. Antimicrob Agents Chemother 25:586–590

Hooper DC, Wolfson JS, Souza KS, Ng EY, McHugh GL, Swartz MN (1989) Mechanisms of quinolone resistance in *Escherichia coli*: characterization of nfxB and cfxB, two mutant resistance loci decreasing norfloxacin accumulation. Antimicrob Agents Chemother 33:283–290

Horowitz DS, Wang JC (1987) Mapping the active site tyrosine of *Escherichia coli* DNA gyrase. J Biol Chem 262:5339–5344

Hoshino K, Kitamura A, Morrisey I, Sato K, Kato J, Ikeda H (1994) Comparison of inhibition of *Escherichia coli* topoisomerase IV by quinolones with DNA gyrase inhibiton. Antimicrob Agents Chemother 38:2623–2627

Howard BMA, Pinney RJ, Smith JT (1993) Function of the SOS process in repair of DNA damage induced by modern 4-quinolones. J Pharm Pharmacol 45:658–662

Howard MT, Neece SH, Matson SW, Kreuzer KN (1994) Disruption of a topoisomerase–DNA cleavage complex by a DNA helicase. Proc Natl Acad Sci USA 91:12031–12035

Hsiang Y-H, Libou MG, Liu LF (1989) Arrest of replication forks by drug-stabilized topoisomerase I–DNA cleavable complexes as a mechanism of cell killing by camptothecin. Cancer Res 49:5077–5082

Hsiung Y, Elsea SH, Osheroff N, Nitiss JL (1995) A mutation in yeast *TOP2* homologous to a quinolone-resistant mutation in bacteria. J Biol Chem 270:20359–20364

Ikeda H, Moriya K, Matsumoto T (1981) In vitro study of illegitimate recombination: involvement of DNA gyrase. Cold Spring Harb Symp Quant Biol 45:399–408

Ikeda H, Aoki K, Naito A (1982) Illegitimate recombination mediated in vitro by DNA gyrase of *Escherichia coli*: structure of recombinant DNA molecules. Proc Natl Acad Sci USA 7:3724–3728

Kato J, Nishimura Y, Imamura R, Niki H, Hiraga S, Suzuki H (1990) New topoisomerase essential for chromosome segregation in *E. coli*. Cell 63:393–404

Kato J, Suzuki H, Ikeda H (1992) Purification and characterization of DNA topoisomerase IV in *Escherichia coli*. J Biol Chem 267:25676–25684

Khodursky AB, Zechiedrich EL, Cozzarelli NR (1995) Topoisomerase IV is a target of quinolones in *Escherichia coli*. Proc Natl Acad Sci USA 92:11801–11805

Kirkegaard K, Wang JC (1981) Mapping the topography of DNA wrapped around gyrase by nucleolytic and chemical probing of complexes of unique DNA sequences. Cell 23:721–729

Kreuzer KN, Cozzarelli NR (1979) *Escherichia coli* mutants thermosensitive for deoxyribonucleic acid gyrase subunit A: effects on deoxyribonucleic acid replication, transcription, and bacteriophage growth. J Bacteriol 140:424–435

Kreuzer KN, Cozzarelli NR (1980) Formation and resolution of DNA catenanes by DNA gyrase. Cell 20:245–254

Krueger JH, Walker GC (1984) *gro*EL and *dna*K genes of *Escherichia coli* are induced by UV irradiation and nalidixic acid in an *htpR*$^+$-dependent fashion. Proc Natl Acad Sci USA 81:1499–1503

Lecomte S, Coupry C, Chenon MT, Moreau NJ (1993) Affinity of sparfloxacin and 4 other fluoroquinolones for magnesium and effect on antibacterial activity. Drugs 45(Suppl 3):148–149

Lecomte S, Baron MH, Chenon MT, Coupry C, Moreau NJ (1994) Effect of magnesium complexation by fluoroquinolones and their antibacterial properties. Antimicrob Agents Chemother 38:2810–2816

Lewin B (1994) Chromatin and gene expression: constant questions, but changing answers. Cell 79:397–406

Lewin CS, Smith JT (1990) DNA breakdown by the 4-quinolones and its significance. J Med Microbiol 31:65–70

Lewin CS, Howard BMA, Ratcliffe NT, Smith JT (1989) 4-Quinolones and the SOS response. J Med Microbiol 29:139–144

Lewis RJ, Singh OMP, Smith CV, Skarynski T, Maxwell A, Wonacott AJ, Wigley DB (1996) The nature of inhibition of DNA gyrase by the coumarins and the cyclothialidines revealed by x-ray crystallography. EMBO J 15:1412–1420

Liu LF (1989) DNA topoisomerase poisons as antitumor drugs. Annu Rev Biochem 58:351–375

Liu LF, Liu C-C, Alberts BM (1980) Type II DNA topoisomerases: enzymes that can unknot a topologically knotted DNA molecule via a reversible double-strand break. Cell 19:697–707

Llorente B, Leclerc F, Cedergreen R (1996) Using SAR and QSAR analysis to model the activity and structure of the quinolone–DNA complex. Bioorg Med Chem 4:61–71

Lockshon D, Morris DR (1985) Sites of reaction of *Escherichia coli* DNA gyrase on pBR322 in vivo as revealed by oxolinic acid-induced plasmid linearization. J Mol Biol 131:63–74

Mamber SW, Kolek B, Brookshire KW, Bonner DP, Fung-Tomc J (1993) Activity of quinolones in the Ames *Salmonella* TA102 mutagenicity test and other bacterial genotoxicity assays. Antimicrob Agents Chemother 37:213–217

Manes SH, Pruss GJ, Drlica K (1983) Inhibition of RNA synthesis by oxolinic acid is unrelated to average DNA supercoiling. J Bacteriol 155:420–423

Maxwell A (1992) The molecular basis of quinolone action. J Antimicrob Chemother 30:409–416

Maxwell A, Gellert M (1984) The DNA dependence of the ATPase activity of DNA gyrase. J Biol Chem 259:14472–14480

Maxwell A, Gellert M (1986) Mechanistic aspects of DNA topoisomerases. Adv Protein Chem 38:69–107

McDaniel LS, Rogers LH, Hill WE (1978) Survival of recombination-deficient mutants of *Escherichia coli* during incubation with nalidixic acid. J Bacteriol 134:1195–1198

McHugh GL, Swartz MN (1977) Elimination of plasmids from several bacterial species by novobiocin. Antimicrob Agents Chemother 12:423–426

Miki T, Park JA, Nagao K, Murayama N, Horiuchi T (1992) Control of segregation of chromosomal DNA by sex factor F in *Escherichia coli*. J Mol Biol 225:39–52

Mizuuchi K, O'Dea MH, Gellert M (1978) DNA gyrase: subunit structure and ATPase activity of the purified enzyme. Proc Natl Acad Sci USA 75:5960–5963

Moreau NJ, Robaux H, Baron L, Tabary X (1990) Inhibitory effects of quinolones of pro- and eucaryotic DNA topoisomerases I and II. Antimicrob Agents Chemother 34:1955–1960

Morrison A, Cozzarelli NR (1979) Site-specific cleavage of DNA by *E. coli* DNA gyrase. Cell 17:175–184

Morrison A, Cozzarelli NR (1981) Contacts between DNA gyrase and its binding site on DNA: features of symmetry and asymmetry revealed by protection from nucleases. Proc Natl Acad Sci USA 78:1416–1420

Morrison A, Higgins NP, Cozzarelli NR (1980) Interaction between DNA gyrase and its cleavage site on DNA. J Biol Chem 255:2211–2219

Nakamura S, Nakamura M, Kojima T, Yoshida H (1989) *gyrA* and *gyrB* mutations in quinolone-resistant strains of *Escherichia coli*. Antimicrob Agents Chemother 33:254–255

Nakamura S, Yoshida H, Bogaki M, Nakamura M, Kojima T (1993) Quinolone resistance mutations in DNA gyrase. In: Andoh T, Ikeda H, Ogura M (eds) Molecular biology of topoisomerases and its application to chemotherapy. CRC, Boca Raton, p 135

Oram M, Fisher M (1991) 4-Quinolone resistance mutations in the DNA gyrase of *Escherichia coil* clinical isolates identified by using the polymerase chain reaction. Antimicrob Agents Chemother 35:387–389

Orphanides G, Maxwell A (1994) Evidence for a conformational change in the DNA gyrase–DNA complex from hydroxyl radical footprinting. Nucleic Acids Res 22:1567–1575

Palù G, Valisena S, Peracchi M, Palumbo M (1988) Do quinolones bind to DNA? Biochem Pharmacol 37:1887–1888

Palù G, Valisena S, Ciarrocchi G, Gatto B, Palumbo M (1992) Quinolone binding to DNA is mediated by magnesium ions. Proc Natl Acad Sci USA 89:9671–9675

Palumbo M, Gatto B, Zagotto Z, Palù G (1993) On the mechanism of action of quinolone drugs. Trends Microbiol 1:233–235

Pedrini AM, Geroldi D, Siccardi A, Falaschi A (1972) Studies on the mode of action of nalidixic acid. Eur J Biochem 25:359–365

Peng H, Marians KJ (1993) *Escherichia coli* topoisomerase IV. J Biol Chem 268:24481–24490

Permana PA, Snapka RM, Shen LL, Chu DTW, Clement JJ, Plattner JJ (1994) Quinobenzoxazines: a class of novel antitumor quinolones and potent mammalian DNA topoisomerase II catalytic inhibitors. Biochemistry 33:11333–11339

Phillips I (1987) Bacterial mutagenicity and the 4-quinolones. J Antimicrob Chemother 20:771–782

Phillips I, Culebras E, Moreno F, Baquero F (1987) Induction of the SOS response by the new 4-quinolones. J Antimicrob Chemother 20:631–638

Piddock LJV (1991) Mechanism of quinolone uptake into bacterial cells. J Antimicrob Chemother 27:399–403

Piddock LJV, Walters RN (1992) Bactericidal activities of five quinolones for *Escherichia coli* strains with mutations in genes encoding the SOS response or cell division. Antimicrob Agents Chemother 36:819–825

Piddock LJV, wise R (1987) Induction of the SOS response in *Escherichia coli* by 4-quinolone antimicrobial agents. FEMS Microbiol Lett 41:289–294

Piddock LJV, Walters RN, Diver JM (1990) Correlation of quinolone MIC and inhibition of DNA, RNA, and protein synthesis and induction of the SOS response in *Escherichia coli*. Antimicrob Agents Chemother 34:2331–2336

Platt DJ, Black AC (1987) Plasmid ecology and the elimination of plasmids by 4-quinolones. J Antimicrob Chemother 20:137–138

Pommier Y, Capranico G, Orr A, Kohn KW (1991a) Distribution of topoisomerase II cleavage sites in simian virus 40 DNA and the effects of drugs. J Mol Biol 222:909–924

Pommier Y, Capranico G, Orr A, Kohn KW (1991b) Local base sequence preferences for DNA cleavage by mammalian topoisomerase II in the presence of amsacrine or teniposide. Nucleic Acids Res 19:5973–5980

Power EGM, Phillips I (1992) Induction of the SOS gene (*umuC*) by 4-quinolone antibacterial drugs. J Med Microbiol 36:78–82

Power EGM, Phillips I (1993a) Correlation between *umuC* induction and *Salmonella* mutagenicity assay for quinolone antimicrobial agents. FEMS Microbiol Lett 112:251–254

Power EGM, Phillips I (1993b) Quinolones and induction of *umuC* and mutagenicity. Drugs 45(Suppl 3):278–279

Ramareddy G, Reiter H (1969) Specific loss of newly replicated deoxyribonucleic acid in nalidixic acid-treated *Bacillus subtilis* 168. J Bacteriol 100:724–729

Rau DC, Gellert M, Thoma F, Maxwell A (1987) Structure of the DNA gyrase–DNA complex as revealed by transient electric dichroism. J Mol Biol 193:555–569

Reece RJ, Maxwell A (1989) Tryptic fragments of the *Escherichia coli* DNA gyrase A protein. J Biol Chem 264:19648–19653

Reece RJ, Maxwell A (1991a) DNA gyrase: structure and function. Crit Rev Biochem Mol Biol 26:335–375

Reece RJ, Maxwell A (1991b) Probing the limits of the DNA breakage–reunion domain of the *Escherichia coli* DNA gyrase A protein. J Biol Chem 266:3540–3546

Reece RJ, Dauter Z, Wilson KS, Maxwell A, Wigley DB (1990) Preliminary crystallographic analysis of the breakage–reunion domain of the *Escherichia coli* DNA gyrase A protein. J Mol Biol 215:493–495

Robinson MJ, Martin BA, Gootz TD, McGuirk PR, Moynihan M, Sutcliffe JA, Osheroff N (1991) Effects of quinolone derivatives on eukaryotic topoisomerase II. A novel mechanism for enhancement of enzyme-mediated DNA cleavage. J Biol Chem 266:14585–14592

Rohatgi K, Courtright JB (1990) Major changes in the structure and morphology of the bacterial nucleoid after treatment of cells with quinolones. In: Siporin C, Heifetz CL, Domagala JM (eds) The new generation of quinolones. Dekker, New York, p 317

Shen LL (1989) A reply: "Do quinolones bind to DNA?" – Yes. Biochem Pharmacol 38:2042–2044

Shen LL, Pernet AG (1985) Mechanism of inhibition of DNA gyrase by analogues of nalidixic acid: the target of the drugs is DNA. Proc Natl Acad Sci USA 82:307–311

Shen LL, Baranowski J, Pernet AG (1989a) Mechanism of inhibition of DNA gyrase by quinolone antibacterials: specificity and cooperativity of drug binding to DNA. Biochemistry 28:3879–3885

Shen LL, Kohlbrenner WE, Weigl D, Baranowski J (1989b) Mechanism of quinolone inhibition of DNA gyrase. Appearance of unique norfloxacin binding sites in enzyme–DNA complexes. J Biol Chem 264:2973–2978

Shen LL, Mitscher LA, Sharma PD, O'Donnell TJ, Chu DTW, Cooper CS, Rosen T, Pernet AG (1989c) Mechanism of inhibition of DNA gyrase by quinolone antibacterials: a cooperative drug–DNA binding model. Biochemistry 28:3886–3894

Siporin C, Heifetz CL, Domagala JM (eds) (1990) The new generation of quinolones. Dekker, New York

Smith JT (1984) Awakening the slumbering potential of the 4-quinolone antibacterials. Pharmacol J 233:299–305

Smith JT (1986) The mode of action of 4-quinolones and mechanisms of resistance. J Antimicrob Chemother 18(Suppl D): 21–29

Snyder M, Drlica K (1979) DNA gyrase on the bacterial chromosome: DNA cleavage induced by oxolinic acid. J Mol Biol 131:287–302

Sonstein SA (1990) Mode of action of quinolones: antibacterial aspects. In: Siporin C, Heifetz CL, Domagala JM (eds) The new generation of quinolones. Dekker, New York, p 63

Staudenbauer WL (1976) Replication of *Escherichia coli* DNA in vitro: inhibition by oxolinic acid. Eur J Biochem 62:491–497

Sugino A, Cozzarelli NR (1980) The intrinsic ATPase of DNA gyrase. J Biol Chem 255:6299–6306

Sugino A, Peebles CL, Kruezer KN, Cozzarelli NR (1977) Mechanism of action of nalidixic acid: purification of *Escherichia coli nalA* gene product and its relationship to DNA gyrase and a novel nicking–closing enzyme. Proc Natl Acad Sci USA 74:4767–4771

Taylor DE, Levine JG (1979) Characterization of a plasmid mutation affecting maintenance transfer and elimination by novobiocin. Mol Gen Genet 174:127–133

Thomsen B, Bendixen C, Lund K, Andersen AH, Sorensen B, Westergaard O (1990) Characterization of the interaction between topoisomerase II and DNA by transcriptional footprinting. J Mol Biol 215:237–244

Tornaletti S, Pedrini AM (1988a) DNA unwinding induced by nalidixic acid binding to DNA. Biochem Pharmacol 37:1881–1882

Tornaletti S, Pedrini AM (1988b) Studies on the interaction of 4-quinolones with DNA by DNA unwinding experiments. Biochim Biophys Acta 949:279–287

van Holde KE, Lohr DE, Robert C (1992) What happens to nucleosomes during transcription? J Biol Chem 267:2837–2840

Vila I, Ruiz J, Goni P, Jimenez de Anta MT (1996) Detection of mutations in *parC* in quinolone-resistant clinical isolates of *Escherichia coli*. Antimicrob Agents Chemother 40:491–493

Walker GC (1984) Mutagenesis and inducible responses to deoxyribonucleic acid damage in *Escherichia coli*. Microbiol Rev 48:60–93

Walker JR, Pardee AB (1968) Evidence for a relationship between deoxyribonucleic acid metabolism and septum formation in *Escherichia coli*. J Bacteriol 95:123–131

Walton L, Elwell LP (1988) In vitro cleavable-complex assay to monitor antimicrobial potency of quinolones. Antimicrob Agents Chemother 32:1086–1089

Wang JC (1985) DNA topoisomerases. Annu Rev Biochem 54:665–697

Watt PM, Hickson ID (1994) Structure and function of type II DNA topoisomerases. Biochem J 303:681–695

Weisser J, Wiedemann B (1985) Elimination of plasmids by new 4-quinolones. Antimicrob Agents Chemother 28:700–702

Weisser J, Wiedemann B (1986) Elimination of plasmids by enoxacin and ofloxacin at near inhibitory concentrations. J Antimicrob Chemother 18:575–583

Wigley DB (1995a) Structure and mechanism of DNA topoisomerases. Annu Rev Biophys Biomol Struct 24:185–208

Wigley DB (1995b) Structure and mechanism of DNA gyrase. In: Eckstein F, Lilley DMJ (eds) Nucleic acids and molecular biology, vol 9. Springer, Berlin Heidelberg New York, p 165

Wigley DB, Davies GJ, Dodson EJ, Maxwell A, Dodson G (1991) Crystal structure of an N-terminal fragment of the DNA gyrase B protein. Nature 351:624–629

Willmott CJR, Maxwell A (1993) A single point mutation in the DNA gyrase A protein greatly reduces the binding of fluoroquinolones to the gyrase–DNA complex. Antimicrob Agents Chemother 37:126–127

Willmott CJR, Critchlow SE, Eperon IC, Maxwell A (1994) The complex of DNA gyrase and quinolone drugs with DNA forms a barrier to transcription by RNA polymerase. J Mol Biol 242:351–363

Yamagishi J, Yoshida H, Yamyoshi M, Nakamura S (1986) Nalidixic acid-resistant mutation of the *gyrB* gene of *Escherichia coil*. Mol Gen Genet 204:367–373

Yoshida H, Kojima T, Yamagishi J, Nakamura S (1988) Quinolone-resistant mutations of the *gyrA* gene of *Escherichia coli*. Mol Gen Genet 211:1–7

Yoshida H, Nakamura M, Bogaki M, Ito H, Kojima T, Hattori H, Nakamura S (1993) Mechanism of action of quinolones against *Escherichia coli* DNA gyrase. Antimicrob Agents Chemother 37:839–845

Ysern P, Clerch B, Castáno M, Gibert I, Barbé J, Llagostera M (1990) Induction of SOS genes in *Escherichia coli* and mutagenesis in *Salmonella typhimurium* by fluoroquinolones. Mutagenesis 5:63–66

Zweerink MM, Edison A (1986) Inhibition of *Micrococcus luteus* DNA gyrase by norfloxacin and 10 other quinolone carboxylic acids. Antimicrob Agents Chemother 29:598–601

CHAPTER 5

The In Vitro Antibacterial Activity of Quinolones: A Review

C. THORNSBERRY

A. Introduction

The age of antibiotics, or "the antibiotic era," began in the 1930s with the sulfonamides. The quinolones were first introduced in the early 1960s and thus have spanned the last 30 years or half of the antibiotic era (MOELLERING 1995; NORRIS and MANDELL 1988). The first of the quinolones to be used substantially was nalidixic acid, an antimicrobial agent that achieves good urinary tract levels but only minimal serum levels and, consequently, was and is still used as a drug for the treatment of urinary tract infections. However, within a short period of time it was recognized that resistance to nalidixic acid readily develops and consequently the drug fails to cure clinical infections. Because of this, nalidixic acid has not been used extensively in recent times as a therapeutic agent. Other quinolone antimicrobials were subsequently developed, for example oxalinic acid and cinoxacin; recently, however, these drugs also have not found a great deal of favor for the treatment of urinary tract infections.

The next significant development came with the introduction of norfloxacin. Norfloxacin is a quinolone with 6-fluorine and 7-piperazine substituents, which give it both gram-negative and gram-positive bacteria (NORRIS and MANDELL 1988). Again, however, the pharmacokinetics were such that norfloxacin was used for treating urinary tract infections and not for other infections involving other tissues. Although norfloxacin has been widely used to treat urinary tract infections, the importance of the introduction of this quinolone was not particularly that it another quinolone antibiotic to be used for the treatment of urinary tract infections, but rather that it signified the introduction of the fluoroquinolones. Fluoroquinolones that have been developed and introduced since then have vastly superior activity against both gram-negative and gram-positive organisms, including *Pseudomonas aeruginosa* and *Staphylococcus aureus*. Numerous fluoroquinolones have been developed since norfloxacin; those that have been used most extensively in the United States are ciprofloxacin and ofloxacin, although others have been approved, such as enoxacin and temafloxacin (the latter has since been withdrawn from clinical use). In some other countries, particularly in Europe, pefloxacin has also been used extensively.

The mechanism of action of quinolones is a selective inhibition of bacterial DNA synthesis (HOOPER and WOLFSON 1993a). This inhibition of DNA synthesis in bacteria is retated to its activity against the enzyme DNA gyrase. One of the purposes of DNA gyrase is to maintain the chromosome in a supercoiled state and also to repair small breaks in DNA that often occur during replication. Because of the activity of these newer fluoroquinolones against gyrase, they are not only efficient bactericidal compounds but also are bactericidal at achievable tissue levels. Bacterial resistance to fluoroquinolones arises as a result of mutations which tend not to occur very frequently and certainly much less frequently than that observed with nalidixic acid and other older quinolones (HOOPER and WOLFSON 1993b). The genes determining resistance were found to be chromosomal rather than plasmid borne.

The development of resistance in clinically important bacteria is obviously of importance because of the wide spread use of fluoroquinolones. Because the resistance to fluoroquinolones is chromosomal, increasing resistance within a species is basically in response to mutations and subsequent selective pressure created by the use of fluoroquinolones. At the moment, the incidence of resistance in most clinical species is relatively low. The incidence of resistance in *P. aeruginosa*, for example, is approximately 10%, but in *Escherichia coli* is <1% (C. Thornsberry, unpublished work). The highest incidence of resistance is in *S. aureus*; resistance in this organism developed very rapidly after the introduction of ciprofloxacin. An examination of the resistance in *S. aureus*, however, shows that almost all fluoroquinolone-resistant strains are also methicillin-resistant (C. Thornsberry, unpublished work). This might have been expected because of the multiple resistance in these isolates (i.e., resistance to most drugs except vancomycin); the selective pressure may be exhibited by all the penicillins and cephalosporins, tetracycline, macrolides, chloramphenicol, and other agents in addition to fluoroquinolones, i.e., any one of these antimicrobials selects for the others. In an institution where there is extensive resistance within a species, there must be concern about the horizontal transfer of resistant isolates and the control of nosocomial infections.

For many gram-positive species, the minimal inhibitory concentrations (MICs) tend to be borderline between susceptible and resistant. Any change in MICs due to methodology could change the results for large numbers of isolates. These borderline MICs are most prominent in *Streptococcus pneumoniae* because of the continuing controversy about the use of fluoroquinolones to treat pneumococcal infections. In our experience, the vast majority of these isolates tend to be either intermediate or barely susceptible, i.e., MICs of 2 or 1 μg/ml for ciprofloxacin.

Another important consideration is that the development of resistance to one fluoroquinolone results in cross-resistance to other fluoroquinolones. There are, however, differences in the activity of different quinolones, i.e., differences in MICs. As a result, much of the developmental work with new fluoroquinolones is directed towards finding new compounds with greater activity (i.e., lower MICs) for gram-positive and anaerobic bacterial species.

The following is a review of the activity of selected fluoroquinolones against a variety of bacterial species. (The values quoted were taken from our own data and from the following publications: APPLEBAUM 1995; BARRY 1988; CHIN et al. 1993; ELIOPOULOS and ELIOPOULOS 1993; ELIOPOULOS and ELIOPOULOS 1989; FELMINGHAM et al. 1993; GOOTZ and MCGUIRK 1994; HOOPER and WOLFSON 1993b; DHOLAKIA et al. 1994; KUWAHARA-AIRAI et al. 1993; MARSHALL et al. 1993; MOELLERING 1995; NAKANE et al. 1993; NORRIS and MANDELL 1988; PHILLIPS et al. 1988; WANG et al. 1993; WEIDEMANN and ATKINSON 1991.)

B. In Vitro Activity

In recent years, many newer fluoroquinolone antibiotics have been introduced. Several of these have been considered by the National Committee for Clinical Laboratory Standards (NCCLS) in one form or another, i.e., either to recommend interpretive breakpoints or to recommend quality control parameters (National Committee for Clinical Laboratory Standards 1993a–b, 1994). Among those that have been considered by the NCCLS are nalidixic acid, cinoxacin, norfloxacin, ciprofloxacin, enoxacin, fleroxacin, lomefloxacin, ofloxacin, levofloxacin, and sparfloxacin. The interpretive breakpoints for

Table 1. Interpretive breakpoints for selected quinolones (from the NCCLS) for organisms other than *Haemophilus* spp., *Neisseria gonorrhoeae*, and *Streptococcus pneumoniae*[a]

Quinolone	Categories by MIC (μg/ml) and disk diffusion (mm)		
	Susceptible	Intermediate	Resistant
Cinoxacin[b]	≤16 μg/ml	NA[c]	≥64 μg/ml
	≥19 mm	15–18 mm	≤14 mm
Ciprofloxacin	≤1 μg/ml	2 μg/ml	≥4 μg/ml
	≥21 mm	16–20 mm	≤15 mm
Enoxacin	≤2 μg/ml	4 μg/ml	≥8 μg/ml
	≥18 mm	15–17 mm	≤14 mm
Fleroxacin	≤2 μg/ml	4 μg/ml	≥8 μg/ml
	≥19 mm	16–18 mm	≤15 mm
Lomefloxacin	≤2 μg/ml	4 μg/ml	≥8 μg/ml
	≥22 mm	19–21 mm	≤18 mm
Nalidixic acid[b]	≤16 μg/ml	NA	≥32 μg/ml
	≥19 mm	14–18 mm	≤13 mm
Norfloxacin[b]	≤4 μg/ml	8 μg/ml	≥16 μg/ml
	≥17 mm	13–16 mm	≤12 mm
Ofloxacin	≤2 μg/ml	4 μg/ml	≥8 μg/ml
	≥16 mm	13–15 mm	≤12 mm

[a] From National Committee for Clinical Laboratory Standards 1993a–b, 1994.
[b] Applies only to isolates from the urinary tract.
[c] NA, not applicable.

Table 2. Interpretive categories (NCCLS) for *Haemophilus influenzae*, *Neisseria gonorrhoeae*, and *Streptococcus pneumoniae*[a]

Organism/drug	Categories (MIC μg/ml; disk diffusion mm)		
	Susceptible	Intermediate	Resistant
H. influenzae			
Ciprofoxacin	≤1[b]; ≥21[b]	–	–
Fleroxacin	≤2[b]; ≥19[b]	–	–
Lomefloxacin	≤2[b]; ≥22[b]	–	–
Ofloxacin	≤2[b]; ≥16[b]	–	–
N. gonorrhoeae			
Ciprofloxacin	≤0.06[b]; ≥36[b]	–	–
Enoxacin	≤0.5[b]; ≥32[b]	–	–
Fleroxacin	≤0.25; ≥33	0.5/28–32	≥1/≤27
Lomefloxacin	≤0.12[b]; ≥36[b]	–	–
Ofloxacin	≤0.25[b]/≥31[b]	–	–
S. pneumoniae			
Ofloxacin	≤2/≥16	4/13–15	≥8/≤12

[a] From National Committee for Clinical Laboratory Standards 1993a–b, 1994.
[b] Only a susceptible category is recommended because of the absence of resistant strains.

some of these agents for which the NCCLS has recommendations are shown in Tables 1 and 2.

In general, the activity of the older quinolones (e.g., nalidixic acid and cinoxacin) is lower than those of the newer fluoroquinolones. Of the fluoroquinolones, norfloxacin is generally the least active, but to some degree that depends upon the organism being tested. As a rule, the activity of fluoroquinolones is greater against gram-negative organisms than against gram-positive organisms. However, even among gram-negative organisms, the activity varies; for example, the activity against many of the Enterobacteriaceae species is much greater than that against *P. aeruginosa* and some of the other non-fermentative gram-negative organisms. The development of resistance to quinolones has been greatest in a few selected species; the most notable example is the rapid development of resistance to quinolones among methicillin-resistant *S. aureus* (MRSA). In contrast, methicillin-susceptible *S. aureus* (MSSA) are, for the most part, susceptible to quinolones. The other organism which has been repeatedly reported to develop resistance is *P. aeruginosa*. However, in a surveillance study carried out in 1991–1992 within the United States, we found that the incidence of resistance to ciprofloxacin among *P. aeruginosa* to be about 10% and that another 5% were intermediate. This indicates that on a national level, approximately 85% of *P. aeruginosa* isolates were susceptible to ciprofloxacin (C. Thornsberry, unpublished work). In certain institutions, however, the incidence of resistant *Pseudomonas* may be higher, but this is likely to be due to larger numbers of the same resistant strain.

In the 1991 surveillance study we demonstrated that essentially all MRSA were resistant to ciprofloxacin and it can be concluded that they are also resistant to all other quinolones because of the cross-resistance that exists among the presently available fluoroquinolones. The in vitro activity of selected fluoroquinolones against various isolates is shown in Tables 3–6. These data are shown as MIC_{90}s or ranges of MIC_{90}s which were taken from the cited references. If the values for the different quinolones are read across for a single species, differences in the activity of the various compounds against that organism can be seen. For example in Table 3 there is a wide range in MIC_{90} from 0.06 to 16 μg/ml for *Acinetobacter species*. On the other hand, the differences for *Aeromonas spp.* are minimal. Likewise, the differences in the activity of a single drug for different organisms can be seen by reading the columns for that drug, e.g., the ciprofloxacin MIC_{90} ranges from 0.008 μg/ml for *Neisseria meningitidis* to >16 μg/ml for *Alcaligenes spp.*

The data for gram-negative species are shown in Table 3. It is apparent that, for the most part, quinolones and particularly fluoroquinolones are very active against gram-negative bacteria. On closer examination, however, it can be seen that there are differences between organisms. *Neisseria spp.*, *Haemophilus spp.*, and *Moraxella catarrhalis* tend to be the more susceptible followed by the Enterobacteriaceae with the non-fermentative gram-negative being the least susceptible of the gram-negative bacteria. The data in this table also shows the markedly increased activity of the newer fluoroquinolones compared with nalidixic acid.

One of the reasons for the early excitement about fluoroquinolones was their activity against *P. aeruginosa*. The data in Table 3 show that ciprofloxacin has the greatest activity with the activity of others being marginal (based on MIC_{90}s). The activity of these compounds against *Burkholderei cepacia* and *Burkholderei pseudomallei* and *Stenotrophomonas maltophilia* is less than that against *P. aeruginosa*.

The activity of fluoroquinolones is quite marked against *Neisseria gonorrhoeae* with MICs in all cases being $\leq$0.12 μg/ml and for ciprofloxacin as low as 0.008 μg/ml. Fluoroquinolones have become a common choice for treatment of patients with gonorrhea. Among other genital pathogens, *Haemophilus ducreyi* was quite susceptible to norfloxacin, ciprofloxacin, ofloxacin, and enoxacin, but *Gardnerella vaginalis* was only marginally susceptible to ciprofloxacin and ofloxacin and resistant to nalidixic acid and enoxacin.

Although resistant strains of Enterobacteriaceae spp. do occur, the incidence is quite low for most of them. The highest MICs for the species of Enterobacteriaceae tend to be those for *Serratia* and *Providencia species* (most of these species still tend to be susceptible to most of the fluoroquinolones).

The data for gram-positive bacteria are shown in Table 4. It is readily apparent that these bacteria, as a group, are less susceptible to quinolones than are gram-negative bacteria. All of the gram-positive species are resistant to nalidixic acid. Although norfloxacin has greater activity (lower MICs) than

Table 3. The activity of various quinolones (as indicated by MIC_{90}s or ranges of MIC_{90}s) against gram-negative species

Species	Ranges of MIC_{90}s (μg/ml)[a]									
	Nalidixic acid	Norfloxacin	Ciprofloxacin	Ofloxacin	Enoxacin	Sparfloxacin	Lomefloxacin	Fleroxacin	Levofloxacin	Pefloxacin
Acinetobacter spp.	8	8–16	0.5–1	0.25–1	4–8	0.06–1	2	0.5–4	0.5–16	2
Aeromonas spp.	0.5	0.06	≤0.06	0.03–0.5	0.06	≤0.12	0.12	0.12–0.25	ND	0.03–0.12
Alcaligenes spp.	ND	64	4–>16	>16	ND	4	ND	4–>16	8	32
Brucella spp.	ND	ND	0.25–1	0.03–0.25	ND	0.25–1	ND	0.5–1	ND	ND
Burkholderei cepacia	ND	32	4	16–32	16	1	16	ND	16	ND
Burkholderei pseudomallei	ND	8–64	8	32	32	ND	ND	ND	ND	ND
Campylobacter spp.	ND	ND	0.12–0.5	ND	ND	0.12–0.25	ND	0.5	0.25	ND
Citrobacter diversus	8	≤0.12	≤0.03	≤0.06	≤0.12	ND	ND	ND	0.12	ND
Citrobacter freundii	8	0.5	0.12	0.5	0.5	0.5	0.5	0.12	0.5	ND
Enterobacter aerogenes	4	0.5	0.06	0.25	0.25	0.12	0.5	0.12	0.12	0.25
Enterobacter agglomerans	ND	0.25	0.06	0.25	0.25	ND	ND	ND	0.12	0.5
Enterobacter cloacae	4	0.5	0.12	0.25	0.5	0.25	0.5	0.25	0.25	0.5
Escherichia coli	4	0.12	≤0.06	0.12	0.25	≤0.06	0.25	0.12	≤0.06	0.12
Gardnerella vaginalis	≥128	16	1–2	2	32	ND	ND	ND	ND	8
Haemophilus ducreyi	≥4	0.12	0.03	0.03	0.12	ND	ND	ND	ND	0.12
Haemophilus influenzae	1–2	0.06–0.12	≤0.03	0.03–0.12	0.12–0.5	≤0.06	0.12	0.06–0.12	0.01	0.06–0.12
Haemophilus parainfluenzae	ND	ND	≤0.03	ND	ND	ND	ND	ND	ND	ND
Hafnia alvei	ND	0.12	0.03–0.12	0.06–0.25	0.12	0.12	0.12	ND	ND	ND
Helicobacter pylori	ND	0.25–8	0.25–8	1	16	0.5–4	2–4	4	0.5	ND
Klebsiella oxytoca	4	0.25	0.12	0.5	0.5	0.06	ND	ND	0.12	1
Klebsiella pneumoniae	32	0.5	0.12	0.25	1	0.12	1	0.5	0.25	2
Moraxella catarrhalis	ND	0.03–0.12	0.01–0.03	0.06–0.5	0.25	0.01–0.12	0.12–0.25	0.25	0.06	0.12–0.25
Morganella morganii	4	0.12	0.06	0.25	0.5	0.5	0.25	0.12	0.5	0.25
Neisseria gonorrhoeae	ND	0.06–0.12	0.008–0.06	0.01–0.12	0.03–0.12	0.01	0.03–0.12	0.01	0.01–0.1	0.12
Neisseria meningitidis	ND	0.03	0.008	0.01–0.5	≤0.06	0.008	0.5	≤0.06	0.008	0.03–1
Pasteurella multocida	ND	ND	≤0.03	0.06	ND	ND	ND	ND	ND	ND
Proteus mirabilis	8	0.12	0.06	0.5	0.5	0.5	1	0.5	0.25	0.25
Proteus vulgaris	4	0.12	0.06	0.25	0.5	0.5	0.5	0.12	0.25	0.25
Proteus rettgeri	ND	2	1	1	1	0.5	4	0.5	1	0.5
Providencia stuartii	≥128	2	0.5	1	2	0.5	1	1	1	4
Pseudomonas aeruginosa	≥128	2–8	0.25–1	2–8	4–8	1–8	4–8	8	4–8	4–16
Salmonella spp.	2–8	0.03–0.12	0.01–0.06	0.06–0.12	0.12–0.25	≤0.06	0.06–0.25	0.06–0.25	0.12	0.12–0.5
Serratia marcescens	8–16	1	0.25	1	2	16	4	4	0.25–16	1
Shigella spp.	2–8	0.03–0.12	0.01–0.03	0.12	0.06–0.12	≤0.06	0.12–0.25	≤0.06	0.12	0.06–0.25
Stenotrophomas maltophilia	64	32–64	8–16	8	16	0.5–>2	8–16	8	4	ND
Vibria cholerae	ND	ND	0.015	0.12	ND	0.002	ND	0.01	0.004	ND
Yersinia enterocolitica	2	0.06–0.12	0.03	0.12–0.25	0.12	≤0.06	ND	≤0.06	0.06	0.25

ND, not determined.

[a]These MIC_{90}s or ranges of MIC_{90}s were taken from Appelbaum 1995, Barry 1988, Chin et al. 1993, Eliopoulos and Eliopoulos 1993, Eliopoulos and Eliopoulos 1989, Felmingham et al. 1993, Gootz and McGuirk 1994, Jones et al. 1993, Kholakia et al. 1994, Kuwahara-Airai et al. 1993, Marshall et al. 1993, Moellering 1995; Nakane et al. 1993; Norris and Mandell 1988, Phillips et al. 1988, Wang et al. 1993, Weidemann and Atkinson 1991, and our own data.

Table 4. The activity of various quinolones (as indicated by MIC_{90}s or ranges of MIC_{90}s) against gram-positive species

Species	Ranges of MIC_{90}s (μg/ml)[a]									
	Nalidixic acid	Norfloxacin	Ciprofloxacin	Ofloxacin	Enoxacin	Sparfloxacin	Lomefloxacin	Fleroxacin	Levofloxacin	Pefloxacin
Bacillus spp.	ND	0.5–1	0.25	0.5	1	0.25	ND	0.5–2	0.25	ND
Corynebacterium spp.	ND	4–8	1	1	8	0.25	16–>16	2	1	8–16
Enterococcus faecalis	>128	8	2	4	8	1	8–16	8	2	8
Enterococcus faecium	>128	>16	4	8	32	1	8	8	4	8
Listeria monocytogenes	ND	8–16	1–2	2	8	2	8	4–8	1	8–16
Staphylococcus, coag neg[b]	64–>128	2–4	0.5	1	2	0.12	2	1–2	0.5	1
Staphylococcus aureus[c]	64	2–16	0.5	0.5–2	2	0.12	2	1	0.25	0.5–1
Streptococcus pyogenes	ND	4	0.5–1	1–2	4–>8	1	8	8	1	8
Streptococcus agalactiae	≥128	8–16	1–2	2–4	8–16	0.5	16	8	2	8–32
Streptococcus bovis	ND	>8	4	4	>8	ND	ND	>8	ND	ND
Streptococcus pneumoniae[d]	≥128	16	1–2	2–4	16	0.5	8	8	2	8–16
Streptococcus, other	≥128	16	2	2–8	16–32	1	8	8	1	>16

ND, not determined.
[a] For data sources, see Table 3.
[b] Coagulase-negative staphylococci.
[c] Most methicillin (oxacillin)-resistant *S. aureus* isolates are resistant to quinolones.
[d] Includes isolates which are susceptible, relatively resistant, or resistant to penicillin. There is no correlation between quinolone resistance and penicillin resistance in this species.

nalidixic acid, its activity is generally lower than the newer fluoroquinolones. Of the fluoroquinolones included here, the most active compounds against gram-positive bacteria are ciprofloxacin, sparfloxacin, and levofloxacin. The fluoroquinolone susceptibility of *S. aureus* isolates is closely tied to methicillin susceptibility – almost all MSSA are susceptible and almost all MRSA are resistant (C. Thornsberry, unpublished work). This is not particularly evident from the data shown in Table 4, probably because the data from which these numbers were derived did not reflect the current level of methicillin resistance. In the United States, the overall national incidence of MRSA is approximately 18% and the incidence of fluoroquinolone-resistant *S. aureus* is also approximately 18% (C. Thornsberry, unpublished). Therefore, in data collected very recently the MIC_{90} would be higher than that shown here, but an MIC_{50} would be low. To some extent this phenomenon also occurs in coagulase-negative staphylococci but there is not a one-to-one correlation as with MRSA.

Much controversy has arisen about the clinical use of fluoroquinolones for patients with *S. pneumoniae* infections. For those fluoroquinolones with the most activity – ciprofloxacin, ofloxacin, sparfloxacin, and levofloxacin – the majority of MICs (and MIC_{90}s) are borderline between susceptible and resistant, e.g., for ciprofloxacin, MICs of 1 μg/ml (susceptible) and 2 μg/ml (intermediate). The data are somewhat similar for enterococci, although they tend to be more resistant than pneumococci.

Data for some anaerobic species are shown in Table 5. The quinolones were considered inadequate in the past for the treatment of anaerobic infections. However, some newer ones have more activity against anaerobic bacteria, e.g., levofloxacin to some degree (MIC_{90}s of 0.5 μg/ml) for *Bacteroides* species and gram-positive cocci, and WIN57273, which has significant activity against all the species included here; neither of these drugs has yet been approved in the US, but levofloxacin is far along in the development process.

The MIC_{90}s for some intracellular pathogens are shown in Table 6. Norfloxacin showed the least activity and sparfloxacin and levofloxacin the most activity. The *Legionella spp* were the most susceptible species to the fluoroquinolones tested, the highest MIC_{90} being 0.12 μg/ml. The least susceptible species was *Ureaplasma urealyticum*. The activity of fluoroquinolones against *Mycobacterium tuberculosis* has created interest in using them as therapeutic agents for tuberculosis. For *Chlamydia* and *Mycoplasma*, sparfloxacin is the most active agent, but several others have marginal, but susceptible MIC_{90}s.

C. The Future

It is clear from these data that if one were developing a new fluoroquinolone, the aim would be to increase activity against gram-positive bacteria and anaerobic bacteria. Several newer compounds have been synthesized which have these properties. Some of these have not survived for safety or other

Table 5. The activity of various quinolones (as indicated by MIC_{90}s or ranges of MIC_{90}s) against anaerobes

Species	Ranges of MIC_{90}s (μg/ml)[a]										
	Nalidixic acid	Norfloxacin	Ciprofloxacin	Ofloxacin	Enoxacin	Sparfloxacin	Lomefloxacin	Fleroxacin	Levofloxacin	Pefloxacin	Win 57273
Bacteroides spp.[b]	≥128	8	1–32	2–32	8	4	8–32	2–64	0.5	8	0.25–1
Bacteroides fragilis	512	128	4–128	2–16	32–128	1–2	8–64	≥16	2–8	16	0.25–1
Clostridium spp.[c]	>512	128	1–16	1–8	32	4	16	2–32	0.25–4	ND	0.12–0.25
Clostridium difficile	≥128	128	8–32	8–16	128	8	≥32	16–32	4	64	ND
Clostridium perfringens	32	2	0.5–8	0.5–8	2	0.5–2	2–8	1–4	ND	1	ND
Fusobacterium spp.	512	16	2–8	2–16	32	1	16	16	ND	ND	ND
Gram-positive cocci	512	8	0.5–4	2–4	8–16	1–4	4–8	2–8	0.5	16	0.03–0.12

ND, not determined.
[a] For data sources, see Table 3.
[b] *Bacteroides* species other than *B. fragilis*.
[c] *Clostridium* species other than *C. difficile* and *C. perfringens*.

Table 6. The activity of various quinolones (as indicated by MIC_{90}s or ranges of MIC_{90}s) against miscellaneous species

Species	Ranges of MIC_{90}s (μg/ml)[a]							
	Norfloxacin	Ciprofloxacin	Ofloxacin	Enoxacin	Sparfloxacin	Lomefloxacin	Fleroxacin	Levofloxacin
Chlamydia pneumoniae	ND	2	1	ND	ND	ND	ND	0.25
Chlamydia trachomatis	≥16	1–4	0.5–2	ND	0.06	2–4	2–8	ND
Legionella spp.	ND	0.01–0.03	0.06	ND	0.03	ND	ND	0.01–0.12
Mycoplasma hominis	8	0.5–2	1–2	ND	≤0.06	2	2	0.25–1
Mycoplasma pneumoniae	16	0.5	2	8	ND	8	ND	0.5
Mycobacterium tuberculosis	2	1	1	2	0.25	ND	2	0.5
Ureaplasma urealyticum	16	1–16	2–4	ND	0.5	4–8	4	0.5

ND, not determined.
[a] For data sources, see Table 3.

reasons. For example, temofloxacin had greater activity against *S. pneumoniae* and was approved for the treatment of pneumococcal infections, but had to be withdrawn for safety reasons. Others have been withdrawn early in their development. Nevertheless, there are a number of these compounds still in development. Examples of these newer compounds are WIN57273, PD117558, PD127391 (CI960 or clinafloxacin), PD131628, PD117596, E4497, E3846, E4868, E5065, E5068, NM394, KB5246, QA241, CP99219 (trevofloxacin), DU6859A, and BAY-Y3118. The MIC_{90}s for some of these can be found in Eliopoulos and Eliopoulos 1993.

Since the data presented here are in vitro data (MICs), it is necessary to point out that there are testing factors which may affect results (Barry 1988; Eliopoulos and Eliopoulos 1993). If MICs are determined in urine (quinolones are used extensively in urinary tract infections) they tend to be considerably higher, i.e., the drug is less active than when tested in broth. Much of this loss of activity is due to magnesium – urines generally have higher contents of magnesium. pH will also affect results; MICs increase with decreasing pH, much higher, i.e., the activity of the drug is lower at a more acidic pH. For example, at pH 4.8 the activity is eight fold less than at pH 6.8 (i.e., the MIC will be eight fold higher). Thus, it is likely that the reduction in activity of quinolones in urine is due to both an increased magnesium concentration and a lower pH. Although the pH effect is evident for the quinolones generally available for clinical use in the US, there are a number of newer quinolones which are not susceptible to these pH changes.

In summary, the fluoroquinolones have a wide spectrum of antimicrobial activity. In general, the activity is greatest against the gram-negative species, especially the Enterobacteriaceae, respiratory pathogens such as *Haemophilus influenzae*, *M. catarrhalis*, and *Neisseria spp*. Among staphylococci, MSSA tend to be susceptible and MRSA are mostly resistant. Streptococci, enterococci, and corynebacteria mostly have MICs that are borderline susceptible. The presently available fluoroquinolones have minimal useful activity against anaerobes. Several of the fluoroquinolones have good to moderate activity against selected intracellular pathogens – the more active compounds were levofloxacin, sparfloxacin, ciprofloxacin, and ofloxacin. For these four drugs, the most susceptible species were *Legionella*, followed by *Mycoplasma*, *Chlamydia*, *M. tuberculosis*, and *U. urealyticum*. Many of the newer fluoroquinolones currently in development have increased activity against gram-positive bacteria and anaerobic species.

References

Appelbaum PC (1995) Comparative in vitro activity of sparfloxacin against organisms causing community-acquired LRTI. International Congress of Chemotherapy, Montreal, 17–20 July 1995

Barry AL (1988) In vitro activity of the fluoroquinolone compounds. Antimicrob Newslett 5:69–76

Chin NX, Huang HB et al. (1993) In vitro activity of DU-6859a, a new fluoroquinolone (Abstr 983; Session 92). Interscience Conference on Antimicrobial Agents and Chemotherapy, New Orleans, 17–20 October 1993

Eliopoulos GM, Eliopoulos CT (1993) Activity in vitro of the quinolones. In: Hooper DC, Wolfson JS (eds) Quinolone antimicrobial agents, 2nd edn. American Society for Microbiology, Washington, pp 161–192

Eliopoulos GM, Eliopoulos CT (1989) Quinolones antimicrobial agents: activity in vitro. In: Wolfson JS, Hooper DC (eds) Quinolone antimicrobial agents. American Society for Microbiology, Washington, pp 35–70

Felmingham D, Robbins MJ, Ghosh G et al. (1993) In vitro studies with DU-6859a, a new fluoroquinolone antimicrobial (abstr 981; Session 92) Interscience Conference on Antimicrobial Agents and Chemotherapy, New Orleans, 17–20 October 1993

Gootz TD, McGuirk PR (1994) New quinolones in development. Exp Opin Invest Drugs 3:93–114

Hooper DC, Wolfson JS (1993a) Mechanisms of quinolone action and bacterial killing. In: Wolfson JS, Hooper DC (eds) Quinolone antimicrobial agents, 2nd edn. American Society for Microbiology, Washington, pp 53–76

Hooper DC, Wolfson JS (1993b) Mechanisms of bacterial resistance to quinolones. In: Wolfson JS and Hooper DC (eds) Quinolone antimicrobial agents, 2nd edn. American Society for Microbiology, Washington, pp 97–118

Jones RN, Erwin ME, Houston AK (1993) In vitro activity of DU-6859a, a novel fluoro-chloroquinolone (Abstr 982; Session 92). Interscience Conference on Antimicrobial Agents and Chemotherapy, New Orleans, 17–20 October 1993

Dholakia N, Rolston KVI, Ho DH et al. (1994) Susceptibilities of bacterial isolates from patients with cancer to levofloxacin and other quinolones. Antimicrob Agents Chemother 38:848–852

Kuwahara-Airai K, Hori S, Hiramatsu K (1993) In vitro antimicrobial activity of a novel quinolone, DU-6859a (Abstr 977; Session 92). Interscience Conference on Antimicrobial Agents and Chemotherapy, New Orleans, 17–20 October 1993

Marshall SA, Jones RN, Murray PR et al. (1993) In vitro comparison of DU-6859a with other quinolones and oral cephalosporins tested against more than 5000 recent clinical isolates in the United States (Abstr 980; Session 92). Interscience Conference on Antimicrobial Agents and Chemotherapy, New Orleans, 17–20 October 1993

Moellering RC (1995) Overview of newer quinolones. International Congress of Chemotherapy, Montreal, 17–20 July 1995

Nakane T, Lyobe S, Mitsuhashi S (1993) Antibacterial activity of DU-6859a, a new quinolone (Abstr 978; Session 92). Interscience Conference on Antimicrobial Agents and Chemotherapy, New Orleans, 17–20 October 1993

National Committee for Clinical Laboratory Standards (1993a) Performance standards for antimicrobial disk susceptibility tests. Approved standard M2-A5. NCCLS, Villanova

National Committee for Clinical Laboratory Standards (1993b) Methods for dilution in antimicrobial susceptibility tests for bacteria that grow aerobically, 3rd edn. Approved standard M7-A3. NCCLS, Villanova

National Committee for Clinical Laboratory Standards (1994) Fifth informational supplement M100S5. NCCLS, Villanova

Norris S, Mandell GL (1988) The quinolones: history and overview. In: Andriole VT (ed) The quinolones. Academic, San Diego, pp 1–22

Phillips I, King A, Shannon K (1988) In vitro properties of the quinolones. In: Andriole VT (ed) The quinolones, Academic, San Diego, pp 83–117

Wang F, Wang Y, Zhang J (1993) In vitro antibacterial activity of DU-6859a. (Abstr 979; Session 92). Interscience Conference on Antimicrobial Agents and Chemotherapy, New Orleans, 17–20 October 1993

Weidemann B, Atkinson BA (1991) Susceptibility to antibiotics: species incidence and trends. In: Lorian V (ed) Antibiotics and laboratory medicine, 3rd edn. Williams and Wilkins, Baltimore, pp 962–1150

CHAPTER 6

Pharmacokinetics of Fluoroquinolones in Experimental Animals

A. DALHOFF and T. BERGAN

A. Introduction

Absorption, distribution, metabolism and excretion (ADME) are important processes which not only characterize the pharmacokinetics of an antibacterial agent but also influence significantly its antibacterial efficacy at the focus of infection. During the developmental process of an antimicrobial agent, infection models in experimental animals bridge the gap between the in vitro and clinical evaluation of an anti-infective agent. However, some restrictions exist, which, if ignored, will compromise the conclusions drawn from the data generated in animal models of infections. One of the major drawbacks in the use of animal models may be the differences in pharmacokinetics of a drug between animals and humans. It is well documented that animals eliminate drugs faster than humans (BOXENBAUM 1982; DEDRICK 1973; MORDENTI 1985, 1986; SAWADA et al. 1984). These differences can be overcome, for example, by repeated fractional dosing or continuous infusion. The species-specific differences in the routes of excretion present difficulties, and in drug metabolism these difficulties cannot be overcome.

Species-specific differences in drug metabolism of fluoroquinolones (FQs) may significantly affect their therapeutic efficacy as exemplified by pefloxacin and fleroxacin. In humans, pefloxacin is mainly metabolized to norfloxacin and pefloxacin-*N*-oxide, whereas in the urine of mice norfloxacin is not detectable; the norfloxacin/pefloxacin ratios were found to be 2–3 in humans and 0 in mice (MONTAY et al. 1984). As norfloxacin is two- to eight-fold more active than pefloxacin against *Pseudomonas aeruginosa*, pefloxacin administered to humans may be more effective against pseudomonal infections that data generated in mice might indicate. Fleroxacin exemplifies interspecies differences in drug metabolism and distribution. In mice and rats unmetabolized drug accounts for approx. 90% of total serum concentrations whereas in rabbits the fraction of intact drug is approx. 50% only. In rabbits, concentrations of unchanged drug at the foci of infection are lower than in mice and rats thus affecting the therapeutic efficacy of fleroxacin in different animal species. These differences in fleroxacin absorption, distribution, metabolism and excretion among the species emphasize the pitfalls inherent in interpreting and extrapolating data generated in various infection models without taking the interspecies differences into account.

Differences in the absolute bioavailability of orally administered FQs in various animal species, on the one hand, and between animals and humans, on the other hand, significantly affect therapeutic efficacies in different settings. For example, the absolute bioavailability of ciprofloxacin is low in rats and monkeys (SIEFERT et al. 1986a), whereas lomefloxacin or fleroxacin are almost completely absorbed (OKEZAKI et al. 1988a; FERNANDES et al. 1989). Thus, without taking these differences into account, data comparing efficacies of different FQs with each other are misleading. Unfortunately, many of the publications on antibacterial efficacies of FQs in experimental animals do not correlate therapeutic efficacies with pharmacokinetics especially of the unmetabolized drug.

In order to put such data into perspective, pharmacokinetic parameters of FQs in animals were compiled maily from the japanese literature. Although the terminology of metabolites of the various FQs is not congruent, the original identification of metabolites as published is quoted for the sake of comparability with original and subsequent publications on a given FQ. Unfortunately, in most of the cases the information on the antibacterial activity of the various metabolites is incomplete or missing. For comparison, only the pharmacokinetic parameters of FQs in healthy men are mentioned in addition; these data are quoted from the review by KARABALUT and DRUSANO (1993). For more detailed information the reader should refer to the references quoted therein and to Chap. 10 in this volume.

B. Norfloxacin

Absorption of norfloxacin after administration of an oral dose to experimental animals was investigated by MURAYAMA et al. (1981) by using the bioassay. Levels peaked in serum 0.5–1 h after administration. Following administration

Table 1. Pharmacokinetic constants of norfloxacin measured by using a bioassay; oral administration

Species (Dose: mg/kg in animals; total mg in man)	Cmax (mg/1)	Tmax (h)	t1/2 (h)	Urinary excretion (%)	Faecal excretion (%)
Mouse (50)	1.0	0.5	NA	6.1[a]	91.4[a]
Rat (50)	3.0	0.5	NA	8.4[a]	85.4[a]
Dog (30)	0.63	1.0	NA	NA	NA
Dog (25)	NA	NA	NA	16.6[a]	73.8[a]
Monkey (25)	NA	NA	NA	17.0[a]	72.4[a]
Man (400)	1.58	1.3	7.4	20–30[b]	28.0[b]

Cmax, maximum serum concentration; Tmax, duration of Cmax; t1/2, elimination half-life; NA, not available.
[a] % of radioactivity.
[b] % of dose.

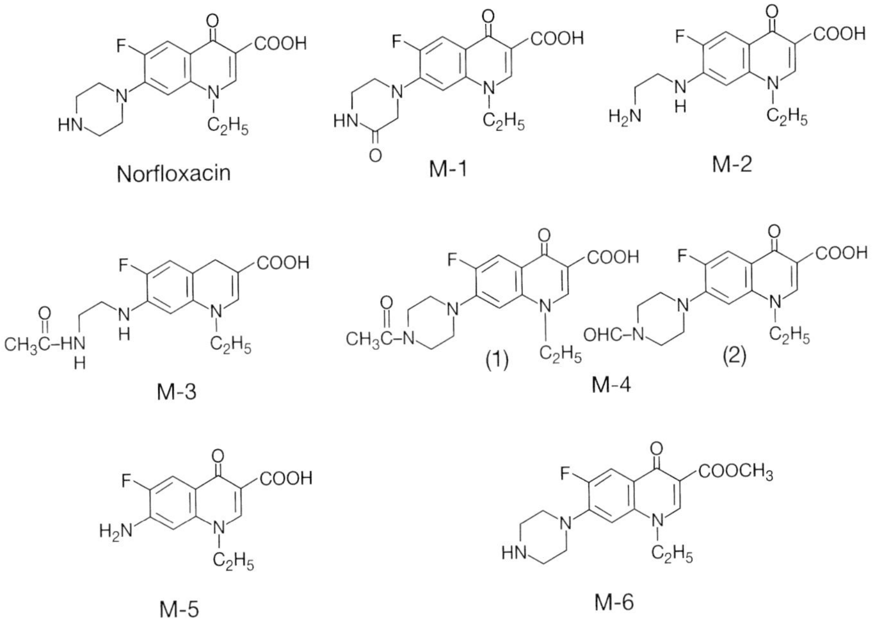

Norfloxacin M-1 M-2 M-3 (1) M-4 (2) M-5 M-6

Fig. 1. Chemical structures of norfloxacin and its metabolites. (From NAGATSU et al. 1981b)

of [^{14}C]norfloxacin to rats, the distribution of radioactivity to most of the tissues studied was rapid (except to the brain), and concentrations in the bladder, submaxillary gland, lymph node, liver, pancreas, spleen, adrenal and kidney significantly surpassed the corresponding serum concentration at 0.25 h after administration (NAGATSU et al. 1981a). Most of the orally administered norfloxacin was excreted in the urine and faeces as unchanged drug (Table 1; NAGATSU et al. 1981b). Seven metabolites were identified (Fig. 1). The main metabolite in rats' and monkeys' urine was M1. The primary metabolite in the bile of rats was M6, which accounted for 57% of the label; M6 was found in the urine of rats to account for 3.5% of the radioactivity. In all the experimental animals studied the major part of the administered dose was excreted via the faeces (Table 2; NAGATSU et al. 1981a). The antibacterial activity of the metabolites was not specified.

C. Pefloxacin

Pefloxacin is well absorbed following oral administration, and the absolute bioavailability of unchanged drug amounted to 64%, 94% and 84% in dogs, monkeys and humans, respectively (Table 3; MONTAY et al. 1984). The major metabolites in rat and dog plasma were pefloxacin glucuronide, pefloxacin-*N*-oxide and norfloxacin. The latter two were also found in monkey and human plasma; no conjugate could be detected.

Table 2. Excretion of norfloxacin (% of total radioactivity either in the urine and/or faeces within 24 h; oral administration)

Species	Unchanged drug		Conjugates of unchanged drug		M1		M2		M4 (1), M4 (2)		M5		M6		Others (unclassified)	
	Urine	Faeces	Urine	Faeces	Urine	Faeces	Urine	Faeces	Urine	Faeces	Urine	Faeces	Urine	Faeces	Urine	Faeces
Rat	74-3	88.3	1.3	1.4	12.4	0.8	NA	2.3	3.7	1.5	1.3	1.2	3.5	n.d.	3.5	4.5
Dog	84-6	83.2	1.5	1.1	0.8	0.7	NA	3.7	0.4	1.7	1.6	2.4	n.d.	n.d.	11.1	7.2
Monkey	81.5	90.6	NA	0.3	9.7	0.4	NA	1.1	2.8	1.1	3.0	0.4	n.d.	n.d.	3.0	6.1
Man	~80.0	NA	NA	NA	NA	NA	NA	NA	NA	NA	NA	NA	NA	NA	NA	NA

n.d., not detectable; NA, not available.

Table 3. Pharmacokinetic constants of pefloxacin; oral administration

Species (Dose: mg/kg in animals; total mg in man)	Cmax (mg/1)	t1/2 (h)	AUC (mg × h/l)
Mouse (50)	5.8	1.9	8.8
Rat (50)	13.0	3.3	56.0
Dog (50)	17.1	3.7	100.6
Monkey (25)	12.4	5.5	108.0
Man (400)	3–8	8.6	48.2

AUC, area under serum-concentration–versus–time curve. See Table 1 for other abbreviations.

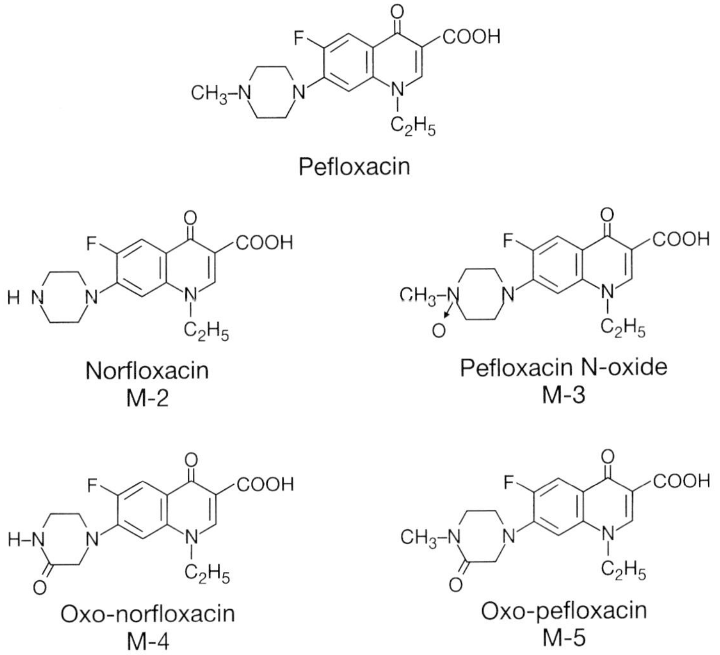

Fig. 2. Chemical structures of pefloxacin and its metabolites. (From Montay et al. 1984)

Pefloxacin was well distributed to tissues and body fluids. In the rat, levels in most of tissues were two to three times higher than the corresponding plasma concentrations; those in liver kidney and spleen were three to six times higher and in the brain four times lower. Metabolism of pefloxacin is significantly different in the various species studied (Table 4; Fig. 2). In urine and bile of mice the only identified metabolites were the glucuronide and *N*-oxide.

Table 4. Renal excretion of pefloxacin (% of dose within 72 h; oral administration)

Species (Dose: mg/kg in animals; total mg in man)	Pefloxacin	Norfloxacin	Pefloxacin glucuronide	Pefloxacin *N*-oxide	Oxafloxacin norfloxacin	Oxafloxacin pefloxacin	Total
Mouse (50)	16.5	n.d.	8.8	4.3	n.d.	Traces	29.5
Rat (25)	6.6	6.4	14.7	10.1	Traces	Traces	37.8
Dog (10)	4.9	4.5	14.9	12.3	Traces	Traces	36.3
Monkey (25)	6.2	9.8	2.5	4.8	2.2	0.9	26.5
Man (400)	9.3	20.2	Traces	23.2	5.4	0.75	58.9

n.d., not detectable.

Table 5. Pharmacokinetic constants of unchanged enoxacin; oral administration

Species (Dose: mg/kg in animals; total mg in man)	Cmax (mg/l)	Tmax (h)	t1/2 (h)	AUC (mg × h/l)
Mouse (20)	0.8	1.0	2.0	1.8
Rat (20)	0.6	0.5	2.8	1.9
Dog (20)	4.2	2.0	5.0	51.8
Monkey (20)	4.2	2.0	4.2	32.0
Man (400)	3.1	1.4	4.9	18.5

See Tables 1 and 3 for abbreviation explanations.

In addition, three unidentified metabolites were detected in the urine. In rat and dog urine and bile the principal metabolites were the glucuronide and the *N*-oxide; norfloxacin and pefloxacin were excreted in minor amounts. Monkey urine contained mainly norfloxacin and pefloxacin and substantially less of pefloxacin-*N*-oxide, oxopefloxacin, oxonorfloxacin, and glucuronide. In humans, mainly norfloxacin and pefloxacin-*N*-oxide were renally excreted. Biliary excretion in monkeys and humans was less than in rats and dogs. The antibacterial activity of the identified pefloxacin metabolites is minimal except for norfloxacin.

D. Enoxacin

Enoxacin was found to be rapidly absorbed in the various animal species studied (Table 5). The metabolite M2, but not other metabolites, could be detected in the plasma of mice, rats and dogs in trace amounts but in monkeys it amounted to about 60% of the unchanged drug (YAMAGUCHI et al. 1984; NAKAMURA et al. 1984). Following repeated b.i.d. dosing of 50 and 40 mg/kg to dogs and monkeys, respectively, for 34 days, the concentration in heart and muscle exceeded the corresponding serum concentrations by approximately a three-fold amount. In the lung a two-fold excess and in the liver and kidney more than a ten-fold excess was measured, whereas levels in the brain were about 7% of those in serum at 3 h after the last dosing. Total urinary excretion varied from approximately 20% to 56% in the various animal species. Most of the drug excreted via the urine was unchanged except in the case of the rat which excreted 6.2% as glucuronide and the monkey which excreted 12% as M2 (Table 6; Fig. 3; YAMAGUCHI et al. 1984; SEKINE et al. 1984; NAKAMURA et al. 1984). Except for the glucuronide all other metabolites had a similar spectrum of antibacterial activity, but their potency was three to ten times lower than that of enoxacin (SEKINE et al. 1984).

The protein binding of enoxacin (5 mg/l) in mice, rats and dogs was 27.6%, 34.5% and 34%, respectively. In humans, protein binding was 35%–40% (NAKAMURA et al. 1984). Following oral administration to human volunteers of either 400 mg or 600 mg, protein binding of the unchanged drug was 48%

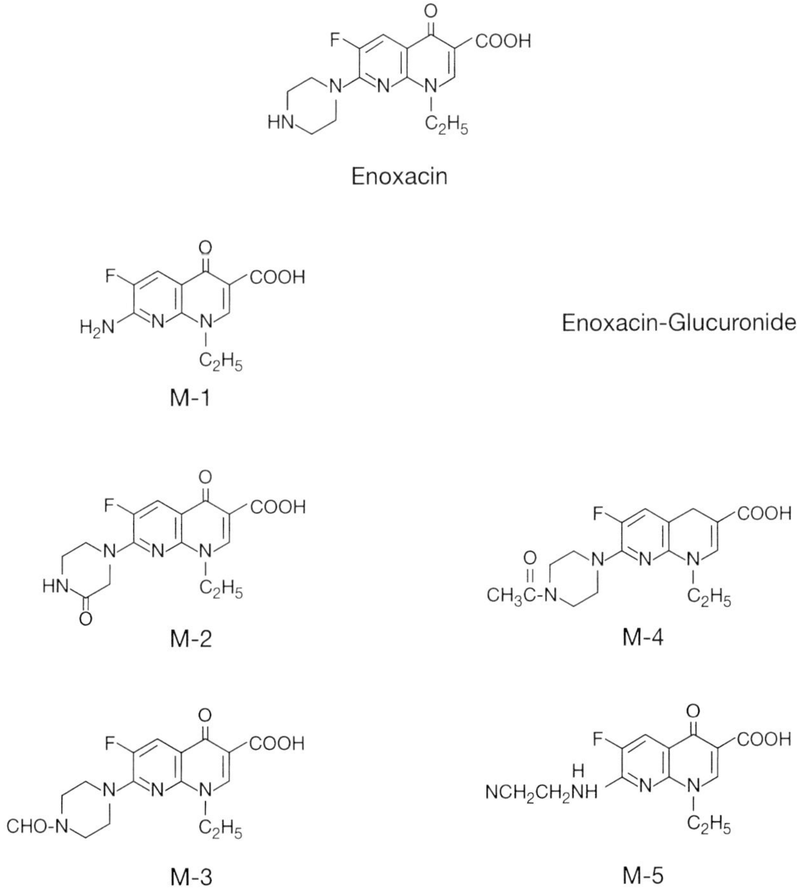

Enoxacin

M-1

Enoxacin-Glucuronide

M-2

M-4

M-3

M-5

Fig. 3. Chemical structures of enoxacin and its metabolites. (From SEKINE et al. 1984)

Table 6. Excretion of enoxacin (% of dose within 24h; oral administration)

Species	Unchanged (Urine)	M1 (Urine)	M2 (Urine)	M3 (Urine)	M4 (Urine)	M5 (Urine)	Glucuronide (Faeces)	Total (Urine/faeces)
Mouse	28.4	0.3	0.7	0.3	0.2	0.9	0	30.8
Rat	12.6	0.1	0.3	0.1	0.2	Traces	6.2	19.4
Dog	27.2	0.1	0.1	0.1	Traces	1.5	0	28.9
Monkey	42.4	0.1	12.0	0.1	0.7	0.7	0	55.9
Man	~50.0	ND	~16.0	ND	ND	ND	ND	~60

ND, not determined.

Table 7. Pharmacokinetic constants of unchanged ofloxacin; oral administration

Species	Cmax (mg/l)	Tmax (h)	t1/2 (h)
Rat	2.6	~2.0	NA
Dog	14.2	1.34	4.37
Monkey	9.3	2.35	2.81
Man (200 mg)	2.2	1.3	5.5

See Table 1 for abbreviation explanations.

and 54%, respectively, and that of the M2 metabolite was 58% and 69%, respectively.

E. Ofloxacin

Upon oral administration to the experimental animals ofloxacin was rapidly absorbed and distributed throughout the extravascular space (Table 7). Radioactively labelled ofloxacin peaked in tissues at 2h postmedication (OKEZAKI et al. 1988a,b; SUDO et al. 1984; TSUMURA et al. 1984). In the pituitary gland, trachea and salivary gland tissue, concentrations exceeded the corresponding serum concentrations by a two- to three-fold amount and in the liver and kidney by a five- to eight-fold amount; in all other tissues studied, levels equalled those in serum.

The main metabolites in rat and dog were the ofloxacin ester glucuronide which was detectable only in low concentrations in the monkey (Table 8; Fig. 4). In man, approximately 4%–5% of the drug is recovered as either desmethyl ofloxacin or ofloxacin *N*-oxide. Demethyl ofloxacin has an antibacterial activity similar to that of enoxacin, whereas *N*-oxide is inactive (WHITE et al. 1987; WHITE 1988, 1993).

Faecal excretion of ofloxacin in rats and dogs was pronounced. The major biliary excretion product in rats was M1 (ofloxacin glucuronide; 76%–85%); biliary excretion of unchanged ofloxacin amounted to 10%–14% (values are not given in per cent of dose but in per cent of total radioactivity in that sample; SUDO et al. 1984). In the rat, approximately 30% of radioactivity excreted via the bile was reabsorbed from the intestine.

Urinary excretion of unchanged ofloxacin varied from 54% to 77% of the administered dose in rats and monkeys, respectively.

Serum protein binding was approximately 50% in the rat and dog and approximately 30% in the monkey following an oral dose of 20 mg/kg and a sampling period of 12 h (OKEZAKI et al. 1988a,b). The authors reported that ofloxacin had a comparably high affinity to erythrocytes.

Table 8. Excretion of ofloxacin (% of dose within 48 h; oral administration)

Species	Unchanged		M1 Glucuronide		M3[a] Demethyl		M4 Oxide	
	Urine	Faeces	Urine	Faeces	Urine	Faeces	Urine	Faeces
Rat	54.3	46.7	4.3–11.4	~0.7	4.6–5.7	4.5–11.9	1.6–2.4	3.8–4.8
Dog	61.5	34.8	3.1–5.0	~1.0	4.9–7.3	4.9–14.9	5.3–8.7	5.0–5.6
Monkey	77.0	1.5	0.8–1.9	0.2–1.5	3.4–9.5	12.3–31.7	1.9–12.9	1.6–9.4
Man	73.6	n.a.	n.d.	n.a.	3.0	n.a.	1.0	n.a.

n.d., not detectable; n.a., not assessed.
[a] % of total radioactivity recovered in each sample.

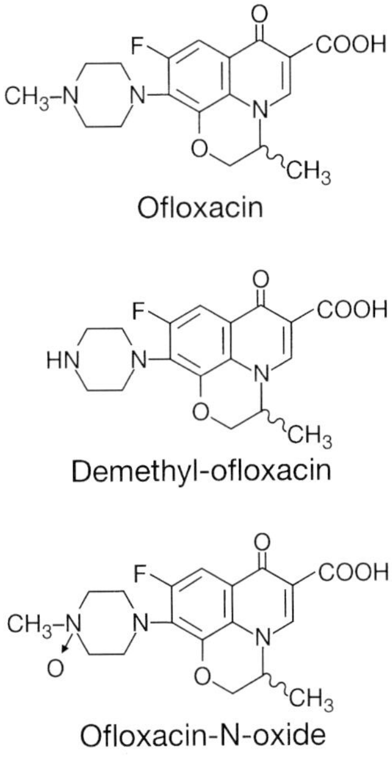

Ofloxacin

Demethyl-ofloxacin

Ofloxacin-N-oxide

Fig. 4. Chemical structures of ofloxacin and its metabolites. (From SUDO et al. 1984)

F. Ciprofloxacin

Following oral administration of [^{14}C]ciprofloxacin to rats, dogs and monkeys, ciprofloxacin was partially absorbed – 30%–44% of the area under the serum-concentration versus time curves (AUCs) obtained with i.v. dosing – and plasma concentrations of unchanged ciprofloxacin peaked at 0.33 h (rats), 1.7 h (dogs) and 3 h (monkeys; Table 9; SIEFERT et al. 1986a; NAKAMURA et al. 1990). The unchanged drug was eliminated with a half-life ranging from 3 h to 4.4 h; the total radioactivity was eliminated from plasma with terminal half-lives ranging from 26 h to 44 h (observation period up to 48 h after dosing).

In the rat, 8.2% of the radioactivity was excreted as unchanged drug via the urine and 81% via the faeces. A further 5% of the dose excreted renally was contributed by a single biotransformation product which was rather unstable. Approximately 5% was excreted as glucuronide. In the monkey, 18% of unchanged ciprofloxacin was excreted renally and 50% via the faeces. Metabolites M1 and M3 (Fig. 5) could be detected in minor amounts in the urine and faeces of monkeys (Table 10).

Autoradiography following i.v. and oral administration to rats revealed that radioactivity was distributed rapidly throughout the body (SIEFERT et al. 1986b). Compared with plasma levels, high ciprofloxacin concentrations were detected in the kidneys, liver, skeleton muscle, pancreas, testes and cartilage; low concentrations were determined in brain and adipose tissue.

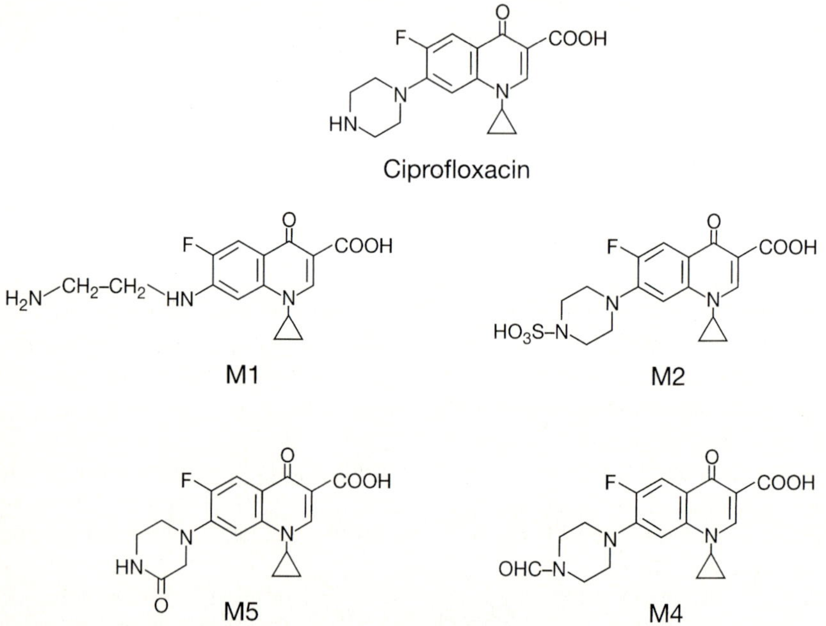

Ciprofloxacin

M1

M2

M5

M4

Fig. 5. Chemical structures of ciprofloxacin. (From ZEILER et al. 1987)

Table 9. Pharmacokinetic constants of unchanged ciprofloxacin; oral administration

Species (Dose: mg/kg in animals; total mg in man)	Cmax (mg/l)	Tmax (h)	t1/2 (h)	AUC (mg × h/l)
Rat (5)	0.25	0.33	3.0	0.78
Dog (5)	0.70	1.7	3.3	4.77
Monkey (30)	0.88	3.0	4.4	7.1
Man (500)	2.5–2.9	1.3	4.1–4.8	11.1–12.7

See Tables 1 and 3 for abbreviation explanations.

Table 10. Excretion of ciprofloxacin in various animals (% of dose within 48 h; oral administration)

Species	Unchanged		M1		M2		M3[a]		M4	
	Urine	Faeces	Urine	Faeces	Urine	Faeces	Urine	Faeces	Urine	Faeces
Rat	8.2	81.0	NA	NA	NA	NA	NA	NA	NA	NA
Monkey	18.0	50.0	0.7	1.8	NA	NA	NA	1.3	NA	NA
Man	44.7	25.0	1.4	0.5	3.7	5.9	6.2	1.1	n.d.	n.d.

NA, not available; n.d., not detectable.
[a] % of total radioactivity recovered in each sample.

The antibacterial activity of the ciprofloxacin metabolite M1 was comparable to that of nalidixic acid, whereas M2 was significantly less active. M4 had an activity similar to norfloxacin against gram-negative bacteria and similar to ciprofloxacin against staphylococci. M3 had a broad spectrum of antibacterial activity which was slightly less marked than that of norfloxacin (ZEILER et al. 1987).

G. Temafloxacin

Absorption, distribution, metabolism and excretion of temafloxacin have been extensively studied in rats and dogs (TAKEDA et al. 1993; ENDO et al. 1993; AIHARA and SHIGEMATSU 1993; KOHNO et al. 1993; YANO et al. 1993). Following oral administration of labelled drug to rats, serum concentrations were found to peak at 0.5 h, and the radioactivity reached its maximum in most tissues at 0.5–1 h after dosing (Table 11). Except for brain, eye and bone, where the radioactivity was low, in all other tissues studied the level of radioactivity was similar or slightly higher than in serum.

In principle, temafloxacin can be metabolized into eight metabolites (Fig. 6). In addition, temafloxacin sulphate was detected in urine and bile of rats. Urinary excretion of temafloxacin was approximately 42% of the administered dose, and the glucuronide and sulphate were renally excreted in amounts equal to 0.6% and 0.8% of the dose respectively, within 24 h. Each of the other three metabolites including their corresponding glucuronides which were excreted amounted to 0.3%–0.6% of the dose.

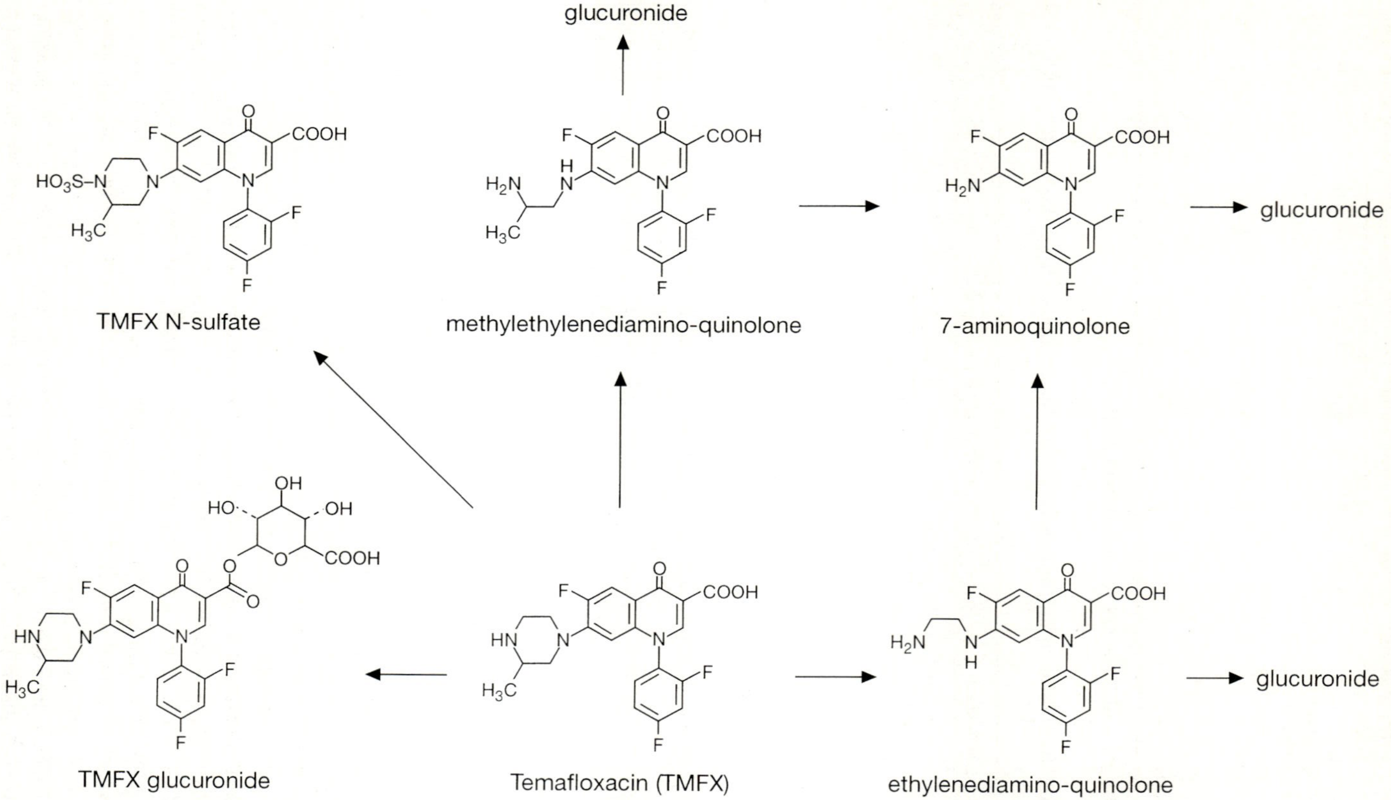

Fig. 6. Chemical structures of temafloxacin and its metabolites. (From KOHNO et al. 1993)

Table 11. Pharmacokinetic constants of unchanged temafloxacin; oral administration

Species (Dose: mg/kg in animals; total mg in man)	Cmax (mg/l)	Tmax (h)	t1/2 (h)	AUC (mg × h/l)	UE (% of dose)
Mouse (20)	8.2	0.4	1.5	12.9	NA
Rat (20)	4.8	0.5	2.1	12.7	41.9
Dog (20)	8.5	1.7	5.0	73.3	38.8
Man (200)	2.3	2–3	7.4	16.9	60

See Tables 1 and 3 for abbreviation explanations.

Similarly, in the dog, 39% of the dose was excreted via the urine, and the ethylenediamine and methylethylenediamine metabolites of temafloxacin were excreted at levels corresponding to 0.2% and 4% of the dose, respectively. In humans, urinary recovery of the metabolites accounted for 6.2% of the dose. The antibacterial activity of the metabolites was not specified. The affinity of temafloxacin to various matrices was determined by ENDO et al. (1993). Sixteen percent of the drug bound to human serum albumin, 6% to α1-acid glycoprotein, 4% to globulin and 6% to bovine β-lipoprotein.

H. Tosufloxacin

Compared with temafloxacin, data on tosufloxacin are scarce. Only one study described the kinetics in mice, rats, rabbits and dogs (YASUDA et al. 1988b). Urinary excretion was highest in mice (21.1%), followed by rabbits (15.2%), rats (11.5%) and dogs (2.7%). Biliary excretion in rats was 0.13% "as the active form". Bioautographic studies revealed that "the main active form in vivo was unchanged tosufloxacin" (YASUDA et al. 1988b). Other methods to identify metabolites of tosufloxacin were not applied. Serum protein binding following oral administration to rats, rabbits and humans were 56.1%, 45.7% and 40.5%, respectively (YASUDA et al. 1988a).

I. Fleroxacin

Following a single oral dose of 10 mg/kg body weight to rats a maximal serum concentration of 3.2 mg/l was recorded, declining with a half-life of 2.3 h; the AUC was 14.8 mg $h^{-1} l^{-1}$ (NAGATSU et al. 1990). Unmetabolized drug accounted for 80%–90% of total serum concentrations in mice and for >90% in rats, whereas the fraction of intact drug in rabbits was 34%–67% (RUBIN et al. 1993). Positron emission tomographic imaging of mice, rats and rabbits revealed that fleroxacin delivery to most tissues was rapid with the notable exception of the brain (FISCHMAN et al. 1992; RUBIN et al. 1993). Within 10 min >90% of the radiolabelled drug was cleared from the blood into tissues.

Within 48 h urinary and faecal recovery of the unchanged drug was 54.2% and 45.7% of the dose, respectively. In rats, 50% of the radioactivity excreted into the bile was reabsorbed. Except for the excretory organs, radioactivity in most other tissues equalled that in serum. Values were lower in the brain, lens, hypophysis and fat. The metabolic pattern in the urine of dogs and rabbits is described by KAWAHARA et al. (1990). Two metabolites were detected in urine of dogs and rabbits: demethyl fleroxacin and fleroxacin *N*-oxide (Fig. 7). In the urine of rabbits, two other metabolites were identified: demethyl-3-oxo fleroxacin and demethyl-4-formyl fleroxacin. The unchanged drug was the predomininant form in the urine from mice and rats, whereas *N*-demethyl fleroxacin was the most abundant form in rabbit urine. Drug metabolism in

Fleroxacin

Demethyl-3-oxo-fleroxacin (M-1)

Formyl-demethylfleroxacin (M-2)

Demethyl-fleroxacin (M-4)

Fleroxacin-N-oxide (M-5)

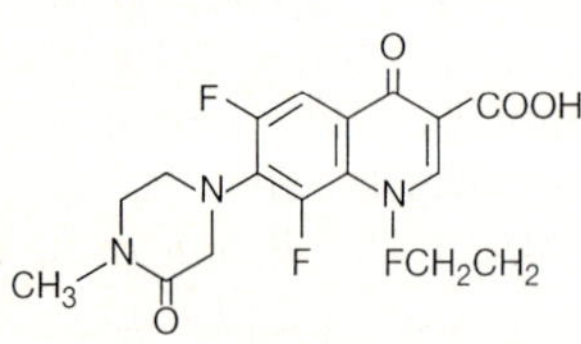

Oxo fleroxacin (M-3)

Fig. 7. Chemical structures of fleroxacin. (From KAWAHARA et al. 1990)

humans was comparable to that in rats and mice (RUBIN et al. 1993). In humans, 50%–65% was excreted as unchanged drug via the urine, with *N*-desmethyl and *N*-oxide metabolites accounting for another 6.5%–11% of the dose.

J. Lomefloxacin

Following oral administration of lomefloxacin to various animal species, peak serum concentrations were reached approximately 2 h after dosing (Table 12). Ratios of the AUCs after oral and intravenous administration to rats (0.85) and dogs (0.92), respectively, indicate nearly complete bioavailability (OKEZAKI et al. 1988a,b). Autoradiography after single oral administration to rats revealed that the radioactivity was equally distributed in serum and most of the organs, with tissue/serum ratios ranging from 0.7 to 1.1 except for spleen (1.5), pancreas (1.8), liver (2.04), kidney (3.7), fat (0.08), eyeball (0.08) and thyroid (0.61; NAGATA et al. 1988b).

Table 12. Pharmacokinetic constants of unchanged lomefloxacin; oral administration

Species (Dose: mg/kg in animals; total mg in man)	Cmax (mg/l)	Tmax (h)	t1/2 (h)	AUC (mg × h/l)
Mouse (20)	5.30	NA	2.9	14.5
Rat (20)	9.73	2	1.83	22.6
Dog (20)	9.79	NA	6.47	60.2
Monkey (20)	4.49	2	2.86	NA
Man (400)	5.22	0.9	8.1	27.9

See Tables 1 and 3 for abbreviation explanations.

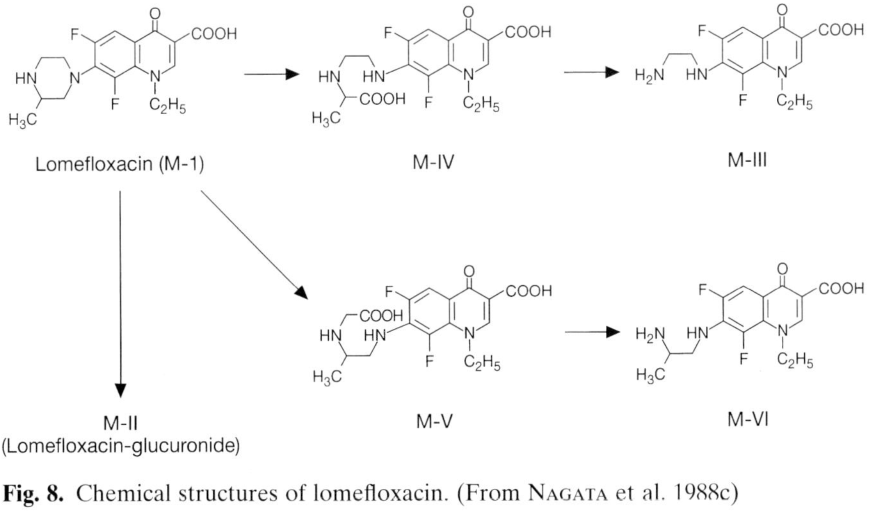

Fig. 8. Chemical structures of lomefloxacin. (From NAGATA et al. 1988c)

Table 13. Excretion of lomefloxacin in various animals; oral administration

Species	Unchanged M1		M2		M3		M4, M5		M6		Others		Total	
	Urine	Faeces	Urine	Faeces	Urine	Faeces	Urine	Faeces	Urine	Faeces	Urine	Faeces	Urine	Faeces
Rat[a]	69.8	16.9	4.3	0.2	0.5	0.1	0.5	0.2	0.8	0.3	n.d.	n.d.	75.9	17.7
Dog[a]	50.9	14.7	1.2	0.5	0.3	0.3	1.4	2.8	0.8	0.7	0.4	0.4	55.0	19.4
Monkey[b]	63.4	5.6	"very small"	"very small"	"very small"	"very small"	"very small"	"very small"	"very small"	"very small"	NA	NA	~63.4	~5.6
Man[a]	~54.1	NA	7.2	NA	NA	NA	NA	NA	NA	NA	NA	NA	61.3	NA

n.d., not detectable; NA, not available.

[a] % of dose within 24 h.

[b] % of dose within 96 h.

In rats, dogs and monkeys lomefloxacin is excreted mostly unchanged (NAGATA et al. 1988b,c; Table 13; Fig. 8). The major metabolite in urine and faeces was M2, the glucuronic acid conjugate of lomefloxacin. Metabolites 4 and 5 were cleaved at the piperazine ring and M3 and 6 were further metabolized through oxidative dealkylation of M4 and M5. Biliary excretion of unchanged lomefloxacin was 4.6% and that of M2 in rats within 24h amounted to 14%–17.3% of the dose; M3 to M6 were excreted in the bile in trace amounts (0.1% to 0.6% of the dose). Bioautography of thinlayer chromatogrammes revealed that no bioactive metabolites were excreted in the urine. Lomefloxacin is mainly renally excreted in all species studied, with urinary recovery ranging from 55% to 76%.

K. Sparfloxacin

Data on sparfloxacin kinetics in experimental animals are scarce. The pharmacokinetic constants vary significantly between the species studied (Table 14; NAKAMURA et al. 1990, 1991). The absolute bioavailability in dogs was found to be 77%. The urinary recoveries within 48h were increased upon β-glucuronidase treatment and amounted to 7.8%, 16.3%, 8.9% and 18.9% of the administered dose in mice, rats, dogs and monkeys, respectively. This indicates that sparfloxacin is first mainly excreted via nonrenal routes and that the acylglucuronide of sparfloxacin is the only metabolite, as in humans.

L. Penetration of Quinolones at Sites of Infection

The pharmacokinetics of antibacterial agents usually involve details of absorption, distribution, metabolism and elimination. Rates of absorption and routes of elimination, whether renal, biliary, transintestinal or metabolic and impacts of diseases such as renal or hepatic impairment, are very important. Tissue distribution, from the amount of attention it receives, would appear to be a minor issue.

Table 14. Pharmacokinetic constants of unchanged sparfloxacin; oral administration

Species (Dose: mg/kg in animals; total mg in man)	Cmax (mg/l)	Tmax (h)	t1/2 (h)	AUC (mg × h/l)	UE (% of dose)
Mouse (5)	0.25	0.3	5.0	0.74	6.7
Rat (5)	0.50	0.4	3.8	2.05	12.9
Dog (5)	1.14	3.9	8.0	17.74	8.6
Monkey (5)	0.49	2.0	11.7	4.61	12.7
Man (200)	0.4–0.7	3.5–6.0	15.8–18.8	10.6–18.8	40

See Tables 1 and 3 for abbreviation explanations.

Serum concentrations of antibacterial agents have been used traditionally to predict or have been correlated to the minimum inhibitory concentrations (MICs) for likely pathogens and to clinical outcome. This approach might be justifiable if bacteraemia is studied experimentally or if a patient is treated for sepsis. However, this approach might be less relevant if the potential sites of infection are within the extravascular site or intracellular. The studies performed by May as early as 1955 and STAMEY et al. (1970) have indicated that local concentrations of antimicrobial agents, being in part higher than those in serum, correlate to clinical outcome in chronic bronchitis or urinary tract infections even if serum concentrations would have predicted a clinical failure. Thus, the concentration of an antimicrobial agent at the site of infection (e.g. in a particular tissue) will contribute to its clinical efficacy. However, in order to exert their potentially benificial effects, the antibacterial agents must penetrate into the tissues or body cavities, i.e. the foci of infection, both in sufficient concentrations and for sufficient periods of time.

Studies by CRAIG and EBERT (1991, 1992), CRAIG et al. (1991), GERBER et al. (1982a,b, 1983, 1986), GERBER and FELLER-SEGESSENMANN (1985) VOGELMANN and CRAIG (1986), and VOGELMANN et al. (1988) have contributed to our understanding of key predicting characteristics of what is needed for adequate activity against an infection. Their pioneering studies have led to the development of concepts correlating pharmacokinetic and pharmacodynamic parameters to therapeutic efficacy. For β-lactams, the neutropenic mouse-thigh model has demonstrated that the essential point is maintenance of the serum concentration above the in vitro MIC of the antibiotic. Serum concentrations exceeding the MIC need to be maintained for a certain period of time, preferably throughout the entire dosage interval, possibly excluding the duration of the postantibiotic effect (PAE). However, the PAE of β-lactams is limited to a few bacterial species only, such as staphylococci and pneumococci or specific strains of gram-negative species. Consequently, the PAE of β-lactams is not of general relevance in vivo. Since β-lactams lack a PAE, serum concentrations should not fall below the MIC as the absence of the PAE cannot prevent organisms from regrowth after β-lactam concentrations fall below the MIC. Furthermore, the bactericidal in vitro efficacy of β-lactams is largely concentration independent. By using a variety of dosing regiments it was convincingly demonstrated that antibacterial in vivo efficacy of various β-lactams correlated best to the period of time during which the serum concentrations exceeded the MIC – ΔT/MIC – but not to elimination half-life (t1/2), AUC or maximal serum concentration (Cmax).

In contrast, aminoglycosides and compounds such as quinolones show a different essential determinative parameter. For these compounds the absolute serum concentrations and the AUC correlate to the outcome of treatment. As there is a strong covariance between serum concentrations and Cmax and between serum concentrations and AUC, both variables (Cmax and AUC) correlate quite well to therapeutic outcome and the suppression of emergence of resistance. Furthermore, quinolones and aminoglycosides exhibit a concentration-dependent bactericidal effect in addition to a PAE, and hence the ΔT (MIC) is not the decisive parameter, but rather the AUC and the Cmax (see chapter by STAHLMANN and LODE, this volume).

The existence of two different classes of antibacterial agents with differing pharmacokinetic/pharmacodynamic (PK/PD) properties might seem surprising. There is no apparent common denominator for *β*-lactams and quinolones/aminoglycosides with respect to their mechanisms of antibacterial action, high/low serum protein binding and host–parasite relationships. However, there is one common denominator for these important preclinical observations, which have been shown to be valid in the clinical arena, that clearly explains the distinction between classes of antibacterial agents. This denominator is tissue penetration ability.

How can pharmacokinetics explain why these two classes of compounds are distinct? That which governs the penetration of antibiotics at sites of infection is the same in both instances and does not depend on their PK/PD properties. The difference which does explain the apparent disparate behaviours of various antibacterials is the end result of penetration. *β*-lactam compounds essentially remain extracellular and reach minute concentrations inside the sessile tissue or parenchymous cells and migrant blood cells with an important phagocytic function, such as polymorphonucelar leukocytes or macrophages.

Compounds remaining extracellularly, such as *β*-lactams which have a limited ability to penetrate, exhibit concentrations in extracellular fluids including peripheral lymph that are on level with the serum concentrations in a true steady state situation after a constant rate infusion. Under these conditions there is an equilibrium between the concentration of non-protein-bound antibiotic in the serum and in extracellular fluids including lymph. However, with a repetitive dosing regimen no true steady state will be achieved and concentrations will fluctuate between peak and trough levels. Similar to the in vitro and in vivo situation, the free, non-protein-bound fraction of an antibiotic is antibacterially active, and only this fraction can diffuse freely between intra- and extravascular compartments; this leaves the free antibiotic in serum and tissue as the relevant fraction for the PK/PD correlations. Because of fluctuation, the concentrations of free drug in plasma and tissue will be different (DERENDORF 1989; DERENDORF et al. 1990; LIMBERG et al. 1990). Only in the true steady-state situation will they be equal. Consequently, under steady-state conditions the pharmacodynamics of antibacterials such as *β*-lactams are governed by the duration for which blood concentrations are maintained above the MIC.

Quinolones and macrolides, in contrast, are compounds which accumulate within tissues – in both sessile and migrant cells – so that tissue concentrations exceed the serum concentrations. Tissue penetration depends on the concentration gradient between plasma and tissue, so that the driving force for the exchange of drug between central and peripheral compartments is the difference of free drug concentrations. Provided that no specific transport or ion-trapping mechanisms mediate drug penetration into specialized sites such as prostatic fluid and lung mucosa (STAMEY et al. 1970; BALDWIN et al. 1992a,b; HONEYBOURNE and BALDWIN 1992), tissue penetration is governed by the Fick equation (BERGAN 1978, 1981). The adherence of the laws of penetration according to Fick's formula has been demonstrated for both β-lactams and cell-accumulating agents in tissue chambers and peripheral human fluid (BERGAN et al. 1987; BERGAN 1978).

These observations corroborate the finding that the total area under the serum concentration curve is the determining factor for the degree of tissue penetration. When a compound with low-level protein binding remains mainly extracellular, e.g. β-lactams such as ampicillin and cefuroxime (RYAN and CARS 1980), balance is almost immediately established between the serum compartment and the extracellular fluid compartment (including peripheral lymph). Consequently, the duration of supra-MIC levels in serum becomes the obvious apparent determinant of the degree of penetration. However, the non-protein-bound fractions, i.e. free plasma and free tissue concentrations need not be equal; on the contrary, they will most frequently differ. This was convincingly demonstrated by DERENDORF (1989) who compared mathematically predicted and measured free drug levels of amoxicillin which exhibits low-level protein binding (15%–20%) and flucloxacillin which exhibits high levels of serum protein binding (96%) in human serum and skin blister fluid. Following an i.v. administration of 1 g each, both drugs showed similar plasma concentration vs. time profiles but significantly different tissue (i.e. skin blister fluid) concentrations: the concentration of amoxicillin was greater and that of flucloxacillin less than the corresponding serum concentrations. Similar findings have been reported for human peripheral lymph (BERGAN et al. 1987). There are limitations to the above-mentioned and DERENDORF's approach, in particular the size of the focus of infection and the limitation to drugs that are distributed by passive diffusion only (LIMBERG et al. 1990).

If penetration beyond the extracellular tissue fluid is achievable, forces of tissue penetration and accumulation are pivotal. Consequently, the covariants Cmax and AUC assume vital importance as the basic and key determinants of penetration to tissues, i.e. the generation of high or low tissue fluid concentrations. However, in specialized sites such as prostatic fluid (STAMEY et al. 1970) or bronchial mucosa and alveolar macrophages (BALDWIN et al. 1992a,b; HONEYBOURNE and BALDWIN 1992), antibacterial agents can be accumulated by mechanisms such as ion-trapping and active transport. For instance, quinolones were shown to be concentrated in tissues by these mechanisms. Thus, quinolone concentrations exceed the corresponding serum concentrations significantly in all tissues studied except fat. This phenomenon has been addressed and confirmed in a large number of studies performed by ourselves and others (for summaries see BERGAN 1978, 1981, 1988; BERGAN et al. 1985, 1987, 1988; DALHOFF 1989). However, determinations of "tissue-concentrations" in total tissue homogenates may be error prone as total tissue homogenate concentrations do not differentiate between, for example, interstitial and cellular fluid (DALHOFF 1985; HONEYBOURNE and BALDWIN 1992; BALDWIN et al. 1992a).

Since the causative pathogens are located in the interstitium – with the exception of intracellular infections – the most relevant information regarding the tissue concentration would be indicated by the amount of drug in the interstitial fluid. Furthermore, since quinolones are concentrated intracellularly, tissue homogenates containing cells and interstitial fluid overestimate quinolone concentrations in the interstitium.

The major sources of criticism are that, first, the most relevant information pertaining to "tissue concentrations" would be indicated by the amount of drug in the interstitial fluid or cellular fluid as – with the exception of intracellular infections – the causative pathogens are located in the interstitium. Second, since quinolones are concentrated intracellularly, tissue homogenates containing cells and interstitial fluid will overestimate quinolone concentrations in the interstitium. In contrast interstitial concentrations of β-lactams, which do not accumulate intracellularly, will be underestimated in tissue homogenates. As methodological problems of quantitating antimicrobial agents at the sites of infection are beyond the scope of this summary, the reader is referred to the above-mentioned reviews for further reading.

By using either appropriate markers or relevant samples, methodological problems can be minimized or even overcome. For example, the epithelial lining fluid (ELF) or peripheral lymph represent pure extracellular sites, while the alveolar macrophage (MØ) or other cell lines such as the J774 macrophage cell line (a continuous reticulosarcoma cell line of murine origin) represent pure intracellular sites. For the sake of conciseness we limit ourselves to these tissues or tissue fluids only to demonstrate that quinolones generate higher concentrations at the sites of infection than corresponding serum concentrations do.

Although serum, bronchial mucosa, ELF and MØ concentrations varied with the drug studied and the dose and dosing schedule employed, the site concentrations of quinolones exceeded the corresponding serum concentrations at any time (Table 15). It is worth mentioning that data summarized in Table 15 were generated by methods which corrected for compromising factors. Furthermore, it is of interest to note that others have measured "lung" concentrations (i.e. lung tissue homogenate concentrations) in samples removed at open lung biopsy (for a summary see DALHOFF 1989). Data obtained by both methods clearly indicate that quinolone concentrations at the potential sites of respiratory tract infection exceed the corresponding serum concentrations significantly. As most of the relevant respiratory tract pathogens exist both intra- and extracellularly, the intracellular accumulation of drugs and their subcellular distribution is also of clinical relevance. We do not intend here to imply that we know the relationship between an MIC, minimum bactericidal concentration or killng rate as it can be determined under standard laboratory conditions in vitro and the drug's activity under in vivo conditions with different environmental constraints, such as protein binding, pH,

Table 15. Mean quinolone concentrations in the potential sites of pulmonary infection

Quinolone	Dose (mg/kg)	Sampling time after last dose (h)	Concentration in serum (mg/l)	Concentration in bronchial mucosa (mg/kg)	Concentration in ELF (mg/l)	Concentration in MØ (mg/l)
Ciprofloxacin[a]	250; p.o., b.i.d., 4d	3–6	1.1	1.8	2.0	14.8
Ciprofloxacin[b]	250; p.o., b.i.d., 2d	1.5–12	0.3–1.4	3.0–29	NA	NA
Ciprofloxacin[a]	100; i.v., once	3–6	1.6	4.4	NA	NA
Ciprofloxacin[c]	200; i.v., b.i.d., 5d	2	1.4	21.6	NA	NA
Lomefloxacin[a]	400; p.o., once, 4d	1–7	3.0	5.6	8.8	54.6
Temafloxacin[a]	400; p.o., b.i.d., 3d	2–6	6–9	12.2	NA	NA
Temafloxacin[a]	600; p.o., b.i.d., 3d	1.5–16	9.6	14.9	26.3	79.2
Rufloxacin[a]	400; p.o., once	4	3.6	5.4	25.0	35.9
Sparfloxacin[a]	400; p.o., once	1–3	0.5	1.3	5.6	9.6
Ofloxacin[d]	200; p.o., b.i.d., 2–4d	1–13	0.9–6.1	1.3–21	NA	NA

ELF, epithelial lining fluid; MØ, macrophage; NA, not available.
[a] Modified from WISE et al. 1991, 1993.
[b] Modified from GARRAFO et al. 1988.
[c] Modified from FABRE et al. 1991.
[d] Modified from DAVEY et al. 1991.

oxygen tension, and the pleiotropic adaptive responses of bacteria to growth at different sites of infection.

Macrolides, for example, are accumulated by phagocytic cells probably by an active process displaying saturation kinetics characteristic of a carrier-mediated transport system (Hand et al. 1987; Carlier et al. 1987; Scorneaux et al. 1991); quinolones are also accumulated intracellularly (Tulkens 1991; Scorneaux and Tulkens 1992a,b). Some of the quinolones which have been studied are accumulated passively: metabolic inhibitors did not affect the uptake of pefloxacin and fleroxacin. Furthermore, pefloxacin, ofloxacin and fleroxacin uptake also occurs at 4°C (Carlier and Tulkens 1988), although ciprofloxacin was not taken up at this temperature (Carlier et al. 1987). High intracellular concentrations, however, do not necessarily correlate with pronounced intracellular activity. A synopsis of the studies performed by Tulkens (1991) (Table 16) reveals that intracellular accumulation per se is not necessarily conducive to antibacterial activity. Although intracellular accumulation of the three quinolones studied differed almost by a factor of three, the quinolone with the lowest intracellular concentrations exhibited the most marked antibacterial effect and vice versa, i.e. the one with the highest concentration showed the least antibacterial activity. Similar findings were obtained with the three macrolides studied.

A comparison of the intracellular activity of macrolides with that of quinolones reveals that the significantly higher degree of intracellular accumulation of macrolides does not translate into significantly greater antibacterial efficacy. This phenomenon does not result from differences in drug disposition, as quinolones, which are not associated with subcellular organelles, exhibited the greatest intrinsic activity against all bacteria studied. In contrast, macrolides, which are almost equally distributed between the cell soluble fraction and lysosomes, were significantly less effective against *Staphylococcus aureus* (escaping host defense by survival within phagolysosomes), *Legionella pneumophila* (surviving in phagosomes) and *Listeria monocytogenes* (surviving in the cytoplasm). Furthermore, these data indicate that antibiotics do not

Table 16. Intracellular activity of quinolones and macrolides in *Staphylococcus aureus*, *Listeria monocytogenes*- and *Legionella pneumophila*-infected J774 macrophages. (Adapted from Tulkens 1991; Scorneaux and Tulkens 1992a,b; Carlier et al. 1987; Scorneaux et al. 1991)

Drug	Extracellular concentration (mg/l)	Intracellular: extracellular concentration	(CFU at time 0 h/CFU at time 24 h)/(Ci/Ce × 10^3)		
			S. aureus	*L. pneumophila*	*L. monocytogenes*
Ciprofloxacin	5.0	5.1	140	330	NA
Pefloxacin	10.0	8.1	85	NA	NA
Sparfloxacin	NA	14	151	255	160
Erythromycin	5.0	10.3	38	140	NA
Roxithromycin	5.0	25.8	30	128	NA
Azithromycin	NA	44	35	89	31

CFU, colony forming units.

necessarily have to be associated with the specific organelles in which bacteria survive or even multiply in order to exert their intracellular antibacterial effect.

The differences in intracellular activity between the quinolones and macrolides may be due to the fact that quinolones moderately bind protein whereas macrolides bind strongly and specifically to α1-acid glycoprotein; an eightfold increase in MICs of roxithromycin against staphylococci and group A, B and D streptococci and pneumococci was observed in 100% human serum compared with conventional bacterial growth media without addition of serum (Andrews et al. 1987).

Sophisticated experimental models have been required to demonstrate that the time period that serum levels exceed the MIC is the only parameter correlating with in vivo efficacy of β-lactams and that the AUC and Cmax are important parameters for quinolone efficacy and prevention of the emergence of resistance (see Craig and Dalhoff, this volume). Controlled prospective clinical studies are required to prove or disprove these preclinical data.

References

Aihara M, Shigematsu A (1993) Absorption, distribution and excretion after multiple oral administrations of [^{14}C]temafloxacin in rats. Chemotherapy (Tokyo) 41:163

Andrews JM, Ashby JP, Wise R (1987) Factor affecting the in vitro activity of roxithromycin. J Antimicrob Chemother 20[Suppl B]:31–37

Baldwin DR, Honeybourne D, Wise R (1992a) Pulmonary disposition of antimicrobial agents: methodological considerations. Antimicrob Agents Chemother 36:1171–1175

Baldwin DR, Honeybourne D, Wise R (1992b) Pulmonary disposition of antimicrobial agents: in vivo observations and clinical relevance. Antimicrob Agents Chemother 36:1176–1180

Bergan T (1978) Kinetics of tissue penetration. Are high plasma peak concentrations or substained levels preferable for effective antibiotic therapy? Scand J Infect Dis Suppl 14:36–46

Bergan T (1981) Pharmacokinetics of tissue penetration of antibiotics. Rev Infect Dis 3:45–66

Bergan T (1988) Pharmacokinetics of fluorinated quinolones. In: Andriole VJ (ed) The quinolones. Academic, London, pp 119–154

Bergan T, Dalhoff A, Thorstensson SB (1985) A review of the pharmacokinetics and tissue penetration of ciprofloxacin. Ciprofloxacin workshop proceedings. Sieber McIntyre, Hongkong, pp 23–36

Bergan T, Engeset A, Olszewski W (1987) Does serum protein binding inhibit tissue penetration of antibiotics? Rev Infect Dis 4:713–718

Bergan T, Dalhoff A, Rohwedder R (1988) Pharmacokinetics of ciprofloxacin. Infection 16 Suppl 1:3–13

Boxenbaum H (1982) Interspecies scaling, allometry, physiological time, and the ground plan of pharmacokinets. J Pharmacokinet Biopharm 10:201–227

Carlier MB, Tulkens PM (1988) Uptake and subcellular localization of lomefloxacin (SC 47111; L) in phagocytic cells in comparison with pefloxacin (P) and fleroxacin (F). Abstracts of the 28th Interscience Conference on Antimicrobial Agents and Chemotherapy, no 52

Carlier MB, Scorneaux B, Zenebergh A, Tulkens PM (1987) Uptake and subcellular distribution of 4 quinolones (pefloxacin, P; ciprofloxacin, C; ofloxacin, O; RO-23-

6240, R) in phagocytes. Abstracts of the 27th Interscience Conference on Antimicrobial Agents and Chemotherapy, no 622

Craig WA, Ebert SC (1991) Killing and regrowth of bacteria in vitro: a review. Scand J Infect Dis Suppl 74:63–70

Craig WA, Ebert SC (1992) Continuous infusion of β-lactam antibiotics. Antimicrob Agents Chemother 36:2577–2583

Craig WA, Leggett JE, Ebert S, Fantin B (1991) Comparative dose-effect relations at several dosing intervals for beta-lactam, aminoglycoside and quinolone antibiotics against gram-negative bacilli in murine thigh-infection and pneumonitis models. Scand J Infect Dis [Suppl] 74:179–184

Dalhoff A (1985) Gewebespiegel von Penicillinen beim Menschen. FAC Fortschr Antimikrob Antineoplast Chemother 4–6:1389–1416

Dalhoff A (1989) A reivew of quinolone tissue pharmacokinetics. In: Fernandes PB (ed) International Telesymposium on quinolones. Prous Science, Barcelona, pp 277–312

Davey PG, Precious E, Winter J (1991) Bronchial penetration of ofloxacin after single and multiple oral dosage. J Antimicrob Chemother 27:335–341

Dedrick RL (1973) Animal scale up. J Pharmacokinet Biopharm 1:435–461

Derendorf H (1989) Pharmacokinetic evaluation of β-lactam antibiotics. J Antimicrob Chemother 24:407–413

Derendorf H, Limberg J, Lebel M (1990) Evaluation of free tissue concentrations of fleroxacin after oral administration. Pharm Res 7:422–424

Endo M, Kohno M, Yamada Y, Otsuka M, Takaiti O (1993) Absorption, distribution and excretion of [^{14}C]temafloxacin in rats. Chemotherapy (Tokyo) 41:140–154

Fabre D, Bressolle F, Gomeni R, Arich C, Lemesle F, Beziau H, Galtier M (1991) Steady-state pharmakokinetics of ciprofloxacin in plasma from patients with nosocomial pneumonia: penetration of the bronchial mucosa. Antimicrob Agents Chemother 35:2521–2525

Fernandes PB, Shipkowitz N, Swanson N (1989) Comparative efficacy of the fluoroquinolones in experimental animal infections: correlation with in vitro potency and pharmacokinetics. In: Fernandes PB (ed) International Telesymposium on quinolones. Prous Science, Barcelone, pp 255–268

Fischman AJ, Livni E, Babich J (1992) Pharmacokinetics of [^{18}F]-labelled fleroxacin in rabbits with Escherichia coli infections, studied with positron emission tomography. Antimicrob Agents Chemother 36:2286–2292

Garrafo R, DE Salvator F, Dellamonica P, Lafallus P (1988) Ciprofloxacin penetration in bronchial mucosa after oral treatment (Abstr 342). International Congress on Infections Diseases, 16–21 April, Rio de Janeiro

Gerber AU, Feller-Segessenmann C (1985) In vivo assessment of in vitro killing patterns of Pseudomonas aeruginosa. J Antimicrob Chemother 15:201–206

Gerber AU, Wiprachtiger P, Stettler-Spichiger U, Lebek G (1982a) Constant infusion versus intermittent doses of gentamicin against Pseudomonas aeruginosa in vitro. J Infect Dis 145:554–560

Gerber AU, Vastola AP, Brandel J, Craig WA (1982b) Selection of aminoglycoside-resistant variants of Pseudomonas aeruginosa in an in vivo model. J Infect Dis 146:691–697

Gerber AU, Craig WA, Brugger HP, Feller C, Vastola AP, Brandel J (1983) Impact of dosing intervals on activity of gentamicin and ticarcillin against Pseudomonas aeruginosa in granulocytopenic mice. J Infect Dis 147:910–917

Gerber AU, Brugger HP, Feller C, Stritzko T, Stalder B (1986) Antibiotic therapy of infections due to Pseudomonas aeruginosa in normal and granulocytopenic mice: comparison of murine and human pharmacokinetics. J Infect Dis 153:90–97

Gerding DN, Peterson LR, Moody JA, Fasching CE (1985) Mezlocillin, ceftizoxime and amikacin alone and in combination against six Enterobacteriaceae in a neutropenic site in rabbits. J Antimicrob Chemother 15:207–219

Hand WE, King-Thomson N, Hohman JW (1987) Entry of roxithromycin (RU 965), imipenem, cefotaxime, trimethroprim and metronidazole into human polymorphonuclear leukocytes. Antimicrob Agents Chemother 31:1553–1557

Honeybourne D, Baldwin DR (1992) The site concentrations of antimicrobial agents in the lung. J Antimicrob Chemother 30:249–260

Karabalut N, Drusano GL (1993) Pharmacokinetics of the quinolone antimicrobial agents. In: Hooper DC, Wolfson JS (eds) Quinolone antimicrobial agents, 2nd edn. American Society for Microbiology, Washington, pp 195–223

Kawahara F, Ooie T, Nagatsu Y, Ulhida H (1991) Metabolism of fleroxacin in humans and various experimental animals. Chemotherapy (Tokyo) 38:122–134

Kohno M, Kodama H, Endo M, Otsuka M, Takaiti O (1993) Absorption, metabolism and excretion of [^{14}C]temafloxacin in dogs. Chemotherapy (Tokyo) 41:177–187

Limberg J, Lebel M, Derendorf H (1990) Evaluation of free tissue concentrations of fleroxacin after oral administration. Pharmac Res 7(4):422–424

May JR (1955) The laboratory background to the use of penicillin in chronic bronchitis and bronchiectasis. Br J Tuberc Dis Chest 49:166–173

Montay G, Goueffon Y, Roquet F (1984) Absorption, distribution, metabolic fate, and elimination of pefloxacin mesylate in mice, rats, dogs, monkeys and humans. Antimicrob Agents Chemother 25:463–472

Mordenti I (1985) Forecasting cephalosporin and monobactam antibiotic half-lives in humans from data collected in laboratory animals. Antimicrob Agents Chemother 27:887–891

Mordenti I (1986) Man versus beast: pharmacokinetic scaling in mammals. J Pharm Sci 75:1028–1040

Murayama S, Hirai K, Ito A, Abe Y, Irikura T (1981) Studies on absorption, distribution and excretion of AM-715 in animals by bioassay method. Chemotherapy (Tokyo) 29:98–104

Nagata O, Yamada T, Takahashi K. Okezaki E (1988a) Disposition and metabolism of NY-198. III. Absorption, metabolism and excretion of NY-198 in monkeys by high-performance liquid chromatography. Chemotherapy (Tokyo) 36:144–150

Nagata O, Yamada T, Yamaguchi T, Okezaki E (1988b) Disposition and metabolism of NY-198. IV. Absorption, distribution and excretion of ^{14}C-NY-198 in rats and dogs. Chemotherapy (Tokyo) 36:151–173

Nagata O, Yamada T, Yamaguchi T, Hasegawa H, Okezaki E (1988c) Disposition and metabolism of NY-198. V. Metabolism of ^{14}C-NY-198 in rats and dogs. Chemoterapy (Tokyo) 36:174–187

Nagatsu Y, Endo K, Irikura T (1981a) Studies on the fate of ^{14}C-labelled AM-715. Chemotherapy (Tokyo) 29:105–118

Nagatsu Y, Endo K, Irikura T (1981b) Studies on the metabolism of ^{14}C-labelled AM-715. Chemotherapy (Tokyo) 29:119–127

Nagatsu Y, Mukai M, Takagi K, Uchida H (1990) Absorption, distribution and excretion of 14C-fleroxacin in rats and rabbits. Chemotherapy (Tokyo) 38:100–114

Nakamura S, Kurobe N, Kashimoto S, Ohue T, Takase Y, Shimizu M (1984) Absorption, distribution, excretion and metabolism of AT-2266 in experimental animals. Chemotherapy (Tokyo) 32:86–94

Nakamura S, Kurobe N, Ohue T, Hashimoto M, Shimizu M (1990) Pharmacokinetics of a novel quinolone, AT-4140, in animals. Antimicrob Agents Chemother 34:89–93

Nakamura S, Kurobe N, Ohue T, Hashimoto M, Shimizu M (1991) Absorption, distribution and excretion of sparfloxacin in animals. Chemotherapy (Tokyo) 39:123–130

Okezaki E, Ohmichi K, Koike S, Takahashi Y, Makino E (1988a) Disposition and metabolism of NY-198. I. Bioassay study of absorption, distribution and excretion in various animals. Chemotherapy (Tokyo) 36:132–137

Okezaki E, Makino E, Ohmichi K, Nagata O, Yamada T, Takahashi K (1988b) Disposition and metabolism of NY-198. II. HPLC and bioassay studies of absorption and excretion in the dog. Chemotherapy (Tokyo) 36:138–143

Rubin RH, Livni E, Babich J, Alpert NM, Liu Y-Y, Tham E, Prosser B, Cleeland R, Callahan RJ, Correia JA, Strauss HW, Fischmann AJ (1993) Pharmacokinetics of fleroxacin as studied by positron emission tomography and [^{18}F]fleroxacin. Am J Med 94 Suppl 3A:31–37

Ryan DM, Cars O (1980) Antibiotic assay in muscle: are conventional tissue levels misleading as indicator of the antibacterial activity. Scand J Infect Dis 12:307–309

Sawada Y, Hanano M, Sugiyama Y, Iga T (1984) Prediction of the disposition of β-lactam antibiotics in humans from pharmacokinetic parameters in animals. J Pharmacokinet Biopharm 12:241–261

Scorneaux B, Tulkens PM (1992a) Activities of macrolides (ML) and fluoroquinolones (FQ) against Legionella pneumophila (L.p.) in a model of J774 macrophages (MØ) in relation to their cellular accumulation. Abstracts of the 32nd International Conference on Antimicrobial Agents and Chemotherapy, no 1701

Scorneaux B, Tulkens PM (1992b) Activities of azithromycin (Az), sparfloxacin (Sp), gentamicin (Gm) and ampicillin (Am) against a series of facultative and obligatory intracellular parasites in a model of J774. Abstracts of the 32nd International Conference on Antimicrobial Agents and Chemotherapy, no 1706

Scorneaux B, Orfila J, Tulkens PM (1991) Activities of erythromycin and Roxithromycin against Chlamydia spp. in the model of J774 MØ in relation to their cellular accumulation. Abstracts of the 17th International Congress of Chemotherapy, no 783

Sekine Y, Yamaguchi T, Miyamoto M, Suzuki R, Yoshida K, Minami A, Nakamura S, Hashimoto M (1984) Pharmacokinetics of a new antibacterial agent AT-2266. I. Metabolism of AT-2266 in rats, dogs, monkeys and man. Chemotherapy (Tokyo) 32:95–102

Siefert HM, Maruhn D, Maul W, Förster D, Ritter W (1986a) Pharmacokinetics of ciprofloxacin. 1st Communication: absorption, concentrations in plasma, metabolism and excretion after a single administration of [^{14}C]ciprofloxacin in albino rats and rhesus monkeys. Arzneimittelforschung 36:1496–1502

Siefert HM, Maruhn D, Scholl H (1986b) Pharmacokinetics of ciprofloxacin. 2nd Communication: distribution to and elimination from tissues and organs following single or repeated administration of [^{14}C]ciprofloxacin in albino rats. Arzneimittelforschung 36:1503–1510

Stamey TA, Meares EM, Winningham DG (1970) Chronic bacterial prostatitis and the diffusion of drugs into prostatic fluid. J Urol 103:187–194

Sudo K, Hashimoto K, Kurata T, Okazaki O, Tsumura M, Tachizawa H (1984) Metabolic disposition of DL-8280. The third report: metabolism of ^{14}C-DL-8280 in various animal species. Chemotherapy (Tokyo) 32:1203–1210

Takeda K, Yano S, Sakuma Y, Koyama Y, Yamaguchi T (1993) Pharmacokinetics of temafloxacin in mice, rats and hamsters. Chemotherapy (Tokyo) 41:128–139

Tsumura M, Sato K, Une T, Tachizawa H (1984) Metabolic disposition of DL-8280. The first report: comparison between absorption and excretion of DL-8280 in the dog and monkey by bioassay and HPLC methods. Chemotherapy (Tokyo) 32:1179–1184

Tulkens PM (1991) Intracellular distribution and activity of antibiotics. Eur J Clin Microbiol Infect Dis 10:100–106

Vogelmann B, Craig WA (1986) Kinetics of antimicrobial activity. J Pediatr 108:835–840

Vogelmann B, Gudmundsson S, Legget J, Turnidge J, Ebert S, Craig WA (1988) Correlation of antimicrobial pharmacokinetic parameters with therapeutic efficacy in an animal model. J Infect Dis 158:831–847

White LO (1988) The pharmacokinetics of ofloxacin, desmethyl ofloxacin and ofloxacin N-oxide in haemodialysis patients with end-stage renal failure. J Antimicrob Chemother 22[Suppl C]:65–72

White LO (1993) The serum concentrations of desmethyl ofloxacin and ofloxacin N-oxide in seriously ill patients and their possible contributions to the antibacterial activity of ofloxacin. J Antimicrob Chemother 34:300–303

White LO, MacGowan AP, Lovering AM, Reeves DS, Mackay IG (1987) A preliminary report on the pharmacokinetics of ofloxacin desmethyl ofloxacin and ofloxacin N-oxide in patients with chronic renal failure. Drugs 34[Suppl 1]:56–61

Wise R, Baldwin DR, Andrews JM, Honeybourne D (1991) Comparative pharmacokinetic disposition of fluoroquinolones in the lung. J Antimicrob Chemother 28[Suppl C]:65–71

Wise R, Andrews J, Imbimbo BP, Greaves I, Honeybourne D (1993) The penetration of rufloxacin into sites of potential infection in the respiratory tract. J Antimicrob Chemother 32:861–866

Yamaguchi T, Suzuki R, Sekine Y (1984) Pharmacokinetics of a new antibacterial agent AT-2266. II. Plasma levels and urinary excretion of AT-2266 and its metabolites in mice, rats, cats, dogs and monkeys. Chemotherapy (Tokyo) 32:103–108

Yano S, Sakuma Y, Takeda K (1993) Absorption, metabolism and excretion of [^{14}C]temafloxacin in dogs. Chemotherapy (Tokyo) 41:177–187

Yasuda T, Watanabe Y, Hayashi T, Kitayama R (1988a) Serum protein binding of T-3262. Chemotherapy (Tokyo) 36:143–148

Yasuda T, Watanabe Y, Minami S, Kumano K, Takagi S, Tsuneda R, Kanayama J (1988b) Absorption, distribution, metabolism and excretion of T-3262 in experimental animals. Chemotherapy (Tokyo) 36:149–157

Zeiler HJ, Petersen U, Gau W, Ploschke HJ (1987) Antibacterial activity of the metabolites of ciprofloxacin and its significance in the bioassay. Arzneimittelforschung 37:131–134

CHAPTER 7

Pharmacodynamics of Fluoroquinolones in Experimental Animals

W. CRAIG and A. DALHOFF

A. Introduction

The efficacy of antimicrobials in experimental infections in animals is dependent on both the pharmacokinetics and pharmacodynamics of the drug. The pharmacokinetics of fluoroquinolones in experimental animals is reviewed in Chap. 5. This chapter reviews the pharmacodynamics of these agents in animal infection models with infections induced by extracellular and intracellular pathogens. The discussion focuses not only on the efficacy of fluoroquinolones, but also on the emergence of resistance to these drugs during therapy.

Pharmacodynamics is concerned with the relationship between drug concentration and the antimicrobial effect. The minimal inhibitory and bactericidal concentrations (MIC and MBC) have been the major parameters used to characterize the antimicrobial effect. While these parameters are useful in describing the potency of the drug, they do not describe the time course of antimicrobial activity. For example, the MIC (i.e., inhibition of visible growth at one particular concentration) does not provide information on the effect of fluctuating drug concentrations as characteristically encountered in a patient following parenteral or oral dosing or whether there are persistent antimicrobial effects that last after antimicrobial exposure. Furthermore, the MBC does not describe the rate of bactericidal activity. Parameters such as the rate of bacterial killing with increasing concentrations, the postantibiotic effect (PAE), and the postantibiotic sub-MIC effect provide a much better description of the time course of antimicrobial activity.

A variety of in vitro studies have demonstrated that the fluoroquinolones exhibit concentration-dependent killing and produce prolonged postantibiotic effects with susceptible gram-positive and gram-negative pathogens (CRAIG and EBERT 1991; CRAIG and GUDMUNDSSON 1991). Further exposure of organisms during the PAE-phase to sub-MIC concentrations can prolong the duration of the postantibiotic effect (ODENHOLT-TORNQVIST et al. 1992). The pattern of antibacterial activity with the fluoroquinolones is markedly different from that of β-lactam antibiotics,which exhibit minimal concentration-dependent killing and produce short-term or no postantibiotic effects (CRAIG and EBERT 1991; CRAIG and GUDMUNDSSON 1991). Confirmation of these pharmacodynamic parameters in animal infection models have been performed in the murine thigh-infection and rat pneumonia models.

B. Bacterial Killing In Vivo

Increasing the dose of ciprofloxacin resulted in faster and more extensive killing of standard strains of both *Klebsiella pneumoniae* and *Staphylococcus aureus* in the thighs of neutropenic mice over the first 2h of therapy. Similar findings have been observed in the same animal model with ofloxacin, fleroxacin, temafloxacin, and BAY y 3118 (Craig et al. 1993; Craig and Watanabe 1992; Ebert et al. 1990; Watanabe et al. 1992a). In comparison with ciprofloxacin, the rates of killing of these two organisms were similar for ofloxacin, fleroxacin and temafloxacin and slightly faster for Bay y 3118. However, the therapeutic significance of a slightly faster rate of in vivo killing is currently unclear. Roosendaal et al. (1987) observed faster and more extensive killing of the same strain of *K. pneumoniae* in the lungs of infected rats.

C. In Vivo Postantibiotic Effects

Neutropenic mouse thigh infection has been the primary experimental infection used to quantify the duration of the in vivo postantibiotic effect with fluoroquinolones (Vogelman et al. 1988a). The infected subcutaneous thread model in mice has also been used to determine the postantibiotic effect with norfloxacin against a few gram-positive and gram-negative organisms (Renneberg and Walder 1989). The duration of the postantibiotic effect with different fluoroquinolones against specific pathogens are summarized in Table 1. In general, fluoroquinolones produce in vivo postantibiotic effects in neutropenic mice, the duration of which varies from 1.8 to 5h. Even longer postantibiotic effects have been observed in normal mice with a standard strain of *K. pneumoniae*. For example, the in vivo postantibiotic effects of ciprofloxacin and temafloxacin increased from 2.4 and 2.9h, respectively, in neutropenic mice to 7.5 and 6.8h respectively, in normal mice (Craig 1993; Craig and Watanabe 1992). The increased duration of the postantibiotic effect in the presence of neutrophils is most likely an in vivo manifestation of the postantibiotic leukocyte enhancement (Macdonald et al. 1981).

Minimal information exists on the postantibiotic effect of fluoroquinolones against intracellular pathogens. Pefloxacin has been shown to exhibit a postantibiotic effect in human macrophages (Vilde et al. 1986) and in an experimental infection with *Legionella pneumophila* in guinea pigs (Dournon et al. 1986).

D. Pharmacodynamic Parameters Determining Efficacy

The pharmacodynamic characteristics of concentration-dependent bacterial killing plus prolonged postantibiotic effects would suggest that peak drug concentrations or the area under the concentration-versus-time curve (AUC)

Table 1. In vivo postantibiotic effects with fluoroquinolones

Organism	Drug	Dose (mg/kg)	Time above MIC (h)	In vivo PAE (h)
K. pneumoniae	Ciprofloxacin	3.12, 6.25	2.4 2.9	2.4 4.7
	Fleroxacin	6.25	2.0	1.8
	Ofloxacin	6.25	2.8	2.7
E. coli	Ciprofloxacin	3.12	3.1	3.8
	Norfloxacin	3.0	1.8	3.4
P. aeruginosa	Ciprofloxacin	50.0	2.2	2.6
	Ofloxacin	25.0	2.5	1.5
	Fleroxacin	25.0	3.0	2.5
	Norfloxacin	3.0	0.9	3.7
S. aureus	Ciprofloxacin	25.0	1.6	1.2
	Fleroxacin	25.0	4.3	1.7
	Ofloxacin	25.0	2.6	1.9
	Norfloxacin	3.0	0.5	4.3
E. faecalis	Norfloxacin	3.0	0.0	3.3

Data on norfloxacin were obtained from the infected subcutaneous thread model in normal mice (RENNEBERG and WALDER 1989); all other data were obtained from the neutropenic murine thigh-infection model (CRAIG and GUDMUNDSSON 1992; CRAIG and WATANABE 1992; CRAIG et al. 1993; EBERT et al. 1990; WANANABE et al. 1992a).

would be the major determinants of in vivo efficacy of the fluoroquinolones. Studies using multiple dosing frequencies at several different total daily doses of drug have been successful in identifying the pharmacodynamic parameter which determines in vivo efficacy in both neutropenic thigh-infection and pneumonitis models (LEGGETT et al. 1989a; VOGELMAN et al. 1988b). Serum concentrations, rather than tissue concentrations, have been used for correlation with efficacy in these experimental infections. RYAN and CARS (1980) have demonstrated that the concentrations of two β-lactams in interstitial fluid closely parallel serum concentrations rather than tissue homogenate concentrations. By implanting cotton threads under the muscle fascia, extracellular muscle tissue fluid was obtained. Under these experimental conditions, the concentrations of ampicillin and cefuroxime in serum and muscle tissue fluid, respectively, were quite similar. Since fluoroquinolones accumulate intracellularly, in constrast to β-lactams, tissue homogenates overestimate the concentration of drug in interstitial fluid. Furthermore, the interstitial fluid is the site of tissue localization of the common extracellular pathogens used in many experimental infection models.

However, drug concentrations at the focus of infection may not always parallel those in serum; antibacterial agents such as quinolones and macrolides accumulate in specialized sites, e.g., prostatic fluid, lung mucosa, and epithelial lining fluid, by active transport, ion-trapping or other mechanisms (STAMEY et al. 1970; HONEYBOURNE and BALDWIN 1992; BALDWIN et al. 1992a,b).

Thus, serum concentrations and tissue fluid concentrations are not necessarily equal. In contrast to β-lactam antibiotics, the site concentrations of quinolones and macrolides, e.g., epithelial lining fluid, alveolar macrophages or histologically defined bronchial mucosa, exceed the corresponding serum concentrations significantly (HONEYBOURNE and BALDWIN 1992). Unfortunately, quinolone concentrations at the focus of infection in experimental animals are poorly defined and were evaluated by various methods. Therefore the pharmacodynamics of quinolones were correlated to serum pharmacokinetics rather than to tissue fluid concentrations. This procedure is most frequently used in the published literature and so the various data sets can be compared with each other.

Dosing studies with different fluoroquinolones against various gram-positive and gram-negative pathogens have been conducted in both neutropenic thigh infection and pneumonitis models (CRAIG and WATANABE 1992; CRAIG et al. 1993; EBERT et al. 1990; LEGGETT et al. 1989a; WATANABE et al. 1992a). Dosing intervals from 1 to 24h at several different total daily doses have been evaluated. The reduction of viable counts at the focus of infection has been correlated to pharmacokinetic parameters and an in vitro parameter to characterize the antimicrobial effect, i.e., the MIC. The MIC has been chosen as a comparator as this parameter is most frequently used to describe the potency of a drug. However, the MIC dose not by definition reflect the bactericidal potency of a drug. By correlating the MIC, MBC, and the killing rate, i.e., reduction of viable counts as a function of time, to the in vivo efficacy in an endocarditis model it was proven that the killing rate correlated best to in vivo efficacy (POTEL et al. 1991). Thus, despite the fact that the MIC may not be the most relevant among the in vitro parameters to be correlated to in vivo efficacy of highly bactericidal drugs, this parameter was nevertheless used for the sake of comparability to other studies mentioned below. Since also in these studies the MIC was used for correlation with in vivo efficacy, we kept this parameter as the common denominator of these studies on the pharmacodynamics of antibacterial drugs.

An example of the results of such studies for ciprofloxacin used against a standard strain of *K. pneumonia* in the pneumonitis model are shown in Fig. 1. The relationship between the change in the number of bacteria (using a logarithmic scale) in the lung after 24h of therapy and the 24-h serum AUC, the serum peak level, and the duration of time that serum concentrations exceed the MIC are shown in Fig. 2a–c, respectively. Since the dotted line represents the number of bacteria in the lung at the start of therapy, negative values (below the line) represent bacterial killing while positive values (above the line) represent bacterial growth.

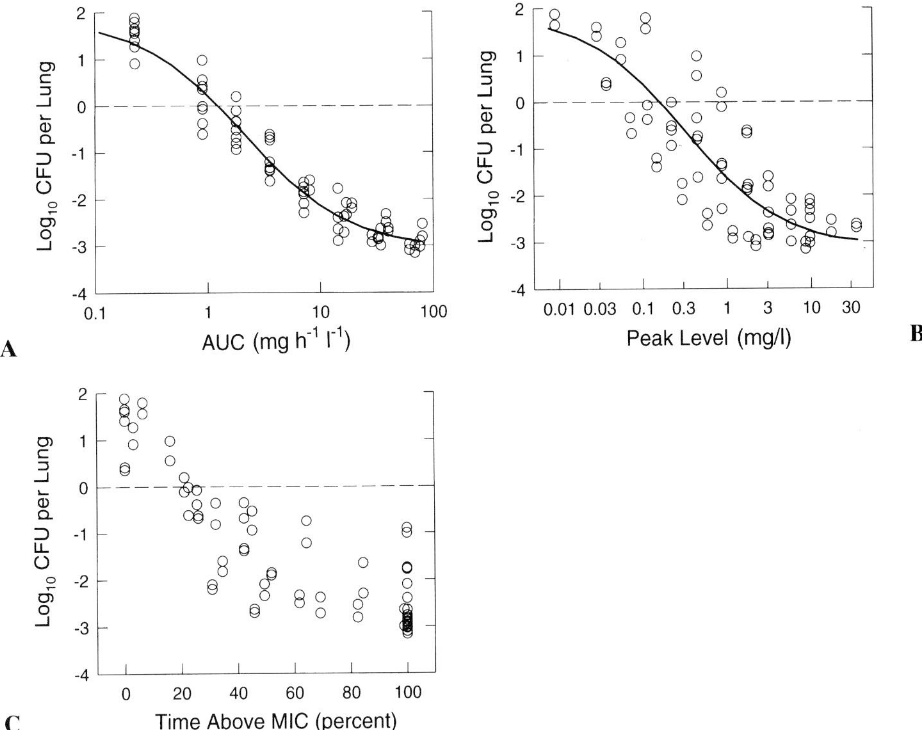

Fig. 1A–C. Relationship between the change colony-forming units (CFU) per lung (logarithmic scale) over 24 h of therapy with ciprofloxacin and the 24-h area under the concentration-versus-time curve (AUC) in serum (**A**), the serum peak level (**B**), and the duration of time serum levels exceed the minimal inhibitory concentration (**C**). Each *point* represents the meant of three to four mice. The *dotted line* represents the CFU per lung (7.35 $\log_{10}$) at the start of therapy

The best overall correlation was observed between efficacy and the 24-h AUC (Fig. 2A). The variation in efficacy for different dosing regiments that provided the same 24-h AUC was only about 1 log of colony-forming units (CFU) per lung. Since the MIC for ciprofloxacin with this organism was 0.06 mg/l, the AUC/MIC ratio required for bactericidal activity was approximately 25. Peak level also correlated with efficacy but the variation in results for a given peak serum level was frequently as much as 3 log CFU per lung (Fig. 2B). A peak level to MIC ratio of at least 10 was required for maximum killing over the 24-h treatment period. With respect to time above MIC, serum concentrations needed to exceed the MIC for about 20% of the time interval to obtain any bacterial killing. However, the extent of bacterial killing at longer durations of time above MIC varied (2–3 log CUF per lung). Somewhat similar correlations have been obtained with other fluoroquinolones and against other bacterial pathogens. AUC (or AUC/MIC) has consistently been the pharmacokinetic (or pharmacodynamic) parameter that best correlates with the antibacterial activity of the fluoroquinolones in both the murine thigh-infection and pneumonitis models. AUC has also been the best parameter for predicting efficacy for aminoglycoside antibiotics, which have many of the same pharmacodynamic characteristics as the fluoroquinolones (VOGELMAN et al. 1988b). However, for β-lactam antibiotics, the duration of time that serum levels exceed the MIC has been the only parameter correlating with in vivo antibacterial activity and efficacy (VOGELMAN et al. 1988b).

DRUSANO et al. (1993) found that peak level was a slightly better predictor of survival than AUC for lomefloxacin in a neutropenic rat model of pseudomonas sepsis. These investigators felt that peak level was the important parameter because peak level/MIC ratios of 10 or more have been shown in vitro to prevent the emergence of resistant bacteria (BLASER et al. 1987; MARCHBANKS et al. 1993). However, the authors did not directly determine in their study whether resistant organisms emerged during therapy. The emergence of resistant bacteria was searched for in the previously cited murine models, but none were found. It is probable that AUC is the major parameter for efficacy for fluoroquinolones for most organisms. However, when the emergence of resistance is likely, the peak level may become the better predictor of in vivo efficacy.

Since the in vitro MIC reflects the amount of drug required to produce a bacteriostatic effect over about 24 h of incubation, the "in vivo MIC" could be defined as the amount of drug required to produce a net bacteriostatic effect at the site of infection over 24 h of therapy. FANTIN et al. (1991) used the neutropenic murine thigh-infection model to demonstrate a linear relationship between the MIC of pefloxacin for a variety of gram-negative pathogens and the 24-h AUC required to produce a net bacteriostatic effect. A similar analysis of data with ciprofloxacin, ofloxacin, temafloxacin, fleroxacin, and Bay y 3118 has also shown a linear relationship between MIC and the 24-h AUC required to produce a net bacteriostatic effect (CRAIG et al. 1993; WATANABE et al. 1992b). Since the slopes of these relationships were approximately 1, the 24-h AUC/MIC ratio was shown to be a parameter that has the potential to correct for differences in activity among fluoroquinolones and among various organisms. By determining the AUC/MIC ratio required to produce a bacteriostatic effect, one can compare the potency of different fluoroquinolones against various bacterial pathogens. A 24-h AUC/MIC value of 24 is equivalent to a constant concentration of 1 times the in vitro MIC for 24 h. In other words a value of about 24 means that the "in-vivo MIC" is very similar to the in vitro MIC.

The impact of the dosing interval on the 24-h AUC/MIC values for six fluoroquinolones against six to 12 different pathogens is shown in Fig. 2. The pathogens in these studies included various strains of *Escherichia coli*, *K. pneumoniae*, enterobacter species, *Serratia marcescens*, *S. aureus*, and *Streptococcus pneumoniae*. Values for 24-h dosing regimens tended to be higher than for those observed at 1- to 12-h dosing intervals. This is not surprising since the

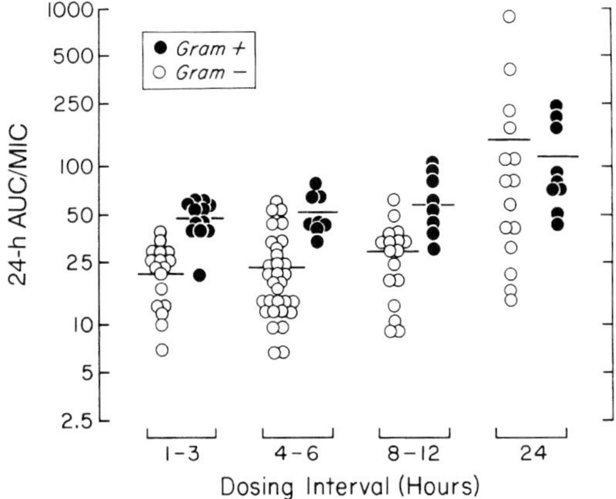

Fig. 2. Relationship between the dosing interval and the 24-h area under the concentration-versus-time curve/minimal inhibitory concentration values required to produce a bacteriostatic effect in the thighs and lungs of neutropenic mice for six fluoroquinolones against gram-negative bacilli and gram-positive cocci

duration of the postantibiotic effect is rarely long enough to cover the duration of time that serum levels are below the MIC with one dose of quinolones per day in mice. If one limits the evaluation to data from 1- to 12-h dosing regimens, the mean 24-h AUC/MIC value for gram-negative bacilli (24.3 ± 14.4) was about twofold less than the value for gram-positive cocci (52.7 ± 18.3). Thus, the in vivo MIC for gram-negative bacilli is very close to the in vitro MIC, while the in vivo MIC for gram-positive cocci is approximately twofold higher than the in vitro MIC. It is worth mentioning that quinolones exhibit their most pronounced bactericidal effect against gram-negative bacteria; gram-positive bacteria are less rapidly killed in vitro. Whether this phenomenon might have an impact on the findings illustrated in Fig. 3 remains unclear.

The mean 24-h AUC/MIC values for six different fluoroquinolones against both gram-negative and gram-positive bacteria are shown in Table 2 (Craig et al. 1993; Watanabe et al. 1992b). The bacteriostatic potencies of the six quinolones were virtually identical. The data are also illustrated for each of the drugs in Fig. 3 to show the varition in results for the different fluoroquinolones. However, when similar organisms were compared, each of the drugs gave very similar results. Enterobacter species and *P. aeruginosa* exhibited significantly lower AUC/MIC values (13.6 ± 5.0 and 20.3 ± 13.3, respectively) than observed for *E. coli*, *K. pneumoniae*, and *S. marcescens* (33.1 ± 17.5, 29.5 ± 8.4, and 32.1 ± 23.0, respectively). The activity of four fluoroquinolones were studied simultaneously in the murine thigh-infection and pneumonitis models against a standard strain of *K. pneumoniae*. The AUC/MIC required for a static effect in each model were similar (28.4 ± 8.0 in the thigh-infection model and 24.8 ± 11.2 in the lung-infection model).

Table 2. Mean 24-h AUC/MIC values required to produce a net bacteriostatic effect in the thighs or lungs of neutropenic mice for six fluoroquinolones against six to 12 different pathogens. (Craig et al. 1993; Watanabe et al. 1992b)

Drug	24-h AUC/MIC (mean ± SD)	*P*
Ciprofloxacin	35.1 ± 21.3	
Fleroxacin	35.5 ± 19.2	NS
Ofloxacin	34.5 ± 24.3	NS
Pefloxacin	29.2 ± 20.3	NS
Temafloxacin	31.4 ± 24.3	NS
Bay y 3118	36.0 ± 15.2	NS

NS, not significantly different to ciprofloxacin; AUC, area under the concentration-versus-time curve; MIC, minimal inhibitory concentration.

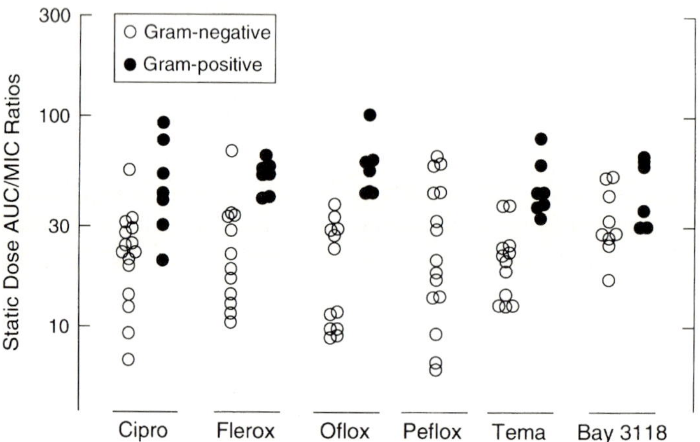

Fig. 3. The 24-h area under the concentration-versus-time curve/minimal inhibitory concentration values for six fluoroquinolones required to produce a bacteriostatic effect in the thighs and lungs of neutropenic mice against gram-negative bacilli and gram-positive cocci

Unfortunately, most other studies of the activity of fluoroquinolones in experimental infections have not used a full dose–response methodology that would allow a calculation of the 24-h AUC/MIC required to produce a static effect. However, Leggett et al. (1989a,b) have shown that the 24-h cumulative dose of drug required to produce a bacteriostatic effect at 24 h (i.e., the "static dose") is very similar to the dose required to protect 50% of the animals from death during a longer course of therapy (i.e., the 50% protective or effective dose). The results of studies with four different fluoroquinolones are shown in Table 3 (Craig and Watanabe 1992; Craig et al. 1993; Leggett et al. 1989a,b;

WATANABE et al. 1992a). In each case the static dose at 24h was an excellent predictor of the 50% protective dose after 4–5 days of therapy. By converting the 24-h cumulative doses used in these dose–response survival studies to 24-h AUC/MIC values, one can evaluate the relationship between the 24-h AUC/MIC ratios and survival. Since survival is a common endpoint used in many experimental infections, this type of analysis allows a comparison of results obtained with different infection models and animal species and with different organisms and fluoroquinolones.

The relationship between the 24-h AUC/MIC values and mortality is illustrated in Fig. 4. The solid points represent the data generated in the neutropenic murine thigh-infection and pneumonitis models (CRAIG and

Table 3. The 24-h cumulative daily doses required to produce a bacteriostatic effect at 24h and protect 50% of mice from death after 4–5 days of therapy in the neutropenic murine thigh-infection and pneumonitis models. (CRAIG and WATANABE 1992; CRAIG et al. 1993; LEGGETT et al. 1989b; WATANABE et al. 1992a)

Drug	Organism	Model	Bacteriostatic dose ($mg\,kg^{-1}\,24\,h^{-1}$)	50% Protective dose ($mg\,kg^{-1}\,24\,h^{-1}$)
Ciprofloxacin	*K. pneumoniae*	Lung	5.22	5.41
	E. coli	Thigh	1.26	1.33
	P. aeruginosa	Thigh	75.3	72.3
Ofloxacin	*E. coli*	Thigh	2.64	3.42
Temafloxacin	*K. pneumoniae*	Thigh	12.6	13.0
Bay y 3118	*S. pnemoniae*	Thigh	11.9	10.8

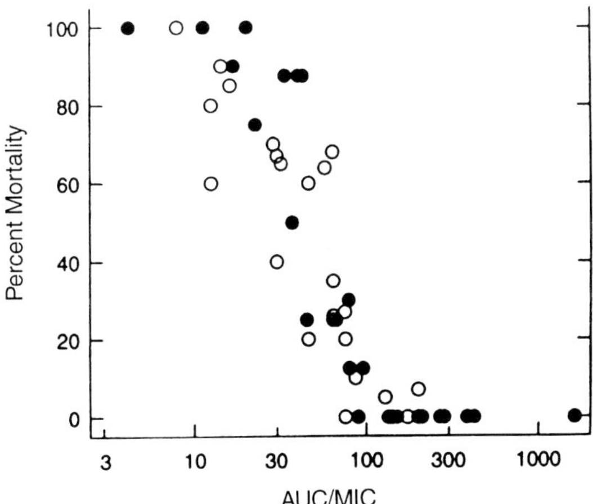

Fig. 4. Relationship between the 24-h area under the concentration-versus-time curve/minimal inhibitory concentration value and mortality in various experimental infection models. The *closed circles* represent data obtained in the neutropenic murine thigh-infection and pneumonitis models. the *open circles* represent data obtained from the literature using pneumonitis, peritonitis and sepsis models in mice, rats, and guinea pigs

WATANABE 1992; CRAIG et al. 1993; LEGGETT et al. 1989b; WATANABE et al. 1992a). The open circles represent data from other experimental infection studies reported in the literature (AZOULAY-DUPUIS et al. 1991, 1992; DRUSANO et al. 1993; GISBY et al. 1991; ROOSENDAAL et al. 1989; SCHIFF et al. 1984; ULRICH et al. 1989). Only those studies that treated animals for at least 2 days, reported survival results at the end of therapy, and provided pharmacokinetic data were used. This included studies of pneumonia, peritonitis and sepsis, performed in mice, rats, and guinea pigs, and using various strains of *P. aeruginosa*, *K. pneumoniae* and *S. pneumoniae*. An excellent correlation between the 24-h AUC and mortality was obtained. In general, AUC/MIC ratios of less than 30 were associated with greater than 50% mortality, whereas AUC/MIC values of 100 or greater were associated with almost 100% survival. A similar relationship between the 24-h AUC/MIC ratio and survival was obtained by SULLIVAN et al. (1993) using only 24h of therapy with ciprofloxacin in an experimental infection of *S.* pneumoniae peritonitis.

Thus, it appears that fluoroquinolone concentrations in serum need to average about four times the MIC for each 24h (i.e., a 24-h AUC/MIC value of 100 divided by 24h yields a value of 4.2) to produce virtually 100% survival in a variety of experimental animal infections. The 24-h AUC/MIC value of 100 or more required for maximum efficacy in experimental infection is very similar to the 24-h AUC/MIC value of 125 recently shown to be associated with satisfactory outcome in seriously ill, infected patients treated with intravenous ciprofloxacin (FORREST et al. 1993). This suggests that the pharmacodynamics of fluoroquinolones in humans may be predictable from experimetnal infection models.

As reported by Dalhoff and Bergan in Chap. 5 of this book, concentrations of fluoroquinolones in animals are two- to fivefold higher in tissues than in serum. Since the majority of the increased amount of drug in tissue is intracellular, one would expect that fluoroquinolones would have an increased potency against intracellular pathogens compared with extracellular pathogens – provided intracellular pathogens are as susceptible to the antibacterial action of drugs as extracellular pathogens are. Thus, one would expect that the 24-h AUC/MIC values in serum required for efficacy would be smaller with intracellular than with extracellular organisms.

Figure 5 illustrates the relationship between 24-h AUC/MIC values for intracellular and extracellular pathogens with mortality. The data points for the extracellular pathogens are the same as those shown in Fig. 5. Data for the intracellular pathogens were obtained from experimental infections in mice or guinea pigs produced by *Chlamydia psittaci*, *L. pneumophila*, *Mycobacterium tuberculosis*, and *Salmonella typhimurium* and treated with seven different fluoroquinolones (BRUNNER and ZEILER 1988; BUTLER et al. 1990; KIMURA et al. 1993; LALANDE et al. 1993; SAITO et al. 1985; TRUFFOT-PERNOT et al. 1991). The relationship between 24-h AUC/MIC values was no different for intracellular organisms than for extracellular organisms. This may suggest that only a fraction of the amount of fluoroquinolone that is intracellular is available for

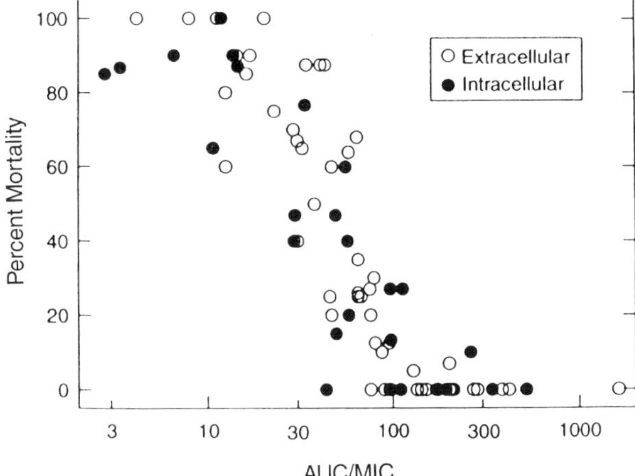

Fig. 5. Relationship between the 24-h area under the concentration-versus-time curve/minimal inhibitory concentration value and mortality for extracellular (*open circles*) and intracellular pathogens (*solid circles*) in various experimental infection models in mice, rats, and guinea pigs

antimicrobial activity. Furthermore, it *may* suggest that serum concentrations are good predictors for the activity of fluoroquinolones against intracellular pathogens as well as extracellular organisms. Alternatively, these data *may* suggest that quinolones might exhibit a moderate concentration-dependent or even concentration-independent killing and/or that quinolones may exhibit a much lower killing rate against these pathogens than against the extracellularly growing gram-negative bacteria.

Furthermore, intracellular bacteria grow and survive in different subcellular compartments, depending on the species studied, where quinolones might not be accumulated; quinolones are nearly exclusively accumulated in the cytosol. In addition, intracellular pathogens such as *Legionella* spp., *Chlamydia* spp., and atypical mycobacteria may cause infections within discrete areas of the lung which are separated from the blood by a variety of barriers to antibiotic diffusion or transport. Thus, the pharmacodynamics of quinolones against intracellularly growing bacteria have still to be examined.

E. Emergence of Resistance to Fluoroquinolones

A large body of evidence has accumulated in the literature that resistance of Enterobacteriaceae to fluoroquinolones emerges in vitro at a frequency of approximately 10^{-9}–10^{-10} (FERNANDEZ et al. 1987; HOOPER and WOLFSON 1993; WIEDEMANN and HEISIG, 1994; see also Chap. 8, this volume). However, fluoroquinolone-resistant mutants of *P. aeruginosa* – and to some extent of *S. aureus* – can be selected 100- to 1000-fold more frequently, i.e., at a rate of

Table 4. Emergence of fluoroquinolone resistance in vivo during fluoroquinolone treatment of experimental infections

Model	Infecting organism	Study drug	Emergence of resistance	MIC (mg/l)[b]		Frequency	Reference
				Pretreatment	Posttreatment		
Pyelonephritis	*E. coli*	Nalidixic acid	No				
		A-56620	No				Fernandez et al. (1987)
		Difloxacin	No				
		Norfloxacin	No				
		Ciprofloxacin	No				
		Pefloxacin	No				Fernandez et al. (1989)
		Enoxacin	No				
		Temafloxacin	No				
		Fleroxacin	No				
Pyelonephritis	*P. aeruginosa*	A-56620	Yes	0.5	0.5–2		Fernandez et al. (1987)
		Difloxacin	Yes	4	4–32	Approx. 3.5×10^{-1}	
		Norfloxacin	Yes	0.5	0.5–8		
		Ciprofloxacin	Yes	0.25	0.12–2		Fernandez et al. (1989)
		Pefloxacin	Yes				
		Enoxacin	Yes				
		Temafloxacin	Yes				
		Fleroxacin	Yes				
Infected chamber	*P. aeruginosa*	Ciprofloxacin	Yes	0.125–1	0.5–1	ND	Bamberger et al. (1986)
	S. marcescens	Ciprofloxacin	Yes	0.125–1	1	ND	Bamberger et al. (1986)
					8	ND	
Osteomyelitis	*P. aeruginosa*	Ciprofloxacin	Yes	0.16	0.6–2.5	ND	Norden and Shinner (1985)
	MRSA	Ciprofloxacin	No				Dworkin et al. (1990)
	MRSA	Pefloxacin	No				
	MRSA	Ciprofloxacin	No				Henry et al. (1987)
Peritonitis	*P. aeruginosa*	Ciprofloxacin	Yes	0.05	8- to 32-fold increase	6.1×10^{-1}	Michea-Hamzehpour et al. (1987)
		Pefloxacin	Yes	1–4	4- to 16-fold increase	7.7×10^{-1}	
		Ciprofloxacin	Yes	ND	ND	20%	Froidefond et al. (1992)
	E. cloacae	Pefloxacin	Yes	0.06	4- to 1024-fold increase	ND	Lucain et al. (1989)
	S. pneumoniae	Sparfloxacin	No				Drugeon et al. (1994)
	S. pneumoniae	Ciprofloxacin	Yes	ND	ND	ND	Drugeon et al. (1994)

Pneumonia	*K. pneumoniae*	Ciprofloxacin	No				Gordin et al. (1985)
	K. pneumoniae	Ciprofloxacin	No				Roosendaal et al. (1987)
	P. aeruginosa	Enoxacin	No				Scribner et al. (1985)
Endocarditis	*P. aeruginosa*	Pefloxacin	Yes	0.19	1.56	~3 × 10^{-1}	Bayer et al. (1988)
	P. aeruginosa	Ciprofloxacin	No				Bayer et al. (1986)
	P. aeruginosa	Ciprofloxacin	No				Thauvin et al. (1989)
	MSSA	Ciprofloxacin	Yes	0.25	5.5–12.5	ND	Kaatz et al. (1987)
	MRSA	Ciprofloxacin	No				Kaatz et al. (1987)
	MRSA	Ciprofloxacin	No				Fernandez-Guerrero et al. (1988)
	MSSA	Ciprofloxacin	No				Perrone et al. (1987)
	MSSA	Ciprofloxacin	No				
	MRSA	Fleroxacin	Yes	0.4	5-fold	ND	Kaatz et al. (1989)
	MSSA	Fleroxacin	Yes	0.5	5- to 10-fold	ND	Kaatz et al. (1991)
	MRSA	Ofloxacin	No				Kaatz et al. (1990)
	MSSA	Ofloxacin	No				Kaatz et al. (1990)
	MRSA	Clinafloxacin	No				Kaatz et al. (1992)
	MSSA	Clinafloxacin	No				Kaatz et al. (1992)
	MRSA	Pefloxacin	Yes	0.25	>4	4%	Thauvin et al. (1988)
	Streptococci	Sparfloxacin	No				Entenza et al. (1994)
	S. adjacens	Temafloxacin	No				Cremieux et al. (1992)
	Enterococcus faecium	Ciprofloxacin	No				Quale (1994)
Pouch	*P. aeruginosa*	Ciprofloxacin	No				Davey et al. (1988)
Salpingitis	*Chlamydia trachomatis*	Clinafloxacin	No				Patton et al. (1993)
Wound infection[a]	MSSA	Ciprofloxacin	No				Kernodle and Kaiser (1994)
		Ofloxacin	No				
	MRSA	Ciprofloxacin	No				
		Ofloxacin	No				
	Staphylococcus epidermidis	Ciprofloxacin	No				
	Staphylococcus epidermidis	Ofloxacin	No				
	Staphylococcus haemolyticus	Ciprofloxacin	No				
	Staphylococcus haemolyticus	Ofloxacin	No				
Abscess	*S. aureus*	Ciprofloxacin	Yes	ND	ND	ND	Doss et al. (1995)

ND, no data; MRSA, methicillin-resistant *S. aureus*; MSSA, methicillin-sensitive *S. aureus*; MIC, minimal inhibitory concentration.
[a] Prophylaxis.
[b] In some cases the increase in MIC is given as the factor of increase.

10^{-6}–10^{-8}. Such numbers of organisms (approx. 10^8 CFU/ml) are found in some wounds, abscesses and the sputum of cystic fibrosis patients. It is therefore tempting to speculate that the higher mutation rates of *P. aeruginosa* and *S. aureus*, on the one hand, and the high bacterial numbers at the focus of infection, on the other, result in treatment failures in infection models in animals. Similarly , fluoroquinolone resistance during therapy of patients developed predominantly although not exclusively in pseudomonal and staphylococcal infections. Therefore this section focuses on these two species. In addition, the majority of preclinical studies have been devoted to these two troublesome species; information on other bacterial species is scarce. Unfortunately , only a few studies in experimental animals have addressed the question of the emergence of fluoroquinolone resistance during treatment, and fewer still the frequency with which the resistant bacteria emerge in vivo. A synopsis of these studies is given in Table 4. In agreement with the above hypothesis fluoroquinolone resistance did not develop in *E. coli*; however, fluoroquinolone resistance did emerge predominantly in *P. aeruginosa* and in staphylococcal infections.

I. *P. aeruginosa* Experimental Infections

FERNANDEZ et al. (1987, 1989) analysed very comprehensively the persisting and fluoroquinolone resistant *P. aeruginosa* populations isolated from kidneys of mice treated for pyelonephritis. In general, resistance to all the fluoroquinolones studied, especially nalidixic acid, developed by the second day of treatment. However, the levels of fluoroquinolone resistance and the frequencies of resistance development differed with the respective quinolones. In addition, within one fluoroquinolone treatment group, the phenotypes, levels and very likely mechanisms of fluoroquinolone resistance were variable. Some of the organisms isolated from the kidneys of treated mice were still susceptible to fluoroquinolones while other strains had either high-level or low-level resistance to the fluoroquinolones.

The level of resistance for the *P. aeruginosa* variants isolated from the kidneys of treated animals ranged from two- to an eightfold increase in MIC. For example, norfloxacin MIC values of the persisters ranged from 0.5 to 8 mg/l; the pretreatment value was 0.5 mg/l (FERNANDEZ et al. 1987, 1989). Some strains with high-level resistance showed no changes in the outer membrane protein pattern whereas other strains showed pleiotropic changes in the cell envelope proteins. The pattern of outer membrane proteins could not be used to predict susceptibility or resistance of *P. aeruginosa* to the fluoroquinolones (FERNANDEZ et al. 1987). The morphology and viability of the persisters varied as well. Although many of the variants were able to survive in the kidneys, they could not grow on MacConkey agar and grew very poorly on subculture to blood agar plates. Among the two morphological variants isolated from the kidneys, the large colonies isolated on blood agar plates with or without the drug were viable upon subculture, whereas, the "pinpoint colonies" were not.

Similarly, resistant organisms rapidly emerged in a model of pseudomonal peritonitis in mice (MICHEA-HAMZEPOUR et al. 1987). Emergence of resistance was obviously dose dependent. A low dose of ciprofloxacin (10 mg/kg) produced more resistance (83%) than a high dose of ciprofloxacin (50 mg/kg; 44%). Treatment of mice with either pefloxacin (25 or 200 mg/kg) or ciprofloxacin (25 mg/kg) yielded similar resistance rates (61%–77%). The fluoroquinolone MIC values for the resistant variants increased 4- to 64-fold, and the imipenem MIC increased two- to eightfold without altering β-lactam and aminoglycoside susceptibility. Some of the fluoroquinolone-resistant variants also showed decreased susceptibility to trimethoprim and chloramphenicol. Either of the fluoroquinolones combined with ceftazidime or amikacin reduced the emergence of resistance, especially with the former antibiotic. Piperacillin in combination with the fluoroquinolones did not reduce the resistance rates of the corresponding monotherapies. Ciprofloxacin produced significantly less resistance and eliminated significantly more bacteria from the infected focus than pefloxacin (MICHEA-HAMZEHPOUR et al. 1987). Therapy of infection of animals with the persisting posttherapy strains of the previous animal experiment resulted in less efficient killing of bacteria; in one of these mice a highly fluoroquinolone-resistant strain was isolated (64-fold MIC increase). The presence of a foreign body in this peritonitis model increased the risk of resistance after therapy.

Another mouse model of experimental *P. aeruginosa* peritonitis was used to evaluate the emergence of resistance during therapy with imipenem, ceftazidime, amikacin, or ciprofloxacin (FROIDEFOND et al. 1992). Resistance emerged in 88% of mice after imipenem, 31% after ceftazidime and 29% after ciprofloxacin. MIC values for the resistant strains were 8- to 512-fold above baseline. The combination of imipenem and ciprofloxacin was followed by lower rates of resistance to each drug (6% and 2%, respectively) than either antimicrobial alone. Combination with amikacin reduced resistant rates for all the antibacterial agents studied. Under the experimental conditions used, the ciprofloxacin-ceftazidime combination provided best results, although the difference with the ciprofloxacin-imipenem combination was not statistically significant (FROIDEFOND et al. 1992).

During pefloxacin therapy of experimental *P. aeruginosa* endocarditis approximately 30% of vegetations sampled from pefloxacin recipients contained bacteria for which pefloxacin MIC values were four- to eightfold higher than the original MIC. The variants also exhibited increased ticarcillin MICs as well as resistance to chloramphenicol, but not to amikacin, ceftazidime, or tetracycline (BAYER et al. 1988). In contrast, resistance to ciprofloxacin was not observed in this model (BAYER et al. 1986). However, in animals receiving amikacin or ceftazidime regimens or both vegetation titers by day 14 of therapy did not differ from those of untreated controls; this was associated with development of amikacin resistance (93%) and ceftazidime resistance (5%). Ceftazidime- resistance increased to 43% by day 5 post-therapy. Both ceftazidime and amikacin-resistant isolates remained highly susceptible to ciprofloxacin (BAYER et al. 1985).

In experimentally induced *Pseudomonas* pneumonia in neutropenic guinea pigs no resistance either to ciprofloxacin or pefloxacin (GORDIN et al. 1985) or to enoxacin (SCRIBNER et al. 1985) emerged. In an infected chamber model a development of resistance was observed in those strains which failed to be eradicated. All of the *P. aeruginosa* strains tested developed greater than eight-fold rises in their posttreatment MIC values (BAMBERGER et al. 1986). Ciprofloxacin treatment of *P. aeruginosa* osteomyelitis in rabbits for 2 weeks resulted in an emergence of resistance (4- to 16-fold) in two of ten treatment failures (NORDEN and SHINNER 1985).

II. Staphylococcal Infections in Experimental Animals

Resistance to fluoroquinolones has rarely emerged in the various staphylococcal infection models studied (Table 4), Especially in experimental endocarditis caused either by methicillin-susceptible *S. aureus* (MSSA) or methicillin-resistant *S. aureus* (MRSA) fluoroquinolones proved effective and were not associated with the development of fluoroquinolone resistance in most of the models. In addition, their in vivo activity was equivalent or superior to that of vancomycin and imipenem; fluoroquinolones were therefore regarded as potential alternatives to therapy with a β-lactam or vancomycin (KAATZ et al. 1987a, 1989, 1990, 1991, 1992; FERNANDEZ-GUERRO et al., 1988, MULLIGAN et al. 1987). However, in common with other antibiotics, fluoroquinolone resistance developed during therapy of human beings. Fluoroquinolone resistance especially among isolates of MRSA was spread by horizontal transmission of resistant clones so that the clinical use of fluoroquinolones in this indication was limited (BALL 1994; DALHOFF 1991).

Because of the magnitude of the problem of surgical wound infections the prophylactic efficacies of ciprofloxacin and ofloxacin were compared to the prophylactic activities of cefazolin and vancomycin in an experimental model of staphylococcal wound infection (KERNODLE and KAISER 1994). Under the experimental conditions studied vancomycin was the most and cefazolin the least effective prophylactic agent. The quinolones provided excellent protection against infection with β-lactamase positive or negative staphylococci as well as MRSA, which were extremely susceptible to quinolone prophylaxis. Fluoroquinolone resistance did not emerge in abscesses which developed despite the quinolone prophylaxis. However, this may merely reflect the observation that the frequency of spontaneous mutation to fluoroquinolone resistance is too low to be observed with a challenge dose of 10^2–10^6 CFU used in their experiments (KERNODLE and KAISER 1994).

III. Miscellaneous Infection Models

After pefloxacin therapy of experimental *Enterobacter cloacae* peritonitis posttherapy strains emerged with decreased susceptibilities to quinolones (4- to 1024-fold), to the structurally unrelated antibiotics chloramphenicol and trimethoprim (4- to 16-fold) and sometimes to tetracyclines and β-lactams (LUCAIN et al. 1989). A recent study of streptococcal endocarditis in rats compared the in vivo efficacy and selection of resistant mutants of streptococci by simulating *parenteral human* kinetics of sparfloxacin and ceftriaxon; therapy lasted either 3 or 5 days. The two drugs exhibited equal in vitro efficacy (MICs and time-kill experiments), but in vivo sparfloxacin generally decreased the bacterial titers in vegetations devoid of phagocytic cells less rapidly than did ceftriaxon (ENTENZA et al. 1994). This difference was quite marked after 3 days of therapy but tended to vanish when treatment was prolonged to 5 days. In contrast, both drugs achieved rapid sterilization of the blood and the spleen, where phagocytic cells are present in high numbers. Therefore the explanation for the differential drug efficacies in the valvular lesion probably relies on the ability of the unbound fraction of a drug to penetrate into the focus of infection, resulting in an adequate in situ bactericidal titer which was greater than or equal to 1000 times the ceftriaxon MIC, despite its high level of protein binding, compared to at best five times the MIC of sparfloxacin (ENTENZA et al. 1994). This finding agrees very well with the hypothesis of KAATZ et al (1987, 1989, 1990, 1991, 1992) discussed below.

IV. Factors Contributing to the Emergence of Resistance In Vivo

A number of preclinical studies have shed some light on factors favoring the in vivo emergence of fluoroquinolone resistance, particularly in *P. aeruginosa* and *S. aureus*. Fleroxacin therapy of experimental MSSA endocarditis cleared effectively bacteremia and reduced bacterial counts in the vegetations. However, resistance to fleroxacin at five- and ten fold the MIC arose to 73% and 27%, respectively (KAATZ et al. 1991). In contrast, only 8% of animals treated with fleroxacin for MRSA endocarditis resistance at five fold the initial MIC arose (KAATZ et al. 1989) It is worth mentioning that the vegations found in the experimental animals having resistant organisms contained more residual CFU and were significantly larger in weight than those found in animals lacking fleroxacin-resistant MSSA. Thus it is conceivable that fleroxacin penetrated these larger vegetations to a lesser extent, resulting in a favorable condition for bacterial survival and the development of resistance by prolonged exposure to subinhibitory concentrations of the drug. This hypothesis is in agreement with all the findings of this group comparing the efficacy of either fleroxacin, ofloxacin, or ciprofloxacin and emergence of fluoroquinolone resistance in the MSSA strain, which produced generally larger vegetations, with the MRSA strain, producing smaller vegetations (KAATZ et al. 1987, 1989, 1990, 1991, 1992).

Similarly, pefloxacin therapy of MRSA endocarditis, was associated with development of resistance in only 1 of 27 animals (4%) despite low rate of sterilization of cardiac vegetations; by comparison, resistance to fosfomycin developed in 5 of 14 animals (36%). The combination of these agents prevented emergence of resistance to either drug alone and was more effective in sterilizing the cardiac vegetations (THAUVIN et al. 1988). Therefore fluoroquinolone efficacy may likely be reduced, and resistance may occur more likely in larger vegetations than in smaller ones because in the larger vegetations drug penetration may be less favorable.

This hypothesis found support in a study comparing the in vitro and in vivo selection of ciprofloxacin-resistant *S. aureus* (DOSS et al. 1995). Mutational frequency was 1.5×10^{-9}–8.4×10^{-10} following an in vitro incubation for 24 h in the presence of 0.5 × MIC and subculture on agar containing 1–5 × MIC ciprofloxacin. A murine subcutaneous abscess model was used for in vivo selection of fluoroquinolone resistance. Mice were treated with such dosages resulting in ciprofloxacin concentrations at the focus of infection equivalent to 0.5 × MIC or 1 × MIC for the pathogen. The number of mice from which mutants were isolated and the mutational frequency were inversely proportional to the ciprofloxacin dose administered and to the duration of treatment. These data are very well in agreement with those of MICHEA-HAMZEHPOUR et al. (1987) on *P. aeruginosa*, indicating that in vivo selection of fluoroquinolone resistance in both species is dose dependent. It is therefore tempting to speculate that drug concentrations at the focus of infection may have a considerable impact on the emergence of resistance. As the selection concentration increases above the MIC the number of resistant mutants decreases (HOOPER and WOLFSON 1993).

In addition, the frequency of selection of fluoroquinolone resitant mutants depends on the bacterial inoculum or challenge dose and the bacterial species. An increase in the bacterial inoculum is associated with an increase in the statistical probability of selecting fluoroquinolone-resistant organisms. In most of the above studies in experimental animals the bacterial counts in the infected foci were as high as or even higher than the mutation frequency, and mutations to fluoroquinolone resistance were therefore very likely not missed. In staphylococcal endocarditis bacterial counts in the vegetations of untreated controls were approximately 8.0–9.0 $\log_{10}$ CFU/g and ranged from 6.0 to 8.0 $\log_{10}$ CFU/g or CFU/ml in the pneumonia or pouch model. However, in *E. coli* pyelonephritis bacterial counts in two kidneys were approximately 4.0 $\log_{10}$ CFU, and in *P. aeruginosa* pyleonephritis the bacterial load ranged from 3.8 to 7.8 $\log_{10}$ CFU/g for both kidneys. The impact of the challenge dose on the in vivo emergence or resistance was confirmed by MICHEA-HAMZEHPOUR et al. (1987). They reported that no resistance occurred in mice challenged with 1.6×10^5 CFU, whereas a higher challenge dose (1.5×10^8 CFU) was associated with the emergence of resistance. Thus, the emergence of in vivo resistance is linked to the bacterial load in the focus of infection.

Furthermore, bacterial species differ in the frequency of selection of mutants. The frequency of development of fluoroquinolone resistance in vitro is low with *E. coli* (approx. 10^{-9}–10^{-11} for low-level resistance, not clinically significant, and $\geq 10^{-27}$ for high-level resistance) but is higher with *P. aeruginosa* (approx. 10^{-6}–10^{-8}; WIEDEMANN and HEISIG 1994; FERNANDEZ et al. 1987; BRYSKIER et al. 1985). Surprisingly, however, mutation frequencies of *P. aeruginosa* were higher by orders of magnitude (Table 1) in in vivo than in in vitro studies; in vivo mutation frequencies ranged from 3 to 7.7×10^{-1}. Because of the discrepancy between in vitro versus in vivo mutation frequencies of *P. aeruginosa* the question was raised whether the emergence of fluoroquinolone-resistant strains in vivo was due to the selection of naturally occurring resistant subpopulations being present even at the very beginning of the experiments rather than to a mutational rise in MICs.

This question was addressed in a series of experiments on the in vivo and in vitro efficacy of ciprofloxacin compared to cefsulodin and sisomicin on *P. aeruginosa* (DALHOFF and DÖRING 1985, 1986; DALHOFF 1991). It was demonstrated that each of the four clinical *P. aeruginosa* isolates and their subpopulations studied were of identical serotype and pyocin type; however, their susceptibilities to fluoroquinolones, β-lactams, and aminoglycosides varied significantly. Single-cell MIC determinations revealed that pretherapy isolates were heterogeneous with respect to MICs. In contrast to conventional MIC testing which yields a single MIC value for the overall population, single-cell MIC testing resulted in individual MIC values up to 64-fold higher than the overall MIC. During exposure of this heterogeneous total population of the challenge strain to ciprofloxacin and the other agents studied, the resistant subpopulations became predominant, and the highly susceptible subpopulations declined in numbers. This finding is very well in agreement with data reported by MICHEA-HAMZEHPOUR et al. (1987). In vitro susceptibility testing of the *P. aeruginosa* strains used in their peritonitis infection model by antibiotic-containing gradient plates revealed that the MIC obtained by the microdilution technique corresponded to the boundary concentrations of the gradient testing. Beyond the boundary concentrations fluoroquinolone-resistant subpopulations grew on the gradient plates up to 12 times higher than the boundary concentration.

MICHEA-HAMZEHPOUR et al. (1987) attributed the rapid selection of resistant variants in their model of pseudomonal peritonitis to the heterogeneity in susceptibilities of the test strains. Resistance rarely occurred in strains containing less resistant variants, but regularly in animals challenged with strains which included the most resistant variants. The in vivo occurrence of resistance in this pseudomonal peritonitis model followed closely the level of resistance of the preexisting variants. Therefore it is tempting to assume that the very frequent decrease in fluoroquinolone susceptibilities of *P. aeruginosa* summarized in Table 1 is due to the selection of naturally occurring subpopulations less susceptible to resistance being present in the challenge dose before the experiments begin. However, it cannot be ruled out that, in addition to the selection of preexisting resistant subpopulation of *P. aeruginosa*, spontaneous mutation contributed to the emergence of resistant variants during fluoroquinolone treatment of experimental animals. Alternatively, resistant subpopulations may have emerged as part of an adaptation process, although this is unlikely as the subpopulations occurred so rapidly and in such high numbers.

The heterogeneity of antibiotic susceptibilities of an otherwise homogeneous (pyocin, phage, genotype) *P. aeruginosa* isolate has been confirmed for the β-lactam and aminoglycoside studied as well (Dalhoff 1991). Similarly, by simulating ciprofloxacin kinetics in an in vitro model subpopulation analysis demonstrated (a) that the inoculum of *P. aeruginosa* is heterogeneous with respect to MICs, and (b) that during exposure to ciprofloxacin the initial decline in CFUs is rapidly followed by the regrowth of resistant subpopulations of *P. aeruginosa* but not those of *E. coli* or *K. pneumoniae* (Dudley et al. 1988). Analogues results were obtained by this group when exposing *P. aeruginosa* but not *E. coli* to netilmicin (Blaser et al. 1985). The presence of aminoglycoside- and quinolone-resistant variants in the initial inoculum and their in vitro selection during exposure to these two classes of antibacterial agents has also been described by others (Nilsson et al. 1987; Radberg et al. 1990). The selection of aminoglycoside-resistant subpopulations has been reported in vivo as well (Gerber and Craig 1982; Gerber et al. 1982).

These data may indicate that emergence of resistance to antibacterial agents in *P. aeruginosa* is due mainly to the heterogeneity of susceptibilities of naturally occurring subpopulations to antibacterial agents and is less likely due to mutational resistance because of drug exposure. This applies for fluoroquinolones as well as for aminoglycosides and β-lactams. A clinical study in patients with cystic fibrosis confirmed that several *P. aeruginosa* subpopulations with varying β-lactam antibiotic susceptibilities were present in each sputum sample; resistant cells comprised at least 10%–20% of all cells in the pretreatment sputum samples (Giwercman et al. 1990). During β-lactam treatment resistance emerged rapidly because of the selection of these resistant subpopulations.

Fluoroquinolone resistance was found to emerge in vivo especially in *P. aeruginosa* and *S. aureus*, perhaps because these species are biologically prone to develop resistance. Resistance frequencies depended on: (a) the mutational frequency of the organism, (b) the presence of resistant subpopulations in the initial inoculum, (c) the number of organisms at the site of infection, (d) the dose administered and thus from the concentration at the site of infection, and (e) the duration of therapy. Emergence of resistance was minimized by adequate dose regimens (i.e., optimal dose, daily dose frequency) and duration of treatment and by combination therapy.

References

Azoulay-Dupuis E, Bedos JP, Vallee E, Hardy DJ, Swanson RN, Pocidalo JJ (1991) Antipneumococcal activity of ciprofloxacin, ofloxacin and temafloxacin in an experimental mouse pneumonia model at various stages of the disease. J Infect Dis 163:319–324

Azoulay-Dupuis E, Vallee E, Veber B, Bedos JP, Bauchet J, Pocidalo JJ (1992) In vivo efficacy of a new fluoroquinolone, sparfloxacin, against penicillin-susceptible and -resistant and multiresistant strains of *Streptococcus pneumoniae* in a mouse model of pneumonia. Antimicrob Agents Chemother 36:2698–2703

Baldwin DR, Honeybourne D, Wise R (1992a) Pulmonary disposition of antimicrobial agents: methodological considerations. Antimicrob Agents and Chemoth 36: 1171–1175

Baldwin DR, Honeybourne D, Wise R (1992b) Pulmonary disposition of antimicrobial agents: in vivo observations and clinical relevance. Antimicrob Agents and Chemother 36: 1176–1180

Ball P (1994) Bacterial resistance to fluoroquinolones: lessons to be learned. Infection 22 Suppl 2:140–147

Bamberger DM, Peterson LR, Gerding DN et al (1986) Ciprofloxacin, azlocillin, ceftizoxime, and amikacin alone and in combination against gram-negative bacilli in an infected chamber model. J Infect Dis 18:51–63

Bayer AS, Norman D, Kim KS (1985) Efficacy of amikacin and ceftazidime in experimental aortic valve endocarditis due to *Pseudomonas aeruginosa*. Antimicrob Agents Chemother 28:781–785

Bayer AS, Blomquist JK, Kim KS (1986) Ciprofloxacin in experimental aortic valve endocarditis due to *Pseudomonas aeruginosa*. J Antimicrob Chemother 17:641–649

Bayer AS, Hirano L, Yin J (1988) Development of beta-lactam resistance and increased quinolone MICs during therapy of experimental *Pseudomonas aeruginosa* endocarditis. Antimicrob Agents Chemother 32:231–235

Blaser J, Stone BB, Zinner SH (1985) Efficacy of intermittent versus continuous administration of netilmicin in a two-compartment in vitro model. Antimicrob Agents Chemother 27:343–349

Blaser J, Stone BB, Groner MC, Zinner SH (1987) Comparative study with enoxacin and netilmicin in a pharmacodynamic model to determine importance of ratio of antibiotic peak concentration to MIC for bacterial activity and emergence of resistance. Antimicrob Agents Chemother 31:1054–1060

Brunner H, Zeiler H-J (1988) Oral ciprofloxacin treatment for *Salmonella typhimurium* infection of normal and immunocompromised mice. Antimicrob Agents Chemother 32:57–62

Bryskier A, Chantot JF (1985) Antibacterial activity of ofloxacin and other 4-quinolone derivatives: in-vitro and in-vivo comparison. J Antimicrob Chemother 16:475–484

Butler T, Cartagenova M, Dunn D (1990) Treatment of experimental *Salmonella typhimurium* infection in mice with lomefloxacin. J Antimicrob Chemother 25:629–634

Craig WA (1993) Post-antibiotic effects in experimental infection models: relationship to in-vitro phenomena and to treatment of infections in man. J Antimicrob Chemother 31 Suppl D:149–158

Craig WA, Ebert S (1991) Killing and regrowth of bacteria in vitro: a review. Scand J Infect Dis Suppl 74:63–70

Craig WA, Gudmundsson S (1992) The postantibiotic effect. In: Lorian V (eds) Antibiotics in laboratory medicine, 3rd edn. Williams and Wilkins, Baltimore, pp 403–431

Craig WA, Watanabe Y (1992) In-vivo pharmacodynamic activity of temafloxacin (Abstr 39). 32nd Interscience Conference on Antimicrobial Agents and Chemotherapy. American Society for Microbiology, Washington, p 117

Craig WA, Ebert S, Moffatt J, Bayer W (1993) Pharmacodynmaic activity of Bay y 3118 in animal infection models (Abstr 1485). 33rd Interscience Conference on Antimicrobial Agents and Chemotherapy. American Society for Microbiology, Washington p 391

Cremieux AC, Saleh-Mghir A, Vallois JM, Maziere B, Muffat-Joly M, Devine C, Bouvet A, Pocidalo JJ, Carbon C (1992) Efficacy of temafloxacin in experimental

Streptococcus adjacens endocarditis and autoradiographic diffusion pattern of ^{14}C-temafloxacin in cardiac vegetations. Antimicrob Agents Chemother 36:2216–2221

Dalhoff A (1991) Clinical perspectives of quinolones resistance in *Pseudomonas aeruginosa*. Antibiotics and Chemotherapy 44:221–239

Dalhoff A (1994) Quinolone resistance in *Pseudomonas aeruginosa* and *Staphylococcus aureus*. Development during therapy and clinical significance. Infection 22[Suppl 2]:111–121

Dalhoff A, Döring G (1985) Interference of ciprofloxacin with the expression of pathogenicity factors of *Pseudomonas aeruginosa*. In Adam D, Hahn H, Opferkuch W (eds) The influence of antibiotics on the host-parasite relationship, II. Springer, Berlin Heidelberg New York, pp 246–255

Dalhoff A, Döring G (1986) Interference of ciprofloxacin with the expression of pathogenicity factors of *Pseudomonas aeruginosa*; in Neu HC Weuta H (eds) 1st International Ciprofloxacin Workshop, Leverkusen. Excerpta Medica, Amsterdam, pp 213–219

Dalhoff A, Döring G (1991) Clinical perspectives of quinolone resistance in *Pseudomonas aeruginosa*. Antibiot Chemother 44:221–239

Davey P, Barza M, Stuart M (1988) Tolerance of *Pseudomonas aeruginosa* to killing by ciprofloxacin, gentamicin and imipenem in vitro and in vivo. J Antimicrob Chemother 21:395–404

Doss SA, Tillotson GS, Brag NL, Amyes SGB (1995) In vitro and in vivo selection of *Staphylococcus aureus* mutants resistant to ciprofloxacin. J Antimicrob Chemother 35:95–102

Dournon E, Rajagopalan P, Vilde J, Pocidalo JJ (1986) Efficacy of pefloxacin in comparison with erythromycin in the treatment of experimental guinea pig legionellosis. J Antimicrob Chemother 17[Supple B]:41–48

Drugeon HB, Drocurt N, Garaffo R (1994) *Streptococcus pneumoniae*: conditions d'apparition des mutants resistans vis à vis de la sparfloxacin et de la ciprofloxacine dans un modile expérimental animal. Colloqúe de la societe francaise de microbiologie Paris, abstract no. 27

Drusano GL, Johnson DE, Rosen M (1993) Pharmacodynamics of a fluoroquinolone antimicrobial agent in a neutropenic rat model of *Pseudomonas* sepsis. Antimicrob Agents Chemother 37:483–490

Dudley MN, Blaser J, Gilbert D, Zinner SH (1988) Bactericidal activity of ciprofloxacin against Pseudomonas aeruginosa and other bacteria in an in vitro two-compartment capillary model. Rev Infect Dis 10 Suppl 1:34–35

Dworkin R, Modin G, Kunz S, Rich R, Zak O, Sande M (1990) Comparative efficacies of ciprofloxacin, pefloxacin and vancomycin in combination with rifampin in a rat model of methicillin-resistant *Staphylococcus aureus* chronic osteomyelitis. Antimicrob Agents Chemother 34[Suppl 6]:1014–1016

Ebert S, Redington J, Rikardsdottir S, Craig WA (1990) In vivo dose-response relationships for fleroxacin versus ciprofloxacin (Abstr 1003) 30th Interscience Conference on Antimicrobial Agents and Chemotherapy. American Society for Microbiology, Washington, p 253

Entenza JM, Blatter M, Glauser MP, Moreillon P (1994) Parenteral sparfloxacin compared with ceftriaxone in treatment of experimental endocarditis due to penicillin-susceptible and resistant streptococci. Antimicrob Agents Chemother 38:2683–2688

Fantin B, Leggett J, Ebert S, Craig WA (1991) Correlation between in vitro and in vivo activity of antimicrobial agents against gram-negative bacilli in a murine infection model. Antimicrob Agents Chemother 35:1413–1422

Fernandez PB, Shipkowitz N, Swanson RN (1989) Comparative efficacy of the fluoroquinolones in experimental animal infections: correlation with in vitro potency and pharmacokinetics. In: Fernandes PBC (Ed) International Telesymposium on Quinolones. Prous Sciences, Barcelona, pp 255–268

Fernandez PB, Hanson CW, Stamm JM et al (1987) The frequency of in vitro resistance development to fluoroquinolones and the use of murine pyelonephritis model to demonstrate selection of resistance in vivo. J Antimicrob Chemother 19:449–465

Fernandez-Guerrero M, Rouse M, Henry N, Wilson W (1988) Ciprofloxacin therapy of experimental endocarditis caused by methicillin-susceptible or methicillin-resistant *Staphylococcus aureus*. Antimicrob Agents Chemother 32: 747–751

Forrest A, Nix DE, Ballow CH, Goss TF, Birmingham MC, Schentag JJ (1993) Pharmacodynamics of intravenous ciprofloxacin in seriously ill patients. Antimicrob Agents Chemother 37:1073–1081

Froidefond S, Saivin S, Lemozy J, Marchou B, Auvergnat JC, Dabernat H (1992) Emergence of resistant strains after antimicrobial therapy of experimental *Pseudomonas aeruginosa* peritonitis. Pathol Biol 40:573–582

Gerber AU, Craig WA (1982) Aminoglycoside-selected subpopulations of Pseudomonas aeruginosa. J Lab Clin Med 100:671–681

Gerber AU, Vastola AP, Brandel J, Craig WA (1982) Selection of aminoglycoside resistant variants of Pseudomonas aeruginosa in an in vivo model. J Infect Dis 146:691–697

Gisby J, Wightman BJ, Beale AS (1991) Comparative efficacies of ciprofloxacin, amoxicillin, amoxicillin-clavulanic acid, and cefaclor against experimental *Streptococcus pneumoniae* respiratory infections in mice. Antimicrob Agents Chemother 35:831–836

Giwercman B, Lambert PA, Rosdahl VT, Shand GH, Hoiby N (1990) Rapid emergence of resistance in *Pseudomonas aëruginosa* in cystic fibrosis patients due to in vivo selection of stable partially depressed β-lactamase producing strains. J Antimicrob Chemother 26:247–259

Gordin FM, Hachbarth CJ, Scott KG, Sande MA (1985) Activities of pefloxacin and ciprofloxacin in experimentally induced Pseudomonas aeruginosa pneumonia in neutropenic guinea pigs. Antimicrob Agents Chemother 27:452–454

Henry NK, Rouse MS, Whitesell AL, McConnell ME, Wilson W (1987) Treatment of methicillin-resistant *Staphylococcus aureus* experimental osteomyelitis with ciprofloxacin or vanomycin alone or in combination with rifampin. Am J Med 82[Suppl 4A]:73–75

Honeybourne D, Baldwin DR (1992) The site concentrations of antimicrobial agents in the lung. J Antimicr Chemother 30:249–260

Hooper DC, Wolfson JS (1993) Mechanisms of bacterial resistance to quinolones. In: Hooper DC, Wolfson JS (eds) Quinolone antimicrobial agents, 2nd edn. American Society for Microbiology, Washington, pp 97–137

Kaatz GW, Barriere SL, Schamberg DR et al. (1987) The emergence of resistance to ciprofloxacin during treatment of experimental *Staphylococcus aureus* endocarditis. J Antimicrob Chemother 20:753–758

Kaatz GW, Seo SM, Barriere SL, Albrecht LM, Rybak MJ (1989) Efficacy of fleroxacin in experimental methicillin-resistant *Staphylococcus aureus* endocarditis. Antimicrob Agents Chemother 33:519–521

Kaatz GW, Seo SM, Barriere SL, Albrecht LM, Rybak MJ (1990) Efficacy of ofloxacin in experimental *Staphylococcus aureus* endocarditis. Antimicrob Agents Chemother 34:257–260

Kaatz GW, Seo SM, Barriere SL, Albrecht LM, Rybak MJ (1991) Development of resistance to fleroxacin during therapy of experimental methicillin-susceptible *Staphylococcus aureus* endocarditis. Antimicrob Agents Chemother 35:1547–1550

Kaatz GW, Seo SM, Lamp KC, Bailey EM, Rybak MJ (1992) CI-960, a new fluoroquinolone, for therapy of experimental ciprofloxacin-susceptible and -resistant *Staphylococcus aureus* endocarditis. Antimicrob Agents Chemother 36:1192–1197

Kernodle DS, Kaiser AB (1994) Comparative prophylactic efficacies of ciprofloxacin, ofloxacin, cefazolin and vancomycin in experimental model of staphylococcal wound infection. Antimicrob Agents Chemother 38:1325–1330

Kimura M, Kishimoto T, Niki Y (1993) In vitro and in vivo antichlamydial activities of newly developed quinolone antimicrobial agents. Antimicrob Agents Chemother 37:801–803

Lalande V, Truffot-Pernot C, Paccaly-Moulin A, Grosset J, Ji B (1993) Powerful bactericidal activity of sparfloxacin (AT-4140) against *Mycobacterium tuberculosis* in mice. Antimicrob Agents Chemother 37:407–413

Leggett JE, Fantin B, Ebert S, Totsuka K, Vogelman B, Calame W, Mattie H, Craig WA (1989a) Comparative antibiotic dose effect relationships at several dosing intervals in murine pneumonitis and thigh-infection models. J Infect Dis 159:281–292

Leggett JE, Ebert S, Fantin B, Craig WA (1989b) A sigmoid dose-response model using bacterial counts predicts dose-survival results for Klebsiella pneumoniae in neutropenic mice (Abstr 313). 29th Interscience Conference on Antimicrobial Agents and Chemotherapy. American Society for Microbiology, Washington, p 153

Leggett JE, Ebert S, Fantin B, Craig WA (1991) Comparative dose-effect relations at several dosing intervals for beta-lactam, aminoglycoside and quinolone antibiotics against gram-negative bacilli in murine thigh-infection and pneumonitis models. Scand J Infect Dis Suppl 74:179–184

Lucain C, Regamy C, Bellido F, Pechere JC (1989) Resistance emerging after pefloxacin therapy of experimental *Enterobacter cloacae* peritonitis. Antimicrob Agents Chemother 33:937–943

MacDonald PJ, Wetherall BL, Pruul H (1981) Postantibiotic leukocyte enhancement: increased susceptibility of bacteria pretreated with antibiotics to activity of leukocytes. Rev Infect Dis 3:38–44

Marchbanks CR, McKiel JR, Gilbert DH, Robillard NJ, Painter B, Zinner SH, Dudley MN (1993) Dose ranging and fractionation of intravenous ciprofloxacin against *Pseudomonas aeruginosa* and *Staphylococcus aureus* in an in vitro model of infection. Antimicrob Agents Chemother 37:1756–1763

Michea-Hamzehpour M, Auckenthaler R, Regamey P et al (1987) Resistance occurring after fluoroquinolone therapy of experimental *Pseudomonas aeruginosa* peritonitis. Antimicrob Agents Chemother 31:1803–1808

Mulligan ME, Ruane PJ, Johnstone L, Wong P, Wheelock JP, MacDonald K, Reinhart JF, Johnson CC, Statner B, Blomquist I, McCarthy J, O'Brien W, Gardner S, Hammer L, Citron DM (1987) Ciprofloxacin for eradication of methicillin-resistant *Staphylococcus aureus* colonization. Am J Med 82 [Suppl 4A]:215–219

Nilsson L, Süren L, Radberg G (1987) Frequencies of variants resistant to different aminoglycosides in *Pseudomonas aeruginosa*. J Antimicrob Chemother 20:255–259

Norden CW, Shinner E (1985) Ciprofloxacin as therapy for experimental osteomyelitis caused by *Pseudomonas aeruginosa*. J Infect Dis 151:291–294

Odenholt-Tornqvist I, Lowdin E, Cars O (1992) Postantibiotic sub-MIC effects of vancomycin, roxithomycin, sparfloxacin, and amikacin. Antimicrob Agents Chemother 36:1852–1858

Patton DL, Cosgrove YT, Kuo CC, Campbell LA (1993) Effects of quinolone analog CI-960 via monkey model of *Chlamydia trachomatis* salpingitis. Animicrob Agents Chemother 37:8–13

Perrone CM, Malinverui R, Glauser MP (1987) Treatment of *Staphylococcus aureus* endocarditis in rats with coumermycin A1 and ciprofloxacin, alone or in combination. Antimicrob Agents Chemother 31:539–543

Potel G, Chau NG, Pangon B, Fantin B, Vallois J-M, Faurisson F, Carbon C (1991) Single daily dosing of antibiotics: importance of in vitro killing rate, serum half-life, and protein binding. Antimicrob Agents Chemother 35:2085–2090

Quale JM, Landman D, Mobarakai N (1994) Treatment of experimental endocarditis due to multidrug resistant *Enterococcus faecium* with ciprofloxacin and novobiocin. J Antimicrob Chemother 34:797–802

Radberg G, Nilsson L, Svensson S (1990) Development of quinolone-imipenem cross-resistance in *Pseudomonas aeruginosa* during exposure to ciprofloxacin. Antimicrob Agents Chemother 34:2142–2147

Renneberg J, Walder M (1989) Postantibiotic effects of imipenem, norfloxacin, and amikacin in vitro and in vivo. Antimicrob Agents Chemother 33:1714–1720

Roosendaal R, Bakker-Woudenberg IAJM, van den Berghe-Van Raffe M, Vink-van den Berg JC, Michel MF (1987) Comparative activities of ciprofloxacin and ceftazidime against *Klebsiella pneumoniae* in vitro and in experimental pneumonia in leukopenic rats. Antimicrob Agents Chemother 31:1809–1815

Roosendaal R, Bakker-Woudenberg IAJM, van den Berghe-Van Raffe M, Vink-van den Berg JC, Michael MF (1989) Impact of the dosage schedule on the efficacy of ceftazidime, gentamicin and ciprofloxacin in Klebsiella pneumoniae pneumonia and septicemia in leukopenic rats. Eur J Clin Microbiol Infect Dis 8:878–887

Ryan DM, Cars O (1980) Antibiotic assay in muscle: are conventional tissue levels misleading as indicator of the antibacterial activity. Scand J Infect Dis 12:307–309

Saito A, Sawatari K, Fukuda Y, Nagasawa M, Koga H, Tomonaga A, Nakazato H, Fujita K, Shigeno Y, Suzuyama Y, Yamaguchi K, Izumikawa K, Hara K (1985) Susceptibility of *Legionella pneumophila* to ofloxacin in vitro and in experimental *Legionella* pneumonia in guinea pigs. Antimicrob Agents Chemother 28:15–20

Schiff JB, Small GJ, Pennington JE (1984) Comparative activities of ciprofloxacin, ticarcillin and tobramycin against experimental *Pseudomonas aeruginosa* pneumonia. Antimicrob Agents Chemother 26:1–4

Scribner RK, Welch DF, Marks MJ (1985) Low frequency of bacterial resistance to enoxacin in vitro and in experimental pneumonia. J Antimicrob Chemother 16:597–603

Stamey TA, Meares EM, Winningham DG (1970) Chronic bacterial prostatitis and the diffusion of drugs into prostatic fluid. J Urol 103:187–194

Sullivan MC, Cooper BW, Nightingale CH, Quintiliani R, Lawlor MT (1993) Evaluation of the efficacy of ciprofloxacin against *Streptococcus pneumoniae* by using a mouse protection model. Antimicrob Agents Chemother 37:234–239

Thauvin C, Lemeland JF, Humbert G, Fillastre JP (1988) Efficacy of pefloxacin-fosfomycin in experimental endocarditis caused by methicillin-resistant *Staphylococcus aureus*. Antimicrob Agents Chemother 32:919–921

Truffot-Pernot C, Ji B, Grosset J (1991) Activities of pefloxacin and ofloxacin against mycobacteria: in vitro and mouse experiment. Tubercle 72:57–64

Ulrich E, Trautmann M, Krause B, Bauernfeind A, Hahn H (1989) Comparative efficacy of ciprofloxacin, azlocillin, imipenem/cilastatin and tobramycin in a model of experimental septicemia due to *Pseudomonas aeruginosa* in neutropenic mice. Infection 17:311–315

Vilde JL, Dournon E, Rajagopalan P (1986) Inhibition of *Legionella pneumophila* multiplication within human macrophages by antimicrobial agents. Antimicrob Agents Chemother 30:743–748

Vogelman B, Gudmundsson S, Turnidge J, Craig WA (1988a) The in vivo postantibiotic effect in a thigh infection in neutropenic mice. J Infect Dis 157:287–298

Vogelman B, Gudmundsson S, Leggett J, Turnidge J, Ebert SE, Craig WA (1988b) Correlation of antimicrobial pharmacokinetic parameters with efficacy in an animal model. J Infect Dis 158:831–847

Watanabe Y, Ebert S, Craig WA (1992a) In-vivo dose-response relationships for ofloxacin versus ciprofloxacin (Abstr 41). 32nd Interscience Conference on Antimicrobial Agents and Chemotherapy. American Society for Microbiology, Washington, p 117

Watanabe Y, Ebert S, Craig WA (1992b) AUC/MIC ratio is unifying parameter for comparison of in vivo activity among fluoroquinolones (Abstr 42) 32nd Interscience Conference on Antimicrobial Agents and Chemotherapy. American Society for Microbiology, Washington, p 117

Wiedemann B, Heisig P (1994) Mechanisms of quinolone resistance. Infection 22 Suppl 2:73–79

CHAPTER 8
Interaction of Quinolones with Host–Parasite Relationship

A. DALHOFF

A. Introduction

The interactions of antibacterial agents with the host–parasite relationship are diverse (Fig. 1). First, the infectious process – and thus the host–parasite relationship – commences by adherence of bacteria to and colonization of epithelial surfaces, followed by penetration into and dissemination within the macroorganism. Antibiosis interferes with these early stages of infection.

Second, the infection as such is a potent immunomodulator causing inflammatory responses and triggering the complex cytokine network. Bacterial exoenzymes, exotoxins and endotoxin released by gram-negative bacteria affect the immune system. Antibiosis may not only reduce bacterial counts but also these products.

Third, the antibacterial agent might directly interact with the immune system. As summarized by SHALIT (1991) and RUBINSTEIN and SHALIT (1993), some quinolones affect immunoglobulin production, synthesis of interleukins, interferons, tumor necrosis factor, and/or colony-stimulating factors (CSFs). Particular attention has been paid to the effect of quinolones on haematopoiesis (for a summary see WANDL et al. 1993).

Fourth, antibacterial agents may indirectly interfere with the host–parasite relationship. They may enhance phagocytosis and/or may make bacteria more vulnerable to intraleukocytic killing by altering the morphology and structure of bacterial surfaces (summarized by MILATOVIC 1983; DASCHNER 1985; SCHUBERT und ULLMANN 1993).

Fifth, antibacterial agents, including quinolones, may indirectly interfere with phagocytic efficacy. Quinolones readily penetrate into phagocytes, are accumulated intracellularly and exhibit intracellular bactericidal activities despite the acidic intracellular pH (for a summary see TULKENS 1991a,b; CARLIER et al. 1990).

Thus, there is a large body of evidence in the literature that quinolones interfere directly and/or indirectly with the humural and cellular immune system. However, some of the results should be interpreted with some reservations since data were generated with different experimental designs (i.e. cells in different stages of differentiation, different stimuli, different and quite frequently unphysiologically high quinolone concentrations). This has resulted in part in contradictory findings and conclusions. Data have been obtained

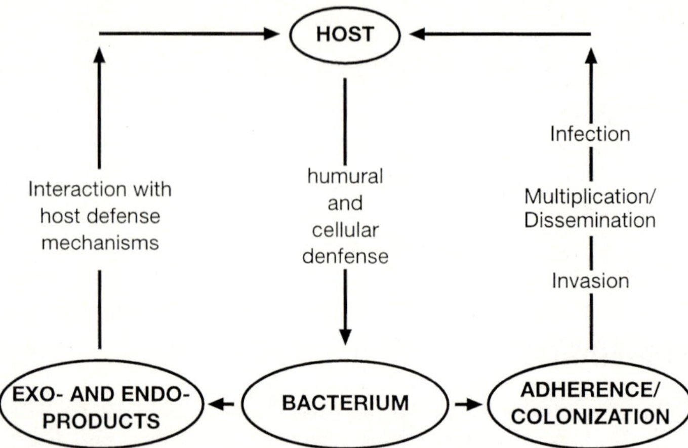

Fig. 1. Interaction of bacteria and bacterial products with the macro-organism and its defence mechanisms

most frequently in vitro or ex vivo, so that comparisons with healthy volunteers or infected patients are lacking and the biological importance of some of these observations is not readily apparent.

Some data, however, are more suggestive. KLETTER et al. (1991, 1994) demonstrated that ciprofloxacin enhanced haematopoiesis and neutrophil function in experimental animals. In cultured murine spleen cells, antimurine granulocyte-macrophage colony-stimulating factor (GM-CSF) and antimurine interleukin-3 (IL-3) antibodies inhibited the stimulation of colony formation by ciprofloxacin, thus indicating that ciprofloxacin-triggered augmentation of haematopoiesis is mediated by colony stimulating factors such as GM-CSF and IL-3. In a clinical study, it was found that in ciprofloxacin-treated patients there seemed to be a higher number of granulocyte-macrophage progenitor cells (CFU-GM) than in the placebo-treated group. Immunocompetent patients undergoing rib resection were treated prophylactically for 2 days prior to operation. In the ciprofloxacin group the numbers of CFU-GM were 80 ± 26 per plate compared with 58 ± 16 per plate in the placebo group. These differences, however, were not statistically significant (BRODIE et al. 1990).

The molecular mechanism(s) of action of physiologically low fluoroquinolone (FQ) concentrations on eukaryotic cell function and the immune system are far from being understood and need to be elucidated.

As the interactions of FQs with the humoral and/or cellular immune system were recently reviewed comprehensively, these will not be summarized again here; the reader is referred to the above-mentioned references. Less attention has been paid to the immunomodulating activities of the infectious agent and the interaction of FQs with these activities. The profound immunomodulating effects of an infection are in its extremes illustrated by the septic syndrome and the progressive states of septic shock ultimately leading to multiple organ failure and death. Antibacterial agents administered to the patients to cope with infection may, on the one hand, reduce bacterial counts and exoenzyme, exotoxin and endotoxin levels. On the other hand, antibiosis may lead to endotoxaemia with all its subsequent deliterious sequelae to the macroorganism. Thus, treatment of severe gram-negative infections involves the elimination of two principally lethal states, bacteraemia and endotoxaemia. In more general terms, antibacterial treatment is aimed at the in vivo growth of the bacterium as well as its pathogenicity and virulence factors (i.e. its exo- and endoproducts). This review summarizes the effect of FQs on adherence and growth in vivo and on exo- and endoproduct synthesis (Fig. 1).

B. Effect on Adherence

The adherence of bacterial pathogens to mammalian mucosa is an essential early step in colonization and ultimately contributes to the development of infection. Several authors have addressed the question of whether FQs might interact with the adhesiveness of bacteria to mucosal cells (Table 1).

By exposing a *Staphylococcus aureus* strain isolated from a bronchial aspirate to subinhibitory concentrations of sparfloxacin, Desnottes and Diallo (1994) demonstrated that 1/2 and 1/4 of the MIC reduced the adherence by more than 50%; 1/8 and 1/16 of the MIC reduced it to 40% and 20%, respectively. No differences to the drug free controls were detectable at 1/32 of the MIC. Similar results were obtained by exposing *Staphylococcus saphrophyticus* to 1/2 MIC of ciprofloxacin (Bohnet et al. 1993). In the presence of sub-MICs of pefloxacin six *S. aureus* strains exhibited a markedly altered capacity for adhesion to buccal cells (Desnottes et al. 1987). The highest significant decrease ranging from 90% to 25% was observed for 1/2 to 1/8 of the MIC, although concentrations as low as 1/512 and 1/1024 of the MIC also decreased attachment.

Fibronectin-mediated adhesion of *S. aureus* to McCoy cells was decreased to 66% and 49%, relative to drug-free controls, during the postantibitoic phase following exposure to ciprofloxacin at 1/2 and 1/1 of the MIC of ciprofloxacin, respectively. During the postantibiotic phase, a ciprofloxacin-dependent decrease of fibronectin receptors on *S. aureus* cells was found (Werk 1991).

Ofloxacin at concentrations of 1/2 to 1/32 of the MIC did not reduce the adherence of *Pseudomonas aeruginosa* to rat alveolar epithelium; in contrast, there was the tendency towards increased adhesion (Yamada et al. 1993).

Preincubation of Mycobacterium avium complex (MAC) with sparfloxacin at concentrations of 1 and 7 mg/l inhibited binding to intestinal mucusal cells by 77%–93% in the case of both drug concentrations (Bermudez et al. 1994). The authors speculate that inhibition of MAC binding to the gastrointestinal mucosa may be one underlying mechanism for the prophylac-

Table 1. Effect of fluoroquinolones on bacterial adherence to eucaryotic cells

Species	Fluoroquinolone	Concentration relative to MIC	Cell type	Method	% Decrease (% increase)	References
S. aureus	Spar	1/2–1/32	Human buccal epithelial cells	Interference contrast microscopy	50–20	Desnottes and Diollo (1994)
S. aureus	Peflo	1/2–1/1024	Human buccal epithelial cells	Interference contrast microscopy	50–90	Desnottes et al. (1987)
S. aureus	Cipro	1/2–1/1	McCoy	Immunofluorescence	66–49	Werk (1991)
S. saprophyticus	Cipro	1/2	3 Human cell lines	MTT cleavage	47	Bohnet et al. (1993)
P. aeruginosa	Oflo	1/2–1/32	Rat alveolar epithelial cells	SEM	Slight increase	Yamada et al. (1993)
P. aeruginosa	Peflo	1/2–1/64	Lymphoblastoid cell line	Plating of non-adherent bacteria	3–79	Marty et al. (1988)
P. aeruginosa	Cipro	1/8–1/32	Vero; uroepithelial cells	Radiometry	20–60	Zahnel et al. (1993)
MAC	Spar	1/8 + 1.2	HT-29 intestinal mucosal cells	Plating of adherent bacteria	77–93	Bermudez et al. (1994)
E. coli	Nal	1/4	Human buccal cells	NS	(38)	Vosbeck et al. (1982)
E. coli	Nal	1/4	Intestine 407 cells	NS	(60)	Vosbeck et al. (1982)
E. coli	Peflo		Uroepithelial cells	NS		Desnottes et al. (1985)
E. coli	Nal	1/4–1/8	Human uroma T24 cells	Haemagglutination; plating	No change	Kovarik et al. (1989)
E. coli	Cipro	1/4–1/8	Human uroma T24 cells	Haemagglutination; plating	No change	Kovarik et al. (1989)
E. coli	Flero	1/4–1/8	Human uroma T24 cells	Haemagglutination; plating		
E. coli	Nor	1/4–1/8	Human uroma T24 cells	Haemagglutination; plating	No change	Kovarik et al. (1989)
E. coli	Enox	1/2–1/8	Uroepithelial cells	SEM	10–90	Burnham (1988)
E. coli	Enox	1/2–1/128	Human buccal; uroepithelial cells	SEM	10–90	Braga and Piatti (1992)
E. coli	Peflo	1/2–1/1024	Uroepithelial cells	Interference contrast microscopy	10–25	Desnottes et al. (1988)

Streptococcus faecalis	Cipro	1/2	Renal and epithelial cells	MTT cleavage	No change	Bohnet et al. (1993)
E. coli	Cipro	1/2	Renal and epithelial cells	MTT cleavage	No change	Bohnet et al. (1993)
E. coli	Peflo	1/2–1/16	Uroepithelial cells	MTT cleavage	47	Loubeyere et al. (1993)
E. coli	Oflo	1/2–1/16	Uroepithelial cells	NS	63	Loubeyere et al. (1993)
E. coli	Cipro	1/2–1/16	Uroepithelial cells	NS	76	Loubeyere et al. (1993)
E. coli	Nal	1/2–1/16	Uroepithelial cells	NS	47	Loubeyere et al. (1993)
E. coli[a]	Enox					
Klebsiella pneumoniae[a]	Nor					
Klebsiella oxytoca[a]	Cipro					
Enterobacter aerogenes[a]	Oflo	1/2–1/3	Small bowel cells	SEM	10–60	Edmunston and Goheen (1989)
Serratia liquiefaciens[a]	Amif					
Citrobacter freundii[a]	Diflo					
Morganella morganii[a]	Nal					
Proteus mirabilis[a]						
Proteus vulgaris[a]						
Serratia marcescens[a]						

Abbreviations: MAC, Mycobactecium avian complex; Spar, sparfloxacin; Peflo, pefloxacin; Cipro, ciprofloxacin; Oflo, ofloxacin; Nal, nalidixic acid; Flero, fleroxacin; Nor, norfloxacin; Enox, enoxacin; Amif, amifloxacin; Diflo, difloxacin; SEM, scanning electron microscopy; NS, not stated; MTT, 3-(4,5-dimethylthiazol-2-yl)-5,5-diphenyltetrazolium bromide; MIC, minimal inhibitory concentration.

[a] Tested against enoxacin, norfloxacin, ciprofloxacin, ofloxacin, amifloxacin, difloxacin and nalidixic acid.

tic effect of quinolones in patients with AIDS; evidence points to the gastrointestinal tract as the major route of infection in these patients.

While nalidixic was found to increase adhesion of *Escherichia coli* to various cells (Vosbeck et al. 1982), some authors have found that FQs at subinhibitory concentrations reduced significantly the ability of *E. coli* to adhere to uroepithelial cells (Desnottes et al. 1985, 1988; Burnham 1988; Sonstein and Burnham 1993; Bohnet et al. 1993; Breines and Burnham 1994). Breines and Burnham (1994) investigated comprehensively the modulation of *E. coli* type 1 fimbrial expression, which mediates the manose-sensitive adherence to mammalian mucosa. Type 1 fimbriae are proteinaceous appendages which are anchored to the outer membrane of many enterobacteriaceae. The expression of type 1 fimbriae in *E. coli* is regulated by phase variation, i.e. the ability to oscillate between on and off expression. Two phase-variation-controlling genes, *fim B* and *fim E*, control the main structural gene, *fim A*. A reduction in *fim A* expression can be mediated either by decreasing *fim B* expression or increasing that of *fim E*.

As FQs affect gene expression they should also affect the expression of fimbrial genes. Expression of *fim A* might be inhibited directly or the promotor could accumulate in the off-orientation. In either case, a fimbriate, non-adherent cell would be the result. However, Breines and Burnham (1994) found that the quinolones studied increased the expression of both *fim B* and *fim E*. Since the *fim B* gene mediates both on to off and off to on inversion while *fim E* mediates on to off inversion exclusively, the effects negate each other. In addition, Hacker et al. (1992) did not find any effect of nalidixic acid or ciprofloxacin at 1/4 of their MICs on the expression of S-fimbriae.

Several other theories have been proposed to explain the effects of FQs at subinhibitory concentrations on bacterial adherence to epithelial cells. First, FQs may induce a partial loss of fimbriae from the surface of bacteria, as has been demonstrated by electron microscopy studies (Burnham 1988; Desnottes et al. 1988; Sonstein and Burnham 1993; Loubeyere et al. 1993). However, doubts are cast on this theory by the finding that the fluorescence assay adopted by Breines and Burnham (1994) to investigate the effect of FQs on the surface expression of type 1 fimbriae showed only small, albeit significant, reductions in the percentage of fimbriate bacteria. Moreover, these bacteria were elongated. Type 1 fimbriae were reduced by only 23%, 21%, 25% and 11% by 0.5 times the MIC of ciprofloxacin, enoxacin, clinafloxacin and PD 131628, respectively. Thus, type 1 fimbriae are present in high numbers on the cell surface, but whether the length and functional properties of these fimbriae remain intact is unknown. Burnham (1988) observed that enoxacin-treated cells had only a few residual fimbriae remaining on their surface and these fimbriae were shorter than those of the control cells.

Second, Eisenstein et al. (1981) suggested that antibiotics interfering with protein biosynthesis may affect fimbrial production and/or assembly resulting in a loss of their lectin-like properties. Streptomycin, for example, which causes misread proteins, was found to lead to the synthesis of aberrant fimbriae. However, DESNOTTES et al. (1988) could not detect any changes in the mobility of P-fimbriae during polyacrylamide gel electrophoresis, thus indicating that the fimbriae were not aberrant. Similarly, KOVARIK et al. (1989) reported that *E. coli* expressed morphologically and functionally intact P-fimbriae following exposure to nalidixic acid, ciprofloxacin, fleroxacin, and norfloxacin. Adhesion to human uroma cells was not changed following exposure to the quinolones.

Thirdly – and independent of specific interactions mediated by fimbriae – FQs may alter surface charge density and hydrophobicity, leading to an increased repulsion between bacteria and epithelial cells (LOUBEYERE et al. 1993; SONSTEIN and BURNHAM 1993; YAMADA et al. 1993). This hypotehesis points to phenotypic changes of bacteria following exposure to FQs. LOUBEYERE et al. (1993) assessed bacterial cell surface properties such as charge and hydrophobicity by determining partition in a two-phase polyethyleneglycol/dextran system. In this system, pefloxacin, ofloxacin, norfloxacin and ciprofloxacin increased the negative charges on the bacterial surface to 88%, 64%, 30% and 80%, respectively. Surprisingly, nalidixic acid did not increase the charge on *E. coli* although it inhibited adhesion by 66% at 1/8 of the MIC. For the fluorinated quinolones a correlation was found between inhibition of adhesion and increase in negative surface charge ($r = 0.65$). The antiadherence properties of pefloxacin were lost when magnesium was added to the test system (LOUBEYERE et al. 1993). For a given pefloxacin concentration, increasing concentrations of magnesium decreased adhesion. All of the quinolones studied were found to increase hydrophobicity; however, there was no correlation between the increase in hydrophobicity and the inhibition of adhesion ($r = -0.21$).

With reference to the work of CHAPMAN and GEORGOPAPADAKOU (1988), LOUBEYERE et al. (1993) speculate that FQs chelate with magnesium which is associated with lipopolysaccharide (LPS) and maintains the integrity of the outer membrane. Chelation of magnesium creates hydrophobic patches in the outer membrane of gram-negative bacteria through which quinolones diffuse (self-promoted pathway or non-porin pathway); quinolones also penetrate through porins.

Although the hypothesis of self-promoted uptake by quinolones has been disputed by MARSHALL and PIDDOCK (1994) they and LECOMTE et al. (1992, 1994) could demonstrate that selected quinolones were chelated by magnesium and that the extent of chelation varied with the quinolone. Chelation – as demonstrated in a fluorescence assay with only the quinolone and magnesium being present in the test system – was greatest with nalidixic acid and enoxacin, intermediate with ciprofloxacin and pefloxacin, and low with ofloxacin. The strength of the magnesium – quinolone complex was weakest for nalidixic acid and enoxacin and strongest for ciprofloxacin and pefloxacin (MARSHALL and PIDDOCK 1994). By using nuclear magnetic resonance, fluorescence- and infrared spectroscopy, LECOMTE et al. (1992) found that affinity increased from sparfloxacin to ofloxacin, ciprofloxacin, norfloxacin and pefloxacin (Table 2).

Table 2. Effect of quinolones on the adhesion of *E. coli* to uroepithelial cells (1/8 MIC), the negative surface charge of *E. coli* (1/8 MIC) and complex formation with divalent cations (20 m*M* Mg^{2+}; 0.05 mg/l quinolone). (LOUBEYERE et al. 1993; MARSHALL and PIDDOCK 1994; and LECOMTE et al. 1992, 1994)

Quinolone	Inhibition of adhesion (%)	Increase in negative charge (%)	Change in fluorescence (%)	Dissociation constant
Nalidixic acid	66	24	+395	n.d.
Pefloxacin	47	15	+104	0.50
Norfloxacin	8	14	+93	0.83
Ofloxacin	63	44	+16	0.95
Ciprofloxacin	76	30	+177	0.87

n.d., not done; MIC, minimal inhibitory concentration.

Thus, chelation of individual quinolones by magnesium varies significantly; for any given quinolone, affinity to magnesium may depend on the methods applied. CHAPMAN and GEORGOPAPADAKOU (1988) working with fleroxacin proposed that magnesium ions bind to the carboxyl group at position 4 and the carboxylic acid group at position 3 of the quinolone nucleus. This was confirmed by LECOMTE et al. (1992) in the case of lomefloxacin and by HELENA et al. (1995) in the case of ciprofloxacin.

Consequently, it is tempting to speculate that quinolones compete for the LPS-associated divalent cations thereby increasing the negative surface charge of gram-negative bacteria. This hypothesis, however, could not be proven by others (SEYDEL 1995, personal communication). In studying the relationships between adherence and charge in bacteria *not* exposed to antibacterial agents it was found, that highly charged strains adhered less well than those carrying little charge (COLLEEN et al. 1979; MOZES et al. 1988). Assuming that the same correlation holds true for quinolone-exposed bacteria, those quinolones being most strongly complexed by magnesium should increase the negative surface charge most markedly and should thus reduce the adherence of gram-negative bacteria to various cells most dramatically. However, no correlation was found between the strength of the quinolone–magnesium complex, the increase in negative surface charge and inhibition of adhesion (Table 2).

Thus, none of the three hypotheses, (1) gene expression of adhesins, (2) surface expression or assembly of adhesins and (3) surface charge and hydrophobicity, can provide a conclusive explanation of the mechanism(s) underlying the proven phenomenological effects of FQs on the adhesion of bacteria to mucosal or epithelial cells. These phenomena have been studied so far in well-defined, in vitro systems; now, the study of these phenomena in the complex in vivo situation is warranted.

Bacterial adherence may be mediated not only by fimbriae but also by extracellular exopolysaccharides, i.e. slime. Adherence of *Staphylococcus epidermidis* to artificial devices is mediated by its slime production. A few studies have yielded unequivocal results on the effect of sub-MIC of FQs on the adherence of *S. epidermidis* and slime production (Schmitt et al. 1989; Perez-Giraldo et al. 1994). In general, only a slight but not statistically significant alteration in slime production was found. This effect was strain dependent in so far as some strains increased the amount of slime production in the presence of FQs while others produced less. When the most productive strains were analysed separately, 1/2 of the MIC resulted in the greatest decrease in slime production. Wilcox et al. (1991) did not analyse the effect of subinhibitory levels on slime production by coagulase-negative staphylococci but quantitated directly the adherence of ten strains to silicone rubber and polystyrene. Strains exposed to 1/4 of the MIC of ciprofloxacin were compared with those exposed to 1/4 of the MIC of cefuroxime, vancomycin and teicoplanin. In general, cefuroxime caused a reduction in adherence, whereas vancomycin exposure frequently appeared to result in increased adherence; ciprofloxacin caused changes midway between these two extremes. There were marked interstrain variations for all the antibacterials tested.

C. Effect Against Slowly Growing Bacteria

Bacteria growing in vivo multiply much more slowly than in vitro, irrespective of whether they are freely floating in body fluids or adhere to cell surfaces. A reduction in their growth rate is a characteristic response of bacteria to environmental changes. Slowly growing bacteria generally survive better than those replicating quickly – spore formation is an extreme example (Brown et al. 1990; Brown and Williams 1985a,b; Williams 1988; Gilbert et al. 1990; Dalhoff 1985).

In the infected foci, bacteria may be walled off by leukocytes and fibrin deposits, thus limiting access to nutrition. For example, generation times of *E. coli* are 0.4 h under optimal in vitro culture conditions. However, the replacement of broth by a kidney homogenate prolonged generation times to 0.6 h. In experimental animals, the same *E. coli* strain, causing pyelonephritis, multiplies every 2.9–3.5 h. In a sepsis model, the generation time of this *E. coli* strain ranges from 0.8 to 1.0 h. Under optimal in vitro conditions, *S. aureus* multiplies every 0.4 h. In experimental animals suffering from osteomyelitis, generation times are prolonged to 8–24 h in the infected host (Zak and Sande 1982; Dalhoff 1985). Many additional environmental changes, such as alterations in temperature, do not support rapid growth of bacteria (Brown et al. 1990; Small et al. 1986; Sheldon 1988).

It is well known that β-lactam antibiotics exert their maximal effect against rapidly growing bacteria; slowly growing bacteria were killed proportionally more slowly. Hence, the bactericidal effect of β-lactam antibiotics is directly proportional to the growth rate. A constant proportion of the population is killed per generation (Cozens et al. 1986; Tuomanen et al. 1986; Driehuis and Wouters 1987). Therefore, it is of clinical importance to determine the bactericidal effect of quinolones – as with any other antibacterial agent – on optimally growing bacteria and their slowly growing counterparts.

ENG et al. (1991) studied the bactericidal effects of antimicrobial agents of various chemical classes and different modes of action on optimally growing, slowly growing and nongrowing bacteria in batch culture. By limiting nutrient supply, the growth rate was controlled. As, however, nongrowing bacteria were killed even in drug-free media under the experimental conditions studied, the results on nongrowing bacteria generated by ENG et al. (1991) are not mentioned here. By using the same quinolone concentration (4 mg/l) for all the different bacteria tested – without considering their MICs – rapidly growing bacteria were killed more effectively than their slowly growing counterparts (ENG et al. 1991; Table 3). However, the authors did not analyse their data, strain and drug specifically. The data summarized in Table 3 clearly indicate that ciprofloxacin is generally more bactericidal than ofloxacin and that slowly growing *P. aeruginosa* and *Enterobacter cloacae* are even more effectively killed by ciprofloxacin than their rapidly growing counterparts. Killing of *Klebsiella pneumoniae* seems to be independent of growth rate, and slowly growing *E. coli* and *S. aureus* are less effectively killed than the rapidly growing ones.

These phenomena having been observed although not mentioned explicitly by ENG et al. (1991) were studied in greater detail by DALHOFF et al. (1995). In batch cultures, generation times of bacteria were controlled by using media of different strengths, and bacteria were exposed to quinolone concentrations at multiples of their MICs, i.e. bioequivalent concentrations were used. Since the MICs for the quinolones tested differed significantly, relative bactericidal potencies were determined by calculating the "killing rate" (k; by analogy to the growth rate) and dividing the k-values by the corresponding MIC. By comparing k/MIC ratios all data are normalized for identical antibacterial activity.

Table 3. Bactericidal activities (log_{10} CFU/ml killed within 24 h; inoculum 10^8 CFU/ml) of ciprofloxacin and ofloxacin against bacteria growing under optimal and subobtimal conditions. (Modified according to ENG et al. 1991)

Species	Control		Ciprofloxacin		Ofloxacin	
	OG	SG	OG	SG	OG	SG
E. coli	–1.8*	–0.1	>7.0	5.8	5.8	5.9
K. pneumoniae	–0.8	–0.2	5.9	5.7	4.8	4.8
P. aeruginosa	–1.2	–0.5	4.9	6.1	4.8	4.8
E. cloacae	–1.1	–0.4	2.9	4.0	2.7	2.7
S. aureus	–1.1	–0.8	4.5	3.9	4.4	2.9

OG, optimal growth; SG, suboptimal growth; growth rates were not specified; CFU, colony-forming unit.
[a] Negative numbers indicate growth.

In general, reduced growth rates affected the bactericidal activities of the quinolones against the individual test strains and species studied differently. A modest increase in generation times of the *E. coli* strains resulted in an augmented bactericidal effect of the quinolones, whereas the slowly growing strains (generation time of 1.4h) were killed less effectively. On average,

Table 4. Relative bactericidal potencies of quinolones against *E. coli*

Quinolone	Strain no./MIC	Killing rates		
		GT = 0.38 h	GT = 0.78 h	GT = 1.41 h
Ciprofloxacin	I/0.031	–70.6	–90.0	–60.0
	II/0.015	–92.9	–82.9	–61.9
	III/0.031	–34.2	–56.5	–55.5
Fleroxacin	I/0.125	–12.9	–16.1	–12.4
	II/0.25	–8.8	–10.5	–8.1
	III/0.25	–7.0	–6.8	–6.8
Norfloxacin	I/0.25	–13.1	–8.3	–7.1
	II/0.125	–17.6	–21.5	–17.4
	III/0.125	–8.9	–13.6	–13.7
Ofloxacin	I/0.125	–17.6	–20.6	–15.4
	II/0.125	–31.4	–29.7	–20.5
	III/0.125	–15.6	–14.5	–12.7

GT, generation time; MIC, minimal inhibitory concentration; –, reduction in colony-forming units.

Table 5. Relative bactericidal potencies of quinolones against *P. aeruginosa*

Quinolone	Strain no./MIC	Killing rates		
		GT = 0.49 h	GT = 0.83 h	GT = 1.26 h
Ciprofloxacin	I/0.062	–23.4	–15.9	–31.5
	II/0.031	–4.2	–15.1	–58.7
	III/0.125	–12.9	–16.3	–21.6
Fleroxacin	I/1.0	–1.6	–2.2	–3.3
	II/1.0	–1.5	–1.4	–1.6
	III/2.0	–0.8	–1.1	–0.9
Norfloxacin	I/0.5	–3.8	–4.6	–4.2
	II/0.25	–3.7	–6.0	–6.8
	III/0.5	–3.6	–3.5	–4.1
Ofloxacin	I/2.0	–2.4	–2.6	–3.2
	II/0.5	–2.7	–2.7	–4.5
	III/1.0	–2.1	–2.2	–5.8

GT, generation time; MIC; minimal inhibitory concentration; –, reduction in colony-forming units.

killing rates were reduced by 11.5% for ciprofloxacin, 4.3% for fleroxacin, 3.9% for norfloxacin and 28.7% for ofloxacin (Table 4).

In contrast, the bactericidal activities of the quinolones against *P. aeruginosa* (Table 5) and *S. aureus* (Table 6) were augmented by reducing the growth rates of the bacteria. The bactericidal effects of quinolones against slowly growing *P. aeruginosa* (generation time of 1.3 h) was increased by 176% for ciprofloxacin, 48% for fleroxacin, 36% for norfloxacin and 86% for ofloxacin, compared with their rapidly growing counterparts (generation time of 0.49 h). The increased bactericidal effect of quinolones against slowly growing *S. aureus* was not as marked. These data indicate, in agreement with the results reported by others, that quinolones exert a bactericidal activity especially against gram-negative bacteria, irrespective of whether they were cultured under optimal or suboptimal in vitro conditions, allowing rapid or slow growth or whether they caused infections in experimental animals (CHALKLEY and KOORNHOF 1985; ROOSENDAAL et al. 1987; ZEILER 1985; ZEILER and ENDERMANN 1986).

Unexpectedly, in batch cultures, viable counts of *P. aeruginosa* were even more effectively and more rapidly reduced when growing slowly compared with the rapidly growing organisms. Hypothetically, this phenomenon is due to changes in the outer membrane protein (omp) pattern of *P. aeruginosa* when growing under conditions of limited nutrient supply. As a result of adaptation to nutrient-deficient media, ompH is overexpressed in *P. aeruginosa* (YOUNG et al. 1992, YOUNG and HANCOCK 1992). As quinolone uptake by *P. aeruginosa* is mediated by ompH, overexpression of this porin results in FQ hypersusceptibility. Overexpression of ompH let to an eight- to

Table 6. Relative bactericidal potencies of quinolones against *S. aureus*

Quinolone	Strain no./MIC	Killing rates		
		GT = 0.40 h	GT = 1.04 h	GT = 3.87 h
Ciprofloxacin	I/0.5	–1.8	–2.4	–1.9
	II/0.25	–3.0	–4.7	–3.2
	III/0.25	–3.2	–4.9	–4.6
Fleroxacin	I/0.5	–1.3	–1.8	–1.9
	II/0.5	–1.6	–2.7	–1.8
	III/0.5	–1.5	–2.5	–2.2
Norfloxacin	I/1.0	–0.79	–1.23	–1.01
	II/1.0	–0.92	–1.29	–0.96
	III/1.0	–0.69	–1.33	–1.36
Ofloxacin	I/1.0	–1.13	–1.34	–1.2
	II/1.0	–1.08	–1.80	–1.25
	III/0.25	–2.76	–4.4	4.0

GT, generation time; MIC, minimal inhibitory concentration; –, reduction in colony-forming units; positive numbers indicate growth.

32-fold increase in susceptibility to nalidixic acid, norfloxacin, fleroxacin and ciprofloxacin (YOUNG and HANCOCK 1992). However, the precise mechanisms causing increased FQ susceptibility of *P. aeruginosa* and especially *S. aureus* under the batch culture conditions remain to be elucidated.

The relative independence of the bactericidal effect of quinolones of the bacterial growth rate might be mirrored by clinical findings. In patients suffering from severe pneumonia, ciprofloxacin cleared the causative pathogens from the lungs (lung aspirate) within 24h in 60% of the patients studied (SCHENTAG 1991; FORREST et al. 1993). The median time to eradication was 1.9 days – provided an adequate dosage regime was used – based on the in vitro bacterial susceptibility. β-lactams, however, cleared significantly less rapidly the pathogens from the lungs of these severely ill patients. However, these comparisons were done historically (SCHENTAG 1991).

D. Effect on Exoenzyme Production

I. *E. coli*

Verotoxin-producing *E. coli* (VTEC) are the causative agents of haemorrhagic colitis and haemolytic uremic syndrome. VTEC strains produce one or several verotoxins which are also termed "shiga-like toxins" because of their structural and biological similarities.

Exposure of a shiga-like-toxin-producing *E. coli* strain to 1/16 the MIC of ciprofloxacin for 12h resulted in a dramatic reduction of toxin production (KARCH et al. 1985); the yield of intracellular and extracellular shiga-like toxin was reduced to 1/16 and 1/8, respectively, of the drug free control values.

The effect of enoxacin and ciprofloxacin on haemolysin activity during the postantibiotic phase was studied by GUAN and BURNHAM (1992). A 1-h exposure to 0.5 times the MIC of the FQs produced a postantibiotic effect of 0.7–0.9h. Following FQ removal the exposed *E. coli* did not exhibit normal haemolysin activity for at least 2h; extracellular haemolysin activity during the postantibiotic phase was about 65% of that of unexposed control cells.

II. *P. aeruginosa*

P. aeruginosa is an opportunistic pathogen causing severe septicaemia in immunocompromised patients and burn wound infections. It is also the principal pulmonary pathogen in patients with cystic fibrosis and diffuse panbronchiolitis. The replication of *P. aeruginosa* at the pulmonary sites of infection is not as important as the production of extracellular enzymes. The exoenzymes exotoxin A, phospholipase C, proteases – in particular elastase and alkaline protease – and exoenzyme S are important pathogenicity factors (DÖRING et al. 1984; NICAS and IGLEWSKI 1985).

Antipseudomonal therapy results in clinical improvement because of multiple actions of the antibiotics employed: (1) reduction of bacterial numbers at the site of infection; (2) reduction of virulence of the causative organism; and (3) interaction with the host immune mechanism(s) to protect host tissues against an overexuberant immune response.

The in vitro effects of ciprofloxacin and those of the less comprehensively studied ofloxacin upon exoenzyme production by *P. aeruginosa* following a 24-h period of growth in broth and exposure to subinhibitory concentrations of FQs was examined by three groups of investigators (Table 7). All the data generated agree very well and clearly demonstrate that FQ concentrations sometimes as low as 1/20 of the MIC were capable of significantly reducing exoenzyme production.

Ciprofloxacin at 1/10 its MIC completely inhibited phospholipase production and elastase activity was as low as 2% of the control (GRIMWOOD et al.

Table 7. In vitro effect of fluoroquinolones on exoenzyme production by *P. aeruginosa* after 24 h of growth in broth culture

Exoenzyme	FQ	Concentration (× MIC)	Exoenzyme production (% of control)	Reference
Elastase	CPX	0.5	5	DALHOFF and DÖRING (1985, 1987)
	CPX	0.2	1	GRIMWOOD et al. (1989b)
		0.1	2	
		0.05	16	
	OFX	0.2	14	MIZUKANE et al. (1991)
Alkaline protease	CPX	0.5	7	DALHOFF and DÖRING (1985, 1987)
Total protease	CPX	0.2	2	GRIMWOOD et al. (1989a,b)
		0.1	14	
		0.05	43	
	OFX	0.05	29	MIZUKANE et al. (1992)
Exotoxin A	CPX	0.5	0	DALHOFF and DÖRING (1985, 1987)
	CPX	0.2	34	GRIMWOOD et al. (1989b)
		0.1	43	
		0.05	88	
Exoenzyme S	CPX	0.2	25	GRIMWOOD et al. (1989b)
		0.1	35	
		0.05	35	
Phospholipase C	CPX	0.2	0	GRIMWOOD et al. (1989b)
		0.1	0	
		0.05	74	

FQ, fluoroquinolone; CPX, ciprofloxacin; OFX, ofloxacin; MIC, minimal inhibitory concentration.

1989a). Similar data were generated by GOVAN and DOHERTY (1985) and HOSTACKA and MAJTAN (1994). During longitudinal in vitro and in vivo experiments a relative resistance to the bactericidal effects of FQs developed quickly in *P. aeruginosa* (DALHOFF and DÖRING 1985, 1987; DALHOFF 1991). Therefore, the test organisms were exposed for 5 and 10 days to a constant subinhibitory concentration of ciprofloxacin (DALHOFF and DÖRING 1985, 1987; GRIMWOOD et al. 1989a; DALHOFF 1991). As anticipated, there was a ten- to 33-fold increase in the MICs of ciprofloxacin by day 3–5 of exposure, but this remained stable until the end of the experiments. Despite steadily increasing MIC values of ciprofloxacin throughout the study period, resulting on day 5 in concentrations as low as 1/320 of the initial MIC, the suppression of exoenzyme production was maintained throughout the total exposure period.

Exoenzyme production remained inhibited up to 4 days and four passages in antibiotic-free medium. This phenomenon suggests that ciprofloxacin induced transient changes in the phenotype of the *P. aeruginosa* strains studied (GRIMWOOD et al. 1989a,b).

The effects of ciprofloxacin on exoenzyme production were compared to those of β-lactams (cefsulodin and ceftazidime) and aminoglycosides (sisomicin and tobramycin) (DALHOFF and DÖRING 1985; GRIMWOOD et al. 1989a). All of the antibacterial agents studied were able to suppress the production of exoenzymes, but this effect was most consistent for ciprofloxacin.

The in vivo relevance of these in vitro studies was confirmed by adopting the rat granuloma pouch model for chronic local *P. aeruginosa* infection (DALHOFF and DÖRING 1985) and a rat lung model of *P. aeruginosa* infection (GRIMWOOD et al. 1989a,b). Treatment of the pouch animals with suboptimal ciprofloxacin doses resulted in a significant reduction of elastase, alkaline protease and exotoxin A concentrations in the pouch exudate. The concentrations of these pathogenicity factors were determined directly from the exudate without subculturing the pathogens in vitro. The general wellbeing of the animals improved in direct correlation to the interference with the biosynthesis of proteases and exotoxin A.

Treatment of rats with chronic *P. aeruginosa* lung infections with suboptimal doses of ciprofloxacin did not affect bacterial numbers in lung homogenates. However, the lungs from ciprofloxacin-treated rats had significantly less histological damage than those from control rats. Furthermore, *P. aeruginosa* isolates from treated rats produced significantly less exoenzyme S (66%) elastase (78%) and total protease (71%) than those from control animals upon subculture in vitro. The effect on exoenzyme production was less than expected considering the degree of protection from histological injury. It may be that subculturing of the rat lung isolates in the absence of ciprofloxacin enabled the bacteria to revert to their original phenotype.

Although the results obtained in the rat pouch model reflect the direct effects of FQs on exoenzyme production by *P. aeruginosa*, the protective effects seen in the rat lung model may be related to factors other than exoenzyme production. It is known that FQs inhibit the adherence of various bacterial species to epithelial surfaces, which underscores the complexity of interactions between subinhibitory FQ concentrations and the microorganisms. The mechanism(s) underlying the effect of FQs on exoenzyme production have not yet been elucidated. It is speculated that in particular the transcription of genes which are controlled by catabolite repression is highly dependent on an interaction with RNA polymerase. FQs and novobiocin affect the supercoiling of circular DNA and thereby modify the interaction of RNA with some promotors. As summarized by DALHOFF and DÖRING (1987), FQs and/or novobiocin do not affect the constitutively expressed genes but rather those which are controlled by catabolite repression only.

E. Quinolone-Induced Endotoxin Release

The endotoxin LPS is released from gram-negative bacteria during exponential growth and cell death into culture media (in vitro) or the bloodstream (in vivo). LPS is known to be bioreactive, affecting the dynamic interplay of the cytokine network. Thus, the release of LPS into the infected macroorganism is a major contributing factor to the pathogenicity of the infecting bacterium.

Concern was raised that a rapid decrease in viable bacterial counts as mediated by quinolones, for example, may be paralleled by an early and sustained increase in endotoxin release. Several in vitro studies indicate that the increase in LPS levels is a direct result of bacterial killing. In contrast, LPS release in vivo did not necessarily correlate with the rate of bacterial killing but depended on the class of antibiotic studied. Treatment of animals with either an aminoglycoside (gentamicin) or a β-lactam (moxalactam) resulted in equal killing of bacteria; however, a seven-to 20-fold higher endotoxin release was observed with moxalactam treatment than with gentamicin treatment. β-lactams with different affinities to their biochemical target, the penicillin-binding proteins (PBPs), exhibit different levels of endotoxin release (for summary see PRINS et al. 1994).

Only among the β-lactams are the differences in their endotoxin-releasing abilities likely to be due to their different affinities to their target, resulting in different modes of antibacterial activities. The significant differences in endotoxin-liberating potencies of various other antibacterial agents of different chemical classes can neither be attributed to their different modes of action nor to their differences in bacterial killing. In general, the conclusion that an increase in LPS levels results directly from bacterial cell wall disintegration seems to be too simple.

Since FQs exert their bactericidal effect rapidly, they may trigger a rapid rise in endotoxin levels. On the other hand, FQs interact with bacterial gyrase, and hence, they may have no potential for endotoxin-liberation. However, in vitro studies demonstrated that FQs caused more LPS release than gentamicin or even the β-lactam amoxicillin (VAN DEN BERG et al. 1992; COHEN and

McConnel 1985). Although no differences in bactericidal activity between the FQs at the different concentrations studied were observed, they differed significantly in their potential to release endotoxin from an in vitro culture of *E. coli* (McConnell and Cohen 1986; Table 8). However, there is no conclusive evidence that this effect is related to differences in clinical outcome between these agents.

Because of the clinical relevance of antibiotic-induced endotoxin release, studies on LPS release were performed during antibacterial treatment of experimental bacteraemia in animals (Nitsche et al. 1994, 1996). Such studies may have more clinical relevance than in vitro data as they integrate pharmacokinetics, pharmacodynamics and the physiological interaction between LPS release and endotoxin clearance mechanisms. In a peritonitis-associated *E. coli* sepsis model in rats, bacterial counts in plasma and peritoneal cavity, plasma endotoxin activity and mean arterial pressure (MAP, as a pathophysiological indicator for endotoxaemia) were monitored. Animals were treated intravenously with doses of ciprofloaxacin, cefotaxime, imipenem and gentamicin recommended for clinical use.

In general, antibacterial treatment resulted in a transient and very modest bacteraemia and a rapid clearance of bacteria from the plasma of treated animals. Differences in antibacterial activity of the agents studied were evident at the focus of infection (Table 9). In the peritoneal cavity cefotaxime was the least effective drug. Doubling the dose of imipenem did not result in an augmented bactericidal effect, whereas doubling the dose of ciprofloxacin resulted in an approximately ten-fold greater reduction of viable counts in the peritoneal lavage. Gentamincin treatment was equally effective as the high-dose ciprofloxacin treatment (Table 9).

Plasma endotoxin activity (*Limulus* amoebocyte lysate; LAL) did not correlate to antibacterial efficacy. Compared with the other agents studied cefotaxime treatment resulted in the most pronounced increase in plasma

Table 8. Release of endotoxin in vitro from *E. coli* by quinolones. Bacterial counts and endotoxin concentrations were determined following an incubation time of 2h

Quinolone	Concentration (mg/l)	MIC (mg/l)	Log_{10} CFU/ml	Endotoxin (mg/ml)	Reference
Control	0	NA	n.d.	140	Cohen and McConnell (1985)
Ciprofloxacin	5	n.d.	n.d.	470	
Control	0	NA	+1.0	956	McConnell and Cohen (1986)
Ciprofloxacin	2.3	0.02	−3.4	956	
Pefloxacin	4.3	0.08	−3.4	1.478	
Norfloxacin	1.5	0.04	−3.4	1.543	
Ofloxacin	5.3	0.02	−4.0	2.043	
Enoxacin	3.7	0.16	−2.9	2.152	

NA, not applicable; n.d., no data; CFU, colony-forming units; MIC, minimal inhibitory concentration.

Table 9. In vivo efficacy of antibacterial therapy of a peritonitis-associated *E. coli* sepsis in rats and its effect on endotoxaemia. (Modified according to NITSCHE et al. 1994, 1995)

Drug	Dose (mg/kg)	Bacterial counts (CFU/ml)		Endotoxin activity in plasma (Eu/ml)
		Plasma	Peritoneum	
Control	NA	6.5×10^3	1.9×10^7	1.18
Gentamicin	5	$\leq 10^1$	7.1×10^3	3.10
Cefotaxime	40	2.5×10^1	2.0×10^5	4.28
Imipenem	7	$\leq 10^1$	4.4×10^4	3.37
Imipenem	14	$\leq 10^1$	3.2×10^4	2.95
Ciprofloxacin	3	$\leq 10^1$	2.3×10^4	3.47
Ciprofloxacin	6	$\leq 10^1$	2.7×10^3	2.48

NA, not applicable; CFU, colony-forming unit.

endotoxin activity although it exhibited the weakest bactericidal efficacy. Compared with cefotaxime, imipenem and, particularly, ciprofloxacin decreased bacterial counts at the focus of infection up to 100-fold more effectively. However, endotoxin activity was significantly lower in the ciprofloxacin group than in the cefotaxime group. The effect of ciprofloxacin on endotoxaemia was dose dependent whereas imipenem exhibited a dose-independent effect (Table 9).

The differential effects on endotoxin release are mirrored by differences in MAP. The decrease in MAP was greatest in the cefotaxime group and lowest in the high-dose ciprofloxacin group. In this group MAP was not different from that measured in the sham-infected group (Fig. 2). In conclusion, antibacterial activity in vivo does not correlate with endotoxaemia under the experimental conditions studied. Ciprofloxacin being the most effective antibacterial drug induced the lowest endotoxin release, in contrast to cefotaxime, which was the least effective agent but triggered the highest endotoxin release.

Hypotension is closely linked to endotoxaemia. Based on these findings the authors speculate that not only plasma endotoxin levels, but also endotoxin-mediated cytokine release at the focus of infection may contribute to adverse sequelae such as hypotension. Low endotoxin levels resulted in a limited local cytokine release and a minor decrease in MAP (NITSCHE et al. 1995).

The beneficial effect of high-dose ciprofloxacin (6 mg/kg) on endotoxaemia might be due to its greater potential to reduce the bacterial load at the focus of infection than low-dose ciprofloxaxin (3 mg/kg). Furthermore, ciprofloxacin exhibits concentration-dependent antiendotoxin properties in vitro which are similar to those known for polymyxin. By incubating in vitro LPS extracted from *Salmonella abortus equi* and *E. coli* EC-5 with equimolar concentrations of polymyxin, ciprofloxacin and gentamicin, ciprofloxacin re-

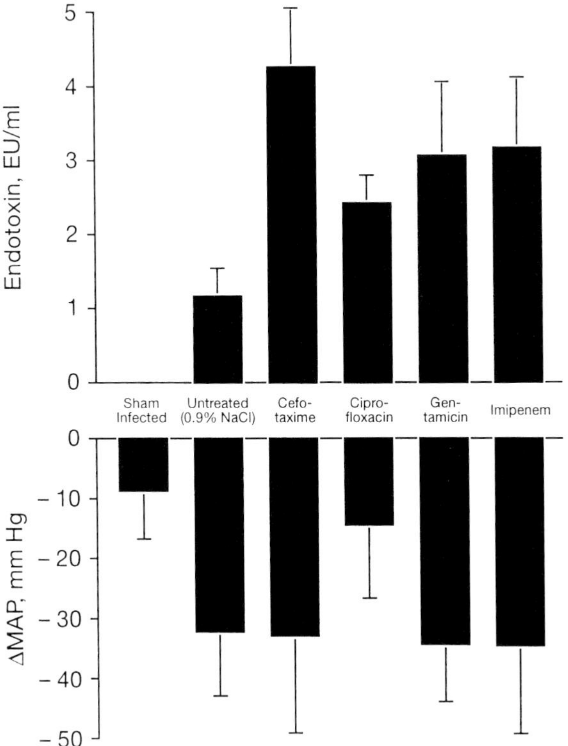

Fig. 2. Plasma endotoxin activity and mean arterial pressure decrease (*m.a.p.*) 5 h after bacterial challenge and antimicrobial treatment of an intra-abdominal *E. coli* infection. Dose: cefotaxime 40 mg/kg ($n = 10$), imipenem 14 mg/kg ($n = 10$), gentamicin 5 mg/kg ($n = 10$), ciprofloxacin 6 mg/kg ($n = 12$). The animals in the untreated group ($n = 9$) received 0.9% NaCl instead of an antibiotic. Neither *E. coli* nor antibiotics were administered to the sham-infected group ($n = 8$)

duced LAL activity approximately 0.25-fold less effectively and gentamicin, 3-fold less effectively than polymyxin. In vitro incubation of drugs at concentrations determined in the plasma of experimental animals in the course of the above study with LPS (3000 endotoxin units/dl) resulted in a reduction in LAL activity of 82% and 92% for low- and high-dose ciprofloxacin, respectively, 54% for gentamicin and 8% for imipenem. A physiological concentration of polymyxin (3.6 mg/l) reduced endotoxin activity to 59%. The reduction in LAL activity was dependent on the pH of the medium and was noted only within the physiological pH range (pH = 7.2; Nitsche et al. 1994). As summarized by Prins et al. (1994) no endotoxin binding could be demonstrated for chloramphenicol, ampicillin, carbenicillin, cefixin, aztreonam, imipenem, netilmicin and ofloxacin.

Thus, antibacterial agents differ in their endotoxin-releasing abilities in vivo and their endotoxin neutralizing effects in vitro. Both phenomena may contribute to clinical outcome. Based on the differences in endotoxin-liberating potential of the various quinolones studied by MCCONNELL and COHEN (1986) and the endotoxin-neutralizing abilities of ciprofloxacin and gentamicin demonstrated by NITSCHE et al. (1994), the latter two agents may be particularly suitable for treatment of gram-negative infections. However, it remains to be demonstrated that modest endotoxin-liberating abilities and marked endotoxin-neutralizing activities contribute to a positive clinical outcome. It also remains to be demonstrated that antibiotic-mediated endotoxin release may contribute to an adverse clinical outcome.

F. Summary

A synopsis of the various interactions of FQs with the host–parasite relationship is presented in Fig. 3. FQs reduce adherence to various epithelial cells and may thus interfere with the early phases of infection. By decreasing or completely inhibiting the synthesis or the amount of exo- and endoproducts released from bacteria the deleterious effects triggered by these bioreactive products will also be decreased.

As FQs exhibit a marked bactericidal effect against bacteria growing slowly in vivo, they reduce the bacterial burden at the focus of infection significantly. The elimination of bacteria from the focus of infection is supported by humural and cellular defense mechanisms which are also augmented by FQs, as reviewed by others (RUBENSTEIN and SHALIT 1993). In addition, the favourable pharmacokinetics of FQs, i.e. marked intracellular penetration and intracellular antibacterial effect in addition to high concentrations at the focus

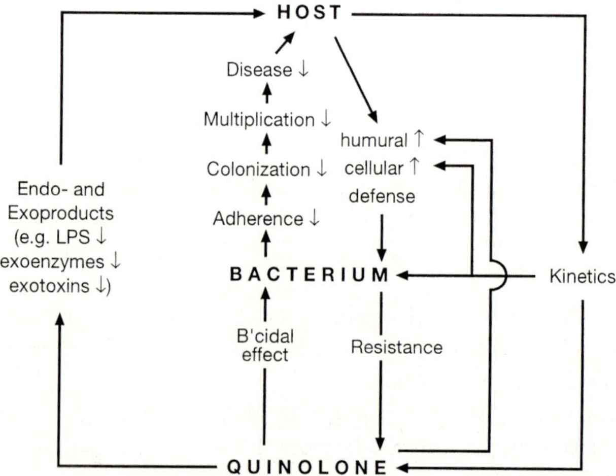

Fig. 3. Pleiotropic interactions of quinolones with host–parasite relationship. ↑ increase; ↓ decrease

of infection, cooperate with the host defense mechanisms. In conclusion, despite the difficulties in defining the mechanisms of interactions with the host–parasite relationship and the relevance of experimental data to the clinical situation, fluoroquinolones may combine a direct antibacterial action with a direct or indirect immunoenhancing activity.

References

Bermudez LE, Young LS, Inderlied CB (1994) Rifabutin and sparfloxacin but not azithromycin inhibit binding of mycobacterium avium complex to HT-29 intestinal mucosal cells. Antimicrob Agents Chemother 38:1200–1202

Bohnet S, Kreft B, Marre R (1993) Influence of ciprofloxacin on the adherence of several uropathogenic bacteriae to eucaryotic cell lines (Abstr 476). 18th International Congress on Chemotherapy, Stockholm

Braga PC, Piatti G (1992) Influence of enoxacin sub-mics on the adherence of Staphylococcus aureus and Escherichia coli to human buccal and urinary epithelial cells. Chemotherapy 38:261–266

Breines DM, Burnham JC (1994) Modulation of Escherichia coli type 1 fimbrial expression and adherence to uroepithelial cells following exposure of logarithmic phase cells to quinolones at subinhibitory concentrations. J Antimicrob Chemother 34:205–221

Brodie E, Duerr D, Shaked N, Yellin A, Liebermann Y, Rosen N, Seger S, Rubinstein E (1990) The effect of ciprofloxacin, ceftriaxone and placebo on peripheral WBC and marrow-derived-granulocytes-marophage-progenitor cells in humans (Abstr 437). 3rd International Symposium on New Quinolones, Vancouver

Brown MRW, Williams P (1985a) Influence of substrate limitation and growth phase on sensitivity to antimicrobial agents. J Antimicrob Chemother 15 Suppl A:7–14

Brown MRW, Williams P (1985b) The influence of environment on envelope properties affecting survival of bacteria in infections. Annu Rev Microbiol 39:527–556

Brown MRW, Collier PJ, Gilbert P (1990) Influence of growth rate on susceptibility to antimicrobial agents: modification of the cell envelope and batch and continuous culture studies. Antimicrob Agents Chemother 34:1623–1628

Burnham JC (1988) Mediation by enoxacin of adherence of Escherichia coli to uroepithelial cells. Rev Infect Dis 10:175

Carlier M-B, Scorneaux B, Zenebergh A et al (1990) Cellular uptake, localization and activity of fluoroquinolones in uninfected and infected macrophages. J Antimicrob Chemother 26 Suppl B:27–39

Chalkley IJ, Koornhof HJ (1985) Antimicrobial activity of ciprofloxacin against Pseudomonas aeruginosa, Escherichia coli and Staphylococcus aureus determined by the killing curve method: antibiotic comparisons and synergistic interactions. Antimicrob Agents Chemother 28:331–342

Chapman JS, Georgopapadakou NH (1988) Routes of quinolone permeation in Escherichia coli. Antimicrob Agents Chemother 32:438–442

Cohen J, McConnell JS (1985) Antibiotic induced endotoxin release. Lancet 8463:1069–1070

Cohen J, McConnell JS (1986) Release of endotoxin from bacteria exposed to ciprofloxacin and its prevention with polymyxin B. Eur J Clin Microbiol 5:13–17

Colleen STIG, Hovelius B, Wieslander A, Mardh P-A (1979) Surface properties of Staphylococcus saprophyticus and Staphylococcus epidermindis as studied by adherence tests and two-polymer, aqueious phase systems. Acta Pathol Microbiol Scand [B] 87:321–328

Cozens RM, Tuomanen E, Tosch W et al (1986) Evaluation of the bactericidal activity of β-lactam antibiotics on slowly growing bacteria cultured in the chemostat. Antimicrob Agents Chemother 29:797–802

Dalhoff A (1985) Differences between bacteria grown in vitro and in vivo. J Antimicrob Chemother 15 Suppl A:175–195

Dalhoff A (1991) Clinical perspectives of quinolone resistance in Pseudomonas aeruginosa. In: Homma IJ, Taminoto H, Holder IA, Hoiby N, Döring G (eds) Pseudomonas aeruginosa in human disease. Karger, Basel, pp 221–239

Dalhoff A, Döring G (1985) Interference of ciprofloxacin with the expression of pathogenicity factors of Pseudomonas aeruginosa. In: Neu HC, Weuta H (eds) 1st International Ciprofloxacin Workshop, Leverkusen. Excerpta Medica, Amsterdam, pp 213–219

Dalhoff A, Döring G (1987) Action of quinolones on gene expression and bacterial membranes. Antibiot Chemother 39:205–214

Dalhoff A, Matutat S, Ullmann U (1995) Effect of quinolones against slowly growing bacteria. Chemotherapy 41:92–99

Daschner FD (1985) Antibiotics and host defence with special reference to phagocytosis by human polymorphonuclear leukocytes. J Antimicrob Chemother 16:135–141

Desnottes JF, Diallo N (1994) Effect of sparfloxacin on Staphylococcus aureus adhesiveness and phagocytosis. J Antimicrob Chemother 33:737–746

Desnottes JF, Diallo N, Santonja R (1985) Effect of subminimal inhibitory concentrations of pefloxacin on the haemagglutination and adhesion of pyelonephritogenic Escherichia coli strains. Chemioterpie 4:574–576

Desnottes JF, Diallo N, Moret G et al (1987) Effects of subinhibitory concentrations of pefloxacin on the adherence of Staphylococcus aureus to human cells. Drugs Exp Clin Res 13(2):69–73

Desnottes JF, Le R, Diallo N (1988) Effect of subminimal inhibitory concentrations of pefloxacin on the piliation and adherence of E. coli. Drugs Exp Clin Res 14(10):629–634

Döring G, Dalhoff A, Vogel O, Brunner H, Dröge U, Botzenhart K (1984) In vivo activity of proteases of Pseudomonas aeruginosa in a rat model. J Infect Dis 149:532–537

Driehuis F, Wouters JTM (1987) Effect of growth rate on the penicillin binding proteins of Escherichia coli. FEMS Lett 48:89–92

Edmunston CE, Goheen MP (1989) Impact of subinhibitory concentrations of quinolones on adherence of enterobacteriaceae to cells of the small bowel. Rev Infect Dis Suppl 5 11:948–949

Eisenstein BI, Ofek I, Beachey EH (1981) Loss of lectin-like activity in aberrant type 1 fimbriae of Escherichia coli. Infect Immun 31:792–797

Eng RHK, Padberg FT, Smith SM et al (1991) Bactericidal effects of antibiotics on slowly growing and nongrowing bacteria. Antimicrob Agents Chemother 35:1824–1828

Forrest A, Nix DE, Ballow CH et al (1993) Pharmacodynamics of intravenous ciprofloxacin in seriously ill patients. Antimicrob Agents Chemother 37:1073–1081

Gilbert P, Collier PJ, Brown MRW (1990) Influence of growth rate on susceptibility to antimicrobial agents: biofilms, cell cycle, dormancy, and stringent response. Antimicrob Agents Chemother 34:1865–1868

Govan JRW, Doherty C (1985) Influence of antibiotics, alginate biosynthesis and hypersensitivity mutations on Pseudomonas virulence factors associated with respiratory infections in patients with cystic fibrosis. In: Ishigani J (ed) Recent advances in chemotherapy. Antimicrobial section 1. University of Tokyo Press, Tokyo, pp 357–358

Grimwood K, To M, Rabin HR, Woods DE (1989a) Subinhibitory antibiotics reduce Pseudomonas aeruginosa tissue injury in the rat lung model. J Antimicrob Chemother 24:937–945

Grimwood K, To M, Rabin HR, Woods DE (1989b) Inhibition of Pseudomonas aeruginosa exoenzyme expression by subinhibitory antibiotic concentrations. Antimicrob Agents Chemother 33:41–47

Guan L, Burnham JC (1992) Postantibiotic effect of CI-960, enoxacin and ciprofloxacin on Escherichia coli: effect on morphology and haemolysin activity. J Antimicrob Chemother 29:529–538

Hacker J, Wallner U, Straube E, Hof H (1992) Einfluβ von Antibiotika auf die Expression Virulenz-assoziierter Gene. Z Arztl Fortbild 86:271–276

Helena M, Teixeira SF, Vilas-Boas LF, Gil VMS, Tetixeira F (1995) Complexes of ciprofloxacin with metal ions contained in antacid drugs. J Chemother 7:126–132

Hostacka A, Majtan V (1994) Elastase and proteinase of P. aeruginosa after treatment with sub-mics of trobramycin, netilmicin and ofloxacin (Abstr 199). 6th International Congress on Infections Diseases, Prague

Karch H, Goroncy-Bermes P, Opferkuch W, Kroll HP, O'Brien A (1985) Subinhibitory concentations of antibiotics modulate amount of shiga-like toxin produced by Escherichia coli. In: Adam B, Hahn H, Opferkuch W (eds) The influence of antibiotics on the hostparasite relationship II. Springer, Berlin Heidelberg New York, pp 239–245

Kletter Y, Riklis I, Shalit I et al. (1991) Enhanced repopulation of murine hematopoietic organs in sublethally irradiated mice after treatment with ciprofloxacin. Blood 78:1685–1691

Kletter Y, Singer A, Nagler A (1994) Ciprofloxacin enhances hematopoiesis and the peritoneal neutrophil function in lethally irradiated, bone marrow-transplanted mice. Exp Hematal 22:360–365

Kovarik JM, Hoepelman IM, Verhoef J (1989) Influence of fluoroquinolones on expression and function of P fimbriae in uropathogenic Escherichia coli. Antimicrob Agents Chemother 33:684–688

Lecomte S, Coupry C, Chenon MT et al (1992) Accurate determination of the affinity of sparfloxacin for mg++, and comparison with that of other fluoroquinonlones (FQ), with regard to antibacterial activity, entry into the cell and resistance. 32nd Interscience Conference on Antimicrobial Agents and Chemotherapy, abstract 785

Lecomte S, Baron MH, Chenon MT, Coupry C, Moreau NJ (1994) Effect of magnesium complexation by fluoroquinolones on their antibacterial properties. Antimicrob Agents Chemother 38:2810–2816

Loubeyere C, Desnottes JF, Moreau N (1993) Influence of sub-inhinbitory concentrations of antibacterials on the surface properties and adhesion of Escherichia coli. J Antimicrob Chemother 31:37–45

Marshall AJH, Piddock LJV (1994) Interaction of divalent cations, quinolones and bacteria. J Antimicrob Chemother 34:465–483

Marty N, Lapchine L, Dournes JL, Agueda L, Chebanon G (1988) Effects of five antibiotics on adhesion and haemagglutinating properties of Pseudomonas aeruginosa isolated from cystic fibrosis patients. Drugs Exp Clin Res 14:635–643

McConell JS, Cohen J (1986) Release of endotoxin from Escherichia coli by quinolones. J Antimicrob Chemother 18:765–766

Milatovic D (1983) Antibiotics and phagocytosis. Eur J Clin Microbiol 2:414–425

Mizukane R, Kaku M, Matsumoto T, Ishida K, Koga H, Kohno S, Hara K (1992) The effect of ofloxacin on production of exoenzymes by Pseudomonas aerugminosa. Fourth International Symposium on New Quinolones, Munich, abstract 72

Mizukane R, Hirakata Y, Kaku M, Ishii Y, Furuya N, Ishida K, Koga H, Kohno S, Yamaguchi K (1994) Comparative in vitro exoenzyme-suppressing activities of azithromycin and other macrolide antibiotics against Pseudomonas aeruginosa. Antimcrob Agents Chemother 38:528–533

Mozes N, Leonard AJ, Rouxhet PG (1988) On the relations between the elemental surface composition of yeasts and bacteria and their charge and hydrophobicity. Biochim Biophys Acta 945:324–334

Nicas TI, Iglewski BH (1985) The contribution of exoproducts to virulence of Pseudomonas aeruginosa. Can J Microbiol 31:387–392

Nitsche D, Schulze C, Oesser S, Dalhoff A, Sack M (1994) The effects of different types of antimicrobial agents on plasma endotioxin activity in gram-negative bacterial infection. In: Garrard C (ed) Ciprofloxacin i.v., defining its role in serious infections. Springer, Berlin Heidelberg New York, pp 21–36

Nitsche D, Schulze C, Oesser S et al (1996) Impact of different classes of antimicrobial agents on plasma endotoxin activity. Arch Surg 131:192–199

Pérez-Giraldo C, Rodriguez-Benito A, Moran FJ et al (1994) In-vitro slime production by Staphylococcus epidermidis in presence of subinhibitory concentrsations of ciprofloxacin, ofloxacin and sparfloxacin. J Antimicrob Chemother 33:845–848

Prins JM, VAN Deventer SJH, Kuijper EJ et al (1994) Clinical relevance of antibiotic-induced endotoxin release. Antimicrob Agents Chemother 38:1211–1218

Roosendaal R, Bakker Woudenberg IAJ, van den Berghe RM et al (1987) Comparative activities of ciprofloxacin and ceftazidime against Klebsiella pneumoniae in vitro and in experimental pneumonia in leukopenic rats. Antimicrob Agents Chemother 31:1809–1815

Rubinstein E, Shalit I (1993) Effects of the quinolones on the immune system. In: Hooper DC, Wolfson JS (eds) Quinolone antimicrobial agents, 2nd edn. American Society for Microbiology, Washington, pp 519–526

Schentag JJ (1991) Correlation of pharmacokinetic parameters to efficacy of antibiotics: relationships between serum concentrations, MIC values, and bacterial eradication in patients with gram-negative pneumonia. Scand J Infect Dis Suppl 74:218–234

Schmitt DD, Bandyk DF, Edmunston CE, Levy MF, Seabrook GR, Towne JB (1989) The in vitro effect of subinhibitory concentrations of quinolones and vancomycin on adherence of slime-producing Staphylococcus epidermidis to vascular prothesis. Rev Infect Dis 11 Suppl 5:947–948

Schubert S, Ullmann U (1993) Effect of antibiotics on chemotaxis, phagocytosis of polymorphonuclear leukocytes and lymphocyte transformation with special reference to modern beta-lactams and fluoroquinolones. In: Ullmann U, Dalhoff A (eds) Significance of cytokines in the treatment of infectious diseases. Fischer, Stuttgart, pp 111–136

Shalit I (1991) Immunological aspects of new quinolones. Eur J Clin Microbiol Infect Dis (Special issue) 10:262–266

Sheldon H (1988) Boyd's introduction to the study of disease 10th edn. Lea and Febiger, Philadelphia, pp 129–155

Small PM, Tauber MG, Hackbarth CJ et al (1986) Influence of body temperature on bacterial growth rates in experimental pneumococcal meningitis in rabbits. Infect Immun 52:484–487

Sonstein SA, Burnham JC (1993) Effect of low concentrations of quinolone antibiotics on bacterial virulence mechanisms. Diagn Microbiol Infect Dis 16:277–289

Tulkens PM (1991a) Intracellular pharmacokinetics and localization of antibiotics as predictors of their efficacy against intragphagocytic infections. Scand J Infect Dis Suppl 74:209–217

Tulkens PM (1991b) Intracellular distribution and activity of antibiotics. Eur J Clin Microbiol Infect Dis 10:100–106

Tuomanen E, Cozens R, Tosch W et al (1986) The rate of killing of Escherichia coli by β-lactam antibiotics is strictly proportional to the rate of bacterial growth. J Gen Microb 132:1297–1304

Van Berg C, DE Neeling AJ, Schot CS, Hustiux WNM, Wemer J, DE Wildt DJ (1992) Delayed antibiotic-induced lysis of Escherichia coli in vitro is correlated with enhancement of LPS release. Scand J Infect Dis 24:619–627

Vosbeck K, Mett H, Huber U et al (1982) Effects of low concentration of antibiotics on Escherichia coli adhesion. Antimicrob Agents Chemother 21:864–869

Wandl U, Dalhoff A, Anders CU (1993) Effect of antiinfective agents on human haematopoietic cell growth. In: Ullmann U, Dalhoff A (eds) Significance of cytokines in the treatment of infectious diseases. Fischer, Stuttgart, pp 159–169

Werk R (1991) Ciprofloxacin influences the fibronectin mediated adhesion of Staphylococcus aureus in the postantibiotic phase (Abstr 649). 17th International Congress of Chemotherapy, Berlin

Wilcox MH, Finch RG, Smith DGE et al (1991) Effects of carbon dioxide and sublethal levels of antibiotics on adherence of coagulase-negative staphylococci to polystyrene and silicone rubber. J Antimicrob Chemother 27:577–587

Williams P (1988) Role of the cell envelope in bacterial adaptation to growth in vivo infections. Biochimie 70:987–1011

Yamada Y, Kojima Y, Nakamura A et al (1993) Effects of antibiotics on the adherence of Pseudomonas aeruginosa to rat alveolar and tracheal epithelia. International Congress of Chemotherapy, Berlin

Young ML, Hancock REW (1992) Fluoroquinolone supersusceptibility mediated by outer membrane protein OprH overexpression in Pseudomonas aeruginosa: evidence for involvement of a nonproin pathway. Antimicrob Agents Chemother 36:2365–2369

Young ML, Bains M, Bell A, Hancock REW (1992) Role of Pseudomonas aeruginosa outer membrane protein OprH in polymyxin and gentamicin resistance: isolation of an OprH-deficient mutant by gene replacement techniques. Antimicrob Agents Chemother 36:2566–2568

Zahnel GG, Kim SO, Davidson RJ, Hoban DJ, Nicolle LE (1993) Effect of subinhibitory concentration of ciprofloxacin and gentamicin on the adherence of Pseudomonas aeruginosa to Vero cells and voided uroepithelial cells. Chemotherapy 39:105–111

Zak O, Sande MA (1982) Correlation of in vitro antimicrobial activity of antibiotics with results of treatment in experimental animal models and human infection. In: Sabath LD (ed) Action of antibiotics in patients. Huber, Bern, pp 55–67

Zeiler H-J (1985) Evaluation of the in vitro bactericidal action of ciprofloxacin on cells of Escherichia coli in the logarithmic and stationary phases of growth. Antimicrob Agents Chemother 28:524–527

Zeiler H J, Endermann R (1986) Effect of ciprofloxacin on stationary bacteria studied in vivo in a murine granuloma pouch model infected with Escherichia coli. Chemotherapy 32:468–472

CHAPTER 9

Mechanisms of Resistance to Fluoroquinolones

M.J. EVERETT and L.J.V. PIDDOCK

A. Introduction

In the early 1980s, when fluoroquinolones were an area of rapid development and considerable pharmaceutical company interest, it was commonly believed that resistance to fluoroquinolones would not become a problem, as it had with previous classes of antibiotics, as fluoroquinolone-resistant *Escherichia coli* could only be selected with great difficulty under laboratory conditions (SMITH 1984). The very low minimum inhibitory concentrations (MICs) of these agents for many pathogens, combined with good pharmacokinetics and hence high achievable concentrations at the site of infection, also convinced many that resistance was unlikely to arise in the clinical setting. Furthermore, those mutations which were selected were chromosomally located and recessive to the wild type; thus interspecies spread of resistance alleles was not expected.

Fifteen years later, fluoroquinolones are available in both oral and intravenous formulations. Because of their high in vitro activity against many common gram-negative pathogens they are often used as an empiric treatment against a wide variety of bacterial infections, including gastroenteritis, urinary tract infections and opportunistic infections associated with chronic lung disorders. Some of the fluoroquinolones also exhibit good activity against gram-positive organisms such as *Staphylococcus aureus*, and have been used to combat infections by methicillin-resistant *S. aureus* (MRSA) in hospitals and multiply drug resistant *Mycobacterium tuberculosis*. New agents under development further expand the spectrum of activity to include anaerobes. Despite early optimism, fluoroquinolone-resistance emerged in a number of clinically important organisms and this trend is likely to increase in the future. The mechanism(s) of resistance in fluoroquinolone-resistant isolates have consequently been the subject of intense research, and in recent years dramatic advances have been made, both in our understanding of these mechanisms and in the range of resistant organisms that have been characterised. This chapter reviews the current state of research in this exciting area.

Bacterial resistance to toxic chemicals such as antibiotics can occur by a number of well-recognised mechanisms. These are alteration of the target site, exclusion of the antibiotic from the target site and enzymatic destruction or modification of the antibiotic so as to render it inactive. Resistance to quinolone antibiotics has been shown to occur via the first two of these mechanisms but not by enzyme-mediated destruction or modification of the antibiotic. The failure to identify systems employing this latter mechanism may be due to the fact that quinolones are totally synthetic chemicals and thus have not previously been involved in biological processes.

B. Target Site Modification

Fluoroquinolone antibiotics exert their antibacterial effect by inhibition of certain bacterial topoisomerase enzymes, namely DNA gyrase (bacterial topisomerase II) and topoisomerase IV (reviewed in Chap. 4). Topoisomerases are a class of enzymes which are able to alter the topological state of DNA molecules. A number of different topoisomerases have been described from both prokaryotic and eukaryotic organisms all of which have the ability to relax supercoiled DNA, or otherwise introduce positive supercoils into covalently closed double-stranded DNA. DNA gyrase is unique among topoisomerases in that it is able to introduce negative supercoils into DNA (GELLERT et al. 1976), and thus can also reduce the linkage number of positively supercoiled DNA. This is achieved by creating a double-stranded staggered break in the DNA, passing a portion of DNA though the resulting gap and resealing the break to leave an intact DNA molecule with two extra negative supercoils. Purification of DNA gyrase from *E. coli* revealed that it is a heterotetrameric protein composed of two subunits, designated A and B (PEEBLES et al. 1978). The genes encoding the A and B subunits, *gyrA* and *gyrB*, were shown to map at 48 min and 83 min in *E. coli*. The mechanism of supercoiling involves the strand breakage of double-stranded DNA by the A subunits, whilst the B subunits facilitate passage of a segment of DNA through the gap. This latter reaction requires the hydrolysis of two molecules of ATP for each strand passage. The break in the DNA is then resealed by the A subunit. The activity of DNA gyrase is essential for chromosome replication and decatenation of circular DNA molecules. DNA gyrase was first implicated as a target of quinolone action by GELLERT et al. (1977), who showed that its activity was sensitive to nalidixic acid; this was later demonstrated to be due to inhibition of the A subunit (PEEBLES et al. 1978).

I. Mutations in *gyrA*

The first quinolone resistance gene described in *E. coli*, *nalA*, was shown by a biochemical approach to encode an A subunit of DNA gyrase with decreased sensitivity to nalidixic acid. It was found that the purified A subunits from a resistant strain (A^r) when mixed with B subunits from a sensitive strain (B^s) formed a DNA gyrase which was refractory to normally inhibitory concentrations of nalidixic acid (PEEBLES et al. 1978). Similar observations have been made with fluoroquinolone-resistant *E. coli* (HOOPER et al. 1986;

SATO et al. 1994), *Pseudomonas aeruginosa* (ROBILLARD and SCARPA 1988), *Campylobacter jejuni* (GOOTZ and MARTIN 1991), *S. aureus* (NAKANISHI et al. 1991a), *Enterococcus faecalis* (NAKANISHI et al. 1991b) and *Shigella dysenteriae* (RAHMAN et al. 1994). Furthermore, it has been shown that if gyrase is reconstituted from a mixture of A^r and A^s subunits the resulting enzyme is sensitive to quinolones indicating that the A^s subunit is dominant to the A^r subunit (SUGINO et al. 1977). This finding agrees with that of HANE and WOOD (1969) who previously demonstrated by genetic means that the wild-type *gyrA* allele is dominant over the resistant allele. Complementation studies in which the cloned wild-type allele of *gyrA*, carried on a multicopy plasmid, has been introduced into resistant strains have been performed on a number of different species, including *E. coli* (NAKAMURA et al. 1989; YOSHIDA et al. 1990a), *P. aeruginosa* (ROBILLARD 1990; YOSHIDA et al. 1990b), *Salmonella typhimurium* (PIDDOCK et al. 1993a), *Klebsiella pneumoniae* (HEISIG and WIEDEMANN 1991; PIDDOCK and ZHU 1991), *Providencia stuartii*, *Acinetobacter baumanii* (HEISIG and WIEDEMANN 1991), *Enterobacter cloacae* (PIDDOCK and ZHU 1991) and *Serratia marcescens* (MASECAR and ROBILLARD 1991; PIDDOCK and ZHU 1991). These studies showed that many gram-negative fluoroquinolone-resistant organisms contained a *gyrA* mutation and could be complemented to quinolone-susceptibility by introduction of the wild-type allele. The wild-type *gyrA* gene of *S. aureus* has recently been subcloned into a gram-negative–gram-positive shuttle plasmid, enabling an analogous assay to be carried out in *S. aureus* (TANKOVIC et al. 1994).

The susceptibility of DNA gyrase activity to quinolone action can be measured by determining the concentration of quinolone (in micrograms per millilitre) required to inhibit supercoiling of plasmid DNA in vitro by 50% (IC_{50}). The extent of plasmid supercoiling can be visualised by agarose gel electrophoresis; the greater the number of supercoils in the plasmid the more compact the molecule, and hence the faster it will migrate through the agarose matrix. Alteration of the A subunit of DNA gyrase from a wide variety of organisms has been found to increase the IC_{50} 25- to 500-fold (CAMBAU and GUTMANN 1993). A problem with the "supercoiling" assay is that it does not distinguish between quinolone-mediated inhibition of DNA gyrase and that caused by non-GyrA inhibitors or experimental artifacts resulting in false positive data. A better in vitro measure of 4-quinolone activity is the "cleavable complex" assay, which does not produce false positive data and is easier to measure (DOMAGALA et al. 1986; BARRETT et al. 1993).

The concentrations required to inhibit DNA gyrase in vitro, however, do not correlate well with the MICs of fluoroquinolones for the organism, with approximately ten to 30 times more drug required to inhibit in vitro reactions than to exert the bactericidal effect (HALLETT and MAXWELL 1991). A better measurement of fluoroquinolone activity is the determination of the concentration of quinolone required to inhibit in vivo DNA synthesis by 50% (IC_{50}). This figure has been found to correlate well with the MIC of a fluoroquinolone for an organism (PIDDOCK et al. 1990).

The first molecular characterisation of a quinolone-resistance mutation in *gyrA* was reported in 1988 when Yoshida et al. (1988) cloned and sequenced the *gyrA* gene from a number of quinolone-resistant mutants of *E. coli* (Table 1). In comparison with the wild-type sequence all four carried point mutations within a small region near the N-terminus of the GyrA polypeptide. These mutations were situated in a relatively hydrophilic region of the polypeptide and close to the tyrosine residue at amino acid 122 which has been shown to be the site covalently bound to the DNA (Horowitz and Wang 1987). Since then, a number of quinolone-resistant *E. coli* mutants, selected in vitro and in vivo, have been examined and all carry point mutations within the same small region from codon 67 to 106, designated the quinolone resistance determining region (QRDR; Table 1; Fig. 1). In particular, many mutations result in the

Table 1. Alterations in DNA gyrase subunit A conferring quinolone resistance

Organism	Amino acid substitution	Reference
Acinetobacter baumanni	Gly 81 → Val	Vila et al. (1995)
	Ser 83 → Leu	Vila et al. (1995)
Aeromonas salmonicida	Ser 83 → Ile	Oppegaard and Sorum (1994)
	Ser 83 → Ile; Ala 67 → Gly	Oppegaard and Sorum (1994)
Coxiella burnetii	Glu 87 → Gly	Vila et al. (1995)
Campylobacter jejuni	Ala 70 → Thr	Wang et al. (1993)
Campylobacter lari	Thr 86 → Ile	Wang et al. (1993)
	Asp 90 → Ala, Asn	Wang et al. (1993); Charvalos et al. (1996)
	Ser 83 → Arg	Korten et al. (1994)
	Glu 87 → Lys, Gly	Korten et al. (1994)
	Thr 86 → Ile; Pro 104 → Ser	Everett and co-workers, unpublished results
Enterobacter cloacae	Ser 83 → Leu	Everett and co-workers, unpublished results
Escherichia coli	Ala 67 → Ser	Yoshida et al. (1988)
	Gly 81 → Cys, Asp	Yoshida et al. (1990a); Cambau et al. (1993)
	Ser 83 → Leu, Trp, Ala	Yoshida et al. (1990a); Oram and Fisher (1991); Cullen et al. (1989); Hallett and Maxwell (1991)
	Ala 84 → Pro	Yoshida et al. (1990a)
	Asp 87 → Asn, Val, Thr, Gly, His	Yoshida et al. (1990a); Oram and Fisher (1991); Heisig et al. (1993); Ouabdesselam et al. (1995); Everett et al. (1996)

Table 1. *Continued*

Organism	Amino acid substitution	Reference
Enterococcus faecalis	Ser 83 → Ile	Korten et al. (1994)
	Gln 106 → His, Arg	Yoshida et al. (1988); Hallett and Maxwell (1991)
Helicobacter pylori	Asn 87 → Lys	Moore et al. (1995)
	Ala 88 → Val	Moore et al. (1995)
	Asp 91 → Gly, Asn, Tyr	Moore et al. (1995)
	Asp 91 → Asn; Ala 97 → Val	Moore et al. (1995)
Mycobacterium avium	Ala 90 → Val	Cambau et al. (1994)
Mycobacterium smegmatis	Ala 90 → Val	Revel et al. (1994)
	Asp 94 → Gly	Revel et al. (1994)
Mycobacterium tuberculosis	Gly 88 → Cys	Takiff et al. (1994)
	Ala 90 → Val	Takiff et al. (1994)
	Ser 91 → Pro	Takiff et al. (1994)
	Asp 94 → Asn, His, Gly, Tyr, Ala	Takiff et al. (1994)
Neisseria gonorrhoeae	Ser 83[a] → Phe	Deguchi et al. (1995)
	Ser 83 → Phe; Asp 87 → Asn	Deguchi et al. (1995)
Pseudomonas aeruginosa	Thr 83 → Ile	Kureishi et al. (1994); Pumbwe et al. (1996)
	Asp 87 → Tyr, Asn, Gly, His	Kureishi et al. (1994); Yonezawa et al. (1995a)
Shigella dysenteriae	Ser 83 → Leu	Rahman et al. (1994)
Salmonella typhi	Ser 83 → Phe	Brown et al. (1996)
Salmonella typhimurium	Ser 83 → Phe, Tyr	Reyna et al. (1995); Griggs et al. (1996)
	Asp 87 → Gly, Tyr, Asn	Griggs et al. (1996)
	Ala 119 → Glu	Griggs et al. (1996)
	Ala 67 → Pro; Gly 81 → Ser	Reyna et al. (1995)
	Ser 83 → Ala; Asp 87 → Asn	Heisig et al. (1995)
Staphylococcus aureus	Ser 84 → Leu, Ala, Phe	Sreedharan et al. (1990); Goswitz et al. (1992); Tokue et al. (1994); Ito et al. (1994)
	Ser 85 → Pro	Ito et al. (1994)
	Glu 88 → Lys, Gly	Goswitz et al. (1992); Ito et al. (1994); Tokue et al. (1994)
Staphylococcus epidermidis	Ser 84 → Phe	Sreedharan et al. (1991)

[a] Codon position based on *E. coli* sequence.

Fig. 1. Quinolone resistance determining regions of DNA gyrase A subunits from different bacteria. Each *box* represents a single amino acid. Sites of known fluoroquinolone resistance mutations are shaded

substitution of Ser83 with a more hydrophobic residue, with a concomittant increase in the MIC of ciprofloxacin from 0.03 to 0.5 mg l^{-1}. Hence, Ser83 would appear to be important in gyrase–quinolone interactions. Site-directed mutagenesis of codon 83 has shown that it is the replacement of serine with a hydrophobic residue, rather than a loss of the hydroxyl group, which is important for resistance (YONEZAWA et al. 1995b). High-level fluoroquinolone-resistant mutants (MIC ciprofloxacin ≥8 mg l^{-1}) have been shown to require two point mutations, one resulting in the Ser83 to Leu substitution, and the other in the Asp87 to Gly substitution (HEISIG and TSCHORNY 1994; VILA et al. 1994). These double mutants can be selected in a stepwise manner in the laboratory with the Ser83 to Leu mutants arising first, suggesting that this mutation may be required to stabilise a second Asp87 to Gly mutation (VILA et al. 1994). A similar double mutant has also been isolated from a patient known to have been previously infected with an *E. coli* strain containing a mutation at Ser83 of *gyrA*, indicating that the serial acquisition of *gyrA* mutations may also be a clinical phenomenon (SOUSSY et al. 1994). However, OUABDESSELAM et al. (1995) have demonstrated that Asp87 to Gly mutations can arise alone in clinical isolates, and that this mutation results in similar increases in MICs of fluoroquinolones as Ser83 to Leu. Site-directed mutagenesis of codon 87 has shown that it is the loss of the negative charge of the aspartate residue which confers resistance (YONEZAWA et al. 1995c).

Similar mutations have been characterised in *S. aureus*, except that in *S. aureus* the equivalent of Ser83 in *E. coli* is found at position 84. Again, substitution of Ser84 with hydrophobic amino acids results in a decreased quinolone susceptibility (MIC of ciprofloxacin = 16–64 mg l^{-1}) compared with the wild type (MIC of ciprofloxacin = 0.5 mg l^{-1}). Some mutants have been found to carry double mutations at Ser84 and either Ser85 or Glu88 which can lead to MICs in excess of 256 mg ciprofloxacin l^{-1}. Similar *gyrA* mutations have recently been characterised in a number of other organisms (Table 1; Fig. 1). In all cases, mutations have been found in amino acids equivalent to Ser83 in *E. coli* suggesting that this residue is the most common site for mutations in *gyrA* (Fig. 1). In *P. aeruginosa* and *C. jejuni*, the serine residue is replaced by a threonine residue at positions 83 and 86, respectively (WANG et al. 1993; KUREISHI et al. 1994). Threonine also contains a hydroxyl group suggesting that this side group is important in fluoroquinolone binding. In *M. tuberculosis* and other mycobacteria, most mutations analysed to date occur at Ala90 which is at the equivalent position as Ser83 in *E. coli* (Table 1; Fig. 1); however mutations have also been found in the adjacent amino acid Ser91 (TAKIFF et al. 1994). Both *S. aureus* and *E. faecalis*, which are gram-positive organisms, contain a glutamic acid residue at the equivalent position of Asp87 in *E. coli* and other gram-negative organisms. This suggests a possible difference in the quinolone-binding site between GyrA of gram-positive and gram-negative bacteria (Fig. 1).

In a number of organisms, mutations at Ser83 result in the loss of a *Hinf*I restriction site. This enables the presence of such a mutation to be detected by cutting a suitable fragment of *gyrA* DNA, amplified by the polymerase chain reaction (PCR), with *Hinf*I and fractionating the products by agarose or polyacrylamide gel electrophoresis (SREEDHARAN et al. 1990). This technique, termed restriction fragment length polymorphism (RFLP; Fig. 2), has been used to detect *gyrA* mutations in *E. coli*, *S. aureus*, *Staphylococcus epidermidis* (SREEDHARAN et al. 1991) and *S. typhimurium* (GRIGGS et al. 1996); however, in other species no *Hinf*I or other suitable restriction enzyme site is present at this position. All *gyrA* mutations so far characterised in resistant isolates have been within the QRDR, except for a single mutation of Ala119 to Glu in *S. typhimurium* (GRIGGS et al. 1996); however, it has yet to be proved that this mutation is responsible for the resistance phenotype.

In many cases the recent increase in the numbers of strains in which *gyrA* mutations have been characterised has been facilitated by the introduction of single-stranded conformational polymorphism analysis (SSCP; TOKUE et al. 1994; KUREISHI et al. 1994). This method allows the detection of point mutations in short lengths of DNA produced by PCR. Denatured single-stranded DNA molecules will attain different conformational structures dependent on nucleotide sequence. Such differences can be detected by fractionating the various fragments on a non-denaturing polyacrylamide gel. A number of factors influence the sensitivity of the method, including gel temperature, polyacrylamide concentration, and presence or absence of mild denaturing agents such as glycerol in the gel. Following electrophoresis, the DNA can be visualised by staining with silver or ethidium bromide, or alternatively by autoradiography of radiolabelled fragments (Fig. 2). Fragments of DNA with different mutations give distinct SSCP patterns compared with the wild type, thus enabling large numbers of PCR products to be screened for the presence of novel mutations which can subsequently be sequenced. By optimisation of

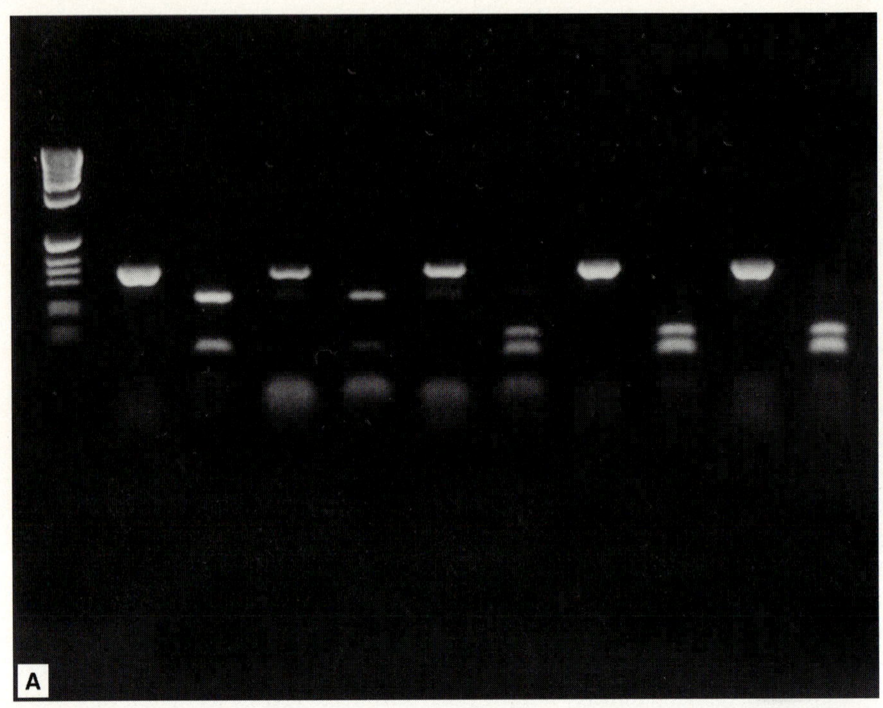
A

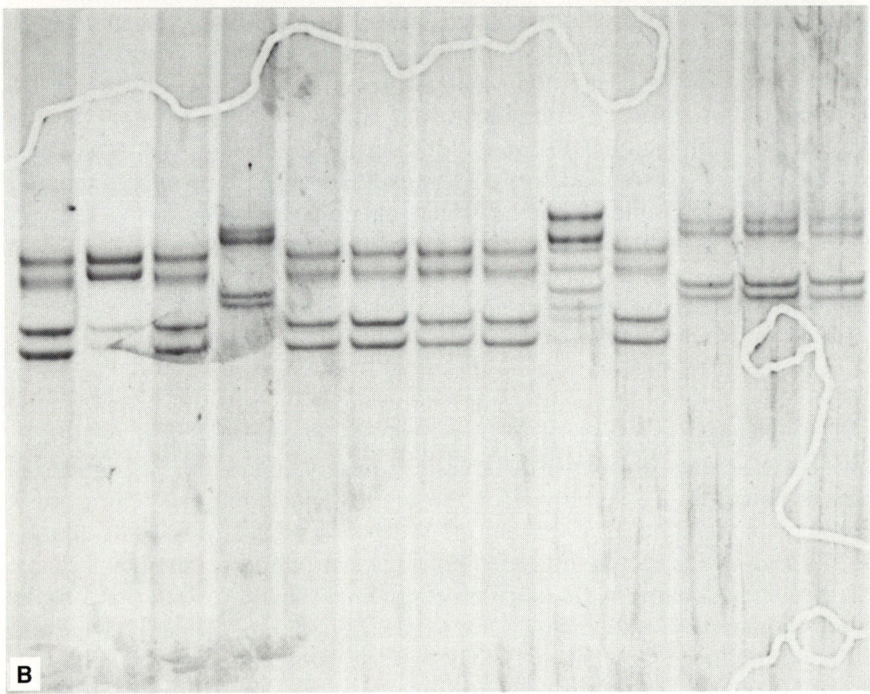
B

the conditions the majority of mutations can be differentiated in fragments up to 400 bp in length; however, it is rare that any one set of conditions is able to detect all possible mutations (M.J. Everett, unpublished observations).

II. Mutations in *gyrB*

Although quinolones are thought to interact with the A subunit of DNA gyrase, mutations have been discovered in the B subunit which also confer quinolone resistance. This phenomenon was first described by YAMAGISHI et al. (1981) who characterised two quinolone-resistant mutants of *E. coli* designated *nalC* (*nal-31*) and *nalD* (*nal-24*). Both were subsequently reported to contain mutations in *gyrB* resulting in the substitution of Asn for Asp426 and Glu for Lys447, respectively (YAMAGISHI et al. 1986). Both mutations conferred resistance to acidic quinolones such as nalidixic acid (MIC = 32–128 mg l^{-1}). However, the Lys447 (*nalC*) mutant was four-fold more susceptible than the wild type to fluoroquinolones containing a piperazine substituent at carbon-7, such as ciprofloxacin (SMITH 1984). NAKAMURA et al. (1989) examined the proportion of *gyrA* and *gyrB* mutations in quinolone-resistant isolates of *E. coli* by introduction of cloned wild-type *gyrA* or *gyrB* genes. In 25 mutants selected in vitro using four times the MIC of nalidixic acid or enoxacin, 13 had *gyrA* mutations and 12 had *gyrB* mutations. Thus, in *E. coli* it would appear that *gyrA* and *gyrB* mutants arise at similar frequencies under laboratory conditions, whilst *gyrA* mutants are more common amongst clinical isolates. A subsequent study of 13 *gyrB* mutants of *E. coli* revealed that all contained mutations in *gyrB* resulting in changes at Asp426 or Lys447. This suggests that these positions delineate a QRDR analogous to that specified for *gyrA* (YOSHIDA et al. 1991). Recently, a novel *gyrB* mutation has been discovered in *gyrB* from a fluoroquinolone-resistant post-therapy strain of *S. typhimurium* at codon 464, resulting in substitution of serine to tyrosine; however, it has yet to be shown that this confers the resistance phenotype (GENSBERG et al. 1995).

The frequency of *gyrB* mutations, compared to *gyrA*, has been shown to be relatively low with a recent survey of 16 clinical *E. coli* isolates identifying only one *gyrB* mutant compared with 15 *gyrA* mutants (VILA et al. 1994). No

Fig. 2A,B. Detection of *gyrA* mutations. **A** Restriction fragment length polymorphism analysis of *Salmonella typhimurium gyrA* fragments. *Lane 1*, DNA size markers; *lanes 2, 4, 6, 8* and *10*, uncut DNA; *lanes 3, 5, 7,* and *9*, DNA cut with *Hinf*I. Mutations at Ser83 result in loss of a *Hinf*I restriction site and hence an altered restriction fragment pattern (*lanes 3* and *5*) compared to non-Ser83 mutants (*lanes 7, 9* and *11*; Griggs et al. 1996). **B** Single-stranded conformational polymorphism analysis of *Campylobacter jejuni gyrA* fragments, denatured and fractionated on a nondenaturing polyacrylamide gel. Point mutations result in changes to the conformation of single-stranded DNA and altered migration through the gel. (M.J. EVERETT, unpublished results)

other *gyrB* mutations have been described in *E. coli*, although *gyrB* mutations have been reported in quinolone-resistant isolates of *P. aeruginosa* (YOSHIDA et al. 1990b) and *Neisseria gonorhoea* (STEIN et al. 1991). Recently, ITO et al. (1994) examined the mechanisms of resistance of 35 fluoroquinolone-resistant laboratory mutants of *S. aureus*. The majority of these carried *gyrA* mutations; however two mutants were identified with a mutation in *gyrB*, resulting in the substitution of Asn for Asp437 in one mutant and Gln for Arg458 in the other, suggesting that in *S. aureus* these residues define a similar resistance determining region as in *E. coli*.

No *gyrB* mutations have been reported that result in cross-resistance between quinolones and the B subunit inhibitors, coumermycin and novobiocin. This is consistent with evidence which suggests that the GyrB protein comprises two distinct domains: an N-terminal domain containing the ATP-hydrolysing and coumermycin binding sites, and a C-terminal domain containing the QRDR of GyrB (REECE and MAXWELL 1991).

It has been proposed that the QRDR of the A subunit of DNA gyrase constitutes the binding site for quinolone antibiotics and that the quinolone–gyrase–DNA complex acts as a cellular poison (KREUZER and COZZARELLI 1979; HALLETT and MAXWELL 1991). This would explain the dominance of susceptible *gyrA* over the resistant allele, as even small amounts of the sensitive A subunit could form a "poisonous" complex with the quinolone. Mutations in the QRDR, such as the replacement of the small hydroxyl side group of serine with bulkier hydrophobic groups, are thought to block access to this site. YOSHIDA et al. (1991) propose that the QRDR of the B subunit also forms part of this quinolone binding site, and that the addition of a negatively charged side group, as in the Lys447 mutation, results in attraction of the positively charged piperzinyl group of the newer fluoroquinolones; hence the increased sensitivity to such antibiotics. Clinical isolates of *S. aureus* and *S. typhimurium* have been described in which fluoroquinolone resistance has been attributed to mutations in both *gyrA* and *gyrB* (ITO et al. 1994; HEISIG 1993). This suggests that both subunits may work in concert. However, the effect of *gyrB* mutations on quinolone–gyrase interactions could also be explained by an indirect effect of the B subunit change on the A subunit.

III. Mutations in Other Topoisomerase Genes

KATO et al. (1990) reported the finding of a second type II topoisomerase enzyme in *E. coli*, designated topoisomerase IV, which was shown to be involved in chromosomal partitioning. Topoisomerase IV, like DNA gyrase, is composed of two protein subunits, ParC and ParE, which show strong homology to GyrA and GyrB, respectively, especially in those domains known to be involved in quinolone action on DNA gyrase (KATO et al. 1990). Subsequently, the *parC* and *parE* genes were cloned and sequenced in *S. typhimurium* (LUTTINGER et al. 1991). *gyrA*-like and *gyrB*-like genes have also been reported in *Rickettsia*, *Chlamydia*, *Borrelia*, *Neisseria*, *Legionella* and *Rhodobacter* (HUANG 1992). In *E. coli*, topoisomerase IV was thought not to be readily inhibited by fluoroquinolones, with 30 times more antibiotic required to inhibit topoisomerase IV than to inhibit DNA gyrase (PENG and MARIANS 1993). However, HOSHINO et al. (1994) subsequently showed that fluoroquinolones may have significant activity against topoisomerase-IV-mediated deconcatenation of DNA, particularly in *gyrA* mutants.

There is now good evidence that topoisomerase IV is a secondary target for fluoroquinolone action in *E. coli* in the absence of a sensitive DNA gyrase and that mutations in *parC* result in further decreased susceptibility (HEISIG 1996; KUMUGAI et al. 1996; KHODURSKY et al. 1995). These mutations have been shown to occur at Ser80 and Glu84, analogous to codons Ser83 and Asp87 of the *E. coli gyrA*, and to be common in fluoroquinolone-resistant clinical isolates of *E. coli* (VILA et al. 1996; EVERETT et al. 1996). A mutation has also been reported in the *parE* gene which results in decreased fluoroquinolone susceptibility in a *gyrA* background (SOUSSY et al. 1996). The amino acid change of leucine 445 to histidine is only two codons away from a fluoroquinolone resistance mutation in *gyrB*; however, as is the case with *gyrB*, such mutations would appear to be rare in clinical isolates (EVERETT et al. 1996).

In *S. aureus*, topoisomerase IV is a primary target of fluoroquinolones. FERRERO et al. (1994) examined eight fluoroquinolone-resistant isolates of *S. aureus* and demonstrated that *gyrA* mutations were only present in the three isolates exhibiting resistance to high levels of ciprofloxacin ($16\,mg\,l^{-1}$). In contrast, all isolates carried mutations in *grlA* (the gene encoding the A subunit of topoisomerase IV), suggesting that a mutation in *grlA* may be a prerequisite for the occurrence of mutations in *gyrA* (FERRERO et al. 1994). Furthermore, these mutations all resulted in changes at Ser80 of GrlA, which is analogous to Ser83 in GyrA, indicating that fluoroquinolones may interact with GrlA in a similar manner as with GyrA.

Mutations in *grlA* may account for the uncharacterised *flq* mutation in *S. aureus*, obtained after single-step fluoroquinolone selection in the laboratory (TRUCKSIS et al. 1991), as well as similar non-*gyrA* mutants described by other investigators (HORI et al. 1993; TANKOVIC et al. 1994). HORI et al. (1993) found that two genetic events were necessary for high-level ofloxacin resistance in both MRSA and methicillin-susceptible *S. aureus* (MSSA). It was concluded that the increased incidence of fluoroquinolone resistance in MRSA, compared with MSSA (see Sect. E.II), was due to the increased prevalence of low-level resistance in these strains, as represented by first stage mutants, enabling high-level resistance to be achieved in a single subsequent step. It was suggested that, in Japan at least, this may be due to previous use of quinolones against MRSA.

Topoisomerase IV mutations have now been characterised in several other organisms. In *Streptococcus pneumoniae* topoisomerase IV appears to be a primary target for fluoroquinolone action (TANKOVIC et al. 1996), whilst in *N. gonorrhoeae* (DEGUCHI et al. 1996) it is a secondary target.

C. Reduced Intracellular Accumulation

In the absence of mechanisms to destroy, modify or sequester antibiotic, the intracellular concentration of active antibiotic can be reduced by exclusion of the antibiotic from the cell. This may be achieved by decreasing the penetration (uptake) of antibiotic into the cell or by increasing efflux of antibiotic from the cell (Fig. 3). Both mechanisms have been shown to contribute to fluoroquinolone-resistance in bacteria (Tables 2, 3).

A number of investigators have measured the accumulation of fluoroquinolones in bacterial cells (reviewed by PIDDOCK 1991). It has been shown that fluoroquinolone accumulation increases linearly with increasing external drug concentrations, suggesting that fluoroquinolones are taken up by a process of passive diffusion rather than via any specific uptake mechanism. Initially, fluoroquinolones are taken up rapidly into bacteria until a steady-state concentration is reached after a few minutes. It has been proposed that this steady-state concentration reflects an equilibrium between the passive diffusion of the antibiotic into the cell and active efflux of the antibiotic from the cell (PIDDOCK 1991; Fig. 4A). Such active efflux mechanisms utilise energy derived from metabolism to pump the antibiotic against a concentration gradient. In the absence of an active efflux mechanism an intracellular concentration equivalent to the extracellular concentration will eventually be reached; this can be demonstrated by the addition of a metabolic inhibitor, such as carbonyl cyanide *m*-chlorophenyl-hydrazone (CCCP), to the growth medium which results in inhibition of the efflux mechanism and a concomitant increase in drug accumulation (Fig. 3). Active efflux systems have now been

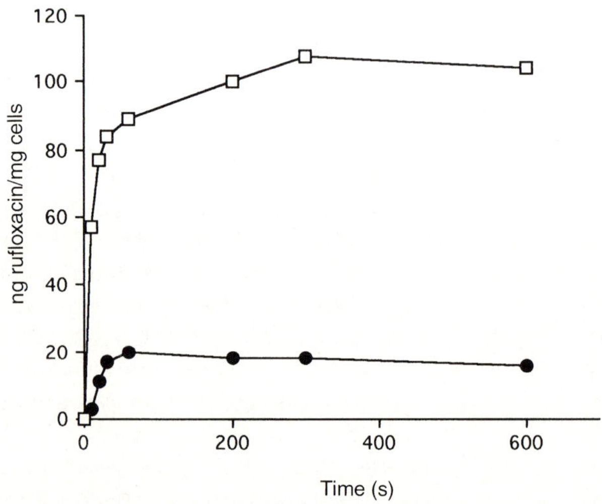

Fig. 3. Typical accumulation of fluoroquinolones in gram-positive bacteria in the presence (-□-) and absence (-●-) of the efflux inhibitor CCCP

Table 2. Mutations and phenotypes associated with decreased quinolone permeability

Organism	Locus	Phenotype	Reference
Escherichia coli	*nalB*	Unknown	HANE and WOOD (1969)
	nfxB, 19 min	Decreased OmpF	HOOPER et al. (1986)
	marR (*cfxB*, *norB*, *nfxC*), 34 min	MAR, decreased OmpF	HOOPER et al. (1987); HOOPER et al. (1992); HIRAI et al. (1986)
	norC, 8 min	Decreased OmpF, altered LPS	HIRAI et al. (1986)
	nalD, 89 min	Changes in OM lipid bilayer	HREBENDA et al. (1985)
Enterobacter cloacae	Unknown	Decreased 37-kDa OMP	GUTMANN et al. (1985); PIDDOCK et al. (1991); LUCAIN et al. (1989)
Klebsiella pneumoniae	Unknown	MAR, decreased 41-kDa OMP	GUTMANN et al. (1985)
Serratia marsescens	Unknown	39 kDa	PIDDOCK et al. (1991)
	Unknown	Decreased 41-kDa OMP, Increased 40-kDa OMP	GUTMANN et al. (1985)
Salmonella typhimurium	Unknown	Decreased OmpF	PIDDOCK et al. (1993a)
Pseudomonas aeruginosa	Unknown	MAR, decreased OprF	PIDDOCK et al. (1992)
	Unknown	Decreased OprD2	HAMZEHPOUR et al. (1991)

OM, outer membrane; OMP, outer membrane protein.

Table 3. Mutations and phenotypes associated with increased quinolone efflux

Organism	Locus	Phenotype	Reference
Escherichia coli	*emrAB*, 57.5 min	MAR, hydrophobic efflux system	LOMOVSKAYA and LEWIS (1992)
	acrAB,	MAR, hydrophilic efflux system	MA et al. (1995)
Proteus vulgaris	Unknown	Loss of 37-kDa OMP	ISHII et al. (1991)
S. aureus	*norA*	MAR, efflux protein	OHSHITA et al. (1990); YOSHIDA et al. (1990c); KAATZ et al. (1993)
Pseudomonas aeruginosa	*mexAB-oprM* (*cfxB nalB*), 20 min	MAR, increased OprM (50 kDa)	ROBILLARD and SCARPA (1988); RELLA and HAAS (1982); POOLE et al. (1996)
	nfxB	MAR, increased OprJ (54 kDa)	OKAZAKI and HIRAI (1992); MASUDA et al. (1996)
	nfxC, 46 min	MAR, increased 50-kDa OMP Decreased OprD	FUKUDA et al. (1990)
Bacillus subtilis	*bmr*	MAR, efflux protein	NEYFAKH (1992)

OMP, outer membrane protein.

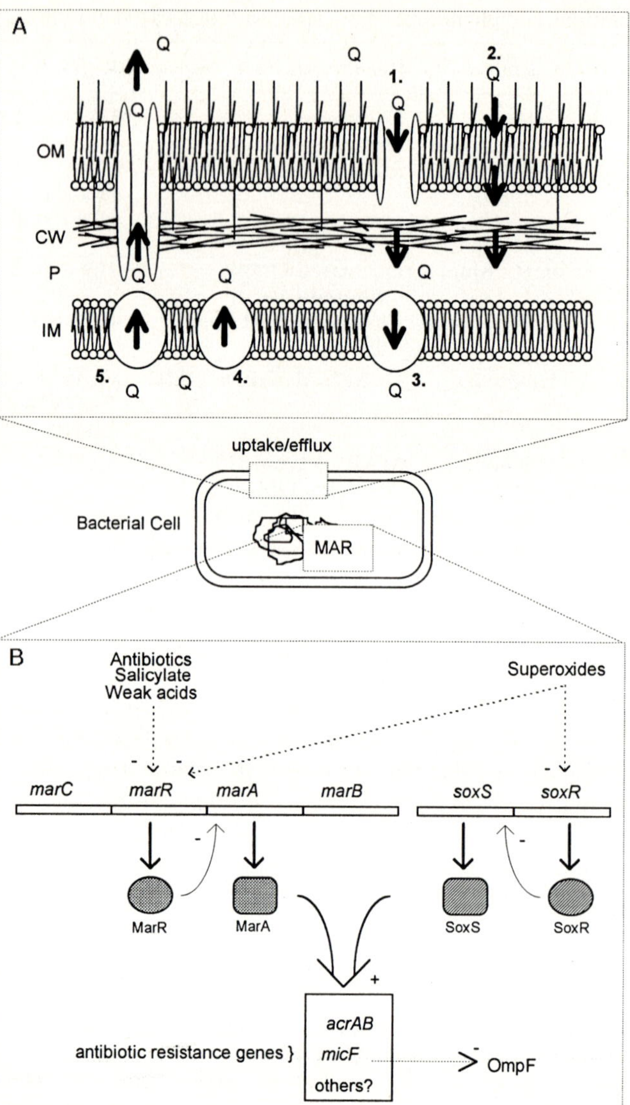

Fig. 4. **A** Possible routes of quinolone (*Q*) flux across a gram-negative cell envelope (*CW*, cell wall): *1*. influx via outer membrane (*OM*) porin(s); *2*. influx directly across the outer membrane; *3*. uptake across the inner membrane (IM); *4*. efflux across the inner membrane into the periplasm (*P*); *5*. efflux across both inner and outer membranes. **B** Regulation of *mar*- and *sox*-mediated multiple antibiotic resistance

demonstrated in many bacteria and it is likely that other organisms in which reduced accumulation as a mechanism of fluoroquinolone resistance has been implicated also possess such a mechanism.

I. Decreased Uptake

DNA gyrase and topoisomerase IV are both located in the cytoplasm of the bacterial cell. Thus, in order to reach their target(s), fluoroquinolone antibiotics must traverse the cell envelope. In gram-positive bacteria this consists of the cell wall and a single membrane, whereas in gram-negative bacteria the fluoroquinolone must first cross the outer membrane. That this acts as an additional barrier to fluoroquinolone penetration has been shown by HOOPER et al. (1989), who demonstrated that in *E. coli* the permeability of the cell envelope is increased by 50% after removal of the outer membrane. Changes in the cell envelope of gram-negative bacteria, particularly in the outer membrane, have been associated with decreased uptake and increased resistance to fluoroquinolones (GUTMANN et al. 1985; PIDDOCK et al. 1991; MORTIMER and PIDDOCK 1993). In contrast, decreased uptake has not been demonstrated to be a mechanism of resistance in gram-positive bacteria.

It is thought that the major route through the outer membrane for hydrophilic fluoroquinolones, such as ciprofloxacin, is through water-filled protein pores called porins. In *E. coli* the major porins are OmpF and OmpC with pore widths of 1.2 and 1 nm, respectively. These can be detected by SDS-PAGE analysis of purified outer membranes. A number of fluoroquinolone-resistant laboratory mutants and clinical isolates have been reported which show complete or partial loss of OmpF from the outer membrane. OmpF also acts as a pore for a number of antibiotics other than fluoroquinolones, such as chloramphenicol, tetracycline and some β-lactams. Thus loss of OmpF often results in cross-resistance to a number of unrelated antibiotics. As well as mutations in the structural gene for OmpF, mutations have also been reported which interfere with the complex regulation of *ompF* in response to environmental changes. These include mutations in the *crp* (cAMP receptor protein) and *cya* (adenyl cyclase) genes which affect cAMP-mediated gene regulation (KUMAR 1976; SCOTT and HARWOOD 1980). Other mutations in *E. coli*, namely *nfxC*, *cfxB*, and *norB* have been shown to be alleles of the *mar* operon, mutations which confer multiple antibiotic resistance (MAR; see Sect. C.III).

The role of OmpF in resistance to antibiotics, including fluoroquinolones, is unclear. CHAPMAN et al. (1989) originally showed that loss of the *ompF* structural gene results in a 50% reduction in the intracellular concentration of hydrophilic fluoroquinolones. MORTIMER and PIDDOCK (1993) found that norfloxacin accumulation was 70% of wild-type levels in an OmpF-deficient strain of *E. coli*, and that this was associated with increased MICs (from 0.06 to 0.5 $mg\,l^{-1}$) Accumulation was further reduced in the absence of OmpC and PhoE. However, even in the absence of all three porins norfloxacin accumulation was still 40% of that in the wild type, indicating that fluoroquinolones must also gain entry by other non-porin pathways. Recently a number of reports have cast doubt on the contribution of *ompF* downregulation to the resistance phenotype (COHEN et al. 1988b; GAMBINO et al. 1993). The most recent of these demonstrates that repression of OmpF synthesis by another regulatory protein, SoxS, which is involved in the superoxide stress response, does not confer antibiotic resistance (MILLER et al. 1994). Thus, it would appear that the fluoroquinolone resistance phenotype of many OmpF deficient mutants cannot be explained solely in terms of loss of OmpF, and that other as yet unidentified factors must be involved.

Changes in the outer membrane protein profiles associated with decreased fluoroquinolone susceptibility have also been reported for a number of other species. These include loss of OmpF in *S. typhimurium* (PIDDOCK et al. 1993a), decreased expression of a 39-kDa protein in *S. marsescens* (PIDDOCK et al. 1991) and of a 37-kDa protein in *E. cloacae* (GUTMANN et al. 1985; PIDDOCK et al. 1991; LUCAIN et al. 1989), *K. pneumoniae* (GUTMANN et al. 1985; PIDDOCK et al. 1991) and *Proteus vulgaris* (ISHII et al. 1991). Some of these changes may be due to the effect of quinolones and/or *gyrA* mutations on differential expression of outer membrane proteins as it has been shown that *gyrA*-mediated changes in supercoiling of DNA can affect expression of porin genes (HIGGINS et al. 1988).

The low outer membrane permeability of *P. aeruginosa* has previously been thought to be the most important factor contributing to the high levels of intrinsic resistance of this species to many antimicrobial agents. However, it has been calculated, using an experimentally determined permeability coefficient for tetracycline, that even very rapid growth (generation time of 20 min) decreases the intracellular drug concentration by less than 0.2% from the equilibrium (LI et al. 1994a). Thus decreased permeability alone cannot produce significantly decreased susceptibility. Many fluoroquinolone-resistant isolates of *P. aeruginosa* have been reported with changed outer membrane profiles; whereas these were originally thought to be permeability mutants, it is now clear that the resistance phenotype of many of these is due to enhanced efflux (see Sect. C.II).

However, the mechanism of resistance of other mutants cannot be explained simply in terms of efflux. HAMZEHPOUR et al. (1991) showed that loss of OprD, the porin for imipenem, confers cross-resistance between imipenem and sparfloxacin. Cross-resistance between fluoroquinolones and carbapenems has also been shown in clinical isolates of *P. aeruginosa* (AUBERT et al. 1992; MOUTON et al. 1995). PIDDOCK et al. (1992) described a multiply antibiotic resistant isolate of *P. aeruginosa* which lacked OprF and had other outer membrane changes. The role of OprF in the outer membrane of *P. aeruginosa* has been the matter of some speculation. Complementation of an OprF-deficient multiply antibiotic resistant mutant with the cloned wild-type *oprF* gene resulted in changes to cellular morphology and increased susceptibility to β-lactams, chloramphenicol and tetracycline, but very little, if any, increased resistance to fluoroquinolones (ZHANEL et al. 1995). These data suggest that OprF in *P. aeruginosa* has a dual function, both as a structural protein and as a porin, but that fluoroquinolones do not enter via this route. This indicates that fluoroquinolone resistance in such mutants is due to another, as yet unidentified, mechanism.

Increased resistance to nalidixic acid in *E. coli* has been associated with changes in the outer membrane lipid bilayer (HREBENDA et al. 1985). Consequently, it has been suggested that hydrophobic quinolones, such as nalidixic acid, are able to pass directly throught the lipid bilayer of the outer membrane. It was suggested that other less hydrophobic quinolones may also pass directly through the lipid bilayer, via a process of self-promoted uptake (CHAPMAN and GEORGOPAPADAKOU 1988; VALISENA et al. 1990). This hypothesis is based on the finding that fluoroquinolones can chelate divalent cations, such as are found in the lipid bilayer. It is proposed that this disrupts the integrity of the lipid bilayer and thus facilitates the passage of the fluoroquinolone through transient gaps in the membrane. The antagonistic effect of divalent cations on quinolone activity was investigated by MARSHALL and PIDDOCK (1994). They confirmed that magnesium forms complexes with quinolones and increases the MICs of a number of quinolones between two- and eightfold for a variety of gram-negative and gram-positive organisms. However, antagonism of quinolone activity was shown not to be due to inhibition of the self-promoted uptake pathway as, unlike other antimicrobial agents known to enter gram-negative bacteria via such a pathway, quinolones clearly did not permeabilise the outer membrane.

II. Increased Efflux

Increased efflux as a mechanism of fluoroquinolone resistance was first reported in fluoroquinolone-resistant isolates of *S. aureus* (YOSHIDA et al. 1990c). The *norA* mutation, which confers a 32-fold decrease in susceptibility to hydrophilic fluoroquinolones, was first thought to be an allele of *gyrA* (UBUKATA et al. 1989). However, further characterisation (YOSHIDA et al. 1990c; OHSHITA et al. 1990) revealed that it encodes a protein with sequence similarity to a family of export proteins, including the Bmr protein from *Bacillus subtilis* and the Tet efflux proteins (LEVY 1992). The *norA* gene is chromosomally encoded and is naturally occuring in *S. aureus*. Similar sequences have been shown by Southern hybridisation to occur in *S. epidermidis* but not in *E. coli*, *K. pneumoniae* or *E. faecalis* (KAATZ et al. 1993). The NorA protein has a hydropathic amino acid profile consistent with a location in the cytoplasmic membrane and exhibits a low level of fluoroquinolone efflux in wild-type cells, with a preference for hydrophilic fluoroquinolones. Efflux is an active process and is inhibited by protonophores, such as CCCP. The natural function of NorA has yet to be elucidated. However, it is known that Bmr mediates the efflux of a number of toxic chemicals, such as chloramphenicol and ethidium bromide as well as fluoroquinolones (NEYFAKH et al. 1991). Thus the possession of nonspecific efflux mechanisms may be a general feature of gram-positive bacteria. Increased efflux of fluoroquinolones in resistant strains was originally thought to be due to a point mutation in *norA* resulting in a higher affinity efflux system (OHSHITA et al. 1990); however, it is now apparent that resistance is due to overexpression of the wild-type gene (KAATZ et al. 1993). NG et al. (1994) recently demonstrated that overexpression of *norA* in a *flqB* mutant was due to a single nucleotide change in the *norA* promoter. This is in contrast to overexpression of *bmr* which is due to gene amplification (NEYFAKH 1992).

In *P. aeruginosa*, resistance to fluoroquinolones as well as to a number of other antimicrobial agents has often been associated with decreased accumulation and increased expression of outer membrane proteins, often with concomitant increases in cytoplasmic membrane proteins (OKAZAKI and HIRAI 1992; MASUDA and OHYA 1992; FUKUDA et al. 1990; LEGAKIS et al. 1989). There is now good evidence to suggest that these changes reflect overexpression of one or more efflux systems capable of exporting fluoroquinolones, tetracycline, chloramphenicol and β-lactams, including meropenem. POOLE et al. (1993) characterised the mechanism of multiple antibiotic resistance in a siderophore-deficient mutant of *P. aeruginosa* which overproduced a 50-kDa outer membrane protein, designated OprK. The gene encoding OprK was identified as being part of an operon composed of three genes previously implicated in the secretion of the siderophore pyoverdine (POOLE et al. 1993). This operon, designated originally *mexA–mexB–oprK*, was shown to encode two cytoplasmic membrane proteins of 40 and 108kDa, as well as OprK, all of which were overexpressed in the mutant. However, recently, insertion of a transposon in *oprK* was discovered to knock out expression of a homologous outer membrane protein, OprM, previously described by MASUDA and OHYA (1992). Thus the *mexAB-oprK* operon has been redesignated *mexAB-oprM* (GOTOH et al. 1995). It is hypothesised that the natural function of the *mexAB-oprM* efflux system may be to export pyoverdin and/or its breakdown products, but that the nonspecificity of the system allows it to recognise a variety of cyclical compounds. Expression of the *mexAB-oprM* operon is under the control of an upstream regulator gene *mexR*. A mutation in this gene (*nalB*: Arg 69 to Trp) results in the activation of the operon and overproduction of OprM (POOLE et al. 1996).

It is clear that there is at least one other efflux system in *P. aeruginosa*, the overexpression of which is associated with a different antibiotic susceptibility profile (HAMZEHPOUR et al. 1995; LI et al. 1994a). MASUDA et al. (1996a) have analysed another class of multidrug resistant strains with mutations in a regulator gene, *nfxB*. These show resistance to many antibiotics but, unlike *nalB* mutants, remain susceptible to meropenem. Such strains overexpress a 54-kDa outer membrane protein, designated OprJ, which has been proposed to be identical to OprK (HAMZEHPOUR et al. 1995).

The glycopeptide vancomycin has been shown to exhibit synergy with ciprofloxacin against *P. aeruginosa* (DAY et al. 1993). As a possible explanation it was suggested that fluoroquinolone efflux may be mediated by the C_{55}

bacterioprenol lipid carrier which is inhibited by binding of vancomycin to UDP-*N*-acetylmuramylpentapeptide (for a review of vancomycin mode of action see NAGARAJAN 1991). Furthermore, O-antigen biosynthesis also utilises the C_{55} lipid carrier; therefore, fluoroquinolones and O-antigen biosynthesis may compete for the same export mechanism. It is suggested that this may explain why fluoroquinolone resistant mutants have been observed to have shorter O-antigen sidechains on their LPS (LEGAKIS et al. 1989).

E. coli, which is intrinsically more susceptible to quinolones than *P. aeruginosa*, has also been shown to possess efflux systems, notably EmrAB (LOMOVSKAYA and LEWIS 1992) and AcrAB (MA et al. 1995). The *emr* operon is located at 57.5 min on the *E. coli* chromosome and encodes two proteins: EmrA, which is mostly hydrophilic but thought to be anchored to a membrane by a hydrophobic domain, and EmrB which has 14 membrane-spanning domains and is thought to be an integral membrane protein (LOMOVSKAYA and LEWIS 1992). Overexpression of *emr* on a multicopy plasmid results in decreased susceptibility to nalidixic acid as well as various hydrophobic energy uncouplers such as CCCP, but not to hydrophilic agents such as fluoroquinolones. Thus, the Emr system of *E. coli* differs from those described in *S. aureus* and *P. aeruginosa* as it strictly exports only hydrophobic compounds. COHEN et al. (1988a) proposed active efflux of norfloxacin in *E. coli*; furthermore, they showed that this was able to mediate fluoroquinolone resistance when combined with a twofold decrease in permeability in an OmpF-deficient mutant. Such an efflux system has now been identified in AcrAB, which is expressed as part of the MAR phenotype in *E. coli* (see Sect. C.III). The *acrAB* operon is at least partly regulated by the repressor AcrR and encodes two proteins: AcrB, which is a membrane efflux protein, and AcrA, which is thought to connect AcrB physically to an outer membrane channel (MA et al. 1994).

III. The *mar* Operon

It is now clear that *E. coli* and a number of other organisms possess mechanisms which provide intrinsic protection against a wide range of chemically unrelated toxic substances, including quinolone antibiotics. Multiple antibiotic resistance (MAR) in *E. coli* has been shown to be the result of mutations in the *mar* locus at 34 min on the *E. coli* chromosome (COHEN et al. 1993a), and studies suggest that a homologue of the *mar* locus exists in other members of the Enterobacteriacae, as well as in other bacteria (COHEN et al. 1993a). In *E. coli*, the *mar* locus consists of two divergently expressed operons, *marC* and *marRAB*, both of which are required for full expression of the MAR phenotype (COHEN et al. 1993a; GOLDMAN et al. 1996). The *marA*, *marB*, and *marC* genes are under the transcriptional control of MarR which is a DNA-binding protein. Mutations which disrupt the MarR protein result in overexpression of the MarA, MarB and MarC proteins (Fig. 4B) and an approximate two- to four-fold increase in resistance to chloramphenicol, rifampicin, tetracycline, ampicillin and nalidixic acid. MarA is itself a transcriptional regulator protein which activates expression of a number of different operons, including *micF* (COHEN et al. 1988b), *inhA* (ROSNER and SLONCZEWSKI 1994) and *acrAB* (OKUSU et al. 1996), all of which contribute to the MAR phenotype. The *micF* gene encodes a short length of mRNA complementary to the 5' end of the *ompF* transcript. Thus expression of *micF* antisense mRNA, due to MarA activation, inhibits the translation of *ompF* mRNA, and hence the production of OmpF porin. The function of *inhA* is unknown; *acrAB* encodes a drug efflux pump with homology to other efflux systems (MA et al. 1995). A deletion in *acrAB* results in hypersusceptibility to antibiotics, whether or not there is also a mutation in *marR*, suggesting that the AcrAB pump is the major effector of the MAR phenotype (OKUSU et al. 1996). The functions of MarB and MarC have not been determined; however, only MarC appears to be required for the MAR phenotype (COHEN et al. 1993a; SULAVIK et al. 1994; GOLDMAN et al. 1996).

In wild-type cells the *mar* regulon can be induced by certain antibiotics, such as tetracycline and chloramphenicol, as well as by salicylates and other related compounds (COHEN et al. 1993b; ROSNER 1985). This phenomenon, known as phenotypic antibiotic resistance (PAR), confers resistance to a number of antibiotics, including nalidixic acid, and is associated with increased *micF* expression and reduced amounts of OmpF. The MAR phenotype can also be activated by mutations in the *soxRS* operon or by induction with certain superoxide-generating agents, indicating an overlap between the superoxide stress reponse and phenotypic antibiotic resistance (ARIZA et al. 1994; Fig. 4B).

Expression of the MAR phenotype, whether by induction or mutation, has recently been shown to protect the cells from fluoroquinolone killing (see Sect. D) at up to four times the MIC (GOLDMAN et al. 1996). This may be more clinically important than the relatively modest increases in MIC associated with MAR, as cells which escape death would have the potential to mutate to higher levels of fluoroquinolone resistance. In support of this, *mar* mutants have been reported among clinical isolates of fluoroquinolone-resistant *E. coli* (GOLDMAN et al. 1996).

D. Reduced Killing

The killing mechanism of quinolones is unclear and remains open to debate; however, current evidence suggests that these drugs kill bacteria via a combination of processes. The concentration required to inhibit supercoiling in vitro is about tenfold higher than the MIC of the quinolone for the organism from which the enzyme was isolated. In contrast, there is very good correlation between the concentration required to inhibit DNA synthesis by 50% (IC_{50}) and the MIC, suggesting that the primary cause of cell death is the consequence of DNA gyrase-mediated inhibition of DNA synthesis, rather than inhibition of supercoiling per se.

Mutations which interfere with induction of the SOS response or cell division have been shown to reduce the lethality of fluoroquinolones in *E. coli* (Table 4; PIDDOCK et al. 1990). These include mutations in the SOS genes *recA* and *lexA*, and the cell division genes *sfiA* and *ftsZ* (*sulB*, *sfiB*; PIDDOCK and WALTERS 1992; Table 4). Such mutants are still killed by fluoroquinolones but more slowly than would otherwise be the case. Likewise, it has been found that the rate of fluoroquinolone killing is reduced in the presence of protein or RNA synthesis inhibitors, such as chloramphenicol or rifampicin (RATCLIFFE and SMITH 1985). It is thought that fluoroquinolone-mediated damage to DNA causes induction of the SOS response via activation of the RecA protein. This is normally a fine-tuned cell process which is downregulated once the inducing signal, in this case nicked DNA, is removed. However, if the nicks cannot be repaired due to the permanent nature of the DNA–DNA gyrase–quinolone complex then runaway activation of the SOS response may occur. A consequence of induction of the SOS response is inhibition of cell division and activation of autolysins and this is believed to contribute to the killing action of fluoroquinolones (PIDDOCK and WALTERS 1992; VINCENT et al. 1991). Thus, it seems that the final stages of fluoroquinolone killing are similar to those of β-lactams which also involve the action of autolytic enzymes. Therefore, it is not surprising that *hipA* mutants, identified as having reduced killing by β-lactams, also exhibit reduced killing by fluoroquinolones; and vice versa, a *hipQ* mutant, identified by reduced fluoroquinolone killing, also exhibits reduced killing by β-lactams (SCHERRER and MOYED 1988; WOLFSON et al. 1990).

Because fluoroquinolones can still kill in the absence of protein or RNA synthesis, albeit more slowly, another mechanism of killing has been postulated which has been termed "mechanism B" (RATCLIFFE and SMITH 1985; LEWIN et al. 1991). This appears to be a direct result of fluoroquinolone action on the DNA itself. One of the paradoxes of fluoroquinolone action is that they are rapidly bactericidal up to a specific concentration, known as the optimal bactericidal concentration (OBC). Above this concentration protein and RNA synthesis is inhibited, presumably by the increasing number of DNA–DNA gyrase–quinolone complexes causing obstruction of RNA polymerase, and there is a corresponding reduction in the kill rate.

Mutations which confer resistance via interference with SOS-mediated cell death have so far only been discovered in vitro, probably because they interfere with essential cell functions, thus placing themselves at a selective disadvantage. However, such mutants are very useful in elucidating how fluoroquinolones kill cells, which is still not fully understood. As described above, *mar* mutants also show reduced killing by fluoroquinolones and these have been identified among clinical isolates (see Sect. B.III).

Various other mutations in *E. coli* have been selected after mutagenesis which result in decreased susceptibility to nalidixic acid (Table 4). These are

Table 4. Mutations conferring decreased susceptibility to quinolone antibiotics in *E. coli*

Locus	Position	Function of gene	MIC ($mg\,l^{-1}$) NAL	CIP	NOR	FOX	CAM	TET
wt			4	0.03	0.016	4	2	2
gyrA	48	DNA gyrase A subunit	128–2048	0.5–64	0.5–64	4	4	2
gyrB	83	DNA gyrase B subunit	32–128	0.06[a]	0.12[a]	4	4	2
ompF	21	Outer membrane porin	16	0.12	0.12	16	8	4
marR	34	Regulator protein of *mar* operon	8–16	0.25	0.12	32	16	8
soxR	92	Regulator protein of superoxide stress response	NA	0.5	NA	NA	NA	4
emr[b]	57.5	Inner membrane efflux protein	8	NA	0.5	NA	2	2
nalB	58	*ompF* expression	8	NA	NA	NA	NA	NA
norC	8	Unknown	1	0.06	0.25	8	4	NA
nalD	89	Outer membrane lipid composition	24	NA	NA	4	4	4
nfxB	19	Expression of *ompF*	16	0.5	NA	>32	32	16
cya	85	Adenyl cyclase; cAMP synthesis	>8	NA	NA	NA	NA	NA
crp	74	cAMP receptor protein	>8	NA	NA	NA	NA	NA
recA	58	DNA repair and recombination; induction of SOS Response	16	0.03	0.03	4	8	1
lexA	92	Regulator protein; mediator of SOS response	16	0.016	0.03	8	8	4
sfiA	22	Regulation of cell division	8	0.03	0.12	8	8	4
ftsZ	2	Cell division apparatus	32	0.03	0.06	8	8	4
hipA	34	Cell wall metabolism	10	NA	NA	NA	NA	NA
hipQ	2	Cell wall metabolism	NA	0.12	0.25	16	NA	NA
icd	25	Isocitrate dehydrogenase	>10	NA	NA	NA	NA	NA
purB	25	Adenosuccinate lyase	>10	NA	NA	NA	NA	NA

NAL, nalidixic acid; CIP, ciprofloxacin; NOR, norfloxacin; FOX, cefoxitin; CAM, chloramphenicol; TET, tetracycline; wt, wild type; NA, data not available; MIC, minimal inhibitory concentration.

[a] *gyrB* Lys447 mutant MIC = 0.008.

[b] MICs obtained after introduction of *emr* on multicopy plasmid into wild-type strain.

all located in genes involved in central metabolism, such as *icd* (isocitrate dehydrogenase; HELLING and KUKORA 1971) and *purB* (adenosuccinate lyase; HELLING and ADAMS 1970). The reasons for the decreased susceptibility of these and other such mutants remain to be elucidated.

E. Prevalence of Fluoroquinolone Resistance

I. Community Acquired Pathogens

Since the introduction of fluoroquinolones into clinical practice in the mid-1980s, a number of national and international surveillance programmes have monitored the emergence of fluoroquinolone resistance in clinical isolates (ACAR et al. 1993; KRESKEN et al. 1994; THORNSBERRY 1994). In general, fluoroquinolone resistance is still rare among common pathogens, such as *E. coli*, *Haemophilus influenzae*, *N. gonorrhoeae* and *Salmonella* and *Shigella* spp., with 97%–100% of strains remaining susceptible (ACAR et al. 1993). This is despite the fact that the bulk of fluoroquinolones (approximately 75% in Europe) are used in the community (KRESKEN et al. 1994). Reports of resistance in *Streptococcus pneumoniae* or *N. gonorrhoeae* are sporadic and is suggested to be the consequence of improper dosage or inadequate penetration of antibiotic to the target tissue (BALL 1994). In a survey of 109 strains of *N. gonorrhoeae* 97.2% were susceptible to ciprofloxacin (MIC $\leq 0.06\,mg\,l^{-1}$; RICE and KNAPP 1994). It has been observed that the early discharge of hospital patients colonised with fluoroquinolone resistant MRSA does not lead to dissemination of such strains within the community (WITTE et al. 1994), suggesting that the clonal spread of resistance may not be as common in the community as it is in hospitals.

Despite the low incidence of resistance in the community, there is evidence to suggest that the use of fluoroquinolones in this environment does impose a significant selective pressure on the emergence of fluoroquinolone-resistant strains. In a survey of fluoroquinolone resistance among gut intestinal flora isolated from the stools of hospital patients, RICHARD et al. (1994) found that administration of fluoroquinolones within the month preceding admission was a significant risk factor in the acquisition of resistance. It was suggested that the use of fluoroquinolones in the community favours the establishment of a faecal reservoir of fluoroquinolone-resistant isolates and that the selective pressure of fluoroquinolones in the community is probably underestimated.

In the southern hemisphere, the incidence of fluoroquinolone resistance among community acquired pathogens differs greatly between "developed" countries, such as Australia and New Zealand, and "developing" countries, such as Korea, China and the Phillipines (TURNIDGE 1994). For example, the percentage of non-urine clinical isolates of *E. coli* exhibiting fluoroquinolone resistance in Australia and New Zealand in 1992 was 0%–0.2%, whereas in Korea and China the percentage was 22% and 21%, respectively. Likewise, the percentage of resistant non-typhoidal salmonella in the Philipines was

11%, compared with 0.1% in Australia. It has been suggested that the high prevalence of resistance in such countries is a consequence of inadequate control of fluoroquinolone use and the illegal marketing of substandard imitations (TURNIDGE 1994).

II. Nosocomial Pathogens

Among nosocomial pathogens, the incidence of fluoroquinolone resistance has increased in recent years, especially in *S. aureus* and *P. aeruginosa*, although wide variations exist between different institutions and between different countries. In a survey of fluoroquinolone resistance in 12 European countries (excluding Spain and Scandinavia) over the period 1983–1990 the overall percentage of *P. aeruginosa* isolates resistant to 4 mg ciprofloxacin l^{-1} increased from 3.1% to 16.1%, and of *S. aureus* from 0% to 6.8% (KRESKEN et al. 1994). Southern European countries generally showed higher levels of resistance compared to those in northern Europe; in particular, high rates of ciprofloxacin resistance in *P. aeruginosa* were observed in Italy (43.3%), France (25.6%) and Greece (19.0%) compared to the other countries surveyed (0%–11.2%; KRESKEN et al. 1994). In the US similar rates of resistance among nosocomial pathogens to those found overall in Europe have been reported (Fig. 5; THORNSBERRY 1994).

A number of factors favour the emergence and dissemination of resistant strains in hospitals. Nosocomial pathogens such as *P. aeruginosa* and staphylococci are intrinsically less susceptible to fluoroquinolones than many commu-

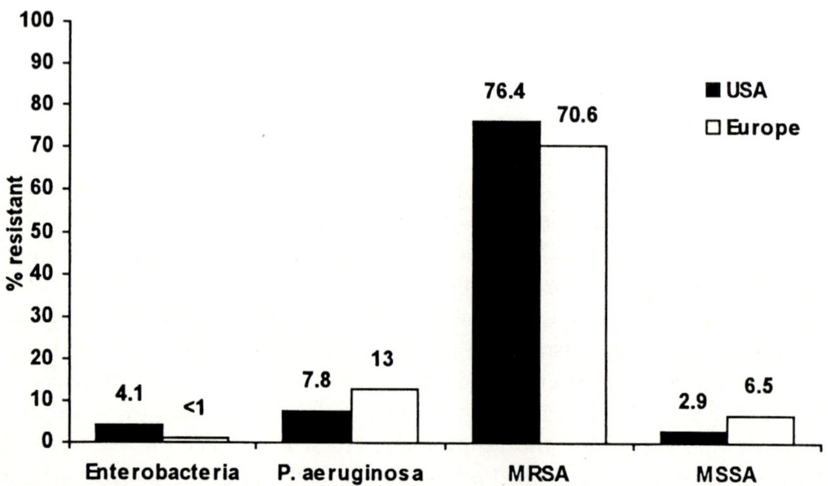

Fig. 5. Prevalence of nosocomial fluoroquinolone resistance in USA and Europe (excluding Spain and Scandinavia) in 1990. Resistant isolates were defined as exhibiting MICs > 4 mg ciprofloxacin l^{-1}. (Data from THORNSBERRY 1994; KRESKEN et al. 1994)

nity acquired pathogens (MIC of approximately 0.5 vs. <0.06 mg ciprofloxacin ml^{-1}), and infections often occur in sites with poor antibiotic accessibility, such as the lungs of cystic fibrosis patients or around indwelling devices (BALL 1994). Thus, even small decreases in susceptiblity can lead to clinical resistance. Additionally, the prophylactic use of fluoroquinolones has been shown to provide selective pressure for the emergence of resistance in neutropenic or immunosuppressed patients (BALL 1994). A recent survey of fluoroquinolone-resistant *E. coli* from ten cancer centres across Europe and the Middle East identified significant strain diversity even within individual centres, with fluoroquinolone resistance commonly associated with multiple antibiotic resistance (OETHINGER et al. 1996).

The high rates of fluoroquinolone resistance revealed by surveys can sometimes be misleading as the figures may be skewed by the occurrence of epidemics of resistant nosocomial pathogens in particular institutions. Analysis of plasmid profiles from resistant *S. aureus* isolates indicate that such outbreaks are often due to the emergence and clonal spread of a single resistant strain within a hospital or ward, rather than repeated selection of resistance in individual patients (WITTE et al. 1994; KOTILAINEN et al. 1994). Thus, the true incidence of fluoroquinolone resistance is probably lower than reported. Such outbreaks are usually localised to one hospital but have been shown to spread to other hospitals via transfer of patients (WITTE et al. 1994). Horizontal spread of resistant organisms is facilitated by contact between contaminated patients and staff. The use of catheters and endoscopic examination increases the risk of transmission between patients and high rates of resistance have been found in intensive care units where the use of such procedures is common (LINGNAU and ALLERBERGER 1994).

A number of reports highlight the high incidence of fluoroquinolone resistance among MRSA in certain hospitals compared with methicillin-susceptible strains. Within 1 year of the introduction of ciprofloxacin for the treatment of *S. aureus* infections in one hospital in the US the percentage of ciprofloxacin resistant isolates of MRSA increased from zero to 79% (BLUMBERG et al. 1991). National surveys in both the US and Europe show a high incidence of fluoroquinolone resistance among MRSA (Fig. 5); however, this varies dramatically from one institute to another with some hospitals as low as 7% and others as high as 90% (JONES et al. 1992). In 16 hospitals surveyed in France in 1990, the incidence of fluoroquinolone-resistant MRSA varied between 65% and 98% (ACAR et al. 1993). A correlation between methicillin resistance and fluoroquinolone resistance among *S. aureus* has also been reported in three South American countries (Argentina, Chile, Venezuela) and three Asian countries (Pakistan, Malaysia, Singapore; ACAR et al. 1993). In Europe, the overall incidence of fluoroquinolone resistance in methicillin-resistant coagulase-negative staphylococci (51.2%) is also higher than in methicillin-susceptible strains (15.9%; KRESKEN et al. 1994). The high rate of fluoroquinolone resistance among methicillin-resistant staphylococci probably reflects the greater use of fluoroquinolones against such strains, rather than any intrinsic difference between methicillin-resistant and susceptible organisms.

Recent years has seen the emergence of enterococci as significant nosocomial pathogens. The incidence of fluoroquinolone resistance among enterococci rose from zero in 1985/1986 to as high as 27% in 1989/1990, 24% of which occurred in isolates of *E. faecalis* (SCHABERG et al. 1992). This occurred despite the fact that fluoroquinolones are not usually recommended for the treatment of enterococcal infections due to the intermediate MIC of fluoroquinolones for enterococci (ciprofloxacin MIC of approximately $1\,mg\,l^{-1}$).

III. Impact of Fluoroquinolone Use in Agriculture

The licensing of fluoroquinolones for use in veterinary medicine and in animal feedstuffs has prompted fears that this may lead to the zoonotic transfer of resistant bacteria from animals to man (ENDTZ et al. 1991). Notably, the incidence of fluoroquinolone resistance in *Campylobacter* strains which cause acute diarrhoea has increased in certain countries over recent years. In Spain, for instance, the percentage of isolates exhibiting fluoroquinolone resistance has risen from 2.3% in 1988 to 32% in 1991 (REINA et al. 1992). The major source of infection with *Campylobacter* spp. is from contaminated chicken carcasses, and there are concerns that the prophylactic use of fluoroquinolones in poultry feed is providing the selective pressure for the emergence and spread of fluoroquinolone-resistance among poultry flocks (ENDTZ et al. 1991). In the UK, which has only recently licensed the veterinary use of enrofloxacin, poultry flocks remain virtually free of fluoroquinolone-resistant *Campylobacter* strains, and there is evidence that the incidence of fluoroquinolone-resistant *Campylobacter* isolates from humans is associated with the consumption of imported poultry (GAUNT and PIDDOCK 1996).

In contrast, there is less evidence of transfer of fluoroquinolone-resistant salmonella from animals to man giving rise to an infection resistant to fluoroquinolone therapy. In a study of resistant isolates of salmonella from veterinary sources, the mechanism of fluoroquinolone resistance was found to be similar as for human isolates (GIGGS et al. 1994; PIDDOCK et al. 1993a). Whilst the veterinary isolates required higher concentrations of fluoroquinolone for inhibition than a wild-type strain, it was concluded that such strains would still be inhibited by therapeutically achievable concentrations of fluoroquinolone (GRIGGS et al. 1994). However, in Germany, fluoroquinolone-resistant isolates of *S. typhimurium* have recently been isolated from both man and cattle, which carry a double mutation in *gyrA* and are resistant to 32 mg ciprofloxacin l^{-1} (HEISIG et al. 1995). Furthermore, the genotypic fingerprinting was identical for all strains, suggesting that the use of fluoroquinolones in animals may indeed lead to the zoonotic transfer of fluoroquinolone-resistant salmonella as well as of *Campylobacter* spp.

F. Distribution of Resistance Mechanisms

Much information is now available regarding the prevalence of fluoroquinolone resistance in clinical bacterial isolates, but rarely does this data distinguish between different resistance mechanisms; hence, the clinical relevance of particular resistance mechanisms is hard to gauge. However, several studies suggest that multiple mechanisms of fluoroquinolone resistance are common amongst clinical, veterinary and laboratory isolates, and can be selected with a variety of fluoroquinolone agents (PIDDOCK et al. 1993a; PIDDOCK and ZHU 1991; GRIGGS et al. 1994).

PIDDOCK et al. (1993a) examined the mechanisms of fluoroquinolone resistance in one pretreatment isolate and a further eleven isolates of *S. typhimurium* obtained over a period of 4.5 months from a patient on a course of ciprofloxacin. The pretreatment isolate was highly susceptible to ciprofloxacin (MIC = $0.03\,\mathrm{mg\,l^{-1}}$); however all subsequent isolates were less susceptible (MIC ≥ 0.12), and six displayed cross-resistance to other antibiotics. On the basis of *gyrA*-plasmid complementation experiments, the five isolates which were resistant to fluoroquinolones only were identified as *gyrA* mutants, as was one of the multiply resistant strains. Multiply resistant strains accumulated less fluoroquinolones than the pretreatment isolate. In a parallel study of 27 nalidixic-acid-resistant isolates of salmonella from veterinary sources, all showed decreased susceptibility to fluoroquinolones and six were cross-resistant to one or more unrelated antibiotics (GRIGGS et al. 1994). DNA synthesis IC_{50} values for enrofloxacin showed good correlation with MICs for 23 isolates, indicating a likely alteration in DNA gyrase, and five of the 27 isolates showed reduced levels of enrofloxacin accumulation. None of the outer membrane protein profiles differed from control strains, although three isolates exhibited differences in their lipopolysccharide profiles. These results illustrate that whilst mutations in DNA gyrase are probably the most frequent cause of fluoroquinolone resistance in *Salmonella* species mutations affecting fluoroquinolone accumulation, which are often associated with cross-resistance to other antibiotics, are also important. In contrast, fluoroquinolone resistance in *Campylobacter* species is predominantly *gyrA*-mediated (M.J. Everett and L.J.V. Piddock, unpublished results).

A number of surveys have studied the relative prevalence of *gyrA* and *gyrB* mutants in clinical isolates of fluoroquinolone-resistant bacteria (NAKAMURA et al. 1989; VILA et al. 1994; ITO et al. 1994). These indicate that, whereas *gyrA* and *gyrB* mutants arise at similar frequencies in the laboratory, *gyrB* mutants are far less common amongst clinical isolates. The reasons for this are unclear; however, *gyrA* mutants tend to be more resistant to fluoroquinolones than *gyrB* mutants, some of which are hypersusceptible to certain fluoroquinolones (Sect. B.II), and therefore may have a selective advantage over *gyrB* mutants. Since the identification of topoisomerase IV as a target for fluoroquinolones in bacteria, several studies have investigated the prevalence of *parC* mutations in clinical isolates. VILA et al. (1996) found *parC* mutations in 15/27 clinical isolates of *E. coli*; similarly, EVERETT et al. (1996) found that 24/36 isolates from humans and poultry carried *parC* mutations. However no *parE* mutations were identified. Both studies recognised a correlation between possession of a *parC* mutation and increased fluoroquinolone resistance. A recent report suggests that clinical *E. coli* isolates may also possess *mar* mutations and that this may be a common first step in a multistep process which leads to high-level fluoroquinolone resistance (GOLDMAN et al. 1996). It is now clear that fluoroquinolone resistance in clinical isolates can and does involve mutations at multiple loci. What is less clear is the source of such mutants, the process by which they accumulate multiple mutations, and the selective pressures which drive that process.

G. Summary

In conclusion, advances in recent years have increased our understanding of fluoroquinolone resistance in a number of important areas:

- The number of species in which fluoroquinolone resistance has been studied has increased dramatically. Mutations in *gyrA* have been confirmed as an important determinant of fluoroquinolone resistance for a wide variety of pathogens.
- Topoisomerase IV has been identified as a primary target of fluoroquinolones in *S. aureus* and as a secondary target in *E. coli*.
- Multiple antibiotic resistance in *E. coli* has been shown to involve the expression of a number of genes, and to be regulated by a complex control system involving both the *mar* and *sox* regulons.
- Permeability of the cell envelope has been recognised as playing a less important role in fluoroquinolone resistance than previously thought, with resistance requiring both a permeability barrier and a functional efflux system. Such efflux systems have now been identified in a variety of organisms.
- The use of fluoroquinolones in animal food stuffs has been implicated in the emergence of fluoroquinolone resistance in bacteria, and in particular in *Campylobacter* species.
- The prevalence of fluoroquinolone resistance amongst common community acquired pathogens remains low; however, in certain hospitals the incidence of fluoroquinolone resistance has increased due the emergence and clonal spread of resistant nosocomial pathogens, in particular MRSA.

References

Acar JF, O'Brien TF, Goldstein FW, Jones RN (1993) The epidemiology of bacterial quinolone resistance. Drugs 45(Suppl 3):24–28

Ariza RR, Cohen SP, Bachhawat N, Levy SB, Demple B (1994) Repressor mutations in the *marRAB* operon that activate oxidative stress genes and multiple antibiotic resistance in *Escherichia coli*. J Bacteriol 176:143–148

Aubert G, Pozzetto B, Dorche G (1992) Emergence of quinolone–imipenem cross-resistance in *Pseudomonas aeruginosa* after fluoroquinolone therapy. J Antimicrob Chemother 29:307–312

Ball P (1994) Bacterial resistance to fluoroquinolones: lessons to be learned. Infection 22(Suppl 2):S140–S147

Barrett JF, Bernstein JI, Krause HM, Hilliard JJ, Ohemeng KA (1993) Testing potential gyrase inhibitors of bacterial DNA gyrase: a comparison of the supercoiling inhibition assay and "cleavable complex assay". Anal Biochem 214:313–317

Blumberg HM, Rimland D, Carroll DJ, Terry P, Wachsmuth IK (1991) Rapid development of ciprofloxacin resistance in methicillin-susceptible and methicillin-resistant *Staphylococcus aureus*. J Infect Dis 163:1279–1285

Brown JC, Shanahan PMA, Jesudason MV, Thomson CJ, Amyes SGB (1996) Mutations responsible for reduced susceptibility to 4-quinolones in clinical isolates of multiresistant *Salmonella typhimurium* in India. J Antimicrob Chemother 37:891–900

Cambau E, Gutmann L (1993) Mechanisms of resistance to quinolones. Drugs 45(Suppl 3):15–23

Cambau E, Bordon F, Collatz E, Gutmann L (1993) Novel *gyrA* point mutation in a strain of *Escherichia coli* resistant to fluoroquinolones but not to nalidixic acid. Antimicrob Agents Chemother 37:1247–1252

Cambau E, Sougakoff W, Jarlier V (1994) Amplification and nucleotide sequence of the quinolone resistance-determining region in the *gyrA* gene of mycobacteria. FEMS Microbiol Lett 116:49–54

Chapman JS, Georgopapadakou NH (1988) Routes of quinolone penetration in *Escherichia coli*. Antimicrob Agents Chemother 32:438–442

Chapman JS, Bertasso A, Georgeopapadakou NH (1989) Fleroxacin resistance in *Escherichia coli*. Antimicrob Agents Chemother 33:239–242

Charvalos E, Peteinaki E, Spyridaki I, Manetas S, Tselentis Y (1996) Detection of ciprofloxacin resistance mutations in *Campylobacter jejuni gyrA* by non-radioisotopic single-strand conformation polymorphism and direct sequencing. J Clin Lab Anal 10:129–133

Cohen SP, Hooper DC, Wolfson JS, Souza KS, McMurray LM, Levy SB (1988a) Endogenous active efflux of norfloxacin in susceptible *Escherichia coli*. Antimicrob Agents Chemother 32:1187–1191

Cohen SP, McMurray LM, Levy SB (1988b) *marA* locus causes decreased expression of OmpF porin in multiple-antibiotic resistant (Mar) mutants of *Escherichia coli*. J Bacteriol 170:5416–5422

Cohen SP, McMurray LM, Hooper DC, Wolfson JS, Levy SB (1989) Cross-resistance to fluoroquinolones in multiple-antibiotic resistant (Mar) *Escherichia coli* selected by tetracycline or chloramphenicol: decreased drug accumulation associated with membrane changes in addition to OmpF reduction. Antimicrob Agents Chemother 33:1318–1325

Cohen SP, Hachler H, Levy SB (1993a) Genetic and functional analysis of the multiple antibiotic resistance (*mar*) locus in *Escherichia coli*. J Bacteriol 175:1484–1492

Cohen SP, Levy SB, Foulds J, Rosner JL (1993b) Salicylate induction of antibiotic resistance in Escherichia coli: activation of the mar operon and a mar-independent pathway. J Bacteriol 175:7856–7862

Cohen SP, Yan W, Levy SB (1993c) A multidrug-resistance regulatory locus is widespread among enteric bacteria. J Infect Dis 168:484–488

Cullen ME, Wyke AW, Kuroda R, Fisher LM (1989) Cloning and characterization of a DNA gyrase A gene from *Escherichia coli* that confers clinical resistance to 4-quinolones. Antimicrob Agents Chemother 33:886–894

Day CA, Marceau-Day ML, Day DF (1993) Increased susceptibility of *Pseudomonas aeruginosa* to ciprofloxacin in the presence of vancomycin. Antimicrob Agents Chemother 37:2506–2508

Deguchi T, Yasuda M, Asano M, Tada K, Iwata H, Komeda H, Ezaki T, Saito I, Kawada Y (1995) DNA gyrase mutations in quinolone-resistant clinical isolates of *Neisseria gonorrhoeae*. Antimicrob Agents Chemother 39:561–563

Deguchi T, Yasuda M, Nakano M, Ozeki S, Ezaki T, Saito I, Kawada Y (1996) Quinone-resistant *Neisseria gonorrhoae* – correlation of alterations in the GyrA subunit of DNA gyrase and the ParC subunit of topoisomerase IV with antimicrobial susceptibility profiles. Antimicrob Agents Chemother 40:1020–1023

Domagala JM, Hanna LD, Heifetz CL, Hutt MP, Mich TF, Sanchez JPS, Solomon M (1986) New structure-activity relationships of the quinolone antibacterials using the target enzyme. J Med Chem 29:394–404

Endtz HP, Ruijs GJ, Van Klingeren B, Jansen WH, Van Der Reyden T, Mouton RP (1991) Quinolone resistance in campylobacter isolated from man and poultry following the introduction of fluoroquinolones in veterinary medicine. J Antimicrob Chemother 27:199–208

Everett MJ, Jin Y-F, Ricci V, Piddock LJV (1996) Contribution of individual mechanisms to fluoroquinolone resistance in 36 *Escherichia coli* isolates from humans and animals. Antimicrob Agents Chemother 40:2380–2386

Ferrero L, Cameron B, Manse B, Lagneaux D, Crouzet J, Famechon A, Blanche F (1994) Cloning and primary structure of *Staphylococcus aureus* DNA topoisomerase IV: a primary target of fluoroquinolones. Mol Microbiol 13:641–653

Fukuda H, Hosaka M, Hirai K, Iyobe S (1990) New norfloxacin resistance gene in *Pseudomonas aeruginosa* PAO. Antimicrob Agents Chemother 34:1757–1761

Furet YX, Deshusses J, Pechere JC (1992) Transport of pefloxacin across the bacterial cytoplasmic membrane in quinolone-susceptible *Staphylococcus aureus*. Antimicrob Agents Chemother 36:2506–2511

Gambino LS, Gracheck SJ, Miller PF (1993) Overexpression of the MarA positive regulator is sufficient to confer multiple antibiotic resistance in *Escherichia coli*. J Bacteriol 175:2888–2894

Gaunt PN, Piddock LJV (1996) Ciprofloxacin resistant campylobacter in humans: an epidemiological and laboratory study. J Antimicrob Chemother 37:747–757

Gellert M, Mizuuchi K, O'Dea MH, Itoh T, Tomizawa J-I (1976) DNA gyrase: an enzyme that introduces superhelical turns into DNA. Proc Natl Acad Sci USA 73:4772–4776

Gellert M, Mizuuchi K, O'Dea MH, Itoh T, Tomizawa J (1977) Nalidixic acid resistance: a second genetic character involved in DNA gyrase activity. Proc Natl Acad Sci USA 74:4772–4776

Gensberg K, Jin YF, Piddock LJV (1995) A novel mutation in a fluoroquinolone-resistant clinical isolate of *Salmonella typhimurium*. FEMS Microbiol Lett 132:57–60

Goldman JD, White DG, Levy SB (1996) Multiple antibiotic resistance (MAR) locus protects *Escherichia coli* from rapid killing by fluoroquinolones. Antimicrob Agents Chemother 40:1266–1269

Gootz TD, Martin BA (1991) Characterization of high-level quinolone resistance in *Campylobacter jejuni*. Antimicrob Agents Chemother 35:840–845

Goswitz JJ, Willard KE, Fasching CE, Peterson LR (1992) Detection of *gyrA* gene mutations associated with ciprofloxacin resistance in methicillin-resistant *Staphylococcus aureus*: analysis by polymerase chain reaction and automated direct DNA sequencing. Antimicrob Agents Chemother 36:1166–1169

Gotoh N, Tsujmoto H, Poole K, Yamagishi JI, Nishino T (1995) The outer membrane protein OprM od *Pseudomonas aeruginosa* is encoded by *oprK* of the *mexA–mexB–oprK* multidrug-resistance operon. Antimicrob Agents Chemother 39:2567–2569

Griggs DJ, Hall MC, Jin YF, Piddock LJV (1994) Quinolone resistance in veterinary isolates of salmonella. J Antimicrob Chemother 33:1173–1189

Griggs DJ, Gensberg K, Piddock LJV (1996) Mutations in *gyrA* gene of quinolone-resistant salmonella serotypes isolated from humans and animals. Antimicrob Agents Chemother 40:1009–1013

Gutmann L, Williamson R, Moreau N, Kitzis MD, Collatz E, Acar JF, Goldstein FW (1985) Cross-resistance to nalidixic acid, trimethoprim and chloramphenicol associated with alterations in outer membrane proteins of *Klebsiella*, *Enterobacter* and *Serratia*. J Infect Dis 151:501–507

Hallett P, Maxwell A (1991) Novel quinolone resistance mutations of the *Escherichia coli* DNA gyrase A protein: enzymatic analysis of the mutant proteins. Antimicrob Agents Chemother 35:335–340

Hamzehpour MM, Furet YX, Pechere JC (1991) Role of protein D2 and lipopolysaccharide in diffusion of quinolones through the outer membrane of *Pseudomonas aeruginosa*. Antimicrob Agents Chemother 35:2091–2097

Hamzehpour MM, Pechere JC, Pleisat P, Kohler T (1995) OprK and OprM define two genetically distinct multidrug efflux systems in *Pseudomonas aeruginosa*. Antimicrob Agents Chemother 39:2392–2396

Hane MW, Wood T (1969) *Escherichia coli* K-12 mutants resistant to nalidixic acid: genetic mapping and dominance studies. J Bacteriol 99:238–241

Heisig P (1993) High-level fluoroquinolone resistance in a *Salmonella typhimurium* isolate due to alterations in both *gyrA* and *gyrB* genes. J Antimicrob Chemother 32:367–377

Heisig P (1996) Genetic evidence for a role of *parC* mutations in development of high-level fluoroquinolone resistance in *Escherichia coli*. Antimicrob Agents Chemother 40:879–885

Heisig P, Tschorny R (1994) Characterization of fluoroquinolone resistant mutants of *Escherichia coli* selected in vitro. Antimicrob Agents Chemother 38:1284–1291

Heisig P, Wiedemann B (1991) Use of a broad-host-range *gyrA* plasmid for genetic characterization of fluoroquinolone-resistant gram-negative bacteria. Antimicrob Agents Chemother 35:2031–2036

Heisig P, Schedletzky H, Falkenstein Paul H (1993) Mutations in the *gyrA* gene of a highly floroquinolone-resistant clinical isolate of *Escherichia coli*. Antimicrob Agents Chemother 37:696–701

Heisig P, Kratz B, Halle E, Graser Y, Atwegg M, Rabsch W, Faber JP (1995) Identification of DNA gyrase A mutations in ciprofloxacin-resistant isolates of *Salmonella typhimurium* from men and cattle in Germany. Microb Drug Resist Mech Epidemiol Dis 1:211–218

Helling RB, Adams BS (1970) Nalidixic acid-resistant auxotrophs of *Escherichia coli*. J Bacteriol 104:1027–1029

Helling RB, Kukora JS (1971) Nalidixic acid-resistant mutants of *Escherichia coli* deficient in isocitrate dehydrogenase. J Bacteriol 105:1224–1226

Higgins CF, Dorman CJ, Stirling DA, Waddel L, Booth IR, May G, Bremer E (1988) A physiological role for DNA supercoiling in the osmotic control of gene expression in *Salmonella typhimurium* and *Escherichia coli*. Cell 52:569–584

Hirai K, Aoyama H, Suzue S, Irikura T, Iyobe S, Mitsuhashi S (1986) Isolation and characterization of norfloxacin-resistant mutants of *Escherichia coli* K-12. Antimicrob Agents Chemother 30:248–253

Hooper D, Wolfson J, Souza KS, Tung C, McHugh G, Swartz M (1986) Genetic and biochemical characterisation of norfloxacin resistance in *Escherichia coli*. Antimicrob Agents Chemother 29:639–644

Hooper D, Wolfson J, Ng E, Swartz M (1987) Mechanisms of action of and resistance to ciprofloxacin. Am J Med 82:12–20

Hooper DC, Wolfson JS, Souza KS, Ng EY, McHugh GL, Swartz MN (1989) Mechanisms of quinolone resistance in *Escherichia coli*: characterization of *nfxB* and *cfxB*, two mutant resistance loci decreasing norfloxacin accumulation. Antimicrob Agents Chemother 33:283–290

Hooper DC, Wolfson JS, Bozza MA, Ng EY (1992) Genetics and regulation of outer membrane protein expression by quinolone resistance loci *nfxB*, *nfxC*, and *cfxB*. Antimicrob Agents Chemother 36:1151–1154

Hori S, Ohshita Y, Utsui Y, Hiramatsu K (1993) Sequential acquisition of norfloxacin and ofloxacin resistance by methicillin-resistant and -susceptible *Staphylococcus aureus*. Antimicrob Agents Chemother 37:2278–2284

Horowitz DS, Wang JC (1987) Mapping the active site tyrosine of *Escherichia coli* DNA gyrase. J Biol Chem 262:5339–5344

Hoshino K, Kitamura A, Morrissey I, Sato K, Kato J, Ikeda H (1994) Comparison of inhibition of *Escherichia coli* topoisomerase IV by quinolones with DNA gyrase inhibition. Antimicrob Agents Chemother 38:2623–2627

Hrebenda J, Heleszko H, Brzostek K, Bielechi H (1985) Mutation affecting resistance of *Escherichia coli* K12 to nalidixic acid. J Gen Microbiol 131:2285–2292

Huang WM (1992) Multiple DNA gyrase-like genes in Eubacteria. In: Andoh T, Ikeda H, Oguro M (eds) Molecular biology of DNA topoisomerases and its application to chemotherapy. CRC, London, pp 39–48

Ishii H, Sato K, Hoshino K, Sato M, Yamaguchi A, Sawai T, Osada Y (1991) Active efflux of ofloxacin by a highly quinolone-resistant strain of *Proteus vulgaris*. J Antimicrob Chemother 28:827–836

Ito H, Yoshida H, Bogaki-Shonai M, Niga T, Hattori H, Nakamura S (1994) Quinolone resistance mutations in the DNA gyrase *gyrA* and *gyrB* genes of *Staphylococcus aureus*. Antimicrob Agents Chemother 38:2014–2023

Jones RN, Reller LB, Rosati LA, Erwin ME, Sanchez ML, The Ofloxacin Surveillance Group (1992) Ofloxacin, a new broad-spectrum floroquinolone. Results from a multicenter, national comparative activity surveillance study. Diagn Microbiol Infect Dis 15:425–434

Kaatz GW, Seo SM, Ruble CA (1993) Efflux-mediated fluoroquinolone resistance in *Staphylococcus aureus*. Antimicrob Agents Chemother 37:1086–1094

Kato JI, Nishimura Y, Imamura R, Niki H, Hiraga S, Suzuki H (1990) New topoisomerase essential for chromosome separation. Cell 63:393–404

Khodursky AB, Zechiedrich EL, Cozzarelli NR (1995) Topoisomerase IV is a target of quinolones in *Escherichia coli*. Proc Natl Acad Sci USA 92:11801–11805

Korten V, Huang WM, Murray BE (1994) Analysis by PCR and direct sequencing of *gyrA* mutations associated with fluoroquinolone resistance in *Enterococcus faecalis*. Antimicrob Agents Chemother 38:2091–2094

Kotilainen P, Huovinen S, Jarvinen H, Aro H, Huovinen P (1995) Epidemiology of the colonization of inpatients and outpatients with ciprofloxacin resistant coagulase negative staphylococci. Clin Infect Dis 21:685–687

Kresken M, Hafner D, Mittermayer H, Verbist L, Bergogne-Gerezin E, Giamarellou H, Esposito S, Van Klingeren B, Kayser FH, Reeves DS, Wiedemann B, The "Bacterial Resistance" Study Group of the Paul-Ehrlich-Society for Chemotherapy (1994) Prevalence of fluoroquinolone resistance in Europe. Infection 22(Suppl 2):S90–S98

Kreuzer KN, Cozzarelli NR (1979) *Escherichia coli* mutants thermosensitive for deoxyribonucleic acid gyrase subunit A: effects on deoxyribonucleic acid replication, transcription and bacteriophage growth. J Bacteriol 140:424–435

Kumar S (1976) Properties of adenyl cyclase and cyclic adenosine 3′,5′-monophosphate receptor protein-deficient mutants of *Escherichia coli*. J Bacteriol 125:545–555

Kumugai Y, Kato J, Hoshino K, Akasaka T, Sato K, Ikeda H (1996) Quinolone-resistant mutants of *Escherichia coli* DNA topisomerase IV *parC* gene. Antimicrob Agents Chemother 40:710–714

Kureishi A, Diver JM, Beckthold B, Schollaardt T, Bryan LE (1994) Cloning and nucleotide sequence of *Pseudomonas aeruginosa* DNA gyrase *gyrA* gene from strain PAO1 and quinolone-resistant clinical isolates. Antimicrob Agents Chemother 38:1944–1952

Legakis NJ, Tzouvelekis LS, Makris A, Kotsifiki H (1989) Outer membrane alterations in multiresistant mutants of *Pseudomonas aeruginosa* selected by ciprofloxacin. Antimicrob Agents Chemother 33:124–127

Levy SB (1992) Active efflux mechanisms for antimicrobial resistance. Antimicrob Agents Chemother 36:695–703

Lewin CS, Morrissey I, Smith JT (1991) The mode of action of quinolones: the paradox in activity of low and high concentrations and activity in the anaerobic environment. Eur J Clin Microbiol Infect Dis 10:240–248

Li X, Livermore D, Nikaido H (1994a) Role of efflux pump(s) in intrinsic resistance of *Pseudomonas aeruginosa*: resistance to tetracycline, chloramphenicol, and norfloxacin. Antimicrob Agents Chemother 38:1732–1741

Li XZ, Ma D, Livermore DM, Nikaido H (1994b) Role of efflux pump(s) in intrinstic resistance of *pseudomonas aeruginosa*: active efflux as a contributing factor to beta-lactam resistance. Antimicrob Agents Chemother 38:1742–1752

Lingnau W, Allerberger F (1994) Control of an outbreak of methicillin-resistant *Staphylococcus aureus* (MRSA) by hygenic measures in a general intensive care unit. Infection 22(Suppl 2):S135–S139

Lomovskaya O, Lewis K (1992) *emr*, an *Escherichia coli* locus for multidrug resistance. Proc Natl Acad Sci USA 89:8938–8942

Lucain C, Regamey P, Bellido P, Pechere J (1989) Resistance emerging after after pefloxacin therapy of experimental *Enterobacter cloacae* peritonitis. Antimicrob Agents Chemother 33:937–943

Luttinger AL, Springer AL, Schmid MB (1991) A cluster of genes that affects nucleoid segregation in *Salmonella typhimurium*. New Biol 3:687–397

Ma D, Cook DN, Hearst JE, Nikaido H (1994) Efflux pumps and drug resistance in gram-negative bacteria. Trends Microbiol 2:489–493

Ma D, Cook DN, Alberti M, Pon NG, Nikaido H, Hearst JE (1995) Genes *acrA* and *acrB* encode a stress-induced efflux system of *Escherichia coli*. Mol Microbiol 16:45–55

Marshall AJH, Piddock LJV (1994) Interaction of divalent cations, quinolones and bacteria. J Antimicrob Chemother 34:465–483

Masecar BL, Robillard NJ (1991) Spontaneous quinolone resistance in *Serratia marcescens* due to a mutation in *gyrA*. Antimicrob Agents Chemother 35:898–902

Masuda N, Ohya S (1992) Cross-resistance to meropenem, cephems, and quinolones in *Pseudomonas aeruginosa*. Antimicrob Agents Chemother 36:1847–1851

Masuda N, Gotoh N, Satoshi O, Nishino T (1996) Quantitative correlation between susceptibility and OprJ production in NfxB mutants of *Pseudomonas aeruginosa*. Antimicrob Agents Chemother 40:909–913

Miller PF, Gambino LF, Sulavik MC, Gracheck SJ (1994) Genetic relationship between *soxRS* and *mar* loci in promoting multiple antibiotic resistance in *Escherichia coli*. J Bacteriol 38:1773–1779

Moore RA, Beckthold B, Wong S, Kureishi A, Bryan LE (1995) Nucleotide sequence of the *gyrA* gene and characterization of ciprofloxacin-resistant mutants of *Helicobacter pylori*. Antimicrob Agents Chemother 39:107–111

Mortimer PG, Piddock LJV (1993) The accumulation of five antibacterial agents in porin-deficient mutants of *Escherichia coli*. J Antimicrob Chemother 32:195–213

Mouton JW, Den-Hollander J, Horrevorts AM (1995) Emergence of antibiotic resistance amongst *Pseudomonas aeruginosa* isolates from patients with cystic fibrosis. J Antimicrob Chemother 31:919–926

Nagarajan R (1991) Antibacterial activities and modes of action of vancomycin and related glycopeptides. Antimicrob Agents Chemother 35:605–609

Nakamura S, Nakamura M, Kojima T, Yoshida H (1989) *gyrA* and *gyrB* mutations in quinolone-resistant strains of *Escherichia coli*. Antimicrob Agents Chemother 33:254–255

Nakanishi N, Yoshida S, Wakebe H, Inoue M, Yamaguchi T, Mitsuhashi S (1991a) Mechanisms of clinical resistance to fluoroquinolones in *Staphylococcus aureus*. Antimicrob Agents Chemother 35:2562–2567

Nakanishi N, Yoshida S, Wakebe H, Inoue M, Mitsuhashi S (1991b) Mechanisms of clinical resistance to fluoroquinolones in *Enterococcus faecalis*. Antimicrob Agents Chemother 35:1053–1059

Neyfakh AA (1992) The multidrug efflux transporter of *Bacillus subtilis* is a structural and functional homolog of the *Staphylococcus* NorA protein. Antimicrob Agents Chemother 36:484–485

Neyfakh AA, Bidnenko VE, Chen LB (1991) Efflux-mediated multidrug resistance in *Bacillus subtilis*: similarities and dissimilarities with the mammaliam system. Proc Natl Acad Sci USA 88:4781–4785

Ng EYW, Trucksis M, Hooper DC (1994) Quinolone resistance mediated by *norA*: physiologic characterization and relationship to *flqB*, a quinolone resistance locus on the *Staphylococcus aureus* chromosome. Antimicrob Agents Chemother 38:1345–1355

Oethinger M, Conrad S, Kaifel K, Cometta A, Bille J, Klotz G, Glauser MP, Marre R, Kern WV (1996) Molecular epidemiology of fluoroquinolone-resistant *Escherichia coli* blood-stream isolates from patients admitted to European cancer centres. Antimicrob Agents Chemother 40:387–392

Ohshita Y, Hiramatsu K, Yokota T (1990) A point mutation in *norA* is responsible for quinolone resistance in *Staphylococcus aureus*. Biochem Biophys Res Commun 172:1028–1034

Okazaki T, Hirai K (1992) Cloning and nucleotide sequence of the *Pseudomonas aeruginosa nfxB* gene, conferring resistance to new quinolones. FEMS Microbiol Lett 97:197–202

Okusu H, Ma D, Nikaido H (1996) AcrAB efflux pump plays a major role in the antibiotic resistance phenotype of *Escherichia coli* multiple-antibiotic-resistance (Mar) mutants. J Bacteriol 178:306–308

Oppegaard H, Sorum H (1994) *gyrA* mutations in quinolone-resistant isolates of the fish pathogen *Aeromonas salmonicida*. Antimicrob Agents Chemother 38:2460–2464

Oram M, Fisher LM (1991) 4-Quinolone resistance mutations in the DNA gyrase of *Escherichia coli* clinical isolates identified by using the polymerase chain reaction. Antimicrob Agents Chemother 35:387–389

Ouabdesselam S, Hooper DC, Tankovic J, Soussy CJ (1995) Detection of *gyrA* and *gyrB* mutations in quinolone resistant clinical isolates of *Escherichia coli* by SSCP analysis and determination of levels of resistance conferred by two different single *gyrA* mutations. Antimicrob Agents Chemother 39:1667–1670

Peebles CL, Higgins NP, Krenzer KN, Morrison A, Brown PO, Sugino A, Cozzarelli NR (1978) Structure and activities of *Escherichia coli* DNA gyrase. Cold Spring Harb Symp Quant Biol 43:41–52

Peng H, Marians KJ (1993) *Escherichia coli* topoisomerase IV. Purification, characterization, subunit structure, and subunit interactions. J Biol Chem 268:24481–24490

Piddock LJV (1991) Mechanism of quinolone uptake into bacterial cells. J Antimicrob Chemother 27:399–403

Piddock LJV, Walters RN (1992) Bactericial activities of five quinolones for *Escherichia coli* strains with mutations in genes encoding the SOS response or cell division. Antimicrob Agents Chemother 36:819–825

Piddock LJV, Zhu M (1991) Mechanism of action of sparfloxacin against and mechanism of resistance in Gram-negative and Gram-positive bacteria. Antimicrob Agents Chemother 35:2423–2427

Piddock LJV, Walters RN, Diver JM (1990) Correlation of quinolone MIC and inhibition of DNA, RNA, and protein synthesis and induction of the SOS response in *Escherichia coli*. Antimicrob Agents Chemother 34:2331–2336

Piddock LJV, Hall MC, Walters RN (1991) Phenotypic characterization of quinolone-resistant mutants of Enterobacteriaceae selected from wild type, *gyrA* type and multiply-resistant (*marA*) type strains. J Antimicrob Chemother 28:185–198

Piddock LJV, Hall MC, Bellido F, Bains M, Hancock RE (1992) A pleiotropic, posttherapy, enoxacin-resistant mutant of *Pseudomonas aeruginosa*. Antimicrob Agents Chemother 36:1057–1061

Piddock LJV, Griggs DJ, Hall MC, Jin YF (1993a) Ciprofloxacin resistance in clinical isolates of *Salmonella typhimurium* obtained from two patients. Antimicrob Agents Chemother 37:662–666

Piddock LJV, Panchal S, Norte V (1993b) Comparison of the mechanism of action and resistance of two new fluoroquinolones, rufloxacin and MF961 with those of ofloxacin and fleroxacin in Gram-negative and Gram-positive bacteria. J Antimicrob Chemother 31:855–863

Poole K, Krebes K, McNally C, Neshat S (1993) Multiple antibiotic resistance in *Pseudomonas aeruginosa*: evidence for involvement of an efflux operon. J Bacteriol 175:7363–7372

Poole K, Tetro K, Zhao Q, Neshat S, Heinrichs DE, Bianco N (1996) Expression of the multidrug resistance operon *mexA-mexB-oprM* in *Pseudomonas aeruginosa*: *mexR* encodes a regulator of operon expression. Antimicrob Agents Chemother 40:2021–2028

Pumbwe L, Everett MJ, Hancock REW, Piddock LJV (1996) Role of *gyrA* mutation and loss of OprF in the multiple antibiotic resistance (MAR) phenotype of *Pseudomonas aeruginosa* G49. FEMS Microbiol Lett 143:25–28

Rahman M, Mauf G, Levy J, Couturier M, Pulverer G, Glasdorff N, Butzler JP (1994) Detection of 4-quinolone resistance mutations in the *gyrA* gene of *Shigella dysenteriae* Type 1 by PCR. Antimicrob Agents Chemother 38:2488–2491

Ratcliffe NT, Smith JT (1985) Norfloxacin has a bactericidal mechanism of action unrelated to that of other 4-quinolones. J Pharm Pharmacol 37:92P

Reece RJ, Maxwell A (1991) DNA gyrase – structure and function. Crit Rev Biochem Mol Biol 26:335–375

Reina J, Borrell N, Serra S (1992) Emergence of resistance to erythromycin and fluoroquinolones in thermotolerant *Campylobacter* strains isolated from feces. Eur J Clin Microbiol Infect Dis 11:1163–1166

Rella M, Haas D (1982) Resistance of *Pseudomonas aeruginosa* PAO to nalidixic acid and low levels of β-lactam antibiotics: mapping of chromosomal genes. Antimicrob Agents Chemother 22:242–249

Revel V, Cambau E, Jarlier V, Sougakoff W (1994) Characterization of mutations in *Mycobacterium smegmatis* involved in resistance to fluoroquinolones. Antimicrob Agents Chemother 38:1991–1996

Reyna F, Huesca M, Gonzalez V, Fuchs LY (1995) *Salmonella typhimurium gyrA* mutations associated with fluoroquinolone resistance. Antimicrob Agents Chemother 39:1621–1623

Rice RJ, Knapp JS (1994) Antimicrobial susceptibilities of *Neisseria gonorrhoeae* strains representing five distinct resistance phenotypes. Antimicrob Agents Chemother 38:155–158

Richard P, Delangle M, Raffi F, Richet H (1994) Impact of fluoroquinolone administration on the emergence of fluoroquinolone-resistant organisms from the GI flora. Proceedings of the 34th Interscience Conference on Antimicrobial Agents and Chemotherapy C40

Robillard NJ (1990) Broad-host-range gyrase A gene probe. Antimicrob Agents Chemother 34:1889–1894

Robillard NJ, Scarpa AL (1988) Genetic and physiological characterization of ciprofloxacin resistance in *Pseudomonas aeruginosa* PAO. Antimicrob Agents Chemother 32:535–539

Rosner JL (1985) Nonheritable resistance to chloramphenicol and other antibiotics induced by salicylates and other chemotactic repellants in *Escherichia coli* K-12. Proc Natl Acad Sci USA 82:8771–8774

Rosner JL, Slonczewski JL (1994) Dual regulation of *inhA* by the multiple antibiotic resistance (Mar) and superoxide (SoxRS) stress response systems of *Escherichia coli*. J Bacteriol 176:6262–6269

Sato K, Inoue Y, Fujii T, Aoyama H, Inoue M, Mitsuhashi S (1994) Purification and properties of DNA gyrase from a fluoroquinolone resistant strain of *Escherichia coli*. Antimicrob Agents Chemother 30:777–780

Schaberg DR, Dillon WI, Terpenning MS, Robinson KA, Bradley SF, Kauffman CA (1992) Increasing resistance of enterococci to ciprofloxacin. Antimicrob Agents Chemother 36:2533–2535

Scherrer R, Moyed HS (1988) Conditional impairment of cell division and altered lethality in *hipA* mutants of *Escherichia coli* K-12. J Bacteriol 170:3321–3326

Scott NW, Harwood CR (1980) Studies on the influence of the cyclic AMP system on major outer membrane proteins of *Escherichia coli*. FEMS Microbiol Lett 9:95–98

Smith JT (1984) Mutational resistance to 4 quinolone antibacterial agents. Eur J Clin Microbiol Infect Dis 3:347–350

Soussy CJ, Truong QC, Ouabdesselam S, Hooper DC, Moreau NJ (1994) Sequential mutations of *gyrA* in *Escherichia coli* associated with quinolone therapy. Proceedings of the 34th Interscience Conference on Antimicrobial Agents and Chemotherapy C82

Soussy CJ, Ng NY, Oubdesselam S, Tankovic J, Hooper DC (1996) Contribution of mutation in the *parE* gene to quinolone resistance in *Escherichia coli* harbouring a mutation in the *gyrA* gene. Proceedings of the 96th ASM General Meeting A-112

Sreedharan S, Oram M, Jenson B, Peterson LR, Fisher LM (1990) DNA gyrase *gyrA* mutations in ciprofloxacin-resistant strains of *Staphylococcus aureus*: close similarity with quinolone resistance mutations in *Escherichia coli*. J Bacteriol 172:7260–7262

Sreedharan S, Peterson LR, Fisher LM (1991) Ciprofloxacin resistance in coagulase-positive and -negative staphylococci: role of mutations at serine 84 in the DNA gyrase A protein of *Staphylococcus aureus* and *Staphylococcus epidermidis*. Antimicrob Agents Chemother 35:2151–2154

Stein DC, Danaher RJ, Cook TM (1991) Characterization of a *gyrB* mutation responsible for low-level nalidixic acid resistance in *Neisseria gonorrhoeae*. Antimicrob Agents Chemother 35:622–626

Sugino A, Peebles CL, Kreuzer KN, Cozzarelli NR (1977) Mechanism of action of nalidixic acid: purification of *Escherichia coli nalA* gene product and its relationship to DNA gyrase and a novel nicking–closing enzyme. Proc Natl Acad Sci USA 74:4767–4771

Sulavik MC, Gambino LF, Miller PF (1994) Analysis of the genetic requirements for inducible multiple-antibiotic resistance associated with the *mar* locus in *Escherichia coli*. J Bacteriol 176:7754–7756

Takiff HE, Salazar L, Ruerrero C, Philipp W, Huang WM, Kreiswirth B, Cole ST, Jacobs WR JR, Telenti A (1994) Cloning and nucleotide sequence of *Mycobacterium tuberculosis gyrA* and *gryB* genes and detection of quinolone resistance mutations. Antimicrob Agents Chemother 38:773–780

Tankovic T, Courvalin P, Duval J (1994) Construction of a *gyrA* plasmid for genetic characterization of fluoroquinolone-resistant *Staphylococcus aureus*. Proceedings of the 34th Interscience Conference on Antimicrobial Agents and Chemotherapy C86

Tankovic J, Perichon B, Courvalin P (1996) Contribution of *gyrA* and *parC* mutations to fluoroquinolone resistance of in vivo and in vitro mutants of *Streptococcus pneumoniae*. Proceedings of the 96th ASM General Meeting A-114

Thornsberry C (1994) Susceptibilty of clinical isolates to ciprofloxacin in the United States. Infection 22(Suppl 2):S80–S89

Tokue Y, Sugano K, Saito D, Noda T, Ohkura H, Shimosato Y, Sekiya T (1994) Detection of novel mutations in the *gyrA* gene of *Staphylococcus aureus* by

nonradioisotopic single-strand conformation polymorphism analysis and direct DNA sequencing. Antimicrob Agents Chemother 38:428–431
Trucksis M, Wolfson J, Hooper DC (1991) A novel locus conferring fluoroquinolones resistance in *Staphylococcus aureus*. J Bacteriol 173:5854–5860
Turnidge J (1994) Epidemiology of quinolones resistance: Southern hemisphere. Drugs 49[Suppl 2]
Ubukata K, Itoh-Yamashita N, Konno M (1989) Cloning and expression of the *norA* gene for fluoroquinolone resistance in *Staphylococcus aureus*. Antimicrob Agents Chemother 33:1535–1539
Valisena S, Palumbo M, Parolin C, Palu G, Meloni GA (1990) Relevance of ionic effects on norfloxacin uptake by *Escherichia coli*. Biochem Pharm 40:431–436
Vila J, Ruiz J, Marco F, Barcelo A, Goni P, Giralt E, Jimenez De Anta T (1994) Association between double mutation in *gyrA* gene of ciprofloxacin-resistant clinical isolates of *Escherichia coli* and MICs. Antimicrob Agents Chemother 38:2477–2479
Vila J, Ruiz J, Goni P, Marcos A, Deanta TJ (1995) Mutation in the *gyrA* gene of quinolone-resistant clinical isolates of *Acinetobacter baumanii*. Antimicrob Agents Chemother 39:1201–1203
Vila J, Ruiz J, Goni P, Marcos A, Deanta TJ (1996) Detection of mutations in parC in quinolone-resistant clinical isolates of *Escherichia coli*. Antimicrob Agents Chemother 40:491–493
Vincent S, Glauner B, Gutmann L (1991) Lytic effect of two fluoroquinolones, ofloxacin and pefloxacin, on *Escherichia coli* W7 and its consequences on peptidoglycan composition. Antimicrob Agents Chemother 35:1381–1385
Wang Y, Huang WM, Taylor DE (1993) Cloning and nucleotide sequence of the *Campylobacter jejuni gyrA* gene and characterization of quinolone resistance mutations. Antimicrob Agents Chemother 37:457–463
Witte W, Braulke D, Heuck D, Cuny C (1994) Analysis of nosocomial outbreaks with multiply and methicillin-resistant *Staphylococcus aureus* (MRSA) in Germany. implications for hospital patients. Infection 22(Suppl 2):S128–S134
Wolfson JS, Hooper DC, Mchugh GL, Bozza MA, Swartz MN (1990) Mutants of *Escherichia coli* K-12 exhibiting reduced killing by both quinolone and β-lactam antimicrobial agents. Antimicrob Agents Chemother 34:1938–1943
Yamagishi J, Furutani Y, Inoue S, Ohue T, Nakamura S, Shimizu M (1981) New nalidixic acid resistance mutations related to deoxyribonucleic acid gyrase activity. J Bacteriol 148:450–458
Yamagishi J, Yoshida H, Yamyoshi M, Nakamura S (1986) Nalidixic acid-resistant mutations of the *gyrB* gene of *Escherichia coli*. Mol Gen Genet 204:367–373
Yonezawa M, Takahata M, Matsubara N, Watanbe Y, Narita H (1995a) DNA gyrase *gyrA* mutations in quinolone-resistant clinical isolates of *Pseudomonas aeruginosa*. Antimicrob Agents Chemother 39:1970–1972
Yonezawa M, Takahata M, Banazawa N, Matsubara N, Watanbe Y, Narita H (1995b) Analysis of the NH2-terminal 83rd amino-acid of *Escherichia coli* GyrA in quinolone-resistance. Microbiol Immunol 39:243–247
Yonezawa M, Takahata M, Banazawa N, Matsubara N, Watanbe Y, Narita H (1995c) Analysis of the NH2-terminal 87th amino-acid of *Escherichia coli* GyrA in quinolone-resistance. Microbiol Immunol 39:517–520
Yoshida H, Kojima T, Yamagishi J, Nakamura S (1988) Quinolone-resistant mutations of the *gyrA* gene of *Escherichia coli*. Mol Gen Genet 211:1–7
Yoshida H, Bogaki M, Nakamura M, Nakamura S (1990a) Quinolone resistance-determining region in the DNA gyrase *gyrA* gene of *Escherichia coli*. Antimicrob Agents Chemother 34:1271–1272
Yoshida H, Nakamura M, Bogaki M, Nakamura S (1990b) Proportion of DNA gyrase mutants among quinolone-resistant strains of *Pseudomonas aeruginosa*. Antimicrob Agents Chemother 34:1273–1275

Yoshida H, Bogaki M, Nakamura S, Ubukata K, Konno M (1990c) Nucleotide sequence and characterization of the *Staphylococcus aureus norA* gene, which confers resistance to quinolones. J Bacteriol 172:6942–6949

Yoshida H, Bogaki M, Nakamura M, Yamanaka LM, Nakamura S (1991) Quinolone resistance-determining region in the DNA gyrase *gyrB* gene of *Escherichia coli*. Antimicrob Agents Chemother 35:1647–1650

Zhanel GG, Karlowsky JA, Saunders MH, Davidson RJ, Hoban DJ, Hancock REW, McLean I, Nicolle LE (1995) Development of multiple antibiotic resistance (MAR) mutants of *Pseudomonas aeruginosa* after serial exposure to fluoroquinolones. Antimicrob Agents Chemother 39:489–495

CHAPTER 10

Toxicology and Safety Pharmacology of Quinolones

E. VON KEUTZ and W. CHRIST

A. Introduction

Quinolones have proven to be valuable drugs with interesting microbiological, pharmacokinetic, and therapeutic properties. The aim of this article is to give an overview of the toxicological profile of this class of compounds. Based on current knowledge, it appears that the pattern of preclinical adverse reactions is comparable for all quinolones, although there are some differences in both the incidence and the type of reaction induced by certain compounds. The published toxicological profile for the quinolones (Table 1) includes arthropathy, Achilles tendinitis and rupture, nephropathy, effects on the central nervous system (CNS), ocular toxicity, impairment of spermatogenesis, cardiovascular effects, possible mutagenic and carcinogenic effects, phototoxicity, photocarcinogenicity, photomutagenicity, drug interactions, and metabolic and nutritional events. The aim of this paper is to review and discuss the data derived from animal toxicology studies of the effects of quinolones at these target sites.

B. Arthropathy

When administered to immature animals (e.g., rats, rabbits, dogs, marmosets), all the quinolones studied – older and newer derivatives – demonstrated arthropathogenic properties in synovial joints. These experimental findings raised concern about the use of quinolones in pediatric patients. Hence, they are contraindicated in children and adolescents in the growing phase, except in cases of specific infections or life-threatening conditions such as cystic fibrosis. Use of these drugs has also been restricted in women during pregnancy and lactation.

There appears to be clear-cut species differences with regard to the quinolone effect on cartilage (Table 2). None of any mice treated with nalidixic acid (1000 mg/kg given orally for 7 days) experienced cartilage damage, whereas rats and rabbits showed joint toxicity at 250 mg/kg and 500 mg/kg, respectively. Of the animal species tested, dogs were by far the most sensitive species, with 100% of juvenile beagles exhibiting cartilage damage at a dose of 100 mg nalidixic acid per kilogram body weight per day (CHRIST and LEHNERT 1990). In addition, also juvenile nonhuman primates, i.e., marmosets, are

Table 1. Toxicological profile of quinolones

Arthropathy
Achilles tendinitis and rupture
Nephropathy
CNS effects
Ocular toxicity
Impairment of spermatogenesis
Cardiovascular effects
Possible mutagenic and carcinogenic effects
Phototoxicity
Photocarcinogenicity, photomutagenicity
Drug interactions
Metabolic and nutritional adverse events

Table 2. Incidence of arthropathies caused by nalidixic acid in various species (CHRIST and LEHNERT 1990)

Species	Dose (mg/kg per day, orally for 7 days)				
	50	100	250	500	1000
Mouse	ND	0/10	0/10	0/10	0/10
Rat	ND	ND	2/8	4/8	7/8
Rabbit	ND	0/4	0/4	3/4	4/4
Dog	0/2	4/4	4/4	4/8	ND

ND, not determined.

susceptible to quinolone-induced arthropathy, as has been shown for ofloxacin at oral doses of 200 mg/kg twice daily for 5 days (STAHLMANN et al. 1990).

The younger the animal, the worse the lesions and the shorter the interval between drug administration and lesion development. It is, however, noteworthy that individual quinolones may also cause arthropathy in adult dogs as has been demonstrated for pefloxacin at an oral dose of 140 mg per kilogram body weight per day for 12 months (RIBARD and KAHN 1991). For most of the quinolones, analogous studies in adult dogs are missing.

The arthropathy evolves within days to weeks and is characterized by cartilage damage accompanied by noninflammatory joint effusion. Histopathologic lesions are characterized by the formation of blisters, fissures, and erosions in joint cartilage. Clustering of chondrocytes has been noted, which is indicative of cartilage repair. Electron microscopic studies showed that ofloxacin toxicity to cartilage results in chondrocyte necrosis followed by dissolution of the matrix throughout the whole articular cartilage and a loss of proteoglycans. Since the cartilage consists of a functional unit of different types of collagens, proteoglycans, and glycoproteins, the loss of any of these constituents may lead to a disturbance of the whole system. In this way the collagen architecture which is stabilized by proteoglycans, largely breaks down at the joint surface. Consequently, dehiscenses may occur under static stress, and, in the extreme case, large blisters may develop.

All of these quinolone-induced cartilage lesions are usually irreversible and thus may promote degeneration leading to arthropathia deformans. If clinically symptomatic this arthropathy manifests in animals as acute arthritis, as evidenced by limping and swelling. On the basis of early investigations with pipemidic acid, it was assumed that joint damage would involve only those joints subjected to static stress during therapy. Subsequent studies, however, have shown that lesions can also develop in immobilized joints, although to a less severe degree.

Most articles published so far on quinolone-induced arthropathy describe effects induced by the older quinolones, e.g., pipemidic acid or nalidixic acid. INGHAM et al. (1977) first noticed lameness in immature dogs after daily administration of nalidixic, oxolinic and pipemidic acid. They described lesions in the joints of the animals noticed at autopsy after 6 months administration, in spite of the fact that the animals had been clinically normal except during the first month of drug administration. Microscopically, hypertrophy of the chondrocytes with large cell nests and multinucleate cells was observed. GOUGH et al. (1979) confirmed these results in another study with oxolinic and pipemidic acid in juvenile dogs. Both compounds induced gross alterations of the articular cartilage in major synovial joints.

Japanese authors (TATSUMI et al. 1978) reported similar effects with pipemidic acid at lower doses: daily oral administration of two doses of 50 mg pipemidic acid per kilogram body weight caused lameness in juvenile beagle dogs which was most pronounced 3–7 days after the start of treatment. After administration of two doses of 15 mg/kg per day cartilage erosions were still detectable but without clinical symptoms. No effect was seen after two doses of 10 mg pipemidic acid per kilogram body weight per day. At autopsy performed 30 days after a treatment period of 1 week (twice 500 mg/kg per day), blisters were still visible in the cartilage, and 4 years after treatment the investigators detected scars in the joint cartilage where blisters and erosions occurred under treatment. Mechanical pressure was shown to enhance pipemidic-acid-induced cartilage damage in dogs.

With the newer quinolones, i.e., ofloxacin and ciprofloxacin, the typical joint lesions are dose dependent, but occur at different doses in different species. For example, damage to the weight-bearing joints occurred in 3- to 4-month-old dogs treated orally for 7–28 days with 20 mg ofloxacin per kilogram body weight per day or 30 mg of ciprofloxacin per kilogram body weight per day. In 4- to 5-week-old rats, however, the doses required for the production of similar effects were 100–300 and 500 mg/kg per day, respectively, for 7–10 days (Table 3) (CHRIST and LEHNERT 1990). Fleroxacin was given orally to male and female juvenile dogs in doses of 8, 10, 20, 30, and 50 mg/kg per day for a period of 2–13 weeks. In the 20 and 50 mg/kg per day groups, arthropathy was noted soon after the start of treatment (CHRIST and LEHNERT 1990).

Table 3. Incidence of quinolone-induced articular cartilage degeneration in juvenile rats (Christ and Lehnert 1990)

Daily dose (mg/kg)	Ciprofloxacin		Norfloxacin		Nalidixic acid		Ofloxacin	
	Male	Female	Male	Female	Male	Female	Male	Female
100	0/10	0/10	5/10	0/10	5/10	2/10	1/10	0/10
250	0/10	0/10	6/10	2/10	3/9	3/10	1/10	0/10
500	1/10	0/10	5/10	4/10	7/10	1/10	0/10	0/10

The apparent difference in toxicity for animal cartilage between quinolone compounds might be attributed, at least in part, to species-specific pharmacokinetic properties of the drugs. Stahlmann et al. (1990) suggested that the high bioavailability of ofloxacin may explain why this quinolone in comparison with other quinolones can induce cartilage alterations even at comparably low dosages. Importantly, in the two animal species investigated (rats and marmosets) the peak plasma concentrations leading to cartilage alterations were approximately ten times higher than those reached under therapeutic conditions in man.

Although not studied intensively so far, cartilage tissue seems to be a deep compartment for quinolones. For example, relatively high concentrations of ciprofloxacin have been measured in skeletal muscle, cartilage, liver, and kidney of rats treated with radioactively labelled material (Siefert et al. 1986a). Since it is known that ciprofloxacin is not metabolized to a great extent in the rat (Siefert et al. 1986b), it can be assumed that the original compound achieves high concentrations in cartilage. The unusually high volume of distribution determined in man may be explained by high concentrations reached in different tissues such as bronchial mucosa and cartilage. Similar results have been published with ofloxacin. During a 3-week treatment period, on the 3rd, 7th, and 14th day of treatment the concentrations in tracheal tissue were significantly higher than in all other organs except for liver and kidney. A high volume of distribution has also been shown for ofloxacin and other fluoroquinolones (e.g., enoxacin, lomefloxacin, pefloxacin, temafloxacin, norfloxacin, fleroxacin; Cullmann et al. 1993).

In an attempt to investigate the direct effect of quinolone derivatives on cultured chondrocytes an in vitro model has been established by Hildebrand et al. (1993). In this model, cell viability, mitochondrial dehydrogenase activity, and proteoglycan content was determined using chondrocytes from the articular cartilage of juvenile dogs. The results obtained with different quinolone derivatives indicate that one major reason for quinolone-induced arthropathy is inhibition of proteoglycan synthesis in chondrocytes. This model can be used to identify strongly arthropathogenic quinolones without the need of performing animal studies.

The pathogenesis of quinolone-induced arthropathy in animals remains speculative. The primary target of all quinolones is bacterial gyrase (topoisomerase II), an essential bacterial enzyme for DNA replication and certain aspects of transcription, DNA repair, recombination, and transposition. Although it is generally accepted that today's quinolones are specific for nuclear prokaryotic enzymes, there are reports of in vitro experiments indicating some inhibition of eukaryotic DNA replication (HUSSY et al. 1986; GOOTZ et al. 1990). It has been found that in immature rats dosed with ofloxacin, chondrocytes of the intermediate zone exhibited a transient decrease and subsequent increase in uptake of tritiated thymidine (KATO and ONODERA 1988). This observation could reflect early depression of DNA synthesis by the drug followed by reparative reaction of the chondrocytes.

Some recent papers (STAHLMANN et al. 1993a,b, 1995) postulate that effects on integrin receptors on chondrocytes and in the matrix might be the primary event leading to quinolone-induced arthropathy in juvenile animals. Integrins are important cell surface receptors which regulate cell–cell as well as cell–matrix interaction. Little is known about chondrocytic integrins, but they have been identified in hyaline cartilage. Since magnesium is an essential requirement for the activity of some integrins it is hypothesized that quinolones damage cartilage by forming chelate complexes with magnesium, thus decreasing the level of ions available for chondrocyte–matrix interaction. The function of specific integrins in cartilage tissue is thereby impaired, and subsequently the integrity of the tissue is lost. The organ specificity of this effect may be explained by the fact that quinolones reach high concentrations in cartilage tissue and the articular cartilage is not supplied with nutritients since it contains no blood vessels and is dependent upon synovial fluid for nutrition. Therefore, disturbances in the electrolyte concentrations can not be as quickly compensated as in other tissues.

To test the hypothesis that quinolone-induced arthropathy is caused (or aggravated) by magnesium deficiency in cartilage, STAHLMANN et al. (1995) induced magnesium deficiency by feeding juvenile rats a magnesium-deficient diet for 9 days and treated the rats with single oral doses of ofloxacin (0, 100, 300, 600, or 1200 mg/kg) during this period. Additional groups of juvenile rats on a normal diet were treated in parallel with the same ofloxacin dosage regimen. Typical cartilage lesions were found in knee joints of all magnesium-deficient rats, including those without ofloxacin treatment. Lesions in these groups were not distinguishable from lesions induced by a single dose of 600 mg ofloxacin per kilogram or higher in rats on a normal diet. It was concluded that quinolone-induced arthropathy is probably caused by a deficit of available magnesium in joint cartilage due to the formation of quinolone–magnesium chelate complexes. Experiments to further prove this hypothesis are underway.

As mentioned earlier, the animal data raised concern about the use of quinolones in children. However, meanwhile there is an increasing amount of data on the efficacy and safety of quinolones in pediatric patients, mostly with serious infections with pathogens resistant to standard antibiotic therapy. By early 1992 published data on the use of fluoroquinolones in children included approximately 1000 prepubertal patients, the majority of which was suffering from cystic fibrosis. These studies – in which ciprofloxacin was preferably used – reported good to excellent efficacies and usually mild and reversible adverse reactions which are not substantially different from those in adults, with the exception of the musculoskeletal system and skin effects (CHYSKY et al. 1991).

The high proportion of pediatric patients with acute exacerbations of cystic fibrosis could be the reason for the higher incidence of arthralgia in the pediatric group. In older patients with cystic fibrosis, arthropathies are estimated to occur in 7%–8%. The majority of these joint diseases are either caused by immune mechanisms (so-called cystic fibrosis arthropathy) or by hypertrophic pulmonary osteoarthropathy. A definite diagnosis of these arthropathies is often difficult. Therefore, the relation of joint symptoms to drug therapy in patients with cystic fibrosis requires cautious interpretation (SCHAAD et al. 1991).

Early diagnosis of cartilage damage remains difficult; magnetic resonance imaging (MRI) appears to be useful. Although many studies have stressed the excellent sensitivity of MRI in detecting early and discrete inflammatory, degenerative, or traumatic lesions of articular cartilage under both clinical and experimental conditions, the most sensitive method to detect or rule out any cartilage damage is the post mortem diagnosis.

In this context, the results of comprehensive morphological studies of potential cartilage toxicity after repeated ciprofloxacin therapies in two pediatric patients with cystic fibrosis which were reported by SCHAAD and WEDGWOOD (1992) are most encouraging. Both patients had repeatedly received oral courses of ciprofloxacin (30 mg/kg per day in two equal doses for 2–12 weeks) amounting to a total of 9–10 months of quinolone treatment during their last 3 years of life. In both patients, autopsy of the left knee revealed macroscopically regular structures and microscopical and electron microscopical examinations showed normal cartilage. Although these observations are limited to individual patients, they certainly are in accordance with other studies on quinolone use in prepubertal patients in which neither clinical, nor radiological, nor magnetic resonance imaging monitoring pointed to quinolone-induced arthropathy. The therapeutic use of quinolones in children may thus be justified in cases where the pediatric benefits associated with the marked antibacterial activity of this class of drugs outweigh the possible risks associated with arthralgia, i.e. whenever the condition of the patient is serious and vital decisions for a favorable prognosis for the sick child are necessary.

C. Achilles Tendinitis and Rupture

Unilateral and bilateral Achilles tendinitis which can be complicated by rupture of the tendon (also bilateral) has been associated with the use of fluoroquinolones. Since the first cases of tendinitis and Achilles tendon rupture induced by norfloxacin and ciprofloxacin were reported (BAILEY et al.

1983; McEwan and Davey 1988), a large number of papers dealing with Achilles tendon disorders induced by these quinolones and also enoxacin, pefloxacin, and ofloxacin have appeared, particularly in France (Bailey and Peddie 1985; Burkhardt et al. 1993; Cohen and Rosenberg 1992; Franck et al. 1991; Jorgensen et al. 1991; Kahn and Carbon 1993; Perrot et al. 1991; Ribard et al. 1991, 1992; Rose et al. 1990; Tin Lee and Collins 1992).

A survey performed by Royer et al. (1994) in France from 1985 to 1992 reported 100 cases of fluoroquinolone-related tendinopathy consisting of 31 cases of Achilles tendon rupture and 69 of tendinitis. About 75% of these cases were related to treatment with pefloxacin. Fluoroquinolones can cause tendinitis and rupture of Achilles tendons even after short-term use. The average time of onset of tendon disorders from the commencement of treatment was 13 days. The time of onset ranged from 1 to 90 days; in seven cases tendon disorders appeared within 1–2 days after the start of treatment and in 14 cases after the completion of treatment. The favored site was the Achilles tendon, while several cases were found in the tendons of the long head of biceps and the long extensor muscle of thumb. Recovery usually occurred within 30 to 60 days after the disorders occurred, except in 13 patients in whom recovery took more than 2 months. The risk of tendon rupture appeared to be higher in males, in persons over the age of 60, and in those playing a great deal of sport.

All patients had received quinolones for infection during the course of treatment with a large variety of drugs for their underlying condition. Due to this situation, drugs concurrently administered with quinolones have been suspected as contributing to the disorder. Seventeen of the 100 patients mentioned above were receiving concomitant steroid therapy. Since long-term corticotherapy is known to predispose to tendon rupture, some authors have focused on the contribution of steroid therapy to the Achilles tendon disorders induced by quinolones. However, one has to acknowledge that even if partly caused by simultaneous long-term therapy with corticoids, these effects evidently occurred in cases in which corticoid treatment was given over a longer period of time without such adverse drug reactions. Only with additional fluoroquinolone therapy did ruptures of Achilles tendons occur.

In an attempt to examine the effects of quinolones on the Achilles tendon in an experimental animal model, Kato et al. (1995) treated juvenile and adult Sprague-Dawley rats with pefloxacin and ofloxacin alone and in combination with cyclosporin A and hydrocortisone. The quinolones were given orally at doses of 300 mg/kg and 900 mg/kg once or as a single daily dose for 2 weeks. In the acute experiment, pefloxacin (300 mg/kg and 900 mg/kg) or ofloxacin (900 mg/kg) induced edema with mononuclear cell infiltration mainly in the inner sheath of the inner Achilles tendon just proximal to the tuber calcanei in rats killed on the next day. Cell infiltration was also seen in the adjacent synovial membrane and joint space. With progression of severity, the lesions extended to the surface tendon tissue, wherein irregularly arranged collagen bundles were detached from each other and nuclei of fibroblasts were pyknotic and fragmented. After a 2-week repeated administration, these lesions were replaced by fibrotic foci with regenerated tendon fibroblasts; the incidence and severity were reduced in the ofloxacin but not the pefloxacin group. Coadministration of cyclosporin A with ofloxacin (300 mg/kg) induced these lesions despite the fact that neither induced lesions alone. The tendon lesions were induced in juvenile rats (4 weeks old) but not in young adults (12 weeks old). The articular cartilage of juvenile rats showed focal degeneration and/or cavitation in the tarsal joints after a single and 2-week administration of pefloxacin or ofloxacin. Hydrocortisone slightly increased the incidence of ofloxacin-induced lesions in both the tendon and cartilage after a 2-week administration. The authors conclude that the occurrence of the tendon lesions is different from that of the Achilles tendon disorders reported in older humans, but they are thought to be a useful model for them.

Kempka et al. (1996) established an in vitro model to investigate potential cytotoxic effects of fluoroquinolones (ciprofloxacin, pefloxacin, sparfloxacin) and triamcinolone acetonide on cultivated tendon cells from humans, dogs, mini-pigs, rats, and marmosets at different ages by measuring mitochondrial dehydrogenase activity (MTT assay), viability, proteoglycan content and proliferation. The following results were obtained: (1) No marked differences were found between various species after fluoroquinolone treatment; (2) No age-dependent effects of fluoroquinolones and triamcinolone acetonide were found in cultivated human tendon cells from patients of different ages; (3) The simultaneous administration of fluoroquinolones and triamcinolone acetonide resulted in a stronger decrease of cell viability than these substances alone. The authors conclude that the combined use of fluoroquinolones and glucocorticoids may favor the occurrence of tendon ruptures.

The pathogenesis of quinolone-induced tendon lesions remains speculative. Christ and Esch (1994) presented the hypothesis that the effect on topoisomerase II, the enzyme corresponding to the bacterial gyrase, may play a causal role in tendinopathies. They assumed that quinolones may inhibit DNA of mitochondria of eukaryotic cells because of its structural similarity to bacterial DNA. Tendons as part of the firm connective tissue have only a low metabolic activity, so that a reduced energy supply can be expected if the few mitochondria are impaired. This could cause degenerative signs that manifest themselves in tendinopathies and even ruptures. The association between tendon rupture and fluoroquinolones has prompted a revision of the labeling stating that fluoroquinolones are contraindicated in persons with a history of tendinitis or tendon rupture associated with the use of any member of the quinolone group of antimicrobial agents.

D. Nephropathy

There is no evidence that quinolones are primarily nephrotoxic. Effects on renal function which have been described for some derivatives of the class are related to foreign body reactions caused by crystals. Such changes include mild interstitial nephritis, crystalluria, occult blood in urine, decreased renal function, and increased kidney weight.

The doses needed to produce nephropathological changes vary greatly with the specific quinolone and the animal species involved. In rats, the necessary doses range from 20–30 mg/kg intravenously to 500–1000 mg/kg orally. In dogs, oral doses as low as 50–100 mg/kg can cause nephropathies (CHRIST and LEHNERT 1990).

Two different studies with ciprofloxacin were performed in rhesus monkeys. A 22-day study utilized intravenous doses of 40 and 80 mg/kg per day. A 28-day study utilized a maximum dose of 18 mg/kg per day. All of these doses were associated with histological evidence of ciprofloxacin crystallization and a subsequent slight to moderate inflammatory response in the distal tubule or the renal medulla and cortex, which in some cases penetrated into the collecting ducts of the medulla. White blood cells and multinucleated giant cells were also observed in these regions. Animals that received 40 mg/kg per day and 80 mg/kg per day and then allowed a 4-week recovery period had normalization of their previously elevated blood urea nitrogen (BUN) and creatinine values. Histologically, these animals showed slight intertubular inflammation and few phagocytes, indicating that recovery from the insult was occurring. Intravenous doses of up to 10 mg/kg per day were not associated with renal toxicity (SCHLÜTER 1987a,b).

Beagle dogs given up to 80 mg ciprofloxacin per kilogram body weight per day orally for 4 weeks suffered no renal abnormalities, and only one of four weanling pigs given 50 mg/kg per day orally for 16 days had elevated BUN and casts in the urine (SCHLÜTER 1987a).

Renal tolerance of quinolones is linked to the pH-dependent solubility of the compounds. Ciprofloxacin is only very slightly soluble at neutral pH. Under alkaline pH conditions up to 9.0, the solubility continues to be slight, but it increases considerably once the pH becomes even slightly acidic. A similar solubility behavior has also been described for norfloxacin, and this compound also causes nephropathological alterations. The kidney damage is thought to develop as a result of the precipitation of acicular shaped crystals with an inflammatory foreign body reaction as a secondary event. Extensive investigations with ciprofloxacin have shown that the observed crystal-like structure represents a complex of ciprofloxacin and/or its metabolites, magnesium, and protein. Magnesium is always found in crystals observed under such conditions (SCHLÜTER 1987a).

Since the solubility of quinolones is pH dependent it is not surprising that crystalluria occurs primarily in animal species with urinary pH in the alkaline range, conditions under which quinolones are least soluble. The rat and the monkey – unlike man – are predisposed to such precipitation. This assumption was proven in a rat study with intravenously administered ciprofloxacin; the acidification of urine via the drinking water resulted in a distinctly reduced amount of acicular crystals excreted with the urine (SCHLÜTER 1987a).

To investigate the question of whether the infusion speed of parenterally administered quinolones may influence the occurrence of intratubular crystallization, a study was performed in rhesus monkeys. Ciprofloxacin (20 mg/kg) was administered either by intravenous injection or by intravenous infusion over a period of 1 h. Tubular damage was only found after bolus injection of 20 mg/kg but not after administration of the same dose in a 60-min infusion (SCHLÜTER 1987a).

In summary, and with respect to the relevance of the adverse effects observed in animal species, there is no evidence that quinolones are primarily nephrotoxic. Nephropathological changes do not seem to occur in the absence of crystalluria. Crystalluria in animals appears to take place under urinary pH conditions which do not generally occur in man. The fact that drug-related crystalluria has been observed only rarely in humans during the clinical investigation of the quinolones seems to support this hypothesis. Nevertheless, patients receiving high doses of quinolones should be well hydrated, and alkalinity of the urine should be avoided.

E. Effects on Central Nervous System

Quinolones have been reported to induce CNS reactions in humans with an incidence of 0.9%–2.1%. Severe reactions such as hallucinations, depressions, nightmares, confusion, and manic reactions are mostly very rare, amounting to only 0.5% of the cases (CHRIST 1990). The CNS reactions of quinolones can be divided into psychiatric and neurological events (Table 4).

The wide range of symptoms makes it difficult to explain the possible targets in the CNS. Comparison of the four quinolones used most often in anti-infective therapy (norfloxacin, ciprofloxacin, ofloxacin, and lomefloxacin) has shown that lomefloxacin induced the CNS symptoms most often, predominantly anxiety, seizures, tremor, and dizziness, while ofloxacin provoked the greatest number of sleep disorders and acute organic psychoses (PATON and REEVES 1991; ZAUDIG et al. 1989). Ciprofloxacin and norfloxacin also induced CNS effects, but to a much smaller extent.

Table 4. Neurotoxic symptoms related to quinolones

Psychiatric events	Delirium, delusions, depersonalization, depression, psychotic depression, hallucinations, manic reaction, manic depression, paranoid reaction, psychosis, schizophrenic reaction, agitation, hostility, anxiety, nervousness, neurosis, insomnia, nightmares, somnolence
Neurologic events	Convulsive seizures, tremor, ataxia, paresthesias, neuritis, vertigo, dizziness, headache, pseudotumor cerebri

Convulsive seizures are rarely reported after quinolone therapy. They have been observed mainly in elderly patients or patients with a history of epilepsy, cerebral trauma, or alcohol abuse (CHRIST 1990). Studies with animal models have shown that the route of administration is also of importance. Investigations with systemic administration were normally done in mice; convulsions occurred very rarely after administration of quinolones. In mice, only very high quinolone concentrations led to convulsions and deaths. For flumequine the LD_{50} was 1300–2500 mg/kg body weight and for fleroxacin, 800 mg/kg. In contrast, equivalent doses of ofloxacin, enoxacin, ciprofloxacin, pefloxacin, and norfloxacin failed to trigger convulsions (CHRIST 1990). The effects on behavior and motor activity were minimal.

Seizure models in animals gave no predictive ranking for the various quinolones comparable to humans. DBA/2 mice, a strain genetically susceptible to sound-induced seizures, and normal mice were used to test the convulsive potential of quinolones in animals (DE SARRO et al. 1994). It was found that quinolones had a certain proconvulsive action in DBA/2 mice. The sequence in this epilepsy model was pefloxacin > enoxacin > ofloxacin > nalidixic acid > rufloxacin > norfloxacin > ciprofloxacin > cinoxacin > temafloxacin. This ranking can to some extent be explained by the degree of lipophilicity of these compounds. It has been found that pefloxacin and ofloxacin are more lipophilic than norfloxacin and ciprofloxacin and may thus reach higher concentrations in the central nervous system.

In other models of seizures provoked by PTZ, picrotoxin, strychnine or electric shock in male mice it was found that neither nalidixic nor oxolinic acid altered the convulsion threshold of PTZ, strychnine, or picrotoxin, but both lowered the threshold for seizures induced by electric shock (WILLIAMS and HELTON 1991). The proconvulsive action of both compounds was blocked by MK-801 and AP-5 (2-amino-4-phosphonobutyric acid), two excitatory amino acid receptor antagonists. These effects are in contrast to former investigations in the PTZ model (MORIKAWA et al. 1987).

Another route of quinolone administration used to provoke seizures is injection into the right lateral ventricle in rats (weighing 180–190 g), thereby achieving higher concentrations in the brain. Pronounced and long-lasting (40 min) clonicotonic convulsions were induced by 100 μg per ventricle in the case of enoxacin and ciprofloxacin, whereas pefloxacin induced convulsions only at 400 μg per ventricle. Single doses of 100 μg and 200 μg of pefloxacin, respectively, had no effect (CHRIST 1990).

The only available in vitro model in this context is the determination of evoked potentials, measured by extracellular recordings in hippocampus slices of the rat (DIMPFEL et al. 1991). Such investigations have identified enoxacin, nalidixic acid, and ofloxacin as the most potent convulsants, whereas ciprofloxacin, norfloxacin, and pefloxacin were less active. With the exception of pefloxacin, these results are in accordance with the results of the study in DBA/2 mice (DE SARRO et al. 1994) and also with a review of the frequency of

CNS side effects in patients in relation to the numbers of prescriptions: fleroxacin (1.86%) ≫ enoxacin (0.99%) > lomefloxacin (0.86%) > sparfloxacin (0.61%) > ofloxacin (0.56%) > ciprofloxacin (0.43%) = tosufloxacin (0.43%) > norfloxacin (0.22%; HORI and SHIMADA 1993).

In the context of the possible side effects of quinolones, the absorption and bioavailability of these drugs and their penetration into the brain are of major interest. Most quinolones appear to be readily absorbed with an absolute bioavailability of 50%–100%. In the cerebrospinal fluid (CSF) only some 2%–10% of the ciprofloxacin concentration in plasma could be found (DALHOFF 1989), and this was also true for most other quinolones. Also in the CSF, 25% of the plasma concentration was reached by ofloxacin, and about 10% by temafloxacin and sparfloxacin (PERCIVAL 1991). In a blood–brain barrier (BBB) in vitro model using bovine cerebrovascular endothelial cells, JAEHDE et al. (1993) have shown that quinolones cross the BBB by passive diffusion. There was no evidence for the operation of saturable transport mechanisms or of any significant metabolism of quinolones at the BBB.

Besides the CNS effects which arise from the direct action of the quinolones, another group of CNS events is mediated by coadministration of other drugs. The additive effect of combinations of quinolones (0.25 mmol/l) and theophylline (0.5 mmol/l) on the CNS has been investigated in vitro using hippocampal slices. Electrical recordings of these slices have shown that aminophylline produced spontaneous excitation, which in vivo might induce epileptiform discharges after reaching a certain threshold. Additive enhancement of both compounds was in fact documented, decreasing in the sequence nalidixic acid > enoxacin > ofloxacin > pefloxacin > ciprofloxacin (DIMPFEL et al. 1994).

In contrast to these in vitro results, DOMAGALA (1994) and POLK (1989) have reported a different sequence of interaction: enoxacin (very high) > ciprofloxacin (high) = norfloxacin (high) > pefloxacin (moderate) > ofloxacin (low). This confirms the pharmacokinetic influence of quinolones on the clearance of theophylline (CHRIST 1990; DAVEY 1988; NIX et al. 1987).

The adenosine receptor might be a candidate for impairment by quinolones. Adenosine receptors are abundant in the brain, especially in the hippocampus, and like the γ-aminobutyric acid (GABA) receptors have mostly an inhibitory function. Adenosine completely suppresses the transmission in the hippocampal CA3 area (ALZHEIMER et al. 1991), and it has also been found to exert anticonvulsive action after electrically or chemically induced seizures. This effect can be antagonized by theophylline (BARABAN et al. 1991).

Quinolones have been shown to inhibit the specific binding of ^{3}H-labeled *N*-ethylcarboxamidoadenosine and L-N^6-phenylisopropyladenosine, two adenosine receptor ligands (DODD et al. 1989). Although the concentration of quinolones in the cerebral fluid is high, their affinity to the adenosine receptors is low, and therefore the impact on the cerebral adenosine receptors due to concomitant treatment with theophylline and quinolones is caused by theophylline.

In November 1986 the Japanese Ministry of Health and Welfare issued a warning against the combined use of enoxacin and fenbufen or its metabolite biphenylacetic acid (BPAA). The warning was issued after several patients with no history of epilepsy or other underlying cerebrovascular dysfunction experienced clonic or acute convulsive seizures after this combination (Ministry of Health and Welfare, Japan, 1986 cited by HORI et al. 1987a). Following this warning animal studies documented that quinolones in combination with fenbufen or other nonsteroidal anti-inflammatory drugs (NSAIDs) provoked convulsions within 120 min in the following ranking: enoxacin = lomefloxacin > ciprofloxacin > norfloxacin > pefloxacin > fleroxacin. The question of whether fenbufen or its major metabolites change the penetration of fluoroquinolones across the blood/brain barrier cannot be answered definitively to date. The published and unpublished data do not give a uniform picture. This may be due to different experimental approaches.

After coadministration of NY-198 (lomefloxacin) and fenbufen (200 mg/kg of each drug; oral administration), serum protein binding and pharmacokinetic parameters in mouse serum and brain, such as the concentration of lomefloxacin and fenbufen, were not found to be influenced by this combination. In contrast, at 60 min after coadministration of ^{14}C-labeled lomefloxacin and fenbufen, the total radioactivity in the brain of mice was somewhat lower compared to control animals (YAMAMOTO et al. 1988).

When rats were pretreated with fenbufen (500 mg/kg, orally) 1 h prior to treatment with ^{14}C-labeled ciprofloxacin (10 mg/kg i.v.), concentrations of total radioactivity both in the CSF and in the brain were much higher in pretreated animals (235% in CSF and 155% in brain, respectively) than in control animals. This was only true for the 2-h time point after administration of ciprofloxacin. At earlier time points, comedication with fenbufen resulted only in slight changes in ciprofloxacin pharmacokinetics (SIEFERT et al. 1988, unpublished data, Bayer AG).

In another study, rats were given fenbufen (100 mg/kg orally) and 10 min later enoxacin (100 mg/kg orally). Animals were killed when they experienced convulsions. Enoxacin was measured in blood and brain by a sensitive HPLC method. The ratio of concentrations in brain and plasma was substantially unaffected by pretreatment with fenbufen (CHRIST 1990).

In a study by Ichikawa et al. after the intravenous administration of norfloxacin or ofloxacin (10 mg/kg) with or without fenbufen (20 mg/kg i.v.), the concentration of the fluoroquinolones in serum, CSF and in the whole brain were determined at various time points. Coadministered fenbufen raised the brain concentrations of both quinolones at 60 and 120 min after administration of the drugs. Brain and CSF concentrations of ofloxacin were higher than those of norfloxacin. Ofloxacin penetrates into the CNS more readily than norfloxacin (ICHIKAWA et al. 1992).

Immediately after coadministration of ciprofloxacin (10 mg/kg i.v., bolus injection) and fenbufen (20 mg/kg i.v.) to male Wistar rats, brain and CSF levels of ciprofloxacin were raised by about 15 to 70%, respectively. In contrast, no elevation of sparfloxacin concentrations were observed in either brain or CSF after coadministration with fenbufen (IWAMOTO et al. 1993).

Investigating the situation in closer detail, AKAHANE et al. (1989) found some structural similarities between quinolones related to position 7 of the quinolone structure. They found that clonic convulsions and deaths were induced by quinolones in the order enoxacin > norfloxacin > ciprofloxacin ≫ pipemidic acid, four compounds which have an unsubstituted piperazine moiety at position 7. The unsubstituted aminopyrrolidine moiety (AM-1091 = BAY v 3545; T 3262 = tosufloxacin) also produced convulsions and deaths, but only at higher concentrations. In contrast, ofloxacin, AT 4140 (sparfloxacin), and nalidixic acid, in which the piperazine is substituted with one or more methyl groups or in which there is no piperazine moiety at position 7, never induced convulsions. On the other hand lomefloxacin, which has a 3-methylpiperazine group, provoked convulsions whereas pyrrolidine moieties had no effect (DOMAGALA 1994).

This in vivo activity has been correlated with the $GABA_A$ receptors. Receptor-binding studies carried out with the aid of ^{3}H-labeled muscimol or ^{3}H-labeled GABA have shown that all quinolones have only a weak affinity for the $GABA_A$ receptors, usually $>1 \times 10^{-4} M$ (TSUI et al. 1988; HORI et al. 1987b). Using ^{3}H-labeled GABA, the authors established the ranking enoxacin > norfloxacin > cinoxacin > ofloxacin > pipemidic acid = lomefloxacin. However, in combination with BPAA the affinity for the receptors increased to 3.5×10^{-8} mol/l in the case of enoxacin and norfloxacin, followed by lomefloxacin (4.7×10^{-7} mol/l). Ciprofloxacin and pipemidic acid, compounds which were even weaker or not convulsive at all, have only a weak affinity for the $GABA_A$ receptors (10^{-5} to 10^{-4} mol/l). Ofloxacin showed no affinity at all. These data are more or less in accordance with the above-outlined structural similarities, where unsubstituted piperazine moieties had the strongest affinity. The ranking of these compounds varies between different authors, but according to both in vivo and in vitro studies these quinolones reached therapeutic range concentrations when they were given concomitantly with NSAIDs (AKAHANE et al. 1989; JANKNEGT 1990; DOMAGALA 1994).

Neither the pharmacokinetics nor the protein binding was changed in concomitantly dosed animals. However, muscimol, a $GABA_A$ agonist, and diazepam antagonized the convulsions, whereas baclofen, a $GABA_B$ receptor agonist, had no influence. These findings lead to the conclusion that the $GABA_A$ receptors are involved (AKAHANE et al. 1989). In vitro binding studies do not provide information on drug-induced functional consequences because both agonists and antagonists replace ^{3}H-labeled muscimol, though with opposite physiological consequences (HALLIWELL et al. 1993). For a more detailed answer whole-cell patch clamp techniques could clarify the primary functional response (GABA-evoked chloride current). Ciprofloxacin and ofloxacin are relatively weak inhibitors of GABA-evoked currents recorded from neurons, but BPAA is able to potentiate this effect (AKAIKE et al. 1989; HALLIWELL et al. 1989, 1991; SHIRASAKI et al. 1991). These authors reported the same ranking as seen in vivo and in receptor-binding studies: norfloxacin > enoxacin > ciprofloxacin > ofloxacin. The electrophysiological and radioligand assays have shown that in combination with NSAIDs quinolones may interact with the agonist recognition site of the $GABA_A$ receptor protein complex. According to this, only agonists of the benzodiazepine binding site diminish the convulsive effect, whereas antagonists such as flumazenil are probably ineffective (HALLIWELL et al. 1993; SHIRASAKI et al. 1991).

According to some authors ofloxacin given concomitantly with NSAIDs does not affect the $GABA_A$ receptors and is not convulsive in vivo (AKAHANE et al. 1989; DAVIES and MAESEN 1989; HALLIWELL et al. 1993), while others have found distinct effects (CHRIST 1990; DOMAGALA 1994; DIMPFEL et al. 1991; NOZAKI et al. 1989). These discrepancies could possibly be due to the different behavior of the two ofloxacin enantiomers. The (+)*R*-enantiomer DR-3354 shows more pronounced side effects relating to the CNS and inhibits the sprouting efficacy of human IMR-32 neuroblastoma cells to a greater extent than the (–)*S*-enantiomer DR-3355 (YOKOTA and KANDA 1989).

The proconvulsive action of quinolones used concomitantly with NSAIDs has also been documented in vitro on hippocampus slices (DIMPFEL et al. 1991). In their studies the authors used very low (0.25 mmol/l) concentrations of quinolones and fenbufen which did not induce any increases in the evoked potentials when used separately. The results did not in fact lead to a clear ranking of the quinolones tested because the differences in potential increases were not sufficiently large, but the tendency was enoxacin > nalidixic acid > pefloxacin > norfloxacin = ofloxacin ≫ ciprofloxacin. This ranking did not correspond exactly to the in vivo findings in rodents (enoxacin = lomefloxacin > norfloxacin = ciprofloxacin > pefloxacin and fleroxacin; CHRIST 1990).

It has been demonstrated in vivo that quinolones do not stimulate the cerebral dopaminergic system (Ungerstedt model; CHRIST 1990), but haloperidol is able to reduce convulsions produced by fenbufen and quinolones. CHRIST et al. (1988a) suggest that dopamine receptors may be involved in the convulsant action of quinolones in combination with opiates and NSAIDs. The cerebral concentrations of catecholamines in rats dosed with ciprofloxacin were determined by microdialysis. All catecholamines were found to be increased dose-dependently, whereas monoamine oxidase was reduced (RAASCH et al. 1992). Ciprofloxacin therefore should increase the excitatory stimuli in the CNS by increasing catecholamine concentrations.

The antinociceptive activity of tilidine, tramadol, morphine, and pentazocine tested in the hot-plate test is antagonized by enoxacin lomefloxacin and other quinolones (CHRIST 1990). On the other hand, naloxone and tramadol potentiate quinolone-induced seizures (CHRIST et al. 1988a; NOZAKI et al. 1989).

The similarities of quinolones to kynurenic acid and to quinoline acids, both endogenous ligands of the glutamate receptors, suggest a possible interaction of quinolones with ligand-gated glutamate receptors. Receptor-binding studies with ^{3}H-labeled glutamate, ^{3}H-labeled kainate, ^{3}H-labeled AMPA, and ^{3}H-labeled *N*-methyl-D-aspartate (NMDA) have shown that there is no specific affinity of quinolones for the ion- or ligand-gated glutamate receptors (Dodd et al. 1989; Halliwell et al. 1993).

Antagonistic properties with respect to the proconvulsive action of quinolones are displayed by AP-5 or AP-7, selective antagonists of the glutamate-binding site of the NMDA receptor (Nozaki et al. 1989; Williams and Helton 1991).

Like the $GABA_A$ receptors, the NMDA receptors belong to the ligand-gated receptor superfamily, and therefore possess two binding sites: one for the agonist glutamate and one which binds glycine. The glycine site is comparable to the benzodiazepine site in the $GABA_A$ receptor and cannot be inhibited by strychnine. Using electrical hippocampus slice recordings, Dimpfel et al. (1995) have found that one agonist of this glycine site, HA 966, lowered the evoked potentials induced by quinolones in this system, whereas another agonist, D-serine, increased the pop-spike amplitude in combination with selected quinolones. Although this finding must be clarified further, it would appear that quinolones are partial or full agonists of the glycine-binding site, since antagonists of this site such as 7-chlorokynurenic acid abolished the evoked potentials completely in these slices (Watanabe et al. 1992a).

According to Williams and Helton (1991), MK-801 and phenylcyclidine, two selective channel blockers of the NMDA receptors, antagonize the proconvulsive activity of quinolones in male mice. This could perhaps be connected with the absence of the Mg^{2+} block in this channel, produced by quinolones. Mg^{2+} blocks the glutamate channel during the inactive stage to individual Ca^{2+} ions. Activation of the channel by glutamate delivers Mg^{2+} ions for increasing Ca^{2+} current, and low Mg^{2+} concentrations therefore enhance the response of synaptic activation in hippocampus slices and lead to epileptic activity (Psarropoulou and Kostopoulos 1990). This theory fits perfectly with the proconvulsive action of quinolones and with the above-mentioned effect of channel blockers which replace missing Mg^{2+} ions (Dimpfel et al. 1994; Schürmann 1996).

In conclusion, quinolones exhibit distinct CNS effects in man and in animals, and many investigations have consequently been carried out to explain the mechanisms by which these effects are brought about. In addition, quinolones potentiate the action of proconvulsants such as methylxanthines, NSAIDs, or beta-lactam antibiotics, and lower the threshold of electrically or chemically induced seizures. Various possibilities have been put forward for quinolone targets or receptors, such as the NMDA and $GABA_A$ receptors. In both cases variations in their normal receptor activity could be responsible for the convulsive and other CNS effects. It may be that chelation of Mg^{2+} or of other ions is a possible cause of these observed effects, most explainable in the case of the NMDA receptors.

The potential to induce CNS events is addressed in the package insert of all quinolones on the market. It is, however, noteworthy that on September 23, 1993, the FDA's Anti-Infective Drugs Advisory Committee stated, after a review of the frequency and spectrum of CNS adverse events due to marketed fluoroquinolone drugs, that differences within the class of fluoroquinolones exist which warrant disparate labeling.

F. Ocular Toxicity

A potential target organ for quinolones in laboratory animals is the eye. There are two possibilities for ocular damage. One is related to the lens and the other is related to the melanin-containing tissues (retina, iris, ciliary body) of pigmented eyes.

Effects on the lens have been described for some of the older quinolones. Rosoxacin caused slight lenticular opacities in rats when administered orally for 2 months at a dose of 400 mg/kg per day (CHRIST and LEHNERT 1990), and the fluoroquinolone pefloxacin induced lenticular opacities in dogs given 100 mg/kg per day orally for 12 months (ANONYMOUS 1988). Due to these findings, the newer quinolones have been intensively tested by using special investigations besides routine ophthalmoscopical and histological lens examinations to exclude such effects.

For ciprofloxacin these investigations were done within the framework of a chronic parenteral toxicity study in monkeys. In this study, rhesus monkeys were treated intravenously with doses of up to 20 mg/kg. At the end of the 6-month treatment period the lenses were specifically examined for even discrete opacities using the special evaluation technique of Scheimpflug photography. No lens densifications were found in the animals treated with doses up to and including 20 mg/kg. Further investigations were performed on the lens of one eye from all of the animals involved in this study to establish the lens weight and water and protein contents, as well as electrophoretic separations of the lens proteins. These investigations also failed to give any evidence of lens alterations caused by ciprofloxacin: in particular, there was no evidence of any shift in the lens protein pattern towards insoluble protein components. To answer the question of whether the lens and the aqueous humor of the eye act as deep compartments for ciprofloxacin, pharmacokinetic investigations were performed in which ciprofloxacin plasma concentrations were compared with the concentrations in the aqueous humor and the lens (capsule and nucleus). It was found that the maximum concentrations of ciprofloxacin detected in the lens by HPLC were equivalent to those found in the aqueous humor and the plasma. There was no evidence of any accumulation of the test substance in the lens tissue. To summarize, the investigations failed to yield any evidence of ciprofloxacin-induced cataractogenic effect, nor was there any evidence to suggest that the substance accumulates in the lens. In addition, a study on syn- and co-cataractogenic potential of ciprofloxacin in rats did not give any evidence for adverse effects on the eye of this compound (SCHLÜTER 1987a).

Since it is known that quinolones strongly bind to melanin, the melanin-containing tissues in the eye, i.e., retina, iris, and ciliary body have to be regarded as potential target organs for toxicity. In his review of pharmacological characteristics of fluoroquinolones, Mitsui (1992) published results of animal experiments showing that after an oral dose of 20 mg levofloxacin per kilogram body weight the amount of drug taken up into the iris–ciliary body was several times greater in pigmented rabbits compared to albino rabbits. In addition, almost no reduction in drug concentration was observed; after 24 h iridociliary concentrations were more than ten times higher than in albino rabbits.

The strong affinity of quinolones for melanin was also demonstrated by an in vitro experiment in which purified melanin obtained from bovine uvea was added to solutions of levofloxacin, ofloxacin, and norfloxacin. The results of this experiment are shown in Table 5 (Mitsui 1992). About 70%–90% of ofloxacin, levofloxacin, and norfloxacin is captured by melanin in the melanin solution at a concentration of 0.1 m*M*, indicating that these drugs are concentrated and accumulated in the iris or ciliary body of the eye.

In order to investigate the possibility that the accumulation of quinolones in the melanin-containing tissues of the eye is associated with ocular toxicity, specific toxicological experiments were performed in pigmented animals, in which the visual function was tested by recording electroretinograms (ERGs). In a study using pigmented Long Evans rats, ciprofloxacin, levofloxacin (the (–)*S*-isomer of ofloxacin), norfloxacin, and nalidixic acid were given at an oral dose of 100 mg/kg for 2 weeks (Nomura et al. 1992). Examinations of ERGs revealed a decrease in the amplitudes of the a- and b-waves, a prolongation of the latency, and a diminution or disappearance of oscillatory potential waves in rats treated with nalidixic acid. Similar but milder changes were also noted in the norfloxacin-treated rats. ERGs from levofloxacin- or ciprofloxacin-treated rats were normal. In a 26-week dog study with temafloxacin (100 mg/kg per day given orally), examinations of ERGs revealed a decrease in the amplitudes of the a- and b-waves (Yamamura et al. 1994).

Table 5. In vitro uptake of fluoroquinolones by melanin[a] (Mitsui 1992)

Drug	Amount taken up by melanin (μmol/g)	
	1 m*M* Melanin	0.1 m*M* Melanin
Lomefloxacin	423.4 ± 1.0	69.1 ± 1.9
Ofloxacin	485.3 ± 6.2	81.0 ± 0.2
Norfloxacin	ND	92.4 ± 0.4

ND, not determined.
[a] Obtained from bovine uvea.

In pigmented rabbits, an intravitreal injection of 500 μg norfloxacin induced changes of the ERG, i.e., the amplitude of the oscillatory potentials was decreased and their peak latencies were delayed 3 h after injection; however, these changes were reversed within 7 days. A marked suppression of the c-wave was noted in one animal. An intravitreal injection of 50 μg norfloxacin produced no significant change in the ERG. Neither 50 nor 500 μg caused any apparent changes in the visual evoked potential and in the retinal histology 7 days after the injection (TANAHASHI et al. 1992).

Toxicity of ciprofloxacin to the retina was evaluated following an intravitreal injection of 100, 250, 500, and 1000 μg into the midvitreous cavity of one eye of pigmented rabbits. Ophthalmoscopical examinations showed focal areas of retinitis following injections of 500 and 1000 μg ciprofloxacin, but none in the case of 250 μg. In the ERG, b-wave amplitudes were reduced following the 1000-μg dose, but not at the lower doses (MARCHESE et al. 1993).

Retinal toxicity of norfloxacin and ofloxacin was evaluated after injection of 0.1 ml of 0.3% aqueous solutions into the vitreal cavity of the eyes of pigmented rabbits. There were no abnormal histological findings in the retina, and only norfloxacin caused temporary changes in the ERG, which returned to normal within 3 days (OOMOMO 1991).

Retinal toxicity of ofloxacin was studied by recording ERGs before and after intravitreal injection of ofloxacin into the eyes of pigmented rabbits. A dose of 200 μg ofloxacin did not cause any deterioration of the b-wave, the c-wave, or the oscillatory potentials throughout the follow-up period of 8 weeks (MOCHIZUKI et al. 1991).

The relevance of the ocular findings observed in preclinical animal studies to long-term therapy in humans remains unclear. It is, however, advisable in cases where a long-term application of a quinolone is intended and in which the oculopharmacokinetic investigations have shown a drug penetration into the eye that additional investigations – besides routine ophthalmological methods – be performed. In such cases, detailed measurements of lens transparency and testing of visual function by applying sophisticated techniques should be carried out.

G. Impairment of Spermatogenesis

Impaired spermatogenesis and/or testicular damage (e.g., reduced spermatogenesis, azoospermia, reduced weight of the prostate, epididymis, and seminal vesicle, and testicular atrophy) have been described for some quinolone-derivatives such as pipemidic acid, rosoxacin, norfloxacin, pefloxacin, enoxacin, and fleroxacin. The investigations included studies in rats, dogs, and monkeys using oral doses ranging from 50 to 3000 mg/kg per day administered over a period of 2–52 weeks.

Pefloxacin caused azoospermia (200 mg/kg per day for 13 weeks) and testicular damage with impaired spermatogenesis (100 mg/kg per day for 52 weeks) in dogs. In rats, a 2-week treatment with 50 mg/kg per day was reported to induce morphological alterations of spermatozoa. A ten-fold higher dose caused a reduction in accessory organ weights (seminal and prostate glands) and testicular atrophy. In standard tests for hormonal activity using mice, rats, and rabbits, pefloxacin (oral doses of 125 and 500 mg/kg per day for up to 7 days) had no estrogen-, progesterone- or antiandrogen-like effects (Anonymous 1988). Enoxacin impaired fertility and induced testicular atrophy in rats after long-term (26 weeks) administration of oral doses of between 1000 and 3000 mg/kg per day. In dogs, reduced spermatogenesis was observed at a lower dose (Stahlmann 1990).

Fleroxacin (320 mg/kg per day orally for 61 days before mating and throughout mating period) affected spermatogenesis and resulted in a reduced weight of the prostate gland and the epididymis in male rats. At histological examination, a reduced number of spermatozoa was present in the epididymis. A dose of 80 mg/kg was considered to be the dose at which fleroxacin had no effect (Suzuki et al. 1990). The relevance of these findings for humans remains unclear. It is probable that the impairment of spermatogenesis in animals by fleroxacin and some other quinolones has no relevance for the therapeutic situation in men. When healthy male volunteers were treated for 3 weeks with high fleroxacin doses (600 mg daily), no effect on sperm function or morphology was observed up to 3 months following the end of the 21-day treatment period.

Sparfloxacin did not impair male fertility of rats which were treated with oral doses of up to 500 mg/kg per day for 91 days including 64 days before mating (Terada et al. 1991). Extensive studies which have been performed with ofloxacin, levofloxacin, and ciprofloxacin in rats, dogs, and monkeys have failed to reveal any adverse impact on the male reproductive capacity (Christ and Lehnert 1990; Mayer 1987; Watanabe et al. 1992b).

Impaired male reproductive function by antimicrobial agents is not a new phenomenon, and antibacterial drugs belonging to other chemical classes such as aminoglycosides and cephalosporins are also known to cause altered spermatogenesis and testicular damage in different animal species. The mechanisms underlying these antifertility effects are still not elucidated. Neuroendocrine mechanisms (possibly the pituitary gland and gonadotropin release) or antiandrogenic action of the quinolones may be involved in these effects.

H. Cardiovascular Effects

After rapid intravenous administration of 10–30 mg/kg to anesthetized cats and dogs, most of the quinolones produce systolic and diastolic hypotension (Christ and Lehnert 1990; Takayama et al. 1995). It has been proposed that these cardiovascular effects are not caused directly by the quinolones but are mediated by histamine release. The intensity and duration of the effects are dose related. Thus, quinolones intended for intravenous use should be administered as an infusion.

Besides this immunologically mediated cardiovascular effect which may be regarded as a class effect, there seems to be a possibility that individual members of the quinolone class of drugs exhibit effects on the heart which need to be assessed differently. Sparfloxacin, one of the newer quinolones, was reported to induce a transient prolongation of the Q-T interval in the electrocardiogram (ECG) of beagle dogs which were orally dosed with 45 mg/kg per day for 4 weeks (IIDA et al. 1991). Increases in the Q-T interval have also been observed in healthy volunteers treated with sparfloxacin. A mean maximum increase of 19 ms was observed at the recommended dose of 400/200 mg. In clinical trials involving 813 patients, the average prolongation was about 3%, and 1.2% of patients developed Q-T intervals greater than 500 ms (prolongation of 100 ms in 0.3%), but with no arrhythmic effects. As a consequence, the use of sparfloxacin in patients with known Q-T prolongation and the concomitant use of drugs known to produce an increase in the Q-T interval and/or torsade de pointes is contraindicated. In light of these findings, it is advisable to ensure that adequate investigations are performed in preclinical and clinical studies in order to exclude a cardiotoxic potential of new quinolones under development.

I. Possible Mutagenic and Carcinogenic Effects

Fluoroquinolones act as inhibitors of bacterial gyrase. This enzyme catalyzes the conversion of relaxed DNA into negatively supercoiled DNA and is therefore important in the regulation of the superhelical state of chromosomal DNA. In combination with the topoisomerase I it is required to correct DNA topology during replication and transcription. Inhibition of bacterial gyrase by quinolones leads to replication arrest and cell death (DRLICA and FRANCO 1988).

Although considered specific for the bacterial enzyme, it has been recognized that fluoroquinolones may also inhibit the functionally related topoisomerase II which is the mammalian version of bacterial gyrase (ELSEA et al. 1992; BARRETT et al. 1989). It has also been shown that inhibition of cellular topoisomerase II correlates with cytotoxicity in those cells (OOMORI et al. 1988; OSHEROFF et al. 1992). In this context, it is interesting to note that some fluoroquinolones proved to be so cytotoxic to mammalian cells that they might become useable as antitumor agents; CP-67015 (FORT 1992) and CP-67804 (ROBINSON et al. 1991) are examples of such fluoroquinolones. For antibacterial quinolones, the need for a sufficiently differential cytotoxicity against bacterial versus mammalian cells is self-evident. Accordingly, for quinolones used as antibacterial drugs, inhibition of the mammalian enzyme is lower by several orders of magnitude than that of bacterial gyrase.

As a class, the quinolone antibacterials tend to give positive results in various tests for genetic toxicity, mainly in vitro. Whether such genotoxic effects prohibit the use of a specific fluoroquinolone in antimicrobial therapy requires careful evaluation. It is apparent that the "classic" structure–activity relationships which have been described, mainly on the basis of experience with DNA-reactive and electrophilic chemicals, will not hold for gyrase (topoisomerase II) inhibitors such as the fluoroquinolones. Furthermore, it is obvious that extrapolation of the bacterial results to mammalian organisms has to take into account the different affinity of the drugs for the bacterial versus the mammalian enzyme.

Several reviews on the mutagenicity and carcinogenicity of fluoroquinolones and mammalian topoisomerase-interactive agents have appeared recently (Fort 1992; Anderson and Berger 1994; Ferguson and Baguley 1994; Albertini et al. 1995; Herbold et al. 1995). These papers provide an excellent basis to summarize and discuss the activity of quinolone antibacterial compounds in a variety of assay systems for the detection of genotoxicity. The test systems used include gene mutation assays, chromosomal aberration assays and DNA damage assays, thus covering all important endpoints of genetic toxicity. For most quinolones, gene mutation assays have been performed in bacterial systems (e.g., *Salmonella typhimurium*), in lower eukaryotic systems (e.g., *Saccharomyces cerevisiae*), and in mammalian cells.

In assays for detecting gene mutations in bacteria, negative results were obtained for fluoroquinolones if DNA-repair deficient *S. typhimurium* strains such as TA 1535, TA 1537, TA 97, TA 98, and TA 100 were used (Fort 1992; Albertini et al. 1995; Herbold et al. 1995). However, all quinolones tested, including nalidixic acid, oxolinic acid, cinoxacin, norfloxacin, lomefloxacin, fleroxacin, ofloxacin, pefloxacin, and ciprofloxacin, were positive in *S. typhimurium* strains which were DNA-repair proficient, such as TA 102 (Albertini et al. 1995). The genotoxic activity increased strictly in parallel with the bacteriotoxic activity from the parent compounds nalidixic acid and oxolinic acid to the new generation of quinolones. It was concluded that intact excision repair and error-prone repair systems are required for the induction of a positive response in this test system (Ysern et al. 1990; Clerch et al. 1992).

In a lower eukaryotic system, i.e., *S. cerevisiae* D7 cells, Albertini et al. (1995) reported a dose-dependent increase in the frequencies of gene conversion after exposure to nalidixic acid, ofloxacin, pefloxacin, enrofloxacin, ciprofloxacin, and marbofloxacin. Structurally related quinolones such as oxolinic acid, cinoxacin, enoxacin, norfloxacin, lomefloxacin, and fleroxacin showed no increase in the induction of gene conversion. Herbold et al. (1995) tested ciprofloxacin in *S. cerevisiae* 6126/16c, Ade^-, Arg^-, and Tyr^- and found no indication of a mutagenic effect.

In mammalian systems such as the hypoxanthine phosphoribosyltransferase (HPRT) forward mutation assay in Chinese hamster V79 cells, ciprofloxacin, ofloxacin, levofloxacin, fleroxacin, nalidixic acid, and norfloxacin were negative (Fort 1992; Shimada et al. 1992). Most of these results were confirmed by Albertini et al. (1995) who showed that the majority of the tested quinolones did not increase the frequency of gene mutations even at cytotoxic doses. In contrast to FORT (1992) and HERBOLD et al. (1995), ALBERTINI et al. (1995), however, reported that ciprofloxacin induced an increase of gene mutations in Chinese hamster V79 cells with 100 μg/ml being the lowest effective dose.

Another mammalian gene mutation assay is the mouse lymphoma assay which can be used to evaluate mutation induction either at the thymidine kinase (TK) or at the HPRT locus. The mouse lymphoma test is believed to be more sensitive than the HPRT test. This may be due to the specific position of the reporter gene in the mouse genome which may allow larger deletions to be tolerated more easily by the affected cells. It is assumed that for this reason clastogenic effects, often resulting in deletions, are also detectable with this system. Therefore, positive results in this test system may be accounted for by chromosome aberrations as well as point mutations. FORT (1992) and HERBOLD et al. (1995) reported that ciprofloxacin, enrofloxacin, norfloxacin, ofloxacin, pefloxacin, and nalidixic acid yielded positive results in the mouse lymphoma thymidine kinase forward mutation assay. In addition, ALBERTINI et al. (1995) evaluated eight gyrase inhibitors including nalidixic acid, norfloxacin, ofloxacin, ciprofloxacin, and enrofloxacin in the mouse lymphoma forward assay at the TK and HPRT loci. They found that enrofloxacin and ciprofloxacin increased the mutation frequency, but only at the TK locus, while nalidixic acid, norfloxacin, and ofloxacin did not give rise to any increase in mutation frequency at either locus, even at the highest doses tested.

Norfloxacin, fleroxacin, cinoxacin, and nalidixic acid were negative for induction of chromosomal aberrations in vitro in cultured cells, i.e., Chinese hamster ovary (CHO) cells and human lymphocytes (FORT 1992; ALBERTINI et al. 1995). For ciprofloxacin and ofloxacin, contradictory results are reported in the literature. Whereas according to FORT (1992) both compounds yielded negative results under in vitro assay conditions, ALBERTINI et al. (1995) found that high concentrations of ciprofloxacin (200 μg/ml) and ofloxacin (400 μg/ml) induced a significant increase in chromosomal aberrations in CHO cells. A dose-dependent increase in sister chromatid exchanges and chromosomal aberrations in Chinese hamster lung cells was also seen with levofloxacin, the (−)*S*-isomer of ofloxacin (SHIMADA et al. 1992). In this context, the already-mentioned positive result of the mouse lymphoma forward assay for ciprofloxacin and ofloxacin could also be accounted for by a possible clastogenic effect in vitro.

The ability of quinolones to induce the formation of micronuclei in vitro was assessed in CHO cells (ALBERTINI et al. 1995). Nalidixic acid and norfloxacin did not induce micronuclei formation even at the highest doses tested, whereas ofloxacin and ciprofloxacin induced significant increases in micronuclei formation, but only at high concentrations (400 μg/ml). For structure–activity studies, CIARAVINO et al. (1993) developed a high capacity in vitro micronucleus assay with Chinese hamster V79 cells. After having tested a large number of quinolone structures with modifications at the 7- or 8-

position they concluded that most of the quinolones tested appeared to induce micronuclei formation, irrespective of the substituent at position 7 or 8.

Under in vivo conditions, ciprofloxacin, cinoxacin, ofloxacin, and levofloxacin gave negative results in the mouse micronucleus test using oral doses of up to 4000 mg/kg (Carlin 1975; Mayer 1987; Shimada et al. 1992; Herbold et al. 1995). In addition, the results of tests for chromosomal aberrations using rat or hamster bone marrow were negative for norfloxacin (Irikura et al. 1981), cinoxacin (Shiratori and Takase 1980), fleroxacin (Albertini et al. 1995), levofloxacin (Shimada et al. 1992), and ciprofloxacin (Herbold et al. 1995). Importantly, ciprofloxacin and ofloxacin did not cause chromosomal aberrations in lymphocytes from patients treated with the drugs (Mayer 1987; Mitelman et al. 1988).

In order to exclude the possibility of chromosomal damage in germ cells, a number of quinolones including ciprofloxacin, norfloxacin, ofloxacin, levofloxacin, and nalidixic acid were tested in the mouse dominant lethal test (Fort 1992; Shimada et al. 1992; Herbold et al. 1995). This test is conducted by treating males with the test compound and then mating them to untreated females over several periods using a different group of females each period. The result is determined by the number of early postimplantation deaths in pregnant females. An increase in the number of such deaths is evidence for germ line chromosomal damage in males. Importantly, all of the tested quinolones gave negative results in the mouse dominant lethal test.

The ability of quinolones to induce DNA damage was assessed in bacteria, yeast and in mammalian cells. Norfloxacin, fleroxacin, and ofloxacin gave a positive result in a DNA repair assay in *Bacillus subtilis*. Nalidixic acid and oxolinic acid gave positive results in *Escherichia coli*, *S. typhimurium*, and *B. subtilis* DNA repair tests, while ciprofloxacin did not induce DNA damage in an *E. coli* pol A_1^- test (Fort 1992; Herbold et al. 1995). In the *E. coli* TK104 SOS repair test, all tested quinolones including nalidixic acid, oxolinic acid, cinoxacin, enoxacin, norfloxacin, lomefloxacin, fleroxacin, ofloxacin, pefloxacin, and ciprofloxacin gave positive results (Albertini et al. 1995). A test for mitotic recombination in yeast (*S. cerevisiae*) with ciprofloxacin provided no evidence of DNA damage (Herbold et al. 1995).

A test for unscheduled DNA synthesis (UDS) in vitro with primary rat hepatocytes provided evidence of increased DNA repair after exposure to ciprofloxacin, pefloxacin, norfloxacin, and ofloxacin (McQueen et al. 1991). However, norfloxacin has also been reported not to induce UDS in human and mouse skin fibroblasts and rat hepatocytes (Hosomi et al. 1988). Similarly, ofloxacin gave a negative result when tested for UDS in human diploid fibroblasts in culture (Mayer 1987; Shimada et al. 1984); the result with levofloxacin (the (–)*S*-isomer of ofloxacin) was also negative in a UDS test on rat primary hepatocytes (Shimada et al. 1992). For ciprofloxacin, Bredberg et al. (1989) reported positive results for UDS with stimulated peripheral human lymphocytes. However, there was no evidence of mutagenic effects when a shuttle vector plasmid test was employed. In the same study, ciprofloxacin-induced DNA strand breaks were detected with an alkaline elution protocol. DNA strand breaks were no longer detectable when cells were allowed to recover within 15 min at 37°C in the absence of ciprofloxacin. It seems likely that the observed strand breaks reflect highly reversible cleavage complexes which are arrested when cells are treated with high concentrations of ciprofloxacin in vitro.

In order to evaluate the biological relevance of the positive result in the UDS in vitro assay, ciprofloxacin was also tested in the UDS test ex vivo. After single subcutaneous applications of 30 and 190 mg/kg to male Fischer 344 rats, no evidence for increased DNA repair was detected in the isolated hepatocytes (Herbold et al. 1995). A UDS ex vivo assay was also performed with hepatocytes from animals that had been exposed repeatedly (three times every 24 h) to ciprofloxacin at concentrations of 40 and 80 mg/kg. Also in this test there was no evidence for DNA repair (McQueen et al. 1991).

In addition, McQueen et al. (1991) reported the results of a ^{32}P-postlabeling study performed with ciprofloxacin. There was no evidence for DNA-adduct formation in cultured hepatocytes after ciprofloxacin treatment. This supports the notion that the high level of DNA repair in the UDS test in vitro cannot be explained by DNA-adducts but rather is due to other phenomena.

As mentioned earlier the mode of action of the quinolones as gyrase inhibitors (topoisomerase II inhibitors) is a plausible explanation for their genotoxic activity. The fluoroquinolone antibiotics generally have an affinity to the mammalian topoisomerase which is lower than their affinity to bacterial gyrase by several orders of magnitude. This specificity is paralleled by the total absence of, or presence of comparatively weak genotoxic effects in mammalian systems. The mechanism of action postulated to give rise to positive results in the mutagenicity assays, i.e., interference with enzyme action rather than a direct DNA-damaging effect, has significance in terms of assessment of risk to patients exposed to quinolone antibiotics. Hence, any effects leading indirectly to DNA or chromosome damage should depend on the concentration of compound reaching the target enzyme within the nucleus of the cell and on the affinity of the compound for the enzyme. This means that, although most of the quinolones yielded positive results in one of the assays for genetic toxicity, the significance of these findings for humans should be estimated on the basis of the concentrations of drug required to elicit the positive results compared with the therapeutic drug levels. For example, ciprofloxacin, norfloxacin, nalidixic acid, and ofloxacin inhibit calf thymus topoisomerase I and topoisomerase II at concentrations at least 100 times those required to inhibit bacterial DNA gyrase (Hussy et al. 1986).

Further estimates of the significance of positive mutagenicity results for patients comes from carcinogenicity studies completed with several quinolones. According to the literature (Christ and Esch 1994; Fort 1992), carcinogenicity studies have been completed for nalidixic acid, flumequine, norfloxacin, ciprofloxacin, enrofloxacin, and fleroxacin.

Under the conditions used in 2-year feeding studies (nalidixic acid administered at dietary concentrations ranging from 1000 to 16000ppm), there was clear evidence of carcinogenic activity of nalidixic acid for Fischer 344 rats, as indicated by increased incidences of preputial gland neoplasms in males and clitoral gland neoplasms in females (Morrissey et al. 1991). In CFD_1 mice, nalidixic acid yielded a negative result but in $B_6C_3F_1$ mice, nalidixic acid gave an equivocally positive result, inducing an increase in the incidence of subcutaneous fibromas or fibrosarcomas in male mice (Fort 1992). In an 18-month feeding experiment with flumequine at a high dose (800mg/kg per day), a tumorigenic response in mice was seen. Both benign and malignant tumors of the liver developed without metastases, with a much higher incidence in males than in females. The same high dose administered for 2 years to rats only induced a mild toxic effect on the liver and no tumorigenic response. An increased incidence of liver tumors in mice, especially in male mice, constitutes insufficient evidence to classify a compound as carcinogenic (Christ and Esch 1994).

An increase in the number of benign tumors and hyperplasia in the kidney without an increase in the number of malignant tumors was seen at the end of a 2-year rat carcinogenicity study with pefloxacin at oral doses of 96 and 384mg/kg per day (Anonymous 1988). No evidence for a carcinogenic activity could be detected with the newer quinolones norfloxacin, ciprofloxacin, and fleroxacin in long-term studies performed with rats and mice (Christ and Lehnert 1990; Cullmann et al. 1993), and with levofloxacin in a multiple organ carcinogenesis model using male rats (Kajimura et al. 1992).

From the results of the mutagenicity and carcinogenicity assays one can conclude that the most likely mechanism giving rise to positive genetic toxicity results with quinolones is interference with enzyme (topoisomerase) action and not a direct effect on DNA. Since interference with enzyme action is a concentration-dependent phenomenon, any risk to humans of genetic toxicity from these compounds depends on the concentrations achieved in vivo. The negative carcinogenicity results for the newer quinolones that have been tested suggest that these compounds at therapeutic concentrations do not present a risk for humans.

J. Phototoxicity

The potential of quinolones to induce phototoxicity is a well-documented side effect in man and has been demonstrated in a variety of experimental studies. Quinolone-induced phototoxicity is almost always nonimmunogenic and appears to be a class effect since there is a dose and an exposure level that will cause an effect for most of the known agents. Quinolones do, however, differ from one another substantially in their level of phototoxic risk both in man and in laboratory studies. Nalidixic acid and – of the newer quinolones – pefloxacin, lomefloxacin, fleroxacin, and sparfloxacin appear to be associated with the highest incidences of phototoxic side effects in patients (CHRIST and ESCH 1994; DOMAGALA 1994).

According to the data generated by a pharmacovigilance inquiry in France in 1995, the phototoxic reactions have a higher frequency of occurrence after sparfloxacin therapy than with other fluoroquinolones for a shorter treatment duration. The frequencies of photosensitivity reactions in patients in relation to the number of prescriptions with all fluoroquinolones marketed in France were as follows: sparfloxacin (one case in 4000) > pefloxacin (one case in 18200) ≫ ciprofloxacin (one case in 387000) = norfloxacin (one case in 403000) > ofloxacin (one case in 1640000).

Among the new fluoroquinolones, sparfloxacin seems to be associated with a new quality of phototoxicity. The degree of seriousness of the phototoxic reactions are high: second degree burns reported in 15.6% of cases, hospitalization in 8.2% of cases, and slow recovery sometimes leaving sequelae. The recurrence of symptoms with renewed sun exposure several weeks after sparfloxacin discontinuation has not been observed with other fluoroquinolones. The phototoxic reaction can occur after a moderate sun exposure (driving a car, shopping, or behind a window) or without sun exposure.

The phototoxic response, which usually occurs in the long-wave, i.e., longer than 320 nm, ultraviolet A (UV-A) range is similar to a severe sunburn with or without severe blistering. The reaction is dependent on the concentration of the phototoxic agent in the skin and the amount of light to which the patient is exposed. The concentrations of fluoroquinolones reached in the skin can be equal to those in blood after ingestion or parenteral administration. The quinolones may require metabolic or physicochemical transformation before they become photoactive. When a quinolone is exposed to UV-A radiation, there are three potential molecular outcomes: the quinolone can gradually decompose (MATSUMOTO et al. 1992); the quinolone can pass the light energy on to oxygen forming singlet oxygen, which is a strong oxidizing agent and radical generator (ROBERTSON et al. 1991); or the quinolone can return to its ground stage unchanged.

It can be assumed that quinolones which are degraded to a large extent may also be capable of inducing phototoxicity. Photoreactivity is mostly influenced by the substituent at the 8-position of the fluoroquinolone nucleus. MATSUMOTO et al. (1992) and MARUTANI et al. (1993) demonstrated that quinolones in which the 8-position is substituted with fluorine are very unstable against UV-A irradiation and may develop toxic degradation products whereas the introduction of a methoxy group at position 8 significantly increases the stability of fluoroquinolones against irradiation by UV light and consequently reduces phototoxicity. DOMAGALA (1994) confirmed these results by demonstrating that the highest level of phototoxicity of quinolones was seen when the 8-position was substituted by halogens. Moreover, he established a ranking order for substituents at the 8-position with the following order of decreasing phototoxicity: CF ≫ CCl > N > CH > CCF_3 > COR.

According to this structure–side effect relationship, the presence of fluorine in the 8-position may explain the comparatively high incidences of phototoxic events reported in man for lomefloxacin, fleroxacin, and sparfloxacin.

Quinolone-induced phototoxicity can be estimated by measuring rates of degradation, by measuring cellular damage in vitro, or through in vivo models such as the mouse phototoxicity model.

Przybilla et al. (1990) screened nine quinolones including the first-generation quinolones cinoxacin, nalidixic acid, oxolinic acid, pipemidic acid, and rosoxacin in addition to the new-generation quinolones ciprofloxacin, enoxacin, fleroxacin, norfloxacin, and ofloxacin in an in vitro photohemolysis test. They found that all first-generation quinolones as well as three of the newer agents (enoxacin, fleroxacin, ofloxacin) caused photohemolysis at a concentration of 10^{-3} mol/l, which was most prominent after exposure to UV-A-rich irradiation. At lower concentrations phototoxic effects were less or absent. Ciprofloxacin did not exhibit phototoxic activity in the test system. Due to hemolysis without irradiation, norfloxacin could not be evaluated at 10^{-3} mol/l; lower concentrations did not cause phototoxic effects.

Johnson et al. (1989) established an in vitro battery of five assays (photohemolysis, photosensitized killing of *Candida albicans*, photosensitized destruction of histidine, photosensitized inhibition of phytohemagglutinin – stimulated lymphocytes, and photosensitized killing of mouse peritoneal macrophages) to assess the phototoxic potential of nalidixic acid, ofloxacin, and ciprofloxacin. It was concluded that although all quinolones tested exhibited some phototoxic effects in vitro, the phototoxicity of ofloxacin and ciprofloxacin was considerably lower than that of nalidixic acid.

Under in vivo conditions, Wagai et al. (1990) investigated the phototoxic potential of quinolone antibacterial agents in Balb/c mice. The mice were orally administered nalidixic acid, enoxacin, ofloxacin, DR-3355 (levofloxacin, the (–)*S*-isomer of ofloxacin), ciprofloxacin, and lomefloxacin, and immediately exposed to UV-A for 4 h (21.6 joules/cm^2). Their ears were examined for overt damage, as a major phototoxic parameter, 0, 24 and 48 h after irradiation ended. At doses of 200 mg/kg, lomefloxacin, nalidixic acid, and enoxacin caused marked cutaneous phototoxic reactions on the ears, whereas ciprofloxacin, ofloxacin and DR-3355 did not. At 800 mg/kg, however, ciprofloxacin, ofloxacin, and DR-3355 also caused phototoxic reactions on the ears. These phototoxic reactions were characterized grossly by edema, and histopathologically by edema and infiltration of inflammatory cells, especially neutrophils, into the connective tissue surrounding the cartilage. Based on the calculation of the 50% erythema-inducing doses (Table 6), the phototoxic potency of the quinolones was ranked as lomefloxacin > enoxacin, nalidixic acid > ofloxacin, DR-3355 (levofloxacin), ciprofloxacin. The same ranking order was established when the doses of lomefloxacin, enoxacin, ofloxacin, ciprofloxacin, and DR-3355 (Table 7) which induced an increase in ear thickness of 50% were calculated by measuring the ear swelling reactions in Balb/c mice (Wagai and Tawara 1991).

Table 6. The 50% erythema-inducing dose (EID_{50}) of quinolone derivates in mice (WAGAI et al. 1990)

Quinolone	EID_{50} (mg/kg)	95% Confidence limit (mg/kg)
Lomefloxacin	19	10–37
Enoxacin	102	89–117
Nalidixic acid	143	107–190
Ofloxacin	553	483–634
DR-3355	619	430–890
Ciprofloxacin	741	522–1052

EID_{50} was calculated by the Litchfield and Wilcoxon method.

Table 7. Doses of quinolones inducing 50% increment of ear thickness ($ETID_{50}$) (WAGAI and TAWARA 1991)

Quinolone	$ETID_{50}$ (mg/kg)	95% Confidence limit (mg/kg)
Lomefloxacin	24.8	22.8–26.6
Enoxacin	81.9	73.7–89.0
Ofloxacin	428.0	394.6–455.6
Ciprofloxacin	457.9	425.6–486.3
DR-3355	526.6	485.4–562.2

In a study using Balb/c mice, the phototoxic effects of sparfloxacin, levofloxacin, and enoxacin on auricular skin were examined histopathologically. The quinolones were orally administered and the animals were then exposed to UV-A for 4h. In the auricle, degeneration of basal epidermal cells was sporadically observed at 2h (during the irradiation). Foci of slight edema with degenerated fibroblasts were seen in the dermis at 4h. Edema and neutrophil infiltration in the dermis became severe at 96h. The effects were induced by sparfloxacin (50 and 100mg/kg), levofloxacin and enoxacin (400 and 800mg/kg). The authors conclude that the combination of orally administered quinolones and UV-A irradiation, which never caused apparent morphological changes alone, was able to induce phototoxic lesions in albino mice (SHIMODA et al. 1993).

In a study using guinea pigs, HORIO et al. (1994) investigated the phototoxic potential of nalidixic acid, norfloxacin, ofloxacin, enoxacin, ciprofloxacin, lomefloxacin, and tosufloxacin. The quinolones were administered orally once and the animals were subsequently exposed to UV-A at a dose of 30 joules/cm^2. The non-phototoxic doses were 6mg/kg (enoxacin, lomefloxacin), 10mg/kg (ofloxacin), 30mg/kg (nalidixic acid, tosufloxacin), and 100mg/kg (norfloxacin, ciprofloxacin). Thus the phototoxic potency was ranked in the order enoxacin, lomefloxacin > ofloxacin, nalidixic acid, tosufloxacin > norfloxacin, ciprofloxacin.

Mechanistically, UV-A-induced quinolone phototoxicity is most likely mediated by reactive oxygen species such as singlet oxygen, OH radicals, superoxide anions and hydrogen peroxide. Robertson et al. (1991) demonstrated that PD 117596, a quinolone with strong phototoxic properties, was a more efficient producer of singlet oxygen than rose bengal and that it produces a strong photohemolytic effect. This photohemolytic action was initiated rapidly, was oxygen dependent, enhanced by D_2O replacement, and effectively inhibited by the antioxidants β-hydroxybutyrate, butyl hydroxytoluene, and α-tocopherol.

The possible direct causal role of reactive oxygens in cutaneous phototoxicity was also investigated by Wagai and Tawara (1992) using cultured mouse 3T3 fibroblast cells and Balb/c mice. In the in vitro study, the cultured cells were exposed to UV-A in the presence of the five quinolones lomefloxacin, enoxacin, ciprofloxacin, ofloxacin, and levofloxacin (the (–)*S*-isomer of ofloxacin). Cytotoxicity after irradiation was assayed by the neutral red and MTT assay methods, both of which revealed dose-dependent phototoxicity for all five quinolones. Phototoxicity was inhibited by the addition of catalase, and was augmented by the addition of superoxide dismutase. Dimethylurea (a hydroxyl radical scavenger) protected against phototoxicity induced by four of these quinolones, but not against that induced by enoxacin.

These results indicate that superoxide anions, hydrogen peroxide, and hydroxyl radicals were generated in solutions of these quinolones under UV-A irradiation. In the in vivo study, mice were injected in the auricle with hydrogen peroxide. Ear swelling reactions appeared in a dose-dependent manner. When irradiated, these reactions were significantly augmented. These data suggest that cutaneous phototoxicity in Balb/c mice is initiated by the generation of reactive oxygen species in the target tissue, especially hydroxyl radicals.

Rosen et al. (1995) used the formation of 8-Oxo-dG in DNA as an indicator system for the production of singlet oxygen and/or OH radicals in cultured rat epithelial cells (ARL-18) exposed to lomefloxacin and UV-A. A dose-dependent increase from 1.1- to four-fold in 8-Oxo-dG levels was found with increasing concentrations of lomefloxacin (50 to 400 μmol/l) together with an irradiation of 20 joules/cm^2 compared to UV-A alone at 20 joules/cm^2. The authors suggest that a potential mechanism for UV-A-induced phototoxicity by lomefloxacin is through an oxidative pathway.

The potential to induce phototoxic events is addressed in the package insert of the quinolones on the market. Patients undergoing therapy should avoid exposure to sun or artificial UV light. It should be noted that the FDA's Anti-Infective and Dermatologic Drugs Advisory Committees met on March 31, 1993, to discuss reports of human phototoxicity thought to be associated with the use of lomefloxacin. After reviewing the clinical data it was concluded that there are differences between lomefloxacin and the other marketed quinolones relating to phototoxic reactions which warrant disparate labeling.

K. Photocarcinogenicity and Photomutagenicity

The aforementioned FDA's Advisory Committees (March 31, 1993) met also to discuss the results of an animal photocarcingenicity study performed by Hoffmann–La Roche (Roche) in Switzerland in which lomefloxacin appeared to act as a tumor promoter.

In this study, albino hairless mice (SKH-1) were treated over a long period of time with fleroxacin, ciprofloxacin, lomefloxacin, ofloxacin, and nalidixic acid. 8-Methoxypsoralen (8-MOP) was used as a positive control. The drugs were administered orally by gavage five times every 2 weeks up to 78 weeks. The doses of quinolones were chosen to give plasma levels (within 2 and 4h after gavage) similar to that reached by 0.2mg/mouse per day of fleroxacin. Approximately 1.5h after drug administration the animals were exposed to UV-A radiation of subtoxic doses (25 joules/cm^2 for the quinolones; initially 3.5 joules/cm^2 and then 2 joules/cm^2 for 8-methoxypsoralen).

The result of the study was that all quinolones tested caused the development of skin tumors. There were, however, large differences in the latent period prior to the onset of skin tumors and in the median latent period, i.e., the duration of time required for just over one half of the mice in a group to

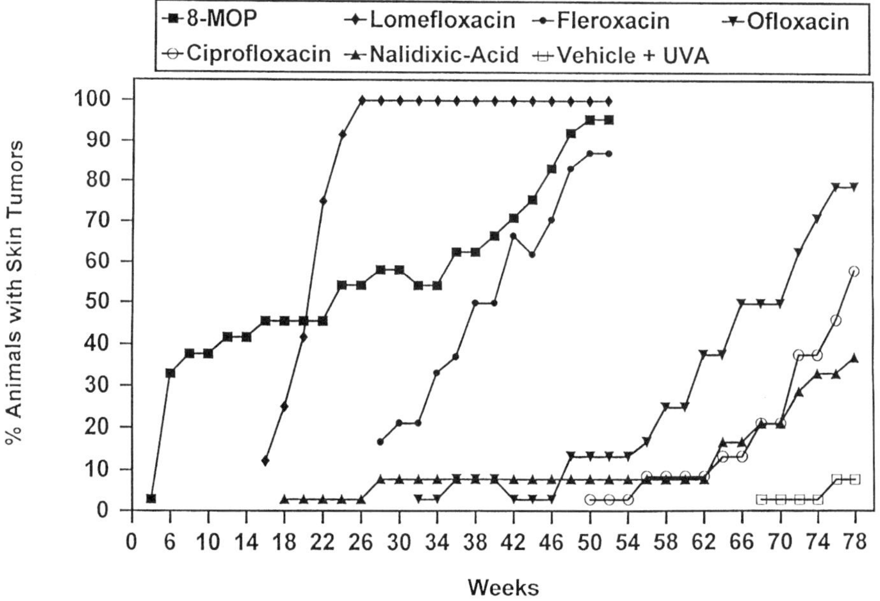

Fig. 1. Photocarcinogenicity of quinolones

develop a skin tumor of at least 1 mm in diameter, between 8-MOP (positive control) and the quinolones, and between the quinolone derivatives themselves (Fig. 1). The shortest latent period prior to skin tumor onset occurred with 8-MOP (4 weeks), followed by lomefloxacin (16 weeks) and fleroxacin (38 weeks). For ofloxacin, nalidixic acid, and ciprofloxacin a much longer latent period (>50 weeks) was observed. Another difference between the quinolone derivatives was related to the type of skin tumors induced: lomefloxacin induced 29 malignant tumors, i.e., squamous cell carcinomas, compared with 12 due to 8-MOP and one due to fleroxacin. For the other quinolones (ciprofloxacin, ofloxacin, nalidixic acid) all skin tumors were of the benign type and of the type expected to be produced by long-term UV-A irradiation of hairless mice. As expected from the UV-A doses used, the combination of vehicle plus UV-A also produced a few benign skin tumors.

At the Advisory Committee Meeting it was suggested that the carcinogen responsible for the skin tumors was UV-A; the quinolones were considered to be enhancers. It was further noted that the compounds tested in the Roche study can be classified into two groups: fast responders (8-MOP, lomefloxacin, fleroxacin) and slow responders (ofloxacin, ciprofloxacin, nalidixic acid). The committee unanimously agreed that for lomefloxacin a bold-type section including the results of the Roche study suggesting possible enhanced animal photocarcinogenicity be added to the precautions section of the package labeling.

Since then, FORBES et al. (1995) published results of a study in hairless mice, in which lomefloxacin was given orally at doses of 20 or 200 mg/kg (once daily, 5 days weekly for 2 weeks). In addition the mice were exposed to simulated sunlight. In this study, signs of erythema, flaking, and epidermal melanization which subsequently resolved were observed. However, at 10 weeks after this 2-week regimen, no mouse and six mice in the 20 and 200 mg/kg dosage groups, respectively, developed cutaneous papillomas.

The results of the photocarcinogenicity study have prompted the investigation of the photomutagenic properties of fluoroquinolone derivatives in several genotoxicity assays under concomitant irradiation with simulated solar light (so-called photomutagenicity assays). CHETELAT et al. (1995) found that ciprofloxacin, fleroxacin, and lomefloxacin induced the UV-light-dependent occurrence of chromosomal aberrations in Chinese hamster V79 cells and DNA strand breaks in mouse lymphoma L5178Y cells (Comet assay). Marginal induction of point mutations was seen in the Ames test using the *S. typhimurium* strains TA 100 and TA 104, while no induction of gene conversion was apparent in *S. cerevisiae* D7. Relative photogenotoxic potencies conformed with previously reported phototoxic activities of the three fluoroquinolones (lomefloxacin > fleroxacin > ciprofloxacin). Preirradiation of the test compounds did not lead to degradation products causing genotoxic effects. The photoclastogenic activity was reduced by addition of scavengers (superoxide dismutase, catalase, dimethylthiourea). The authors conclude that light-induced formation of short-lived active oxygen species is responsible for the photogenotoxicity of fluoroquinolones, a mechanism not connected to the light-independent genotoxicity of the topoisomerase inhibitors.

Brendler-Schwaab and Herbold (1995) and Brendler-Schwaab et al. (1995) investigated the photomutagenic potential of the experimental fluoroquinolone BAY y 3118, a compound with demonstrated strong phototoxic properties, in a photo–HPRT assay (irradiation with simulated sunlight) using Chinese hamster V79 cells. They found that BAY y 3118 was a photomutagen which revealed its photomutagenic property only when it was present inside the cell during irradiation. Furthermore, they found that the underlying mechanism of the photomutagenicity of BAY y 3118 was not an inhibition of the repair of UV-induced lesions.

In conclusion, it can be stated that all quinolones with a demonstrated photo-instability should be candidates for photocarcinogenicity testing. In this context, the use of data from relevant short-term photomutagenicity assays may be predictive of a photocarcinogenic potential. In vitro photomutagenicity tests can aid in understanding the mechanisms involved in photocarcinogenicity by detecting those skin carcinogens which can act as genotoxic (initiating) agents. For photocarcinogens which are also photomutagens, a key question in risk assessment is the relative contribution of genotoxicity compared to promoting events, such as hyperplasia due to phototoxicity. The genetic portion can be assessed by comparing drug concentrations which are mutagenic in vitro with plasma/skin concentrations in animal photocarcinogenicity studies, and ultimately with plasma concentrations in humans exposed to the drugs. The effect of hyperplasia may depend upon the degree of erythema seen in patients. Finally, it should be stressed again that the results of the photocarcinogenicity and photomutagenicity studies clearly indicate that humans undergoing antibiotic therapy with quinolones should avoid extensive exposure to sunlight or artificial UV-light.

L. Drug Interactions

It is becoming clear that quinolones interact with a number of other agents, with the potential for clinical disturbances. Because these interactions are reviewed in numerous papers (Wolfson and Hooper 1989; Paton and Reeves 1991, 1992; Christ and Lehnert 1990; Cullmann et al. 1993; Edwards et al. 1988; Lomaestro and Bailie 1991; von Rosenstiel and Adam 1994) they are only briefly mentioned here. The most important interactions concern drugs and foodstuffs containing Ca^{+2}, Mg^{+2}, Fe^{+2}, Zn^{+2}, and Al^{+3}, H_2-receptor antagonists, sucralfate, theophylline, caffeine, fenbufen, and to a minor degree, warfarin, cyclosporin, cimetidine, and oral diabetic drugs. A comprehensive overview on drug interactions is given in the chapter entitled Clinical Pharmacology.

M. Metabolic and Nutritional Effects

Possible metabolic and nutritional effects related to quinolones are summarized in Table 8 (Christ and Esch 1994). It should be mentioned that most of the findings are discussed here with respect to side effects concerning specific organ systems, e.g., renal, hepatic, and blood disorders. In general, among the monitored laboratory parameters, there does not seem to be important variations among the different fluoroquinolones. Profound lactic acidosis was reported in a 16-year-old girl with hepatic dysfunction while on nalidixic acid (2 g q.i.d.) (Paton and Reeves 1991). Hyperglycemia and hypoglycemia were only rare events with older and newer quinolones. Under temafloxacin therapy, however, some patients developed severe hypoglycemia. The multisystem disease caused by temafloxacin which finally prompted the withdrawal of the drug from the market is characterized by acute renal failure, hemolytic anemia/hemolysis, thrombocytopenia, liver dysfunction, and disseminated intravascular coagulation. As a secondary effect of temafloxacin therapy, a significant but clinically irrelevant decrease in serum immunoglobulins was found in humans. With lomefloxacin, a decrease in serum immunoglobulins was observed in rats (after oral administration) and in the dog (after intravenous administration; Christ and Esch 1994).

N. Conclusion

The toxicological profile of quinolones has been well characterized in a large number of preclinical studies. Despite the impression that in principle the overall pattern of potential toxicities is comparable for all quinolones, it is obvious that there are marked differences in both the incidence and the type of reaction induced by certain compounds. Very recently, new toxicological aspects have come to light: a new quality of phototoxicity associated with sparfloxacin and effects on the conduction system of the heart (prolongation of Q-T interval), also associated with sparfloxacin. This demonstrates that –

Table 8. Metabolic and nutritional effects related to quinolones (Christ and Esch 1994)

Elevated	Transaminases
	Alkaline phosphatase
	γ-Glutamyl transpeptidase
	LDH
	Serum creatinine
	Uric acid and blood urea nitrogen
	Cholesterol
Decreased	Serum albumin (temafloxacin)
	Serum globulin (in rats only)
Hypernatremia, hypercalcemia	
Hyperglycemia, hypoglycemia	

although the toxicological profile of a new fluorquinolone can be predicted to a large extent – one must always be aware of the possibility of unexpected surprises.

The preclinical findings (e.g., arthrotoxicity in juvenile animals, nephropathy, CNS toxicity, phototoxicity) have lead to restrictions in the clinical use of quinolones. Under these restrictions, today's quinolones have been proven to be safe drugs which, due to their exciting microbiological and pharmacokinetic properties, provide a useful antibiotic therapy for many patients.

Acknowledgement. The authors thank Dr. G. Schmuck for her contribution to the section on the effects of quinolones on the central nervous system.

References

Akahane K, Sekiguchi M, Une T, Osada Y (1989) Structure relationship of quinolones with special reference to their interaction with γ-aminobutyric acid receptor sites. Antimicrob Agents Chemother 33:1704–1708

Akaike N, Shirasaki T, Yakushiji T (1989) Quinolones and fenbufen interact with $GABA_A$ receptors in dissociated hippocampal cells of rat. J Neurophysiol 66:497–504

Albertini S, Chételat AA, Miller B, Muster W, Pujadas E, Strobel R, Gocke E (1995) Genotoxicity of 17 gyrase- and four mammalian topoisomerase II-poisons in prokaryotic and eukaryotic cell systems. Mutagenesis 10(4):343–351

Alzheimer C, Röhrenbeck J, ten Bruggencate G (1991) Adenosine depressed induction of LTP at the mossy-fiber-CA3 synapse in vitro. Brain Res 543:163–165

Anderson RD, Berger NA (1994) Mutagenicity and carcinogenicity of topoisomerase-interactive agents. Mutat Res 309:109–142

Anonymous (1988) Pefloxacin. Rhone-Poulenc, Antony, France (Product monograph)

Bailey RR, Peddie BA (1985) Enoxacin for the treatment of urinary tract infection. N Z Med J 98:268–288

Bailey RR, Kirk JA, Peddie BA (1983) Norfloxacin-induced rheumatic disease. N Z Med J 96:590

Baraban JM, Cole AJ, Stratton KR, Pritchett J, Alvaro J, Abushakra S, Saffen DW, Worley PF (1991) Neuronal excitability: focus on second messenger systems. In: Fischer RS, Coyle JT (eds) Neurotransmitters and epilepsy. Wiley-Liss, New York, pp 35–37

Barrett JF, Gootz TD, McGuirk PR, Farrell CA, Sokolowski SA (1989) Use of in vitro topoisomerase II assays for studying quinolone antibacterial agents. Antimicrob Agents Chemother 33:1697–1703

Bredberg A, Brant M, Riesbeck K, Azou Y, Forsgren A (1989) 4-Quinolone antibiotics: positive genotoxic screening tests despite an apparent lack of mutation induction. Mutat Res 211:171–180

Brendler-Schwaab SY, Herbold BA (1995) BAY y 3118: a photomutagenic quinolone. 7th Congress of Toxicology, July 2–6, Seattle

Brendler-Schwaab SY, von Keutz E, Schlüter G, Herbold BA (1995) A new approach to screen for photomutagenicity with the HPRT-assay. Workshop "Photocarcinogenesis: Mechanisms, Models and Human Health Implications", Oct 27–28, Washington

Burkhardt JE, Hill MA, Lamar CH, Smith GN, Carlton WW (1993) Effects of difloxacin on the metabolism of glycosamino-glycans and collagen in organ cultures of articular cartilage. Fundam Appl Toxicol 20:257–263

Carlin H (1975) Pharmacology review of cinoxacin. Food and Drug Administration, Rockville

Chételat A, Albertini S, Gocke E (1995) Photogenotoxicity of fluoroquinolone antibiotics. Workshop "Photocarcinogenesis: Mechanisms, Models and Human Health Implications", Oct 27–28, Washington

Christ W (1990) Central nervous system toxicity of quinolones: human and animal findings. J Antimicrob Chemother 26(B):219–225

Christ W, Esch B (1994) Adverse reactions to fluoroquinolones in adults and children. Infect Dis Clin Pract 3(3):168–176

Christ W, Lehnert T (1990) Toxicity of the quinolones. In: Siporin C, Heifetz CL, Domagala JM (eds) The new generation of quinolones. Dekker, New York, pp 165–186

Christ W, Gindler K, Gruene S, Hecker W, Jacobsen M, Junge H, Park H-H (1988a) Interactions of quinolones with opioids and fenbufen, a nonsteroidal antiinflammatory drug: involvement of dopaminergic neurotransmission. Rev Infect Dis 11(5):1393–1394

Christ W, Lehnert T, Ulbrich B (1988b) Specific toxicologic aspects of the quinolones. Rev Infect Dis 10:141–146

Chysky V, Kapila K, Hullmann R, Arcieri G, Schacht P, Echols R (1991) Safety of ciprofloxacin in children: worldwide experience based on compassionate use. Emphasis on joint evaluation. Infection 19:289–296

Ciavarino V, Suto MJ, Theis JC (1993) High capacity in vitro micronucleus assay for assessment of chromosomal damage: results with quinolone/naphthyridone antibacterials. Mutat Res 298:227–236

Clerch B, Barbé J, Llagostera M (1992) The role of the excision and error-prone repair systems in mutagenesis by fluorinated quinolones in Salmonella typhimurium. Mutat Res 281:207–213

Cohen A, Rosenberg F (1992) Trois nouveaux de tendiopathie achilléenne après traitement par fluoroquinolones (Abstr). Rev Rhum Mal Osteoartic 59:E23

Cullmann W, Geddes AM, Weidekamm E, Urwyler H, Braunsteiner A (1993) Fleroxacin: a review of its chemistry, microbiology, toxicology, pharmacokinetics, clinical efficacy and safety. Int J Antimicrob Agents 2:203–230

Dalhoff A (1989) In: Fernandes PB (ed) Quinolones. A review of quinolone tissue pharmacokinetics. Praus, pp 277–312

Davey PG (1988) Overview of drug interactions with the quinolones. J Antimicrob Chemother 22C:97–107

Davies BI, Maesen FPV (1989) Drug interactions with quinolones. Rev Infect Dis 11(5):1083–1090

De Sarro A, Ammendola D, De Sarro G (1994) Effects of some quinolones on imipenem-induced seizures in DBA/2 mice. Gen Pharmacol 25:369–379

Dimpfel W, Spüler M, Dalhoff A, Hofmann W, Schlüter G (1991) Hippocampal activity in the presence of quinolones and fenbufen in vitro. Antimicrob Agents Chemother 35:1142–1146

Dimpfel W, Dalhoff A, Hofmann W, Schlüter G (1994) Electrically evoked potentials in the rat hippocampus slice in the presence of aminophylline alone and in combination with quinolones. Eur Neuropsychopharmacol 4:151–156

Dimpfel W, Dalhoff A, von Keutz E (1995) In vitro modulation of hippocampal pyramidal cell response by quinolones. Effect of HA 966 and γ-hydroxybutyric acid J Antimicrob Chemother 40(11):2573–2576

Dodd PR, Davies LP, Watson WEJ, Nielsen B, Deyer JA, Wong LS, Johnston GAR (1989) Neurochemical studies on quinolone antibiotics: effects on glutamate, GABA and adenosine systems in mammalian CNS. Pharmacol Toxicol 64:404–411

Domagala JM (1994) Structure–activity and structure–side-effect relationship for the quinolone antibacterials. J Antimicrob Chemother 33:685–706

Drlica K, Franco RJ (1988) Inhibitors of DNA topoisomerase. Biochemistry 27:2253–2259

Edwards DJ, Bowles SK, Svensson CK, Rybak MJ (1988) Inhibition of drug metabolism by quinolone antibiotics. Clin Pharmacokinet 15:194–204

Elsea SH, Osheroff N, Nitiss JL (1992) Cytotoxicity of quinolones towards eukaryotic cells. J Biol Chem 267(19):13150–13153

Ferguson LR, Baguley BC (1994) Review article: topoisomerase II enzymes and mutagenicity. Environ Mol Mutagen 24:245–261

Forbes PD, Sambuca CP, Arocena MA, D'Aloisio LC, Hoberman AM (1995) Fluoroquinolone antibiotics (FQA) induce photosensitized skin reactions in hairless mice exposed to simulated sunlight (Abstr 1712). Toxicologist 15:319

Fort FL (1992) Mutagenicity of quinolone antibacterials. Drug Saf 7:214–222

Franck JL, Bouteiller G, Chagnaud P, Sapene M, Gautier D (1991) Rupture des tendons d'achille chez deux adultes traités par péfloxacine dont un cas bilatéral. Rev Rhum Mal Osteoartic 58:904

Gootz TD, Barrett JF, Sutcliffe JA (1990) Inhibitory effects of quinolone antibacterial agents on eucaryotic topoisomerases and related test systems. Antimicrob Agents Chemother 34:8–12

Gough A, Barsoum NJ, Mitchell L, McGuire EJ, De la Iglesia FA (1979) Juvenile canine drug-induced arthropathy: clinicopathological studies on articular lesions caused by oxolinic and pipedimic acids. Toxicol Appl Pharmacol 51:177–187

Halliwell RF, Labert JJ, Davey PG (1989) Actions of quinolones and nonsteroidal antiinflammatory drugs on γ-aminobutyric acid currents of rat dorsal root ganglion neurons. Rev Infect Dis 11(5):1398–1399

Halliwell RF, Davey PG, Lambert JJ (1991) The effect of quinolones and NSAIDs upon GABA-evoked currents recorded from dorsal root ganglia neurons. J Antimicrob Chemother 27:209–218

Halliwell RF, Davey PG, Lambert JJ (1993) Antagonism of $GABA_A$ receptors by 4-quinolones. J Antimicrob Chemother 31:457–462

Herbold BA, Gahlmann R, Schlüter G (1995) Ciprofloxacin, a potent gyrase inhibitor: Studies on genotoxicity. Mutat Res (in press)

Hildebrand H, Kempka G, Schlüter G, Schmidt M (1993) Chondrotoxicity of quinolones in vivo and in vitro. Arch Toxicol 67:411–415

Hori S, Shimada J (1993) Effects of quinolones on the central nervous system. In: Hooper DC, Wolfson JS (eds) Quinolone antimicrobial agents, 2nd edn. American Society for Microbiology, Washington

Hori S, Shimada J, Saito A, Miyahara T, Kurioka S, Matasuda M (1987a) A study on enhanced epileptogenicity of new quinolones in the presence of antiinflammatory drugs (Abstr 30). 27th Interscience Conference on Antimicrobial Agents and Chemotherapy, Oct 4–7, New York. American Society for Microbioloy, Washington, p 101

Hori S, Shimada J, Saito A, Matsuda M, Miyahara T (1987b) Comparison of the inhibitory effects of new quinolones on γ-aminobutyric acid receptor binding in the presence of antiinflammatory drugs. Rev Infect Dis 11(5):1397–1398

Horio T, Miyauchi H, Asada Y, Aoki Y, Harada M (1994) Phototoxicity and photoallergenicity of quinolones in guinea pigs. J Dermatol Sci 7(2):130–135

Hosomi J, Maeda A, Oomori Y, Irikura T, Yokota T (1988) Mutagenicity of norfloxacin and AM-833 in bacteria and mammalian cells. Rev Infect Dis 10(1):148–149

Hussy P, Maass G, Tummler BL, Grosse FL, Schomburg U (1986) Effect of 4-quinolones and novobiocin on calf thymus DNA polymerase α primase complex, topoisomerases I and II, and growth of mammalian lymphoblasts. Antimicrob Agents Chemother 29:1073–1078

Ichikawa N, Naora K, Hayashibara M (1992) Effect of fenbufen on the entry of new quinolones, norfloxacin and ofloxacin, into the central nervous system in rats. J Pharm Pharmacol 44:915–920

Iida M, Yasuba M, Nakajima F, Maeda T, Matsuoka N, Ohnishi K (1991) Four week oral subacute toxicity study of sparfloxacin in beagle dogs. Chemotherapy (Tokyo) 39(4):195–202

Ingham B, Brentnall DW, Dale EA, McFadzean JA (1977) Arthropathy induced by antibacterial fused *N*-alkyl-4-pyridone-3-carboxylic acids. Toxicol Lett 1:21–26

Irikura T, Suzuki H, Sugimoto T (1981) Mutagenicity studies of AM-715 in animals. Chemotherapy (Tokyo) 29(4):932–937

Iwamoto K, Naora K, Katagiri Y, Ichigawa N, Hayashibara M, Tanaka K, Yamaguchi T, Sekine Y (1993) Comparative neurotoxicity study of ciprofloxacin and sparfloxacin after coadministration with fenbufen in rats. Drugs 45(3):290–291

Jaehde U, Goto T, de Boer AG, Breimer DD (1993) Blood–brain transport rate of quinolone antibacterials evaluated in cerebrovascular endothelial cell cultures. Eur J Pharm Sci 1:49–55

Janknegt R (1990) Drug interactions with quinolones. J Antimicrob Chemother 26(D):7–29

Johnson BE, Walker EM, Ferguson J (1989) Quinolone-induced photosensitivity. Rev Infect Dis 11(5):1396–1397

Jorgensen C, Anaya JM, Didry C, Canovas F, Serre I, Baldet P, Ribard P, Kahn MF, Sany J (1991) Arthropathies et tendinopathie achilléenne induites par la péfloxacine. A propos d'une observation. Rev Rhum Mal Osteoartic 58:623–625

Kahn MF, Carbon C (1993) Tendinopathies et fluoroquinolones. Concours Med 115:819–823

Kajimura T, Tojo H, Kudo G, Yamada M, Domon S, Nomura M, Takayama S (1992) Effect of the new quinolone antibacterial agent levofloxacin on multiple organ carcinogenesis initiated with wide-spectrum carcinogens in rats. Arzneimittelforschung 42(3a):390–395

Kato M, Onodera T (1988) Morphological investigation of cavity formation in articular cartilage induced by ofloxacin in rats. Fundam Appl Toxicol 11:110–119

Kato M, Takada S, Kashida Y, Nomura M (1995) Histological examination on Achilles tendon lesions induced by quinolone antibacterial agents in juvenile rats. Toxicol Pathol 23(3):385–392

Kempka G, Ahr HJ, Rüther W, Schlüter G (1996). The effects of fluoroquinolones and glucocorticoids on cultivated tendon cells in vitro. Toxicol In Vitro 10:743–754

Lomaestro BM, Bailie GR (1991) Quinolone–cation interactions: a review. DICP Ann Pharmacother 25:1249–1258

Marchese AL, Slana VS, Holmes EW, Jay WM (1993) Toxicity and pharmacokinetics of ciprofloxacin. J Ocul Pharmacol 9(1):69–76

Marutani K, Matsumoto M, Otabe Y et al (1993) Reduced phototoxicity of a fluoroquinolone antibacterial agent with a methoxy group at the 8 position in mice irradiated with long-wavelength UV light. Antimicrob Agents Chemother 37:2217–2223

Matsumoto M, Kojima K, Nagano H, Matsubara S, Yokota T (1992) Photostability and biological activity of fluoroquinolones substituted at the 8 position after UV-irridation. Antimicrob Agents Chemother 36:1715–1719

Mayer DG (1987) Overview of toxicological studies (ofloxacin). Drugs 34(1):150–153

McEwan SR, Davey PG (1988) Ciprofloxacin and tendosynovitis. Lancet 2:900

McQueen CA, Way BM, Queener SM, Schlüter G, Williams GM (1991) Study of potential in vitro and in vivo genotoxicity in hepatocytes of quinolone antibiotics. Toxicol Appl Pharmacol 111:255–262

Mitelman F, Kolnig A-M, Strömbeck B, Norrby R, Kromann-Andersen B, Sommer P, Wadstein J (1988) No cytogenetic effects of quinolone treatment in humans. Antimicrob Agents Chemother 32:936–937

Mitsui Y (1992) Pharmacological characteristics of fluoroquinolones. J Eye 9(2):215–223

Mochzuki K, Torisaki M, Wakabayashi K (1991) Effects of vancomycin and ofloxacin on rabbit – ERG in vivo. Jpn J Ophthalmol 35(4):435–445

Morikawa K, Nagata O, Kubo S, Kato H (1987) Unusual CNS toxic action of new quinolones. 27th Interscience Conference of Antimicrobial Agents and Chemotherapy, October 4–7, New York. American Society for Microbiology, Washington

Morrissey ER, Eustis S, Haseman JK, Huff J, Bucher JR (1991) Toxicity and carcinogenicity studies of nalidixic acid in rodents. Drug Chem Toxicol 14:45–66
Nix DE, De Vito JM, Whitbread MA, Schentag JJ (1987) Effect of multiple dose oral ciprofloxacin on the pharmacokinetics of theophylline and indocyanine green. J Antimicrob Chemother 19:263–269
Nomura M, Yamada M, Yamamura H, Kajimura T, Takayama S (1992) Ophthalmotoxicity and ototoxicity of the new quinolone antibacterial agent levofloxacin in long evans rats. Arzneimittelforschung 42(3A):398–403
Nozaki M, Takeda N, Tanaka K, Tsurumi K (1989) GABA-ergic mechanism on convulsions induced by combination with new quinolones and NSAIDs; true or not. 16th International Congress of Chemotherapy, Jerusalem
Oomomo A (1991) Intravitreal injection of norfloxacin or ofloxacin. II Retinal toxicity (in Japanese). Atarashii Ganka 8(4):651–654
Oomori Y, Yasue T, Aoyama H, Hirai K, Suzue S, Yokota T (1988) Effects of fleroxacin on HeLa cell functions and topoisomerase II. J Antimicrob Chemother 22(D):91–97
Osheroff N, Elsea SH, Nitiss JL (1992) Cytotoxicity of quinolones towards eukaryotic cells. J Biol Chem 267:13150–13153
Paton JH, Reeves DS (1991) Clinical features and management of adverse effects of quinolone antibacterials. Drug Saf 6:8–27
Paton JH, Reeves DS (1992) Adverse reactions to the fluoroquinolones. Adverse Drug React Bull 153:575–578
Percival A (1991) Impact of chemical structure on quinolone potency, spectrum and side effects. J Antimicrob Chemother 28(C):1–8
Perrot S, de Bourran-Cauet G, Lanchand AT, Desplaces N, Kaplan G, Ziza JM (1991) Nouvelle complication liée aux quinolones: la rupture du tendon d'achille (Abstr). Rev Rhum Mal Osteoartic 58:I25
Polk RE (1989) Drug–drug interactions with ciprofloxacin and other fluoroquinolones. AM J Med 87(5A):5A–76S
Przbilla B, Georgii A, Bergner T, Ring J (1990) Demonstration of quinolone phototoxicity in vitro. Dermatology 181:98–103
Psarropoulou C, Kostopoulos G (1990) Long-term potentiation of postsynaptic excitability after brief exposure to Mg^{++}-free medium in normal and epileptic mice. Brain Res 508:70–75
Raasch W, Grobecker H, Kees F (1992) Influence of ciprofloxacin on brain metabolism of biogenic amines in the rat. Pharm Pharmamacol Lett 2:153–156
Ribard P, Kahn MF (1991) Rheumatological side-effects of quinolones. Baillieres Clin Rheumatol 15:175–191
Ribard P, Audisio F, DeBandt M, Kahn MF, Palazzo E, Meyer O (1991) 5 cas de tendinite d'achille, dont 2 avec rupture, au cours de traitememts par les fluoroquinolones (Abstr). Rev Rhum Mal Osteoartic 58:I26
Ribard P, Audisio F, Kahn MF, DeBandt M, Jorgensen C, Hayem G, Meyer O, Palazzo E (1992) Seven Achilles tendinitis including 3 complicated by rupture during fluoroquinolone therapy. J Rheumatol 19:1479–1481
Robertson DG, Epling GA, Kiely JS, Bailey DL, Song B (1991) Mechanistic studies of the phototoxic potential of PD 117595, a quinolone antibacterial compound. Toxicol Appl Pharmacol 111:221–232
Robinson MJ, Martin BA, Gootz TD, McGuirk PR, Moynihan M, Sutcliffe JA, Osheroff N (1991) Effects of quinolone derivative on eukaryotic topoisomerase II. J Biol Chem 266:14585–14592
Rose TF, Bremner DA, Collins J, Ellis-Pegler R, Isaaca R, Richardson R, Small M (1990) Plasma and dialysate levels of pefloxacin and its metabolites in CAPD patients with peritonitis. J Antimicrob Chemother 25:657–664
Rosen JE, Prahalad AK, Williams GM (1995) Quinolone antibacterial phototoxicity as measured by oxidative DNA damage (Abstr 157). Toxicologist 15:29
Royer RJ, Pierfitte C, Netter P (1994) Features of tendon disorders with fluoroquinolones. Therapie 49:75–76

Schaad UB, Wedgwood J (1992) Lack of quinolone-induced arthropathy in children. J Antimicrob Chemother 30:414–416

Schaad UB, Stoupis C, Wedgwood J, Tschaeppeler H, Vock P (1991) Clinical, radiologic and magnetic resonance monitoring for skeletal toxicity in pediatric patients with cystic fibrosis receiving a three-month course of ciprofloxacin. Pediatr Infect Dis J 10:723–729

Schlüter G (1987a) Toxicology of quinolones. In: Naber KG, Adam D, Grobecker H (eds) Gyrase-Hemmer II. FAC (Fortschritte der antimikrobiellen und antineoplastischen Chemotherapie), vol 6–10. Futuramed, Munich, pp 1631–1642

Schlüter G (1987b) Ciprofloxacin: review of potential toxicologic effects. Am J Med 82(4A):91–96

Schürmann A (1995) Potentielle neurotoxische Wirkung von Chinolonen im zentralen Nervensystem. PhD thesis, Ruhr Universit, Bochum

Shimada H, Yutaka E, Kurusawa Y, Arauchi T (1984) Mutagenicity studies of DL-8280, a new antibacterial drug. Chemotherapy (Tokyo) 32(1):1162–1170

Shimada H, Itoh S, Hattori C, Tada S, Matsuura Y (1992) Mutagenicity of the new quinolone antibacterial agent levofloxacin. Arzneimittelforschung 42(3A):378–385

Shimoda K, Yoshida M, Wagai N, Takayama S, Kato M (1993) Phototoxic lesions induced by quinolone antibacterial agents in auricular skin and retina of albino mice. Toxicol Pathol 21(6):554–561

Shirasaki T, Harata N, Nakaye T, Akaike N (1991) Interaction of various non-steroidal antiinflammatories and quinolone antimicrobials on GABA response in rat dissociated hippocampal pyramidal neurons. Brain Res 562:329–331

Shiratori O, Takase S (1980) Mutagenic activity tests on cinoxacin in vitro and in vivo cytogenetic tests in mammalian cells. Chemotherapy (Tokyo) 28(4):523–529

Siefert HM, Maruhn D, Maul W, Förster D, Ritter W (1986a) Pharmacokinetics of ciprofloxacin, 1st communication: absorption, concentrations in plasma, metabolism and excretion after a single administration of [^{14}C]ciprofloxacin in albino rats and rhesus monkeys. Arzneimittelforschung 36:1496–1502

Siefert HM, Maruhn D, Scholl H (1986b) Pharmacokinetics of ciprofloxacin, 2nd communication: distribution to and elimination from tissues and organs following single or repeated administration of [^{14}C]ciprofloxacin in albino rats. Arzneimittelforschung 36:1503–1510

Stahlmann R (1990) Safety profile of the quinolones. J Antimicrob Chemother 26(D):31–44

Stahlmann R, Merker HJ, Hinz N, Chahoud I, Webb J, Heger W, Neubert D (1990) Ofloxacin in juvenile, non-human primates and rats. Arthropathia and drug plasma concentrations. Arch Toxicol 64:193–204

Stahlmann R, Förster C, Shakibaei M (1993a) Decrease of β 1-integrins in cartilage from juvenile rats after ofloxacin treatment. Naunyn Schmiedebergs Arch Pharmacol 348(Suppl):R170

Stahlmann R, Förster C, Van Sickle D (1993b) Quinolones in children. Are concerns over arthropathy justified? Drug Saf 9:397–403

Stahlmann R, Förster C, Shakibaei M, Vormann J, Stürje H (1995) Ofloxacin-induced joint cartilage lesions in immature rats are identical with lesions induced by magnesium-deficient diet (Abstr. 888). Toxicologist 15:167

Sutcliffe JA, Osheroff N (1991) Effects of quinolone derivatives on eucaryotic topoisomerase. II. J Biol Chem 266:14585–14592

Suzuki H, Takahashi T, Sato Y, Abe Y (1990) Fertility study on fleroxacin in rats. Chemotherapy (Tokyo) 38(2):261–271

Takayama S, Hirohashi M, Kato M, Shimada H (1995) Toxicity of the quinolone antimicrobial agents (review). J Toxicol Environ Health 45:1–45

Tanahashi T, Mochizuki K, Torisaki M, Yamashita Y, Komatsu M, Higashide T, Ogata M (1992) Effect of intravitreal injection of norfloxacin on the retina in pigmented rabbits. Lens Eye Toxicol Res 9(3–4):493–503

Tatsumi H, Senda H, Yatera S, Takemoto Y, Yamoyoshi M, Ohnishi K (1978) Toxicological studies on pipemidic acid. V. Effect on diarthrodial joints of experimental animals. J Toxicol Sci 3:357–367

Terada Y, Aoki Y, Mukumoto K, Shigematsu K, Nishimura K, Ohnishi K (1991) Reproductive and developmental toxicity studies of sparfloxacin. I. Fertility study in rats (in Japanese). Yakuri to Chiryo 19(4):1241–1255

Tin Lee W, Collins JT (1992) Ciprofloxacin associated bilateral Achilles tendon rupture. Aust N Z J Med 22:500

Tsuji A, Sato H, Kume Y, Tamai I, Okezaki E, Nagata O, Kato H (1988) Inhibitory effects of quinolone antibacterial agents on γ-aminobutyric acid binding to receptor sites in the rat brain membranes. Antimicrob Agents Chemother 32:190–194

Von Rosenstiel N, Adam D (1994) Quinolone antibacterials. An update of their pharmacology and therapeutic use. Drugs 47:872–901

Wagai N, Tawara K (1991) Quinolone antibacterial-agent-induced cutaneous phototoxicity: ear swelling reactions. Toxicol Lett 58:215–223

Wagai N, Tawara K (1992) Possible direct role of reactive oxygens in the cause of cutaneous phototoxicity induced by five quinolones in mice. Arch Toxicol 66:392–397

Wagai N, Yamaguchi F, Sekiguchi M, Tawara K (1990) Phototoxic potential of quinolone antibacterial agents in Balb/c mice. Toxicol Lett 54:299–308

Watanabe Y, Himi T, Saito H, Abe K (1992a) Involvement of glycin site associated with the NMDA receptor in hippocampal long-term potentiation and acquisition of spatial memory in rats. Brain Res 582:58–64

Watanabe T, Fujikawa K, Harady S, Ohura K, Sasaki T, Takayama S (1992b) Reproductive toxicity of the new quinolone antibacterial agent levofloxacin in rats and rabbits. Arzneimittelforschung 42(3A):374–377

Williams PD, Helton DR (1991) The proconvulsive activity of quinolone antibiotics in an animal model. Toxicol Lett 58:23–28

Wolfson JS, Hooper DC (1989) Fluoroquinolone antimicrobial agents. Clin Microbiol Rev 2:378–424

Yamamoto K, Naitoh Y, Inoue Y, Yoshimura K, Morikawa K, Nagata O, Hashimoto S, Yamada T, Kubo S (1988) Seizure discharges induced by the combination of a new quinolinecarboxylic acid antimicrobial drug and non-steroidal anti-inflammatory drugs. Chemotherapy (Tokyo) 36(2):300–324

Yamamura T, Kuse H, Susami M, Kawai Y, Hori M (1993) A 26-week oral toxicity study of temafloxacin in dogs. Chemotherapy (Tokyo) 41(5):214–224

Yokota T, Kanda K (1989) Influence of new quinolones on the extension of nerve fibers in redifferentiated human neuroblastoma IMR-32 cells. Rev Infect Dis 11(5):1399–1400

Ysern P, Clerch B, Castano M, Gibert I, Barbe J, Llagostera M (1990) Introduction of SOS genes in Escherichia coli and mutagenesis in Salmonella typhimurium by fluoroquinolones. Mutagenesis 5:63–66

Zaudig M, von Bose M, Weber MM, Bremer D, Zieglgänsberger W (1989) Psychotic effects of ofloxacin. Pharmacopsychiatry 22:11–15

CHAPTER 11
Clinical Pharmacology

J. KUHLMANN, H.-G. SCHAEFER, and D. BEERMANN

A. Introduction

In drug development clinical pharmacology forms not only the all-important link between preclinical and clinical research providing the necessary prerequisites for targeted clinical trials, but also the link between clinical efficacy and drug concentration. With the aid of experimental and biochemical pharmacology, toxicology and clinical chemistry, clinical pharmacology deals with the safety and tolerance, working mechanism, pharmacokinetics and metabolism of a drug in the human organism under physiological and pathological conditions. The aim must be to increase knowledge in the field of pharmacotherapy in order to make it work in an optimal way.

Knowledge of the relationships between dose, drug concentration in blood and clinical response (effectiveness and undesirable effects) is important to identify an appropriate starting dose, to adjust dosage to the needs of a particular patient and to define a dose beyond which increases are unlikely to provide added benefit. Once a concentration–effect relationship has been defined, pharmacokinetic changes can be used to predict the change in intensity of a specific therapeutic effect or side effect. Although the effect must be assessed ultimately in patients, early clinical pharmacology studies in healthy volunteers can contribute enormously to drug development as demonstrated in the development of anti-infective drugs.

The antibacterial activity combined with the pharmacokinetic properties are key criteria determining the therapeutic and prophylactic potential of drugs against infections. On the basis of such information, the relevance of results from animal experiments for humans can be determined, recommendations may be made regarding dosage and predictions made about the likely therapeutic index. When prescribing antimicrobials, several factors may have an impact on the proper choice of a specific antibacterial agent. The physician must take into account the antibiotics' activity against the infecting organisms, the potential toxicity of the agent of choice in relation to the patient's health status and the pharmacokinetic parameters which describe disposition in the body. In contrast to most other drugs, where noninvasive and invasive methods are necessarily used to measure the pharmacodynamic effects in healthy volunteers and patients, most information regarding antibacterial activity of anti-infective drugs is gleaned from in vitro studies of antibiotic action on bacteria. While in vitro killing rate studies can explore bacterial killing in relation to an array of static concentrations, the study of antibiotic killing actions on bacteria in animal models needs to focus on the relationship between animal pharmacokinetics and metabolism-changing serum drug concentrations.

The reliable extrapolation of animal data to human treatment requires careful consideration of the pharmacokinetic differences between animals and humans. Even in the infection models simulating human drug time concentration curves, drug penetration from serum to tissues or fluids is different due to species differences in serum protein binding. These differences must be considered when making a comparison of drug concentrations in target tissues/fluids in animals and humans. Determination of the correlation between pharmacokinetics in humans, the killing of bacteria in animals and the antibacterial activity in vitro may provide valuable information on efficacy earlier and with higher precision and can lead to more economical drug development.

B. Pharmacokinetics of Fluoroquinolone Antibiotics

In the course of drug development initial information on pharmacokinetic properties of a new drug is gained in healthy subjects. Key parameters such as half-life and clearance are evaluated from results of phase I studies performed in young healthy male volunteers. Later on, during phases II and III of clinical drug development, pharmacokinetic parameters are assessed in the target population, i.e. in male and female patients of all ages who are suffering from illnesses against which the new drug is directed.

All fluoroquinolones, including enoxacin, tosufloxacin (both derived from naphthyridine), ofloxacin, levofloxacin (derived from benzoxacine) and rufloxacin (derived from benzothiazine) share the same basic structural features (see Chap. 2, this volume) which determine their physicochemical characteristics. Ofloxacin has an asymmetric center at the methyl substitution of the oxacine ring making it a racemic compound. The (–)- and (+)-isomers differ by as much as 100-fold in activity against Gram-positive and Gram-negative bacteria (LAMP et al. 1992; HAYAKAWA 1986). The more active (–)-isomer has been developed as a separate drug, namely levofloxacin. Tosufloxacin is also a racemic mixture of enantiomers due to the asymmetric center at the amino substitution of the pyrrolidine ring; however, the (+)- and (–)-enantiomers differ by only two- to four-fold in their in vitro activities (MINAMI et al. 1993b). Chiral centers are also present in lomefloxacin and sparfloxacin. Whereas lomefloxacin is used as the racemic compound with no available data on single enantiomer activities (FREEMAN et al. 1993; MITSCHER et al. 1989), sparfloxacin is used as a pure enantiomer (SHIMADA et al. 1993).

Fluoroquinolones exist as zwitterionic compounds at physiological pH values (SÖRGEL and KINZIG 1993a). This is due to their carboxyl groups having

Table 1. Isoelectric points (pI) and distribution coefficients (D) of quinolones (SÖRGEL and KINZIG 1993a)

Drug	pI	D
Fleroxacin	6.78	0.173
Amifloxacin	6.84	n.a.
Difloxacin	6.85	4.001
Pefloxacin	6.95	1.538
Ofloxacin	7.14	0.318
Rufloxacin	7.18	0.55
Norfloxacin	7.34	0.042
Ciprofloxacin	7.42	0.058
Enoxacin	7.50	0.039
Sparfloxacin	7.53	0.4
Lomefloxacin	7.56	n.a.*

* n.a. data not available.

Table 2. Pharmacokinetic parameters of ciprofloxacin in healthy subjects

Dose (mg)	Administration	C_{max} (mg/l)	AUC (mg h^{-1}l^{-1})	$t_{1/2}$ (h)	Reference
50	p.o.	0.28	1.00	3.40	HÖFFKEN et al. (1985a)
100	p.o.	0.49	1.90	4.10	KARABALUT and DRUSANO (1993)
250	p.o.	1.45	6.37	3.97	BRITTAIN et al. (1985)
500	p.o.	2.56	11.10	4.15	KARABALUT and DRUSANO (1993)
750	p.o.	2.65	12.20	4.75	HÖFFKEN et al. (1985a)
1000	p.o.	3.38	16.60	6.30	TARTAGLIONE et al. (1986)

pKa values in the range of 5.5–6.3 and to their heterocyclic substituents such as piperazine rings with pKa values between 7.4 and 9.3 (dependent on substitution of the basic nitrogen). At their isoelectric points (pI), which range between 6.78 and 7.56 (see Table 1), the solubility of the quinolones is lowest in aqueous media and highest in a lipophilic environment.

The pharmacokinetic behaviour of the individual drugs greatly depends on the relationship of their pI to the different pH values of body fluids and tissues and pH gradients across membranes, which become involved in absorption, distribution and secretion processes. Thus, although being structurally closely related, the fluoroquinolones vary in their pharmacokinetic properties.

I. Healthy Subjects

Basic pharmacokinetic parameters of ciprofloxacin, enoxacin, fleroxacin, levofloxacin, lomefloxacin, norfloxacin, ofloxacin, pefloxacin, rufloxacin, sparfloxacin and tosufloxacin are given in Tables 2–22. The vast majority of the data in these tables have been obtained from evaluation of serum or plasma concentrations, which were measured by high-performance liquid chromatog-

Table 3. Pharmacokinetic parameters of ciprofloxacin in elderly subjects

Dose (mg)	Administration	Age (years)	C_{max} (mg/l)	$t_{1/2}$ (h)	CL_{tot} (ml/min)	CL_{ren} (ml/min)	Reference
100	p.o., s.d.°	74	0.8	3.3	443.0	88.0	BALL et al. (1986)
250	p.o., s.d.	67	1.7	3.4	560.8	244.8	BAYER et al. (1987)
500	p.o., s.d.	75.4	3.2	6.8	394.0	152.0	LEBEL et al. (1986)
750	p.o., s.d.	78.1[a]	5.9	5.2	415.0	100.0	GUAY et al. (1987)
750	p.o., st.st.*	78.1[b]	8.6	4.8	336.0	172.0	GUAY et al. (1987)
750	p.o., st.st.	78	7.6	4.2	n.a.	n.a.	HIRATA et al. (1989)
200	i.v., s.d.	74	4.2	4.2	441.0	n.a.	KEES et al. (1989)
200	i.v., s.d.	57–84	4.2	4.3	407.0	n.a.	NABER et al. (1989)
200	i.v., st.st.	77.1	3.5	5.8	289.1	164.3	HIRATA et al. (1989)
200	i.v., st.st.	78	3.6	5.7	274.5	157.9	HIRATA et al. (1989)

CL_{tot}, total clearance; CL_{ren}, renal clearance; n.a., data not available. ° s.d. single dose; * st.st. steady state.
[a] Ill patients.
[b] Convalescent patients.

Table 4. Pharmacokinetic parameters of enoxacin in healthy subjects

Dose (mg)	C_{max} (mg/l)	AUC ($mg\,h^{-1}l^{-1}$)	$t_{1/2}$ (h)	Reference
200	1.02	4.67	3.16	CHANG et al. (1988)
400	3.09	18.47	4.90	WOLF et al. (1984)
600	4.18	27.30	4.74	WOLF et al. (1984)
800	3.70	26.99	5.44	WOLF et al. (1984)
800	3.80	25.75	4.93	CHANG et al. (1988)

Table 5. Pharmacokinetic parameters of enoxacin in elderly subjects

Dose (mg)	Administration	Age (years)	C_{max} (mg/l)	$t_{1/2}$ (h)	CL_{tot} (ml/min)	CL_{ren} (ml/min)	Reference
200	p.o., s.d.	81.8	1.9	6.1	199.6	n.a.	WISE et al. (1987)
400	p.o., s.d.	70.8	3.6	7.3	177.3	82.0	NABER et al. (1986)
600	p.o., s.d.	74.8	6.9	6.8	207.3	91.2	DOBBS et al. (1987)

See Table 3 for abbreviations.

raphy (HPLC). Since the conjugated heterocyclic ring systems of quinolones provide an optimal basis for fluorescence detection, very sensitive and selective HPLC assays have been developed employing this detection principle (SÖRGEL et al. 1987).

Metabolites of several quinolones have been reported to have antibacterial activity. Since results of microbiological assays relate to the microbiologically active material in the sample, these assays may not reflect the true concentration of the parent drug. Hence, results from HPLC assays are usually preferred for the calculation of pharmacokinetic parameters.

Table 6. Pharmacokinetic parámeters of fleroxacin in healthy subjects

Dose (mg)	Administration	C_{max} (mg/l)	AUC (mg $h^{-1}l^{-1}$)	$t_{1/2}$ (h)	Vd/f (l/kg)	CL/f (ml/min)	CL_{ren} (ml/min)	Reference
100	p.o.	1.6	18.3	10.6	91	n.a.	n.a.	WEIDEKAMM et al. (1987)
100	i.v.	n.a.	10.2	8.6	1.6	168	105	WEIDEKAMM and PORTMANN (1989)
200	p.o.	2.33	20.9	8.9	159	n.a.	n.a.	WEIDEKAMM et al. (1987)
400	p.o.	4.36	48.3	9.2	138	n.a.	n.a.	WEIDEKAMM et al. (1987)
600	i.v.	n.a.	80.7	9.6	1.3	131	76	WEIDEKAMM and PORTMANN (1989)
800	p.o.	7.04	106.1	10.3	126	n.a.	n.a.	WEIDEKAMM et al. (1987)

Vd/f, volume of distribution devided by bioavailability; CL/f, total clearance devided by bioavailability; see Table 3 for other abbreviations.

Table 7. Pharmacokinetic parameters of oral fleroxacin in elderly subjects

Dose (mg)	Age (years)	C_{max} (mg/l)	$t_{1/2}$ (h)	CL/f (ml/min)	CL_{ren} (ml/min)	Reference
400	60–74	5.2	10.6	106.0	60.0	RICHER and LEBEL (1993)
800	75	15.6	16.0	43.0	17.0	TABURET et al. (1990)

See Tables 3 and 6 for abbreviations.

Table 8. Pharmacokinetic parameters of lomefloxacin in healthy subjects

Dose (mg)	Administration	C_{max} (mg/l)	t_{max} (h)	AUC (mg $h^{-1}l^{-1}$)	$t_{1/2}$ (h)	CL/f (ml/min)	CL_{ren} (ml/min)	Reference
100	p.o., s.d.	1.10	0.80	6.40	7.10	265	210	MORRISON et al. (1988)
100	p.o., s.d.	0.90	0.86	6.20	7.60	269	175	MORSE (1990)
200	p.o., s.d.	2.50	0.73	13.20	7.30	256	187	MORRISO et al. (1988)
200	p.o., s.d.	1.60	1.14	12.00	7.80	286	192	MORSE (1990)
200	p.o., o.d.*	2.00	0.80	11.60	7.70	291	221	FREEMAN et al. (1993)
400	p.o., s.d.	3.00	1.27	27.40	7.80	247	189	MORRISON et al. (1988)
400	p.o., s.d.	3.50	1.14	28.90	8.00	233	140	MORSE (1990)
400	p.o., o.d.	2.80	1.50	25.90	7.80	259	167	FREEMAN et al. (1993)
400	p.o., b.i.d.°	5.40	0.90	30.60	8.20	219	146	FREEMAN et al. (1993)
600	p.o., s.d.	4.80	1.63	35.50	8.00	290	162	MORRISON et al. (1988)
600	p.o., s.d.	5.90	1.39	48.10	7.80	209	140	MORSE (1990)
600	p.o., o.d.	5.30	1.00	40.50	8.20	248	140	FREEMAN et al. (1993)
600	p.o., b.i.d.	7.00	1.20	46.30	7.80	219	135	FREEMAN et al. (1993)
800	p.o., s.d.	7.50	0.97	57.00	8.00	238	161	MORRISON et al. (1988)
800	p.o., s.d.	7.00	1.14	62.10	8.10	218	143	MORSE (1990)
800	p.o., o.d.	6.00	0.90	47.00	7.50	290	165	FREEMAN et al. (1993)

For abbreviations, see Tables 3 and 6. * o.d. once daily; ° b.i.d. twice daily.

1. Absorption

The site of absorption has been investigated for ciprofloxacin by topical administration of a drug solution to defined areas of the gastrointestinal tract (HARDER et al. 1990; STAIB et al. 1989b). The extent of absorption (in relation to an orally administered solution) decreases from nearly 100% in the duode-

Table 9. Pharmacokinetic parameters of oral norfloxacin in healthy subjects (SWANSON et al. 1983)

Dose (mg)	C_{max} (mg/l)	AUC ($mg\,h^{-1}l^{-1}$)	$t_{1/2}$ (h)
200	0.75	3.56	7.3
400	1.58	6.26	7.4
800	2.41	11.4	6.2
1200	3.15	16.1	5.7
1600	3.87	19.7	6.8

Table 10. Pharmacokinetic parameters of norfloxacin in elderly subjects

Dose (mg)	Administration	No. of subjects	Age (years)	C_{max} (mg/l)	$t_{1/2}$ (h)	Reference
400	p.o., st.st.	8	81	1.5	5.0	MACGOWAN et al. (1988)
400	p.o., st.st.	8	83.8	n.a.	5.2	KELLY et al. (1988)

See Table 3 for abbreviations.

Table 11. Pharmacokinetic parameters of ofloxacin in healthy subjects

Dose (mg)	Administration	C_{max} (mg/l)	AUC ($mg\,h^{-1}l^{-1}$)	$t_{1/2}$ (h)	Vd/f (l/kg)	CL/f (ml/min)	CL_{ren} (ml/min)	Ae ur* %	Reference
100	i.v.	n.a.	7.30	4.5	1.3	235	185	73	LODE et al. (1987)
200	p.o.	2.19	14.60	5.6	1.6	230	197	74	LODE et al. (1987)
200	i.v.	n.a.	14.0–14.4	4.3–5.4	1.2–1.5	235–280	140–190	66–77	FARINOTTI et al. (1988)
400	p.o.	3.51	28.00	4.9	1.5	240	202	73	LODE et al. (1987)

For abbreviations, see Tables 3 and 6. * Ae ur amount excreted in urine.

Table 12. Pharmacokinetic parameters of ofloxacin in elderly subjects

Dose (mg)	Administration	Age (years)	C_{max} (mg/l)	$t_{1/2}$ (h)	CL_{tot} (ml/min)	Reference
200	p.o., s.d.	77	3.60	13.30	82.80	VEYSSIER et al. (1986)
200	p.o., s.d.	75.2	2.90	6.10	189.40	GRABER et al. (1988)
300	p.o., s.d.	85	2.70	6.20	83.30	MOLINARO et al. (1992)

See Table 3 for abbreviations.

num and jejunum to 25% in the ileum and less than 5% in the descending colon. Rapid uptake from the upper gastrointestinal tract was also demonstrated by 19-F nuclear magnetic resonance spectroscopy for fleroxacin (SÖRGEL et al. 1988d; JYNGE et al. 1990). It can be assumed that other

Table 13. Pharmacokinetic parameters of oral pefloxacin in healthy subjects. (BARRE et al. 1984)

Dose (mg)	C_{max} (mg/l)	AUC ($mg\,h^{-1}\,l^{-1}$)	$t_{1/2}$ (h)
200	1.5	25.7	11.7
400	3.2	54.5	10.5
600	5.5	87.9	11.3
800	7.0	105.0	12.6

Table 14. Pharmacokinetic parameters of rufloxacin in healthy subjects. (SEGRE et al. 1992)

Dose (mg)	Administration	C_{max} (mg/l)	t_{max} (h)	AUC ($mg\,h^{-1}\,l^{-1}$)	$t_{1/2}$ (h)	Ae ur (%)	CL_{ren} (ml/min)
100	po, s.d.	1.1	9.33	63.8	58.5	37.8	10.0
200	po, s.d.	2.0	4.00	96.6	70.4	53.5	20.0
400	po, s.d.	4.0	5.70	186.8	82.8	49.3	18.3
800	po, s.d.	10.2	13.20	484.0	54.9	15.8	4.5
150	po, m.d.* day 1	4.7	4.91	65.5[b]	n.a.	38.4	11.5
200	po, m.d. day 1	4.7	6	63.6[b]	n.a.	35.9	9.9
200[a]	po, m.d. day 6	4.06	n.a.	71.40	n.a.	n.a.	n.a.
300	po, m.d. day 6	5.1	4.36	69.8[b]	33.6	42.2	11.3
400	po, m.d. day 6	6.2	8	92.2[b]	36.2	57.1	11.5

See Tables 3 and 11 for abbreviations. *m.d. multiple dose.
[a] MATTINA et al. (1991b).
[b] AUC(0–24).

Table 15. Pharmacokinetic parameters of sparfloxacin in healthy subjects (from SHIMEDA et al. 1993)

Dose (mg)	Administration	C_{max} (mg/l)	t_{max} (h)	AUC ($mg\,h^{-1}\,l^{-1}$)	$t_{1/2}$ (h)
200	po, s.d.	0.70	4.0	18.75	20.8
400	po, s.d.	1.18	5.0	32.73	18.2
600	po, s.d.	1.65	5.0	46.21	20.3
800	po, s.d.	1.97	5.0	57.45	20.0
100	po, s.d.	0.39	2.3	8.5	15.8
200	po, s.d.	0.62	3.5	14.7	15.8
400	po, s.d.	1.36	4.3	34.7	16.9
200	po, s.d.	0.65	4.7	17.64	18.2
400	po, s.d.	1.48	4.7	41.6	19.6
200	po, b.i.d.	0.89[a]	n.a.	14.9[a]	n.a.
200	po, b.i.d.	1.64[b]	n.a.	15.8[b]	14.9
300	po, o.d.	1.00[a]	n.a.	21.8[a]	n.a.
300	po, o.d.	1.40[b]	n.a.	21.9[b]	16.8

For abbreviations, see Tables 3 and 8.
[a] C_{max} and $AUC_{(0-oo)}$ values for first dose presented.
[b] C_{max} and $AUC_{(0-tau)}$ values for steady state presented.

Table 16. Pharmacokinetic parameters of tosufloxacin enantiomeres in healthy subjects. (MINAMI et al. 1993)

Dose (mg)	Enantiomer	C_{max} (mg/l)	t_{max} (h)	AUC ($mg\,h^{-1}l^{-1}$)	$t_{1/2}$ (h)	Ae ur (%)	CL_{ren} (ml/min)
204 (rac)	(+)	0.40	2.60	2.78	3.61	35.4	226
	(–)	0.44	2.40	2.87	3.49	32.4	202

For abbreviations, see Tables 3 and 11.

quinolones have the same characteristics, since the positively charged protonated moieties present in the acidic stomach are less likely to be absorbed, whereas the neutral molecules – which are formed by rapid equilibration with the zwitterionic form present in the slightly alkaline environment of the duodenum and jejunum with pH values of 7–8 – are effectively and nearly completely absorbed (SÖRGEL and KINZIG 1993a). In line with this assumption are findings which have been reported for ciprofloxacin: absorption following administration by nasoduodenal tube was slightly better than by nasogastric tube and both were not significantly different to that following oral administration (YUK et al. 1990).

The absorption of quinolones is rapid, and peak concentrations in serum or plasma (C_{max}) are reached between less than 1 h and up to 3 h. The rate and extent of absorption is independent of gender (HÖFFLER et al. 1984; FRYDMAN et al. 1986; GROS and CARBON 1990; SÖRGEL et al. 1991a). Following oral dosing within the recommended therapeutic range, the C_{max} values and the areas under the concentration–vs–time curves (AUC) for plasma or serum are linearily related to increasing doses of amifloxacin (COOK et al. 1990), ciprofloxacin (BERGAN 1986; BERGAN et al. 1988; BORNER et al. 1986; CAMPOLI-RICHARDS et al. 1988; DRUSANO et al. 1986; HÖFFKEN et al. 1985; HÖFFLER et al. 1984; WINGENDER et al. 1984), enoxacin (TOOTHAKER 1989; ZINNER 1989), fleroxacin (STUCK et al. 1992; NIGHTINGALE 1993; WEIDEKAMM et al. 1987), levofloxacin (DAVIS and BRYSON 1994; NAKASHIMA et al. 1992), lomefloxacin (FREEMAN et al. 1993), norfloxacin (HOLMES et al. 1985; SWANSON et al. 1983), ofloxacin (VERHO et al. 1986; LAMP et al. 1992), pefloxacin (FRYDMAN et al. 1986), rufloxacin (KISICKI et al. 1992; SEGRE et al. 1992), sparfloxacin (SHIMADA et al. 1993), temafloxacin (SÖRGEL et al. 1991b) and tosufloxacin (LATHIA et al. 1991; MINAMI et al. 1993b). In line with linear pharmacokinetics, steady-state plasma levels attained after multiple dosing of fluoroquinolones are elevated compared to first dose levels with the increase depending on the relation of dosing interval to biological half-life; accumulation beyond steady-state concentrations that are predictable from single dose data, has not been reported for healthy subjects (BERGAN 1988).

2. Intravenous Administration

Several fluoroquinolones are available in parenteral formulation. Intravenous infusion of different doses of ciprofloxacin (BEERMANN et al. 1987; BORNER et al. 1986; WINGENDER et al. 1984), enoxacin (ZINNER 1989), fleroxacin (STUCK et al. 1992), ofloxacin (LAMP et al. 1992; LODE et al. 1987a) and pefloxacin (FRYDMAN et al. 1986; SÖRGEL et al. 1988b) confirm the linear pharmacokinetic behaviour of these fluoroquinolones.

Pharmacokinetic model parameters from single dose ciprofloxacin infusions were used for simulation of plasma concentration vs time courses of ciprofloxacin following two-step intravenous infusion regimens (high-rate start phase followed by low-rate maintenance phase). The measured steady-state concentrations at the end of a 6-h infusion (150 mg/h for 2 h followed by 50 mg/h for 4 h in six healthy male subjects) matched exactly the calculated level of 1.0 mg/l, thereby proving the validity of the linear pharmacokinetic model (BEERMANN et al. 1987; WINGENDER et al. 1988).

3. Bioavailability

The systemic availability of orally administered fluoroquinolones was assessed by the quotient of AUC values after oral administration to AUC values after intravenous dosing. Almost complete bioavailability has been reported for fleroxacin (96%–100%; STUCK et al. 1992), levofloxacin (approx. 100%; DAVIS and BRYSON 1994), lomefloacin (>95%; FREEMAN et al. 1993), ofloxacin (95%–100%; LODE et al. 1987) and pefloxacin (98%; FRYDMAN et al. 1986; SÖRGEL et al. 1988b). Two other quinolones have shown good absolute bioavailability: ciprofloxacin (70%–85%; BERGAN et al. 1986, 1988; BORNER et al. 1986; DRUSANO et al. 1986; WINGENDER et al. 1984) and enoxacin (87%–92%; TOOTHAKER 1989; ZINNER 1989). Moderate to poor bioavailability has been estimated for norfloxacin (30%–70%; HOLMES et al. 1985; LODE et al. 1989a) from excretion of unchanged drug into urine and faeces. However, figures for the absolute bioavailability are not known for norfloxacin, rufloxacin, sparfloxacin and tosufloxacin, as intravenous studies with these drugs in man have not yet been reported.

Comparison of bioavailability data and plasma levels between different quinolones has only limited clinical implication. Distribution into infected tissue and bactericidal activity may differ vastly between compounds. In a crossover study, ciprofloxacin, ofloxacin and norfloxacin were compared with respect to their plasma levels and serum bactericidal activities. Whereas concentration profiles of ofloxacin in plasma were about twice as high as those of ciprofloxacin and norfloxacin, the bactericidal activity of ciprofloxacin was higher than that of ofloxacin and norfloxacin by a factor of 4 (BEERMANN et al. 1984; ZEILER et al. 1988).

4. Distribution

Fluoroquinolones bind to plasma proteins only to a relatively small degree, i.e. between 15% and 40% (see Table 17), which has no clinical implications (SÖRGEL and KINZIG 1993b). Distribution in body fluids and tissues is fast and extensive as can be concluded from the volume of distribution ranging between 1.5 and 4 liter per kilogram body weight.

Table 17. Binding of quinolones to plasma proteins. (SÖRGEL and KINZIG 1993b)

Drug	Percentage binding
Norfloxacin	14
Pefloxacin	20–30
Lomefloxacin	21
Fleroxacin	23
Ofloxacin	25
Temafloxacin	26
Tosufloxacin[a]	37
Ciprofloxacin	40
Difloxacin	42
Sparfloxacin	44
Enoxacin	51
Rufloxacin	60

[a] MINAMI et al. (1993b).

Table 18. Penetration of quinolones into skin blister fluid

Drug	Route	Dose (mg)	C_{max} (mg/l)	t_{max} (h)	AUC (mg $h^{-1} l^{-1}$)	Ratio[f]
Pefloxacin[a]	i.v.	400	3.3	1.8	38.8	0.69
Fleroxacin[b]	po	400	3.8	4.0	70.4	0.90
Rufloxacin[c]	po	400	3.2	3.5	91.6	0.90
Lomefloxacin[d]	po	400	3.5	2.7	n.a.	1.00
Norfloxacin[a]	po	400	1.0	2.3	5.8	1.07
Enoxacin[a]	po	600	2.9	3.7	32.8	1.14
Sparfloxacin[e]	po	400	1.3	5.0	n.a.	1.17
Ciprofloxacin[a]	po	500	1.4	2.6	11.6	1.17
Ciprofloxacin[a]	i.v.	100	0.6	1.25	3.4	1.21
Ofloxacin[a]	po	600	5.2	5.3	71.8	1.25
Enoxacin[a]	i.v.	400	2.2	0.5	23.1	1.30

n.a., data not available.
[a] WISE et al. (1986).
[b] WISE et al. (1987).
[c] WISE et al. (1991).
[d] STONE et al. (1988).
[e] JOHNSON et al. (1992).
[f] Ratio = $AUC_{serum}/AUC_{blister\ fluid}$.

An approach to compare the distribution characteristics of various fluoroquinolones is the assessment of their penetration into inflammatory blister fluid. Wise and co-workers have investigated a wide spectrum of compounds using comparable protocols (WISE et al. 1986, 1987, 1991; JOHNSON et al. 1992; STONE et al. 1988). Their technique is to raise skin blisters by taping cantharides plasters to the anterior aspect of the forearm on the night before dosing the subjects. The results of these studies (see Table 18) demonstrate excellent penetration for most quinolones with AUC ratios of between 90% and 130%. An appraisal of attained levels in various tissues and their relevance for therapy is given in Chap. 12 of this volume.

5. Disposition

The disposition of fluoroquinolones and their metabolites occurs via four basic routes: (a) renal excretion into urine by both glomerular filtration and tubular secretion, (b) hepatic metabolism, (c) biliary secretion and faecal excretion and (d) transintestinal secretion across the colon mucosa into the faeces. Because of the different degrees to which these pathways are involved, the various quinolone drugs behave differently with respect to their elimination characteristics. Moreover, impairment of renal and/or liver function changes these characteristics to different degrees and requires drug-specific alterations of dosing regimens in the affected patients (see below).

The biological half-life indicates the time course characteristic for the disposition from plasma or serum. In order to balance the considerable variation recorded within studies and between studies, BERGAN (1988) calculated mean values for half-life by using the number of individual time courses studied as weighting factors (see Table 19).

Three groups of quinolones can easily be distinguished:

1. Ciprofloxacin, norfloxacin, enoxacin and ofloxacin are eliminated with terminal half-lives of 4–5 h; these drugs are recommended for twice daily dosing regimens.
2. Pefloxacin and fleroxacin have biological half-lives of 9–12 h and are thus candidates for once daily administration schemes.
3. Rufloxacin has a half-life of 26 h which requires a 50% reduction after the first dose for continuation on a once daily dosing regimen (e.g. 400 mg on day 1 followed by 200 mg on day 2 and subsequent days).

Table 19. Half-lives of quinolones

Drug	Number of time courses	$t_{1/2}$[a] (h)	Reference
Ciprofloxacin	826	4.0	BERGAN (1988)
Norfloxacin	205	4.6	BERGAN (1988)
Enoxacin	105	4.7	BERGAN (1988)
Ofloxacin	479	4.8	BERGAN (1988)
Fleroxacin	54	9.6	BERGAN (1988)
Pefloxacin	59	11.9	BERGAN (1988)
Difloxacin	23	25.7	BERGAN (1988)
Rufloxacin	45	34–83	SEGRE et al. (1992)
Rufloxacin	32	28–54	KISICKI et al. (1992)

[a] Mean half-life given for all drugs except rufloxacin.

a) Renal Elimination

Fluoroquinolones undergo glomerular filtration and most of them are also actively secreted by the proximal tubules as either acids or bases (SÖRGEL and KINZIG 1993b). The anionic secretion mechanism was proven for ciprofloxacin (JAEHDE et al. 1989) and fleroxacin (SHIDA et al. 1989) by blockage using probenecid. The cationic transport system may be inhibited by cimetidine; this was shown for ciprofloxacin (WINGENDER et al. 1985), fleroxacin (PORTMANN 1992) and temafloxacin (SÖRGEL 1992).

Pefloxacin, difloxacin, rufloxacin, sparfloxacin and – to a small extent – fleroxacin show tubular reabsorption. This was concluded from the observation that their renal clearance is lower than that calculated from the glomerular filtration rate and unbound fraction in serum (SÖRGEL and KINZIG 1993b). These different mechanisms of renal elimination of the quinolones give rise to potential drug interactions (see below).

b) Metabolism

With the exception of sparfloxacin, which is largely glucuronidated, and pefloxacin, which can be regarded as a prodrug of norfloxacin (identical to pefloxacin's major metabolite desmethyl pefloxacin) and which is excreted unchanged only to an extent of 5%–10% (FRYDMAN et al. 1986), the fluoroquinolones are metabolized to an extent of less than 25%. Detailed data are summarized in Table 20.

The quinolone antibiotics are mainly metabolized by oxidation at the heterocyclical substituent ring system to form oxo derivatives, *N*-oxides, *N*-desmethyl, *N*-sulphonyl and *N*-formyl compounds, and products resulting from dealkylation or cleavage of the piperazine ring (desethylene compounds). Glucuronidation seems to be limited to a few compounds (difloxacin and sparfloxacin). Less than 5% of ofloxacin is excreted in urine as metabolites. Even in renal failure, metabolite concentrations do not achieve clinical relevance in plasma or urine (LAMP et al. 1992). A first pass effect had been suggested for ciprofloxacin (BORNER et al. 1984), but based on data from a study with ^{14}C-labelled ciprofloxacin it was estimated to be less than 5% (BEERMANN et al. 1986).

II. The Elderly

Pharmacokinetic parameters usually show a larger variation in the elderly population (including both genders) than in young healthy males. Age-related changes of the body composition (muscle to fat ratio) and the main organ functions (cardiac output, renal clearance, metabolic capacity) may affect the pharmacokinetics of a drug (MAYERSOHN 1992). The age-related changes of lean body mass exert an influence on distribution and hence on pharmacokinetic parameters in elderly subjects.

Table 20. Metabolism of fluoroquinolones (excretion into urine and faeces in percent of dose). (SÖRGEL and KINZIG 1993b)

Drug	Urine		Faeces	
	Parent drug	Metabolites	Parent drug	Metabolites
Ciprofloxacin	61.5	9.5	15.2	2.6
Desethylene		1.3–2.1		
Oxo		3.9–6.2		
Difloxacin	9.8	18.6	n.a.	n.a.
Glucuronide		11.2–12.4		
N-oxide		2.1–3.2		
Desmethyl		3.1–4.2		
Desethylene		0.1–0.2		
Enoxacin	53.6	12.5	n.a.	n.a.
Oxo		11–12		
Fleroxacin	61	15	15	n.a.
Glucuronide		3.6		
Desmethyl	4.7–7.4			
N-oxide		2.9–6.3		
Lomefloxacin	71.8	10	n.a.	n.a.
Glucuronide		5.8–8.0		
Desethylene		0.1		
Norfloxacin	20–40	10	n.a.	n.a.
Desethylene		1.5		
Oxo		5.1		
Ofloxacin	90	5–10	n.a.	n.a.
Desmethyl		3.0–3.2		
N-oxide		1.0–1.1		
Pefloxacin		35	n.a.	n.a.
Desmethyl		16.2–23.2		
N-oxide		15.8–20.2		
Oxo		0.75		
Rufloxacin	30–50	2	n.a.	n.a.
Desmethyl		2		
Sparfloxacin	10–15	25–35	n.a.	n.a.
Glucuronide		28.2 + –6.8		
Temafloxacin	70	n.a.	n.a.	n.a.
Glucuronide		2–5		

n.a., data not available.

Due to alterations in the gastrointestinal tract which occurs with increasing age, a decrease in drug absorption may be expected (LOI and VESTAL 1988). Whereas no differences were observed between young and elderly subjects for the time required to reach maximal serum or plasma levels (t_{max}) of quinolones (SCHENTAG and GOSS 1992), the peak levels attained (C_{max}) were consistently higher in the elderly. Consequently, the area under the concentration vs. time curve (AUC) was larger for elderly subjects than for young subjects. The respective parameters for ciprofloxacin (BALL et al. 1986; BAYER et al. 1987; GUAY et al. 1987; LEBEL et al. 1986; LEBEL and BERGERON 1987;

NABER et al. 1989), enoxacin (DOBBS et al. 1987), fleroxacin (WEIDEKAMM 1987, cited by RICHTER and LEBEL 1993), ofloxacin (MOLINARO et al. 1992), norfloxacin (NABER and BARTOSIK-WICH 1984), lomefloxacin (CROME and MORRISON 1991) and rufloxacin (COGO et al. 1992; MATTINA et al. 1991) are shown in Tables 3, 5, 7, 10, 12, 21, and 22.

These observations can be explained by several factors: (1) a reduced volume of distribution as a consequence of less lean body mass and a higher proportion of fat tissue; (2) a reduced clearance as a consequence of decreasing function of eliminating organs (liver, kidneys); (3) an increased absorption due to decreased gastric emptying time and different gastric pH.

Another possible explanation for smaller volumes of distribution could be serum protein binding. It is, however, too low for the fluoroquinolones – ranging from 14 to 60% – to cause a change in distribution patterns with decreasing serum albumin levels in the elderly. An additional possible reason for increased levels of quinolones in the serum of elderly patients is the decrease in hepatic blood flow associated with age as well as the reduced activity of liver microsomal enzyme systems (MAYERSOHN 1992), which are responsible for metabolism of several quinolones. A corresponding reduction in nonrenal clearance was reported for ciprofloxacin (BALL et al. 1986; BAYER et al. 1987; LEBEL et al. 1986) and enoxacin (DOBBS et al. 1987; NABER et al. 1986). Since the most important route of elimination of fluoroquinolones is via the kidneys, compromised renal function – which is frequently seen in elderly patients – has a marked influence on the pharmacokinetics of many fluoroquinolones and, hence, on dosing recommendations.

Table 21. Pharmacokinetic parameters of lomefloxacin in elderly subjects (from CROME and MORRISON 1991)

Dose (mg)	Administration	Age (years)	C_{max} (mg/l)	t_{max} (h)	$t_{1/2}$ (h)	CL_{tot}/f (ml/min)
200	p.o., s.d.	65–85	2.40	0.96	8.54	146.9
400	p.o., s.d.	65–85	4.48	1.31	8.75	143.0

Table 22. Pharmacokinetic parameters of rufloxacin in elderly subjects

Dose (mg)	Administration	Age (years)	C_{max} (mg/l)	t_{max} (h)	$t_{1/2}$ (h)	CL_{tot}/f (ml/min)
400[a]	p.o., s.d.	55–75	3.17	4.2	n.d.	n.d.
400[a]	p.o., st. st.	55–75	7.26	5.4	38.2	49.0
400/200[b]	p.o., st. st.	63–80	6.46	4.3	28.7	35.0

[a] Data from RIMOLDI et al. 1992.
[b] Dose on first day 400 mg, thereafter 200 mg once daily; data from COGO et al. 1992.

III. Patients with Various Degrees of Renal Failure

Renal elimination plays a predominant role in the disposition of quinolone drugs and their metabolites. Thus, in absence of alternate routes of elimination, a dosage adjustment in renal impairment is absolutely necessary for most quinolones. This rule is exemplified by ofloxacin, which is largely excreted unchanged by the kidneys of healthy volunteers (HÖFFLER and KOEPPE 1987). Other quinolones either follow the renal route of elimination to a smaller extent or are subject to alternate mechanisms of elimination. An exception to the rule is pefloxacin; the majority of a pefloxacin dose is excreted only as metabolites in the urine (FRYDMAN et al. 1986).

1. Ciprofloxacin

Neither absorption (PLAISANCE et al. 1987) nor distribution (WEBB et al. 1986) of ciprofloxacin are significantly changed by renal dysfunction. Since renal clearance of ciprofloxacin in healthy subjects greatly exceeds creatinine clearance, indicating a significant contribution of tubular secretion (WINGENDER et al. 1984; DRUSANO et al. 1987), a distinct effect of renal dysfunction on its disposition may be assumed. Indeed, renal excretion of a parent drug and its metabolites are reduced threefold in patients with severe renal impairment (creatinine clearance CL_{crea}, of less than 20ml/min) (ROHWEDDER et al. 1990); however, the elimination half-life is only increased to 10–12h (BOELAERT et al. 1985; WEBB et al. 1986; DRUSANO et al. 1987; GASSER et al. 1987a,b; SINGLAS et al. 1987; ROHWEDDER et al. 1990). The reason for this apparent contradiction is transintestinal elimination, which becomes a major route of elimination of ciprofloxacin in these patients and thus serves as a safety valve of excretion which compensates for reduced renal function (see Table 23; ROHWEDDER et al. 1990; BERGAN et al. 1989b). Thus, ciprofloxacin is not accumulated significantly in patients with renal failure, who in general do not require adjustment of the ciprofloxacin dosage. Only in cases of terminal renal impairment is a dose reduction to 50% or doubling of the dosing interval recommended. Only 2% of a ciprofloxacin dose is removed during haemodialysis sessions

Table 23. Excretion of ciprofloxacin and metabolites after 200mg intravenous infusion in healthy subjects and patients with severe renal impairment (in percent of dose). (ROHWEDDER et al. 1990)

	Urine		Faeces		Total
	Parent drug	Metabolites	Parent drug	Metabolites	
Normal renal function	65.3 ± 10.71	12.2 ± 2.3	11.4 ± 2.6	7.3 ± 1.6	96.3 ± 14.1
Renal failure	19.0 ± 15.9	5.8 ± 5.1	37.2 ± 12.5	26.2 ± 6.5	88.1 ± 20.9

Data given as mean ± SD; n = 5.

(BOELAERT et al. 1985) and even less during continuous ambulatory peritoneal dialysis (CAPD) treatment (SHALIT et al. 1986).

2. Norfloxacin

Elimination half-lives of norfloxacin do not increase beyond twofold, even in patients with glomerular filtration rates of less than 20 ml/min (ARRIGO et al. 1985; EANDI et al. 1983). Thus, similar dosage adjustments are recommended for severe renal impairment as for ciprofloxacin.

3. Ofloxacin

Approximately 75%–80% of an oral ofloxacin dose is eliminated via the kidneys (LODE et al. 1987a; VERHO et al. 1985). Mild to moderate renal impairment leads to four- to ten-fold prolonged elimation half-lives which require altered dosing regimens (FILLASTRE et al. 1987; HÖFFLER and KOEPPE 1987). Ofloxacin dosage must be carefully adapted to the degree of renal impairment with one doubling of the dosing interval to 24h for moderate renal failure (CL_{crea} > 20 ml/min) and a second doubling to 48h for patients with severe renal impairment (CL_{crea} < 20 ml/min; FILLASTRE et al. 1987). Approximately 20% of an ofloxacin dose is removed by haemodialysis (KAMPF et al. 1990), leading to a recommendation of additional postdialysis dosing, at least after the first session (DÖRFLER et al. 1987).

4. Enoxacin

In patients with severe renal failure (CL_{crea} < 15 ml/min), the clearance of both enoxacin and its oxo metabolite is reduced, resulting in an elimination half-life of up to 20h for the parent drug (VAN DER AUWERA et al. 1990; NIX et al. 1988). It is therefore recommended to reduce the daily dose by 50% for patients with CL_{crea} < 30 ml/min (VAN DER AUWERA et al. 1990). Clearance by haemodialysis is not significant; thus, supplemental doses after dialysis are not needed (WHITE et al. 1988).

5. Fleroxacin

Two studies investigating the pharmacokinetics of fleroxacin in renally impaired patients have led to the recommendation to modify the dosage regimen in patients with CL_{crea} < 30 ml/min per 1.73 m^2, keeping the first dose as a loading dose and reducing subsequent doses by 50% (maintenance doses; SINGLAS et al. 1990; STUCK et al. 1989). Patients on haemodialysis may require additional postdialysis dosing (SINGLAS et al. 1990; STUCK et al. 1989).

6. Pefloxacin

Although plasma clearance and volume of distribution are not changed by severe renal impairment, dosage adjustment by 50% is recommended in patients with moderate and severe renal impairment (HÖFFLER et al. 1988). The reason is a possible accumulation of pefloxacin metabolite concentrations in plasma, i.e. *N*-oxide and norfloxacin, by decreased renal function (HÖFFLER et al. 1988; JUNGERS et al. 1987).

7. Lomefloxacin

In several studies involving patients with various degrees of renal impairment, single doses of 400 mg were excreted with half-lives of up to 44 h (BLUM 1992; LEROY et al. 1990; NILSEN et al. 1992). In patients with severe renal impairment ($CL_{crea} < 30$ ml/min per 1.73 m^2) a decrease of the dose by 50% is recommended (BLUM et al. 1990; NILSEN et al. 1992). Patients undergoing haemodialysis do not need additional doses after dialysis treatment, since the half-life is not significantly changed from approximately 30 h in the predialysis period (LEROY et al. 1990).

8. Summary

The overall elimination profile of the quinolones renders them primary candidates for the treatment of infections in patients with various degrees of renal dysfunction. Their use in renal failure patients is safe, as long as the dose or the dosing interval is adjusted to compensate for the reduced elimination capacity of the kidneys. However, possible accumulation of metabolites as result of reduced renal function and its implication for the safety of the newer fluoroquinolones need further study.

IV. Patients with Hepatic Failure

Compared with the extensive knowledge available on the effect of renal impairment, there is only little information available on the effect of hepatic failure on the pharmacokinetics of fluoroquinolones. This fact may relate to the smaller prevalence of patients with liver failure and to the lack of clearly defined parameters for assessing liver function in relation to drug pharmacokinetics.

From a study with fleroxacin in cirrhotic patients with and without ascites it was recommended to reduce the dose by 50% in cirrhotics with ascites (BLOUIN et al. 1992). Clearance of pefloxacin is markedly decreased in cirrhotic patients (DANAN et al. 1985), which warrants an increased dosing interval (RICHER and LEBEL 1993). Since ofloxacin is almost exclusively eliminated via the renal route, no dosing adaptation should be required in hepatic insufficient patients. However, in a study with 12 cirrhotic patients with normal serum creatinine levels, a doubling of elimination half-life was observed (SILVAIN et al. 1989), which may be due to an impaired tubular secretion of ofloxacin in the cirrhotic patients. In order to prevent accumulation, individualized dosage adjustments may be required (RICHER and LEBEL 1993).

Studies with lomefloxacin (LEBREC et al. 1992) and ciprofloxacin (FRAISE and SMITH 1990; FROST et al. 1989) in cirrhotic patients gave no evidence of altered elimination kinetics. Thus, cirrhotic patients may be treated with the same dose regimens as used in patients without hepatic failure.

V. Fluoroquinolones in Pediatric Patients

With the exception of norfloxacin which was approved in 1991 for use in children in Japan (FUJII 1992), none of the fluoroquinolones has yet been licensed for use in children in any major market of the world due to cartilage damage induced in young animals (CHRIST et al. 1988; see also Chap. 10, this volume). However, there seems to be no clinical correlate to this experimental finding (KUBIN 1993). Recently a consensus report was published on the use of fluoroquinolones in pediatrics (SCHAAD et al. 1995) which indicates that ciprofloxacin is safe and efficacious for use in selected pediatric patients, for example, those suffering from cystic fibrosis (CF). Although a considerable number of successful treatments of children have been published (KUBIN 1993), only very limited pharmacokinetic data are available which could serve as a basis for dosage recommedations in children of different age groups.

Norfloxacin was investigated in 76 out of 406 patients aged 5 years and older and in 23 children of 2–4 years of age who suffered from infections considered to be an indication for norfloxacin (FUJII 1992). Dose-proportional pharmacokinetics were demonstrated with C_{max}, AUC and $t_{1/2}$ values similar to those reported in adults. Based on these data and on clinical outcome, the following dosing recommendations for norfloxacin use in children were derived:

- Administration to pediatric patients only if other antibiotics are not effective
- Dosage of 2–4 mg per kilogram body weight orally three times daily
- Dose adjustment according to symptoms
- Duration of treatment usually 1 week
- No concommitant use of phenyl acetic or propionic acid NSAIDs

For ciprofloxacin, several reports on pharmacokinetic studies in children were published during the past few years. From the treatment of seven infants (aged 9–14 weeks) and nine young children (aged 1–5 years) with a dose of 15 mg/kg ciprofloxacin, PELTOLA et al. (1992) concluded that elimination of ciprofloxacin is faster in young children ($t_{1/2} = 1.28 \pm 0.52$ h) than in infants (2.73 ± 0.28 h), requiring a dosing scheme of 10–15 mg/kg three times daily in this age group (1–5 years). In another study, sequential therapy of 11 children (aged 6–12 years) with intravenous ciprofloxacin (10 mg/kg three times daily) followed by oral ciprofloxacin (20 mg/kg every 12 h) gave rise to the conclusion that ciprofloxacin at 20 mg/kg twice daily would be effective to eradicate *Pseudomonas aeruginosa* from young CF patients (RUBIO et al. 1994).

The first detailed study to investigate the pharmacokinetics of ciprofloxacin in children was undertaken in ten CF patients aged 6–16 years (SCHAAD et al. 1996; SCHAEFER et al. 1996). Ciprofloxacin was administered twice daily by intravenous infusion (10 mg/kg) for one day followed by oral dosing (15 mg/kg). Plasma concentration vs time profiles had similar shapes for young and adult CF patients and healthy adult volunteers. Population pharmacokinetic calculations led to a formula which allows a calculation of the daily dose for children, as a function of their body weight, which is necessary for achieving steady-state levels in plasma similar to those obtained in adults after administration of 750mg ciprofloxacin twice daily orally or 400mg three times daily intravenously. The calculated oral dosage regimens suggest that younger children (14–28kg body weight) should receive oral ciprofloxacin at 20–28mg/kg twice daily whereas older children (28–42kg) should be treated with 15–20mg/kg twice daily. For intravenous administration, doses of 10–15mg/kg twice daily are sufficient to achieve steady-state plasma concentrations which have been shown to be effective in adults.

A similar dosage recommendation based on body weight was given by STASS et al. (1996) who investigated a new liquid formulation of ciprofloxacin for oral treatment in 16 non-CF infected children (aged < 1–12 years). Despite a strong correlation between apparent clearance of ciprofloxacin and age, a dose regimen of 10mg/kg administered to children three times daily was considered equivalent to daily treatment with three intravenous 400mg doses or two 750mg oral doses in adults, assuming that AUC/MIC is predictive for clinical and microbiological outcome.

C. Interactions of Fluoroquinolone Antibiotics with Other Drugs

The availability of intravenous and oral dosage forms of fluoroquinolones allows the sequential therapy of patients. Starting with intravenous treatment in the hospital the patient can be put on oral treatment as soon as improvement has been shown, resulting in shorter hospital stays. With the use of fluoroquinolones in outpatients the likelihood of concomitant treatment with other drugs, especially with OTC drugs, is increased and drug–drug interactions can be expected. Pharmacokinetic interactions can occur during absorption, metabolism and excretion. Protein binding of fluoroquinolones ranges from 20%–60% and therefore drug–drug interactions by displacement from protein binding are unlikely. In addition, pharmacodynamic interactions can occur due to similar receptor affinity or antagonistic/agonistic effects (Table 24). This chapter summarises the current knowledge of drug–drug interactions for fluoroquinolones.

I. Interactions During the Absorption Process

Ciprofloxacin is the only fluoroquinolone for which absorption along the gastrointestinal tract was investigated using the technique of the high frequency capsule (HARDER et al. 1990). This study showed that ciprofloxacin was primarily absorbed in the duodenum and to a smaller extent in the jejunum (relative bioavailability of 37%), whereas in lower parts of the gastrointestinal tract (colon) the bioavailability is low (5%–7%). In the more distal part of the gastrointestinal tract, ciprofloxacin is actively secreted from the blood into the lumen (JAEHDE et al. 1989; MAHR et al. 1989). This is most likely also the case

Table 24. Drug–drug interactions of fluoroquinolone antibiotics

Interactions during the absorption process
Food and dairy products
Al^{3+}-, Ca^{2+}-, Mg^{2+}-containing antacids
Sucralfate
Didanosine
Other metal cations
Chemotherapeutic agents
Activated charcoal
Interactions of fluoroquinolones due to alterations in metabolism
Theophylline, caffeine and structurally, closely related substances
Antipyrine
Phenytoin
H_2-receptor antagonists
K^+/Na^+-ATPase inhibitors
Warfarin
Cyclosporine
Rifampin
Oral contraceptive steroids
Benzodiazepines (diazepam, temazepam)
Alterations in renal excretion
Probenecid
β-Lactam antibiotics
Pharmacodynamic interactions
NSAIDs
Metronidazole

for other fluoroquinolones, as shown by SÖRGEL and KINZIG (1993a) using the charcoal model. WINGENDER et al. (1986) investigated the effect of gastric emptying on drug absorption using intravenously administered *N*-butyl-scopolamine bromide (40 mg) and metoclopramide (10 mg), which delay or accelerate gastric motility, respectively. It was shown that *N*-butyl-scopolamine bromide delayed the absorption of ciprofloxacin, prolonging the time required to reach the maximal concentration and slightly increased the bioavailability (+14%), whereas metoclopramide accelerated the absorption of ciprofloxacin (shorter time required to reach maximum concentration) and slightly reduced the bioavailability (–10%). Similar results were obtained for pirenzepine, an anticholinergic drug for the treatment of gastritis and gastric/duodenal ulcers, which reduces gastric motility and the stomach emptying rate. The absorption of ofloxacin and ciprofloxacin was prolonged, whereas the bioavailability was not changed (DEPPERMANN and LODE 1993). Therefore, all influences which result in a decrease in gastroduodenal transit time or prevent the absorption of fluoroquinolones in the upper gastrointestinal tract can result in a relevant interaction.

1. Food and Dairy Products

The administration of fluoroquinolones (ciprofloxacin, fleroxacin, pefloxacin) with food does not reduce the amount absorbed, but changes the rate of absorption, resulting in reduced maximal drug concentrations in plasma and a prolongation of the time required to reach the maximal concentration. For tosufloxacin, the extent of absorption increased by about 40% after administration with a normal breakfast (TAI et al. 1988; SÖRGEL and KINZIG 1993a). Only for rufloxacin was a decrease in the extent of absorption reported (SEGRE et al. 1992). However, this study was performed as a group comparison and did not have a crossover design, which is more appropriate for investigating the effect of food. Overall, the observed food effect for the fluoroquinolones has no clinical significance. Also the intake of ciprofloxacin or ofloxacin together with enteral feeding does not influence the extent of absorption (YUK et al. 1989; MUELLER et al. 1993). There are no data for other quinolones.

In several of the morning food effect studies for ciprofloxacin, a high calcium meal or a standard meal with milk was given, and no reduction in the extent of absorption was observed. However, the simultaneous intake of ciprofloxacin and 300 ml milk or 300 mg yoghurt in a fasting state reduced the bioavailability by 33% and 36%, respectively (NEUVONEN et al. 1991). The maximum ciprofloxacin plasma concentration decreased by 36% and 47%, whereas the time required to reach the maximum concentration increased by 8% and 42%, respectively. For norfloxacin (KIVISTOE et al. 1992) the same study design resulted in a 50% reduction in the extent of absorption and in maximum norfloxacin plasma concentrations. Also 200 ml milk can reduce the bioavailability of norfloxacin (MINAMI et al. 1993a). No interaction was found when ofloxacin was given 0.5 h after 8 oz milk (DUDLEY et al. 1991), and only a minor, clinically irrelevant reduction in bioavailability of 6% and 9%, when given together with 300 ml milk or 300 ml yoghurt (NEUVONEN and KIVISTOE 1992). Similarly, the bioavailability of enoxacin was not influenced by milk (LEHTO and KIVISTOE 1995). Data for other fluoroquinolones are lacking. The effect of milk is most likely caused by Ca^{2+}-ions contained in the dairy products. Therefore, the results reported on Ca^{2+} in the following paragraph (no. 2) can be extrapolated to the situation when dairy products are taken. The simultaneous intake of fluoroquinolones with milk and dairy products should be avoided, with the exception of enoxacin and ofloxacin.

2. Al^{3+}-, Ca^{2+}- and Mg^{2+}-Containing Antacids

Fluoroquinolones can form chelation complexes via the 3-carboxyl and 4-oxo functional groups with various polyvalent metal ions such as Al^{3+}, Mg^{2+}, Ca^{2+} and Fe^{2+} (GUGLER and ALLGAYER 1990; TUNCEL and BERGISADI 1992; ZUPANCIC et al. 1991). It is thought that these complexes cannot be absorbed, which leads to a reduced bioavailability of the fluoroquinolones. The complex is formed in a ratio (metal ion/fluoroquinolone) of up to 1:3 (KAKANO et al. 1978; OKAZAKI et al. 1988). For ciprofloxacin HÖFFKEN et al. (1985) demonstrated that the bioavailability in healthy volunteers is reduced when given simultaneously with aluminium–magnesium antacids (Maalox: 10 × 600 mg $Mg(OH)_2$ and 900 mg $Al(OH)_3$). The amount of ciprofloxacin excreted in urine decreased by 91%. In subsequent reports (FLEMING et al. 1986; PREHEIM et al. 1986) it was shown that this interaction has a high clinical relevance, especially in patients undergoing CAPD who received ciprofloxacin and aluminium-containing phosphate-binding antacids.

In order to find the appropriate time interval between antacid and ciprofloxacin dosing, three seperate studies have been performed (NIX et al. 1989b,c). As an antacid, 30 ml Maalox were used (17.3 mmol of aluminium ion and 20.6 mmol of magnesium ion per 30 ml) and administered 2, 4 and 6 h before, simultaneously with and 2 h after administration of 750 mg ciprofloxacin. The mean relative bioavailabilities, compared to fasting state, decreased by 77%, 30%, 0%, 85% and 0%, respectively. Therefore it was concluded that antacids should be given either 6 h before or 2 h after ciprofloxacin administration. Also aluminium- or magnesium-containing antacids alone can reduce the bioavailability of ciprofloxacin (FROST et al. 1992). After administration of 750 mg ciprofloxacin 5 min after 4 × 600 mg $Al(OH)_3$ tablets the relative bioavailability was decreased by 84%. Relative to the control, 100 ml of a magnesium citrate solution (8 g active compound) reduced ciprofloxacin's bioavailability to 21% (BROUWERS et al. 1990a).

With regard to calcium the situation remains unclear: FLEMING et al. (1986) found no difference between C_{max} and C_{min} in patients receiving calcium carbonate capsules (Titralac) and the control group. However, the time interval was not specified. LOMAESTRO and BAILIE (1991) showed that a single dose of calcium carbonate given 2 h prior to ciprofloxacin does not change the extent of absorption but increases C_{max} (1.98 vs 2.42 mg/l). SAHAI et al. (1993a,b) investigated the effect of a calcium supplement (Oscal – 500 mg elemental calcium) given for 5 days three times daily prior to 500 mg ciprofloxacin. The relative bioavailability compared with ciprofloxacin intake without Oscal was 59%. The intake of ciprofloxacin together with a high calcium meal reduced its bioavailability only by 10% (CARLSON et al. 1988). 4 × 850 mg $CaCO_3$ tablets given 5 min prior to 750 mg ciprofloxacin reduced the bioavailability by 42% (FROST et al. 1992). LOMAESTRO and BAILIE (1993) determined the effect on the relative bioavailability of a staggered single dose of ciprofloxacin given 2 h after a morning dose of calcium carbonate given three times daily over the previous days. The relative bioavailability was 87%, which was not significantly different to the control. When Maalox TC (1.8 g $Mg(OH)_2$ and 3.6 g $Al(OH)_3$), was given 0.5, 2 and 8 hours before 400 mg enoxacin, the bioavailability of enoxacin was reduced by 73%, 49% and 9%, respectively (GRASELA et al. 1989).

For lomefloxacin, the simultaneous intake with Maalox resulted in a 52% decrease in bioavailability, whereas the antacid given 12 h before or 4 h after did not result in a change in bioavailability compared to the control (KUNKA et al. 1988). FOSTER and BLOUIN (1991) showed that the bioavailability of lomefloxacin is only slightly reduced, when given 2 or 4 h after Maalox (–20%, –10%) or 2 h before Maalox (–12%). The effect of the simultaneous administration of Kolantyl (400 mg $Al(OH)_3$, 200 mg MgO, 5 mg dicyclonune HCl) and the nonmetallic gastrointestinal agent Ulgut (200 mg benexate HCl-cyclodextrin clathrate) on lomefloxacin's bioavailability was investigated in

Japanese volunteers. Ulgut had no effect on bioavailability, whereas Kolantyl led to a 40% reduction (SHIMADA et al. 1992).

For levofloxacin, the *S*-(−)-enantiomer of ofloxacin, the concomitant administration of aluminium hydroxide (1 g) resulted in a significant reduction in AUC and C_{max} to 44% and 65% of control, respectively. After the concomitant administration of 500 mg magnesium oxide, a significant decrease in AUC and C_{max} was observed (78% and 62% of control, respectively) but lower than that for aluminium hydroxide coadministration. Coadministration of 500 mg calcium carbonate did not impair the pharmacokinetics of levofloxacin (SHIBA et al. 1992, 1993).

The reduction in norfloxacin urinary recovery was 90% when given together with 1632 mg $Mg(OH)_2$ (milk of magnesia United States Pharmacopeia) and 86% with $Al(OH)_3$ (10 ml Amphojel) indicating that norfloxacin bioavailability is reduced by an amount similar to that for ciprofloxacin (CAMPBELL et al. 1992). When Maalox (30 ml) was given 5 min before norfloxacin, the urinary excretion was reduced by more than 90%. Norfloxacin administration 2 h before Maalox (30 ml) resulted in a 30% reduction. For calcium carbonate (30 ml Titralac) given as a 5-min pretreatment, a 62% reduction was observed (NIX et al. 1989a, 1990).

ALBRECHT et al. (1992) reported that the ofloxacin maximal serum concentration was reduced by about 60% when given for 3 days concomitantly with Maalox 70. FLOR et al. (1990) investigated the influence of 15 ml Maalox and 5 ml Titrilac ($CaCO_3$) on ofloxacin bioavailability. No effect was seen for calcium carbonate and Maalox given 24 h before or 2 h after ofloxacin intake, whereas the administration of Maalox 2 h prior to ofloxacin intake lead to a 20% reduction in bioavailability. CARBARGA et al. (1991) administered ofloxacin together with 1 g colloidal aluminium phosphate (Fosfalumina). The total amount excreted in urine was only slightly reduced (−7%) compared to the control. The simultaneous intake of ofloxacin with Maalox (600 mg $Mg(OH)_2$ and 900 mg $Al(OH)_3$) resulted in a 73% decrease in bioavailability (HÖFFKEN et al. 1988). Five-hundred milligrams of magnesium trisilicate and calcium carbonate had no effect on the bioavailability of ofloxacin, whereas aluminium hydroxide reduced the saliva concentration vs time AUC by about 20% (AKERELE and OKHAMAFE 1991). The interaction between Maalox and fleroxacin resulted in an approximately 26% decrease in bioavailability (LODE et al. 1987b).

The clinical importance of the interaction between fluoroquinolones and antacids is significant, since a decrease or blockage in absorption prevents access of the drugs to the infection site and thus leads to therapeutic failure. In addition this situation might favour the selection of resistant strains.

3. Sucralfate

Sucralfate is a complex of aluminium hydroxide and sulphated sucrose. The sucrose binds to proteins in ulcerated tissue forming a barrier against acid diffusion across the gastrointestinal mucosa. One gram sucralfate contains

190 mg aluminium. Two- and 6-h pretreatments with sucralfate resulted in a 30% decrease in ciprofloxacin absorption (Nix et al. 1989a,b,d). Concurrent administration of 1 g sucralfate reduced the bioavailability by 87.5% (Garrelts et al. 1990), whereas 2 g sucralfate reduced the bioavailability by 96% compared with the control (Brouwers et al. 1990a). Similar results were obtained by Van Slooten et al. (1991) when 2 g sucralfate were given concurrently, 2 h and 6 h after 750 mg ciprofloxacin. The relative bioavailabilities were 4.3%, 82.9% and 96.5%, respectively. Administration of 1 g sucralfate 2 h prior to or concomitantly with norfloxacin resulted in a 43% and 98% reduction in bioavailability (Nix et al. 1989a). The relative bioavailability of fleroxacin given with 1 g sucralfate was 76% compared to the control (Lubowski et al. 1992). Also lomefloxacin bioavailability is reduced by 51% when given simultaneously with 1 g sucralfate (Lehto et al. 1994a).

4. Didanosine

Due to didanosine's extreme acid lability at pH < 3, the marketed formulation (Videx, Bristol Meyer Sqibb) is a chewable tablet. It contains dihydroxy aluminium sodium carbonate, magnesium hydroxide and sodium citrate. In order to evaluate the magnitude of interaction of this formulation with ciprofloxacin, two didanosine placebo tablets were administered simultaneously with ciprofloxacin (Sahai et al. 1993b). The relative bioavailability of ciprofloxacin was reduced by 98%. A later study found that 500 mg ciprofloxacin 2 h after two didanosine-placebo tablets (i.e. all antacid additives but didanosine) reduced the ciprofloxacin serum levels below minimal inhibitory concentrations, but ciprofloxacin 2 h before gave normal blood levels (Sahai 1995). Therefore, ciprofloxacin should be given at least 2 h before or 6 h after didanosine. Other quinolone antibiotics interacting with antacids are expected to interact similarly but so far reports are lacking.

5. Other Metal Cations

Besides Al^{3+}, Mg^{2+} and Ca^{2+}, other cations can form chelate complexes with fluoroquinolones and reduce their bioavailability. The administration of 200 mg Fe^{2+} as a Fe^{2+}-glycine-sulphate complex with ciprofloxacin and ofloxacin resulted in a 48% and 36% reduction in bioavailability of the fluoroquinolones, respectively (Lode et al. 1989b). Simultaneous administration of 500 mg ciprofloxacin and 200 mg ferrous fumarate as a suspension reduced the bioavailability of ciprofloxacin by 70% (Brouwers et al. 1990b), whereas the bioavailability of 200 mg ofloxacin was only reduced by 10.9%, when given together with 1050 mg of ferrous sulphate (Fero-Gradumet; Polk et al. 1989). However, this formulation is a sustained release formulation of ferrous sulphate and hence, ferrous sulphate is not released instantaneously. This might explain the low magnitude of interaction in this study as suggested by Nix (1993). Lehto et al. (1994b) showed that the bioavailabilities of norfloxacin, ciprofloxacin and ofloxacin were decreased by the simultaneous adminsitration of ferrous sulphate (Duroferon, 100 mg ferrous sulphate) by

73%, 57% and 25%, respectively. A 7-day pretreatment of 325 mg ferrous sulphate (three times daily) reduced ciprofloxacin bioavailability by 64% (Polk et al. 1989). The bioavailability of 100 mg levofloxacin was reduced by 19%, when given concomitantly with 160 mg ferrous sulphate (Shiba et al. 1992).

Also a multivitamin preparation (Stresstabs 600 with zinc) containing trace amounts of metal ions (23.9 mg zinc, 4 mg copper) can reduce the bioavailability of fluoroquinolones: a 7-day once daily pretreatment of the multivitamin preparation resulted in a reduction of ciprofloxacin's bioavailability by 24% (Polk et al. 1989). The urinary excretion of norfloxacin was measured after coadministration with a variety of OTC products containing different metal ions. The 24-h urinary excretion was reduced by 55% with ferrous sulphate (300 mg) and by 56% with zinc sulphate (200 mg; Campbell et al. 1992).

Bismuth subsalicylate (528 mg, 30 ml Pepto-Bismol) did not change the amount excreted in urine for norfloxacin (Campbell et al. 1992). The same dose of bismuth subsalicylate had only a slight effect on ciprofloxacin bioavailability, which was reduced by about 13% (Rambout et al. 1993).

The conclusion of a study (Akerele and Okhamafe 1991) investigating the effect of sodium bicarbonate (500 mg), potassium citrate (3 g), ferrous sulphate (200 mg), magnesium trisilicate (500 mg), calcium carbonate (500 mg) and aluminium hydroxide (500 mg) on the saliva and urine concentrations of ofloxacin has to be questioned due to methodological problems. Only for aluminium hydroxide was a slight decrease in saliva AUC(0–9) and maximal urine concentration reported, whereas all other comedications including ferrous sulphate did not show any effect. However, there was no information given with regard to saliva sampling methodology and the saliva pH. In addition, the total urinary excretion was not determined. Urinary concentrations are highly dependent on the urine volume and therefore not directly comparable from one study period to another.

6. Chemotherapy Treatment

Infections in neutropenic patients are commonly caused by endogenous Gram-negative bacilli from the patients' alimentary tract and occasionally by oropharyngeal flora. Chemoprophylaxis with fluoroquinolones over 3–5 days eliminates Enterobacteriaceae from the alimentary tract of patients and thus results in a reduced frequency of Gram-negative infections. In some haematology units all patients who are undergoing remission-induction therapy for haematological malignancy and who are likely to be neutropenic for more than 7 days are routinely given ciprofloxacin 500 mg b.i.d. orally. Cytotoxic drugs have a significant effect on rapidly dividing cells of the intestinal tract mucosa which might result in a reduced absorption of orally administered drugs. Kuhlmann et al. (1981, 1982) reported for digitoxin, which is absorbed in upper and lower parts of the gastrointestinal tract, no change in the extent of absorption during chemotherapy, but for digoxin, which is mainly absorbed in the upper small intestine, the extent of absorption was reduced by 20%–30% due to chemotherapy. Thus, for fluoroquinolones such as ciprofloxacin, which are mainly absorbed in the upper parts of the small intestine, a reduced bioavailability during chemotherapy can be expected. JOHNSON et al. (1990) investigated in six patients with a newly diagnosed haematological malignancy the absorption of ciprofloxacin (500 mg twice daily) prior to, during and for at least 21 days after chemotherapy. Serum ciprofloxacin levels were measured before, and 1, 2, 3, and 4 h after the morning dose of ciprofloxacin. In five out of six patients the C_{max} was higher before treatment than after. Mean 2-h C_{max} values (95% confidence intervals) were 3.3 mg/l (1.6–5.0 mg/l; prechemotherapy) and 1.8 mg/l (1.5–2.1 mg/l; 8–13 days postchemotherapy). The AUC(0–4) was reduced from 10.7 mg $h^{-1} l^{-1}$ (6.2–15.2 mg $h^{-1} l^{-1}$; prechemotherapy) to 5.7 mg $h^{-1} l^{-1}$ (4.7–7.0 mg $h^{-1} l^{-1}$; 13 days postchemotherapy).

The authors concluded that ciprofloxacin was absorbed less well following cytotoxic chemotherapy in neutropenic patients. However, the levels achieved are probably adequate for treatment of most infections; caution should be exercised when treating systemic infections with oral ciprofloxacin in this group of patients. MARCHBANKS et al. (1989) found in six patients that ciprofloxacin absorption is only slightly (~17%) reduced when given 5–7 days after courses of induction chemotherapy (nadic of neutropenia). In another study conducted by BROWN et al. (1993) 10 patients with non-Hodgkin's lymphoma or acute mylloid leukaemia were given 500 mg ofloxacin at breakfast time, the maximum serum ofloxacin levels were reduced significantly 2–3 days after chemotherapy (by 18%) but none of the other measurements (AUCs) were changed by the cytotoxic treatment. The authors suggest that these changes are of limited clinical significance and dosage alterations are not necessary. Nothing appears to be documented about any of the other quinolone antibiotics.

7. Activated Charcoal

In a randomised two-way cross-over study TORRE et al. (1989) investigated in six healthy volunteers the effect of 1 g charcoal given soon after the intake of 500 mg ciprofloxacin. The AUC of ciprofloxacin was reduced by about 10% and C_{max} by 3% compared with 500 mg ciprofloxacin without charcoal. The finding that charcoal does not reduce ciprofloxacin absorption must be regarded with caution since only one dose of charcoal was given soon (not further specified) after intake of ciprofloxacin. SÖRGEL and KINZING (1993a) showed that sufficient pretreatment with activated charcoal can absorb all drug that is secreted into the gut lumen and blocks the reabsorption of the compound back into the bloodstream when the fluoroquinolone is given as an intravenous infusion. Thus, it is likely that the results of the study done by TORRE et al. (1988) are due to the special design and that higher doses of charcoal given before or during fluoroquinolone treatment can reduce the bioavailability of the fluoroquinolone.

II. Interactions of Fluoroquinolones Due to Alterations in Metabolism

Drug–drug interactions at the metabolic level are most often due to induction or inhibition of hepatic microsomal enzyme systems. Hepatic metabolism can be reduced by inhibition of hepatic enzymes, resulting in a decrease in intrinsic clearance of the drug, and by changes in hepatic blood flow, resulting in changes in clearance of high clearance drugs. Fluoroquinolone antibiotics inhibit hepatic enzymes and do not change hepatic blood flow, as shown for ciprofloxacin (Nix et al. 1987).

Early clinical observations of drug–drug interactions between enoxacin and antipyrine (Logamann and Ohnhaus 1986) and between enoxacin and methylxanthines (Wijnands et al. 1984) provided evidence that fluoroquinolones inhibit a subclass of the P450 enzyme system. Several studies showed that different isoforms of cytochrome P450 are responsible for the 1- and 3-demethylation and the 8-hydroxylation pathways of methylxanthine metabolism. A clear association of one P450 enzyme system with definite metabolic pathways remains unclear because different methods (e.g. human microsomes, cell lines) resulted in different associations (Ha et al. 1995). However, the 1- and 3-demethylation steps are mainly attributed to the CYP1A2 isoform, whereas the 8-hydroxylation step is mainly attributed to the CYP2E1 and CYP3A4 isoforms. That the CYP1A2 subclass is selectively inhibited by fluoroquinolones is supported by the fact that no interaction was observed between ciprofloxacin and quinidine (Bleske et al. 1990), ciprofloxacin and cyclosporine (van Buren et al. 1990; Tan et al. 1989) or ciprofloxacin and clarithromycin (Polk 1993). All drugs are primarily metabolized by the CYP3A4 subclass. For most investigated fluoroquinolones the in vitro inhibitory potency measured by using the 3-demethylation rate of caffeine in human liver microsomes correlated with the in vivo inhibitory potency (Fuhr et al. 1992). This allows the determination of a possible clinically relevant interaction between fluoroquinolones and methylxanthines already during the preclinical development phase (Schaefer et al. 1995).

Extensive investigations to characterise the relationship between the structure of the fluoroquinolone and the in vitro inhibition of the human cytochrome P450 isoform CYP1A2 have been performed recently (Fuhr et al. 1990, 1993; Sarkar et al. 1990). It was found that the 3′-oxo metabolites in general had a reduced activity compared with their parent compounds, which was in contrast to the findings of earlier investigations (Wijnands et al. 1985). The activity after piperazine ring cleavage was generally increased in comparison with the parent compound. A formula to predict the inhibition potency from the chemical structure was developed (Fuhr et al. 1993).

1. Theophylline, Caffeine and Structurally Closely Related Substances

Theophylline is extensively metabolized (85%–90% of the dose) by 1-, and 3-demethylation to form 3-methylxanthine and 1-methylxanthine, and by 8-

hydroxylation to form 1-, 3- and 1,3-dimethyluric acid. These metabolites are eliminated predominantly by the kidneys. Caffeine, which is structurally related to theophylline, is also cleared by N-demethylation and hydroxylation resulting in di- and monomethylxanthines in addition to mono-, di- and trimethyluric acids. In vitro investigations using human hepatic microsomes (SARKAR et al. 1990) demonstrated that the inhibition of the N-demethylation step is more pronounced than that of the hydroxylation step. In addition, there is a huge difference in inhibitory potency between the various fluoroquinolones.

a) *Enoxacin*

WIJNANDS et al. reported already in 1984 severe adverse reactions such as nausea, vomiting, headache and tachycardia in eight out of ten patients who received concurrent treatment with enoxacin and theophylline. These observations were confirmed by MAESEN et al. (1984), who observed serious unwanted drug effects in 12 of 15 patients receiving enoxacin. Nine of them had to stop taking enoxacin. Three patients receiving enoxacin and theophylline had no complains at all. In patients treated with pefloxacin and theophylline or ciprofloxacin and theophylline no adverse events were observed. In a study with six patients with chronic obstructive lung disease (COPD), after receiving a continuous theophylline infusion until steady-state was achieved enoxacin (400–600 mg/day) treatment was initiated. Three days later theophylline plasma concentrations were increased from 8.4 to 15.1 mg/l. Theophylline clearance was reduced by 42% (WIJNANDS et al. 1985). In a subsequent trial in patients with COPD receiving maintenance theophylline treatment as a controlled-release tablet (600–1200 mg/day), concomitant treatment with 400 mg enoxacin b.i.d. resulted in a 63.6% reduction in theophylline clearance (WIJNANDS et al. 1986).

Further controlled studies in healthy volunteers were done (BECKMANN et al. 1987; NIKI et al. 1987). A significant decrease in the urinary excretion of 1-methyluric acid and 3-methylxanthine, but no change in excretion of 1,3-dimetyluric acid was observed (WIJNANDS et al. 1988a). After discontinuation of the enoxacin administration theophylline plasma concentrations dropped back to normal within 3–5 days (NIKI et al. 1988). STAIB et al. (1987) investigated the effect of multiple doses of enoxacin (400 mg b.i.d.) on the pharmacokinetics of a single caffeine dose in 12 healthy volunteers. Total body clearance was decreased by about 80% and terminal half-life increased from 3.3 to 11.8 h. These results are not surprising, given the similarity between theophylline and caffeine in structure and metabolism.

b) *Ciprofloxacin*

Several controlled studies in patients and healthy volunteers in addition to case reports have been documented for a ciprofloxacin–theophylline interaction. The magnitude of the clearance reduction is substantially lower than that caused by enoxacin. MAESEN et al. (1984) reported the occurrence of halucinations in one out of six patients receiving 750 mg ciprofloxacin b.i.d. together with theophylline. However, theophylline levels were not increased. No adverse events were reported in 20 patients taking 500 mg ciprofloxacin b.i.d.

During a clinical trial with 33 hospitalised patients suffering from respiratory tract infections, who received intravenous theophylline, 750 mg oral ciprofloxacin b.i.d. was given. Significant (>4 mg/l) increases in theophylline levels occurred in 20 patients (61%). These patients were in general older than the remaining 13 patients (mean age 65 ± 12 vs 46 ± 16 years). The increase in serum theophylline concentrations ranged from 4.0 to 30 mg/l. Symptoms of theophylline toxicity did not always correlate with the change in serum theophylline concentrations (RAOFF et al. 1987). The influence of subject age on the inhibition of the oxidative metabolism by ciprofloxacin was investigated in 13 young (23–34 years) and nine elderly (65–82 years) volunteers using antipyrine. The results suggest that elderly volunteers are not more sensitive to the inhibitory effect on antipyrine metabolism. Since the enzymes involved in theophylline and antipyrine metabolism are the same these results might be valid also for theophylline (WAITE et al. 1991).

The effect of ciprofloxacin on theophylline pharmacokinetics has also been investigated in healthy young volunteers. The mean total body clearance of theophylline was in general reduced by about 17.8%–31.0% across several studies with wide inter-individual variability in the magnitude of the interaction (SCHWARTZ et al. 1988; BACHMANN et al. 1988; NIX et al. 1987; PRINCE et al. 1989; ROBSON et al. 1990a). In a recent study (BATTY et al. 1995) involving nine healthy volunteers the mechanism of the ciprofloxacin–theophylline interaction was investigated in great detail. The mean reduction in theophylline clearance after a 5-day pretreatment with 500 mg ciprofloxacin b.i.d. was 19%. However, in four volunteers (group A) a little decrease was observed, whereas the other 5 volunteers (group B) showed a mean decrease of 30%, supporting the significant inter-individual variability. Group A subjects showed only a slight inhibition of 1-demethylation (–12.8%), while group B subjects showed a significantly greater inhibition of 1-demethylation (–49.9%), 3-demethylation (–44.8%) and 8-hydroxylation (–27%). This significant inhibition of the 8-hydroxylation is in contrast to in vitro findings using human microsomes (FUHR et al. 1990), where only a weak inhibition of 8-hydroxylation was found.

Caffeine clearance was reduced by about 33% when given after a 5-day pretreatment with ciprofloxacin (250 mg b.i.d.) (STAIB et al. 1987). Thus, the effect is similar to that reported for theophylline. The results were confirmed in several subsequent studies (HEALY et al. 1989; HARDER et al. 1988) in which a similar reduction in caffeine clearance with higher doses of ciprofloxacin was observed. PARKER et al. (1994) used the caffeine breath test to correlate the inhibition of caffeine metabolism by ciprofloxacin in children with cystic fibrosis to a certain metabolic pathway. Caffeine was labelled using a nonradioactive stable isotope (^{13}C on the 3-methyl group of caffeine). The caffeine given orally undergoes 3-N demethylation, which is a cytochrome-P4501A2-dependent reaction. After N demethylation, the labelled methyl group enters the carbon pool as it is converted to formaldehyde, formate and bicarbonate, and is then exhaled as carbon dioxide. There was a significant decrease in the 2-h cumulative labelled CO_2 exhaled during ciprofloxacin treatment suggesting that at least the 3-N demethylation of caffeine is inhibited by ciprofloxacin.

Pentoxifylline (1-(5-Oxohexyl)theobromin) was reported to show a significant drug interaction with ciprofloxacin (CLEARY 1992). Six healthy male volunteers received 400 mg pentoxifylline on day 1 and then 500 mg ciprofloxacin for 3 days with pentoxifylline given with the final ciprofloxacin dose. Peak pentoxifylline serum concentrations increased from 114.5 μg/l to 179.5 μ/l (+57%) and the area under the curve from 954.7 to 1097.0 $\mu g\, h^{-1} l^{-1}$ (+15%). All patients reported frontal headaches on the combination which required therapy with an analgesic. Doxofylline, a novel methylxanthine bronchodilator, showed no interaction with ciprofloxacin (250 mg b.i.d.; O'CONNELL et al. 1994). However, basic details with regard to the study design are lacking and therefore the conclusion of this study is questionable.

c) *Fleroxacin*

Five healthy male volunteers received a slow-release theophylline preparation (200 mg) twice daily for 9 days. From the fifth day onwards, 200 mg of fleroxacin were given. No effect was seen when theophylline levels on days 4, 7 and 9 were compared (SOEJIMA et al. 1989). Also a once-daily 400-mg oral dose of fleroxacin was shown to have no effect on theophylline pharmacokinetics in young and elderly male volunteers (PARENT et al. 1990).

d) *Lomefloxacin*

Eight volunteers received a 7-day treatment of 400 mg lomefloxacin twice daily. An intravenous dose of 6 mg/kg theophylline before and 7 days later during lomefloxacin treatment was administered. There was no effect on theophylline metabolism noted (WIJNANDS et al. 1989). STAIB et al. (1989a) investigated in 12 healthy volunteers the effect of a 5-day treatment with 400 mg lomefloxacin per day on a single oral dose of 260 mg theophylline. Again, no interaction was found and results were confirmed by NIX et al. (1989e), LEBEL et al. (1990) and ROBSON et al. (1990a). Also in vitro studies in human liver microsomes demonstrated that lomefloxacin does not inhibit the cytochrome P-450 mediated drug metabolism of tolbutamide, ethinylestradiol and mianserin (WINN et al. 1989). Similarly with caffeine there was no interaction found (HEALY et al. 1991).

e) *Norfloxacin*

The effect of norfloxacin on theophylline metabolism was investigated in three separate studies (SANO et al. 1988; HO et al. 1988; BOWLES et al. 1988). All studies demonstrated that pretreatment with norfloxacin reduces the theophylline clearance only slightly, from 7.4% to 14.9%, resulting in a marginal increase in the mean C_{max} value of theophylline of 5%–11%. HARDER et al. (1988) found that caffeine clearance is only slightly increased when norfloxacin (400 mg b.i.d.) is given as a 4-day pretreatment. Only the change in terminal half-life was statistically significant (3.2 h vs 3.7 h), but this was not clinically relevant. At higher doses (800 mg b.i.d.) CARBO et al. (1989) observed a 35% decrease in caffeine clearance after a 1-day pretreatment.

f) Ofloxacin

Eight patients with COPD on maintenance treatment with a controlled-release theophylline formulation (Theolin retard, 300–600 mg b.i.d.) received 400 mg ofloxacin twice daily for 5.5 days. No effect on theophylline clearance was observed compared with the placebo group (WIJNANDS et al. 1987). This result was confirmed in subsequent studies which found only a clinically nonrelevant increase in mean maximum theophylline concentrations of less than 10% (NIKI et al. 1988; SANO et al. 1988; GREGOIRE et al. 1987). Ofloxacin did not affect the pharmacokinetics of caffeine (HARDER et al. 1988; STAIB et al. 1987).

g) Pefloxacin

In a study comparing the relative potency of the effect of pefloxacin and ciprofloxacin on theophylline clearance, it was found that both drugs are comparable with regard to this interaction: 400 mg pefloxacin given twice daily for 5.5 days to patients with chronic obstructive lung disease receiving 300–600 mg theophylline twice daily (Theolin retard) reduced the theophylline clearance by 29.4% compared with 30.4% for 500 mg ciprofloxacin twice daily (WIJNANDS et al. 1987). In young healthy volunteers given 200 mg pefloxacin twice daily, theophylline serum concentrations increased by 17%, while the area under the serum concentration–time curve increased by 19% (NIKI et al. 1987). The reduction in caffeine clearance by about 45% in the case of pefloxacin is slightly higher than that observed for ciprofloxacin (MAHR et al. 1990).

h) Rufloxacin

The effect of a single dose of 400 mg rufloxacin on a single 300 mg theophylline dose given 30 min earlier in healthy volunteers demonstrated that rufloxacin does not change the pharmacokinetics of theophylline relative to the control (CESANA et al. 1991).

i) Sparfloxacin

The effect of a 1-week coadministration of 200 mg sparfloxacin once daily on the pharmacokinetics and metabolism of theophylline was investigated in six asthmatic patients receiving chronic theophylline therapy (a sustained-release theophylline tablet of 200–300 mg twice daily). No change in theophylline clearance was observed (Takagi et al. 1991).

j) Tosufloxacin

Twelve healthy male volunteers received theophylline orally twice daily for 10 days to produce theophylline plasma concentrations of about 10 mg/l. Tosufloxacin (450 mg twice daily) was given from day 6 to day 10. A comparison of the pharmacokinetic parameters of theophylline on days 5 and 10 showed a significant decrease in the apparent oral clearance and a significant increase in C_{max} and AUC (Muralidharan et al. 1992). However, no information about the magnitude of the effect was given and thus no dosing recommendations can be given.

2. Antipyrine

Antipyrine is a frequently used substrate for assessing the effect of various factors on hepatic oxidative drug metabolism. It is metabolized by at least two different isoenzymes of cytochrome P-450. Ludwig et al. (1988) reported a 35% reduction in antipyrine clearance when antipyrine (15 mg/kg intravenously) was given after 8–10 days of 500 mg ciprofloxacin b.i.d. treatment compared with the administration before ciprofloxacin treatment. Single doses of ofloxacin (400 mg) and ciprofloxacin (500 mg) had no effect on antipyrine clearance (Boettcher and Wolter 1989). In contrast, a pronounced inhibitory effect on the hydroxylation of antipyrine was found after a single dose of 400 mg enoxacin (Boettcher and Wolter 1989). Also a 7-day pretreatment of 200 mg b.i.d. ofloxacin had no effect on the antipyrine metabolism (Graber et al. 1989). No effect of age on the inhibition of antipyrine metabolism by ciprofloxacin was seen (Waite et al. 1991).

3. Phenytoin

Phenytoin is metabolized by the cytochrome P450 enzyme system. Thus an interaction between fluoroquinolones and phenytoin can be considered. Gardner et al. (1990) found no significant differences between the pharmacokinetics of phenytoin (14 days 200 mg po o.d.) alone or after concomitant ciprofloxacin administration (500 mg b.i.d.) on days 10–14 in healthy volunteers. This was confirmed by Job et al. (1994). A modest increase in serum phenytoin concentrations by an average of 24% was observed by Schroeder et al. (1991) in seven epiletic patients on maintenance therapy with phenytoin when ciprofloxacin (500 mg b.i.d.) was given for 10 days in addition to the usual phenytoin dose. No change in seizure activity or significant adverse reactions have been noted. In contrast, in a case report by Dillard and Fink (1992) it was speculated that a decrease in phenytoin serum concentration seen in a 78-year-old man may be attributed to simultaneous administration of ciprofloxacin. One volunteer in a study done by JOB et al. (1994) also experienced a significant decrease in phenytoin plasma levels while receiving ciprofloxacin, whereas HULL (1993) reported that the concurrent use of ciprofloxacin and phenytoin in an 87-year-old woman may have contributed to the increased phenytoin blood concentrations in this patient. Further well-designed clinical trials and data on other quinolones are lacking. Thus, until the contradictory findings are resolved, special care should be taken when fluoroquinolones and phenytoin are combined.

4. H_2-Receptor Antagonists

Cimetidine, a nonselective inhibitor of the cytochrome P450 enzyme system, has been shown to alter the pharmacokinetics of a variety of drugs that undergo hepatic biotransformation. In contrast, ranitidine has no significant inhibitor properties and exhibits fewer interactions with other drugs. Treatment with H_2-receptor antagonists increases the gastric pH to a range (5–6) in which the solubility of fluoroquinolones is considerably reduced.

The effect of cimetidine on the pharmacokinetics of ciprofloxacin was investigated by PRINCE et al. (1990) in 20 healthy male volunteers taking 500mg ciprofloxacin b.i.d. for 10 days. On days 4–7, 800mg cimetidine was given together with the evening dose of ciprofloxacin. On day 7, ciprofloxacin's apparent oral clearance was reduced by about 20%. Thus, dose adjustment is not necessary. SÖRGEL et al. (1988a) found a decrease in the total clearance when a single intravenous dose of 400mg pefloxacin was given 6 days after oral pretreatment with cimetidine. PORTMANN (1992) investigated the effect of cimetidine on the pharmacokinetics of fleroxacin and found a 20% reduction in fleroxacin total body clearance. NIX et al. (1989c) and HÖFFKEN et al. (1988) demonstrated the lack of interaction between ranitidine and ciprofloxacin and ranitidine and ofloxacin in healthy volunteers. The same was found to hold true for lomefloxacin (NIX and SCHENTAG 1989) and fleroxacin (LODE et al. 1987a).

Intravenous ranitidine administered 2h prior to oral intake of enoxacin led to a 40% reduction in enoxacin bioavailability (GRASELA et al. 1989). However, the mechanism of this interaction is due to the increase in the gastric pH leading to a decrease in enoxacin's absorption (LEBSACK et al. 1992). DAVIS et al. (1992) investigated the effect of a second enzyme inhibitor (ciprofloxacin) on hepatic microsomal enzyme activity in subjects already maximally inhibited by cimetidine. In a randomised crossover study, six subjects received 5mg theophylline per kilogram body weight on day 6 of therapy with cimetidine (2.4mg/day), ciprofloxacin (1g/day), both drugs, or neither drug.

Theophylline clearance was greatest when patients received neither drug (100%), less during ciprofloxacin therapy (69%), lesser with cimetidine therapy (60%) and least with combined cimetidine/ciprofloxacin therapy (55%). Thus, the addition of a second enzyme inhibitor in subjects receiving maximally inhibiting doses of cimetidine can produce a further decrease in the hepatic metabolism of drugs that are metabolized by the cytochrome P-450 microsomal enzyme system. This effect is independent of age and gender (LOI et al. 1993a,b).

5. K^+/Na^+-ATPase Inhibitors

Omeprazole is an irreversible inhibitor of the K^+/Na^+-ATPase ("proton pump inhibitor") with much more efficiency than the established H_2-receptor antagonists. Omeprazole acts as an inhibitor of cytochrome P-450 isoenzymes. It reduces the clearance of both diazepam and phenytoin. In addition to its inhibitory action, omeprazole causes induction of the cytochrome P-450 system. Administration of omeprazole at therapeutic doses for 4 days caused induction of cytochrome P-450 1A2. FASSBENDER et al. (1993) excluded an interaction between omeprazole and three fluoroquinolones, ciprofloxacin, oflocxacin and lomefloxacin.

6. Warfarin

Warfarin is a racemic mixture of the enantiomers (*R*)-warfarin and (*S*)-warfarin. The more pharmacologically active (*S*)-enantiomer is oxidized via (*S*)-7-hydroxy warfarin to the (*S*, *S*)-warfarin alcohol. Two hypotheses for a mechanism of a warfarin–fluoroquinolone interaction are (a) inhibition of warfarin metabolism and thus increasing warfarin plasma levels and (b) suppression of vitamin K producing gut bacteria, and thus potentiation of the warfarin effect. Due to the low plasma protein binding of fluoroquinolones (20%–60%) displacement of warfarin from the binding sites is very unlikely.

Several case reports have been published which demonstrate a possible interaction between warfarin and fluoroquinolones (e.g. norfloxacin, ciprofloxacin, ofloxacin) resulting in increased prothrombin times and bleeding (LINVILLE and MATANIN 1989; MOTT et al. 1989; KAMADA 1990; RENZI and FINKBEINER 1991; DUGONI-KRAMER 1991; BACIEWICZ et al. 1993; JOHNSON et al. 1991; LEOR and MATETZKI 1988).

However, the subsequent evaluation of this interaction in healthy volunteer studies for norfloxacin (ROCCI et al. 1990), fleroxacin (SÖRGEL and KINZIG 1993b), enoxacin (TOON et al. 1987) and studies in patients on warfarin for ciprofloxacin (RINDONE et al. 1991; BIANCO et al. 1992; POLK et al. 1994) could not confirm the case reports. No changes in prothrombin time have been observed. Therefore, the question of whether a warfarin–fluoroquinolone interaction occurs in infected patients on warfarin treated with fluoroquinolones remains largely unanswered. A close monitoring of patients receiving warfarin and fluoroquinolones is mandatory.

7. Cyclosporine

Cyclosporine is an immunosuppressive agent used to prevent rejection in organ transplant recipients. Cyclosporine is mainly metabolized by CYP3A isoenzymes and not by CYP1A, which is mainly inhibited by ciprofloxacin.

Therefore from a theoretical point of view an interaction with the metabolism is unlikely. However, altered pharmacokinetics of cyclosporine due to concurrent fluoroquinolone therapy with the consequence of enhanced nephrotoxicity have been reported as case reports for ciprofloxacin (AVENT et al. 1988; NASIR et al. 1991). Others who reported no interaction (HOOPER et al. 1988). ELSTON and TAYLOR (1988) speculated that the interaction (enhanced nephrotoxicity) is due to a direct local effect in the kidney rather than pharmacokinetic alterations of cyclosporine. Well-designed pharmacokinetic studies in healthy volunteers (VAN BUREN et al. 1990; TAN et al. 1989) and transplant patients (KRÜGER et al. 1990; LANG et al. 1989a; EHNINGER and KRÜGER 1989; ROBINSON et al. 1990) have clearly demonstrated that ciprofloxacin does not interact with cyclosporine. In these studies no increase in cyclosporine plasma concentrations or nephrotoxicity was observed. Thus, ciprofloxacin may be confidently prescribed to patients receiving cyclosporines as also recommended by HOLY and LAKE (1994). This may be also apply to other fluoroquinolones such as ofloxacin (WYNCKEL et al. 1991) and pefloxacin (LANG et al. 1989b); fewer data are available on these quinolones. In contrast, results recently reported by MCLELLAN et al. (1995) strongly suggest that norfloxacin does inhibit the metabolic clearance of cyclosporine in children. Patients receiving cyclosporine and norfloxacin must be monitored closely to maintain levels within the target range for effective immunosuppression and the prevention of adverse side effects such as nephrotoxicity (MCLELLAN et al. 1995).

8. Rifampin

Rifampin and fluoroquinolones are sometimes combined in order to treat infections with fluoroquinolone-resistant staphylococci. Rifampin is a well-recognized (relatively nonspecific) inducer of hepatic microsomal enzymes, and the pharmacokinetics of a variety of drugs are altered by concomitant rifampin intake. CHANDLER et al. (1990) investigated in a group comparison in 12 elderly patients colonized with methicillin-resistant *Staphylococcus aureus* a 14-day therapy with ciprofloxacin (750 mg b.i.d.; group A; n = 6) versus a 14-day therapy with ciprofloxacin (750 mg b.i.d.) and oral rifampin (300 mg b.i.d.; group B; n = 6). No significant differences in ciprofloxacin pharmacokinetics were noted, and rifampin pharmacokinetics were not different from historical control data. Also DEETER et al. (1989) found no effect of rifampin on ciprofloxacin pharmacokinetics.

HUMBERT et al. (1991) investigated the effect of a 10-day induction by rifampin (900 mg/day) on pefloxacin pharmacokinetics (400 mg b.i.d. for 3 days; on day 4 400 mg; intravenous administration) in healthy volunteers. The total plasma clearance increased by 35%, thus indicating a significant influence of rifampin on pefloxacin in healthy volunteers. Due to the modest intensity of this effect no a priori dose modification is necessary, but pefloxacin plasma assays may be advised for long-term treatments as a drug monitoring help in situations involving patients at risk.

SÖRGEL and KINZIG (1993b) reported that when fleroxacin was given concomitantly with 600 mg of rifampin for 7 days, the metabolic clearance of fleroxacin increased by 27%, the half-life decreased by 12% and AUC decreased by 14%. However, these changes are clinically not relevant. For long-term treatment the same advice as for pefloxacin seems appropriate.

9. Oral Contraceptive Steroids

The possibility of a link between antibiotic usage and oral contraceptive steroid failure was first reported by REIMERS and JEZEK (1971). While in the case of rifampin this interaction is related to rifampin's ability to induce hepatic microsomal enzymes responsible for oral contraceptive steroid metabolism, further reports on different antibacterial drugs such as penicillins, cephalosporins and tetracyclines were published, suggesting a different mechanism of interaction. After absorption from the gastrointestinal tract, oral contraceptive steroids form their glucuronide and sulphate conjugates in the liver and are then biliary excreted. In the gut lumen, bacterial enzymes hydrolyse the conjugates and the steriods are reabsorbed so that effective plasma levels are maintained. Suppressing the intestinal flora by antibiotics results in a loss of the enterohepatic circulation of the steroids and thus decrease plasma concentrations with the consequence of a possible failure of contraceptive protection. Fluoroquinolones are able to affect the gut flora (ENZENSBERGER et al. 1985). However, MAGGIOLO et al. (1991) showed that a 7-day treatment with ciprofloxacin did not affect steroid treatment outcome (measured by FSH, LH and estrogen blood levels). A very comprehensive study of ciprofloxacin was performed by DROPPERT et al. (1992), including parameters like luteal progesterone, follicular and preovulatory estradial blood levels, follicular diameter, and sex-hormone-binding globulin. It was concluded that ciprofloxacin does not interfere with ovulatory inhibition by oral contraceptives containing at least 30 μg ethinylestradiol. Data for other quinolones are not published.

10. Benzodiazepines (Diazepam, Temazepam)

Diazepam is a commonly used anxiolytic drug and is mainly metabolized by N-demethylation to *N*-desmethyldiazepam and by hydroxylation to 3-hydroxydiazepam (temazepam), which is then mainly excreted by glucuronidation. WIJNANDS et al. (1990) investigated in 10 healthy volunteers the effect of a 3-day pretreatment of ciprofloxacin (500 mg b.i.d.) on a single intravenous dose of 10 mg diazepam. There was a nonsignificant decrease in total body clearance of about 10% when diazepam was given after pretreatment.

In contrast, KAMALI et al. (1993) demonstrated a significant decrease of about 37% in diazepam clearance when given as a single 5 mg intravenous dose after a 7-day pretreatment with ciprofloxacin (500 mg b.i.d.) compared to the control. The terminal half-life increased from 36.7 h to 71.1 h. However, no significant changes were detected in psychometric tests such as digit symbol substitution, tapping rate and short memory, in addition to levels of concentration, vigilance and tension measured by visual analogue scales. Data for other quinolones are not available. Thus, the clinical relevance of this interaction remains questionable. Careful monitoring of central nervous system effects when diazepam and fluoroquinolones are given concomitantly is warranted. Temazepam pharmacokinetics and pharmacodynamics are not affected by ciprofloxacin as recently shown by KAMALI et al. (1994).

III. Alterations in Renal Excretion

Pharmacokinetic studies have shown that most of the fluoroquinolones are excreted in the urine. By comparing the renal clearance with the creatinine clearance the relative proportion of glomerular filtration versus active tubular secretion and reabsorption can be estimated. For most fluoroquinolones and their metabolites tubular secretion is involved in the renal excretion process.

1. Probenecid

Probenecid is known to be actively secreted by the renal tubules and to block the active secretion of many organic acids by its high affinity to the carrier for anion transport. WINGENDER et al. (1985, 1986) already found a reduction in ciprofloxacin renal clearance from $4.6\,ml\,min^{-1}\,kg^{-1}$ to $2.3\,mg\,min^{-1}\,kg^{-1}$, when 1 g probenecid was given orally prior to a 500-mg ciprofloxacin tablet, but no significant changes in AUC and $t_{1/2}$ occurred; thus a corresponding increase in extrarenal elimination occurred in compensation as proven by BERGAN et al. (1989b) and ROHWEDDER et al. (1990).

For other fluoroquinolones such as enoxacin (WIJNANDS et al. 1988a), norfloxacin (SHIMADA et al. 1983) and fleroxacin (SHIBA et al. 1989) a significant reduction in renal clearance was also observed when probenecid was given simultaneously. It can be expected that also other fluoroquinolones primarily excreted in urine with active tubular secretion being involved (e.g. ofloxacin, lomefloxacin) will show this interaction.

2. β-Lactam Antibiotics

Ciprofloxacin and azlocillin may potentially be used in combination for the treatment of Gram-negative bacterial infections, mainly those caused by *P. aeruginosa*. Both drugs are eliminated by renal and hepatic routes and both are organic acids and, therefore, may compete for the carrier for anion transport in the renal tubules or in other tissues.

BARRIERE et al. (1990) studied the effect of single intravenous doses of ciprofloxacin, azlocillin, and the two drugs simultaneously on separate occasions in six healthy volunteers. The pharmacokinetic disposition of azlocillin was unchanged despite the simultaneous administration of ciprofloxacin. The total renal and nonrenal clearance and the volume of distribution at steady-state of ciprofloxacin were reduced by 35%, 39%, 31% and 27%, respectively.

The effect on the nonrenal clearance can be explained by effects on the metabolism of ciprofloxacin, inhibition of hepatic uptake and/or by interference in the direct transintestinal secretion of ciprofloxacin into the gut lumen, a process by which about 15% of the dose was excreted after intravenous administration (BEERMANN et al. 1986). Surprisingly, a study with six healthy volunteers, to whom amoxycillin (3g) and ofloxacin (400mg) were given orally on separate occasions and simultaneously (PAINTAUD et al. 1993), showed no significant differences in C_{max}, AUC, $t_{1/2}$ and mean residence time of ofloxacin when combined with amoxycillin compared with the control. Data on renal excretion are not available for that study. Data on other fluoroquinolones are lacking.

IV. Pharmacodynamic Interactions

Pharmacodynamic interactions very often occur at the receptor level and result in an enhanced or diminished effect, depending on whether the drugs involved exhibit agonistic or antagonistic properties. Pharmacodynamic interactions are difficult to detect in clinical trials due to the fact that precise methods to quantify such effects are lacking or due to the difficulty in separating the overall variability observed in clinical trials into the portion associated with pharmacokinetics and that associated with pharmacodynamics. However, the methodology used to investigate pharmacokinetic/pharmacodynamic relationships in drug development has been improved significantly during the last decade (CUTLER et al. 1994).

1. Nonsteroidal Anti-inflammatory Drugs

In 1986 the Japanese Welfare Ministry issued a warning to avoid prescribing the combination of enoxacin and fenbufen, based on the observation that seven patients which had received the combination experienced central nervous effects (convulsion). Sixty-one patients were given ofloxacin and fenbufen simultaneously prior to 1986, but no side effects were observed which involved convulsion (SAITO et al. 1987). Because of the lack of sensitive and specific measurable endpoints in humans predictive for the clinical situation, the interaction between NSAIDs and fluoroquinolones has been primarily investigated in in vitro models using mouse and rat brain synaptic membranes (HORI et al. 1989; TSUJI et al. 1988b) and frog neurons (YAKUSHIJI et al. 1992). It was shown that fluoroquinolones inhibit the binding of γ-aminobutyric acid (GABA), an inhibitory neurotransmitter, in a concentration-dependent manner (HORI et al. 1989; TSUJI et al. 1988; YAKUSHIJI et al. 1992). The concentration required to reduce GABA binding by 50% (IC50) was 10^{-5} *M* for norfloxacin, 10^{-4} *M* for enoxacin, 10^{-3} *M* for ofloxacin, 10^{-4}–10^{-5} *M* for ciprofloxacin and 10^{-3}–10^{-4} *M* for fleroxacin. However, it should be noted that these concentrations are equivalent to about 10–300mg/l. In the presence of biphenylacetate (10^{-4} *M*), an active metabolite of fenbufen, the inhibitory activities of norfloxacin, enoxacin, ofloxacin and ciprofloxacin were remarkably enhanced (up to 1000 times).

With fleroxacin, only a week enhancement of inhibitory activity was observed in the presence of biphenylacetate. A lower enhancement was achieved with fenbufen, flubiprofen and indomethacin. Acetylsalicylic acid had no effect. The extrapolation of these results to humans is difficult due to interspecies differences in GABA receptor affinity and the fact that differences may exist in patients with predisposing neurological conditions (HEALY and SMALL 1989; ROLLOF and VINGE 1993).

A recent study in healthy volunteers supported the notion that the interaction between NSAIDs and fluoroquinolones is pharmacodynamic in nature by demonstrating the lack of any changes in pharmacokinetics of pefloxacin and ofloxacin when given simultaneously with ketoprofen (FILLASTRE et al. 1992). POLK (1994) concluded that in light of the high frequency with which fluoroquinolones and NSAIDs are administered concurrently, this interaction must be exceedingly uncommon. Agents from these two groups of drugs can probably be coadministered safely, although clinical data regarding their interaction are sparse.

2. Metronidazole

To achieve a reliable anaerobic coverage with fluoroquinolones they have to be combined with antianaerobic agents such as clindamycin and metronidazole. Pharmacokinetic interactions between several fluoroquinolones (ciprofloxacin, ofloxacin, enoxacin, fleroxacin) and clindamycin and metronidazole have been excluded (BOECKH et al. 1990; SHAH et al. 1995). Case reports of increased CNS toxicity have been published for ciprofloxacin–metronidazole and pefloxacin–metronidazole combinations (LUCET et al. 1988; SEMEL and ALLEN 1989). The symptoms of increased CNS toxicity were agitation, confusion, disorientation, involuntary movement and slurred speech. Patients receiving metronidazole and fluoroquinolone combinations should be closely monitored for CNS toxicity.

V. Conclusions

In spite of the wide application of fluoroquinolones these drugs can be administered safely with a huge variety of different drugs without causing serious drug–drug interactions. However, clinically relevant drug–drug interactions can be avoided if a few rules are followed:

1. The simultaneous intake with antacids containing aluminium and magnesium salts, iron or zinc preparations and with sucralfate should be avoided. The fluoroquinolone should be given 2 h prior to or 6 h after antacid intake or acid suppression should be done using for example ranitidine (although not with enoxacin) or omeprazole together with ciprofloxacin, ofloxacin or lomefloxacin.
2. The patient should be informed about the potential hazard of other metal cations, which are very often in OTC multivitamin supplements since they impair the bioavailability of the fluoroquinolone.

3. In HIV-patients the simultaneous administration of didanosine and fluoroquinolones should be avoided.
4. Fluoroquinolones inhibit theophylline metabolism to different degrees. Enoxacin had the largest effect on theophylline clearance, whereas fleroxacin, lomefloxacin, norfloxacin, ofloxacin, rufloxacin and sparfloxacin had no or only a small effect. Ciprofloxacin and pefloxacin had an intermediate effect. The clinical importance of this interaction depends on the initial theophylline concentration (e.g. whether it is at the low or high end of the therapeutic range for theophylline) and on the type and dose of the fluoroquinolone used. For enoxacin, ciprofloxacin and pefloxacin, theophylline plasma concentrations should be closely monitored. The metabolism of other methylxanthines such as caffeine and pentoxiphylline is also inhibited. However, due to the larger therapeutic range of these drugs plasma concentration monitoring is not necessary. If adverse events with central nervous system symptoms are observed, caffeine consumption should be reduced.
5. If the fluoroquinolone is given together with warfarin, prothrombin times should be monitored. Further studies investigating this drug–drug interaction are necessary.
6. Due to the conflicting results reported, the careful monitoring of central nervous system effects is necessary when fluoroquinolones are combined with diazepam.
7. β-Lactam antibiotics and probenecid reduce the tubular secretion of fluoroquinolones resulting in a reduced renal clearance. For azlocillin, also a reduction in the nonrenal clearance was observed. However, dose adjustments are not necessary.
8. Fluoroquinolones have a potential to cause various central nervous effects, which can be potentiated by other drugs (e.g. theophylline, caffeine, metronidazole). Also fenbufen, a nonsteroidal anti-inflammatory agent, can enhance the epileptogenic effect of fluoroquinolones by inhibition of GABA receptors in the brain. A clinical relevance of this interaction was observed only for enoxacin. Thus, other fluoroquinolones and NSAIDs can be administered together. Patients predisposed to seizures should be carefully observed.

D. Adverse Reactions of Fluoroquinolones

The original quinolones, nalidixic acid, oxolinic acid, pipemidic acid and cinoxacin had limited clinical use because of their poor absorption, narrow antibacterial spectrum, the rapid development of bacterial resistance and high incidence of, sometimes severe, side effects. The newer fluoroquinolones, about which we have the most information, include norfloxacin, ciprofloxacin, ofloxacin, enoxacin and pefloxacin.

In general, the tolerability of these agents has been good and, compared with other antimicrobial agents, they can be considered relatively save agents (Ball and Tillotson 1995; Wilton et al. 1996). More severe adverse effects, however, have been seen with some more recently developed compounds, e.g. temafloxacin, lomefloxacin, sparfloxacin and fleroxacin. Temafloxacin was voluntarily withdrawn worldwide on account of its high frequency of serious adverse reactions including severe hypoglycaemia, hepatic failure, haemolytic anaemia, coagulopathy, nephrotoxic reactions, central nervous system complications and severe anaphylaxis (Blum et al. 1994).

The incidence and severity of adverse effects can only be evaluated from phase III and IV of clinical trials and postmarketing surveillance data. It is not the purpose of this section to review all these data in detail, but rather to give a general overview of the most common adverse effects of the newer fluoroquinolones. The emphasis is on some severe adverse events which must be considered individually for the different members of the quinolone class. A comprehensive overview of the toxicological profile of the different quinolones is given in Chap. 10 of this book.

I. Gastrointestinal Tract

Gastrointestinal side effects are the most frequent adverse effects of all fluoroquinolones. There does not appear to be any major difference in frequency and severity of gastrointestinal side effects between the fluoroquinolones. The most observed side effects are nausea, vomiting, dyspepsia, abdominal pain, anorexia, diarrhoea, flatulence, looseness of stools, and dry mouth (Wolfson 1989; Wolfson and Hooper 1989; Paton and Reeves 1991, 1992; Schacht et al. 1988, 1989; Cullmann et al. 1993; Chysky et al. 1991; Stahlmann and Lode 1988; Corrado et al. 1987; Jüngst and Mohr 1988; Sawada et al. 1991; Rubinstein and Carbon 1994; Simon et al. 1993; Halkin 1988; Janknegt 1989). Antibiotic-associated gastrointestinal bleeding or colitis have been seen but only extremely rarely (Adam 1989a). The cause of the gastrointestinal symptoms is postulated to be a combination of gastric irritation and CNS-mediated effects (Norrby 1991; Shimada and Hori 1992).

II. Central Nervous System

Central nervous system side effects are the second most commonly observed adverse reactions of quinolones and can be devided into mild reactions and severe neurotoxic side effects that require interruption of therapy. Mild reactions include headache, dizziness, tiredness, insomnia, faintness, agitation, listlessness, restlessness, abnormal vision, sleep disorders, tremors, and catatonic syndrome.

Severe reactions are rare and include hallucinations, depressions, psychotic reactions and in some cases convulsions (Wolfson 1989; Wolfson and Hooper 1989; Paton and Reeves 1991, 1992; Schacht et al. 1988, 1989;

CULLMANN et al. 1993; CHYSKY et al. 1991; MODAI 1988; CORRADO et al. 1987; JÜNGST and MOHR, 1988; SAWADA et al. 1991; RUBINSTEIN and CARBON 1994; SIMON et al. 1993; HALKIN 1988; JANKNEGT 1989). Convulsions have usually been associated with predisposing factors such as a previous history of epilepsy or pre-existing brain lesions or concomitant treatment with methylxanthines such as theophylline, fenbufen and other NSAIDs or β-lactam antibiotics. Patients with known or suspected CNS disorders should be observed carefully while on fluoroquinolones (British National Formulary 1996).

The exact mechanism of central nervous system toxicity associated with quinolone therapy has not been defined. It has been suggested that the dose-dependent inhibition of GABA binding to receptors in the brain leading to CNS stimulation may be involved in the induction of epileptogenic neurotoxicities (HORI et al. 1985, 1986, 1989; TSUJI et al. 1988a,b; HALLIWELL et al. 1993). The concentrations of quinolones required for GABA binding are generally high relative to therapeutically achievable concentrations in plasma. In vivo penetration to the brain may also be an important variable. Unfortunately, data on the concentrations of quinolones in the brain are not widely available and a correlation with the differential lipophilicities of the quinolones cannot be made. Predicting direct CNS effects for the quinolones is, therefore, still difficult (HALLIWELL et al. 1993; RIETBROCK and STAIB 1987). In a controlled, double-blind study HERRMANN et al. (1984) compared the CNS effects of ciprofloxacin at doses of 500 mg and 1000 mg with that of 2000 mg nalidixic acid and placebo in 12 healthy young volunteers (ten men and two women). Electroencephalograms and psychological tests were performed before and after dosing in all subjects. After nalidixic acid dosing, a significant difference to placebo was detected in terms of alpha-slow-wave index (ASI) in the confirmatory statistical tests. This was interpreted as an indicator of a possible increase in the vigilance tone. On the basis of the psychological tests, some dysphoria was present along with heightened anxiety. A rise in the flicker fusion frequency also suggested CNS stimulation. After dosing with 500 mg or 1000 mg ciprofloxacin, no significant difference was found on the confirmatory statistical tests. Explorative tests indicated a higher ASI after both doses of ciprofloxacin than after placebo. The other changes observed for 500 mg ciprofloxacin were always weak and gave no indication of significant CNS activity. For the 1000 mg dose some modification of CNS parameters could not be excluded, the changes being interpreted as low degree CNS excitation. The psychological tests gave no indication of any systematic CNS effects. On the basis of the results obtained in this study, the investigators concluded that 1000 mg ciprofloxacin is likely to cause less undesirable CNS effects than 2000 mg of nalidixic acid.

The question of whether fluoroquinolones should be labelled disparately for CNS toxicity was discussed by FDA's Anti-Infective Drugs Advisory Committee at its September 23 meeting in 1993. After hearing all the data, the committee decided that class labelling should remain but the agency and pharmaceutical sponsors should have some options about addressing differences in side effects, which is appropriate because there are differences between the compounds.

III. Skin and Allergic Reactions

Skin and allergic reactions are the third most commonly observed side effects and include hypersensitivity reactions and photosensitivity reactions of skin surfaces exposed to sunlight. Allergic skin reactions to the quinolones are uncommon and include erythema nodosum, urticaria, rash, pruritus, fever and edema (WOLFSON 1989; WOLFSON and HOOPER 1989; PATON and REEVES 1991; SCHACHT et al. 1988, 1989; STAHLMANN and LODE 1988; ADAM et al. 1987; CORRADO et al. 1987). Anaphylactic reactions have been reported rarely.

1. Photosensitivity

Drug-induced photosensitivity is an adverse reaction of the skin which results from simultaneous exposure to certain drugs and to ultraviolet radiation (UVR) or visible light. There are two types of reactions: phototoxic, which can occur in all individuals and is essentially an exaggerated sunburn response, and photoallergic, which involves an immunological response.

a) Phototoxicity

Photoxic drug reactions are more frequent than photoallergic sensitivity and are described for different chemical classes which may be therapeutic, cosmetic, industrial or agricultural (ALLEN 1993; JOHNSON and FERGUSON 1990; ROSEN 1989; EPSTEIN and WINTROUB 1985; THE MEDICAL LETTER 1995). Photoxic reactions are said to occur in all individuals who are exposed to a high dose of both the drug and the appropriate wavelengths of radiation.

Quinolone phototoxicity is well known and was described for the first time with nalidixic acid (RAMSAY and OBRESHKOVA 1974). It appears to be a class effect, since there is a dose and an exposure level that will cause an effect for most of the known agents (PATON and REEVES 1991). Quinolones do differ from one another substantially in their level of phototoxic risk both in man and in laboratory studies (NORRBY 1991; DOMAGALA 1994). Of all the new quinolones, pefloxacin and lomefloxacin have a clearly established photosensitizing effect in the UVA range (CHRIST and LEHNERT 1990).

The unusual high incidence and severity of phototoxic reactions related to treatment with lomeflocaxin was the reason for a FDA meeting held on March 31, 1993, during which the phototoxic potential of lomefloxacin and other quinolones was assessed by linking the number of prescriptions for the particular drug with the number of reports of phototoxic reactions. Table 25 displays the compelling evidence that quinolones, while all are capable of inducing phototoxicity, differ substantially in their phototoxic potential. Other examples are fleroxacin and sparfloxacin with overall higher incidences of phototoxic reactions (GEDDES 1993; CARBON and RUBINSTEIN 1994). Clearly,

Table 25. Phototoxic potential of fluoroquinolones

Drug	Year of launch (US)	R × 1000	Reports of phototoxic reactions (n)
Norfloxacin	1986	10.226	19
Ciprofloxacin	1987	32.369	29
Ofloxacin	1990	3.161	13
Lomefloxacin	1992	259	183

R, number of prescriptions for drug.

only if the phototoxic potential of the drug is high enough does the problem of phototoxicity become serious enough to alert the medical community. In order to use quinolones safely we have to be aware of their differential phototoxic potential. The assessment of the phototoxic potential can be done retrospectively estimating the incidence and severity of reported phototoxic cases. A difficulty with this approach is that such assessment depends on both dose and clinical setting. The dose dependency can be illustrated by results of one study (BOWIE et al. 1989) in which the incidence of phototoxic reactions associated with fleroxacin treatment was 0%, 11% and 19% following a 400-, 600- and 800-mg daily dose of fleroxacin, respectively. In two different studies employing the same dose of lomefloxacin (400 mg once daily) the numbers of reported phototoxic cases were as different as two of 324 patients in one study (YERNAULT and RUSSEL 1992) and ten of 235 patients in the other study (NERINGER et al. 1992). This clearly indicates that factors other than merely dose influence the incidence of phototoxic events. Another obstacle with this approach which is even more important is that the information on phototoxicity of the given agents is gained relatively late in the drug development after many patients has already experienced adverse events. Sparfloxacin, for example, was launched in France in September 1994. A pharmacovigilance survey has shown a recent and significant increase in phototoxicity cases, sometimes severe, with a frequency which appears to be higher than with other fluoroquinolones. These symptoms are similar to erythema, mainly on the uncovered parts of the body and can be more severe as second-degree burns. This has led to restrictions being imposed on the indication for this agent – sparfloxacin – may now be used only for community-acquired pneumonia and sinusitis when other antibacterials are inappropriate (ANONYMOUS 1996a).

In order to circumvent the problems described above, the assessment of the phototoxic potential of quinolone antibiotics can be performed preclinically in vitro and in vivo by measuring rates of degradation, by measuring cellular damage in vitro or by using in vivo models (MATSUMOTO et al. 1992; ROBERTSON et al. 1991; WAGAI and TAWARA 1992; SESNIE et al. 1990; BRITISH PHOTODERMATOLOGY GROUP 1990). The method of choice for estimating and predicting the phototoxic potential is however phototesting in subjects/patients exposed to the drug in question. The primary parameter of such investigations is the determination of the minimal erythema dose (MED) which is the minimum dose of energy required to produce a defined erythema and abnormal skin reactions (FERGUSON and JOHNSON 1990).

The following is a review of the literature on quinolones with respect to phototesting results. The first published study of this kind was an open study in which standard phototesting procedures with an irradiation monochromator were used to determine the phototoxic potential of ciprofloxacin in 12 patients taking the drug (FERGUSON and JOHNSON 1990). The dosage regimen ranged from 250 mg ciprofloxacin orally b.i.d. to 750 mg b.i.d. taken for a minimum of 7 days. MED at different wave lengths (UVA and UVB) was determined at baseline, and after at least 1 week of treatment, with repeated phototesting at 2-week intervals for 4 weeks if a persistent abnormality, i.e. a decrease in MED was detected. The maximum decrease in MED was observed at 335 ± 30 mm from 4.04 to 1.92 J/cm^2 (52.5%). In no instance was any abnormal immediate erythema observed, nor was there any evidence of an abnormal sensation, urticaria or blistering. In all patients, MED determined 2 weeks after stopping the drug was within normal limits, indicating reversibility.

In another, this time randomised, double-blind, placebo-controlled study groups of eight healthy subjects received either 200 mg or 400 mg norfloxacin, placebo or 500 mg ciprofloxacin (FERGUSON and JOHNSON 1993). Both drugs and placebo were administered orally twice daily for 7 days. The phototesting results revealed only marginal phototoxicity of norfloxacin without any evidence of dose dependency. An overall comparison between the ciprofloxacin and placebo groups showed again a decrease in MED following ciprofloxacin administration at the wavelengths 335 ± 30 nm and 365 ± 30 nm. However, only one subject reached a level beyond the normal data values. Again, within the 2 weeks there was a normalization of MED and no abnormal skin reactions were observed.

In a recent study with lomefloxacin, a morning and an evening dose regimen (400 mg) was investigated in terms of the phototoxic potential in a randomized, double-blind crossover trial (LOWE et al. 1994). In contrast to the studies described above, MED was determined in this study using the whole spectrum of UVA or UVB light. In the lomefloxacin morning dosing group (2-h interval between dosing and phototesting), the mean immediate and delayed MED_{UVA} values were significantly reduced (24% and 26%, respectively) compared with their mean baseline ($p < 0.05$). In contrast, in the lomefloxacin evening dosing group (16-h interval between dosing and phototesting), no difference was observed. This supports what had already been suspected, namely a relationship between the drug concentrations and the amount of energy required to cause a signficant change in MEDs.

While the mechanisms of quinolone phototoxicity are not known, it could be speculated that quinolones which are degraded to a large extent in vitro after UVA irradiation may also be more likely to produce phototoxicity (PATON and REEVES 1991; CULLMANN et al. 1993). In general, excessive sunlight or artificial UV light should be avoided by patients on fluoroquinolone therapy, and therapy should be discontinued if phototoxicity occurs.

2. Photoallergy

In photoallergic drug photosensitivity the chemical agent present in the skin absorbs photons and forms a photoproduct. This photoproduct then binds to a soluble or membrane-bound protein to form an antigen. To be considered a photoallergic reaction, an immune-mediated mechanism must be demonstrated. Because of the role of specific immunity, only a small percentage of people exposed to the potentially sensitizing drug and the culpable wavelengths will be affected.

IV. Nephropathy and Crystalluria

Nephrotoxicity and crystalluria are rare quinolone adverse reactions and very difficult to judge. The quinolones are probably not primarily nephrotoxic. Potential changes include interstitial nephritis, occult blood in urine and decreased renal function (PATON and REEVES 1991, 1992; SCHACHT et al. 1988, 1989). No signs of nephrotoxicity were found in a study on the renal tolerance of ciprofloxacin using monoclonal antibodies and by measurement of brush border enzymes (FALKENBERG et al. 1987).

In animals crystalluria may cause marked renal damage. This phenomenon is related to the solubility of the quinolones which is lowest at pH 7–9 for most of the derivatives. Crystalluria does not appear to be a cause of renal damage in humans, and renal function returns to normal after discontinuing the drug (PATON and REEVES 1991; THORSTEINSSON et al. 1987b; BERGAN et al. 1989b). In healthy volunteers, with higher than therapeutic doses, no crystals were observed after nalidixic acid or ofloxacin. After cooling the urine to room temperature for 24 h, crystals were observed after ofloxacin, ciprofloxacin and norfloxacin.

THORSTEINSSON et al. (1986) administered 500 mg or 1000 mg ciprofloxacin to six healthy adult volunteers (three men and three women) who had been given ammonium chloride to acidify urine, sodium bicarbonate to alkalize urine or no dietary supplement. Freshly voided urine kept at 37°C was examined microscopically with polarized light. When urine was alkaline ($pH > 7.3$), crystals of ciprofloxacin were observed in samples from five of six subjects given 1000 mg and three of six given 500 mg. Under acidic conditions ($pH < 5.8$), crystals were not observed in any sample at either dose. In subjects on a normal diet (urine $pH \leq 6.9$), one of six subjects given 1000 mg and none of the six given 500 mg had ciprofloxacin crystals in their urine. Thus, this study indicates that under alkaline conditions it is possible for crystals of ciprofloxacin to form in the urine. Analytical studies on the composition of the crystals indicated that they consisted of ciprofloxacin in combination with magnesium. Only small amounts of one metabolite (M_2) were also shown to be present (Holm et al. 1985; Weber et al. 1985). In man, the urine is typically acidic, generally ranging in pH from 5.5–6.5. This probably explains why crystalluria clearly related to ciprofloxacin has rarely been observed during clinical investigation. Similar changes may in principle be induced by all the fluorinated quinolones, but with different probabilities of occurrence (Thorsteinsson et al. 1987b). Patients receiving quinolones should be well hydrated, and alkalinity of the urine should be avoided (Bergan et al. 1989; Thorsteinsson et al. 1986).

V. Arthropathy and Musculoskeletal Disorders

Fluoroquinolones are not recommended for use in pregnant women, nursing mothers or in children and adolescents in the growing phase. The recommendation has to a large extent been based on the finding that administration of high doses of quinolones to juvenile animals may produce cartilage damage in weight-bearing joints (Paton and Reeves 1991; Christ and Lehnert 1990; Gough et al. 1992; Ribard and Kahn 1991). It is very difficult to apply the animal findings to the human situation. To date there is little evidence of any quinolone-induced arthropathy in man (Adam et al. 1989a; Schaad and Wedgwood 1992; Schaad et al. 1991, 1992; Schaad 1992) although the risk of such effects may increase with very prolonged usage (Chevalier et al. 1992). Therefore, carefully controlled prospective clinical trials should be undertaken to evaluate whether fluoroquinolones can be used in children (Christ and Lehnert 1990; Norrby 1991; Adam 1989b). Fluoroquinolones should be used during pregnancy only if the potential benefit justifies the potential risk for the fetus. Unilateral and bilateral tendinitis and rupture of the tendon have been recently published (Jorgensen et al. 1991; Perrot et al. 1992; Ribard et al. 1992; Chaslerie et al. 1992; Lee and Collins 1992). The pathogenesis of these musculoskeletal side effects of quinolones is still poorly understood. The onset of tendon pain is sudden, and symptoms occur a mean of 13 days after starting treatment. The risk of rupture appears to be greatest when patients previously confined to bed start to walk again (Anonymous 1996b).

VI. Body Systems

Only in very rare cases have adverse reactions involving cardiovascular, respiratory, haematopoetic and lymphatic systems or the liver been observed (Wolfson 1989; Wolfson and Hooper 1989; Paton and Reeves 1991, 1992; Cullmann et al. 1993; Stahlmann and Lode 1988; Adam et al. 1987; Norrby 1991; Schacht et al. 1989; Corrado et al. 1987; Jüngst and Mohr 1988; Sawada et al. 1991; Rubinstein and Carbon 1994; Simon et al. 1993; Halkin 1988; Janknegt 1989). In animals testicular and ocular toxicity have been described, especially for pefloxacin, but in none of the clinical investigations were therapeutic doses of the quinolones found to be related to these animal toxicity finding (Stahlmann and Lode 1988; Schacht et al. 1988, 1989;

ARCIERI et al. 1987; Hullmann et al., personal communication). In ophthalmological tests no subjective and/or objective symptoms related to ofloxacin eye drops were noted after the drug was applied to ten healthy volunteers for 2 weeks (HARA 1985).

The tolerability of intravenous ciprofloxacin appears to be similar to that of oral ciprofloxacin. However, the major difference between the two formulations is the occurrence of local reactions at the site of intravenous infusion (THORSTEINSSON et al. 1987a, 1989; ARCIERI et al. 1989). Local i.v. site reactions of ciprofloxacin are more frequent if infusion time is 30 min or less or if small veins of the hand are used. Therefore, ciprofloxacin should be infused into large veins slowly and if possible the infusion site should be put in a higher position than the rest of the body in order to allow a faster flow from the site of infusion (ROHWEDDER, personal communication).

VII. Others

Quinolones can suppress the senses of smell and taste which can last for several months and is reversible. Patients should be advised accordingly (SCHACHT et al. 1989).

References

Adam D (1989a) Klinische Bedeutung der neueren Chinolonderivate. MMW 131:1–2

Adam D (1989b) Use of quinolones in pediatric patients. Rev Infect Dis 11(Suppl 5):1113–1116

Akerele JO, Okhamafe AO (1991) Influence of oral co-administered metallic drugs on ofloxacin pharmacokinetics. J Antimicrob Chemother 28:87–94

Albrecht D, Zahorsky R, Krausse R, Wittke J-W, Ullmann U, Niedermayer W (1992) Influence of magnesium- and aluminiumhydroxide on pharmacokinetic data of ofloxacin in patients with prophylaxis of peptic ulcer. Arch Pharmacol 345:R1

Allen JE (1993) Drug-induced photosensitivity. Clin Pharm 12:580–587

Anonymous (1996a) Sparfloxacin: phototoxicity warrants restriction on use in the community. Presc Int 5(27):6–8

Anonymous (1996b) Tendon rupture on fluoroquinolones. Presc Int 5(27):20

Arcieri G, Griffith E, Gruenwaldt G, Heyd A, O'Brien B, Becker N, August R (1987) Ciprofloxacin: an update on clinical experience. Am J Med 82(Suppl 4A):381–386

Arcieri GM, Becker N, Esposito B, Griffith E, Heyd A, Neumann C, O'Brien B, Schacht P (1989) Safety of intravenous ciprofloxacin. A review. Am J Med 87(Suppl 5A):92–97

Arrigo G, Cavaliere G, D'Amico G, Passarella E, Broccali G (1985) Pharmacokinetics of norfloxacin in chronic renal failure. Int J Clin Pharmacol Ther Toxicol 23:491–96

Avent CK, Krinsky D, Kirklin JK, Bourge RC, Figg WD (1988) Synergistic nephrotoxicity due to ciprofloxacin and cyclosporine. Am J Med 85:452–453

Bachmann K, Schwartz JI, Jauregui L (1988) Predicting the ciprofloxacin–theophylline interaction from single plasma measurements. Br J Clin Pharmacol 26:191–194

Baciewicz AM, Ashar BH, Locke TW (1993) Interaction of ofloxacin and warfarin. Ann Intern Med 199:1223

Ball AP, Fox C, Ball ME, Brown IRF, Willis JV (1986) Pharmacokinetics of oral ciprofloxacin, 100 mg single dose, in volunteers and elderly patients. J Antimicrob Chemother 17:629–635

Ball P, Tillotson G (1995) Tolerability of fluoroquinolone antibiotics: past, present and future. Drug Saf 13(6):343–358

Barre J, Houin G, Tillement JP (1984) Dose dependent phamacokinetic study of pefloxacin, a new antibacterial agent, in humans. J Pharm Sci 73:1379–1382

Barriere SL, Catlin DH, Orlando PL, Noe A, Frost RW (1990) Alteration in the pharmacokinetic disposition of ciprofloxacin by simultaneous administration of azlocillin. Antimicrob Agents Chemother 34:823–826

Batty KT, Davis TME, Ilett KF, Dusci LJ, Langton SR (1995) The effect of ciprofloxacin on theophylline pharmacokinetics in healthy subjects. Br J Clin Pharmacol 39:305–311

Bayer A, Gajewska A, Stephens M, Stark JM, Pathy J (1987) Pharmacokinetics of ciprofloxacin in the elderly. Respiration 51:292–295

Beckmann J, Elsäßer W, Gundert-Remy U, Hertrampf R (1987) Enoxacin – a potent inhibitor of theophylline metabolism. Eur J Clin Pharmacol 33:227–230

Beermann D, Wingender W, Zeiler HJ, Förster D, Graefe KH, Schacht P (1984) Comparative pharmacokinetics of three new quinolone carboxylic acid antibiotics after oral administration in healthy volunteers. J Clin Pharmacol 24:403

Beermann D, Scholl H, Wingender W, Förster D, Beubler E, Kukovetz WR (1986) Metabolism of ciprofloxacin in man. In: Neu H, Weuta H (eds) Proceedings of the First International Ciprofloxacin Workshop, Leverkusen 1985. Exerpta Medica, Amsterdam, pp 141–46

Beermann D, Wingender W, Horstmann R (1987) Intravenous infusion regimens for rapidly achieving steady state levels of ciprofloxacin. Am J Med 83 (Suppl 4A):360–362

Bergan T (1988) Pharmacokinetics of fluorinated quinolones. In: Andriole VT (ed) The quinolones. Academic, New York, pp 119–154

Bergan T, Thorsteinsson SB (1986) Pharmacokinetics and bioavailability of ciprofloxacin. In: Neu H, Weuta H (eds) Proceedings of the First International Ciprofloxacin Workshop, Leverkusen 1985. Excerpta Medica, Amsterdam, pp 111–121

Bergan T, Thorsteinsson SB, Kolstad IM, Johnsen S (1986) Pharmacokinetics of ciprofloxacin after intravenous and increasing oral doses. Eur J Clin Microbiol 5:187–192

Bergan T, Dalhoff A, Rohwedder R (1988) Pharmacokinetics of ciprofloxacin. Infection 16(Suppl 1):3–13

Bergan T, Rohwedder R, Thorsteinsson SB (1989a) Significance of crystalluria caused by quinolones. Rev Infect Dis 11(Suppl 5):1395–1396

Bergan T, Thorsteinsson SB, Rohwedder R, Scholl H (1989b) Elimination of ciprofloxacin and three major metabolites and consequences of reduced renal function. Chemotherapy 35:395–405

Bianco TM, Bussey HI, Farnett LE, Linn WD, Roush MK, Wong J (1992) Potential warfarin–ciprofloxacin interaction in patients receiving long-term anticoagulation. Pharmacotherapy 12:435–439

Bleske BE, Carver PL, Annesley TM, Bleske JR, Morady F (1990) The effect of ciprofloxacin on the pharmacokinetic and ECG parameters of quinidine. J Clin Pharmacol 30:911–915

Blouin RA, Hamelin BA, Smith DA, Foster TS, John WJ, Welker HA (1992) Fleroxacin pharmacokinetics in patients with liver cirrhosis. Antimicrob Agents Chemother 36:632–638

Blum RA (1992) Influence of renal function on the pharmacokinetics of lomefloxacin Compared with Other Fluoroquinolones. Am J Med 92(Suppl 4A):18–21

Blum RA, Schultz RW, Schentag JJ (1990) Pharmacokinetics of lomefloxacin in renally compromised patients. Antimicrob Agents Chemother 34:2364–2368

Blum MD, Graham DJ, McCloskey CA (1994) Temafloxacin syndrome. Clin Infect Dis 18:946–950

Boeckh M, Lode H, Deppermann KM, Grineisen S, Shokry F, Held R, Wernicke K, Koeppe P, Wagner J, Krasemann C, Borner K (1990) Pharmacokientics and serum bactericidal activities of quinolones in combination with clindamycin, metronidazole, and ornidazole. Antimicrob Agents Chemother 34:2407–2414

Boelaert J, Valcke Y, Schurgers M, Daneels R, Rosseneu M, Rosseel MT, Bogaert MG (1985) The pharmacokinetics of ciprofloxacin in patients with impaired renal function. J Antimicrob Chemother 16:87–93

Boettcher J, Wolter M (1989) Influence of new quinolones on drug metabolism (Abstr 247). J Chemother Infect Dis Malig (Suppl 1)

Borner K, Höffken G, Prinzing C, Lode H (1984) Renal execretion of ciprofloxacin and several metabolites following single oral or intravenous administration of 50 mg. Fortschr Antimikrob Antineoplast Chemother 3:695–699

Borner K, Höffken G, Lode H, Koeppe P, Prinzing C, Glatzel P, Wiley R, Olschewski P, Sievers B, Reinitz D (1986) Pharmacokinetics of ciprofloxacin in healthy volunteers after oral and intravenous administration. Eur J Clin Microbiol 5:179–186

Borner K, Borner E, Lode H (1993) A metabolite of sparfloxacin in urine. Drugs 45(Suppl 3):303–304

Bowie WR, Willetts V, Jewesson PJ (1989) Adverse reactions in a dose-ranging study with a new long-acting fluoroquinolone, fleroxacin. Antimicrob Agents Chemother 1778–1782

Bowles SK, Popovski Z, Rybak MJ, Beckman HB, Edwards DJ (1988) Effect of norfloxacin on theophylline pharmacokinetics at steady-state. Antimicrob Agents Chemother 32:510–512

British National Formulary (1996) No. 31. Pharmaceutical, London, pp 257–259

British Photodermatology Group (1992) Workshop report: diagnostic phototesting in the United Kingdom. Br J Dermatol 127:297–299

Brittain DC, Scully BE, McElrath MJ, Steinman R, Labthavikul P, Neu H (1985) The pharmacokinetics and serum and urine bactericidal activity of ciprofloxacin. J Clin Pharmacol 25:82–88

Brouwers JRBJ, Van Der Kam HJ, Sijtsma J, Prost JH (1990a) Important reduction of ciprofloxacin absorption by sucralfate and magnesium citrate solution. Drug Invest 2:197–199

Brouwers JRBJ, Van Der Kam HJ, Sijtsma J, Prost JH (1990b) Decreased ciprofloxacin absorption with concomitant administration of ferrous fumarate. Pharm Weekbl [Sci] 12:182–183

Brown NM, White LO, Blundell EL, Chown SR, Slade RR, MacGowan AP, Reeves DS (1993) Absorption of oral ofloxacin after cytotoxic chemotherapy for haematological malignancy. J Antimicrob Ag Chemother 32:117–122

Campbell NRC, Kara M, Hasinoff BB, Haddara WM, McKay DW (1992) Norfloxacin interaction with antacids and minerals. Br J Clin Pharmacol 33:115–116

Campoli-Richards DM, Monk JP, Price A, Benfield P, Todd PA, Ward A (1988) Ciprofloxacin – a review of its antibacterial activity, pharmacokinetic properties and therapeutic use. Drugs 35:373–447

Carbarga MM, Navarro AS, Gandarillas CIC, Dominguez-Gil A (1991) Effects of two cations on gastrointestinal absorption of ofloxacin. Antimicrob Agents Chemother 35:2102–2105

Carbó M, Segura J, De la Torre R, Badenas JM, Cami J (1989) Effects of quinolones on caffeine disposition. Clin Pharmacol Ther 45:234–240

Carbon C, Rubinstein E (eds) (1994) Sparfloxacin monograph. ADIS, Chester

Carlson JD, Dietz AJ, Hertsgaard DM, Heyd A, Frost RW, Lettieri JT (1988) Ciprofloxacin bioavailability after a standard or high-calcium meal. Pharm Res 5(Suppl 10):S166

Cesana M, Broccali G, Imbimbo BP, Crema A (1991) Effect of single doses of rufloxacin on the disposition of theophylline and caffeine after single administration. Int J Clin Pharmacol Ther Toxicol 29:133–138S

Chandler MHH, Toler SM, Rapp RP, Muder RR, Korrick JA (1990) Multiple-dose pharmacokinetics of concurrent ciprofloxacin and rifampin therapy in elderly patients. Antimicrob Agent Chemother 34:442–447
Chang T, Black A, Dunky A, Wolf R, Sedman A, Latts J, Welling PG (1988) Pharmacokinetics of intravenous and oral enoxacin in healthy volunteers. J Antimicrob Chemother 21(Suppl B):49–56
Chaslerie A, Bannwarth B, Landreau JM, Yver L, Begaud B (1992) Ruptures tendineuses et fluoroquinolones: un effect indésirable de classe (Lettre). Rev Rhum Mal Osteoartic 59:297–298
Chevalier X, Albengres E, Voisin MC, Tillement JP, Larget-Peirt B (1992) A case of destructive polyarthropathy in a 17-year-old youth following pefloxacin treatment. Drug Saf 7:310–314
Christ W, Lehnert T (1990) Toxicity of quinolones. In: Spiron L, Heifetz CL, Domagala JM (eds) The new generation of quinolones. Dekker, New York, pp 165–187
Christ W, Lehnert T, Ulbrich B (1988) Specific toxicologic aspects of the quinolones. Rev Infect Dis 10(Suppl 1):S141–S146
Chrome P, Morrison PJ (1991) Pharmacokinetics of a single oral dose of lomefloxacin in healthy elderly volunteers. Drug Invest 3:183–187
Chysky V, Kapila K, Hullmann R, Arcieri G, Schacht P, Echols R (1991) Safety of ciprofloxacin in children: world-wide clinical experience based on compassionate use. Emphasis on joint evaluation. Infection 19:289–296
Cleary JD (1992) Ciprofloxacin and pentoxifylline: a clinically significant drug interaction. Pharmacotherapy 12:259–260
Cogo R, Rimoldi R, Mattina R, Imbimbo BP (1992) Steady-state pharmacokinetics of rufloxacin in elderly patients with lower respiratory tract infections. Ther Drug Monit 14:36–41
Cook JA, Silverman MH, Schelling DJ, Nix DE, Schentag JJ, Brown RR, Stroshane RM (1990) Multiple dose pharmacokinetics and safety of oral amifloxacin in healthy volunteers. Antimicrob Agents Chemother 34:974–979
Corrado ML, Struble WE, Peter C, Hoagland V, Sabbaj J (1987) Norfloxacin: review of safety studies. Am J Med 82:22–26
Crome P, Morrison PJ (1991) Pharmacokinetics of a single oral dose of lomefloxacin in health elderly volunteers. Drug Invest 3:183–187
Cullmann W, Geddes AM, Weidekamm E, Urwyler H, Braunsteiner A (1993) Fleroxacin: a review of its chemistry, microbiology, toxicology, pharmacokinetics, clinical efficacy and safety. Int J Antimicrob Agents 2:203–230
Cutler NR, Sramek JJ, Narang PK (eds) (1994) Pharmacodynamics and drug development: perspectives in clinical pharmacology. Wiley, New York
Danan G, Montay G, Cunci R, Erlinger S (1985) Pefloxacin kinetics in cirrhosis. Clin Pharmacol Ther 38:439–442
Davis R, Bryson HM (1994) Levofloxacin – a review of its antibacterial activity, pharmacokinetics and therapeutic efficacy. Drugs 47:677–700
Davis RL, Quenzer RW, Kelly HW, Powell JR (1992) Effect of the addition of ciprofloxacin on theophylline pharmacokinetics in subjects inhibited by cimetidine. Ann Pharmacother 26:11–13
Deeter RG, Weinstein MP, Swanson KA, Gross JS, Hildebrant A (1989) Comparative pharmacokinetics of oral ciprofloxacin alone and in combination with rifampin in healthy elderly volunteers (Abstr 201). Program and abstracts of the 28th Interscience Conference on Antimicrobial Agents and Chemotherapy, September 1989, Houston
Deppermann K-M, Lode H (1993) Fluoroquinolones: interaction profile during enteral absorption. Drugs 45(Suppl 3):65–72
Dillard ML, Fink MR (1992) Ciprofloxacin phenytoin interaction. Ann Pharmocother 26:263
Dobbs BR, Gazely LR, Campbell AJ, Edwards IR (1987) The effect of age on the pharmacokinetics of enoxacin. J Clin Pharmacol 33:101–104

Domagala JM (1994) Structure–activity and structure–side effect relationships for the quinolone antibacterials. J Antimicrob Chemother 33:685–706
Dörfler A, Schulz W, Burkhardt F, Zichner M (1987) Pharmacokinetics of ofloxacin in patients on haemodialysis treatment. Drugs 34(Suppl 1):62–70
Dow J, Frydman AM, Djebbar F, Gaillot J (1988) Single- and multiple-dose pharmacokinetics of pefloxacin in elderly patients. Rev Infect Dis 10(Suppl 1):107
Doyle GD, Donohoe J, Kelly JG, Laher MS (1988) Single- and multiple-dose pharmacokinetics of norfloxacin in renal impairment. Rev Infect Dis 10 (Suppl 1):111–112
Droppert RM, Scholten PC, Zwinkels M, Hoepelman JM, Te Velde ER (1993) Lack of influence of ciprofloxacin on the effectiveness oral contraceptives. Drugs 45(Suppl 3):286–287
Drusano GL, Standiford HC, Plaisance K, Forrest A, Leslie J, Caldwell J (1986) Absolute oral bioavailability of ciprofloxacin. Antimicrob Agents Chemother 30:444–446
Drusano GL, Weir M, Forrest A, Plaisance K, Emm T, Standiford HC (1987) Pharmacokinetics of intravenously administered ciprofloxacin in patients with various degrees of renal function. Antimicrob Agents Chemother 31:860–864
Drusano GL, Johnson DE, Rosen M, Lockatell V, Standiford HC (1991) Pharmacodynamics of the flouroquinolone lomefloxacin (Abstr 1926). 17th ICC, June 23–28, Berlin
Dudley MN, Marchbanks CR, Flor SC, Beals B (1991) The effect of food or milk on the absorption kinetics of ofloxacin. Eur J Clin Pharmacol 41:569–571
Dugoni-Kramer BM (1991) Ciprofloxacin–warfarin interaction. Ann Pharmacother 25:1397
Eandi M, Viano I, DiNola F, Leone L, Genazzani E (1983) Pharmacokinetics of norfloxacin in healthy volunteers and patients with renal and hepatic damage. Eur J Clin Microbiol 2:253–259
Ehninger G, Krüger HU (1989) No severe drug interaction of the 1-quinolone derivative ciprofloxacin with cyclosporin (Abstr 252). J Chemother Infect Dis Malig (Suppl 1)
Elston RA, Taylor J (1988) Possible interaction of ciprofloxacin with cyclosporin A. J Antimicrob Chemother 21:679–680
Enzensberger R, Shah PM, Knothe H (1985) Impact of oral ciprofloxacin on the fecal flora of healthy volunteers. Infection 13:273–275
Epstein JH, Wintroub BU (1985) Photosensitivity due to drugs. Drugs 30:42–57
Falkenberg FW, Mondorf AW, Dalhoff A (1987) Untersuchungen über die Nierenverträglichkeit von Ciprofloxacin mit Hilfe monoklonaler Antikörper. FAC 6–10:2265–2280
Farinotti R, Trouvin JH, Bocquet V, Vermerie N, Carbon C (1988) Pharmacokinetics of ofloxacin after single and multiple intravenous infusions in healthy subjects. Antimicrob Agents Chemother 32:1590–1592
Fassbender M, Lode H, Stuht H, Wischmann L, Borner K, Koeppe P (1993) Influence of omeprazole on the pharmacokinetics of three quinolones and a macrolide-antibiotic (Abstr 1700). 33rd Interscience Conference on Antimicrobial Agents and Chemotherapy, Oct 17–20, New Orleans
Ferguson J, Johnson BE (1990) Ciprofloxacin-induced photosensitivity: in vitro and in vivo studies. Br J Dermatol 123:9–20
Ferguson J, Johnson B (1993) Clinical laboratory studies of the photosensitizing potential of norfloxacin, a 4-quinolone broad-spectrum antibiotic. Br J Dermatol 128:285–295
Fillastre JP, Singlas E (1991) Pharmacokinetics of newer drugs in patients with renal impairment (part 1). Clin Pharmacokinet 20:293–310
Fillastre JP, Leroy A, Humbert G (1987) Ofloxacin pharmacokinetics in renal failure. Antimicrob Agents Chemother 31:156–160
Fillastre JP, Leroy A, Borsa-Lebas F, Etienne I, Gy C, Humbert G (1992) Effects of ketoprofen (NSAID) on the pharmacokinetics of pefloxacin and ofloxacin in healthy volunteers. Drugs Exp Clin Res 18:487–492

Fillastre JP, Montay G, Bruno R, Etienne I, Dhib M, Vivier N, LeRoux Y, Guimart C, Gay G, Schott D (1994) Pharmacokinetics of sparfloxacin in patients with renal impairment. Antimicrob Agents Chemother 38:733–737

Fischman AJ, Livni E, Babich J, Alpert NM, Liu YY, Thom E, Cleeland R, Prosser BL, Correia JA, Strauss HW, Rubin RH (1993) Pharmacokinetics of [18-F]-fleroxacin in healthy human subjects studied by using positron emission tomography. Antimicrob Agents Chemother 37:2144–2152

Fleming LW, Moreland TA, Stewart WK, Scott AC (1986) Ciprofloxacin and antacids. Lancet 8501:294

Flor S, Guay DRP, Opsahl JA, Tack K, Matzke GR (1990) Effects of magnesium-aluminium hydroxide and calcium carbonate antacids on bioavailability of ofloxacin. Antimicrob Agents Chemother 34:2436–2438

Foster TS, Blouin R (1991) The effect of antacid timing on lomefloxacin bioavailability. Pharmacotherapy 11:101

Fraise AP, Smith SP (1990) Ciprofloxacin in combined renal and hepatic impairment. J Antimicrob Chemother 25:297–303

Freeman CD, Nicolau DP, Belliveau PP, Nightingale CH (1993) Lomefloxacin clinical pharmacokinetics. Clin Pharmacokinet 25:6–19

Frost RW, Lettieri JT, Krol G, Shamblen EC, Lasseter KC (1989) The effect of cirrhosis on the steady-state pharmacokinetics of oral ciprofloxacin. Clin Pharmacol Ther 45:608–616

Frost RW, Lasseter KC, Noe AJ, Shamblen EC, Lettieri JT (1992) Effects of aluminium hydroxide and calcium carbonate antacids on the bioavailability of ciprofloxacin. Antimicrob Agents Chemother 36:830–832

Frydman AM, LeRoux Y, Lefebvre MA, Djebbar F, Fourtillan JB, Gaillot J (1986) Pharmacokinetics of pefloxacin after repeated intravenous and oral administration (400 mg bid) in young healthy volunteers. J Antimicrob Chemother 17(Suppl B):65–79

Fuhr U, Wolff T, Harder S, Schymanski P, Staib AH (1990) Quinolone inhibition of cytochrome P450 dependent caffeine metabolism in human liver microsomes. Drug Metab Dis 18:1005–1010

Fuhr U, Anders EM, Mahr G, Sörgel F, Staib AH (1992) Inhibitory potency of quinolone antibacterial agents against cytochrome P450IA2 activity in vivo and in vitro. Antimicrob Agents Chemother 36:942–948

Fuhr U, Strobl G, Manaut F, Anders EM, Sörgel F, Lopez-de-Brinas E, Chu DTW, Pernet AG, Mahr G, Sanz F, Staib AH (1993) Quinolone antibacterial agents: relationship between structure and in vitro inhibition of the human cytochrome P450 isoform CYPIA2. Mol Pharmacol 43:191–199

Fujii R (1992) The use of norfloxacin in children in Japan. Adv Antimicrob Antineoplast Chemother 11(2):219–230

Gardner K, Job ML, Strom JG, Jacobs NF, Dsoura J (1990) Effect of ciprofloxacin on the pharmacokinetics of steady-state phenytoin serum concentrations. Annual ASHP Midyear Clin Meet 25:PP-426E

Garrelts JC, Godley PJ, Peterie JD, Gerlach EH, Yakshe CC (1990) Sucralfate significantly reduces ciprofloxacin concentrations in serum. Antimicrob Agents Chemother 34:931–933

Gasser TC, Ebert SC, Graversen PH, Madsen PO (1987a) Pharmacokinetic study of ciprofloxacin in patients with impaired renal function. Am J Med 82 (Suppl 4A):139–141

Gasser TC, Ebert SC, Graversen PH, Madsen PO (1987b) Ciprofloxacin pharmacokinetics in patients with normal and impaired renal function. Antimicrob Agents Chemother 31:709–712

Geddes AM (1993) Safety of fleroxacin in clinical trials. Am J Med 94(Suppl 3A):201–203

Goodwin SD, Gallis HA, Chow AT, Wong FA, Flor SC, Bartlett JA (1994) Pharmacokinetics and safety of levofloxacin in patients with human immunodeficiency virus infection. Antimicrob Agents Chemother 38:799–804

Gough AW, Kasali OB, Sigler RE, Baragi V (1992) Quinolone arthropathy – acute toxicity to immature articular cartilage. Toxicol Pathophysiol 20:436–450
Graber H, Ludwig E, Arr M, Lanyi P (1988) Pharmacokinetics of ofloxacin in young and elderly patients. Rev infect Dis 10 (Suppl 1):106
Graber H, Ludwig E, Magyar T, Csiba A, Szekely E (1989) Ofloxacin does not influence antipyrine metabolism. Rev Infect Dis 11(Suppl 5):1094–1095
Grasela TH, Schentag JJ, Sedman AJ, Wilton JH, Thomas DJ, Schultz RW, Lebsack ME, Kinkel AW (1989) Inhibition of enoxacin absorption by antacids or ranitidine. Antimicrob Agents Chemother 33:615–617
Gregoire SL, Grasela TH, Freer JP, Tack KJ, Schentag JJ (1987) Inhibition of theophylline clearance by coadministered ofloxacin without alteration of theophylline effects. Antimicrob Agents Chemother 31:375–378
Gros I, Carbon C (1990) Pharmacokinetics of lomefloxacin in healthy volunteers: comparison of 400 milligrams once daily and 200 milligrams twice daily given orally for 5 days. Antimicrob Agents Chemother 34:150–152
Grözinger KH, Beermann D, Elsas S (1989) Biliary kinetics of ciprofloxacin in humans. Rev Infect Dis 11 (Suppl 5):1132–1133
Guay DRP, Auni WM, Peterson PK, Obaid S, Breitenbucher R, Matzke GR (1987) Pharmacokinetics of ciprofloxacin in acutely ill and reconvalescent elderly patients. Am J Med 82(Suppl 4A):124–129
Gugler R, Allgayer H (1990) Effects of antacids on the clinical pharmacokinetics of drugs – an update. Clin Pharmacokinet 18:210–219
Ha HR, Chen J, Freiburghaus AU, Follath F (1995) Metabolism of theophylline by cDNA-expressed human cytochromes P-450. Br J Clin Pharmacol 39:321–326
Halkin H (1988) Adverse effects of the fluoroquinolones. Rev Infect Dis 10:S258–S261
Halliwell RF, Davey PG, Lambert JJ (1993) Antagonism of $GABA_A$ receptors by 4-quinolones. J Antimicrob Chemother 31:457–462
Hara J (1985) Phase I study on DE-055 eye drops. Jpn Rev Clin Ophthalmol 79:1712–1717
Harder S, Staib AH, Beer C, Papenburg A, Stille W, Shah PM (1988) 4-Quinolones inhibit biotransformation of caffeine. Eur J Clin Pharmacol 35:651–656
Harder S, Fuhr U, Beermann D, Staib AH (1990) Ciprofloxacin absorption in different regions of the human gastrointestinal tract. Investigations with the HF-Capsule. Br J Clin Pharmacol 30:35–39
Hayakawa I, Atarashi S, Yokohama S, Imamura M, Sakano K, Furukawa M (1986) Synthesis and antibacterial activities of optically active ofloxacin. Antimicrobial Agents Chemother 29:163–164
Healy DP, Small RE (1989) Possible diclofenac–quinolone interaction. Clin Pharm 8:837
Healy DP, Polk RE, Kanamati L, Rock DT, Mooney ML (1989) Interaction between oral ciprofloxacin and caffeine in normal volunteers. Antimicrob Agents Chemother 33:474–478
Healy DP, Schoenle JR, Stofka J, Polk RE (1991) Lack of interaction between lomefloxacin and caffeine in normal volunteers. Antimicrob Agents Chemother 35:660–664
Herrmann W, Andressen C, Rohmel J, Aigner B, Rosener B, Kern U (1984) Combined pharmaco-EEG and pharmacopsychological study for the investigation and characterization of possible CNS effects of the broad spectrum chemotherapeutic ciprofloxacin 500 mg and 1000 mg versus nalidixic acid 2000 mg and a placebo. Bayer, Leverkusen (Internal Bayer R-report no. 3121 (P))
Hirata CA, Guay RP, Awni WM, Stein DJ, Peterson PK (1989) Steady-state pharmacokinetics of intravenous and oral ciprofloxacin in elderly patients. Antimicrob Agents Chemother 33:1927–1931
Ho G, Tierney MG, Dales RE (1988) Evaluation of the effect of norfloxacin on the pharmacokinetics of theophylline. Clin Pharmacol Ther 44:35–38
Höffken G, Lode H, Prinzing C, Borner K, Koeppe P (1985a) Pharmacokinetics of ciprofloxacin after oral and parenteral administration. Antimicrob Agents Chemother 27:375–379

Höffken G, Borner K, Glatzel PD, Koeppe P, Lode H (1985b) Reduced enteral absorption of ciprofloxacin in the presence of antacids. Eur J Clin Microbiol 4:345
Höffken G, Lode H, Wiley R, Glatzel TD, Sievers D, Olschewski T, Borner K, Koeppe T (1988) Pharmacokinetics and bioavailability of ciprofloxacin and ofloxacin: effect of food and antacid intake. Rev Infect Dis 10(Suppl 1):S138–S139
Höffler D, Koeppe P (1987) Pharmacokinetics of ofloxacin in healthy subjects and patients with impaired renal function. Drugs 34(Suppl 1):51–55
Höffler D, Dalhoff A, Au W, Beermann D, Michl A (1984) Dose- and sex-independent disposition of ciprofloxacin. Eur J Clin Microbiol 3:363–366
Höffler D, Schäfer I, Koeppe P, Sörgel F (1988) Pharmacokinetics of pefloxacin in normal and impaired renal function. Drug Res 38:739–743
Höffler D, Waetcke K, Koeppe P, Metz R, Sörgel F (1989) Pharmacokinetics of lomefloxacin in normal and impaired renal function. Acta Theriol 15:321–336
Holm R, Rohwedder R, Rühl C, Wünsche C (1985) Morphology and composition of crystalline concretion in the kidney and urine after ciprofloxacin administration. Bayer, Leverkusen (Internal Bayer PB-report no 13668)
Holmes B, Brogden RN, Richards DM (1985) Norfloxacin – a review of its antibacterial activity, pharmacokinetic properties and therapeutic use. Drugs 30:482–513
Holy LL, Lake KD (1994) Does ciprofloxacin interact with cyclosporine? Ann Pharmacother 28:93–96
Hooper TL, Gould FK, Swinburn CR, Featherstone G, Odom NJ, Corris PA, Freeman R, McGregor CGA (1988) Ciprofloxacin: a preferred treatment for legionella infections in patients receiving cyclosporin A. J Antimicrob Chemother 22:952–953
Hori S, Shimada J, Saito A, Miyahara T, Kurioka S, Matsuda M (1985) Effect of new quinolones on gamma-aminobutyric acid receptor binding (Abstr 396). Proceedings of the 25th Interscience Conference on Antimicrobial Agents and Chemotherapy, Minneapolis. American Society for Microbiology, Washington
Hori S, Shimada J, Saito A, Miyahara T, Kurioka S, Matsuda M (1986) Inhibitory effect of quinolones on gamma-aminobutyric acid receptor binding. Structure activity relationship (Abstr 438). Proceedings of the 26th Interscience Conference on Antimicrobial Agents and Chemotherapy, New Orleans. American Society for Microbiology, Washington
Hori S, Shimada J, Saito A, Matsuda M, Miyahara T (1989) Comparison of the inhibitory effects of new quinolones on γ-aminobutyric acid receptor binding in the presence of antiinflammatory drugs. Rev Infect Dis 11(Suppl 5):S1397–1398
Hull RL (1993) Possible phenytoin–ciprofloxacin interaction. Ann Pharmacother 27:1283
Humbert G, Brumpt I, Montay G, Le Liboux A, Frydman A, Borsa-Lebas F, Moore N (1991) Influence of rifampin on the pharmacokinetics of pefloxacin. Clin Pharmacol Ther 50:682–687
Jaehde U, Sörgel F, Naber KG (1989) Gastrointestinal secretion of ciprofloxacin (CIP) in healthy volunteers (Abstr 202). 29th Interscience Conference on Antimicrobial Agents and Chemotherapy. American Society for Microbiology, Washington
Janknegt R (1989) Fluoroquinolones. Pharm Weekbl [Sci] 11:124–127
Job LM, Arn SK, Strom JG, Jacobs NF, D'Souza MJ (1994) Effect of ciprofloxacin on the pharmacokinetics of multiple-dose phenytoin serum concentrations. Ther Drug Monit 16:427–431
Johnson BE, Ferguson J (1990) Drug and chemical photosensitivity. Semin Dermatol 9:39–46
Johnson EJ, MacGowan AP, Potter MN, Stockley RJ, White LO, Slade RR, Reeves DS (1990) Reduced absorption of oral ciprofloxacin after chemotherapy for haematological malignancy. J Antimicrob Chemother 25:837–842
Johnson KC, Joe RH, Self T (1991) Drug interaction. J Fam Pract 33:338
Johnson JH, Cooper MA, Andrews JM, Wise R (1992) Pharmacokinetics and inflammatory fluid penetration of sparfloxacin. Antimicrob Agents Chemother 36:2444–2446

Jorgensen C, Anaya JM, Didry C, Canovas F, Serre I, Baldet P, Ribard P, Kahn M-F, Sany J (1991) Arthropathies et tendinopathie achilléene induites par la péfloxacine. A propos d'une observation. Rev Rhum Mal Osteoartic 58:623–625

Jungers P, Ganeval D, Hannedouche T, Prieur B, Montay G (1987) Steady-state levels of pefloxacin and its metabolites in patients with severe renal impairment. Eur J Clin Pharmacol 33:463–467

Jüngst G, Mohr R (1988) Overview of postmarketing experience with ofloxacin in Germany. J Antimicrob Chemother 22(Suppl C):167–175

Jynge P, Skjetne T, Gribbestad I, Kleinbloesem CH, Hoogkamer HFW, Antonsen O, Krane J, Bakoy OE, Furuheim KM, Nilsen OG (1990) In vivo tissue pharmacokinetics by fluorine magnetic resonance spectroscopy: a study of liver and muscle disposition of fleroxacin in humans. Clin Pharmacol Ther 48:481–489

Kakano M, Yamamoto M, Arita T (1978) Interaction of Al, Mg, calcium ions with nalidixic acid. Chem Pharm Bull (Tokyo) 26:1505–1510

Kamada AK (1990) Possible interaction between ciprofloxacin and warfarin. Pharmacother 24:27–28

Kamali F, Thomas SHL, Edwards C (1993) The influence of steady-state ciprofloxacin on the pharmacokinetics and pharmacodynamics of a single dose of diazepam. Eur J Clin Pharmacol 44:365–367

Kamali F, Herd B, Edwards C, Nicholson E, Wynne H (1994) The influence of ciprofloxacin on the pharmacokinetics and pharmacodynamics of a single dose of temazepam in the young and elderly. J Clin Pharm Ther 19:105–109

Kampf D, Borner K, Pustelnik A (1990) Pharmacokinetics of ofloxacin and adequacy of maintenance dose for patients on haemodialysis. J Antimicrob Chemother 26(Suppl D):61–68

Karablut N, Drusano GL (1993) Pharmacokinetics of the quinolone antimicrobial agents. In: Hooper DC, Wolfson JS (eds) Quinolone antimicrobial Agents, 2nd edn. American Society for Microbiology, Washington, pp 195–223

Kazmierczak A, Pechinot A, Duez JM, Haas O, Favre JP (1987) Biliary tract excretion of ofloxacin in man. Drugs 34(Suppl 1):39–43

Kees F, Naber KG, Meyer GP, Grobecker H (1989) Pharmacokinetics of ciprofloxacin in elderly patients. Drug Res 39:523–527

Kelly JG, Deaney NB, Lavan J, Noel J (1988) Chronic dose urinary and serum pharmacokinetics of norfloxacin in the elderly. Br J Clin Pharmacol 26:787–790

Kisicki JC, Griess RS, Ott CL, Cohen GM, McCormack RJ, Troetel WM, Imbimbo BP (1992) Multiple-dose pharmacokinetics and safety of rufloxacin in normal volunteers. Antimicrob Agents Chemother 36:1296–1301

Kivistö KT, Ojala-Karlsson P, Neuvonen PJ (1992) Inhibition of norfloxacin absorption by dairy products. Antimicrob Agents Chemother 36:489–491

Kljucar S, Heimesaat M, Von Pritzbuer E, Timm J, Scholl H, Beermann D (1989) Efficacy and safety of higher-dose intravenous ciprofloxacin in severe hospital-acquired infections. Am J Med 87(Suppl 5A):52–56

Krüger HU, Schuler U, Proksch B, Göbel M, Ehninger G (1990) Investigation of potential interaction of ciprofloxacin with cyclosporine in bone marrow transplant recipients. Antimicrob Agents Chemother 34:1048–1052

Kubin R (1993) Safety and efficacy of ciprofloxacin in paediatric patients – review. Infection 21:413–421

Kuhlmann J, Zilly W, Wilke J (1981) Effects of cytostatic drugs on plasma level and renal excretion of β-acetyldigoxin. Clin Pharmacol Ther 30:518–527

Kuhlmann J, Wilke J, Rietbrock N (1982) Cytostatic drugs are without significant effect on digitoxin plasma level and renal excretion. Clin Pharmacol Ther 32:646–651

Kunka RL, Wong YY, Lyon JL (1988) Effect of antacid on the pharmacokinetics of lomefloxacin. Pharm Res 10:S-165

Lamp KC, Bailey EM, Rybak MJ (1992) Ofloxacin clinical pharmacokinetics. Clin Pharmacokinet 22:32–46

Lang J, De Villaine JF, Garraffo R, Touraine JL (1989a) Cyclosporine (cyclosporin A) pharmacokinetics in renal transplant patients receiving ciprofloxacin. Am J Med 87(Suppl 5A):82–85

Lang J, De Villain JF, Guemei A, Touraine JL, Faucon C (1989b) Absence of pharmacokinetic interaction between pefloxacin and cyclosporin A in patients with renal transplants. Rev Infect Dis 11(Suppl 5):S1094

Lathia C, Bansal S, Batra V, Czernik B, Faulkner R, Greene D, Kinzig M, Kuye O, Pascente V, Sörgel F, Yacobi A (1991) Single-dose pharmacokinetics of tosufloxacin at three dose levels in normal, healthy volunteers. Pharm Res 8(Suppl):303

LeBel M, Bergeron MG (1987) Pharmacokinetics in the elderly, studies on ciprofloxacin. Am J Med 82 (Suppl 4A):108–114

LeBel M, Barbeau G, Bergeron MG, Roy D, Valleé F (1986) Pharmacokinetics of cipofloxacin in elderly subjects. Pharmacotherapy 6:87–91

LeBel M, Vallée F, St.-Laurent M (1990) Influence of lomefloxacin on the pharmacokinetics of theophylline. Antimicrob Agents Chemother 34:1254–1256

Lebrec D, Gaudin C, Benhamou JP (1992) Pharmacokinetics of lomefloxacin in patients with cirrhosis. Am J Med 92(Suppl 4A):41–44

Lebsack ME, Nix DE, Ryerson B, Toothaker RD, Welage L, Norman SA, Schentag JJ, Sedman AJ (1992) Effect of gastric acidity on enoxacin absorption. Clin Pharmacol Ther 52:252–256

Lee WT, Collins JF (1992) Ciprofloxacin associated bilateral achilles tendon rupture. Austr N Z J Med 22:500

Lehto P, Kivistoe KT (1995) Effects of milk and food on the absorption of enoxacin. Br J Clin Pharmacol 39:194–196

Lehto P, Kivistö KT, Neuvonen PJ (1994a) Different effects of products containing metal ions on the absorption of lomefloxacin. Clin Pharmacol Ther 56:477–482

Lehto P, Kivistö KT, Neuvonen PJ (1994b) The effect of ferrous sulphate on the absorption of norfloxacin, ciprofloxacin and ofloxacin. Br J Clin Pharmacol 37:82–85

Leor J, Matetzki S (1988) Oflaxacin and warfarin. Ann Intern Med 109:761

Leroy A, Fillastre JP, Humbert G (1990) Lomefloxacin pharmacokinetics in subjects with normal and impaired renal function. Antimicrob Agents Chemother 34:17–20

Lettieri JT, Schaefer HG, Heller AH (1992) Pharmacokinetic determination of dose equivalency between intravenous and oral doses of ciprofloxacin in healthy volunteers. Naunyn Schmiedebergs Arch Pharmacol 345(Suppl):R11S

Linville T, Matanin D (1989) Norfloxacin and warfarin. Ann Intern Med 110:751

Lode H, Höffken G, Olschewski P, Sievers B, Kirch A, Borner K, Koeppe P (1987a) Pharmacokinetics of ofloxacin after parenteral and oral administration. Antimicrob Agents Chemother 31:1338–1342

Lode H, Hampel B, Wachter T, Borner K, Koeppe P (1987b) Single dose pharmacokinetics of fleroxacin in volunteers after fasting, a standard breakfast, and antacids (Abstr 14). 15th International Congress of Chemotherapy, July 19–24, Istanbul

Lode H, Höffken G, Borner K, Koeppe P (1989a) Unique aspects of quinolone pharmacokinetics. Clin Pharmacokinet 16(Suppl 1):1–4

Lode H, Stuhlert P, Deppermann KM, Mainz D, Borner K, Kotvas K, Koeppe P (1989b) Pharmacokinetic interactions between oral ciprofloxacin/ofloxacin and ferro-salts (Abstr 213). 29th Interscience Conference on Antimicrobial Agents and Chemotherapy, 17–20 Sept Houston. American Society for Microbiology, Washington

Logamann C, Ohnhaus EE (1986) Enoxacin, a differential inhibitor of the microsomal liver enzyme system in men (Abstr 481). 26th Interscience Conference on Antimicrobial Agents and Chemotherapy. American Society for Microbiology, Washington

Loi M, Vestal RE (1988) Drug metabolism in the elderly. Pharmacol Ther 36:131–149

Loi CM, Parker BM, Korrapati MR, Cusack BJ, Vestal RE (1993a) Effects of age and gender on inhibition of theophylline metabolism. Clin Pharmacol Ther 53:192

Loi CM, Parker BM, Cusack BJ, Vestal RE (1993b) Individual and combined effects of cimetidine and ciprofloxacin on theophylline metabolism in male nonsmokers. Br J Clin Pharmacol 36:195–200

Lomaestro BM, Bailie GR (1991) Effect of staggered dose of calcium on the bioavailability of ciprofloxacin. Antimicrob Agents Chemother 35:1004–1007

Lomaestro BM, Bailie GR (1993) Effect of multiple staggered doses of calcium on the bioavailability of ciprofloxacin. Ann Pharmacother 27:1325–1328

Lowe NJ, Fakouhi TD, Stern RS (1994) Photoreactions with a fluoroquinolone antimicrobial: evening versus morning dosing. Clin Pharmacol Ther 56:587–591

Lubowski TJ, Nightingale CH, Sweeney K, Quintiliani R (1992) Effect of sucralfate on pharmacokinetics of fleroxacin in healthy volunteers. Antimicrob Agents Chemother 36:2758–2760

Lucet JC, Tilly H, Lerebours G, Gres JJ, Piguet H (1988) Neurological toxicity related to pefloxacin. J Antimicrob Chemother 21:811–812

Ludwig E (1993) Controversies on pharmacokinetics of fluoroquinolones in elderly patients. Int J Antimicrob Agents 3:49–59

Ludwig E, Szekely E, Csiba A, Graber H (1988) The effect of ciprofloxacin on antipyrine metabolism. J Antimicrob Chemother 22:61–67

MacGowan AP, Greig MA, Clarke EA, White LO, Reeves DS (1988) The pharmacokinetics of norfloxacin in the aged. J Antimicrob Chemother 22:721–727

Maesen FPV, Teengs JP, Baur C, Davies BI (1984) Quinolones and raised plasma concentrations of theophylline. Lancet 2:530

Maggiolo F, Puricelli G, Dottorini M, Caprioli S, Bianchi W, Suter F (1991) The effect of ciprofloxacin on oral contraceptive steroid treatments. Drugs Exp Clin Res 17:451–454

Mahr G, Sörgel F, Naber KG, Muth P, Kinzig M, Weigel D (1989) Principles of gastrointestinal secretion of quinolones in man (Abstr 585). Interscience Conference on Antimicrobial Agents and Chemotherapy. American Society for Microbiology, Washington

Mahr G, Seelmann R, Dammeyer J, Seibel M, Muth P, Gottschalk B, Stephan U, Sörgel F (1990) The effect of pefloxacin and enoxacin on the elimination of caffeine in human volunteers (Abstr 425). 3rd International Symposium on New Quinolones. July 12–14, Vancouver

Marchbanks CR, Longest TK, Zinner SH, Dudley MN (1989) Effect of cancer chemotherapy-induced intestinal injury on the oral absorption of ciprofloxacin (Abstr PIII-114). Clin Pharmacol Ther 47:207

Matsumoto M, Kajima K, Nagano H, Matsubara S, Yokota T (1992) Photostability and biological activity of fluoroquinolones substituted at the 8-position after UV irradiation. Antimicrob Agents Chemother 36:1715–1719

Mattina R, Bonfiglio G, Cocuzza CE, Gulisano G, Cesana M, Imbimbo BP (1991) Pharmacokinetics of rufloxacin in healthy volunteers after repeated oral doses. Chemotherapy 37:389–397

Mayersohn MB (1992) Special pharmacokinetic considerations in the elderly. In: Evans WE, Schentag JJ, Jusko WJ (eds) Applied pharmacokinetics: principles of therapeutic drug monitoring. Applied Therapeutics, Vancouver, pp 9.1–9.43

McLellan RA, Drobityh RK, McLellan DH, Caott PD, Crocker JFS, Renton KW (1995) Norfloxacin interferes with cyclosporine disposition in pediatric patients undergoing renal transplantation. Clin Pharmacol Ther 58:322–327

Mignot A, Couet W, Lefebvre MA, Fourtillan JB, Fillastre JP, Humbert G (1988) Pharmacokinetics of ofloxacin in patients with renal failure. Rev Infect Dis 10(Suppl 1):109

Minami R, Inotsume N, Nakano M, Sudo Y, Higashi A, Matsuda J (1993a) Effect of milk on absorption of norfloxacin in healthy volunteers. J Clin Pharmacol 33:1238–1240

Minami R, Inotsume N, Nakamura C, Nakano M (1993b) Stereoselective analysis of the disposition of tosufloxacin enantiomers in man. Eur J Clin Pharmacol 45:489–491

Mitscher LA, Sharma PN, Zavod RM (1989) The influence of optical isomerism on the biological properties of quinolone antimicrobial agents. In: Fernandes PB (ed) Quinilones. Prous, Barcelona, pp 73–83

Modai J (1988) Safety profile of oral quinolones. 6th Congress of Chemotherapy, May 22–27, Taormina

Molinaro M, Villani P, Regazzi MB, Rondanelli R, Doveri G (1992) Pharmacokinetics of ofloxacin in elderly patients and in healthy young subjects. Eur J Clin Pharmacol 43:105–107

Montay G, Gaillot J (1990) Pharmacokinetics of fluoroquinolones in hepatic failure. J Antimicrob Chemother 26(Suppl B):61–67

Morrison PJ, Mant TGK, Norman GT, Robinson J, Kunka RL (1988) Pharmacokinetics and tolerance of lomefloxacin after sequential increasing oral doses. Antmicrob Agents Chemother 32:1503–1507

Morse IS (1990) Pharmacokinetics and safety of single oral doses of lomefloxacin. Biopharm Drug Dispos 11:543–551

Mott FE, Murphy S, Hunt V (1989) Ciprofloxacin and warfarin. Ann Intern Med 111:542–543

Mueller BA, Brierton DG, Abel SR, Bowman L (1993) Effect of enteral feeding with Ensure® on the oral bioavailability of ofloxacin and ciprofloxacin (Abstr 2). Pharmacotherapy 13:674

Muralidharan G, Bansal S, Brocklebank D, Greene D, Hibbord M (1992) Effects of tosufloxacin on the pharmacokinetics of theophylline. Clin Pharmacol Ther 51:155

Naber KG, Bartosik-Wich B (1984) Serum and urine concentrations of norfloxacin, ciprofloxacin, and ofloxacin in elderly urological patients. Fortschr Antimikrob Antineoplast Chemother 3:701–709

Naber KG, Sörgel F, Gutle F, Bartosik-Wich B (1986) In vitro activity, pharmacokinetics, clinical safety and therapeutic efficacy of enoxacin in the treatment of patients with complicated urinary tract infection. Infection 14(Suppl 3):203–208

Naber KG, Sörgel F, Kees F, Jaehde U, Schumacher H (1989) Brief report: pharmacokinetics of ciprofloxacin in young (healthy volunteers) and elderly patients. Am J Med 87(Suppl 5A):57–59

Nakashima M, Uematsu T, Kanamaru M, Okazaki O, Hakusui H (1992) Phase I study of levofloxacin, (S)-(−)-ofloxacin. Jpn J Clin Pharmacol Ther 23:515–520

Nasir M, Rotellar C, Hand M, Kulczcki, Alijani MR, Winchester JF (1991) Interaction between cyclosporin and ciprofloxacin. Nephron 57:245–246

Navarro AS, Lanao JM, Recio MMS, Hurlé ADG, Romo JMT, Sanchez JCG, Delgado MMT (1990) Effect of renal impairment on distribution of ofloxacin. Antimicrob Agents Chemother 34:455–459

Neringer R, Forsgren A, Hansson C, Ode B, South Swedish Lolex Study Group (1992) Lomefloxacin versus norfloxacin in the treatment of uncomplicated urinary tract infections: three-day versus seven-day treatment. Scand J Infect Dis 24:773–780

Neuvonen PJ, Kivistoe KT (1992) Milk and yoghurt do not impair the absorption of ofloxacin. Br J Clin Pharmacol 33:346–348

Neuvonen PJ, Kivistoe KT, Lehto P (1991) Interference of diary products with the absorption of ciprofloxacin. Clin Pharmacol Ther 50:498–502

Nightingale CH (1993) Overview of the pharmacokinetics of fleroxacin. Am J Med 94(Suppl 3A):38–43

Niki Y, Soejima R, Kawane H, Sumi M, Umeki S (1987) New synthetic quinolone antibacterial agents and serum concentration of theophylline. Chest 92:663–669

Niki Y, Sumi M, Hino J, Kawane H, Soejima R (1988) Effects of new quinolones on serum concentrations of theophylline in healthy volunteers. Rev Infect Dis 10(Suppl 1):S140

Nilsen OG, Saltvedt E, Walstad RA, Marstein S (1992) Single-dose pharmacokinetics of lomefloxacin in patients with normal and impaired renal function. Am J Med 92(Suppl 4A): 38–40

Nix DE (1993) Drug–drug interactions with fluoroquinolone antimicrobial agents. In: Hooper DC, Wolfson JS (eds) Quinolone antimicrobial agents, 2nd edn. American Society for Microbiology, Washington
Nix DE, Schentag JJ (1989) Lomefloxacin absorption kinetics when administered with ranitidine and sucralfate (Abstr 1276). 29th Interscience Conference on Antimicrobial Agents and Chemotherapy, September 1989, Houston
Nix DE, DeVito JM, Whitbread MA, Schentag JJ (1987) Effect of multiple dose oral ciprofloxacin on the pharmacokinetics of theophylline and indocyanine green. J Antimicrob Chemother 19:263–269
Nix DE, Schultz RW, Frost RW, Sedman AJ, Thomas DJ, Kinkel AW, Schentag JJ (1988) The effect of renal impairment and haemodialysis on single dose pharmacokinetics of oral enoxacin. J Antimicrob Chemother 21(Suppl B):87–95
Nix DE, Wilton JH, Schentag JJ, Parpia SH, Norman A, Goldstein HR (1989a) Inhibition of norfloxacin absorption by antacids and sucralfate. Rev Infect Dis 2(Suppl 5):S1096
Nix DE, Watson WA, Lener ME, Frost RW, Krol G, Goldstein H, Lettieri J, Schentag JJ (1989b) Interaction of ciprofloxacin with aluminium and magnesium antacids, ranitidine, and sucralfate. In: Symposium papers: ciprofloxacin, major advances in intravenous and oral quinolone therapy. Am J Med 87:5A
Nix DE, Watson WA, Lener ME, Frost RW, Krol G, Goldstein H, Lettieri J, Schentag JJ (1989c) Effects of aluminium and magnesium antacids and ranitidine on the absorption of ciprofloxacin. Clin Pharmacol Ther 46:700–705
Nix DE, Watson WW, Handy L, Frost RW, Rescott DL, Goldstein HR (1989d) The effect of sucralfate pretreatment on the pharmacokinetics of ciprofloxacin. Pharmacotherapy 9:377–380
Nix DE, Norman A, Schentag JJ (1989e) Effect of lomefloxacin on theophylline pharmacokinetics. Antimicrob Agents Chemother 33:1006–1008
Nix DE, Wilton JH, Ronald B, Distlerath L, Williams VC, Norman A (1990) Inhibition of norfloxacin absorption by antacids. Antimicrob Agents Chemother 34:432–435
Norrby SR (1991) Side effects of quinolones: comparisons between quinolones and other antibiotics. Eur J Microbiol Infect Dis 10:378–383
O'Connell K, Sekora D, Kane P, Harning R (1994) A crossover study of the effect of ciprofloxacin HCl, phenytoin sodium and rifampin on doxofylline pharmacokinetics. Can J Physiol Pharmacol 72(Suppl 1):496
Okzaki O, Kurata T, Tachizawa H (1988) Studies on the mechanism of PK interaction of aluminium hydroxide, an antacid, with new quinolones in rats. Xenobiol Metab Dispos 3:387–394
Onishi H, Tanimura H, Ichimiya G, Aoki Y, Ishimoto K, Oka S, Kobayashi Y, Yamaue H, Uchiyama K, Ochiai M, Uenishi M (1993) Excretion of levofloxacin into bile and gallbladder tissue. Drugs 45 (Suppl 3):260–261
Paintaud G, Alvan G, Hellgren U, Nilsson-Ehle I (1993) Lack of effect of amoxycillin on the absorption of ofloxacin. Eur J Clin Pharmacol 44:207–209
Parent M, St-Laurent M, LeBel M (1990) Safety of fleroxacin coadministered with theophylline to young and elderly volunteers. Antimicrob Agents Chemother 34:1249–1253
Parker AC, Preston T, Heaf D, Kitteringham NR, Choonara I (1994) Inhibition of caffeine metabolism by ciprofloxacin in children with cystic fibrosis as measured by caffeine breath test. Br J Clin Pharmacol 38:573–576
Paton JH, Reeves DS (1991) Clinical features and management of adverse effects on quinolone antibacterials. Drug Saf 6:8–27
Paton JH, Reeves DS (1992) Adverse reactions to the fluoroquinolones. Adverse Drug React Bull 153:575–578
Peltola H, Väärälä M, Renkonen O-V, Neuvonen PJ (1992) Pharmacokinetics of single-dose oral ciprofloxacin in infants and small children. Antimicrob Agents Chemother 36:1086–1090
Perrot S, Kaplan G, Ziza JM (1992) 3 cas de tendinite achilléenne sous péfloxacine dont deux avec rupture (Lettre). Rev Rhum Mal Osteoartic 59:162

Perry G, Mant TGK, Morrison PJ, Sacks S, Woodcock J, Wise R, Imbimbo BP (1993) Pharmacokinetics of rufloxacin in patients with impaired renal function. Antimicrob Agents Chemother 37:637–641
Plaisance K, Drusano G, Forrest A, Weir M, Standiford H (1987) The effect of renal function on the bioavailability of ciprofloxacin. Clin Pharmacol Ther 41:195
Polk R (1993) (Abstr 596). In: Program and abstracts of the 32nd Interscience Conference on Antimicrobial Agents and Chemotherapy, October 1992, Anaheim
Polk RE (1994) Drug interactions with fluoroquinolone antibiotics and patient education. Int Dis Clin Pract 3 Suppl 3:S185–S194
Polk RE, Healy DP, Sahai J, Drwal L, Racht E (1989) Effect of ferrous sulfate and multivitamins with zinc on absorption of ciprofloxacin in normal volunteers. Antimicrob Agents Chemother 33:1841–1844
Polk RE, Israel D, Sintek C, Klein C, Swaim WA, Pluhar R, Lettieri J, Heller A (1994) Effect of ciprofloxacin on response to warfarin in anticoagulated adults (Abstr A3). 34th Interscience Conference on Antimicrobial Agents and Chemotherapy, Orlando
Portmann R (1992) Influence of cimetidine on fleroxacin pharmacokinetics (Abstr 134). 4th International Symposium on New Quinolones, Aug 27–29, Munich
Preheim LC, Cuevas TA, Roccaforte JS, Mellencamp MA, Bittner MJ (1986) Ciprofloxacin and antacids. Lancet 8497:48
Prince RA, Casabar E, Adair C, Wexler DB, Lettieri J, Kasik JE (1989) Effects of quinolone antimicrobials on theophylline pharmacokinetics. J Clin Pharmacol 29:650–654
Prince RA, Lion W, Kasik JE (1990) Effect of cimetidine on ciprofloxacin pharmacokinetics. Pharmacotherapy 10:233
Privitera G, Nicastro G, Imbimbo BP, Cesana M, Visconti M, Lombardi F, Tagliabue G, Faleschini E, Colturani F, Franzini P, Nervetti G (1993) Biliary excretion of rufloxacin in humans. Antimicrob Agents Chemother 37:2545–2549
Rambout L, Sahai J, Gallicano K, Garber G (1993) Effect of bismuth subsalicylate on ciprofloxacin absorption in healthy volunteers. Pharmacotherapy 13:647
Ramsay CA, Obreshkova E (1974) Photosensitivity from nalidixic acid. Br J Dermatol 91:523–528
Raoff S, Wollschlager C, Khan FA (1987) Ciprofloxacin increases serum levels of theophylline. Am J Med 82(Suppl 4A):115–118
Reimers D, Jezek A (1971) The simultaneous use of rifampicin and other antitubercular agents with oral contraceptives. Prax Pneumol 25:225
Renzi R, Finkbeiner S (1991) Ciprofloxacin interaction with sodium warfarin: a potentially dangerous side effect. Am J Emerg Med 9:551–552
Ribard P, Kahn MF (1991) Rheumatological side-effects of quinolones. Baillieres Clin Rheumatol 5:175–191
Ribard P, Audisio F, Kahn M-F, De Bandt M, Jorgensen C, Hayem G, Meyer O, Palazzo E (1992) Seven achilles tendinitis including 3 complicated by rupture during fluoroquinolone therapy. J Rheumatol 19:1479–1481
Richer M, LeBel M (1993) Pharmacokinetics of fluoroquinolones in selected populations. In: Hooper DC, Wolfson JS (eds) Quinolone antimicrobial agents, 2nd edn. American Society for Microbiology, Washington, pp 225–244
Rietbrock N, Staib AH (1987) Gyrase-Hemmer: unerwünschte zentralnervöse Wirkungen. Dtsch Med Wochenschr 112:201
Rimoldi R, Fioretti M, Albrici A, Imbimbo BP (1992) Pharmacokinetics of rufloxacin once daily in patients with lower respiratory tract infections. Infection 20:89–93
Rindone JP, Keney CL, Jones WN, Garewal HS (1991) Hypoprothrombinemic effect of warfarin not influenced by ciprofloxacin. Clin Pharm 10:13–138
Ritz M, Lode H, Faßbender M, Borner K, Koeppe P, Nord CE (1994) Multiple-dose pharmacokinetics of sparfloxacin and its influence on fecal flora. Antimicrob Agents Chemother 38:455–459

Robertson DG, Epling GA, Kiely JS, Bailey DL, Song B (1991) Mechanistic studies on the phototoxic potential of PD 117596, a quinolone antibacterial compound. Toxicol Appl Pharmacol 111:221–232

Robinson JA, Venezio FR, Costanzo-Nordin MR, Pifarre R, O'Keefe PJ (1990) Patients receiving quinolones and cyclosporine after heart transplantation. J Heart Transplant 9:30–31

Robson RA (1992) Quinolone pharmacokinetics. Int J Antimicrob Agents 2:3–10

Robson RA, Begg EJ, Atkinson HC, Saunders DA, Frampton M (1990a) Comparative effects of ciprofloxacin and lomefloxacin on the oxidative metabolism of theophylline. Br J Clin Pharmacol 29:491–493

Robson RA, Begg EJ, Saunders DA (1990b) The pharmacokinetics and effects on renal function of lomefloxacin at steady-state in patients with varying degrees of renal insufficiency. Clin Exp Pharmacol Physiol (Suppl 17):66

Robson RA, Bailey RR, Lynn KL (1993) Lomefloxacin pharmacokinetics after intravenous and oral administration in patients with acute pyelonephritis. Drugs 45(Suppl 3):269

Rocci ML, Vlasses PH, Distlerath LM, Gregg MH, Wheeler SC, Zing W, Bjornsso TD (1990) Norfloxacin does not alter warfarin's disposition or anticoagulant effect. J Clin Pharmacol 30:728–732

Rohwedder RW, Bergan T, Thorsteinsson SB, Scholl H (1990) Transintestinal elimination of ciprofloxacin. Diagn Microbiol Infect Dis 13:127–133

Rollof J, Vinge E (1993) Neurologic adverse effects during concomitant treatment with ciprofloxacin, NSAIDs, and chloroquine: possible interaction. Ann Pharmacother 27:1058–1059

Rosen C (1989) Photo-induced drug eruptions. Semin Dermatol 8:149–157

Rubinstein E, Carbon C (1994) Sparfloxacin monograph. ADIS, Chester, pp 35–43

Rubio TT, Miles MV, Church DA, Echols RM, Pickering LK (1994) Pharmacokinetic studies of ciprofloxacin in children with cystic fibrosis. Pediatr Res 35:A195

Sahai J, Healy DP, Stotka J, Polk RE (1993a) The influence of chronic administration of calcium carbonate on the bioavailability of oral ciprofloxacin. Br J Clin Pharmacol 35:302–304

Sahai J, Gallicano K, Oliveras L, Khaliq S, Hawley-Foss N, Garber G (1993b) Cations in the didanosine tablet reduce ciprofloxacin bioavailability. Clin Pharmacol Ther 53:292–297

Sahai J (1995) Avoiding the ciprofloxacin-didanosine interaction. Ann Intern Med 123:394–395

Saito A, Harak K, Yamaguchi K, Kohno S, Furuhama K (1987) On side effects of antimicrobials – adverse drug reaction of ofloxacin and other new quinolones. 15th International Congress of Chemotherapy, July 19–24, Istanbul, pp 1723–1735

Saito A, Oguchi K, Harada Y, Shinoda I, Komeda H, Okano M (1992) Pharmacokinetics of levofloxacin in patients with impaired renal function (in Japanese). Chemotherapy (Tokyo) 40(Suppl 3):188–195

Sano M, Kawakatsu K, Ohkita C, Yamamoto I, Takeyama M, Yamashina H, Goto H (1988) Effects of enoxacin, ofloxacin and norfloxacin on theophylline disposition in humans. Eur J Clin Pharmacol 35:161–165

Sarkar M, Polk RE, Guzelian PS, Hunt C, Karnes HT (1990) In vitro effect of fluoroquinolones on theophylline metabolism in human liver microsomes. Antimicrob Agents Chemother 34:594–599

Sawada M, Nakamura S, Yamada A, Kobayashi T, Okada S (1991) Phase IV study and post-marketing surveillance of ofloxacin in Japan. Chemotherapy 37:134–142

Schaad UB (1992) The role of the new quinolones in pediatric practice. Pediatr Infect Dis J 11:1043–1046

Schaad UB, Wedgwood J (1992) Lack of quinolone-induced arthropathy in children. J Antimicrob Chemother 30:414–416

Schaad UB, Stoupis C, Wedgwood J, Tschaeppeler H, Vock P (1991) Clinical, radiologic and magnetic resonance monitoring for skeletal toxicity in pediatric patients with cystic fibrosis receiving a three-month course of ciprofloxacin. Pediatr Infect Dis J 10:723–729

Schaad UB, Sander E, Wedgwood J, Schaffner T (1992) Morphological studies for skeletal toxicity after prolonged ciprofloxacin therapy in two juvenile cystic fibrosis patients. Pediatr Infect Dis J 11:1047–1049

Schaad UB, Salam MA, Aujard Y, Dagan R, Green SDR, Peltola H, Rubio T, Smith AL, Adam D (1995) Use of fluoroquinolones in pediatrics: consensus report of an International Society of Chemotherapy Commission. Pediatr Infect Dis J 14:1–9

Schaad UB, Wedgwood J, Ruedeberg A, Kraemer R, Hampel B (1996) Ciprofloxacin as anti-pseudomonal treatment in patients with cystic fibrosis. (submitted)

Schacht P, Arcieri G, Branolte J, Bruck H, Chysky V, Griffith E, Gruenwaldt G, Hullmann R, Kanopka CA, O'Brien B, Rham V, Ryoki T, Westwood A, Weuta H (1988) World-wide clinical data on efficacy and safety of ciprofloxacin. Infection 16 (Suppl 1):29–43

Schacht P, Arcieri G, Hullmann R (1989) Safety of oral ciprofloxacin. An update based on clinical trial results. Am J Med 87(Suppl 5A):98–102

Schaefer HG, Ahr G, Kuhlmann J (1995) Overview: pharmacokinetic development of quinolone antibiotics. Int J Clin Pharmacol Ther 33:266–276

Schaefer HG, Stass H, Wedgwood J, Hampel B, Fischer C, Kuhlmann J, Schaad UB (1996) Pharmacokinetics of ciprofloxacin in pediatric cystic fibrosis patients. Antimicrob Agents Chemother 40:29–34

Schentag JJ, Goss F (1992) Quinolone pharmacokinetics in the elderly. Am J Med 92(Suppl 4A):33–37

Schentag JJ, Nix DE, Forrest A (1993) Pharmacodynamics of the fluoroquinolones. In: Hooper DC, Wolfson JS (eds) Quinolone antimicrobial agents, 2nd edn. American Society for Microbiology, Washington, pp 259–271

Schrenzel J, Cerruti F, Herrmann M, Leemann T, Weidekamm E, Portmann R, Hirschel B, Lew DP (1994) Single-dose pharmacokinetics of oral fleroxacin in bacteremic patients. Antimicrob Agents Chemother 38:1219–1224

Schroeder D, Frye J, Alldredge B, Messing R, Flaherty J (1991) Effect of ciprofloxacin on serum phenytoin concentrations in epileptic patients (Abstr 75). Pharmacotherapy 11:275

Schwartz J, Jauregui L, Lettieri J, Bachmann K (1988) Impact of ciprofloxacin on theophylline clearance and steady-state concentrations in serum. Antimicrob Agents Chemother 32:75–77

Segre G, Cerretani D, Moltoni L, Urso R (1992) Pharmacokinetics of rufloxacin in healthy volunteers. Eur J Clin Pharmacol 42:101–105

Semel JD, Allen N (1989) Combination effects of ciprofloxacin, clindamycin, and metronidazole intravenously in volunteers. South Med J 84:465–468

Sesnie JC, Heifetz CL, Joannides ET, Malta TE, Shapiro MA (1990) Comparative phototoxicity of quinolones in a mouse phototolerance model (Abstr 399). 30th Interscience Conference on Antimicrobial Agents and Chemotherapy, Atlanta. American Society for Microbiology, Washington

Shah A, Lettieri J, Heller AH (1995) Pharmacokinetics of ciprofloxacin and metronidazole when given concurrently. In: Selected papers of the 19th International Congress of Chemotherapy, July 1995, Montreal, Canada. Reed Elsevier, Amsterdam

Shalit I, Greenwood RB, Marks MI, Pederson JA, Frederick DL (1986) Pharmacokinetics of single dose oral ciprofloxacin in patients undergoing chronic ambulatory peritoneal dialysis. Antimicrob Agents Chemother 30:152–156

Shiba K, Saito A, Shimada J (1989) Interactions of fleroxacin with dried aluminium hydroxide gel and probenecid. Rev Infect Dis 10(Suppl 1):S137–S138

Shiba K, Sakai O, Shimada J, Okazaki O, Aoki H, Hakusui H (1992) Effects of antacids, ferrous sulfate, and ranitidine on absorption of DR-3355 in humans. Antimicrob Agents Chemother 36:2270–2274

Shiba K, Okazaki O, Aoki H, Shimada J, Sakai O (1993) Influence of antacids and ranitidine on the absorption of levofloxacin in men. Drugs 45(Suppl 3):299–300

Shimada J, Hori S (1992) Adverse effects of quinolones. Prog Drug Res 38:133–144
Shimada J, Yamagii T, Ueda Y, Uchida H, Kusajima H, Irikura T (1983) Mechanism of renal excretion of AM-715, a new quinolonecarboxylic acid derivative, in rabbits, dogs and humans. Antimicrob Agents Chemother 23:1–7
Shimada J, Shiba K, Oguma T, Miwa H, Yoshimura Y, Nishikawa T, Okabayashi Y, Kitagawa T, Yamamoto S (1992) Effect of antacid on absorption of the quinolone lomefloxacin. Antimicrob Agents Chemother 36:1219–1224
Shimada J, Nogita T, Ishibashi Y (1993) Clinical pharmacokinetics of sparfloxacin. Clin Pharmacokinet 25:358–369
Silvain C, Bouquet S, Breux JP, Becq-Giraudon B, Beauchant M (1989) Oral pharmacokinetics and ascitic fluid penetration of ofloxacin. Eur J Clin Pharmacol 37:261–265
Simon C, Stille W, Wilkinson (1993) Antibiotic therapy. Schattauer Stuttgart, pp 231–252
Singlas E, Taburet AM, Landru I, Albin H, Ryckelinck JP (1987) Pharmacokinetics of ciprofloxacin tablets in renal failure: influence of haemodialysis. Eur J Clin Pharmacol 31:589–593
Singlas E, Leroy A, Sultan E et al (1990) Disposition of fleroxacin, a new trifluoroquinolone, and its metabolites. Pharmacokinetics in renal failure and influence of haemodialysis. Clin Pharmacokinet 19:67–79
Soejima R, Niki Y, Sumi M (1989) Effect of fleroxacin on serum concentrations of theophylline. Rev Infect Dis 11(Suppl 5):P1099
Sörgel F, Kinzig M (1993a) Pharmacokinetics of gyrase inhibitors. I. Basic chemistry and gastrointestinal disposition. Am J Med 94(Suppl 3A):44–55
Sörgel F, Kinzig M (1993b) Pharmacokinetics of gyrase inhibitors, II. Renal and hepatic elimination pathways and drug interactions. Am J Med 94(Suppl 3A):56–69
Sörgel F, Muth P, Mahr G, Manoharan M (1987) Pharmacokinetics and analytics of gyrase inhibitors (in German). Fortschr Antimikrob Antineoplast Chemother 6–10:1963–1986
Sörgel F, Mahr G, Koch HV, Stephan U, Wiesemann HG, Malter U (1988a) Effects of cimetidine on the pharmacokinetics of pefloxacin in healthy volunteers. Rev Infect Dis 10(Suppl 1):S137
Sörgel F, Mahr G, Stephan U, Koch HU, Wiesemann HG (1988b) Absolute bioavailability and pharmacokinetics of pefloxacin in healthy volunteers. Rev Infect Dis 10(Suppl 1):93
Sörgel F, Koch HU, Malter U, Metz R, Mahr G, Stephan U (1988c) Metabolism of pefloxacin in humans. Rev Infect Dis 10 (Suppl 1):95–96
Sörgel F, Hentschel D, Weikl A, Seelmann R, Ladebeck R, Muth P (1988d) Nuclear magnetic resonance spectroscopy (NMR), a new non-invasive technique in the assessment of pharmacokinetic processes in the human body. Book of Abstracts, 2nd International Symposium on New Quinolones, Geneva, 25–27 August 1988, p 177
Sörgel F, Naber KG, Jaehde U, Reiter A, Seelmann R, Sigl G (1989a) Brief report: gastrointestinal secretion of ciprofloxacin. Evaluation of the charcoal model for investigations in healthy volunteers. Am J Med 87(Suppl 5A):62–65
Sörgel F, Naber KG, Metz R, Morgenroth (1989b) Enoxacin – pharmacokinetics and tissue penetration in comparison (in German). Infection 17(Suppl 1):14–18
Sörgel F, Naber KG, Mahr G, Gottschalk B, Stephan U, Seelmann R, Metz R, Muth P, Jaehde U, Zürcher J (1991a) Gender related distribution of quinolones. Eur J Clin Microbiol Infect Dis (Special Issue):189–190
Sörgel F, Naber KG, Kinzig M, Mahr G, Muth P (1991b) Comparative pharmacokinetics of ciprofloxacin and temafloxacin in humans: a review. Am J Med 91(Suppl 6A):51–66
Stahlmann R, Lode H (1988) Safety overview: toxicity, adverse effects and drug interactions. In: Andriole VT (ed) The quinolones. Academic, London, pp 201–233

Staib AH, Stille W, Dietlein G, Shah PM, Harder S, Mieke S, Beer C (1987) Interaction between quinolones and caffeine. Drugs 34(Suppl 1):170–174

Staib AH, Harder S, Fuhr U, Wack C (1989a) Interaction of quinolones with the theophylline metabolism in man: investigations with lomefloxacin and pipemidic acid. Int J Clin Pharmacol Ther Toxicol 27:289–293

Staib AH, Beermann D, Harder S, Fuhr U, Liermann D (1989b) Absorption differences of ciprofloxacin along the human gastrointestinal tract determined using a remote-control drug delivery device (HF-capsule). Am J Med 87(Suppl 5A):66–69

Stass H, Kuhlmann J, Peltola H, Rahm V (1996) Single dose and steady state pharmacokinetics of ciprofloxacin in pediatric patients following administration of a new oral suspension. Arch Pharmacol 353(Suppl):R153

Stone JW, Andrews JM, Ashby JP, Griggs D, Wise R (1988) Pharmacokinetics and tissue penetration of orally administered lomefloxacin. Antimicrob Agents Chemother 32:1508–1510

Stroshane RM, Silverman MH, Sauerschell R, Brown RR, Boddy AW, Cook JA (1990) Preliminary study of the pharmacokinetics of oral amifloxacin in elderly subjects. Antimicrob Agents Chemother 34:751–754

Stuck AE, Frey F, Heizmann P, Brandt R, Weidekamm E (1989) Pharmacokinetics and metabolism of intravenous and oral fleroxacin in subjects with normal and impaired renal function and in patients on CAPD. Antimicrob Agents Chemother 33:373–381

Stuck AE, Kim DK, Frey FJ (1992) Fleroxacin clinical pharmacokinetics. Clin Pharmacokinet 22:116–131

Swanson BN, Boppana VK, Vlasses PH, Rotmensch HH, Ferguson RK (1983) Norfloxacin disposition after sequentially increasing oral doses. Antimicrob Agents Chemother 23:284–288

Taburet AM, Devillers A, Thomare P, Fillastre JP, Veyssier P, Singlas E (1990) Disposition of fleroxacin and its metabolites. Pharmacokinetics in elderly patients. Clin Pharmacokinet 19:80–88

Tai M, Sugimoto Y, Katayama Y, Maeda T (1988) Phase I study of T-3262 metabolites in serum, urine and feces. Chemotherapy 36:208–215

Takagi K, Yamaki K, Nadai M, Kuzuya T, Hasegawa T (1991) Effect of a new quinolone, sparfloxacin, on the pharmacokinetics of theophylline in asthmatic patients. Antimicrob Agents Chemother 35:1137–1141

Tan KKC, Trull AK, Shawket S (1989) Co-administration of ciprofloxacin and cyclosporine: lack of evidence for a pharmacokinetic interaction. Br J Clin Pharmacol 28:185–187

Tartaglione TA, Raffalovich AC, Poynor WJ, Espinel-Ingroff A, Kerkering TM (1986) Pharmacokinetics and tolerance of ciprofloxacin after sequential increasing oral doses. Antimicrob Agents Chemother 29:62–66

The Medical Letter (1995) Drugs that cause photosensitivity. Med Lett Drugs Ther 37:35–36

Thorsteinsson SB, Bergan T, Oddsdottier S, Rohwedder R, Holm R (1986) Crystalluria and ciprofloxacin, influence of urinary pH and hydration. Chemotherapy 32:408–417

Thorsteinsson SB, Bergan T, Johannesson G, Thorsteinsson HS, Rohwedder R (1987a) Tolerance of ciprofloxacin at injection site, systemic safety and effect on electroencephalogram. Chemotherapy 33:448–451

Thorsteinsson SB, Rohwedder R, Bergan T (1987b) Urinary crystal formation upon administration of ciprofloxacin, nalidixic acid, norfloxacin and ofloxacin (Apstr). 15th International Congress of Chemotherapy, July 19–24, Istanbul, pp 1748–1749

Thorsteinsson SB, Rahm V, Bergan T (1989) Tolerance of intravenous ciprofloxacin. Scand J Infect Dis 60:116–119

Toon S, Hopkins KJ, Garstang FM, Aarons L, Sedman A, Rowland M (1987) Enoxacin–warfarin interaction: pharmacokinetic and stereochemical aspects. Clin Pharmacol Ther 42:33–41

Toothaker RD (1989) Enoxacin absorption and elimination characteristics. Clin Pharmacokinet 16(Suppl 1):52–58

Torre D, Sampietro C, Rossi S, Bianchi W, Maggiolo F (1989) Ciprofloxacin and activated charcoal: pharmacokinetic data. Rev Infect Dis 11(Suppl 5):1015–1016

Triger DR, Granai F, Woodcock J, Wise R, Imbimbo BP (1993) Multiple-dose pharmacokinetics of rufloxacin in patients with cirrhosis. Hepatology 18:847–852

Tsuji A, Sato H, Kume Y, Tamai I, Okezaki E, Nagata O, Kato H (1988a) Inhibitory effects of quinolone antibacterial agents on gamma-aminobutyric acid binding to receptor sites in rat brain membranes. Antimicrob Agents Chemother 32:190–194

Tsuji A, Sato H, Okezaki E, Nagata O, Kato H (1988b) Effect of the anti-inflammatory agent febufen on the quinolone-induced inhibition of γ-aminobutyric acid receptor binding to rat brain membrane in vitro. Biochem Pharmacol 34:4408–4411

Tuncel T, Bergisadi N (1992) In vitro absorption of ciprofloxacin hydrochloride on various antacids. Pharmazie 47:304–305

Van Buren DH, Koestner J, Adedoyin A, McLune T, MacDonell R, Johnson HK, Carroll J, Nylander W, Richie RE (1990) Effect of ciprofloxacin on cyclosporine pharmacokinetics. Transplantation 50:888–889

Van der Auwera P, Stolear JC, Dudley MN (1990) Pharmacokinetics of enoxacin and its oxometabolites following intravenous administration to patients with different degrees of renal impairment. Antimicrob Agents Chemother 34:1491–1497

Van Slooten AD, Nix DE, Wilton JH, Love JH, Spivey JM, Goldstein HR (1991) Combined use of ciprofloxacin and sucralfate. Ann Pharmacother 25:578–582

Verho M, Malerczyk V, Dagrosa E, Korn A (1985) Dose linearity and other pharmacokinetics of ofloxacin: a new broad spectrum antimicrobial agent. Pharmacotherapeutica 4:376–382

Verho M, Dragosa EE, Malerczyk V (1986) The clinical pharmacology of ofloxacin: a new chemotherapeutic agent belonging to the group of gyrase inhibitors. Infection 14(Suppl 1):47–53

Veyssier P, Fourtillan JB, Modai J (1986) Pharmacokinetics of ofloxacin in elderly subjects (65–85 years) with normal renal function after 200 mg single oral dose. Pathol Biol 34:596–599

Wagai N, Tawara K (1992) Possible direct role of reactive oxygens in the cause of cutaneous photoxicity induced by five quinolones in mice. Arch Toxicol 66:392–397

Waite NM, Rybak MJ, Krakovsky DJ, Steinberg JD, Warbasse LH, Edwards DJ (1991) Influence of subject age on the inhibition of oxidative metabolism by ciprofloxacin. Antimicrob Agents Chemother 35:130–134

Webb DB, Roberts DE, Williams JD, Asscher AW (1986) Pharmacokinetics of ciprofloxacin in healthy volunteers and patients with impaired kidney function. J Antimicrob Chemother 18(Suppl D):83–87

Weber B, Wünsche C, Menold R, Luttge B (1985) Analysis of urinary sediments in rats and humans after application of BAY o 9867 (ciprofloxacin). Bayer, Leverkusen (Internal Bayer PB-report no 13450 (P))

Weidekamm E (1993) Pharmacokinetics of fleroxacin in renal impairment. Am J Med 94(Suppl 3A):70–74

Weidekamm E, Portmenn R (1989) Variation of pharmacokinetic parameters after intravenous and oral administration of fleroxacin. Rev Infect Dis 11(Suppl 5):1023

Weidekamm E, Portmann R, Suter K, Partos C, Dell D, Lücker PW (1987) Single- and multiple-dose pharmacokinetics of fleroxacin, a trifluorinated quinolone, in humans. Antimicrob Agents Chemother 31:1909–1914

Weidekamm E, Stöckel K, Dell D (1988) Single-dose pharmacokinetics of the new fluoroquinolone Ro 23-6240 (AM 833) in humans. Rev Infect Dis 10(Suppl 1):94–95

White LO, MacGowan AP, Lovering AM, Reeves DS, Mackay IG (1987) A preliminary report on the pharmacokinetics of ofloxacin, desmethyl ofloxacin, and ofloxacin N-oxide in patients with chronic renal failure. Drugs 34(Suppl 1):56–61

White LO, MacGowan AP, Mackay IG, Reeves DS (1988) The pharmacokinetics of ofloxacin, desmethyl ofloxacin, and ofloxacin N-oxide in hemodialysis patients with end-stage renal failure. J Antimicrob Chemother 22(Suppl C):65–72

Wijnands WJA, van Herwaarden CLA, Vree TB (1984) Enoxacin raises plasma theophylline concentrations (Letter). Lancet 2:108–109

Wijnands WJ, Vree TB, van Herwaarden CLA (1985) Enoxacin decreases the clearance of theophylline in man. Br J Clin Pharmacol 20:583–588

Wijnands WJA, Vree TB, van Herwaarden CLA (1986) The influence of quinolone derivatives on theophylline clearance. Br J Clin Pharmacol 22:677–683

Wijnands WJA, Vree TB, Baars AM, van Herwaarden CLA (1987) The influence of the 4-quinolones ciprofloxacin, pefloxacin and ofloxacin on the elimination of theophylline. Pharm Weekbl [Sci] 9(Suppl):S72–S75

Wijnands WJA, Janssen TJ, Guelen PJM, Vree TB, DeWitte TMC (1988a) The influence of ofloxacin and enoxacin on the metabolic pathways of theophylline in healthy volunteers. Pharm Weekbl [Sci] 10:272–276

Wijnands WJA, Vree TB, Baars AM, van Herwaarden CLA (1988b) Pharmacokinetics of enoxacin and its penetration into bronchial secretions and lung tissue. J Antimicrob Chemother 21(Suppl B):67–77

Wijnands WJA, Cornel JH, Martea M, Vree TB (1989) Lack of multiple-dose oral lomefloxacin on theophylline metabolism. In: Rubinstein E, Adam D (eds) Recent Advances in chemotherapy. Proceedings of the 16th ICC, Jerusalem 1989. Lewin-Epstein, Jernsalem

Wijnands WJA, Trooster JFG, Teunissen PC, Cats HA, Vree TB (1990) Ciprofloxacin does not impair the elimination of diazepam in humans. Drug Metab Dispos 18:954–957

Wilton LV, Pearce GL, Mann RD (1996) A comparison of ciprofloxacin, norfloxacin, ofloxacin, azithromycin and cefixime examine by observational cohort studies. Br J Clin Pharmacol 41:277–284

Wingender W, Graefe KH, Gau W, Förster D, Beermann D, Schacht P (1984) Pharmacokinetics of ciprofloxacin after oral and intravenous administration in healthy volunteers. Eur J Clin Microbiol 3:355–359

Wingender W, Beermann D, Förster D, Graefe KH, Schacht P, Scharbrodt V (1985) Mechanism of renal excretion of ciprofloxacin (BAY o 9867), a new quinolone carboxylic acid derivative, in humans. Chemioterapie 4-1(Suppl):403

Wingender W, Beermann D, Foerster D, Graefe KH, Kuhlmann J (1986) Interactions of ciprofloxacin with food intake and drugs. Neu HC et al (eds) Proceedings of the First International Ciprofloxacin Workshop, Leverkusen 1985. Excerpta Medica, Amsterdam, pp 136–140

Wingender W, Beermann D, Förster D, Horstmann R (1988) Steady-state pharmacokinetics of ciprofloxacin in healthy volunteers after intravenous infusion. Rev Infect Dis 10(Suppl 1):93–94

Winn MJ, Kitteringham NR, Coleman MD, Newby S, Park BK (1989) The effects of lomefloxacin on cytochrome P-450 mediated drug metabolism in human liver microsomes. Br J Clin Pharmacol 28(2):2389

Wise R, Lister D, McNultry CAM, Griggs D, Andrews JM (1986) The comparative pharmacokinetics of five quinolones. J Antimicrob Chemother 18(Suppl D):71–81

Wise R, Kirkpatrick B, Ashby J, Griggs D (1987) The pharmacokinetics and tissue penetration of RO 23-6240, a new trifluoroquinolone. Antimicrob Agents Chemother 31:161–163

Wise R, Griggs D, Andrews JM (1988) Pharmacokinetics of the quinolones in volunteers: a proposed dosing schedule. Rev Infect Dis 10(Suppl 1):83–89

Wise R, Johnson J, O'Sullivan N, Andrews JM, Imbimbo BP (1991) Pharmacokinetics and tissue penetration of rufloxacin, a long acting quinolone antimicrobial agent. J Antimicrob Chemother 28:905–909

Wolf R, Eberl R, Dunky A, Martz N, Chang T, Goulet JR, Latts J (1984) The clinical pharmacokinetics and tolerance of enoxacin in healthy volunteers. J Antimicrob Chemother 14(Suppl C):63–69

Wolfson JS (1989) Quinolone antimicrobial agents: adverse effects and bacterial resistance. Eur J Clin Microbiol Infect Dis 8:1080–1092

Wolfson JS, Hooper DC (1989) Fluoroquinolone antimicrobial agents. Clin Microbiol Rev 2:378–424

Wolfson JS, Hooper DC (1991) Pharmacokinetics of quinolones: newer aspects. Eur J Clin Microbiol Infect Dis 10:267–174

Wynckel A, Toupance O, Melin JP, David C, Lavaud S, Wong T, Lamiable D, Chanard J (1991) Traitement des légionelloses par ofloxacine chez transplanté rénal–absence d'interférence avec la ciclosporine A. Press Med 20:291–293

Yakushiji T, Shirasaki T, Akaike N (1992) Non-competitive inhibition of $GABA_A$-responses by a new class of quinolones and non-steroidal antiinflammatories in dissociated frog sensory neurons. Br J Pharmacol 105:13–18

Yernault JC, Russel D (1992) Lomefloxacin versus amoxicillin in the treatment of acute exacerbation of chronic bronchitis: results of multinational studies. Int J Antimicrob Agents 2:39–48

Yuk JH, Nightingale CH, Sweeney KR, Quintiliani R, Lettieri JT, Frost RW (1989) Relative bioavailability in healthy volunteers of ciprofloxacin administered through a nasogastric tube with and without enteral feeding. Antimicrob Agents Chemother 33:1118–1120

Yuk JH, Nightingale CH, Quintilliani R, Yeston NS, Orlando R, Dobkin ED, Kambe JC, Sweeney KR, Buonpane EA (1990) Absorption of ciprofloxacin administered through a nasogastric or a nasoduodenal tube in patients receiving enteral nutrition. Diagn Microbiol Infect Dis 13:99–102

Zeiler HJ, Beermann D, Wingender W, Förster D, Schacht P (1988) Bactericidal activity of ciprofloxacin, norfloxacin and ofloxacin in serum and urine after oral administration to healthy volunteers. Infection 16(Suppl 1):19–23

Zinner SH (1989) Enoxacin – pharmacokinetics and clinical experience. In: Fernandes PB (ed) Quinolones. JR Prous Science, pp 383–407

Zupancic T, Marolt-Gomiscek M, Veber M, Komarek J, Durst R, Novak D, Gomiscek S (1991) Some aspects of the interaction of four-quinolones with copper (II) ions. Eur J Clin Microbiol Infect Dis (Special Issue):665–664

CHAPTER 12

Concentration–Effect Relationship of the Fluoroquinolones

R. Stahlmann and H. Lode

A. Introduction

Relevant pharmacodynamic data and the pharmacokinetics of a drug provide the theoretical foundation of any drug therapy. For anti-infective pharmacotherapy the minimum inhibitory concentrations of a compound for a specific pathogen (MIC) and the serum concentrations of the drug are traditionally used as a basis. Both variables are relatively easy to determine, but definitely do not reflect the complex interaction between an antimicrobial agent and a pathogen at the site of infection. Therefore, since the early times of antimicrobial chemotherapy researchers have tried to develop more sophisticated approaches to obtain data which could fill the gap between the very simple level of information (MIC and serum concentration) and the very complex and individual situation in an infected patient. Improved ways of microbiological testing, the use of animal models of infection, determination of tissue concentrations in animal models and man, investigation of the distribution of the drug in cellular compartments and inactivation of antibiotics by bacterial enzymes or human proteins are just a few examples of approaches taken by researchers in order to establish a better basis for antimicrobial chemotherapy with a newly developed drug or to optimize therapy with agents known for a longer time.

Several relationships between pharmacokinetic variables and antimicrobial activity have been evaluated in in vitro studies, in animal models and in clinical trials. Three pharmacokinetic variables are of major importance: (1) peak concentration, (2) the time period during which drug concentrations are above MIC and (3) AUC values ("area under the plasma–drug–concentrations–versus–time curve"). It can be difficult to resolve which of these parameters is the best predictor of clinical efficacy. They are closely related to each other, and increasing one of them usually increases the others.

In this chapter we first briefly describe the situation with β-lactams and aminoglycosides because the comparably solid data basis for these two groups of drugs allows a comparison with the quinolones. We then summarize the current knowledge on the relationships between concentrations of fluoroquinolones achieved in vivo and the effects on the pathogenic organism.

B. Pharmacodynamic Data of Antimicrobials as a Basis for Clinical Use

Data on the MIC or the minimal bactericidal concentration can be useful if a more or less superficial comparison of antibiotics is performed. For a more detailed evaluation, other pharmacodynamic data must be given, such as the kinetics of bacterial killing and the postantibiotic effect; these variables have gained much interest during the last years. The differences in killing curves of antimicrobial agents from different classes (e.g. β-lactams, aminoglycosides, fluoroquinolones) have been investigated in many studies. Principally, the killing rate of antimicrobial agents can be described as relatively concentration *dependent* or relatively concentration *independent.* β-Lactams and aminoglycosides are two examples of thoroughly studied drugs which belong to one of these groups.

I. β-Lactams: Concentration-Independent Killing Rate

β-Lactams are characterized by a steep concentration–effect curve in vitro: a small increase in concentration results in a large increase in killing rate. The range of concentrations between minimal and maximal effective concentrations is rather small and a further increase beyond the maximal effective concentration does not improve the antibacterial action: very high concentrations of β-lactam antibiotics do not result in a more rapid killing of bacteria. In fact, a paradoxical effect ("Eagle effect") was described at the beginning of the antibiotic era: the bactericidal activity at very high concentrations is characterized by a decreasing rate of killing (Eagle and Musselman 1948). However, the clinical significance of this observation is very questionable.

Data from experimental infections indicated that the antibacterial effect of β-lactams in vivo is highly dependent on the time duration that the drug level exceeds the minimal inhibitory concentration. For example, Craig and Ebert (1992) and other researchers found that a continuous infusion of β-lactams was consistently more potent than intermittent administration for Enterobacteriaceae and *Pseudomonas aeruginosa* in various experimental infection models. Continuous infusion or frequent dosing of the various β-lactams was usually at least eightfold more active than intermittent injections. With continuous infusion or hourly dosing, 90%–100% of the maximal efficacy was obtained with antibiotic levels that were only one to four times the MIC. Treatment with β-lactams by continuous infusion has resulted in good clinical efficacy in a variety of patients, including those with neutropenia. However, the number of randomized human trials comparing the two dosage regimens are scarce (for a review see Craig and Ebert 1992).

Because there is a slight concentration dependency of the killing rate for β-lactams up to two to four times the MIC, and because penetration to some infection sites results in lower concentrations than those observed in plasma, it may be wise to have two to four times the MIC of free drug as trough levels for

β-lactams as a clinical goal (DRUSANO 1995). The data published on β-lactams suggest that administration of these drugs by continuous infusion has potential advantages and might be the optimal way to treat patients, especially for gram-negative bacillary infections. The administration of a loading dose prior to continuous infusion would eliminate the only potential pharmacokinetic disadvantage of this concept and ensure the rapid onset of antibacterial activity. However, further clinical trials are needed to clarify the benefits and disadvantages of continuous infusion of β-lactams over currently used intermittent dosing regimens (CRAIG and EBERT 1992).

Studies on detailed aspects of pharmacodynamics are often performed with the goal of improving therapy. This could be by an increase in efficacy, decrease of toxicity or a reduction of costs. The continuous infusion of β-lactam antibiotics could result in a reduction of the total amount of drug used. Since β-lactams exhibit a low toxicity, the main reason to consider this an advantage might be the fact that the costs for therapy would be reduced. Because this approach is still connected with many open questions (and uncertainties), it is very questionable whether or not a reduction in drug costs (which will result only in a slight reduction of total hospital costs) should be considered as an advantage, considering the risks which a new dosing approach might have.

II. Aminoglycosides: Concentration-Dependent Killing Rate

In contrast to the findings with penicillins and cephalosporins, the bactericidal effect of aminoglycosides does increase with increasing concentrations and high peak concentrations are important for the efficacy of aminoglycosides in vivo. Further theoretical reflections and results from toxicological experiments in several animal species led to the recommendation of modifying the established treatment regimens for this group of antibiotics. It is well known that nephro- and ototoxicity are important problems associated with the clinical use of aminoglycosides. Therefore, data from studies which showed that high peak concentrations of aminoglycosides do not increase the toxic effects of these drugs significantly were important for the development of recommendations for new dosing strategies (GILBERT 1991).

Indeed, it has been shown in many clinical trials that for the vast majority of patients the once-daily dosing regimen is at least equal if not superior to the conventional three-times-daily regimen with respect to efficacy and toxicity. The peak concentrations of the antibiotics closely correlated with the ultimate outcome. If all the evidence from in vitro studies, animal infectivity models and clinical experience is taken together, it becomes clear that for aminoglycosides, it is very probably the peak concentration to MIC ratio that is linked to outcome (TER BRAAK et al. 1990; ROZDZINSKI et al. 1993; NICOLAU et al. 1995).

Since there is at least a tendency towards better clinical efficacy and reduced toxicity, the cost reduction effect of the once-daily dosing approach for aminoglycosides is an acceptable further advantage. With this concept, costs of ancillary service time and drug monitoring are significantly reduced. Substantial savings can also be expected by reducing the incidence of nephrotoxicity.

III. Fluoroquinolones

The fluoroquinolones possess many microbiological properties in common with aminoglycosides, e.g. they exhibit a concentration-dependent killing and high potency. Exposure of susceptible bacteria to fluoroquinolones results in rapid cell death; the bactericidal effect of fluoroquinolones occurs more rapidly than with the β-lactams.

Figure 1 shows a comparison of the bactericidal activity of a β-lactam, an aminoglycoside and a fluoroquinolone. These in vitro data indicate the similarity of the in vitro effects of fluoroquinolones and aminoglycosides (CRAIG and EBERT 1991). Analogous data were generated by ROOSENDAAL et al. (1987).

Although the rate of bacterial killing generally increases with drug concentration, it characteristically approaches a maximum. At very high concentrations (e.g. 40 times the MIC which is unphysiologically high in most instances), a reduction in bacterial killing may result from inhibition of RNA synthesis, which appears to be necessary for the bactericidal effect.

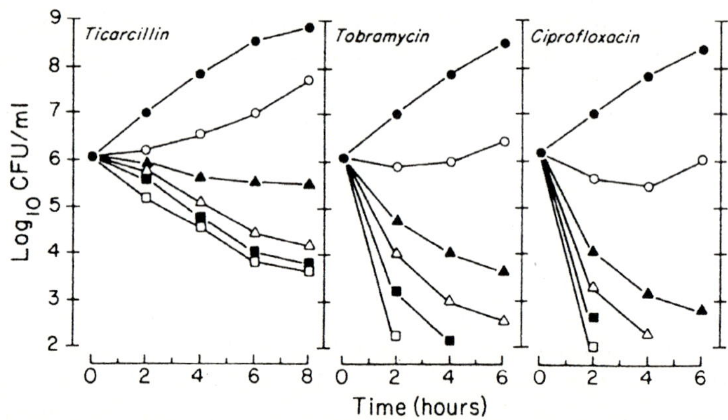

Fig. 1. Time kill curves of *P. aeruginosa* ATCC 27853 with exposure to tobramycin, ciprofloxacin and ticarcillin at concentrations from one fourth to 64 times the MIC. *Closed circle*, control; *open circle*, 1/4 MIC; *closed triangle*, MIC; *open triangle*, four times the MIC; *closed square*, 16 times the MIC; *open square*, 64 times the MIC. Tobramycin or other aminoglycosides are characterized by marked increases in bactericidal activity when concentrations are raised. With penicillins – such as ticarcillin – a rise in the killing rate is demonstrable up to a point of maximum effect (usually about four to five times the MIC). Beyond this point increasing concentrations do not enhance the rate or extent of killing. Increasing concentrations of ciprofloxacin are associated with a continual rise in the rate extent of bactericidal activity. (Modified from CRAIG and EBERT 1991)

1. In Vitro Models

Blaser et al. (1987) examined the killing of *Klebsiella pneumoniae* after exposure to two regimens of enoxacin, one administering the whole dose every 24h and the other administering half the dose every 12h. Administering the whole dose once daily yielded the best outcome. In agreement with the findings with aminoglycosides, the attainment of a peak concentration to MIC ratio of greater than 10:1 resulted in complete sterilization of the system in every instance. The effect was not as convincing at lower ratios which can be explained by resistent mutants in the bacterial population. Fluoroquinolone resistance in gram-negative bacteria is often mediated by a gyrA mutation or by a transport mutation which usually increases the MIC four to eight-fold. Therefore, it is understandable that peak concentration to MIC ratios above 10:1 can often suppress not only the parent strain, but the mutant organism as well.

These findings are in agreement with the data published by Dudley et al. (1987) who investigated the effects of ciprofloxacin on *P. aeruginosa*. Their results also demonstrate the superiority of a single, high-dose administration, and their data indicate that the regrowth phenomenon is linked to emergence of resistent mutants.

2. Animal Models

A comprehensive study on the impact of the dosage schedule of an aminoglycoside, a β-lactam and a quinolone on *K. pneumoniae* pneumonia in rats was performed by Roosendaal et al. (1989). The drugs were administered either intermittently in six hourly intervals or continuously to leukopenic rats. The β-lactam (ceftazidime) was far more effective when given continuously than when given intermittently. The daily doses of ceftazidime that protected 50% of the animals from death (PD50) were 1.5 and 24.4mg/kg given continuously and intermittently, respectively. Gentamicin tended to be more efficacious after intermittent than after continuous administration (PD50 values were 2.8 and 3.8mg/kg, respectively). Similarly, ciprofloxacin when given intermittently was more effective than when administered continuously (PD50 values were 3.3 and 6.5mg/kg, respectively). In the group which was given 7.5mg ciprofloxacin per kilogram body weight in an intermittent dosing schedule, the time to death was 17 days, with nine out of ten animals surviving. In the continuous dosing group, the time to death was 7 days, with four out of ten rats surviving. Evaluation of antibacterial activities in vivo by quantifying the numbers of *K. pneumoniae* in the lungs of treated rats revealed that the therapeutic efficacy of ceftazidime was independent of the doses administered whereas gentamicin and ciprofloxacin exhibited a significant dose-dependent antibacterial efficacy in vivo (Roosendaal et al. 1989).

Leggett et al. (1991) examined the effects of ciprofloxacin against *P. aeruginosa* and *K. pneumoniae* in mouse infection models (pneumonia, thigh infection). Increasing the dose interval (maximum of 12h for *Klebsiella*) had only a minimal effect on the dose which mediated a 50% reduction in bacterial population.

DRUSANO et al. (1993) examined the effects of lomefloxacin in a neutropenic rat model of *Pseudomonas* sepsis. A dose of 80 mg lomefloxacin per kilogram body weight was administerd as a single daily dose, as 40 mg/kg every 12 h, or as 20 mg/kg every 6 h. Concentrations in plasma were monitored. As shown in Fig. 2, the regimen with the highest peak concentration to MIC ratio was associated with a significantly better survival rate compared with the more fractionated regimens (74% vs 32% vs 36%).

At a first glance the results of these two studies seem to be contradictory. However, the inoculum used was different and might explain the different findings. In Leggett's study the inoculum was considerably smaller (10^6 organisms) than in Drusano's study (10^9 organisms). Since the frequency of mutations is usually cited as being between $1/10^7$ and $1/10^8$ it is much more likely that in the study using the higher challenge, the inoculum would contain resistant mutants. The large once-daily dose (80 mg/kg) resulted in peak concentration to MIC ratios of approximately 20:1. The ratios of the lower doses were less than 10:1 and would have a much greater chance of allowing breakthrough growth with a resistant mutant.

Another important outcome of the study with different doses of lomefloxacin in the rat model of *Pseudomonas* sepsis is shown in Fig. 3. Each panel of the figure describes the relationship between survival and different pharmacodynamic variables. Figure 3a shows the relationship between the peak concentration to MIC ratio obtained and the percent survival in an

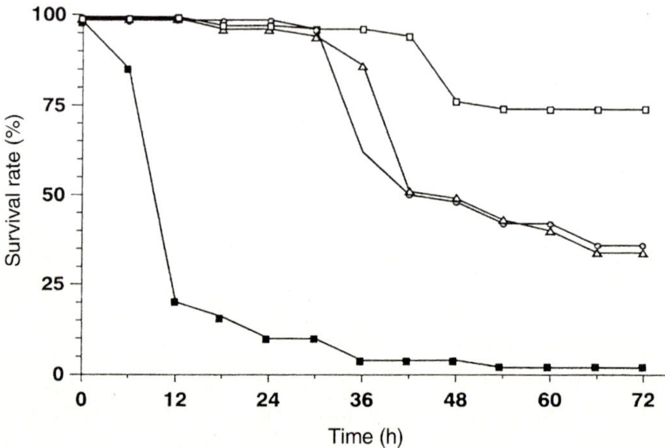

Fig. 2. Dose fractionation experiment with lomefloxacin in a neutropenic rat model of *Pseudomonas* sepsis. The MIC for the challenge organism was 1 mg/l. Regimens of 80 mg/kg every 24 h (*open squares*), 40 mg/kg every 12 h (*open triangles*) and 20 mg/kg every 6 h (*open circles*) were evaluated. Control animals received a saline injection (*closed squares*). There were 50 rats evaluated per group. (Modified from DRUSANO et al. 1993)

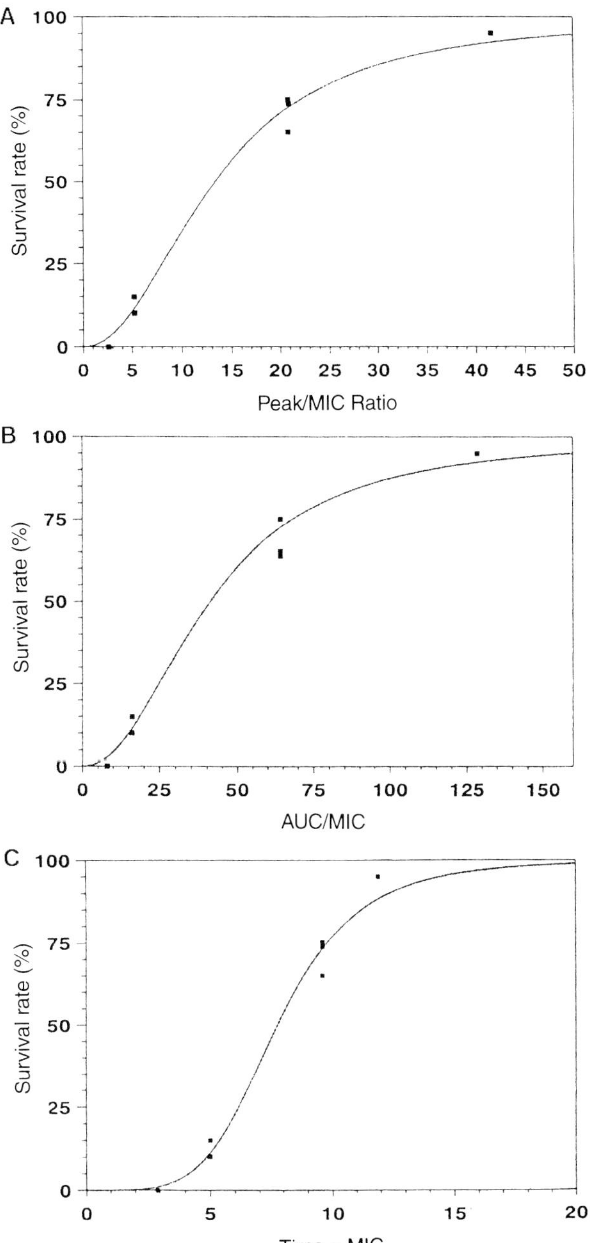

Fig. 3A–C. Modified Hill's (sigmoid Emax) model evaluating survival rate as a function of three different independent variables: the peak concentration to MIC ratio (**A**); the AUC to MIC ratio (**B**); and the time that the concentration in the plasma was above the MIC (**C**). An excellent fit of the model to the data was obtained with r^2-values >0.98 in all cases. This demonstrates that when administration schedule is a constant, there is maximal covariance between the pharmacodynamic variables and outcome survival rate. (Modified from DRUSANO et al. 1993)

animal model. The fit of the model to the data is excellent ($r^2 = 0.99$). Figure 3b shows the same data, but this time the AUC to MIC ratio is examined and obviously a very similar excellent fit is obtained. Finally, also for the time during which plasma concentration is above MIC levels, a good relationship is obtained which is only slightly inferior in comparison with the other two. Obviously, the model with each of the three variables displays an extremely good, statistically significant fit to the data because there is a strong covariance between these three variables (DRUSANO et al. 1993).

Challenges of animals with less susceptible strains and the use of lomefloxacin in doses which yield lower peak concentration to MIC ratios resulted in markedly diminished survival. Overall, when high peak concentration to MIC ratios (10:1 to 20:1) were obtained, this parameter was directly linked to survival. At lower doses producing ratios <10:1, the AUC to MIC ratio was most closely linked to outcome. The time duration that levels in serum exceeded the MIC did not influence the survival. Thus, high peak concentration to MIC ratios strongly reduce the viable counts of the parent strain and suppress the emergence of mutants for which the MIC is higher but limited, i.e. up to eight times that for the parent strain (DRUSANO et al. 1993).

3. Clinical Data

Only very limited clinical data are available which could answer the open questions of which pharmacodynamic variable is most important for the antibiotic effect of fluoroquinolones under clinical conditions and whether antimicrobial therapy with fluoroquinolones can be improved by altered dosage regimens.

In one of the first, larger clinical studies Schentag and colleagues tried to characterize the relationship between plasma ciprofloxacin concentrations, individual MIC values and bacterial eradication from respiratory secretions in critically ill adults in an intensive care unit (PELOQUIN et al. 1989; SCHENTAG 1991). A group of 50 patients with nosocomial lower respiratory tract infections caused by gram-negative bacteria was treated with ciprofloxacin intravenously every 12h and the correlations between clinical outcome and pharmacokinetics were evaluated. Most patients received single doses of 200mg, but ten patients whose pathogen had an initial MIC of 0.5 to 2.0mg/l were treated with 300mg every 12h.

High ratios of the peak ciprofloxacin concentration to the MIC were associated with successful eradication of bacteria from the respiratory tract. These ratios were 112 ± 35 for organisms eradicated, and 35 ± 59 for persisting pathogens. For patients infected with *P. aeruginosa*, the C_{max} to MIC ratio was 5. Significant differences also existed for the "% time > MIC", representing the percentage of the 12-h dosing interval during which ciprofloxacin concentrations exceeded the MIC of the pathogen. However, no significant difference was observed for AUC above MIC. For patients whose pathogens were era-

Table 1. AUIC vs percentage of clinical and microbiological cures (modified from FORREST et al. 1993a)

AUIC range	Total no. of patients	Patients cured (%)	
		Clinical	Microbiological
0–62.5	9	44	22
62.5–125	10	40	30
125–250	16	88	81
250–500	7	71	86
500–5541	22	77	82

AUIC, area under the concentration–versus–time curve to minimum inhibitory concentration ratio.

dicated, the AUC above MIC was $11.6 \pm 8.4\,\mathrm{mg\,h^{-1}\,l^{-1}}$, while this value was $10.8 \pm 9.5\,\mathrm{mg\,h^{-1}\,l^{-1}}$ for patients with persisting pathogens. *P. aeruginosa* proved to be the most difficult pathogen to eradicate, and 70% of the patients with pneumonia who had this pathogen showed bacteriological persistence (PELOQUIN et al. 1989).

Three years later these data were reanalyzed after the addition of 24 new patients. Eleven of these new patients had serious staphylococcal infections, with five receiving the combination of rifampin plus ciprofloxacin. In this reanalysis, the AUC to MIC ratio (AUIC) was the variable most closely linked to outcome in a multivariate analysis by logistic regression. The time required for the lower respiratory tract to be cleared of the pathogen was also significantly related to the AUC to MIC ratio achieved (Table 1). At a ratio of less than 125 (19 patients), the percent probabilities of clinical and microbiological cures were 42% and 26%, respectively. At an AUIC above 125 (45 patients), the probabilities were 80% and 82%, respectively. The authors concluded that an AUIC below 125 represented inadequate activity, one between 125 and 250 was acceptable, and one between 250 and 500 was optimal. They stated that "beyond 500 the patient is receiving more than twice the drug than is associated with optimal response, which adds to the cost of treatment and the risk of toxicity, for little added benefit" (FORREST et al. 1993b).

Pharmacokinetic data from this study indicated that this group of patients had smaller distribution volumes (by 30%) and total clearances (by 44%) than volunteers matched for weight and creatinine clearance. It was shown with the analysis of three or even two well-timed samples (e.g. at 15 to 30 min or 2.5 h postinfusion and at trough) that estimates of total clearance values were precise (FORREST et al. 1993a).

One important aspect in the interpretation of these studies is the fact that fixed dosing intervals were used. On a fixed schedule (twice daily) it is not possible to increase the AUC to MIC ratio without simultaneously increasing the peak concentration to MIC ratio and the time period during which the concentration is above the MIC, as discussed above in Sect. B.III.2.

4. Overall Evaluation of Pharmacodynamics

Overall, the pharmacodynamic pattern of quinolones in vitro resembles that of aminoglycosides more than that of β-lactams. The microbial killing is most likely linked to AUC to MIC ratios, but high peak concentrations are important to suppress resistant mutants. If the peak concentration to MIC ratio is not high enough a mutant subpopulation may emerge. The emergence of resistance in situations of low peak concentration to MIC ratios in vitro or in animal models as well as in a clinical setting has been described. For fluoroquinolones, attainment of high peak concentration to MIC ratios and high AUC to MIC ratios is important. Both of these aims can be reached by administration of single daily doses, but further work is necessary to decide if the currently used dosing strategies for fluoroquinolones could be optimized.

So far, the old therapeutic rule that patients who suffer from infections by less susceptible pathogens (e.g. *P. aeruginosa*) require higher doses than others with less severe infections is still applicable. It is doubtful whether new concepts will be developed in the near future which will abrogate this rule based on clinical experience.

In contrast to the situation with aminoglycosides, fluoroquinolones do not show specific toxicities which would limit their use in critically ill patients. Gastrointestinal reactions (which are usually mild) are the most often observed adverse reactions with fluoroquinolones and some data indicate that they are not significantly reduced by dose reduction. Phototoxicity of fluoroquinolones – which definitely is dose related – does not seem to be of major importance for a severely ill, hospitalized patient and it is questionable whether CNS reactions to fluoroquinolones would occur at lower incidences if dose regimens are changed. Possibly, once-daily dosing regimens would lead to higher incidences of CNS toxicity than the dosing strategies used currently.

For an overall evaluation of quinolone pharmacodynamics it should be considered that a reduction in the amount of drug used bears the risk of decreased efficacy, which might be a fatal risk for some patients. Even if this risk might be small it seems unacceptable if only a small (financial) advantage is gained. So far there is no indication that adverse reactions to fluoroquinolones are reduced by altered dosing strategies.

C. Pharmacokinetic Aspects

If concentration–effect relationships of drugs are evaluated, it is of major importance to consider the differences which exist in vivo in comparison with in vitro studies which provide basic data for the use of the drugs. Only a few pharmacokinetic aspects are mentioned in the following paragraphs, which have been recognized as important factors which complicate the application of in vitro results to the in vivo situation.

I. Protein Binding

Binding to serum proteins may affect several pharmacokinetic and pharmacodynamic variables such as renal elimination, tissue distribution and the activity of antibiotics in blood. Comprehensive studies and reviews on this topic are abundant in the literature (e.g. CRAIG and WELLING 1977; BARZA and CUCHURAL 1985; WISE 1986). In general, the aminoglycosides and quinolones exhibit low plasma protein binding, whereas the β-lactams have low to very high plasma protein binding depending on the individual drug. Thus, protein binding seems to be most important for certain β-lactam antibiotics. The currently used fluoroquinolones are bound less than 50% by serum proteins.

Only the non-protein-bound fraction of a drug can penetrate into a tissue, because protein binding keeps the molecule from passing through the capillary wall. Furthermore, only the unbound form of the antimicrobial is active in vitro (and probably also in vivo); however, the in vivo effects of protein binding are still a matter of controversy. There are a few clinical data showing poor activity of highly protein-bound drugs, but since protein binding is rapidly reversible, even highly protein-bound agents can be efficacious.

Albumin is the protein to which drugs preferentially bind. Binding occurs mainly via ionic, hydrogen, and hydrophobic interactions, and a small alteration in chemical structure may cause significant differences in the extent of protein binding. Usually protein binding depends on drug concentration, type and concentration of the protein, pH and lipid solubility among other factors. It is probably of clinical significance that critical illnesses can shift protein binding, at least of certain antibiotics (NIX et al. 1991a,b).

II. Tissue Concentrations and Volume of Distribution

Antimicrobial activity determined in vitro, protein binding and volumes of distribution are important determinants for the effects of an antibiotic in an infected organism. But it still is difficult to predict the in vivo effects from these variables. One of the most important aspects useful in predicting antimicrobial activity in vivo is the concentration of the drug at the site of infection.

Generally, it is thought that at least the MIC as determined in vitro should be achieved at the site of infection for antimicrobial therapy to be effective. However, it is also known that the achievement of "therapeutic concentrations" of antimicrobial agents at the site of infection may not be sufficient for clinical cure. On the other hand, there is evidence that in some instances subinhibitory concentrations of antibiotics may produce antimicrobial effects in combination with defense mechanisms.

Tissue concentration is most commonly defined as the amount of drug in a certain volume of homogenized tissue. In vivo, however, the constitution of tissue is far from being homogeneous. As a minimum requirement, intracellular and extracellular concentrations should be distinguished. This can be of clinical significance since – although most bacteria are located in the extracel-

Table 2. Intracellular concentrations and activity of antibiotics in macrophages (modified from VAN DER AUWERA 1991)

Antimicrobial agent	Ratio of intra- to extracellullar concentration	Intracellular activity	
		S. aureus	*Legionella* spp.
Azithromycin	>50	n.a.	+++
Erythromycin	>10	(+)	+++
Roxithromycin	>10	++	+++
Fluoroquinolones	2–15	+++	+++
Vancomycin	>10	n.a.	n.a.
Clindamycin	>10	(+)	n.a.
Rifampicin	>3	++	+++
β-Lactam antibiotics	<1	(+)	n.a.
Aminoglycosides	<1	(+)	n.a.

(+) low; ++ moderate; +++ high; n.a., no data available.

lular space – several pathogens are located intracellularly. Table 2 shows the ratios of intra- and extracellular concentrations of antimicrobials in macrophages. With fluoroquinolones, intracellular concentrations can be obtained which are sufficient for the treatment of intracellularly located pathogens, such as staphylococci, chlamydia and legionella species (VAN DER AUWERA 1991).

When tissue is homogenized intracellular enzymes can be released which may bind or inactivate the antibiotic. Also, the concentration of intracellularly located drugs may be diluted in the presence of extracellular fluid. Total tissue concentration of an antimicrobial agent is not an exact indicator of the amount of available active drug, since binding to tissue occurs, and therefore some of it is probably inactive just as protein-bound drug in plasma is. Depending on the physicochemical properties of the agent, it may be released from the binding proteins – which can act as an reservoir – and subsequently exhibit its pharmacological activity.

One typical pharmacokinetic feature of the fluoroquinolones is their high volume of distribution of 100 l, 200 l or more (often exceeding the total body mass). This pharmacokinetic variable, which more precisely should be referred to as an "*apparent* volume of distribution" (because this variable indicates the volume that a dose of drug *appears* to be dissolved in as determined by measuring its concentration in serum) does not provide information on specific tissue concentrations of a drug but it does give useful information on the general distributional properties of a drug and indicates that it is widely distributed in extravascular spaces.

In general, body fluid and tissue concentrations measured after administration of fluoroquinolones are equal to or higher than the concurrent plasma concentrations. Relatively low concentrations are achieved for example in the aqueous humour, lens and CSF (CULLMANN et al. 1993)

D. Summary and Conclusion

Results from a well-designed clinical trial provide the most conclusive information on the efficacy of a drug. There is no way – and probably never will be – to find a substitute for this approach. However, results from clinical trials very often have their shortcomings. To name just a few: in open trials many biases are possible, the number of patients is often too small or the group of patients investigated is too heterogeneous to draw clear conclusions.

Therefore, other approaches besides clinical trials are necessary to provide basic information for therapeutic decisions. Such basic data are essential when a new drug is given to an infected patient for the very first time. However, we should be aware of the fact that many details of the action of fluoroquinolones and other drugs in patients are unknown and therefore it seems prudent to be restrictive with the recommendation of new dosing regimens derived from experimental studies. Certainly, for aminoglycosides the once daily dosing concept is a reasonable one, but it is doubtful if alterations in the dosing recommendations is acceptable for other antimicrobials if possible risks and benefits are carefully estimated. The small advantage of a slight cost reduction alone does not justify new dosing strategies.

References

Barza M, Cuchural G (1985) General principles of antibiotic tissue penetration. J Antimicrob Chemother 15(Suppl A):59–75

Blaser J, Stone BB, Groner MC, Zinner SH (1987) Comparative study with enoxacin and netilmicin in a pharmacodynamic model to determine importance of ratio of antibiotic peak concentration to MIC for bacterial activity and emergence of resistance. Antimicrob Agents Chemother 31:1054–1070

Craig WA, Ebert SC (1991) Killing and regrowth of bacteria in vitro: a review. Scand J Infect Dis (Suppl 74):63–70

Craig WA, Ebert SC (1992) Continuous infusion of β-lactam antibiotics. Antimicrob Agents Chemother 36:2577–2583

Craig WA, Welling PG (1977) Protein binding of antimicrobials: clinical pharmacokinetic and therapeutic implications. Clin Pharmacokinet 2:252–268

Cullmann W, Geddes AM, Weidekamm E, Urwyler H, Braunsteiner A (1993) Fleroxacin: a review of its chemistry, microbiology, toxicology, pharmacokinetics, clinical efficacy and safety. Int J Antimicrob Agents 2:203–230

Drusano GL (1995) Pharmacology of anti-infective agents. In: Mandell GL, Bennett JE, Dolin R (eds) Mandell, Douglas and Bennett's principles and practice of infectious diseases, 4th edn. Churchill Livingstone, Edinburgh, pp 225–233

Drusano GL, Johnson DE, Rosen M, Standiford HC (1993) Pharmacodynamics of a fluoroquinolone antimicrobial agent in a neutropenic rat model of *Pseudomonas* sepsis. Antimicrob Agents Chemother 37:483–490

Dudley MN, Mandler HD, Gilbert D, Ericson J, Mayer KH, Zinner SH (1987) Pharmacokinetics and pharmacodynamics of intravenous ciprofloxacin. Studies in vivo and in an in vitro dynamic model. Am J Med 82(Suppl 4A):363–368

Eagle H, Musselman AD (1948) The rate of bactericidal action of penicillin in vitro as a function of its concentration and its paradoxically reduced activity at high concentrations against certain organisms. J Exp Med 88:99–131

Forrest A, Ballow CH, Nix DE, Birmingham MA, Schentag JJ (1993a) Development of a population pharmacokinetic model and optimal sampling strategies for intravenous ciprofloxacin. Antimicrob Agents Chemother 37:1065–1072

Forrest A, Nix DE, Ballow CH, Goss TF, Birmingham MA, Schentag JJ (1993b) Pharmacodynamics of intravenous ciprofloxacin in seriously ill patients. Antimicrob Agents Chemother 37:1073–1081

Gilbert DN (1991) Once-daily aminoglycoside therapy. Antimicrob Agents Chemother 35:399–405

Leggett JE, Ebert S, Fantin B, Craig WA (1991) Comparative dose–effect relations at several dosing intervals for beta-lactam, aminoglycoside and quinolone antibiotics against gram-negative bacilli in murine thigh-infection and pneumonitis models. Scand J Infect Dis (Suppl 74):179–184

Nicolau DP, Freeman CD, Belliveau PP, Nightingale CH, Ross JW, Quintiliani R (1995) Experience with a once-daily aminoglycoside program administered to 2, 184 adult patients. Antimicrob Agents Chemother 39:650–655

Nix DE, Goodwin SD, Peloquin CA, Rotella DL, Schentag JJ (1991a) Antibiotic tissue penetration and its relevance: models of tissue penetration and their meaning. Antimicrob Agents Chemother 35:1947–1952

Nix DE, Goodwin SD, Peloquin CA, Rotella DL, Schentag JJ (1991b) Antibiotic tissue penetration and its relevance: impact of tissue penetration on infection response. Antimicrob Agents Chemother 35:1953–1959

Peloquin CA, Cumbo TJ, Nix DE, Sands MF, Schentag JJ (1989) Evaluation of intravenous ciprofloxacin in patients with nosocomial lower respiratory tract infections. Arch Intern Med 149:2269–2273

Roosendaal R, Bakker-Woudenberg IAJ, van den Berghe-van Raffe M, Vink-van den Berg JC, Michel MF (1987) Comparative activities of ciprofloxacin and ceftazidime against Klebsiella pneumoniae in vitro and in experimental pneumonia in leukopenic rats. Antimicrob Agents Chemother 31:1809–1815

Roosendaal R, Bakker-Woudenberg IAJ, van den Berghe-van Raffe M, Vink-van den Berg JC, Michel MF (1989) Impact of the dosage schedule on the efficacy of ceftazidime, gentamicin and ciprofloxacin in Klebsiella pneumoniae pneumonia and septicemia in leukopenic rats. Eur J Clin Microbiol Infect Dis 8:878–887

Rozdzinski E, Kern WV, Reichle A, Moritz T, Schmeiser T, Gaus W, Kurrle E (1993) Once-daily versus thrice-daily dosing of netilmicin in combination with β-lactam antibiotics as empirical therapy for febrile neutropenic patients. J Antimicrob Chemother 31:585–598

Schentag JJ (1991) Correlations of pharmacokinetic parameters to efficacy of antibiotics: relationships between serum concentrations, MIC values, and bacterial eradication in patients with gram-negative pneumonia. Scand J Infect Dis Suppl 74:218–234

Ter Braak EW, deVries PJ, Bouter KP, van der Vegt SG, Dorrestein GC, Nortier JW, van Dijk A, Verkooyen RP, Verbrugh HA (1990) Once-daily dosing regimen for aminoglycoside plus β-lactam combination therapy for serious bacterial infections: comparative trial with netilmicin plus ceftriaxone. Am J Med 89:58–66

Van der Auwera P (1991) Interactions between antibiotics and phagocytosis in bacterial killing. Scand J Infect Dis Suppl 74:42–48

Wise R (1986) The clinical relevance of protein binding and tissue concentrations in antimicrobial therapy. Clin Pharmacokinet 11:470–482

CHAPTER 13

Clinical Use of Quinolones

P. SCHACHT

A. Introduction

Since the introduction of the fluorinated quinolone antibiotic norfloxacin in the early 1980s, successor drugs such as ciprofloxacin or ofloxacin represent a significant advance in the history of infection control. Until their introduction, death due to infection by certain organisms, particularly *Pseudomonas aeruginosa*, was a common event. Quinolones have a broad spectrum of activity; however, in vitro they are inactive against anaerobes and show only moderate activity against many gram-positive organisms. Historically, therefore, they have been used mainly to treat gram-negative urinary tract and hospital infections; but it is important to note that in vivo clinical results do not differ significantly from those obtained with β-lactam antibiotics in the treatment of gram-positive infections, including pneumococcal diseases, if the quinolones are dosed appropriately. Today there is, on the one hand, a move towards higher doses of quinolones, e.g. two times 400mg ofloxacin orally for respiratory tract infections or up to three times 400mg ciprofloxacin i.v. for serious infections. On the other hand, single daily doses for urinary and gastrointestinal tract infections or even single shot dosing for uncomplicated urinary tract infections, gonorrhea, chancroid, and gastrointestinal infections are in clinical use today.

Whereas norfloxacin is mainly given to patients with urinary and gastrointestinal tract infections, the other quinolones are used to treat a wide variety of infections. The quinolones continue to make a significant contribution in the fight against community or hospital acquired infections and add a new dimension of antibacterial therapy to the well-established armamentarium of anti-infective drugs.

B. Urinary Tract Infections and Prostatitis

The excellent in vitro activity against almost all urinary pathogens – including *Pseudomonas* species – and the high recovery rate in urine after oral therapy make the quinolones ideal compounds for the treatment of urinary tract infections (SHAH 1985; WOLFSON and HOOPER 1989; NABER 1989). The first comparative clinical study (NABER 1986) demonstrated the high clinical activity of norfloxacin and ciprofloxacin in a controlled trial in which a single daily dose of each drug was administered to hospitalized patients suffering from complicated urinary tract infections. Since then hundreds of clinical studies have been carried out comparing standard therapy with fluoroquinolone therapy (Grubbs et al. 1992) or fluoroquinolones in different dosage regimens or duration of treatment.

I. Acute Uncomplicated Urinary Tract Infection

The introduction of the quinolones made it possible to shorten treatment (Iravani et al. 1995), e.g. 100 mg ciprofloxacin b.i.d. for 3 days or even a single dose treatment with fleroxacin, ciprofloxacin (Iravani 1993; Raz et al. 1989; Karachalios et al. 1991), and pefloxacin (Guibert 1995). These regimens are in clinical use. Recommendations by Stamm and Hooton (1993) are summarized in Table 1.

II. Complicated Urinary Tract Infection

Complicated urinary tract infections (UTIs) chiefly associated with functional, anatomical abnormalities of the urinary tract are often caused by resistant pathogens. Table 2, modified from Stamm and Hooton (1993), reflects the clinical use of quinolones in complicated UTIs, orally as well as parenterally. The results of clinical trials comparing different quinolones are given in Table 3. These results show comparable efficacy rates of the drugs and underline the importance of the quinolones for the treatment of UTIs. Quinolones are a valuable addition to the already available anti-infective agents in this important area of infectious diseases.

III. Prostatitis

Several studies and controlled study programs have shown quinolones to be effective in the treatment of chronic bacterial prostatitis (Stamm and Hooton 1993; Naber 1989; Weidner and Schiefer 1991; Echols and Heyd 1995). The problems associated with the interpretation of the results are due to the different criteria used by investigators around the world. Hence, one can only rely upon studies with a long-term follow-up to identify patients with relapse or late relapse (Table 4) and controlled studies with well-defined criteria (Echols et al. 1994). These so called pivotal studies require the use of techniques to confirm localization of microbiologically documented infection.

Additionally to the data in Table 4 which summarizes long-term follow-up studies, five pivotal studies were conducted: two double-blind studies, one controlled, randomized nonblind, and two prospective uncontrolled studies. Ciprofloxacin at a dose of 500 mg b.i.d. was compared to TMP/SMX and oral carbenicillin (approved drug for prostatitis in the USA). The results of the five studies concerning bacteriological response in the patients may be summa-

Table 1. Treatment regimens for uncomplicated urinary tract infections (modified from STAMM and HOOTON 1993)

Condition	Characteristic pathogens	Complicating circumstances	Recommended empirical treatment[a]
Acute uncomplicated cystitis in women	*Escherichia coli* *Staphylococcus saprophyticus* *Proteus mirabilis* *Klebsiella pneumoniae*	None	3-Day regimens: oral trimetoprim–sulfamethoxazole, trimethoprim, norfloxacin, ciprofloxacin, ofloxacin, lomefloxacin, or enoxacin[a]
		Diabetes, symptoms for >7 days, recent urinary tract infection, use of diaphragm, age >65 years	Consider 7-day regimen: oral trimethoprim-sulfamethoxazole, trimethoprim, norfloxacin, ciprofloxacin, lomefloxacin, or enoxacin
		Pregnancy	Consider 7-day regimen: oral amoxicillin, macrocrystalline, nitrofurantoin, cefpodoxime proxetil, or trimethoprim–sulfamethoxazole
Acute uncomplicated pyelonephritis in women	*E. coli* *P. mirabilis* *K. pneumoniae* *S. saphrophyticus*	Mild-to-moderate illness, no nausea or vomiting – outpatient therapy	Oral trimethoprim–sulfamethoxazole, norfloxacin, ciprofloxacin, ofloxacin, lomefloxacin, or enoxacin for 10–14 days
		Severe illness or possible urosepsis; hospitalization required	Parenteral trimethoprim–sulfamethoxazole, ceftriaxone, ciprofloxacin, ofloxacin, or gentamicin (with or without ampicillin) until fever is gone; then oral trimethoprim–sulfamethoxazole, norfloxacin, ciprofloxacin, ofloxacin, lomefloxacin, or enoxacin for 14 days
		Pregnancy – hospitalization recommended	Parenteral ceftriaxone, gentamicin (with or without ampicillin), aztreonam or trimethoprim–sulfamethoxazole until fever is gone; then oral amoxicillin, a cephalo-sporin, or trimethoprim–sulfamethoxazole for 14 days

[a] Multiday oral regimens for cystitis are as follows: trimethoprime–sufamethoxazole, 160 mg/800 mg every 12 h; trimethoprim, 100 mg every 12 h; norfloxacin, 400 mg every 12 h; ciprofloxacin, 250 mg every 12 h; ofloxacin, 200 mg every 12 h; lomefloxacin, 400 mg every day; enoxacin, 400 mg every 12 h; macrocrystalline nitrofurantoin, 100 mg four times a day; amoxicillin, 250 mg every 8 h; and cefpodoxime proxetil, 100 mg every 12 h. Single-dose treatment of modern quinolones is indicated on the package insert in most countries outside the USA for acute uncomplicated cystitis.

Table 2. Treatment regimens for complicated urinary tract infections (modified from STAMM and HOOTON 1993)

Pathogen	Severity of illness	Recommended empirical treatment
E. coli *Proteus* species *Klebsiela* species *Pseudomonas* species *Serratia* species *Enterococcus* species *Staphylococcus* species	Mild to moderate illness, no nausea or vomiting – outpatient therapy	Oral norfloxacin, ciprofloxacin, ofloxacin, lomefloxacin, or enoxacin for 10–14 days
	Severe illness or possible urosepsis – hospitalization required	Parenteral ampicillin and gentamicin, ciprofloxacin, ofloxacin, ceftriaxone, aztreonam, ticarcillin-clavulanate or imipenem-cilastatin until fever gone, then oral trimethoprim-sulfamethoxazole norfloxacin, ciprofloxacin, ofloxacin, lomefloxacin, or enoxacin for 10–21 days

Multiday parenteral or oral regimens, involvement of, e.g., *Pseudomonas aeruginosa*, requires highly active drugs such as either oral or parenteral ciprofloxacin and prolonged treatment 14–21 days, especially in endangered subpopulations of patients (e.g., paraplegics).

Table 3. Overview of a selection of prospective, randomized, comparative studies with oral quinolones in patients with complicated urinary tract infections

Drug regimen (no. of patients evaluated)	Efficacy (% of patients)		Reference
	Clinical cure (according to guidelines)	Bacteriological eradication	
Ciprofloxacin 500 mg once daily p.o. 7–15 days (30)	29	30	NABER et al. (1986)
Norfloxacin 800 mg once daily p.o. 7–15 days (30)	29	30	
Ciprofloxacin 500 mg twice daily 10–14 days (70)	96	96	COX (1992)
Lomefloxacin 400 mg twice daily 10–14 days (72)	99	97	
Ciprofloxacin 500 mg twice daily 7–14 days (70)	87	96	NABER (1992)
Lomefloxacin 400 mg qd 7–14 days (72)	92	97	
Ciprofloxacin 500 mg twice daily 14–21 days (29)	79	n.a.	SCHAEFFER and ANDERSON (1992)
Norfloxacin 400 mg twice daily 10–21 dyas (29)	72	n.a.	

n.a., data not available.

Table 4. Studies with quinolones in the treatment of chronic bacterial prostatitis with long-term follow-up investigation (modified from NABER 1989)

Drug	Dosage	Duration of treatment (days)	Patients (n)	Bacterial eradication (%)	Follow-up (months)
Norfloxacin	400 mg b.i.d.	28	14	64	6
Ciprofloxacin	500 mg b.i.d.	14	15	60	12
Ciprofloxacin	500 mg b.i.d.	28	19	58	16–36
Ciprofloxacin	500 mg b.i.d.	60–150	7	86	12

Table 5. Pivotal studies: bacteriological response by patient (modified from ECHOLS and HEYD 1995)

	End of therapy[a]				Follow-up[b]		
	CIP	TMP/SMX	IC		CIP	TMP/SMX	IC
Eradication	147 (94%)	45 (76%)	17 (85%)	Continued eradication	125 (85%)	33 (85%)	17 (100%)
Persistence	10 (6%)	14 (24%)	3 (15%)	Recurrence	9 (6%)	4 (10%)	0
Eradication with superinfection	2	5	0	Reinfection	13 (9%)	2 (5%)	0

CIP, ciprofloxacin; TMP/SMX, trimethoprim–sulfamethoxazole; IC, indanyl carbenicillin.
[a] Therapy lasted 4 to 9 days.
[b] 1 month following completion of therapy.

rized as follows (Table 5): *Escherichia coli* 130 isolates (49%), followed by *Enterococcus faecalis* 38 isolates (14%) and *Proteus mirabilis* 29 isolates (11%) were the most frequently isolated pathogens. Ciprofloxacin eradicated 95% of the strains compared with 77% in patients treated with trimethoprim–sulfamethoxazole (TMP/SMX) or 85% in patients treated with indanyl carbenicillin (IC). Ciprofloxacin at a dose of 500 mg b.i.d. orally for 14–28 days has been shown to be consistently effective in these pivotal trials which clearly confirms the conclusion from early phase II studies shown in Table 5 that its clinical use is justified.

C. Gastrointestinal Infections and Traveller's Diarrhea

Gastrointestinal infections are still a major threat to mankind especially to children in developing countries.

I. *Salmonella typhi* – Enteric Fever

Several controlled trials established the value of the quinolones in the treatment of typhoid fever. LIMSON and LITTAVA (1989) compared ciprofloxacin with TMP/SMX. All patients receiving ciprofloxacin were cured, and in each case the salmonella were eradicated at follow-up. In the TMP/SMX group, two patients failed to respond to TMP/SMX treatment; both patients were cured later by ciprofloxacin. MORELLI et al. (1992) studied the efficacy of ofloxacin, pefloxacin, ciprofloxacin, enoxacin, and norfloxacin compared with chloramphenicol and concluded that ofloxacin and ciprofloxacin were more effective than chloramphenicol. Ciprofloxacin was found to be highly efficacious in patients with enteric fever caused by mutidrug resistant *S. typhi* (ALAM et al. 1993; SINGH et al. 1993).

In a review by DUPONT (1993b) the results of controlled trials with orally administered quinolones were summarized (Table 6). In comparative trials, oral ciprofloxacin was as effective as other quinolones (ofloxacin, pefloxacin, fleroxacin, and norfloxacin). In children with typhoid fever caused by multiresistant bacteria ciprofloxacin was highly effective. TAKKAR et al. (1994), CHEW et al. (1992) showed that a 7-day course for enteric fever with ciprofloxacin (500 mg b.i.d.) is highly promising.

II. *Salmonella typhi* Carriers

Ciprofloxacin 750 mg twice daily for 28 days eradicated salmonella carriage in 11 of 12 patients (FERRECIO et al. 1988). Norfloxacin and ciprofloxacin were both effective in eradicating intestinal excretion of *S. typhi* in chronic carriers. It is estimated that between 2% and 4% of enteric fever patients become chronic carriers and excrete *S. typhi* for at least 1 year (DUPONT 1993b; Table 7). These patients are important in the epidemiology of the infection. Chloramphenicol is not effective in preventing typhoid carriage but there is enough evidence that quinolones are useful drugs in this indication.

Table 6. Controlled clinical trials with orally administered fluoroquinolones in the treatment of typhoid fever (modified from DUPONT 1993a)

Drug	Regimen	Duration of therapy (days)	Patients (*n*)	Cure rate (*n*)	Cure rate (%)
Comparisons with chloramphenicol					
Fleroxacin	400 mg o.d.	7	10	20	100
	400 mg o.d.	14	10		
Chloramphenicol	50 mg/kg q.i.d.	14	10	10	100
Ofloxacin	200 mg b.i.d.	14	35	35	100
Chloramphenicol	500–750 mg every 6 h	14	38	28[a]	74
Comparisons with cotrimoxazole					
Ciprofloxacin	500 mg b.i.d.	10	20	20	100
Cotrimoxazole	160/800 mg b.i.d.	14	20	18	90
Pefloxacin	400 mg b.i.d.	14	24	24	100
Cotrimoxazole	160/800 mg b.i.d.	14	18	18	100

[a] Fever persisted in 10 patients for >168 h.

III. Salmonella Gastroenteritis Outbreaks

Most of the cases of salmonellosis are self-limiting in nature with the exception of endangered patient groups such as elderly, young, and immunocompromised (e.g., AIDS) patients (WILCOX and SPENCER 1992).

The benefit of quinolones in the treatment of disease remains still unclear. AHMAD et al. (1991) obtained good results with 500mg ciprofloxacin b.i.d. (7 days) for the treatment of a *Salmonella virchow* outbreak in a psychiatric hospital. Comparable results were reported by DYSON et al. (1995) who controlled an outbreak of *Salmonella enteritidis* in a hospital of mentally handicapped patients with ciprofloxacin. A placebo-controlled comparison between ciprofloxacin and TMP/SMX demonstrated advantages for ciprofloxacin in that it shortened the duration of diarrhea and increased the cured/improved patient number compared with the placebo. These effects could not be shown for TMP/SMX (GOODMAN et al. 1990). Controversial results came from SANCHEZ et al. (1993) who could not detect a difference in a double-blind study between TMP/SMX, ciprofloxacin, and placebo. Unacceptably high failure rates in patients with salmonellosis were reported by NEILL et al. (1991). LEIGH (1992) confirmed high microbiological failure rates (stools were positive up to 14 weeks) but found that clinically the response to treatment was very high. As mentioned above, ciprofloxacin should be administered under sound judgement (LIGHTFOOT et al. 1990) to well-defined patient groups (see Sect. C.VI).

IV. Shigellosis

Two prospective, randomized, double-blind trials (BENNISH et al. 1990, 1992) established the benefit of ciprofloxacin in the treatment of Shigellosis. In the first trial 500mg ciprofloxacin every 12h was compared with standard therapy (ampicillin every 6h); both were adminsterd orally for 5 days to 121 adult males severely infected. Ciprofloxacin proved to be effective in treating patients with shigellosis. The second study compared a single 1-g ciprofloxacin dose with 500mg b.i.d. for 5 days. The single 1-g dose was effective against

Table 7. Clinical trials with orally administered fluoroquinolones in the treatment of chronic intestinal carriage of *Salmonella typhi* (modified from DUPONT 1993a)

Drug	Regimen	Duration of therapy (days)	Patients	Cure rate (*n*)	(%)	Reference
Ciprofloxacin	750mg b.i.d.	28	12	11	92	FERRECCIO et al. (1988)
Norfloxacin	400mg b.i.d.	28	23	18	78	GOTUZZO et al. (1988)

species of *Shigella* other than *Shigella dysenteriae* type 1. The 1-g regimen was inferior to the multiple administration of ciprofloxacin for treating patients infected with *S. dysenteriae* type 1. Clinical use of ciprofloxacin is indicated. The results of a study by MURPHY et al. (1993) suggests the concomitant use of loperamide to shorten the duration of diarrhea.

V. Cholera

Substitution of fluid and electrolytes is the basic standard therapy in patients suffering from cholera. Administered anti-infective agents reduce the severity and duration of diarrhea and can stop excretion of *Vibrio cholerae* 01 within 24h as demonstrated by KHAN et al. (1995) for ciprofloxacin, compared with erythromycin, nalidixic acid, tetracycline, and pivmecillinam, which did not exhibit these effects.

Since tetracycline and nalidixic acid lose their effectiveness due to resistance problems, quinolones may become valuable alternatives as demonstrated in a study in which ciprofloxacin was found to be the most effective therapy when administered as a 3-day course of 500mg b.i.d. orally. GOTUZZO et al. (1995) conducted a double-blind trial comparing a 250-mg daily dose of ciprofloxacin for 3 days with 500mg tetracycline four times a day for 3 days in the treatment of moderate to severe cholera (Table 8). The encouraging results with a low dose of ciprofloxacin may – as stated above – render quinolones a valuable and cost effective alternative in the treatment of cholera. Recently KHAN et al. (1994) showed that single dose treatment with ciprofloxacin was as effective as a single dose of doxycycline in its effect on the duration and volume of watery diarrhea; however, ciprofloxacin was superior in the eradication of *V. cholerae* 0139 infections.

Table 8. Microbiological outcome for patients with cholera who were treated with ciprofloxacin or tetracycline (modified from GOTUZZO et al. 1995)

Outcome variable	Ciprofloxacin (n = 100)		Tetracycline (n = 102)		p value
	(n)	(%)	(n)	(%)	
Microbiological efficacy					
Cure	99	99	97	95	n.a.
Failure	1	1	5	5	0.212
Stool cultures negative for *Vibrio cholera*					
Study day 1	86	86	82	80	0.348
Study day 2	100	100	99	97	0.246
Posttreatment day 1	99	99	97	95	0.212

n.a., data not available.

VI. Traveller's Diarrhea – Prevention

In one of the first studies using a quinolone for prevention of traveller's diarrhea RADEMAKER et al. (1989) demonstrated in a double blind placebo-controlled study that 500 mg ciprofloxacin once daily for 1 week could protect travellers from diarrhea efficiently. Since quinolones can cause serious side effects (SCHACHT et al. 1989), the decision to opt for this kind of prevention should be soundly based on a benefit and risk evaluation which was carefully drawn up by WISTRÖM and NORRBY (1990). They described the possible candidates for quinolone prophylaxis as:

a) Travellers with immunosuppression or immunodeficiency
b) Travellers with reduced gastric acid barrier
c) Travellers with inflammatory bowel disease
d) Travellers with other diseases and taking other medication (digitalis, diuretics)
e) "Travellers on critical missions" (business people, politicians, athletes, artists, soldiers, medical personnel, medical scientists going to a congress).

VII. Traveller's Diarrhea – Therapy

In contrast to the issue of prevention it is quite obvious that bacterial diarrhea is an indication for antimicrobial treatment additional to the basic fluid and electrolyte replacement. More concrete recommendations for the use of quinolones are best summarized by GOODMAN-OBOT and DUPONT (1981; see Table 9), because by SALAM et al. (1994) suggest strongly that a single tablet of 500 mg was an effective empirical treatment. In contrast to the findings with shigellosis, MURPHY et al. (1993) and TAYLOR et al. (1991) could not detect a beneficial effect of ciprofloxacin and loperamide on traveller's diarrhea caused by enterotoxigenic *E. coli* in a double-blind trial.

D. Respiratory Tract Infections

Since early clinical research with the fluoroquinolones there has been a continous concern about their efficacy in chest infections. The concern was mainly raised by clinical microbiologists (DAVIES and MAESEN 1986) and directed to the possibility of the development of resistance, e.g., by *P. aeruginosa*, and clinical failure in *Streptococcus pneumoniae* infections, but also by clinicians (BALL et al. 1995). These reservation were quite understandable because the new fluoroquinolone norfloxacin did not fulfill the expectations raised by in vitro findings.

The quinolones were labelled as good drugs for urinary and enteric infections. Early studies, however, with ciprofloxacin and ofloxacin in the treatment of different types of respiratory tract infections (RTI), e.g., community-acquired pneumonia (CHRYSANTHOPOULOS and BASSARIS 1989; VETTER et al.

Table 9. Quinolones in clinical use for traveller's diarrhea (modified from GOODMAN OBOT and DUPONT 1993)

Drug	Dosage
Norfloxacin	400 mg twice a day for 3 days
or	
Ciprofloxacin	500 mg twice a day for 3 days
or	
Ofloxacin	300 mg twice a day for 3 days
or	
Fleroxacin	400 mg twice a day for 3 days

1985; SCHÖNWALD 1990), and RTI patients treated with ciprofloxacin and cefalexin in a randomized comparison indicated clinical efficacy. Even with serious conditions such as cystic fibrosis, BENDER et al. (1984) reported on improvement in a small group of patients treated with oral ciprofloxacin. Meta-analysis-like evaluations (SCHACHT et al. 1988) revealed high *S. pneumoniae* eradication rates for ciprofloxacin. More than 90% eradication was also reported for ofloxacin (KHAJOTIA et al. 1991). These early findings justified a broad program of clinical studies to establish the usefulness of quinolones in chest infections (Table 10).

I. Acute Exacerbation of Chronic Bronchitis

BASRAN et al. (1990) compared ciprofloxacin and amoxicillin in the treatment of acute exacerbation of chronic obstructive airways disease in 140 elderly patients. Isolated pathogens were *Haemophilus influenzae* (43%), *S. pneumoniae* (25%), *Pseudomonas* spp., *Staphylococcus aureus*, and *Branhamella catarrhalis*. Ciprofloxacin at a dose of 500–750 mg b.i.d. over 7 days produced comparable results to amoxicillin at a dose of 250–500 mg t.i.d. administered for 7 days. An independent (blind) observer analyzed the study results as follows:

Ciprofloxacin produced a success rate of 91.8% (complete success 21.9%, partial success 69.9%) while a 73% success rate was achieved with amoxicillin (complete success 10.4%, partial success 62.7%). None of the isolates were resistant to ciprofloxacin. In another study QUENZER et al. (1990) compared 500 mg ciprofloxacin b.i.d. (28 patients) to 250 mg cefaclor every 8 h (27 patients) for 5 days or longer. The responses to the two anti-infective therapies did not differ statistically; there were no failures in either group with respect to *S. pneumoniae* eradication.

CHODOSH (1991a) compared two quinolones: 500 mg ciprofloxacin twice daily and 600 mg temafloxacin (which has been withdrawn from the market due to unexpected adverse reactions) twice daily for 7 days. Both drugs were administered orally to patients suffering from mild to moderate lower respiratory tract infections, mainly acute exacerbations of chronic bronchitis. The

Table 10. Efficacy of ciprofloxacin against *Streptococcus pneumoniae* respiratory tract infection (modified from KHAN 1990)

Reference	Number of patients[a]	Number of isolates[b]	Mean MIC (mg)	Clinical outcome (*n*)	Bacteriological outcome (N)	Ciprofloxacin dose[c]
WOLLSCHLAGER et al. (1987)	7	8	NA	C (7/7)	E (6/7)	750 mg
KHAN et al. (1989)	9	9	1.1	C (8/9) F (1/9)	E (8/9)	200 mg intravenous b.i.d. for 3 to 5 days, followed by 500 mg b.i.d. orally
GLEADHILL et al. (1986)	7	12	1.56	C (7/7)	E (7/7)	500 mg
KOBAYASHI (1987)	25	42	NA	C (23/25) F (2/25)	E P (13/42)	29/42) 200 to 1200 mg (70% received 600 mg daily)
DAVIES et al. (1986)	11 with single isolates 15 with multiple isolates	48	0.73 pre-treatment; 0.93 post-treatment	I (9/26) F (17/26)	P (21/48)	500 to 1000 mg
ERNST et al. (1986)	6	6	NA	C (6/6)	NA	750 mg
FASS (1987)	7	7	NA	C (7/7)	E (7/7)	500 mg
ESPOSITO et al. (1987)	3	3	0 255	C (1/3) I (2/3)	E (2/3) P (1/3)	250 mg
CHRYSANTHOPOULOS (1989)	12	12	NA	NA	NA	200 mg intravenous b.i.d. for 2 to 10 days, followed by 500 mg b.i.d. orally

C, cured; F, failure; I, improvement; E, eradication; P, persistence; NA, data not available; MIC, minimal inhibitory concentration.
[a] With *Streptococcus pneumoniae* infection.
[b] *Streptococcus pneumoniae* (multiple isolates of same organism in some patients).
[c] All doses given orally, b.i.d., unless otherwise noted.

study was multicenter, double-blind, and randomized in design. Sixty-four patients receiving temafloxacin and 67 patients receiving ciprofloxacin were evaluable for efficacy. The results were not statistically different; 91% of the temafloxacin patients and 94% of the ciprofloxacin patients were cured or improved. All isolates of *H. influenzae*, *B. catarrhalis*, *S. aureus*, *S. pneumoniae*, and *E. coli* were eradicated. Notable failures were reported for two of six *Klebsiella pneumoniae* infections in the temafloxacin group and three of seven *P. aeruginosa* infections in the ciprofloxacin group. Recently CATENA and CESANA (1995) published the results of a randomized multicenter study comparing a 5-day oral course of 200mg rufloxacin daily (the loading dose was twice 200mg) versus a 7-day ciprofloxacin course (500mg b.i.d.) in 179 hospitalized adults suffering from acute exacerbations of chronic bronchitits. At follow-up (2 weeks) the clinical success rates were 92% with rufloxacin and 99% with ciprofloxacin. Failures in both groups ocurred with *P. aeruginosa* infections. *S. pneumoniae* was eradicated in both treatment groups, but the numbers of the pathogens were relatively small, too small to draw final conclusions.

In a similar trial GROSSMAN et al. (1994) found comparable cure rates with TMP/SMX (92.6%) and ciprofloxacin (97.4%) in 65 evaluable patients. Interestingly, of 261 causative organisms isolated from 216 patients four were resistant to ciprofloxacin and 70 to TMP/SMX.

The fluoroquinolones are reported to have good to excellent activity against most common pathogens in this disease (CHODOSH 1991a). Enoxacin, ofloxacin, and ciprofloxacin have yielded high clinical success rates. In comparative clinical trials ciprofloxacin was at least as effective as commonly established standards. A meta-analysis by BYL et al. (1993) supports these results. The authors analyzed 34 studies in which ciprofloxacin was compared with aminopenicillins (12 studies), cephalosporins (12 studies), other quinolones (four studies) SMP/TMX (two studies), doxycycline (two studies), macrolides (one study), and imipenem/cilastatin (one study). In ten studies involving 751 patients with acute exacerbation of chronic bronchitis the difference in efficacy was 92.1% versus 80% in favor of ciprofloxacin. A hundred isolates of *S. pneumoniae* were available for judgement and ciprofloxacin proved to be as effective as the standard therapy.

II. Pneumonia

BYL et al. (1993) did not find any difference in four studies of 266 patients with pneumonia in their meta-analysis comparing ciprofloxacin with standard therapy. THYS et al. (1991) described the benefit of quinolones particularly in community-acquired pneumonia caused by pneumococci as limited. KHAN and BASIR (1990) on the other hand reported good results with the quinolones in pneumococcal infections and confirmed the results of the early explorative study by CHRYSANTHOPOULOS and BASSARIS (1989) in patients suffering from pneumococcal pneumonia. They conclude that the quinolones are not indicated for aspiration pneumonia which was the general opinion from the beginning of the fluoroquinolone era. In contrast quinolones should be suitable for the treatment of nosocomial pneumonias because of their high in vitro activity against gram-negative bacteria. There is some indication that at least ciprofloxacin is clinically effective in the treatment of legionnaires' disease (UNERTL et al. 1989). The general approach in the treatment of moderate to severe pneumonia is to begin with intravenous therapy and, if the patient is able to take oral medication, to continue with the tablet. The clinical practice is to use quinolones if both i.v. and oral preparations are available, e.g., ciprofloxacin or ofloxacin and pefloxacin.

In an open explorative trial FASS (1987) did not detect resistance problems with ciprofloxacin when used at a dose of 500–750 mg b.i.d. orally for gram-negative pneumonia in the hospital.

In three randomized controlled trials in the early phase of quinolone intravenous use, LODE et al. (1987) investigated 201 hospitalized patients with serious bacterial infections including pneumonia. They compared ciprofloxacin (100–200 mg i.v. two or three times followed by 750 mg orally b.i.d. for 2 weeks), imipenem/cilastatin (500–1000 mg i.v. three or four times a day for 2 weeks) ofloxacin (200–300 mg twice a day) and ticarcillin clavulanate (5200 mg three times a day for 2 weeks). Cure and improvement rates were 89% for ciprofloxacin-treated patients, 76% for imipenem/cilastatin-treated patients, 90% for ofloxacin-treated patients and 94% for ticarcillin-clavulanate-treated patients. *P. aeruginosa* resistance rates were 11/24 (46%) for ciprofloxacin, 2/3 (66%) for imipenem/cilastatin, 5/5 (100%) for ofloxacin 1/1 (100%) for ticarcillin clavulanate. An interpretation of these results is difficult because of the historical controls used in some cases and the small numbers of isolates, but one may conclude that quinolones are not worse than other standard anti-infective agents in serious chest infections.

In a well-designed multicenter study, Table 11 FINK et al. (1994) used ciprofloxacin at the appropriate dose of 400 mg i.v. three times a day in contrast to the earlier studies by LODE et al. (1987) who used lower doses because the safety profile of i.v. ciprofloxacin was not fully established at that time. FINK et al. (1994) compared ciprofloxacin with imipenem/cilastatin in patients suffering from severe pneumonia; approximately 79% of the patients were mechanically ventilated at the time of randomization and the mean APACHE II

Table 11. Comparison of ciprofloxacin and imipenem in the treatment of serious respiratory tract infections (modified from FINK et al. 1994)

Drug	Dosage	Duration	Patients (*n*)	Clinical cure/improvement (%)	Bacteriological eradication (%)
Ciprofloxacin	400 mg i.v. three times a day	10.5 days	95	69	69
Imipenem/cilastatin	1 g i.v. three times a day	10.1 days	94	56	59

score was 17.6. The eradication rate for Enterobacteriaceae showed the greatest difference and was significantly higher (41/43; 93%) in the ciprofloxacin-treated patients than in the imipenem/cilastatin group (45/68, 65%; $p = 0.009$). Thirty-three percent of *P. aeruginosa* strains isolated from the ciprofloxacin group were resistant to ciprofloxacin, whereas 53% of strains isolated from patients in the imipenem/cilastatin arm were resistant to imipenem and cilastatin.

This resistance developed earlier than 3 days in either group and requires special treatment strategies. This multicenter, prospective, randomized, blind, controlled trial demonstrated that:

a) Evaluable patients with severe pneumonia had a significantly better response with ciprofloxacin than with imipenem therapy.
b) The bacteriological response by organ system showed a trend towards superiority for ciprofloxacin when compared to imipenem.
c) The organism bacteriological response to Enterobactericeae was significantly better in ciprofloxacin-treated patients than in the imipenem group.
d) *P. aeruginosa* is a problem organism difficult to eradicate (especially in ventilator-dependent patients) with either regimen alone or with subsequent additional combination therapy.
e) Even after considering other prognostic indicators such as ventilation status and *P. aeruginosa* infection, treatment with ciprofloxacin significantly improved patients' response both bacteriologically and clinically.
f) The incidence rate of seizures was significantly higher in imipenem-treated patients.
g) Initial monotherapy for severe pneumonia with ciprofloxacin is safe and effective but may need modification for patients with *P. aeruginosa* infection.

The German Fleroxacin Study Group MARKLEIN et al. (1995) published results from a study comparing the two quinolones ciprofloxacin and fleroxacin. The study was prospective, randomized, multicenter, and controlled, but not blind, and involved the treatment of patients with severe pneumonia (i.v., oral regimen). An interim analysis revealed the results summarized in Table 12. The demographic data and the micriobiological data are comparable in this interim

Table 12. Interim analysis of a study comparing ciprofloxacin and fleroxacin (modified from MARKLEIN et al. 1995)

Drug	Patients (*n*)	Dosage and duration	Cure (%)	Improvement (%)	Failure (%)
Ciprofloxacin	53	400 mg i.v. b.i.d., 2–4 days; 500 mg orally b.i.d. followed up to 14 days	64.2	7.5	18.9
Fleroxacin	49	400 mg i.v. o.d., 2–4 days; initially 400 mg orally o.d. continued for up to 14 days	61.2	6.1	24.5

analysis. The final results of the study may indicate whether a once-daily dose of a quinolone is equivalent to a b.i.d. regimen. Greater numbers of pneumonia patients need to be evaluated (BASSARIS et al. 1995) to prove the better activity of a once-daily regimen of a quinolone in this study comparing ciprofloxacin b.i.d. with a once-daily dose of ofloxacin. There seems to be a trend in favor of the b.i.d. regimen – cure rates of 75% with a b.i.d. regimen versus 63% with an o.d. regimen have been observed for ofloxacin. The studies with ciprofloxacin have clearly shown that it is a useful drug in the treatment of pneumonia when used in a b.i.d. regimen.

III. Recurrent Respiratory Tract Infections in Patients with Cystic Fibrosis

Since their appearance, modern quinolones such as ciprofloxacin and ofloxacin have been candidates for compassionate use to treat exacerbations of gram-negative colonization in the lung of cystic fibrosis patients, even in children (RUBIO 1990; KUBIN 1993; SCHAAD 1994). Ciprofloxacin and ofloxacin represent an important additional therapy for *P. aeruginosa* infections (HOIBY et al. 1993). Patients may be treated at home, follow their usual social activities and avoid intravenous treatment. The disease itself is not curable but the fluoroquinolones improve the condition of the patients (BOSSO 1989).

IV. Sinusitis

The anti-infective regimen in the empirical treatment of sinusitis should have a spectrum of activity against the common causative organisms *H. influenzae*, *S. pneumonia*, and *Moraxella catarrhalis*. In a small study by KLEIN et al. (1993) with acute sinusitis patients 500mg ciprofloxacin b.i.d. were compared with 250mg cefuroxime axetil two times daily for 10–14 days. Ciprofloxacin treatment achieved 92% clinical cure or improvement whereas cefuroxime axetil cured or improved 74% of the patients. Bacterial eradication was 100% versus 74% in favor of ciprofloxacin. Because of the small number of patients, no statistical analysis was performed but the trend favors ciprofloxacin treatment.

A double-blind comparison of ciprofloxacin and amoxicillin/clavulanic acid in the treatment of chronic sinusitis was reported by LEGENT et al. (1994a). They treated 118 patients with 500mg oral ciprofloxacin b.i.d. for 9 days and compared this treatment with 500mg amoxicillin/clavulanic acid three times a day also for 9 days. The study design was double-blind, double-placebo-controlled, prospective, and multicenter. Patients at inclusion had purulent or mucopurulent rhinorrhea. Causative organisms were chiefly *S. aureus* (n = 45), followed by *H. influenzae* (n = 35), *S. pneumoniae* (n = 32), and Enterobacteriaceae (n = 31). The clinical cure and bacterial eradication rates were not statistically different (approximately 90% clinical cure). But at day 40 of follow-up patients with a positive pretreatment culture in the ciprofloxacin group had a significantly higher cure rate than those treated with amoxicillin/clavulanic acid (83.3% versus 67.6%; $p = 0.043$). These results suggest that ciprofloxacin may be a useful therapeutic alternative in this indication.

V. Bacterial Otitis (Chronic Suppurative Otitis Media)

Ciprofloxacin has been studied in both types of disease. In an open, noncontrolled trial FOMBEUR et al. (1994) treated 76 patients with either 500 mg ciprofloxacin b.i.d. orally or 750 mg ciprofloxacin b.i.d. orally for a period of 9 days. An overall analysis of cure (disappearance of otorrhea) revealed higher rates (70%) with the 750-mg dose than with the 500-mg dose.

LEGENT et al. (1994b) performed a controlled, prospective study of oral ciprofloxacin versus amoxicllin/clavulanic acid in the treatment of chronic suppurative otitis media in 76 adults. Five-hundred milligrams of ciprofloxacin b.i.d. was compared with 500 mg amoxicillin/clavulanic acid three times daily, for 9 days. Significantly better clinical results (57.5% cure rate with ciprofloxacin versus 37.1% with amoxicillin/clavulanic acid) and bacteriological results (69.7% versus 27.3%) were obtained with ciprofloxacin. It has to be mentioned that amoxicillin/clavulanic acid is in clinical use for this indication empirically and in this controlled trial *P. aeruginosa* was the main pathogen in the study and all strains were resistant to amoxicillin/clavulanic acid. Ciprofloxacin appears to be a highly active agent and its clinical use is indicated especially if *P. aeruginosa* or Enterobacteriaceae are suspected.

VI. Malignant External Otitis

Malignant external otitis is a serious infection which is most frequently caused by *P. aeruginosa* and is difficult to treat. The disease affects mainly elderly patients; most of them are diabetics. GEHANNO (1994) recently reviewed 84

Table 13. Efficacy of ciprofloxacin monotherapy in patients with malignant external otitis from different studies with >1 patient (modified from GEHANNO 1994)

Study Reference	Ciprofloxacin dosage (mg/l)	Duration of treatment (days)	Duration of post-treatment follow-up (months)	Cases	Eradication (%)	Cure (%)
CHRYSANTHOPOULOS and BASSARIS (1990)	750 mg b.i.d., then 500 mg b.i.d. p.o.	18–50	12	15	100	100
GALANAKIS et al. (1990)	750 mg b.i.d. p.o.	90–172	12	13	100	100
HICKEY et al. (1989)	750 mg b.i.d. p.o.	63–70	5–9	2	100	100
MORRISON and BAILY (1988)	750 mg b.i.d. p.o.	180	6	2	100	100
MURPHY et al. (1993)	750 mg b.i.d. p.o.	60–100	9–12	3	100	100
JOACHIMS et al. (1988)	1500 mg o.d. p.o.	90	7–12	2	100	100
SADÉ et al. (1989)	750 mg b.i.d. or 750 mg t.i.d. p.o.	42–57	9–25	23	100	91.3

clinical case studies collected from 13 publications in which the quinolone ciprofloxacin was administered. Most patients received 750 mg b.i.d. orally for a mean duration of 3 months. This review confirms the high in vitro activity of ciprofloxacin against *P. aeruginosa* in the treatment of patients suffering from a life-threatening disease and proves it to be particularly useful antibiotic for the treatment of malignant external otitis (Table 13).

E. Osteomyelitis

It became evident in the early clinical research phase that oral quinolones may replace parenteral treatment for osteomyelitis; this should drastically reduce the costs of treating patients since they would not have to remain in hospital (SWEDISH STUDY GROUP 1988). In a study initially intended to be randomized and controlled and designed to compare injectable antibiotics with ciprofloxacin, neither the physicians nor the patients accepted this choice; instead, all wanted to try oral ciprofloxacin (NORRBY 1989). Therefore only very few studies are controlled. The available data indicate that 500–750 mg twice daily oral ciprofloxacin for at least 4 weeks is effective. Cure rates ranged from 62% to 76% (SWEDISH STUDY GROUP 1988; YAMAGUTI et al. 1993; MACGREGOR et al. 1990; HOOGKAMP-KORSTANJE et al. 1989; see Table 14). Surgical debridement of the bone combined with ciprofloxacin treatment (750 mg b.i.d. orally for 1–4 months) resulted in higher cure rates (95%; DAN et al. 1990) than either treatment alone. In one controlled trial oral ciprofloxacin was as effective as parenteral ceftazidime or nafcillin plus aminoglycoside. Resistant organisms were not a problem in this study (GENTRY and RODRIGUEZ 1990). These results confirm those obtained by DAN et al. (1990) and underline the importance of surgical intervention. For the vast majority of chronic osteomyelitis patients oral ciprofloxacin offers an attractive alternative to parenteral therapy. In a recent controlled study performed by GENTRY et al. (1994) equally good results were obtained with lomefloxacin and ciprofloxacin. Oral quinolone therapy is strongly indicated for this infection.

Table 14. Bacteriological results[a] in patients with osteomyelitis treated with ciprofloxacin (modified from NORRBY 1989)

Bacterial species	Strains (*n*)	Bacteriological outcome (no. of strains)	
		Eradicated	Persisted[b]
Pseudomonas aeruginosa	85	65	20 (9)
Other gram-negatives	38	35	3 (1)
Gram-positives	15	11	4 (0)
Total	–	111	27 (10)

[a] Data from collated open studies.
[b] Numbers within parentheses represent persisting strains which developed resistance.

F. Skin and Skin Structure Infection

Ciprofloxacin has been investigated because of its broad antimicrobial spectrum in patients with surgical wound infections, infected ulcers, cellulitis, abscesses, and infected burns. FASS (1986) carried out the first explorative investigations and ciprofloxacin seemed to be particularly useful as a substitute for parenteral agents (Table 15). The results of these uncontrolled studies have been substantiated by a large multicenter study comparing 750mg oral ciprofloxacin b.i.d. with 2g intravenous cefotaxime every 8h. The most common pathogens were *S. aureus*, *P. mirabilis*, *E. coli*, *P. aeruginosa*, and *K. pneumoniae*. Evaluable were 432 patients, 217 from the ciprofloxacin group

Table 15. Clinical trials[a] with ciprofloxacin (modified from GENTRY 1991)

Dosage	(*n*)	Success	*S. aureus*	*Enterobacter* spp.	*P. aeruginosa*	Superinfection
500–750 mg p.o., two to three times daily 250 mg i.v., two to three times daily	358	274 (77%)	(83% 48% MRSA)	(100%)	(72%)	(9%)

MRSA, multiresistant *S. aureus*.
[a] A total of 20 noncomparative studies.

Table 16. Response to therapy with oral ciprofloxacin versus parenteral cefotaxime (modified from GENTRY 1989)

	Ciprofloxacin		Cefotaxime	
	(*n*)	(%)	(*n*)	(%)
Clinical response				
Total sites of infection	230		246	
Resolution	187	81	180	74
Improvement	37	16	48	20
Failure	6	3	14	6
Indeterminate	0	0	4	
Bacteriological response				
Total sites of infection	230		246	
Eradication	200	87	205	83
Reduction	16	7	9	3
Recurrence	2	1	4	2
Persistence	11	5	26	11
Indeterminate	1	1	2	
Overall response				
Total patients	217		215	
Completely successful	164	76	162	75
Partially successful	48	22	36	17
Failure	5	2	17	8
Indeterminate	0	0	0	0

and 215 from the cefotaxime group (Table 16). The results presented demonstrate that the clinical use of oral ciprofloxacin is as safe and effective as parenteral cefotaxime. Similar cure rates were reported by PARISH et al. (1988) who evaluated data from 795 patients. Subsequent studies consistently confirmed the good results. Sequential i.v./po administration of ciprofloxacin and ceftazidime (GENTRY et al. 1989) proved to be equally effective. Two studies (McCARTY 1993; MONROE 1993) yielded comparable results for 500 mg ciprofloxacin b.i.d. and 400 mg lomefloxacin od, but compared to the above-mentioned ciprofloxacin studies, the infections were only mild to moderate in nature in both of these studies.

These studies substantiate the efficacy of the quinolones in the treatment of skin structure infections. Initial concerns about these agents (see Sect. A) and their efficacy outside the urinary and gastrointestinal tract have been refuted by the results of a substantial number of clinical trials.

G. Sexually Transmitted Diseases

I. Gonorrhea

Since the start of clinical investigations with fluoroquinolones in the treatment of acute gonorrhoea single doses usually produced 100% cure rate. ECHOLS and HEYD (1995) reported on about 815 patients with 910 infected sites treated with a single 250 mg dose of ciprofloxacin for uncomplicated gonorrhoea. Eradication was achieved in 563 (100%) male urethral, 101 (100%) female cervical, 101 (90%) male and female rectal, and 47 (96%) male and female pharyngeal infections (Table 17).

Compared to the standard regiments, ciprofloxacin was as effective as ceftriaxone, ampicillin, amoxicillin plus probenecid, and spectinomycin. No

Table 17. Bacteriological eradication at various sites in patients treated with a single 250-mg dose of ciprofloxacin for the management of uncomplicated gonorrhea (from ECHOLS et al. 1994)

Site	Number of infections eradicated (%)			Lower limit of a one-sided CI95 (%)
	United States studies	International studies	Total	
Urethra, male	168 (100)	395 (100)	563 (100)	99.4
Cervix/urethra, female	93 (100)	106 (100)	199 (100)	98.2
Total, urogenital			762 (100)	
Rectum, male or female	26 (96.3)	75 (100)	101 (99)	95.0
Pharynx, male or female	11 (100)	36 (94.7)	47 (95.9)	87.0
Total, extragenital			148 (98)	
Total eradication	298 (99.7)	612 (99.7)	910 (99.7)	

CI95, 95% confidence interval.

differences could be detected when compared with other quinolones such as pefloxacin (LEOW et al. 1995) and sparfloxacin (MOI and JEAN 1994). The clinical use of quinolones, especially ciprofloxacin, for the treatment of acute gonorrhea, regardless of localization, has been documented and is indicated.

II. Chancroid

Clinical cure rates of chancroid for ciprofloxacin range from 92% to 100% with a single dose treatment of 500mg orally (BALLARD et al. 1994; BODHIDATTA et al. 1988; TRAISUPA et al. 1988). This disease is a major problem in many developing countries; therefore, the treatment has to be inexpensive and easy to administer. Enoxacin has also been found to be effective, but requires a longer treatment course (NAAMARA et al. 1987). A comparison between ciprofloxacin and ofloxacin (ciprofloxacin: 500mg o.d.; ofloxacin: 400mg o.d. and 400mg b.i.d. for 3 days) by BALLARD et al. (1994) showed that a single daily dose ciprofloxacin was as effective as a multiple dose of ofloxacin; however, statistically there was no difference. The clinical use of fluoroquinolones is indicated in the treatment of chancroid.

III. Nongonococcal Urethritis

Ciprofloxacin is generally not as effective in the treatment of urethritis caused by *Chlamydia trachomatis* or *Ureaplasma urealyticans* as it is in the treatment of gonorrhea and chancroid (FONG et al. 1987). A study by HOOTON et al. (1990) confirmed the ineffectiveness of ciprofloxacin in men even at high doses of 2g per day for 7 days compared with 100mg doxycycline b.i.d., in which case no relapses were observed. There is some evidence that ofloxacin is the most active quinolone available.

H. Intra-abdominal Infections

I. Anaerobic Intra-abdominal Infections

Since most of the intra-abdominal infections are caused by aerobic and anaerobic bacteria a combination therapy with an antianaerobic agent is indicated. Ciprofloxacin with metronidazole was compared with amoxicillin/clavulanic acid. The clinical response of these two treatments was similar (96% versus 90%) in patients with established intra-abdominal infections (YOSHIOKA et al. 1991). Recently, the results of a prospective, randomized, double-blind study comparing ciprofloxacin/metronidazole with imipenem/cilastatin for the treatment of complicated intra-abdominal infections have been reported by SOLOMKIN and REINHART (1995). Patients were randomized to receive ciprofloxacin/metronidazole i.v. or imipenem/cilastatin i.v. or ciprofloxacin/metronidazole i.v. followed by oral treatment with ciprofloxacin plus metronidazole. Some 671 patients were available for intend to treat analysis. The overall success rates for each group of patients in the intend to treat analysis were 82% for the ciprofloxacin/metronidazole arm, 82% for the imipenem/cilastatin arm, and 84% for the ciprofloxacin/metronidazole i.v./oral arm. The results demonstrate statistical equivalence between both of the i.v. groups and indicate that a change to oral therapy guided by the APACHE II score is as effective.

II. Cholangitis

A randomized, controlled study in 100 patients with acute suppurative cholangitis compared 200 mg ciprofloxacin i.v. b.i.d. with a triple therapy comprising ceftazidime (1 g b.i.d. i.v.), ampicillin (500 mg q.i.d.), and metronidazole (500 mg i.v. t.d.s.). The clinical response was 85% with ciprofloxacin versus 77% with the triple therapy. The mean duration of fever and length of hospital stay were both significantly shorter in the ciprofloxacin group. In a randomized multinational study comparing ciprofloxacin twice daily with ofloxacin once daily, both administered intravenously for at least 3 days followed by oral administration Bassaris et al. (1995), both drugs were equally efficient in curing biliary tract infections (100% versus 97.7% cure rates) in 91 patients. All the studies show good results for quinolones in clinical use in intra-abdominal infections.

III. Peritonitis in Chronic Ambulatory Dialysis Patients

Studies with ciprofloxacin in chronic ambulatory peritoneal dialysis (CAPD) patients were mostly performed using intraperitoneally administered ciprofloxacin at a dose of 25 or 50 mg/l for each bag of dialysate for 5–7 days (Ludlam et al. 1990; Dryden et al. 1991; Perez et al. 1993). Cure rates for gram-negative infections were higher (95%) than for gram-positive infections (67%). Ciprofloxacin at a dose of 25 mg/l given intraperitoneally for 5 days is as effective as higher doses. It is a safe and simple regimen for the treatment of CAPD peritonitis, and on the basis of reports in the literature, the intraperitoneal use of ciprofloxacin should be preferred over oral treatment.

IV. Gynecological Infections

Like in other intra-abdominal infections, anaerobic cover should also be provided in the antimicrobial treatment of gynecological infections. Fischbach et al. (1994) studied ciprofloxacin plus metronidazole versus cefoxitin plus doxycycline in women suffering from pelvic inflammatory disease (PID). According to their evaluation criteria, 97% of the ciprofloxacin/metronidazole group was successfully treated whereas the other group achieved 87% cure. Heinonen et al. (1989) underline on the basis of their results (ciprofloxacin i.v./po versus doxycycline plus metronidazole in PID with sussess rates of 94% for ciprofloxacin compared with 70% of the combination) that ciprofloxacin is a promising new alternative for the treatment of acute PID. Noteworthy is the good activity of ciprofloxacin against female chlamydial infections which is in contrast to the findings in men.

I. Bacteremia and Sepsis

Since the most common sources of sepsis are pulmonary, urinary, biliary, and gastrointestinal, most of the published results pertaining to quinolone efficacy in the treatment of sepsis are covered by the different sections on infections in this article. Bouza and the Spanish Group for the Study of Ciprofloxacin (1989) and BASSARIS et al. (1995) reported consistently good results for the treatment of bacteremia. Initial intravenous treatment with 200 mg b.i.d. and 400 mg b.i.d. ciprofloxacin and 400 mg o.d. to 400 mg b.i.d. for ofloxacin is recommended for the treatment of gram-negative bacteremia. A study by GANGJI et al. (1989) compared the parenteral regimen (200 mg b.i.d.) with the sequential i.v./oral regimen (200 mg b.i.d. followed by 750 mg orally)

Table 18. Outcome of treatment of bacteremic patients according to the type and clinical severity of infection (modified from SKOUTELIS et al. 1994)

Type and severity of infection	Patients (*n*)	Outcome of treatment			
		Successful		Unsuccessful	
		(*n*)	(%)	(*n*)	(%)
Biliary tract	63	61	97	2	3
Good	47	47		0	
Fair	14	14		0	
Critical	2	0		2	
Urinary tract	17	17	100	0	
Good	12	12		0	
Fair	4	4		0	
Critical	1	1		0	
Pneumonia	8	7	87	1	13
Good	3	3		0	
Fair	4	4		0	
Critical	1	0		1	
Typhoid fever	4	4	100	0	
Good	4	4		0	
Fair	0	0		0	
Critical	0	0		0	
Soft tissue abscess	2	1	50	1	50
Good	2	1		1	
Fair	0	0		0	
Critical	0	0		0	
Overall	94	90	95	4	4
Good	68	67		0	
Fair	22	22		0	
Critical	4	1		3	

and found equal efficacy in gram-negative bacillary septicemia in a nongranulocytopenic adult population. In an open series of 94 gram-negative bacteremic patients (Table 18) SKOUTELIS et al. (1994) confirmed these results; treatment was successful in 95% of the patients. The i.v. treatment period was 3–10 days (mean 4.5 days), the duration of oral treatment ranged from 3 to 12 days (mean 4.9 days).

In a recently published study by CABALLERO et al. (1995), it was reported that the sequential administration of 400 mg ciprofloxacin b.i.d. i.v. followed by 500 mg oral ciprofloxacin b.i.d. was more effective than 1 g cefotaxime i.v. (or i.m.) every 6 h – 95% versus 87% – in the treatment of 234 sepsis patients. The causative gram-negative organisms in these patients were *E. coli* (63), *Pseudomonas* spp. (11), *Enterobacter* spp. (11), *Proteus* spp. (8), *Klebsiella* spp. (5), *Acinetobacter* spp. (3), and others (7). Gram-positive pathogens were *S. aureus* (31), other staphylococcal spp. (43), *S. pneumoniae* (35), other streptococcal spp. (15), and others (5).

J. Surgical Prophylaxis

Numerous studies with ciprofloxacin and other quinolones have been carried out in different areas of surgical prophylaxis.

I. Transurethral Prostatic Surgery

GRABE et al. (1987) carried out the first study using ciprofloxacin in a controlled trial in which a comparison was made with an untreated group and concluded that oral ciprofloxacin is a valuable alternative antimicrobial for use in conjunction with transurethral resection. GOMBERT et al. (1989) and COX (1989a) compared intravenous ciprofloxacin (300 mg) with intravenous cefotaxime (1 g) and found similar results for both antibiotics.

II. Endoscopic Retrograde Cholangiopancreatography

In two studies (MEHAL et al. 1993; ALWEYN et al. 1991), the usefulness of oral ciprofloxacin was demonstrated for the prevention of sepsis following endoscopic retrograde cholangiopancreatography procedures.

III. Abdominal Surgery

KUJATH (1989) found in his study with 200 mg intravenous ciprofloxacin versus 2 g intravenous ceftriaxone equal efficacy as did GÖRTZ et al. (1990) in a double-blind study comparing ciprofloxacin plus metronidazole with moxalactam all given intravenously to patients undergoing colorectal surgery. Considering the clinical usefulness of quinolones in surgical prophylaxis, the oral formulations should be preferable because of their similar efficacy to intravenously administered ciprofloxacin, more convenient administration, and because of the cost savings.

K. Infections in Neutropenic Patients

I. Empirical Treatment of Febrile Neutropenic Patients

The effectiveness of quinolones in the empirical treatment of febrile neutropenic patients has not yet been established. A definitive study is still lacking and the use of quinolone as monotherapy does not currently appear to be a good approach for treatment (MEUNIER et al. 1991, The European Organization for Research on Treatment of Cancer (EORTC) International Antimicrobial Therapy Cooperative Group). VERHAGEN et al. (1993), in the framework of an EORTC study, found an equal efficacy with ceftazidime and ciprofloxacin in similar populations. In contrast, ciprofloxacin was clearly less effective in the treatment of documented gram-positive infections than ceftazidime (one cure and seven failures vs two cures and one failure). The clinical response with ciprofloxacin even when used in combination with other drugs (PHILPOTT-HOWARD et al. 1990; FLAHERTY et al. 1989; HYATT et al. 1991; CHAN et al. 1989; KELSEY et al. 1990a,b; JOHNSON et al. 1992; WIMPERIS et al. 1991) is below the levels reported by the EORTC. Further studies are required with an appropriate dose of ciprofloxacin, e.g., 400 mg three times daily, intravenously, and with the appropriate combination partner which offers cover for gram-positive bacteria.

II. Prophylaxis in Neutropenic Cancer Patients

The clinical experience with the prophylactic use of quinolones is increasing and has become standard practice in several centers (DEL FAVERO and MENICHETTI 1993). Prophylaxis with norfloxacin and ofloxacin reduced the number of gram-negative infections in comparison with placebo, vancomycin plus polymixin, and TMP/SMX. Similarly, ciprofloxacin was more effective than placebo (LEW et al. 1991) in bone marrow transplantation patients in preventing gram-negative infections. The efficacy of ciprofloxacin was as good as TMP/SMX plus colistin in patients with acute leukemia (DEKKER et al. 1987; ARNING et al. 1990; LEW et al. 1995). A study performed by DONELLY et al. (1992) yielded results in favor of TMP/SMX with respect to infective complications, fever duration and occurrence of fever. Studies comparing quinolones carried out by the Gimema Infection Program Group (1991; Gruppo Italiano Malattie Ematologiche Maligne dell' Adulto) and D'ANTONIO et al. (1994) found ciprofloxacin to be more effective than ofloxacin, pefloxacin, and norfloxacin in preventing gram-negative infections.

It was concluded by DEL FAVERO and MENICHETTI (1993) that enough evidence exists to suggest that the new quinolones are the best choice for prevention of gram-negative infections in neutropenic cancer patients; however, their use should be closely monitored for the emergence of resistance.

L. Use of Quinolones in Pediatrics

Since quinolones cause serious arthropathy in growing animals, the use of these drugs has been restricted to life-threatening or otherwise untreatable cases (KUBIN 1993). A relatively sound data base exists for ciprofloxacin. More than 1500 children were treated and in no instance were toxic effects such as those observed in animals detected. An international expert group came to a consensus for the use of fluoroquinolones in pediatrics after a review of results and internal discussions (SCHAAD and SALAM 1995). This group carefully indicated a possible use in bacterial exacerbation of cystic fibrosis, complicated urinary tract infections, chronic suppurative otitis media with the involvement of *P. aeruginosa* infections in immunocompromised children, neonatal infections, e.g., with multiresistant organisms, serious gastrointestinal infections, and the prevention of meningitis due to *Neisseria meningitidis*. The experts emphasize that they do not promote the use of quinolones in children. Their use should be based on a sound benefit and risk evaluation on an individual basis. Clearly, the quinolones should not be administered in conditions for which other effective antimicrobials with a more established safety profile are available.

M. Remarks

Extensive clinical experience especially with ciprofloxacin but also with norfloxacin and ofloxacin together with a rich body of postmarketing data have confirmed the good tolerability and efficacy of these quinolones. There are still open questions, e.g., concerning the usefulness of ofloxacin and ciprofloxacin in mycobacterial infections. The development of resistance has to be closely monitored and also appropriate dosing has to be defined for special conditions. The future quinolones will also have to undergo rigorous clinical testing. The concept of once-daily dosing is reasonable for aminoglycosides, but is questionable for quinolones, at least in serious infections.

References

Ahmad F, Bray G, Presscott RWG et al (1991) Use of ciprofloxacin to control a salmonella outbreak in a long-stay psychiatric hospital. J Hosp Infect 17:171–178

Alam MN, Haq SA, Mazid MN, Siddique RU, Hasan Z, Khan MAS, Dutta P (1993) Clinical efficacy of ciprofloxacin in multi-drug resistant enteric fever. Clin Res 41:3

Alweyn GG, Robertson DAF, Wright R et al (1991) Prevention of sepsis following endoscopic retrograde cholangiopancreatography. J Hosp Infect 19(Suppl C):65–70

Arning M, Wolf HH, Aul C et al (1990) Infection prophylaxis in neutropenic patients with a acute leukaemia – a randomized, comparative study with ofloxacin, ciprofloxacin and cotrimoxazole/colistin. J Antimicrob Chemother 26(Suppl D):137–142

Ball P, Tillotson G, Wilson R (1995) Chemotherapy for chronic bronchitis: controversies. Presse Med 24:189–194

Ballard RC, Radebe F, Mampuru M et al (1994) A comparison of ofloxacin and ciprofloxacin in the treatment of chancroid (Abstr). Sex Transm Dis 21(Suppl):135

Basran GS, Joseph J, Abbas AMA (1990) Treatment of acute exacerbations of chronic obstructive airways disease – a comparison of amoxicillin and ciprofloxacin. J Antimicrob Chemother 26(Suppl F):19–24

Bassaris H, Akalin E, Calangu S et al (1995) A randomized, multinational study with sequential therapy comparing ciprofloxacin twice daily with ofloxacin once daily. Infection 23:277–233

Bayston KF, Want S, Cohen J (1989) A prospective, randomized comparison of ceftazidime and ciprofloxacin as initial empiric therapy in neutropenic patients with fever. Am J Med(Suppl 5A):269–273

Bender SW, Posselt HG, Wönne R, Stöver B, Strehl R, Shah PM (1984) Oral ciprofloxacin for Pseudomonas bronchopneumonia in CF. 9th International Cystic Fibrosis Congress, Brighton

Bennet-Jones DN, Russel GI, Barrett A (1990) A comparison between oral ciprofloxacin and intraperitioneal vancomycin and gentamycin in the treatment of CAPD peritonitis. J Antimicrob Chemother 26(Suppl F):73–76

Bennish ML, Salam MA, Haider R (1990) Therapy for shigellosis. II. Randomized, double-blind comparison of ciprofloxacin and ampicillin. J Infect Dis 162:711–716

Bennish ML, Salam MA, Khan WA (1992) Treatment of shigellosis. III. Comparison of one- or two-dose ciprofloxacin with standard 5-day therapy. A randomized, blinded trial. Ann Intern Med 117:727–734

Bodhidatta L, Taylor DN, Chitwarakorn A et al (1988) Evaluation of 500- and 1000-mg doses of ciprofloxacin for the treatment of chancroid. Antimicrob Agents Chemother 32:723–725

Bosso JA (1989) Use of ciprofloxacin in cystic fibrosis patients. Am J Med(Suppl 5A):123–127

Bouza E, Diaz-Lopea MD, de Bernaldo QJCL (1989) Ciprofloxacin in patients with bacteremic infections. Am J Med 87(Suppl 5A):228–231

Brown AE, Smith G (1989) Treatment of sepsis in patients with neoplastic diseases with intravenous ciprofloxacin. Am J Med 87(Suppl 5A):266–268

Byl B, Kaufman L, Jacobs F et al (1993) Ciprofloxacin vs comparative antibiotics in LRTI. A meta-analysis (Abstr). Drugs 45(Suppl 3):428–429

Caballero M, Bonora V, Calabuig JR et al (1995) Clinical and microbiological comparison of sequential IV/PO ciprofloxacin vs parenteral cefotaxime in patients with sepsis (Abstr 1123). Fifth International Symposium on New Quinolones. 25–27 August 1994, Singapore

Catena E, Cesana M (1995) Rulfloxacin in AE of CBSG. 5-Day rufloxacin treatment versus 7-day ciprofloxacin treatment in patients with acute exacerbation of chronic bronchitis: a randomized study. Clin Drug Invest 9:334–343

Chan CC, Oppenheim BA, Anderson H et al (1989) Randomized trial comparing ciprofloxacin plus netilmicin versus piperacillin plus netilmicin for empiric treatment of fever in neutropenic patients. Antimicrob Agents Chemother 33:87–91

Cheesbrough JS, Ilunga Mwema F, Green SE et al (1991) Quinolones in children with invasive salmonellosis. Lancet 338:127

Chew SK, Monteiro EHA, Lim YS (1992) A 7-day course of ciprofloxacin for enteric fever. J Infect 25:267–271

Chodosh S (1991a) Use of quinolones for the treatment of acute exacerbations of chronic bronchitis. Am J Med 91:93S–100S

Chodosh S (1991b) Temafloxacin compared with ciprofloxacin in mild to moderate lower respiratory tract infections in ambulatory patients. A multicentre, double-blind, randomized study. Chest 100:1497–1502

Chrysanthopoulos CJ, Bassaris HP (1989) Use of oral ciprofloxacin in communmity-acquired pneumonia. J Chemother 1:103–106

Chrysanthopoulos CJ, Bassaris HP (1990) Treatment of malignant otitis with oral ciprofloxacin (Abstr 366). 3rd International Symposium on New Quinolones, Vancouver

Cordon F, Auquer F, Gorina E et al (1995) Comparison of ciprofloxacin single dose vs. 3 day norfloxacin in non-complicated urinary tract infections in female patients (Abstr). Seventh European Congress of Clinical Microbiology and Infectious Diseases, p 261

Cox CE (1989a) Comparison of intravenous ciprofloxacin and intravenous cefotaxime for antimicrobial prophylaxis in transurethral surgery. Am J Med 87(Suppl 5A):252–254

Cox CE (1989b) Brief report: sequential intravenous and oral ciprofloxacin versus intravenous ceftazidime in the treatment of complicated urinary tract infections. Am J Med 87(Suppl 5A):157–159

Cox CE (1992) A comparison of the safety and efficacy of lomefloxacin and ciprofloxacin in the treatment of complicated or recurrent urinary tract infections. Am J Med 92(Suppl 4A):82–86

D'Antonio D, Piccolomini R, Iacone A et al (1994) Comparison of ciprofloxacin, ofloxacin and pefloxacin for the prevention of the bacterial infection in neutropenic patients with haematological malignants. J Antimicrobial Chemother 33:837–844

Dan M, Siegman-Igra Y, Pitlik S et al (1990) Oral ciprofloxacin treatment of Pseudomonas aeruginosa osteomyelitis. Antimicrob Agents Chemother 34:849–852

Davies BI, Maesen FPV (1986) Quinolones in chest infections. Antimicrob Agents Chemother 18:296–299

Davies BI, Maesen FP, Baur C (1986) Ciprofloxacin in the treatment of acute exacerbations of chronic bronchitis. Eur J Clin Microbiol 5:226–231

Del Favero A, Menichetti F (1993) The new fluorinated quinolones for antimicrobial prophylaxis in neutropenic cancer patients. Eur J Cancer 29A(Suppl 1):2–6

Dekker AW, Rozenberg-Arska M, Verhoef J (1987) Infection prophylaxis in acute leukemia: a comparison of ciprofloxacin with trimethoprim – sulfamethoxazole and colistin. Ann Intern Med 106:7–11

Donelly JP, Maschmeyer G, Daenen S (1992) Selective oral antimicrobial prophylaxis for the prevention of infection in acute leukaemia – ciprofloxacin versus cotrimoxazole plus colistin. Eur J Cancer 29A:873–878

Dryden MS, Wing AJ, Phillips I (1991) Low dose intraperitoneal ciprofloxacin for the treatment of peritonitis in patients receiving continuous ambulatory peritoneal dialysis (CAPD). J Antimicrob Chemother 28:131–139

Dupont HL (1993a) Quinolones in Salmonella typhi infections. Drugs 45[Suppl 3]:119–124

Dupont HL (1993b) Traveller's diarrhoea. Which antimicrobial? Drugs 45:910–917

Dutta P, Rasaily R, Saha MR et al (1993) Ciprofloxacin for treatment of severe typhoid fever in children. Antimicrob Agents Chemother 37:1197–1199

Dyson C, Ribeiro CD, Westmoreland D (1995) Large scale use of ciprofloxacin in the control of a salmonella outbreak in a hospital for the mentally handicapped. J Hosp Infect 29:287–296

Echols RM, Heyd A (1995) Efficacy and safety of ciprofloxacin for chronic bacterial prostatitis. Infect Med 12(6):283–290

Echols RM, Heyd A, O'Keeffe BJ (1994) Single-dose ciprofloxacin for the treatment of uncomplicated gonorrhea: a wordwide summary. Sex Transm Dis 21:345–352

Ernst JA, Sy ER, Colon LH et al (1986) Ciprofloxacin in the treatment of pneumonia. Antimicrob Agents Chemother 29:1088–1089

Esposito S, Galante D, Bianchi W et al (1987) Efficacy and safety of oral ciprofloxacin in the treatment of respiratory tract infections associated with chronic hepatitis. Am J Med 82(Suppl 4A):211–214

Fass RJ (1986) Treatment of skin and soft tissue infections with oral ciprofloxacin. J Antimicrob Chemother 18(Suppl D):153–157

Fass RJ (1987) Efficacy and safety of oral ciprofloxacin in the treatment of serious respiratory infections. Am J Med 82(Suppl 4A):202–207

Ferreccio C, Morris JR Jr, Valdivieso C et al (1988) Efficacy of ciprofloxacin in the treatment of chronic typhoid carriers. J Infect Dis 157:1235–1239

Fink MP, Snydman DR, Niederman MS et al (1994) Treatment of severe pneumonia in hospitalized patients: results of a multicenter, randomized, double-blind trial comparing intravenous ciprofloxacin with imipenem-cilastatin. Antimicrob Agents Chemother 38:547–557

Fischbach F, Deckardt R, Graeff H (1994) Ciprofloxacin/metronidazole vs cefoxitin/doxycycline: comparison of two antibiotic regimens in the treatment of PID (in German). Geburtshilfe Frauenheilkd 54:337–340

Flagerty JP, Waily D, Edlin B et al (1989) Multicenter, randomized trial of ciprofloxacin plus azlocillin versus ceftazidime plus amikacin for empiric treatment of febrile neutropenic patients. Am J Med 20(Suppl 5A):278–282

Fombeur JP, Barrault C, Koubbi G et al (1994) Study of the efficacy and safety of ciprofloxacin in the treatment of chronic otitis. Chemotherapy 40(Suppl 1):29–34

Fong IW, Linton W, Simbul M et al (1987) Treatment of non-gonococcal urethritis with ciprofloxacin. Am J Med 82(Suppl 4A):311–316

Galanakis N et al (1990) Ciprofloxacin in the treatment of external malignant otitis (Abstr 365). 3rd International Symposium on New Quinolones

Gangji D, Jacobs F, De JJ et al (1989) Brief report: randomized study of intravenous versus sequential intravenous/oral regimen of ciprofloxacin in the treatment of gram-negative septicemia. Am J Med 87(Suppl 5A):206–208

Gehanno P (1994) Ciprofloxacin in the treatment of malignant external otitis. Chemotherapy 40(Suppl 1):35–40

Gentry LO (1991) Review of quinolones in the treatment of infections of the skin and skin structure. J Antimicrob Chemother 28(Suppl C):97–110

Gentry LO, Koshdel A (1989) Intravenous/oral ciprofloxacin versus intravenous ceftazidime in the treatmnt of serious gram-negative infections of the skin and skin structure. Am J Med 87(Suppl 5A):132–135

Gentry LO, Rodriguez GG (1990) Oral ciprofloxacin compared with parenteral antibiotics in the treatment of osteomyelitis. Antimicrob Agents Chemother 34:40–43

Gentry LO, Ramirez RE, Rodriquez NE et al (1989) Oral ciprofloxacin vs parenteral cefo-taxime in the treatment of difficult skin and skin structure infections. Arch Intern Med 149:2579–2583

Gentry LO, Rodriguez-Gomez G, Roniker B (1994) The safety and efficacy of lomefloxacin and ciprofloxacin in the treatment of acute and chronic osteomyelitis (Abstr). 5th International Symposium on New Quinolones, p 158

Giamarellou H, Galanakis N (1987) Use of intravenous ciprofloxacin in difficult-to-treat infections. Am J Med 82(Suppl 4A):346–351

Gimema Infection Program (1991) Prevention of bacterial infection in neutropenic patients with hematologic malignancies. A randomized, multicenter trial comparing norfloxacin with ciprofloxacin. Ann Intern Med 115:7–12

Gleadhill IC, Ferguson WP, Lowry RC (1986) Efficacy and safety of ciprofloxacin in patients with respiratory tract infections in comparison with amoxicillin. J Antimicrob Chemoth 18(Suppl D):133–138

Gombert ME, DuBouchet L, Aulicino TM et al (1989) Brief report: intravenous ciprofloxacin versus cefotaxime during transurethral surgery. Am J Med 87(Suppl 5A):250–251

Goodman LJ, Trenholme GM, Kaplan RL et al (1990) Empiric antimicrobial therapy of domestically acquired acute diarrhea in urban adults. Arch Intern Med 150:541–546

Goodman Obot E, DuPont HL (1993) Current approach to the treatment of traveller's diarrhoea. Int J Antimicrob Agents 3:97–103

Görtz G, Boese-Landgraf J, Hopfenmüller W et al (1990) Ciprofloxacin as single-dose anti-biotic prophylaxis in colorectal surgery. Results of a randomized, double-blind trail. Diagn Microbiol Infect Dis 13:181–185

Gotuzzo E, Seas C, Echevarría J et al (1995) Ciprofloxacin for the treatment of cholera: a randomized, double-blind, controlled clinical trial of a single daily dose in Peruvian adults. Clin Infect Dis 20:1485–1490

Grabe M, Forsgren A, Björk T et al (1987) Controlled trial of a short and a prolonged course with ciprofloxacin in patients undergoing transurethral prostatic surgery. Eur J Clin Microbiol 6:11–17

Grossman RF, Beaupre A, LaForge J (1994) A prospective randomized parallel single-blind comparison of oral ciprofloxacin with oral cotrimoxazole in the treatment of respiratory tract infections in patients with chronic obstructive lung disease. Drug Invest 8:110–117

Grubbs NC, Schultz HJ, Henry NK et al (1992) Ciprofloxacin versus trimethoprim – sufamethoxazole: treatment of community-acquired urinary tract infections in a prospective controlled double-blind comparison. Mayo Clin Proc 67:1163–1168

Guibert J (1995) Acute cystitis in women: effectiveness of ciprofloxacin versus pefloxacin given as a single oral dose (in French). Presse Med 24:304–308

Heinonen PK, Teisala K, Miettinen A et al (1989) A comparison of ciprofloxacin with doxy-cycline plus metronidazole in the treatment of acute pelvic inflammatory disease. Scand J Infect Dis suppl 60:66–73

Hickey SA et al (1989) Treating malignant otitis with oral ciprofloxacin. Br Med J 299:550–551

Hoiby N, Pedersen SS, Jensen T et al (1993) Fluoroquinolones in the treatment of cystic fibrosis. Drugs 45(Suppl 3):98 101

Hoogkamp-Korstanje JA, van Bottenburg HA, van Bruggen J et al (1989) Treatment of chronic osteomyelitis with ciprofloxacin. J Antimicrob Chemother 23:427–432

Hooton TM, Rogers ME, Medina TG et al (1990) Ciprofloxacin compared with doxycycline for non-gonococcal urethritis. Ineffectiveness against Chlamydia trachomatis due to relapsing infection. JAMA 264:1418–1421

Hyatt DS, Rogers TRF, McCarthy DM (1991) A randomized trial of ciprofloxacin plus azlocillin versus netilmicin plus azlocillin for the empirical treatment of fever in neutropenic patients. J Antimicrob Chemother 28:324–326

Iravani A (1993) Multicenter study of single-dose and multiple-dose fleroxacin versus ciprofloxacin in the treatment of uncomplicated urinary tract infections. Am J Med 94(Suppl 3A):89–96

Iravani A, Tice AD, McCarthy J (1995) Short-course ciprofloxacin treatment of acute uncomplicated urinary tract infection in women: the minimum effective dose. Arch Intern Med 155:485–494

Johnson PRE, Liu YJA, Tooth JA (1992) A randomized trial of high-dose ciprofloxacin versus azlocillin and netilmicin in the empirical therapy of febrile neutropenic patients. J Antimicrob Chemother 30:203–214

Karachalios GN, Georgiopoulos AN, Nasopoulou-Papadimitriou DD (1991) Value of single-dose ciprofloxacin in the treatment of acute uncomplicated urinary tract infection in women. Drugs Exp Clin Res 17(10–11):521–524

Kelsey SM, Collins PW, Delord C et al (1990a) A randomized study of teicoplanin plus ciprofloxacin versus gentamicin plus piperacillin for the empirical treatment of fever in neutropenic patients. Br J Haematol 76(Suppl 2):10–13

Kelsey SM, Wood ME, Shaw E (1990b) A comparative study of intravenous ciprofloxacin and benzyl-penicillin versus netilmicin and piperacillin for the

empirical treatment in neutropenic patients. J Antimicrob Chemother 25:149–157
Khajotia RR, Vetter N, Harazin H (1991) Ofloxacin in the treatment of lower respiratory tract infections: report of a prospective, comparative trial. Clin Ther 13(4):460–466
Khan FA (1990) The role of quinolones in respiratory tract infections. Int J Clin Pract 6(Suppl 1):57–66
Khan FA, Basir R (1989) Sequential intravenous–oral administration of ciprofloxacin vs ceftazidime in serious bacterial respiratory tract infections. Chest 96:528–537
Khan FA, Basir R, Afzal Q (1989) Controlled, comparative study of low-dose ciprofloxacin vs ampicillin for respiratory tract infections. Rev Infect Dis 11(Suppl 5):23
Khan WA, Seas C, Khan EH et al (1994) Randomized, double-blinded comparison of single dose ciprofloxacin with single dose doxycycline in the treatment of Vibrio cholerae O139 (Bengal) infection (Abstr). 34th Interscience Conference on Antimicrobial Agents and Chemotherapy, p 245
Khan WA, Begum M, Salam MA et al (1995) Comparative trial of five antimicrobial compounds in the treatment of cholera in adults. Trans R Soc Trop Med Hyg 89:103–106
Klein GL, Heyd A, Echols R (1993) Oral ciprofloxacin vs cefuroxime axetil in acute bacterial sinusitis. A pilot study in adults (Abstr). Drugs 45(Suppl 3):324
Kobayashi H (1987) Clinical efficacy of ciprofloxacin in the treatment of patients with respiratory tract infections in Japan. Am J Med 82(Suppl 4A):169–173
Kubin R (1993) Safety and efficacy of ciprofloxacin in paediatric patients – review. Infection 21:413–421
Kujath P (1989) Brief report: antibiotic prophylaxis in biliary tract surgery. Ciprofloxacin versus ceftriaxone. Am J Med 87(Suppl 5A):255–257
Legent F, Bordure P, Beauvillain C et al (1994a) A double-blind comparison of ciprofloxacin and amoxycillin/clavulanic acid in the treatment of chronic sinusitis. Chemotherapy 40(Suppl 1):8–15
Legent F, Bordure P, Beauvillain C et al (1994b) Controlled prospective study of oral ciprofloxacin versus amoxycillin/clavulanic acid in chronic suppurative otitis media in adults. Chemotherapy 40(Suppl 1):16–23
Leigh DA (1992) The treatment of a large outbreack of acute bacterial gastroenteritis with ciprofloxacin. J Antimicrob Chemother 30:733–735
Leow YH, Chan RKW, Cheong LL et al (1995) Comparing the efficacy of pefloxacin and ciprofloxacin in the treatment of acute uncomplicated gonococcal urethritis in male. Ann Acad Med Singapore 24:515–518
Lew MA, Kehoe K, Ritz J et al (1991) Prophylaxis of bacterial infections with ciprofloxacin in patients undergoing bone marrow transplantation. Transplantation 51:630–636
Lew MA, Kehoe K, Ritz J et al (1995) Ciprofloxacin versus trimethoprime/sulfamethoxazole for prophylaxis of bacterial infections in bone marrow transplant recipients: a randomized, controlled trial. J Clin Oncol 13:239–250
Lightfoot NF, Ahmad F, Cowden J (1990) Management of institutional outbreaks of salmonella gastroenteritis. J Antimicrob Chemother 26(Suppl F):37–46
Lim SH, Smith MP, Goldstone AH et al (1990) A randomized prospective study of ceftazidime and ciprofloxacin with or without teicoplanin as an empiric antibiotic regimen for febrile neutropenic patients. Br J Haematol 76(Suppl 2):41–44
Limson BM, Littaua RT (1989) Comparative study of ciprofloxacin versus cotrimoxazole in the treatment of Salmonella enteric fever (Letter). Infection 17:105–106
Lode H, Wiley R, Höffken G et al (1987) Prospective randomized controlled study of ciprofloxacin versus imipenem–cilastatin in severe clinical infections. Antimicrob Agents Chemother 31:1491–1496
Ludlam HA, Barton I, White L et al (1990) Intraperitoneal ciprofloxacin for the treatment of peritonitis in patients receiving coninuous ambulatory peritoneal dialysis. J Antimicrob Chemother 25:843–851

MacGregor RR, Graziani AL, Esterhai JL (1990) Oral ciprofloxacin for osteomlitis. Orthopedics 13:55–60
Mandal BK (1991) Modern treatment of typhoid fever. J Infect 22:1–4
Marklein G, German Fleroxacin Study Group (1995) Fleroxacin vs ciprofloxacin in the empirical treatment of severe pneumonia: iv/oral switch regimen (Abstr). 7th European Congress of Clinical Microbiology and Infectious Diseases, p 139
McCarty J (1993) Lomefloxacin vs ciprofloxacin in the treatment of non-necrotising skin and skin structure infections (Abstr). Drugs 45(Suppl 3):387
Mehal WZ, Culshaw KD, Tillotson GS et al (1993) Antibiotic prophylaxis for ERCP: a randomized clinical trial of ciprofloxacin versus cefuroxime (Abstr). Gastroenterology 104(Suppl):A370
Meunier F, Zinner SH, Gaya H et al (1991) Prospective randomized evaluation of ciprofloxacin versus piperacillin plus amikacin for empiric antibiotic therapy of febrile granulocytopenic cancer patients with lymphomas and solid tumors. Antimicrob Agents Chemother 35:873–878
Moi H, Jean C (1994) Comparative efficacy and safety of sparfloxacin versus ciprofloxacin in the treatment of acute gonococcal urethritis in males: a multicentre, international, double-blind randomized study of 238 patients (Abstr). 5th International Symposium on New Quinolones, p 144
Monroe A (1993) Lomefloxacin vs ciprofloxacin in the treatment of non-necrotising skin and skin structure infections (Abstr). Drugs 45(Suppl 3):390
Morelli G, Mazzoli S, Tortoli E et al. (1992) Fluoroquinolones vs chloramphenicol in the therapy of typhoid fever: a clinical and microbiological study. Curr Ther Res 52:532–542
Moss PJ, Read RC (1995) Empiric antibiotic therapy for acute infective diarrhoea in the developed world. J Antimicrob Chemother 35:903–913
Murphy GS, Bodhidatta L, Echeverria P et al (1993) Ciprofloxacin and loperamide in the treatment of bacillary dysentery. Ann Intern Med 118:582–586
Naamara W, Plummer FA, Greenblatt RM et al (1987) Treatment of chancroid with ciprofloxacin: a prospective randomized clinical trial. Am J Med 82:317 320
Naber KG (1989) Use of quinolones in urinary tract infections and prostatitis. Rev Infect Dis 11(Suppl 5):1321–1337
Naber KG (1992) Lomefloxacin versus norfloxacin versus ciprofloxacin in the treatment of complicated urinary tract infections. Int J Antimicrob Agents 2:33–37
Naber KG, Bartosik-Wich B (1986) Ciprofloxacin versus norfloxacin in the treatment of complicated urinary tract infections: in vitro activity, serum and urine concentrations, safety and therapeutic efficacy. In: Neu HC, Weuta H (eds) 1st International Ciprofloxacin Workshop Excerpta Medica, Amsterdam, pp 314–317
Neill MA, Opal SM Heelan J et al (1991) Failure of ciprofloxacin to eradicate convalescent fecal excretion after acute salmonellosis: experience during an outbreak in health care workers. Ann Intern Med 114:195–199
Norrby SR (1989) Ciprofloxacin in the treatment of acute and chronic osteomyelitis: a review, Scand J Infect Dis suppl 60:74–78
Nye KJ, Gibson SP, Nwosu AC et al (1993) Single-dose intraperitoneal vancomycin and oral ciprofloxacin for the treatment of peritonitis in CAPD patients: preliminary report. Perit Dial Int 13(1):59–60
Obot EG, DuPont HL (1993) Curent approach to the treatment of traveller's diarrhoea. Int J Antimicriob Agents 3:97–103
Parish LC, Witkowski JA, Jungkind DL et al (1988) Treatment of cutaneous infections. Worldwide experience with ciprofloxacin. Int J Dermatol 27:131–133
Pérez FM, Rosales M, Fernandez F et al (1993) Treatment of CAPD peritonitis with ciprofloxacin – long-term results (in Spanish). Nefrologia 13(2):149–154
Philpott-Howard JN, Barker KF, Wade JJ et al (1990) Randomized multicentre study of ciprofloxacin and azlocillin versus gentamicin and azlocillin in the treatment of febrile neutro-penic patients. J Antimicrob Chemother 26(Suppl F):89–99
Pichler HE, Diridl G, Seiberl G et al (1990) Ciprofloxacin in the treatment of gastrointestinal infection. Int J Clin Pract 6(Suppl 1):37–43

Quenzer RW, Davis RL, Neidhart MM (1990) Prospective randomized study comparing the efficacy and safety of ciprofloxacin with cefaclor in the treatment of patients with purulent bronchitis. Diagn Microbiol Infect Dis 13:143–148

Rademaker CMA, Hoepelman IM, Wolfhagen MJHM et al (1989) Results of a double blind placebo-controlled study using ciprofloxacin for prevention of traveller's diarrhea. Eur J Clin Microbiol 8:690–694

Ramirez CA, Bran JL, Mejia CR et al (1985) Open, prospective study of the clinical efficacy of ciprofloxacin. Antimicrob Agents Chemother 28:128–132

Raz R, Rottensterich E, Hefter H (1989) Single-dose ciprofloxacin in the treatment of uncomplicated urinary tract infection in women. Eur J Clin Microbiol 8:1040–1042

Rozenberg-Arska M, Dekker A, Verdonck L et al (1989) Prevention of bacteremia caused by alpha-hemolytic streptococci by roxithromycin (RU 28 965) in granulocytopenic patients receiving ciprofloxacin. Infection 17:240–244

Rubenstein EB, Rolston K, Benjamin RS et al (1993) Outpatient treatment of febrile episodes in low-risk neutopenic patients with cancer. Cancer 71:3640–3646

Rubio TT (1990) Ciprofloxacin in the treatment of Pseudomonas infection in children with cystic fibrosis. Diagn Microbiol Infect Dis 13:153–155

Sadé J, Lang R, Goshen S et al (1989) Ciprofloxacin treatment of malignant external otitis. Am J Med 87(Suppl 5A):138–141

Salam I, Katelaris P, Leigh-Smith S (1994) Randomized trial of single-does ciprofloxacin for traveller's diarrhoea. Lancet 344:1537–1539

Sánchez C, García-Restoy E, Garau J et al (1993) Ciprofloxacin and trimethoprim – sulfamethoxazole versus placebo in acute uncomplicated Salmonella enteritis: a double-blind trial. J Infect Dis 168:1304–1307

Schaad UB (1994) Use of the new quinolones in pediatrics. Isr J Med Sci 30:463–468

Schaad UB, Salam MA (1995) Use of fluoroquinolones in pediatrics: consensus report of an international society of chemotherapy commission. Pediatr Infect Dis J 14: 1–9

Schaad UB, Wedgewood-Krucko J, Guenin K (1989) Antipseudomonas therapy in cystic fibrosis: aztreonam and amikacin versus ceftazidime and amikacin administered intravenously followed by oral ciprofloxacin. Eur J Clin Microbiol 8:858–865

Schacht P, Arcieri G, Branolte J et al (1988) Worldwide clinical data on efficacy and safety of ciprofloxacin. Infection 16(Suppl 1):29–44

Schacht P, Arcieri G, Hullmann R (1989) Safety of oral ciprofloxacin. Am J Med 87(Suppl 5A):98–103

Schaeffer AJ, Anderson RU (1992) Efficacy and tolerability of norfloxacin vs ciprofloxacin in complicated urinary tract infection. Urology 40:446–449

Schönwald S et al (1990) (Abstr 318). 3rd International Symposium on New Quinolones July 12–14, Vancouver

Schulte JM, Schmid GP (1993) Recommendations for treatment of chancroid. Clin Infect Dis 20(Suppl 1):39–46

Shah PM (1985) Clinical experience with quinolones-overview. Quinolones Bulletin, vol I:19–25

Singh CP, Singh N, Brar GK et al (1993) Efficacy of ciprofloxacin and norfloxacin in multidrug resistant enteric fever in adults. J Indian Med Assoc 91:156–157

Skoutelis A, Chrysanthopoulos C, Starakis J et al (1994) Treatment of gram-negative aerobic bacteremia with sequential intravenous to roal ciprofloxacin. Infect Dis Clin Pract 3:352–256

Solomkin J, Reinhart H (1995) Results of a trial of IV+/–PO ciprofloxacin/ metronidazole vs imipenem for complicated intraabdominal infections (Abstr). 7th European Congress of Clinical Microbiology and Infectious Diseases, pp 218–219

Stamm WE, Hooton T (1993) Management of urinary tract infections in adults. N Engl J Med 329:1328–1334

Strandvik B, Hjelte L, Lindblad A et al (1989) Comparison of efficacy and tolerance of intravenously and orally administered ciprofloxacin in cystic fibrosis patients with acute exacerbation of lung infection. Scand J Infect Dis Suppl 60:84–88

Sung JJY, Lyon DJ, Suen R et al (1995) Intravenous ciprofloxacin as treatment for patients with acute suppurative cholangitis: a randomized, controlled clinical trial. J Antimicrob Chemother 35:855–864

Swedish Study Group (1988) Therapy of acute and chronic Gram-negative osteomylitis with ciprofloxacin. J Antimicrob Chemother 22:221–228

Szaff M, Hoiby N, Flensborg EW (1983) Frequent antibiotic therapy improves survival of cystic fibrosis patients with chronic Pseudomonas aeruginosa infection. Acta Paediatr Scand 72:651–657

Takkar VP, Kumar R, Khurana S et al (1994) Comparison of ciprofloxacin versus cephelexin and gentamicin in the treatment of multi-drug resistant typhoid fever. Indian Pediatr 31:200–201

Taylor DN, Sanchez JL, Candler W et al (1991) Treatment of traveller's diarrhea: ciprofloxacin plus loperamide compared with ciprofloxacin alone. A placebo-controlled, randomized trial. Ann Intern Med 114:731–734

Thys JP, Jacobs F, Byl B (1991) Role of quinolones in the treatment of bronchopulmonary infections, particularly pneumococcal and community-acquired pneumonia. Eur J Clin Microbiol Infect Dis 10:304–315

Traisupa A, Wongba C, Tesavibul P (1988) Efficacy and safety of a single dose therapy of a 500 mg ciprofloxacin tablet in chancroid patients. Infection 16(Suppl 1):S44–S45

Unertl KE, Lenhart FP, Forst H et al (1989) Brief report; ciprofloxacin in the treatment of legionellosis in critically ill patients including those cases unresponsive to erythromycin. Am J Med 87(Suppl 5A):128–131

Verhagen C, DE Pauw B, Donelly J et al (1993) Ciprofloxacin versus ceftazidime monotherapy in the empirical treatment of febrile neutropenic patients (Abstr). 33rd Interscience Conference on Antimicrobial Agents and Chemotherapy, p 308

Vetter N, Veist H, Orlicek M, Optual I, Weuta H (1985) RTI patients treated with ciprofloxacin and cefalexin; a randomized comparison. International Chemotherapy Congress, Kyoto, pp 38–95

Wallace MR, Yousif AA, Mahroos GA et al (1993) Ciprofloxacin versus ceftriaxone in the treatment of multiresistant typhoid fever. Eur J Clin Microbiol Infect Dis 12:907–910

Weidner W, Schiefer HG (1991) Chronic bacterial prostatitis: therapeutic experience with ciprofloxacin. Infection 19(Suppl 3):165–166

Wilcox MH, Spencer RC (1992) Quinolones and salmonella gastroenteritis. J Antimicrob Chemother 30:221–228

Wimperis JZ, Baglin TP, Marcus RE et al (1991) An assessment of the efficacy of antimicrobial prophylaxis in bone marrow autografts. Bone Marrow Transplant 8:363–367

Wiström J, Norrby R (1990) Antibiotic prophylaxis of traveller's diarrhoea. Scand J Infect Dis 22(Suppl 70):111–129

Wolfson JS, Hooper DC (1989) Treatment of genitourinary infections with fluoroquinolones: activity in vitro, pharmacokinetics, and clinical efficacy in urinary tract infections and prostatitis. Antimicrob Agents Chemother:1655–1661

Wollschlager CM, Raoof S, Khan FA et al (1987) Controlled comparative study of ciprofloxacin versus ampicillin in treatment of bacterial respiratory tract infections. Am J Med 82(Suppl 4A):164–168

Yamaguti A, Trevisanello C, Lobo IMF et al (1993) Oral ciprofloxacin for treatment of chronic osteomyelitis. Int J Clin Pharmacol Res 13(2):75–79

Yoshioka K, Youngs DJ, Keighly MRB (1991) A randomized prospective controlled study of ciprofloxacin with metronidazole versus amoxicillin/clavulanic acid with metronidazole in the treatment of intraabdominal infection. Infection 19:25–29

CHAPTER 14
Future Aspects

S. SEGEV and E. RUBINSTEIN

A. Introduction

The introduction of the new fluoroquinolones into medical practice some 15 years ago raised great expectations. The fluoroquinolones were regarded as being as close as possible to the "ideal" antimicrobial agent since they possess a broad spectrum of antibacterial activity, induce a low frequency of spontaneous single-step mutations, and exhibit excellent oral absorption and good tissue distribution, achieving remarkable interstitial fluid levels, adequate penetration into macrophages and other phagocytic cells and excellent urinary concentrations.

Animal experiments and a decade of clinical experience raised doubts about the efficacy of the new fluorquinolones in certain clinical indications. Emergence of resistance, particularly among troublesome bacteria, primarily staphylococci and *Pseudomonas aeruginosa*, arthropathy in specific animal species (particularly dog puppies), a handful of serious side effects in humans and unfavorably drug interactions have been detected throughout these years and are described in detail in other chapters of this book. These drawbacks have led to considerable efforts to convert the basically "good compound" into an "ideal" antimicrobial.

The future of the fluoroquinolones will be determined both by the skill of basic scientists and by the prudence of physicians in their use of these valuable compounds. The pharmaceutical industry will doubtlessly direct its efforts toward the synthesis of the "right" product – will that be a target-oriented antibiotic or a broad spectrum antimicrobial drug? Will the new compounds display their in vitro characteristics in clinical setting? Will bacterial resistance in the troublesome species develop less frequently than with the current fluoroquinolone agents? Will the safety of the new products be increased? Will infected newborns be able to enjoy the advantages of these new fluoroquinolones?

For the discussion here, we have selected some aspects of the future of the fluoroquinolones, aspects that seem intriguing and of great importance.

B. Molecular Structure and Mechanism of Action

The molecular structure of the new quinolones is discussed in detail in Chap. 26 of this volume and in other publications (DOMAGALA 1994; HOOPER 1993; HOOPER and WOLFSON 1993a,b).

Fluoroquinolones currently in clinical use are all structurally similar and have an N-1-substituted 1,4-dihydro-4-oxo-pyridine-3-carboxylic acid moiety as the basic nucleus. Chemists have been able to synthesize more than 10000 modifications of the quinolone molecule, modified primarily at the N-1 position and at the C-6, C-7, and C-8 positions (DOMAGALA 1994; NEU 1992; MITSCHER et al. 1993). The accumulated knowledge of the advantageous structure–activity relationships has produced several prominent compounds. The most significant advantage was obtained by substitution of an increased steric bulk at R_7. This bulk, which is usually achieved through alkylation, apart from reducing side effects, improved potency against gram-positive species and improved in vivo efficacy as manifested by the fluoroquinolone PD 138312 (HAGEN et al. 1991), DU 6859a (CHEVALIER 1992), and other newer agents.

Addition of a piperazine group at position C-7 increased activity also against pseudomonads (DOMAGALA 1994). Activity against staphylococci is also achieved by the addition of a fluorine atom at position C-8 (BOUZARD et al. 1989). Adding a cyclopropyl group at position N-1, an amino group at position C-5 and a fluorine group at C-8 resulted in an increased activity against mycoplasma and chlamydia. Improvement of the pharmacokinetics was achieved by the addition of a second fluorine group at position C-8. Absorption was increased and a longer half-life was obtained by substitution of a methyl group for the piperazine group (DOMAGALA 1994). Despite all attempts to correlate the chemical structure of the new fluoroquinolones with activity and/or side-effects, it seems likely that the optimum fluoroquinolone agent cannot be theoretically designed but has to be identified experimentally.

Structure–side effect relationship was even better defined than structure–activity relationship. Some adverse events have no associated structural feature in the fluoroquinolone molecule and are not expected to be modified by chemical changes (DOMAGALA 1994). These include gastrointestinal discomfort, which is the most common side effect; gastrointestinal symptoms might reflect antibacterial activity and transepithelium secretion of fluoroquinolones by the intestinal mucosa. By improving the antibacterial activity, an increase in gastrointestinal disturbances might be anticipated, in addition to arthropathy (GOUGH et al. 1992) and interaction with Ca^{2+}, Mg^{2+}, and Fe^{3+} (LOMAETRO and BAILIE 1991; SHENTAG and NIX 1990). Other adverse events can be modified by chemical alteration, e.g., the increased steric bulk at R_7 is responsible for the favorable effect on CNS side effects, drug interaction with NSAID and theophyllin, and genotoxic risk. Crystalluria can be reduced by increasing the water solubility of the fluoroquinolone molecule; this can be achieved by alkyl substitution of the R_7 substituent and with CF, CCl, CCF or COM as the X_8 substituent (ROSEN et al. 1988).

Skin rashes can be due to allergic reactions, photosensitivity or histamine release phenomena. The latter could be related to the R_7 group, but evidence for this relationship is still lacking (REMUZON et al. 1992). Phototoxicity in animals was reduced in the fluoroquinolone Q-35 (MATSUBARA et al. 1993; MARUTANI et al. 1993; MATSUMOTO et al. 1993) and was related to the X_8 substituent halogen and to alkylated R_7 groups as well as to bulky R_5 side chains. Coincidentally those substituents contributed also to the efficacy and pharmacokinetics of some other derivatives such as fleroxacin, lomefloxacin, CI-938, clinafloxacin, and PD 138312 (DOMAGALA 1994; PATON and REEVES 1991). One can hope that these modifications will not be given up for the sake of the associated, negligible side effects. Central nervous system side effects are the second most common adverse events associated with fluoroquinolones and include headache, insomnia, agitation, and on rare occasions, convulsions. These phenomena have been associated with the binding of fluoroquinolones to the γ-aminobutyric acid ($GABA_A$) receptors; other receptors are also involved. GABA interactions achieved by high fluoroquinolone concentrations are not physiological: the mechanism of action is probably due to the blocking of the natural ligand GABA, thereby stimulating the CNS (HALLIWELL et al. 1993; SEGEV et al. 1988). Under those conditions interaction with dopaminergic opioid and other receptors in the brain are possible. It appears that the bulky side chain in R_7 exhibits less binding to GABA receptors (DOMAGALA 1994). Animal studies showed that convulsion induction was reduced by AM-1155 (SEGEV et al. 1988; HORI et al. 1993) and SYN987 (HORI and SHIMADA 1993). Obviously, penetration into the brain is also important and depends on a high lipophilicity of the compound. Fleroxacin was believed to be associated with a paucity of CNS side effects because of its low lipophilicity; unfortunately the opposite was observed (CHU and FERNANDES 1991). Thus, this field still requires more conclusive data. As stated by MITSCHER et al. (1993):

Although treating each atom of the quinolone nucleus individually and attempting to define the optimum substituent at each center implies that the centers all contribute independently and that the optimum quinolone ("utopiafloxacin") will emerge by a cut-and-paste assembly of the best individual substituents at each center into a single composite molecule, there is ample and increasing evidence that such is not always the case and that the best collection of substituents needs to be found experimentally by tinkering collectively with all of the variables.

Clearly, new fluoroquinolones with improved safety and potency can be designed following the relationships described (DOMAGALA 1994). The case of temafloxacin is a an exceptional example for an unexpected side effect (hemolysis, renal failure, and thrombocytopenia) discovered after the approval of the agent for clinical use (BLUM et al. 1994). The structure of temafloxacin is similar to that of tosufloxacin in having a difluorophenyl substituent at the 1-position; tosufloxacin, however, has been marketed in Japan with no comparable adverse events.

While it is possible that yet other side effects might eventually appear, it is equally likely that extensive knowledge of structure–toxicity relationships will permit medicinal chemists to design new agents that minimize these effects.

Such structure–toxicity relationships force the evolution of quinolones toward a maximum ratio of benefit to risk.

Thus, the quinolone nucleus provides opportunities for future chemical modifications that may lead to:

- More potent derivatives
- Derivatives with specific antibacterial activity
- Derivatives which cause less frequent induction of resistance in the troublesome species
- Improved water solubility for better oral absorption and injectable formulations
- Better central nervous system penetration
- Decreased drug interaction potential
- Better patient safety and tolerance
- Agents which may be administered to newborns and to pregnant women

The newer agents already possess some of these advantages, and one can foresee their future specific applications. Let us assume, for example, that a compound has a high in vitro activity against gram-positive bacteria (as many of the newer agents do). What additional characteristics will be required for the following conditions?

1. Skin and soft tissue infections – for treating abscesses, antibacterial activity that will not be hampered by acidic pH and anaerobic conditions.
2. CNS infections – additional high lipophilicity.
3. Respiratory tract infections – additional high bronchial concentrations. Antimycoplasma activity will extend the efficacy.
4. Prophylaxis for orthopedic or vascular procedures – better tissue penetration with a long half-life and activity against adherent bacteria with their peculiarities (e.g., slime production and decreased respiration).

C. Antimicrobial Activity

We selected organisms for examination which are considered to be relatively resistant to the current fluoroquinolones – gram-positive cocci, anaerobes and *Mycobacterium* species. There are many more organisms of interest that are discussed in other reviews and are not less important, such as *Chlamydia* spp. (KIMURA et al. 1993; ARAI et al. 1992; HAMMERSCHLAG et al. 1992), *Rickettsia* spp. (RAOULT and DRANCOURT 1991; RUIZ-BELTRAN and HERRERO HERRERS 1992), and *Brucella* spp. (LANG and RUBINSTEIN 1989; AL-SIBAI et al. 1992), which seem to be quite susceptible to the newer quinolones (although clinical failure was reported in brucellosis).

New indications such as malaria, nocardiosis, Lyme disease, toxoplasmosis, and pneumocystosis should create a challenge to the chemists in the synthesis of new compounds. These new indications are beyond the scope of this review; it may even be too early for speculation on the use of fluoroquinolones in the treatment of these illnesses.

I. Gram-Positive Bacteria

A few years ago NEU and CHIN (1987) stated that "it is conceivable that agents with markedly superior activity against gram-positive species will be synthesized, but one must analyze whether such agents are truly needed." At that time, resistance of gram-positive bacteria to fluoroquinolones did not pose a major problem. Streptococci were originally not sensitive to the fluoroquinolones as opposed to the majority of staphylococci, which were quite susceptible. About 6 years ago, a rapid development of resistance to fluoroquinolones of *Staphylococcus aureus* was reported worldwide. Much of this resistance developed when these agents were used in settings in which the underlying infection was difficult or impossible to eradicate with antimicrobial agents alone. Whether the fluoroquinolones will continue to have an important role in the treatment of patients infected with gram-positive bacteria will depend on the outcome of clinical trials with the newer fluoroquinolones which have been shown to have greater in vitro activities against these pathogens than currently available agents.

1. Staphylococci

Data from many investigations and from a recent surveillance study indicate that in the United States (PETERSON 1994) nearly all methicillin-resistant *S. aureus* (MRSA) are resistant to fluoroquinolones. However, data from acute care hospitals show that most resistant staphylococci belong to a single clonal type. It is probable that the number of resistant clones worldwide is limited and that horizontal spread of resistant strains is the more likely mechanism of resistance spread. Therefore, efforts to control the spread of fluoroquinolone resistance in MRSA should include not only the rational use of these agents but also infection control measures such as patient isolation and other techniques of hospital hygiene. The number of resistant staphylococci vary widely and depend on geography (country and institution), use of the drugs, and infection control. The widespread indiscriminate use of quinolones has led to high frequencies of resistance in some institutions.

Current efforts are directed toward the production of new molecules with different mechanisms of activity and far greater in vitro efficacy against ciprofloxacin-resistant staphylococci. Most newer fluoroquinolones are at least fourfold more active than ciprofloxacin against staphylococci, including methicillin-resistant strains, regardless of their susceptibility to unrelated drugs (PIDDOCK 1994; KORTEN et al. 1994; PIDDOCK et al. 1994; Table 1). All new agents described by PIDDOCK et al. (1994) inhibited 90% of staphylococci at a concentration of 0.25 μg/ml or less and their optimum bactericidal concentrations (OBC) were ten- to 20-fold higher than their MICs. Recent reports describe enhanced activity of Bay Y 3118, clinafloxacin, PD 131628, DU-6859A, and other derivatives against methicillin-sensitive *S. aureus* (MSSA) and MRSA (PETERSON 1994; PIDDOCK et al. 1994; KORTEN et al. 1994; FUCHS et al. 1991; HOSAKA et al. 1992; IMADA et al. 1992; WAKABAYASHI and MITSUHASI 1994).

Table 1. Activity of newer quinolones against *S. aureus*

Agent	MIC (μg/ml)	
	50%	90%
Ciprofloxacin	0.25–4.0	0.5–4.0 (>64)[a]
Ofloxacin	0.5–4.0 (16.0)[a]	0.5–4.0 (32)[a]
AM-1155	0.1–0.2	0.1–6.25 (6.25)[a]
BAY Y 3118	0.015 (0.25)[a]	0.03 (1.0)[a]
Clinafloxacin	≤0.008	0.03
CP-99,219	0.03 (0.061)	0.06–0.12
DU-6859a	0.03–0.064	0.06–0.125
DU-7751	n.a.	0.2 (6.25)[a]
Levofloxacin	0.4–1.0	2.0–8.0 (16)[a]
PD 131628	0.06	0.06–0.25
Sparfloxacin	0.06–0.25 (8)[a]	0.1–1.25 (32)[a]
WIN 57273	<0.008 (<0.008)	0.03

n.a., data not available; MIC, minimal inhibitory concentration.
[a] Methicillin resistant.

Recent animal experiments demonstrate high therapeutic activity of CI-960 against ciprofloxacin-susceptible and resistant *S. aureus* in the rabbit model of endocarditis (KAATZ et al. 1992). Temafloxacin was also very active in the rat endocarditis model and was more efficient than vancomycin against MRSA (HESSE et al. 1990).

2. Streptococci

Streptococci are less susceptible than staphylococci to the newer fluoroquinolones, but 90% of the strains examined so far are inhibited by 1 μg/ml or less of sparfloxacin, PD 131628, DU 6859a, clinafloxacin, AM-1155, WIN 57273, and some other newer compounds (PIDDOCK 1994; KORTEN et al. 1994; PIDDOCK et al. 1994; LOO et al. 1994). The benefit of fluoroquinolones in the treatment of pneumococcal infections is controversial. The MICs of ciprofloxacin against *Streptococcus pneumoniae* strains in the United States surveillance study were mostly between 0.5 and 1 μg/ml. The breakpoint is ≤1 μg/ml for susceptible strains; thus *S. pneumoniae* can be considered borderline susceptible.

The newer agents are so much more active than ciprofloxacin against pneumococci that clinicians may well have greater confidence in the use of these fluoroquinolones in the treatment of respiratory tract infections due to *S. pneumoniae* strains which are susceptible, intermediately resistant and resistant to penicillin.

Sparfloxacin, PD 127391, and particularly WIN 57273 were shown in a Canadian survey (LOO et al. 1994) to be significantly more active against *S.*

pneumoniae than either ciprofloxacin or ofloxacin (Table 2). These agents demonstrated in vitro activity superior to that of penicillin G against penicillin-resistant strains of *S. pneumoniae* (SPANGLER et al. 1992, 1993). Sparfloxacin demonstrated high efficacy in experimental meningitis caused by highly penicillin-resistant *S. pneumoniae* (DECAZES et al. 1994). Other reports showed activity of Bay y 3118 against *S. pneumoniae* that approaches the activity of penicillin G (PETERSON 1994; PIDDOCK et al. 1994; BAUERNFEIND 1993; ELIOPOULOS 1993).

The need for new agents against *S. pneumoniae* has increased due to the ever-growing resistance of this bacterium against penicillin (SPANGLER et al. 1992; MORI et al. 1993) and to a lesser degree against the macrolides (ALVAREZ et al. 1993). A relatively high rate of in vitro selection of resistance to macrolides was demonstrated, particularly in the case of clarithromycin (THORBURN et al. 1993). This indicates that any increased usage of macrolides will lead to a substantial increase in macrolide resistance among community-acquired pathogens.

So far there is no cross resistance in *S. pneumoniae* with either β-lactam agents or with the macrolides which makes the newer quinolones an attractive alternative for treatment of respiratory tract infections. There is already encouraging clinical experience with sparfloxacin treatment of respiratory infections including pneumonia caused by *S. pneumonia* (GIALDRONI-GRASSI and BRUMPT 1994; ALLEGRA and BRUMPT 1994). Earlier clinical studies with temafloxacin also showed that this drug is highly efficient against pneumococcal pneumonia (CARBON 1993; CARBON et al. 1992). It is thus conceivable that fluoroquinolones will be the therapy of choice in the future for community-acquired infections, in particular in areas in which penicillin-resistant pneumococci are common.

Table 2. Activity of newer quinolones against *Streptococcus pneumoniae*

Agent	MIC μg/ml	
	50%	90%
Ciprofloxacin	1.0–3.13	1.56–4.0
Ofloxacin	1.0–6.25	2.0–12.5
AM-1155	0.39	0.39
BAY Y 3118	0.015	0.03
Clinafloxacin	0.03–0.5	0.06
DU-6859a	0.06–0.03	0.06–0.125
OPC-17116	0.25	0.5
PD 131628	0.12	0.25–0.5
Sparfloxacin	0.2–0.25	0.25–0.78
WIN 57273	0.007	0.01–0.06

MIC, minimal inhibitory concentration.

3. Enterococci

Only 58% of *Enterococcus faecalis* strains were found to be susceptible in the United States surveillance study (PETERSON 1994). *Enterococcus faecium* was found to be even more resistant – only 30% of strains were sensitive to ciprofloxacin. With the currently available quinolones, monotherapy is therefore not recommended (PIDDOCK 1994; KORTEN et al. 1994; PIDDOCK et al. 1994; WAKABAYASHI and MITSUHASI 1994), while combined therapy might be of benefit in some cases (PETERSON 1994).

Only clinafloxacin inhibited 90% of *E. faecium* and *E. faecalis* at a concentration of 1μg/ml or less. However, at the OBC there was no more than one logarithmic decrease in viable bacterial count, suggesting that fluoroquinolones are only bacteriostatic against enterococci (PIDDOCK 1994).

In vitro studies of fluoroquinolones against 200 recent clinical isolates of enterococci (Table 3) revealed the following order of activity: WIN 57273 > sparfloxacin > ciprofloxacin (MURRAY et al. 1993). Fortunately the newer quinolones have identical in vitro activity against strains which have high and strains which have low resistance to gentamicin and more importantly against vancomycin-resistant strains of enterococci as well as against gentamicin- and vancomycin-sensitive strains (NEU and CHIN 1987).

There are no clinical data concerning the use of fluoroquinolones in the treatment of enterococcal endocarditis. The activity of ciprofloxacin was only marginal in the rat endocarditis model and improved when combined with penicillin (INGERMAN et al. 1987). In contrast, sparfloxacin and clinafloxacin alone in combination with gentamicin demonstrated high efficacy in experimental ampicillin-resistant enterococcal endocarditis in rabbits (VAZQUEZ et al. 1993).

Table 3. Activity of newer quinolones against *Enterococcus faecalis*

Agent	MIC μg/ml	
	50%	90%
Ciprofloxacin	0.5–0.1	1.0–4.0
Ofloxacin	0.5–2.0	1.0–4.0
AM-1155	n.a.	0.78
BAY Y 3118	0.12	0.5–1.0
Clinafloxacin	0.25	0.25
CP-99,219	1.0	2.0 (range, 0.12–2.0)
DU-6859a	0.12–0.25	0.39–1.0
PD 131628	n.a.	0.5
Sparfloxacin	0.25–0.5	0.78–2.0
WIN 57273	0.12	2.0 (range, 0.03–8.0)

n.a., data not available; MIC, minimal inhibitory concentration.

II. Anaerobic Bacteria

Some of the newer fluoroquinolones, such as sparfloxacin (COOPER et al. 1990), PD 131628 (COOPER et al. 1992), temafloxacin (FINEGOLD et al. 1991), AM-1155 (LOO et al. 1994), and BAY y 3118 (NORD et al. 1993) are active against anaerobes (Table 4). Other newer agents are no more active than ciprofloxacin or ofloxacin (ELIOPOULOS and ELIOPOULOS 1993). It is possible that anaerobic organisms are inherently less susceptible to fluoroquinolones because of a less susceptible DNA gyrase or because the pH of the environment in which these organisms grow adversely affects the pK_a of fluoroquinolones rendering them less active (AL-SIBAI et al. 1992). Some of the newer derivatives such as clinafloxacin, sparfloxacin, and WIN 57273 show in vitro activity against *Bacteroides fragilis*. However, there is a discrepancy between the in vivo and in vitro activity of these fluoroquinolones in anaerobic infections (THADEPALLI et al. 1993). Of the newly developed fluoroquinolones, clinafloxacin and PD 131628 (the bioactive form of CI-990) were more active in vitro than sparfloxacin and much more active than ciprofloxacin (BARRY et al. 1993). AM-1155 showed a broad antibacterial spectrum against gram-positive and gram-negative anaerobes (UENO et al. 1993), as did PD 131628 (NORD and HAGELBACK 1993; NORD 1992). WIN 57273 (GOLDSTEIN 1993) inhibited 95% of anaerobic strains tested at ≤2 μg/ml. Animal models of anaerobic infections are few, with no evidence of in vivo efficacy. More in vivo and clinical data is necessary to determine the efficacy of the newer quinolones in anaerobic infections.

The newer fluoroquinolones do not disrupt the fecal flora, despite their anaerobic activity, possibly because of their binding to fecal material (GOLDSTEIN 1993). Therefore there may well be a role for these agents in the

Table 4. Activity of newer quinolones against anaerobes

Agent	MIC μg/ml[a]			
	50%		90%	
	Bacteroides fragilis	*Peptostr.*	*Bacteroides fragilis*	*Peptostr.*
Ciprofloxacin	3.13	0.1–1.0	25–50	0.2–6.25
Ofloxacin	3.13	0.39	12.5–25	0.78–8.0
AM-1155	0.78	0.1	1.56–6.25	0.2
Clinafloxacin	0.5	0.03	2.0	0.5
DU-6859a	n.a.	0.1	n.a.	0.2
Sparfloxacin	0.78	0.05–1.0	1.0–6.25	0.39–4.0
WIN 57273	n.a.	n.a.	0.25–1.0	0.03–0.012
CP 99219	0.5	n.a.	2.0	

Peptostr., *Peptostreptococcus* spp; n.a., data not available; MIC, minimal inhibitory concentration.
[a] MICs of Q-35 ranged from 0.025 to 3.13 μg/ml for all anaerobes.

treatment and/or prophylaxis of anaerobic infections that occur outside the intestinal tract.

III. *Mycobacterium tuberculosis*

The fluoroquinolones demonstrate favorable intracellular pharmacokinetics, diffusing and accumulating in the phagocytes without association with the cellular organelles. Additionally, the new fluoroquinolones are active in vitro against some intracellular organisms other than *Mycobacteriae*; their efficacy has already been described in typhoid fever (DUPONT 1993), tularemia (SCHEEL et al. 1992; SYRJALA et al. 1991), Mediterranean spotted fever (DI-LASCIO et al. 1991), legionellosis (PECHERE 1993), and in leprosy (FRANZBLAU and WHITE 1990; PATTYN 1991; GELBER et al. 1992; CHAN et al. 1994). Similarly the new fluoroquinolones are potentially promising for use in the therapy of *M. tuberculosis* infection on the basis of in vitro activity and animal model data. Ciprofloxacin and ofloxacin have already been used in human infections caused by *M. tuberculosis* and *Mycobacterium avium-intracellulare* complexes, and they exhibit synergism with amikacin (GEVAUDAN et al. 1989), rifampin, and isoniazide (GEVAUDAN et al. 1989; CASAL et al. 1987, 1989). Multidrug regimens including fluoroquinolones have been found to be as effective as the commonly used regimens in mycobacterial infections (BERGSTERMANN et al. 1991).

The newer fluoroquinolones seem to have even a greater advantage in the treatment of mycobacterial infections (Table 5). The in vitro activity of sparfloxacin against mycobacteria growing within macrophages is quite striking, and in addition sparfloxacin was found to exhibit an early killing activity against *M. tuberculosis* (PERRONNE et al. 1991; MARTIN-LUENGO and SEMPERE 1993; RASTOGI and GOH 1991; RASTOGI et al. 1991). However, for quinolone-resistant *M. tuberculosis* strains sparfloxacin exhibited only a marginally better activity against these strains than ciprofloxacin or ofloxacin (ALANGADEN et al. 1994). Nevertheless, sparfloxacin seems to be effective against multidrug-

Table 5. Activity of newer quinolones against *M. tuberculosis* and *M. avium-intracellulare*

Agent	MIC μg/ml (range)	
	50%	90%
Ciprofloxacin	0.5–8.0[a]	0.5–16.0[a]
Ofloxacin	12.5–16[a] (0.78)[b]	1.0–32.0[a] (0.2)[b]
DU-6859a	1.56[a] (0.1)[b]	12.5[a] (0.2)[b]
Sparfloxacin	0.25–8[a] (0.1)[b]	4–16.0[a] (0.2)[b]

MIC, minimal inhibitory concentration.
[a] *M. avium-intracellulare*
[b] *M. tuberculosis*

resistant (MDR) clinical isolates of mycobacterium in a murine model (KLEMENS et al. 1993). It has been recommended to include sparfloxacin in a regimen of ultrashort tuberculosis therapy and in the treatment of MDR *M. tuberculosis*.

As for *M. avium* complex, current in vitro data (KLOPMAN et al. 1993; GARCIA-RODREIGUEZ and GOMEZ-GARCIA 1993) are insufficient to predict the clinical efficacy of the newer quinolones. It was reported that fluoroquinolones and ethambutol exhibited synergistic effect against *M. avium* (HOFFNER et al. 1989). Clinical studies in AIDS patients might clarify the effect of combined treatment of the new quinolones with clarithromycin and ethambutol against *M. avium* complex infections.

The following issues regarding the use of fluoroquinolones in mycobacterial diseases have arisen.

1. Possible role of the newer fluoroquinolones as a single drug treatment (in the case of resistance or intolerance to isoniazide). Sparfloxacin and levofloxacin seem to be effective in animals experiments (PETERSON 1994).
2. The potential emergence of resistance among mycobacterial populations exposed to extensive use of quinolones.
3. The newer fluoroquinolones as a potential alternative treatment for tuberculosis when other agents fail.
4. Inclusion of the fluoroquinolones in ultrashort treatments of tuberculosis. A mouse model evaluating this type of treatment using sparfloxacin seems to be promising (KLEMENS et al. 1993).

The quinolones seem to have in the future an important role in treatment of tuberculosis (BERGSTERMANN et al. 1991). Fluoroquinolones might be a part of a multidrug schedule for otherwise untreatable mycobacterial diseases, such as those caused by multiresistant isolates, in cases of failure or adverse events with previous chemotherapy and for infections in compartments which are difficult to reach, such as osteomyelitis.

It is most probably that the major use of fluoroquinolones, at least initially, will be in the treatment of tuberculosis caused by MDR *M. tuberculosis*.

D. Bacterial Resistance

The development and spread of resistance to fluoroquinolones has been discussed previously (HOOPER and WOLFSON 1993a; WIEDEMAN and HEISIG 1994; PETERSON 1994; LIMB et al. 1987). Resistance to this group of drugs has not been found on plasmids but is chromosomally mediated. However, from a theoretical point of view, the of a plasmid-encoded efficient efflux mechanism cannot be excluded. In single-step mutations leading to fluoroquinolone resistance in gram-negative bacteria or staphylococci, the highest level of resistance is always caused by an alteration of the DNA gyrase or, more precisely, its subunit A. In high-level resistant strains one mutation alone is generally not responsible for the levels of resistance observed. For example in ciprofloxacin resistant *S. aureus* increased efflux is often coupled with a gyrA mutation (NAKANISHI et al. 1991).

ACAR et al. (1993) reviewed the epidemiology of resistance in several countries and concluded that among common pathogens in the community, the incidence of resistance is low and stable in contrast to the fairly high incidence of resistance observed among nosocomial pathogens. This implies that the control of antibiotic usage in hospitals should reduce the frequency of resistance. Unnecessary antibiotic treatments should be avoided in circumstances where therapy provokes resistance and does not solve the problem. A recent report illustrated the consequences of ofloxacin treatment in patients with orthopedic infections. When infected prostheses were removed, a high rate of cure was obtained and when prosthesis remained in place, the rate of failure was much higher, with a high percentage of resistant staphylococci (PETERSON 1994; DRANCOURT et al. 1993).

It is to be assumed that the newer oral fluoroquinolones will gain wide usage in the community, particularly because of their efficacy in community-acquired lower and upper respiratory tract infections. Therefore, the spread of resistance seems unavoidable. In the early 1990s, it was forecasted that oral quinolones will be the dominant antibiotic class, displacing the cephalosporins as the market leader (GENESIS REPORT 1992). KUNIN (1985) made the following statement regarding antimicrobial agents: "These are precious drugs which represent one of the most important achievements in modern medicine. How well we use them is a strong indicator of how effective we are in fulfilling our responsibilities as physicians and scientists".

Will the newer fluoroquinolones induce a less frequent emergence of resistance? Should we use antibiotic combinations rather than single drugs and for which conditions? Our view is that better antimicrobial activity does not prevent the development of resistance; vigilance and good clinical judgment should be exercised in each case.

We share the forecast that fluoroquinolones will be used for many indications in the community, particularly the new agents with better anti-gram-positive and antichlamydia/mycoplasma activities.

In the future, the new fluoroquinolones will be administered once a day and will probably not cause too many serious adverse effects. Emergence of resistance and cross-resistance between the current agents and the newer quinolones is, however, a real threat for the future.

E. Future Use of Quinolones in Pediatrics

Arthropathies in young animals of several species (CHRIST and LEHNERT 1990; RIBARD and KAHN 1991) have led to important restrictions in the clinical use of fluoroquinolones in pregnant women, nursing mothers, children, and adolescents. Although nalidixic acid is known to cause extensive damage to cartilage in young animals, no such clinical adverse reactions have been reported in children despite frequent administration (CHRIST and LEHNERT 1990; CHYSKY et al. 1991; SCHAAD and WEDGEWOOD 1992). The clarification of arthropathy in children is expected to require epidemiological studies over a period of 20–40 years (CHRIST and ESCH 1994).

In addition to numerous children having been treated with fluoroquinolones, particularly ciprofloxacin, on a compassionate use basis, approximately 1000 prepubertal patients mainly with cystic fibrosis have been treated with ciprofloxacin for various periods of time (CHYSKY et al. 1991). There was no apparent evidence for drug-induced arthropathy. SCHAAD and WEDGEWOOD (1992) reported no articular histopathological findings at autopsy of two children with cystic fibrosis treated for 9–10 months with ciprofloxacin. SCHAAD (1993) concluded from his experience and that of others: "Magnetic resonance imaging and histopathological and clinical monitoring during ciprofloxacin use together with published experience of other groups suggest that quinolones do not cause arthropathy in humans." He suggested useful guidelines for using quinolones in pediatric patients. CHRIST and ESCH (1994) stressed, however, that therapeutic use of the quinolones in childhood can be justified only in carefully monitored trials on safety and efficacy in specific infections.

Since indisputable data on fluoroquinolone use in newborns and children will not be available in the near future, special attention to side effects ought to be paid, in particular to articular damage. We presume that the minimal age of treatment will continue to be gradually reduced; however, in the next few years physicians will remain conservative. Fluoroquinolones will probably be used even in newborns and pregnant women if no other alternative choice exists.

F. Antitumor Potential

Although it is generally accepted that quinolone are selective for bacterial DNA gyrase, there are in vitro experiments indicating some inhibition of eukaryotic replication due to activity against topoisomerase II (ANDERSSON and KIHLMAN 1989; GOOTZ and OSHERROFF 1993; GOOTZ et al. 1994). Some aspects of their structure–activity relationships have been elucidated (KOHLBERNNER et al. 1992). Thus, exciting potential for the fluoroquinolones may result from the development of new compounds with higher specific affinity for the topoisomerase of malignant human cells.

Congeners derived from two modified quinolone structures, those with a hydroxyphenyl group at position 7 of the quinolone nucleus (CP-67 804, CP-115 953) and those with an isothiazole ring bridging positions 2 and 3 (A-65281, A-65282) exhibited a strikingly enhanced activity against eukaryotic topoisomerase II (GOOTZ and OSHEROFF 1993). CP-67 804 and CP-115 953 exhibited cytotoxicity against wild-type Chinese hamster ovary cells and seem to be of special interest (ROBINSON et al. 1992).

BO-2367 strongly inhibited both mammalian and bacterial topoisomerase II (YOSHINARI et al. 1993). Its activity was more than twofold stronger than that of VP-16. Intraperitoneal injection of this compound increased the life span of CDF1 female mice bearing ascitic L1210 leukemia 2.4-fold, and subcutaneous injection completely inhibited the growth of colon 26 carcinoma implanted subcutaneously. These results suggest that B-2367, a representative of the fluoroquinolone family, is a potent antitumor agent.

In the development of new quinolones the antitumor activity should be the mirror-image of the drug safety profile. Clastogenicity (disruption of the chromosome) is expected in any quinolone to occur at concentrations 30–10000 times above antibacterial concentrations (GOOTZ and OSHEROFF 1993; CIARAVINO et al. 1993). Ciaravino reported a high capacity assay using micronuclei formation as a clastogenic endpoint. GRACHECK et al. (1991) found a very strong correlation between cytotoxicity and measure of clastogenic risk. The most important positions for determining potential clastogenicity are R_1, X_8, and R_7 (DOMAGALA 1994). Several new quinolones with extra rings that extend the aromatic nucleus are also highly cytotoxic and clastogenic (CHU 1992).

It is still too early to predict the possible oncological application of the new quinolones but it seems quite promising as the molecule can be engineered to produce almost any antieukaryotic, antimicrobial, or antiparasitic activity, with the limiting factor being the agent's toxicity and pharmacokinetics.

G. New Attitudes

Obviously, the aim of the pharmaceutical industry is to produce active and safe compounds with long half-lives. Apparently even the activity of the more advanced derivatives of the newer quinolones could be improved by the appropriate combination with other antibiotic agents. As the fluoroquinolones do not exhibit pronounced synergy with most antibacterials by conventional in vitro testing, several compounds consisting of a quinolone and a β-lactam physically linked by a chemical bond have been developed (HAMILTON 1994). These hybrids gave rise to a dual-action antibiotic of high potency. At least one of the compounds described, RO 23-9423 (the ester-linked fleroxacin/desacetylcefotaxime adduct), is reported to have proceeded to clinical trials (CORRAZ et al. 1992).

More stable compounds such as RO 24-4383 and RO 24-81383 are being developed (GEORGOPAPDAKOU and BERTASSO 1993; PFALLER et al. 1993). Their in vitro activity is not clearly superior to currently approved agents but there is hope that the combination can be given at a much higher dose with less toxicity than agents now being used.

The desacetylcephalothin–ciprofloxacin hybrid is more potent on a molar basis than the individual components, suggesting some synergism (DEMUTH et al. 1991). The molecular combination is sometimes more active than the mixture but more often somewhat less active. Thus the question of which alternative is more effective is not settled at the present time, and the clinical future of the cephalosporin–quinolone hybrids is not yet clear. HAMILTON (1994) wonders: "Will these 'quinocephs' or 'cephoquins' come to have a useful clinical role, or are they just gimmicks, diverting resources from the search for truly new chemical entities? Only time and experience will tell".

H. Directions of Future Research on the Quinolones

ANDRIOLE (1993) analyzed the practical approach of community physicians in prescribing oral quinolones based on the GENESIS REPORT (1992). They considered the quinolones to be as effective as other antibiotic agents against gram-positive bacteria and superior against gram-negative organisms. They favored the use of quinolones because they have broad spectrum activity, can be given once or twice a day, and are rarely associated with adverse events.

Andriole invites the scientific community to decide whether we would prefer or need more quinolones with broader antimicrobial activity or whether we should begin to develop new compounds designed for specific targets. Only time will tell which direction of development the drug industry will pursue. The medical community will contribute to the success of the new agents by wise and controlled utilization.

References

Acar JF, O'Bmen TF, Goldstein FW, Jones RN (1993) The epidemiology of bacterial resistance to quinolones. Drugs 45(S3):24–28

Alangaden GJ, Manavathu EK, Lerner SA (1994) Activity of ciprofloxacin (Cip), ofloxacin (Ofl), sparfloxacin (Spl) and five investigational fluoroquinolones against quinolone-resistant strains of mycobacterium tuberculosis (Mtb) (Abstr 29). 5th International Symposium on New Quinolones, Singapore, 25–27 August 1994

Allegra L, Brumpt I (1994) Sparfloxacin (S) in the management of acute exacerbations (AE) of chronic obstructive pulmonary disease (COPD): a comparison with augmentin (AC) (Abstr 139). 5th International Symposium on New Quinolones, Singapore, 25–27 August 1994

Al-Sibai MB, Halim MA, El-Shaker MM, Khan BA, Qadri SMH (1992) Efficacy of ciprofloxacin for treatment of Brucella melitensis infections. Antimicrob Agents Chemother 36:150–152

Alvarez M, Velasco P, Unzaga J, Berdonces P, Busto C, Amezua B, Cisterna R (1993) Antibiotic susceptibilities of microorganisms in respiratory tract infection (1988–1992). Recent advances in chemotherapy. Proceedings of the 18th International Congress of Chemotherapy, Stockholm, 27 June–11 July 1993, pp 322–323

Andersson HC, Kihlman BA (1989) The production of chromosomal alterations in human lymphocytes by drugs known to interfere with the activity of DNA topoisomerase II. I.m-AMSA. Carcinogenesis 10:123–130

Andriole VT (1993) The future of quinolones. Drugs 45(Suppl 3):1–7

Arai S, Gohara Y, Kuwano K, Kawashima T (1992) Antimycoplasmal activity of new quinolones, tetracyclines, and macrolides against Mycoplasma pneumoniae. Antimicrob Agents Chemother 36:1322–1324

Barry AL, Fuchs PC, Citron DM, Allen SD, Wexsler HM (1993) Methods for testing the susceptibility of anerobic bacteria to two fluoroquinolone compounds, PD 131628 and clinafloxacin. J Antimicrob Chemother 31(6):893–900

Baurenfeind A (1993) Comparative in-vitro activities of the new quinolone, Bay 3118, and ciprofloxacin, sparfloxacin, tosufloxacin, CI-960 and CI-990. J Antimicrob Chemother 31:505–522

Bergstermann H, Seifert S, Bauer M, Rifai M (1991) Quinolones for mycobacterial diseases. Eur J Clin Microbiol Infect Dis [Special Issue]:301–302

Blum MD, Graham DJ, McCloskey CA (1994) Temafloxacin syndrome: review of 95 cases. Clin Infect Dis 18:946–950

Bouzard D, Di Cesare P, Essiz M, Jacquet JP, Remuzon A, Weber T, Oki T, Masuyoshi M (1989) Fluoronaphthyidines and quinolones as antibacterial agents. I. Synthesis and structure–activity relationships of new 1-substituted derivatives. J Med Chem 32:537–542

Carbon C (1993) Quinolones in the treatment of lower respiratory tract infections in adult patients. Drugs 45(Suppl 3):91–97

Carbon C, Leophonte P, Petitpretz P, Chauvin JP, Hazebroucq J (1992) Efficacy and safety of temafloxacin versus those of amoxicillin in hospitalized adults with community-acquired pneumonia. Antimicrob Agents Chemother 36:833–839

Casal M, Gutierrez J, Gonzales J, Ruiz P (1987) In vitro susceptibility of M. tuberculosis to ofloxacin and ciprofloxacin in combination with rifampin and isoniazid. Chemioterapia 6:437–439

Casal M, Rodriguez F, Gutierrez J, Ruiz P et al (1989) Promising new drugs in the treatment of tuberculosis and mycobacteriosis. J Chemother 1(1):39–45

Chan G, Garcia-Ignacio B, Chavez V, Livelo J, Jimenez C, Parrilla M, Franzblau S (1994) Clinical trial of sparfloxacin for lepromatous leprosy. Antimicrob Agents Chemother 38:61–65

Chevalier X, Albengres E, Voisin MC, Tillement JP, Larget-Piet B (1992) A case of destructive polyarthropathy in a 17 year old youth following pefloxacin treatment. Drug Saf 7:310–314

Christ W, Esch B (1994) Adverse reactions to fluoroquinolones in adults and children. Infect Dis Clin Pract (Suppl 3):S168–S176

Christ W, Lehnert T (1990) Toxicity of the quinolones In: Siporin C, Heifetz CL, Domagala JM (eds) The new generation of quinolones. Dekker, New York, pp 165–187

Chu DTW (1992) Isothiazoloquinolones: antibacterial and antineoplastic agents. Drugs Fut 17:1101–1109

Chu DTW, Fernandes PB (1991) Recent developments in the field of quinolone antibacterial agents. Adv Drug Res 21:39–144

Chysky V, Kapila K, Hullmann R, Acieri G, Schact P, Echols R (1991) Safety of ciprofloxacin in children: worldwide clinical experience based on compassionate use. Emphasis on joint evaluation. Infection 19:298–296

Ciaravino V, Suto MJ, Thiss JC (1993) High capacity in vitro micronucleus assay for the assessment of chromosome damage: results with quinolone/naphthyridine antibacterials. Mutat Res 298:227–236

Cooper MA, Andrews JM, Ashby JP, Matthews S, Wise R (1990) In vitro activity of sparfloxacin, a new quinolone antimicrobial agent. J Antimicrob Chemother 26:667–676

Cooper MA, Andrews M, Wise R (1992) In vitro activity of PD 131628, a new quinolone antimicrobial agent. J Antimicrob Chemother 29:519–528

Corraz AJ, Dax SL, Dunlap NK, Georgopapadakou NH, Keith DD, Pruess DL, Rossman PL, Then R, Unowsky J, Wei C (1992) Dual action penems and carbapenems. J Med Chem 35(10):1828–1839

Decazes JM, Casin I, Kitzis MD, Kerbou M, Vallois JM, Geslin P, Guttmann I (1994) Sparfloxacin in experimental meningitis due to highly penicillin-resistant streptococcus pneumoniae (HPRSp) (Abstr 39). 5th International Symposium on New Quinolones, Singapore, 25–27 August 1994

Demuth TP, White RE, Tietjen RA, Storrin RJ, Skuster JR, Andersen JA, McOsker CC, Freedman R, Rourke RJ (1991) Synthesis and antibacterial activity of new C-10 quinolonyl-cephemesters. J Antibiot (Tokyo) 44:200–209
Di-Lascio G, Salvini S, Cermola, Grassi C (1991) Use of pefloxacin in boutonneuse fever. Clin Ther 136:101–106
Domagala JM (1994) Structure–activity and structure–side-effect relationships for the quinolone antibacterials. Antimicrob Chemother 33:685–706
Drancourt M, Stein A, Argenson JN et al (1993) Oral rifampin plus ofloxacin for treatment of Staphylococcus-infected orthopedic implants. Antimicrob Agents Chemother 37:1214–1218
DuPont HL (1993) Quinolones in Salmonella typhi infection. Drugs 45(Suppl 3):119–124
Eliopoulos GM, Eliopoulos CT (1993) Activity in vitro of the quinolones. In: Hooper DC, Wolfson JS (eds) Quinolone antimicrobial agents, 2nd edn. American Society for Microbiology, Washington
Eliopoulos GM, Klimm K, Eliopoulos CT et al (1993) In vitro activity of CP-99, 219, a new fluoroquinolone. Antimicrob Agents Chemother 37:366–370
Finegold SM, Molitoris E, Reeves D, Weber HM (1991) In vitro activity of temafloxacin against anerobic bacteria: a comparative study. Antimicrob Chemother 28(Suppl C):25–30
Franzblau SG, White KE (1990) Comparative in vitro activities of 20 fluoroquinolones against Mycobacterium leprae. Antimicrob Agents Chemother 34:229–231
Fuchs PC, Barry AL, Pfaller MA, Allen SD, Gerlach EH (1991) Multicenter evaluation of the in vitro activities of three new quinolones, sparfloxacin, CI-960, and PD 131 628, compared with the activity of ciprofloxacin against 5252 clinical bacterial isolates. Antimicrob Agents Chemother 35:764–766
Garcia Rodreiguez JA, Gomez-Garcia AC (1993) In vitro activities of quinolones against mycobacteria. J Antimicrob Chemother 32:797–800
Gelber RH, Iranmanesh A, Murray L, Siu P, Tsang M (1992) Activities of various quinolones. Antimicrob Agents Chemother 36:2544–2747
Genesis Report (1992) Vol 1(2)
Georgopapadakou NH, Bertasso A (1993) Mechanisms of action of cephalosporin 3′-quinolone esters, carbmates, and tertiary amines in Escherichia coli. Antimicrob Agents Chemother 37:559–565
Gevaudan MJ, Mallet MN, Gulian C et al (1989) Etude de la sensibilité de sept espèces de mycobacteries aux nouvelles quinolones. Pathol Biol (Paris) 36:477–481
Gialdroni-Grassi G, Brumpt I (1994) Sparfloxacin: an empiric therapy in community-acquired pneumonia (CAP). A meta-analysis of two comparative studies (Abstr 138). 5th International Symposium on New Quinolones, Singapore, 25–27 August 1994
Goldstein EJ (1993) Patterns of susceptibility to fluoroquinolones among anerobic bacterial isolates in the United States. Clin Infect Dis 16(Suppl 4):S377–S381
Gootz TD, Osheroff N (1993) Quinolones and eukaryotiuc topoisomerases. In: Hooper DC, Wolfson JS (eds) Quinolone antimicrobial agents, 2nd edn. American Society for Microbiology, Washington
Gootz TD, Barrett JF, Sutcliffe JA (1990) Inhibitory effects of quinolone antibacterial agents on eucaryotic topoisomerases and related test systems. Antimicrob Agents Chemother 34:8–12
Gootz TD, Barrett JF, Sutcliffe JA (1994) Inhibitory effects of quinolone antibacterial agents on eucaryotic topoisomerases and related test systems. Antimicrob Agents Chemother 8–12
Gough AW, Kasali OB, Sigler RE, Baragi B (1992) Quinolone arthropathyacute toxicity to immature articular cartilage. Toxicol Pathol 20:436–450
Gracheck SJ, Mychajlonka M, Gambino L, Cohen MA, Roland G, Ciaravino V, (1991) Correlations of quinolone inhibition endpoints vs bacterial DNA gyrase, mammalian topoisomerase I and II and cell clonogenic survival, an in vitro miconucleus

induction (Abstr 1488). 31st Interscience Conference on Antimicrobial Agents and Chemotherapy, Chicago. American Society for Microbiology, Washington
Hagen S, Domagala JM, Heifetz C, Sanchez J, Sesnie J, Stier M et al (1991) Synthesis and antibacterial activity of new 7-[3-(1-amino-1-methylethyl)-1-pyrolidinyl]-4-quinolones (Abstr 1438). 31st Interscience Conference on Antimicrobial Agents and Chemotherapy, Chicago. American Society for Microbiology Washington
Halliwell RF, Davey PG, Lambert JJ (1993) Antagonism of GABA receptors by 4-quinolones. J Antimicrob Chemother 31:457–462
Hamilton JMT (1994) Dual-action antibiotic hybrids. J Antimicrob Chemother 33:197–202
Hammerschlag MR, Hyman CL, Roblin PM (1992) In vitro activities of five quinolones against Chlamydia pneumoniae. Antimicrob Agents Chemother 36:682–683
Hesse MT, Pitsakis PG, Kaye D (1990) Oral temafloxacin versus vancomycin for therapy of experimental endocarditis caused by methicillin-resistant Staphylococcus aureus. Antimicrob Agents Chemother 354:1143–1142
Hoffner SE, Kratz M, Olsson-Liljequist SB, Svenson SB, Kallenius G (1989) In vitro synergistic activity between ethambutol and fluorinated quinolones against Mycobacterium avium complex. J Antimicrob Chemother 24:317–324
Hooper DC (1993) Introduction. In: Hooper DC, Wolfson JS (eds) Quinolone antimicrobial agents, 2nd edn. American Society for Microbiology, Washington
Hooper DC, Wolfson JS (1993) Mechanisms of bacterial resistance to quinolones. In: Hooper DC, Wolfson JS (eds) Quinolone antimicrobial agents, 2nd edn. American Sotiety of Microbiology, Washington
Hooper DC, Wolfson JF (1993) Mechanisms of quinolone action and bacteria killing. In: Hooper DC, Wolfson JF (eds) Quinolone antimicrobial agents, 2nd edn. American Society of Microbiology, Washington
Hori S and Shimada J (1993) Effect of new fluoroquinolone compounds on the receptor binding of γ-aminobutryic acid. Recent advances in chemotherapy. Proceedings of the 18th International Congress of Chemotherapy, pp 293–294
Hori S, Kanemitsu K, Shimada J (1993) Convulsant activity AM-1155, a new fluoroquinolone: a comparative study of convulsant activities of fluoroquinolones in the presence of anti-inflammatory drugs. Recent advances in chemotherapy. Proceedings of the 18th International Congress of Chemotherapy, Stockholm, 27 June–11 July 1993, pp 283–284
Hosaka M, Yasue T, Fukuda H, Tomizawa H, Aoyama H, Hirai K (1992) In vitro and in vivo antibacterial activities of AM-1155, a new 6-fluor-8-methoxy quinolone. Antimicrob Agents Chemother 36:2108–2117
Imada T, Miyazaki S, Nishida M, Yamaguchi K, Goto S (1992) In vitro and in vivo antibacterial activities of new quinolone, OPC-17116. Antimicrob Agents Chemother 36:573–579
Ingerman M, Pitsakis PG, Rosenberg A, Hessen MT, Abrutyn E, Murray BE et al (1987) Beta-lactamase production in experimental endocarditis due to aminoglycoside-resistant Streptococcus faecalis. J Infect Dis 155:1226–1232
Kaatz GW, Seo SM, Lamp SM, Bailey EM, Rybak MJ (1992) CI-960, a new fluoroquinolone, for therapy of experimental ciprofloxacin-susceptible and resistant Staphylococcus aureus endocarditis. Antimicrob Agents Chemother 36:1192–1197
Kimura M, Kishimoto Y, Hashiguchi K, Niki Y, Soejima R (1993) In vitro and in vivo activities of sparfloxacin against chlamydial species. Drugs 45(Suppl 3):37
Klemens SP, Stefano MS, Cynamon MH (1993) Therapy of multidrug-resistant tuberculosis: lessons from studies with mice. Antimicrob Agents Chemother 37:2344–2347
Klopman G, Wang S, Jacobs MR, Ellner JJ (1993) Antimycobacterium avium activity of quinolones structure activity relationship studies. 37(9):1807–1815
Kohlbrenner WE, Wideburg N, Weigl D, Saldivar A, Chu DTW (1992) Induction of calf thymus topoisomerase I-mediated DNA breakage by the antibacterial isothiazoloquinolones A-65281 and A-65282. Antimicrob Agents Chemother 36:81–86

Korten V, Tomayko JF, Murray B (1994) Comparative in vitro activity of DU-68591, a new fluoroquinolone agent, against gram-positive cocci. Antimicrob Agents Chemother 38:611–615

Kunin CM (1985) The responsibility of the infectious disease community for the optimal use of antimicrobial ágents. J Infect Dis 1561:388–398

Lang R, Rubinstein E (1989) Quinolones for the treatment of brucellosis. J Antimicrob Agents Chemother 33:1–5

Limb DI, Dabbs DJW, Spencer C (1987) In vitro selection of bacteria resistant to the 4-quinolone agents. J Antimicrob Chemother 19:65–71

Lomaetro BM, Bailie GR (1991) Quinolone–cation interactions: a review. DICP Ann Pharmacother 25:1249–1258

Loo VG, Lavellee J, McAlear D, Robson HG (1994) The in vitro susceptibilities of 326 Streptococcus pneumoniae isolates to nine antimicrobial agents including penicillin and newer quinolones. J Antimicrob Chemother 33:641–645

Martin-Luengo F, Sempere MA (1993) Activity of Sparfloxacin and other quinolones against mycobacteria. Drugs 45(Suppl 3):219–220

Marutani K, Matsumoto M, Tanaka K, Nagano H, Matsubara S, Yokota T, Matsumoto F (1993) Reduced phototoxicity of Q-35, a new 8-methoxy fluoroquinolone. II. UVA irradiation of murine ears. Recent advances in chemotherapy. Proceedings of the 18th International Congress on Chemotherapy, Stockholm, 27 June–11 July 1993, pp 270–272

Matsubara S, Matsumoto M, Nagano H, Yokota T, Matsumoto F (1993) Reduced phototoxicity of Q-35, a new 8-methoxy fluoroquinolone 1. Chemical stability in vitro. Recent Advances in Chemotherapy, 18th International Congress on Chemotherapy, Stockholm, pp 269–270

Matsumoto F, Takahashi T, Matsubara S, Ueda Y (1993) Possible relationship between the distribution of fluoroquinolones into sweat and the development of phototoxicity: preliminary study in healthy male volunteers. Recent Advances in Chemotherapy, 18th International Congress on Chemotherapy, Stockholm, pp 271–272

Mitscher LA, Devasthale P, Zavod R (1993) Structure–activity relationships. In: Hooper DC, Wolfson JC (eds) Quinolone antimicrobial agents, 2nd edn. American Society for Microbiology, Washington

Mori T, Ebe T, Takahasi M, Isonuma H, Ikemoto H, Oguri T (1993) Lung abscess: analysis of 66 cases from 1979–1991. Intern med 32(4):287–284

Murray PR,, Bratcher JL, Niles AC, Hampton CM (1993) In vitro activity of nine fluoroquinolone antibiotics against 200 strains of enterococci. Diagn Microbiol Infect Dis 16:83–85

Nakanishi N, Yoshida S, Wakebe H, Inoue M, Yamaguchi T et al (1991) Mechanisms of clinical resistance to fluoroquinolones in Staphylococcus aureus. Antimicrob Agents Chemother 35:2562–2567

Neu HC (1992) Quinolone antimicrobial agents. Annu Rev Med 43:465–486

Neu HC, Chin NX (1987) In vitro activity of two new quinolone antimicrobial agents S-25930 and S-25932 compared with that of other agents. Antimicrob Chemother 19:175–186

Nord CE (1992) Susceptibility of anerobic bacteria to PD 131628 (Abstr A30). Proceedings of the 4th Symposium on New Quinolones, Munich, 25–27 August 1992

Nord CE, Hagelback A (1993) In vitro activity of PD 131628 against anerobic bacterial. Drugs 45(Suppl 3):187

Nord CE, Lindmark A, Persson I (1993) In vitro activity of BAY y3118 against anerobic bacteria. Recent advances in chemotherapy. Proceedings of the 18th International Congress of Chemotherapy. Stockholm, 27 June–11 July 1993, pp 291–293

Paton JH, Reeves DS (1991) Clinical features and management of adverse effects of quinolone antibacterials. Drug Saf 6:8–27

Pattyn SR (1991) Anti-mycobacterium leprae activity of several quinolones studied in the mouse. Int J Lepr Other Mycobact Dis 59:613–617

Pechere JC (1993) Quinolones in intracellular infections. Drugs 45(Suppl 3):29–36

Peterson LR (1994) Quinolone resistance in gram-positive bacteria. Infect Dis Clin Pract 3(Suppl 3):S127–S137

Perronne C, Gikas A, Tuffot-Pennot C, Grosset J, Vilde JL et al (1991) Activities of sparfloxacin, azithromycin, temafloxcin and rifapentine compared with that of clarithromycin against multiplication of Mycobacterium avium complex within human macrophages. Antimicrob Agents Chemother 35:1356–1359

Pfaller MA, Barry AL, PC Fuchs (1993) RO 23-9424, a new cephalosporin 3′-quinolone: in vitro antimicrobial activity and tentative disc diffusion interpretive criteria. J Antimicrob Chemother 31:81–88

Piddock LJV (1994) New quinolones and gram-positive bacteria. Antimicrob Agents Chemother 38:163–169

Piddock LJV, Marshall AJ, Jin YF (1994) Activity of Bay y3118 against quinolone-susceptible and resistant gram-negative and gram-positive bacteria. Antimicrob Agents Chemother 38(3):422–427

Raoult D, Drancourt M (1991) Antimicrobial therapy of rickettsial disease. Antimicrob Agents Chemother 35:2457–2462

Rastogi N, Goh KS (1991) In vitro activity of new difluorinated quinolone sparfloxacin (AT-4140) against Mycobacterium tuberculosis compared with activities of ofloxacin and ciprofloxacin. Antimicrob Agents Chemother 35:1933–1936

Rastogi N, Labrousse V, Goh KS, Carvalho de Sousa JP (1991) Antimycobacterial spectrum of sparfloxacin and its activites alone and in association with other drugs against Mycobacterium avium complex growing extracellularly and intracellularly in urine and human macrophages. Antimicrob Agents Chemother 35:2473–2480

Remuzon P, Bouzard D, Guiol L, Jacquet J (1992) Fluoronaphthyridines as antibacterial agents. VI. Synthesis and structure–activity relationship of new chiral 7-(1-,3-,4-, and 6-methyl-2,5-diazabicyclo[2.2.1]heptan-2-yl)naphthyridine analogues of 7-[1R, 4R)-2-5diazabicyclot[2.2.1]heptan-2-yl]-1-(1,1-dimethylethyl)-6-fluoro-1,4-dihydro-4-oxo-1,8-naph-thyridine-3-carboxylic acid. Influence of the configuration on blood pressure in dogs. A quinolone-class effect. J Med Chem 31:221–232

Ribard P, Kahn MF (1991) Rheumatological side-effects of quinolones. Baillieres Clin Rheumatol 5:175–191

Robinson M, Matin BA, Gootz TD, McGuirk PR, Osheroff N (1992) Effects of novel fluorquinolones on the catalytic activities of eukaryotic topoisomerase II. Influence of the C-8 fluorine group. Antimicrob Agents Chemother. 36:751–756

Rosen T, Chu DTW, Lico IM, Fernandes PB, Marsh K, Shen L et al (1988) Design, synthesis and properties of (4S)-7-(4-amino-2-substituted-pyrrolinyl) quinolone-3-carboxylic acids. J Med Chem 31:1598–1611

Ruiz-Beltran R, Herrero Herrers JI (1992) Evaluation of ciprofloxacin and doxycyline in the treatment of Mediterranean spotted fever. Eur J Clin Microbiol Infect Dis 11:427–431

Schaad UB (1993) Use of quinolones in paediatrics. Drugs 45(Suppl 3):37–41

Schaad UB, Wedgewood J (1992) Lack of quinolone-induced arthropathy in children. J Antimicrob Chemother 30:414–416

Scheel O, Reiersen R, Hoel T (1992) Treatment of tularemia with ciprofloxacin. Eur J Clin Microbiol Infect Dis 11:447–448

Segev S, Rehavi M, Rubinstein E (1988) Quinolones, theophylline, and diclofenac interactions with the γ-aminobutyric acid receptor. Antimicrob Agents Chemother 32:1624–1626

Shentag JJ, Nix DE (1990) Pharmacokinetics and tissue penetration of fluoroquinolones. In: Sanders EE, Sanders CC (eds) Fluoroquinolones in the treatment of infectious disease. Physicians and Scientists, Glenview, pp 29–44

Spangler SK, Jacobs MR, Appelbaum PC (1992) Susceptibilities of penicillin-susceptible and -resistant strains of Streptococcus pneumoniae to RP 59500, vancomycin, erythromcin, PD 131628, sparfloxacin, temafloxacin Win 57273, ofloxacin and ciprofloxacin. Antimicrob Agents Chemother 36:856–859

Spangler SK, Jacobs MR, Appelbaum PC (1993) Susceptibility of 170 penicillin-susceptible and resistant pneumococci to various antimicrobials. Drugs 45 (Suppl 3):194–195

Syrjala H, Schildt R, Raisainen S (1991) In vitro susceptibility of Francisella tularensis to fluoroquinolones and treatment of tularemia with norfloxacin and ciprofloxacin. Eur J Clin Microbiol Infect Dis 10:68–70

Thadepalli H, Chuah SL, Gollapudi S (1993) Quinolones and anaerobic bacterial infections. Recent advances in chemotherapy. Proceedings of the 18th International Congress of Chemotherapy, Stockholm, 27 June–11 July 1993, pp 481–482

Thorburn C, Tyler J, Knott S (1993) Rates of in vitro selection of resistance to 11 oral antibiotics for respiratory tract pathogens. Recent advances in chemotherapy. Proceedings of the 18th International Congress of Chemotherapy, Stockholm, 27 June–11 July 1993, pp 323–325

Ueno K, Watanabe K, Kato N, Kato H, Bandoh K, Tanaka Y (1993) Comparative in vitro activity of AM-1155, a new quinolone, against anaerobic bacteria. Drugs 45(Suppl 3):184–186

Vazquez J, Perri MB, Thal LA, Donabedian SA, Zervos MJ (1993) Sparfloxacin and clinafloxacin alone or in combination with gentamicin for therapy of experimental ampicillin-resistant enterococcal endocarditis in rabbits. J Antimicrob Chemother 32:715–721

Wakabayashi E, Mitsuhasi S (1994) In vitro antibacterial activity of AM-1155, a novel 6-fluor-8-methoxy Quinolone. Antimicrob Agents Chemother 38:594–601

Wiedeman B, Heisig P (1994) Quinolone resistance in gram-negative bacteria. Infect Dis Clin Pract (Suppl 3):S115–S126

Wolfson JS, Hooper DC (1989) Fluoroquinolone antimicrobial agents. Clin Microbiol Rev 2:378–424

Yoshinari T, Mano E, Arakawa H, Kurama M, Iguchi T, Nakagawa S, Tanaka N, Okura A (1993) Stereo (C7) dependent topoiserase II inhibition and tumor growth suppression by a new quinone, BO-2367. J Cancer Res 84:800–806

Subject Index

A-84066 64
A-867191 83–4
A-101211 83
A-104954 64
abdominal surgery 443–4
absorption, distribution, metabolism, excretion (ADME) 179
N-acetyl piperazine 23 (table)
Achilles tendinitis/rupture 302–4
Acinetobacter spp. 171, 172 (table)
Acinetobacter baumanni 262 (table)
acrosoxacin 3, 4
activated charcoal 364
adenosine receptors 308
  ligands 308
ADPNP 132, 154, 156
Aeromonas spp. 171, 172 (table)
Aeromonas salmonicida 262 (table)
agriculture, fluorquinolone use in 284
AIDS 465
L-alanine 81
Alcaligenes spp. 172 (table)
aliphatic aromatic amines, secondary 16
alkaline protease 245, 246 (table)
l-alkyl-(aza)cinnolone carboxylic acids 48 (fig.), 49
alkyl halides 64
alklylsulphonic acid esters 64
alpha-slow-wave index (ASI) 380
AM-1091 90
AM-1155 461 (table), 462 (table), 463 (table)
amifloxacin 64
amine synthesis 78–9
1-aminoazabicycloalkanes 86–7
aminoazetidines 89
γ-aminobutyric acid (GABA) 308, 457
  receptors 457
3-amino-4,4-dimethylpyrrolidines 81
aminoglycosides 6
  concentration-dependent killing rate 409–10
  distribution 6–8
aminomercaptomalonic esters 17
4-amino-3-methoximino-pyrrolidine 80
1-aminomethylazacycloalkanes 86–7
4-aminomethyl-3-methoxoamino-pyrrolidine 80
3-(1-amino-1–methylethyl)pyrrolidine 82–3
5-aminomethylisoxazolidine 89
3-aminomethyl-morpholine derivatives 89
4-amino-4-methylpyrazolidine derivatives 89
aminomethylpyrrolidine 79–83
  derivatives 82
3-aminomethyl-pyrrolidines 82 (fig.)
6-amino-naphthyridone carboxylic acid 77
N-amino-norfloxacin 95
8-amino-ofloxacin 72
aminopiperidines 89
2-amino-propionaldehyde dimethyl acetal 27
4-amino-pyrido[2,3d]pyimidines 72
aminopyrrolidine derivatives 79–83
3-aminopyrrolidine derivatives 80
S-3-aminopyrrolidinyl quinolones 80
3-aminopyrrolidinyl-substituted qunolones 80
3-amino-quinolones 66
5-amino-quinolones 73, 74, 75 (fig.), 75–6
6-amino-quinolones 77
  carboxylic acids 77
ampicillin 278
*m*AMSA 135
amoxicillin 248
ampicillin 251
anaerobic bacteria 463–4
anaerobic intro-abdominal infections 463–4
antacids 359–61
antibacterial agents 233–65
antibiosis 233
antibiotics 167

bacterial resistance 259–60
antimicrobial activity 458–65
antimicrobial agents, bactericidal effects 242
antipseudomonal therapy 245–6
antipyrine 370
antitumor drugs 135
antitumor potential 467–8
AP-5 312
AP-7 312
1-arylcinnolone carboxylic acid 47, 48 (fig.)
1-arylpyridone carboxylic acids 52
arthropia deformans 299
arthropathy 297–302
cystic fibrosis associated 302
juvenile rats 300 (table)
magnesium deficiency 301
5-arylpyridone carboxylic acids 51
1-aryl-quinolones 63, 75
Asp87 267
L-aspartic acid 80, 81
ATPase 136–7
AUC 407
AUIC 415
azabicycloalkanes with amino groups on annelated ring 87–9
8-azaquinolone carboxylic acid 13, 21
aztrenom 251

Bacillus spp. 173 (table)
Bacillus subtilis 121
baclofen 310
bacteremia 442–3
animal 249
bacteria
conjugation inhibition 127
filamentation 125–7, 129
gram-negative 171
gram-positive 171–4
nalA' mutation 230
quinolones effect 129–30
bacterial otitis (chronic suppurative otitis media) 436
bacterial resistance 465–6
Bacteroides spp. 175 (table)
Bacteoides fragilis 175 (table), 463
balfloxacin 37 (table)
Baltz-Schiemann reaction 77
BAY 12-8039 85 (fig.), 90
BAY V 3548 90
BAY Y 3118 85, 177, 459, 460 (table), 461 (table), 462 (table)
benofloxacin 22 (table), 35
benzodiazepines 374–5
benzo[h][1,6]naphthyridines 24
benzo[ij]quinolizinones 21
benzoxacine 17
benzoxazolo[3,2a]quinolone carboxyclic acid ester 31, 32
benzoylperoxide 49
benzthiazole[3,2a]quinolone carboxylic acid ester 31, 32 (fig.)
1-benzyl-3,4-epoxypyrrolidine 81
1-benzyl-3-hydoxy-4-hydroxymethylpyrrolidine 82, 83
Biere and Seelen approach 35
binfloxacin 22 (table)
bioisteric active substances 42
biphenylacetic acid 309
1,6-bisarylpyridine carboxylic acid esters 51 (fig.), 52
bismsuth subsalicylate 363
blood-brain barrier (BBB) 308
BMY 40062 37 (table)
boron complexes 91–2
Borsche synthesis 47
5,6-bridged imidazo-quinolone 77
3,4-bridged pyrrolidien derivatives 83–95
5,6-bridged triazolo-quinolone 77
6-bromine derivatives 78
5-bromo-naphthyridone carboxylic acid 75
5-bromo-quinolone 75
5-bromo-quinolone carboxylic acid 75
8-bromo-quinolone carboxylic acid esters 100
Brucella spp. 172 (table)
brucellosis 458
burn wound infection 245
Burkholderei capacia 171, 172 (table)
Burkholderei pseudomallei 171, 172 (table)
7-(S)-*tert*-butoxycarbonylamino-azaspiro[2,4]heptane 81 (fig.)
N-butyl scopolamine bromide 358

caffeine 366, 367, 378
CI-938 457
cAMP-mediated gene regulation 273
cAMP receptor protein (*crp*) 273
Camps quinolone synthesis 39–41
camptothecin 154
Campylobacter spp. 172 (table)
Campylobacter jejuni 262
Campylobacter lari 262 (table)
carbapenems 66–7, 274
carbenicillin 251
carbonyl cyanide *m*-chlorophenyl-hydrazone (CCCP) 270
carcinogenic effects 317–22

cardiovascular effects 316–17
CcdB-gyrase DNA complex 136
CcdB protein 136
cefixin 251
cefsulodin 247
ceftazidine 247
cell killing mechanism 150–6
central intermediates 17
  cyclocondensation 19 (fig.)
  synthesis 18 (fig.)
central nervous system 306–13, 457
cephalosporins 66–7, 93
  3-hydroxymethyl group 67, 94
cerebral dopaminergic system 311
CH-acid compunds 68
chancroid 440
charcoal, activated 364
chemistry 1–4
chemotherapy 363–4
children *see* pediatrics
clinafloxacin 80
Chinese hamster V79 cells 319
chiral triacyloxyboroydride 17
Chlamydia pneumoniae 176 (table)
Chlamydia trachomatis 176 (table)
chloramphenicol 127, 251, 275, 277, 279
chloranil 24 (fig.), 24
chlorobenzene 96
7-chloro-1,4-dihydro-1-ethyl-4-oxoquinoline 3 carboxylic acid (compound (a)) 1, 2
chlorohydrin 30
7-chlorokynurenic acid 312
4-chloro-3-methoxycarbonyl-4-methylquininium hexaflurophosphate 41, 42 (fig.)
5-chloro-naphthyridone carboxyic acids 75
chloroquine 2
5-chloro-quinoline 75
5-chloro-quinoline carboxybic acid 75
cholangiopancreatography, endoscopic retrograde 443
cholangitis 441
cholera 428
chromosomal aberrations 319
chronic bronchitis, acute exacerbation 430–3
chronic obstructive lung disease (COPD) 366
chronic suppurative otitis media (bacterial otitis) 436
CI-960 460
cimetidine 371
cinnolone carboxylic acid 13
4-cinnolone-3-carboxylic acids 45–9
cinoxacin 13, 21, 46, 47 (fig.)
  interpretive breakpoint 169 (table)
ciprofloxacin 5, 6, 13, 21, 25, 36 (table)
  absorption 357–8, 364
  activity against:
    anaerobes 463
    E. faecalis 462 (table)
    M. tuberculosis 464 (table)
    M avium intracellulare 464 (table)
    S. aureus 464 (table)
    S. pneumoniae 461 (table)
  bactericidal potency against *S; aureus* 21 (table)
  children 356–7
  endotoxin release 250
  high-dose 252
  interactions with other drugs 366–8
  DNA cleavage 136
  DNA gyrase footprinting effect 132
  DNA synthesis inhibition 127
  effects on exoenzyme production 247
  endotoxin-neutralizing ability 252
  E. coli bactericidal potency 243 (table)
  E. coli treatment 125, 126 (fig.)
  N-hydroxylated 94–5
  intermediate 26 (fig.)
  interpretive breakpoint 169 (table)
  methicillin-resistant *S. aureus* resistance 171
  mutagenic effect 128
  P. aeruginosa bactericidal potency 243 (table)
  P. aeruginosa susceptibility 170
  pharmacokinetic parameters 341 (table)
  phospholipase production inhibition 246
  retinal toxicity 315
  S. pneumoniae respiratory tract infection, treatment 431 (table)
  structure 120 (fig.)
  synthesis 25 (fig.)
Citrobacter diversus 172 (table)
Citrobacter freundii 172 (table), 237 (table)
clinafloxacin (PD127391) 5, 36 (table), 90, 177, 391, 459 (table)
  activity against:
    anaerobes 463 (table)
    E. faecalis 462 (table)
    S. aureus 464 (table)
    S. pneumoniae 461 (table)
  6-methylation 77
  synthetic routes 91 (fig.)

clinical pharmacology 39–86
chiral starting materials 80
Clostridium spp. 174 (table)
Clostridium difficile 174 (table)
Clostridium perfringens 174 (table)
colorectal surgery 443
community acquired infections/ pathogens 281–2, 461
Corynebacterium spp. 173 (table)
coumarin drugs 129
Coxiella burnetii 262 (table)
CP67015 4, 149
CP99219 177, 462 (table), 463 (table), 460 (table)
CP115,953 120 (fig.)
crystalluria 456
7-cycloalkyl-quinolone carboxylic acid 98 (fig.)
cycloaracylation procedure 25–36, 64, 75
cyclopropyl 63
1-cyclo-6,8-difluro-1,4-dihydro-7-(2,6-dimethyl-4-pyridinyl)-4-quinolone carboxylic acid 96
N-cyclopropylanilines 21
1-cyclopropyl-6,7-difluroquinolone-3-carboxylic acid 39
1-cyclopropyl-6,8-difluroquinolone 80
l-cyclopropyl-1,8-naphthridone carboxylic acid 24
7-cyclopropyl-quinolone carboxylic acid 97 (fig.)
8-cyclopropyl-pyridol[2,3-d]pyrimidine-6-carboxylic acid 24
l-cyclopropylquinolones 72
cyclosporine 372–3
CYP1A2 365
CYP2E1 365
CYP384 365
cystic fibrosis 245, 356, 445
  recurrent respiratory tract infections 435

danofloxacin 36 (table)
2,3-dehydro-ofloxacin 28, 29
3-deoxy-α-methyl-17α-acetoxyprogesterone 68 (fig.)
desacetylcephalothin-ciprofloxacin hybrid 468–9
L-2,4-diaminobutyric acid 80
5,6-diamino-quinolones 77
diazabicycloalkanes 84–6
2,5-diazabicyclo[2,2,1]heptane system 79 (fig.)
diazaquinolone carboxylic acid derivatives 21
diazepam 310, 374–5
2,4-dichloro-5-fluorobenzoate 24
2,4-dichloro-5-flurobenzoyl chloride 25
2,6-dichloro-5-fluoronicotinic acid 21
2,4-dichloro-5-nitrobenzoate 24
didanosine 362
Dieckmann cyclization of diesters 21–4
diethyl malonate 68
difluro-isothiazole[5,4-b]quinoline-3,4-diones 33, 34 (fig.)
9.10-difluro-3-methylene-pyridobenzoxazine carboxylic acid ester 30
2,4-diflurophenyl 63
N(2,4-diflurophenyl aniline 21
difluroquinolone carboxylic acid 26
4,5-dihydrofuran-2-carbonitrile 87
1,4-dihydro-1-methyl-6-nitro-4-oxo-quinoline carboxylic acid 64
1,4-dihydro-4-oxo-quinoline-3-carboxylic acids 13
dihydro-pyrrolopyridines 85 (fig.)
2,5-dimethoxytetrahydrofuran 76
3-dimethylaminoacrylonitrile 51
5-dimethyl-amino-naphthyridone carboxylic acid 75
5-dimethyl-amino-quinolone carboxylic acid 75
dimethylformamide (DMF) 16, 23 (fig.), 27 (fig.), 91 (fig.) 92 (fig.)
  synthesis 25 (fig.), 57 (fig.), 71
dimethylsulfoxide (DMSO) 25 (fig.) 73
2,4-dinitro-chlorobenzene 20, 64
2,4-dinitro-fluorobenzene 64
dioxane 2
1,1-dioxides benzothiazine-2-carboxylic acid 42
diphenyl ether 16
diothiocarbamate 19
DMSO 73
  synthesis 73 (fig.)
DNA 6, 119
  binding 131–3
  biosynthesis inhibition 130
  break 260
  breakage-reunion 119, 130–1
  cleavage 133–6
    requirement for quinolone binding to gyrase-DNA complex 141–2
    site 135
  cooperative quinolone-DNA binding model 140–1
  damage, quinolone-induced 127–8
  decatenation 15
  different topological forms 138
  drug-dependent cleavage 135

gyrase *see* DNA gyrase
magnesium ions in quinolone binding 142–4
norfloxacin binding 6, 138
8-Oxo-dg formation 326
pBR322 139
polymerase III replication complex blocking 156
protein complex 131
quinbenzoxazine binding 44–5
quinolones binding to 138–9
quinolone-induced damage 320
relaxation 151
relaxed double-stranded 138
shorter 137
SOS repair system 128
strand breaks 321
supercoiling 119, 121, 151, 260
inhibition 6
negative 260
relaxation 130
synthesis inhibition 127
topoisomerase(s) 119, 122
topoisomerase HH′ relaxation reaction 131
unknotting 131
unscheduled synthesis 320
DNA gyrase 6, 66, 119–33, 130–58, 260
alterations in subunit A conferring quinolone resonance 262–3 (table)
A protein 131
ATPase 136–7
A subunits, quinolone resistance determining regions 264 (fig.)
binding mode 137–50
B protein 131
catenation/decatenation reactions catalysed 131
cleavage-religation equilibrium 136
current problems 149–50
DNA binding 131–2
DNA cleavage 133–6
effects on 130–7
Escherichia coli 121 (tables), 122
footprinting 132
ciprofloxacin effect 132
hydroxyl radical 132
genes 121
GyrA subunit 119
GyrB subunit 119
gyrase-drug complex effect 129
illegitimate recombination 137
quinolone binding effect 139–40
quinolone-resistant mutants, quinolone binding 145–6
reactions 130–1
topisomerase II-targeting drugs binding 146–8
transient electric dichroism 132
DnaK 128
DNase I footprints 132
Dowtherm A 16
doxorubicin 135
DR-3354 311
drug interactions 329
DU-6859 37 (table), 81
DU-6859a 177, 456, 460 (table), 461 (table), 462 (table), 463 (table), 464 (table)

E3846 177
E4497 177
E4868 177
E5065 177
E5068 177
"Eagle effect" 408
elastase 246 (table)
endoscopic retrograde cholangiopancreatography 443
endotoxemia 250
endotoxin release, fluroquinoline-induced 248–52
enoxacin 5, 13, 21, 24 (fig.), 309
adverse reactions 366
+ fenbufen 309
interactions with othe drugs 366
interpretive breakpoints 169 (table)
pharmacokinetic patterns in healthy subjects 42 (table)
enrofloxacin 25, 36 (table)
Enterobacteriaceae 171
Enterobacter aerogenes 172 (table), 174 (table)
Enterobacter agglomerans 172 (table)
Enterobacter cloacae 172 (table), 242, 262 (table)
bactericidal activity 242 (table)
enterococci 284, 462
Enterococcus faecalis 265, 172 (table), 263 (table), 462
newer quinolones against 462 (table)
Enterococcus faecium 172 (table)
3,4-epoxypyrrolidine derivatives 82
esafloxacin 22 (table)
Escherichia coli 5, 236 (table), 237 (table)
bactericidal activity 242 (table)
ciprofloxacin treatment 125, 126 (fig.)
DNA gyrase 121 (table)
subunits 121 (table)

efflux systems 277
enzyme 119
fluroquinolone-resistant 259, 260, 283
KL16 cells 126 (fig.)
multiple antibiotic resistance 277
mutations conferring decreased susceptibility to quinolones 280 (table)
nalidix acid resistance 275
nalidix acid treated 125–7
norfloxacin treatment 125
quinolone activity 172 (table)
quinolone bactericidal potencies 243 (table)
quinolone resistance 262 (table)
mutations 123–4
SOS response 128
type 1 fimbriae 238
verotoxin-producing 245
Eschweiler-Clarke conditions 76
ethidium bromide 275
3-ethoxy-2-benzolyacrylic acid esters 28, 30
3-ethylamino-propionic acid 21
N-ethylcarboxamidoadenosine 308
ethyl 7-chloro-1-cyclopropyl-6-fluoro-1,4-dihydro-5-methyl-4-oxo-1,8-naphthyridine-3-carboxyate 78 (fig.)
flurobenzoylacetate 26, 31
ethyl-3(2,6-dichloro-5-fluoro-3-pyridyl)-3-oxopropionate 31
ethyl-3-cyclopropylaminoacrylate 25
ethyl-2,4-dichloro-5-fluorobenzoylacetate 26, 31
ethyl iodide 84
5-ethyl-naphthyridone carboxylic acid 175
ethyl nicotinoylacetate 41
1-ethyl piperazine 25
1-ethyl quinolone 72
5-ethyl-quinolone carboxylic acid 75
exoenzyme production 245–8
exoenzyme S 245, 246 (table)
exotoxin A 245, 246 (table)
experimental animals, pharmacodynamics of fluorquinolones 179–02, 207–27
bacterial killing *in vivo* 208
ciprofloxcin 188–9, 234
excretion 189
pharmacokinetic constants 19 (table)
structure 188 (fig.)
enoxacin 184–6
excretion 186 (table)
pharmacokinetic constants 184 (table)
structure 185 (fig.)
fleroxacin 191–3
structure 192 (fig.)
lomefloxacin 193–5
excretion 194 (table)
pharmacokinetic constants 193 (table)
structure 193 (fig.)
minimal bactericidal concentration 207
minimal inhibitory concentration 207
norfloxacin 180–1, 182 (table)
ofloxacin 186–7
excretion 187 (table)
pharmacokinetic constants 186 (table)
structure 187 (fig.)
perfloxacin 181–4
pharmacokinetic constants 183 (table)
renal excretion 184 (table)
structure 183 (fig.)
pharmacodynamic parameters determining efficacy 208–17
area under concentrations 211
AUC/MIC ratio 211–16
postantibiotic effects *in vivo* 208, 209 (table)
resistance of fluroquinolones 217–27
factors contributing to emergence of resistance *in vivo* 223–7
miscellaneous infection models 223
P aeruginosa experimental infection 202–2
staphylococcal infections 222
sparfloxacin 195
pharmacokinetic constants 195 (table)
temafloxacin 189–91
pharmacokinetic constants 191 (table)
structure 190 (fig.)
tosufloxacin 191
quinolone penetration at sites of infection 195–22
alveolar macrophage 199
area under concentration 196, 198
epithelial lining fluid 199
J774 macrophage cell line 199
lung concentration 199
macrolides 201–2

maximum serum concentration 196, 198
mean concentrations in potential sites of pulmonary infection 200 (table)
minimum inhibitory concentration 196
peripheral lymph 199
postantibiotic effect 196
tissue concentration 199

fenbufen 309, 378
ferrous sulphate 362–3
fleroxacin 5, 23 (table)
bactericidal potency against *E. coli* 243 (table)
bactericidal potency against *S. aureus* 244 (table)
interactions with other drugs 368
interpretive breakpoint 169 (table)
pharmacokinetic parameters in:
elderly subjects 343 (table)
healthy subjects 343 (table)
flumequin 2, 19, 35
fluorine C-6 2
2-fluoroethyl tosylate 64
3-fluoro-9-(4-flurophenyl)-2-(4-methylpiperazin-1-yl)-6H-6-oxo-pyrido[1,2-a]pyrimidine-7-carboxlic acid 43, 44 (fig.)
9-fluro-3-methylene-10-(4-methyl-1-pieperazine)-7-oxo-2,3-dihydro-7H-pyrido[1,2,3-de][1,4]benzoxamine-6-carboxylic acid
fluoroquinolone(s) 13, 14–15, 35, 77, 127, 167
adolescents 385
adverse reactions 378–86
allergic 381–4
arthropathy 385
body systems 385–6
central nervous system 379–81
crystalluria 384–5
gastrointestinal tract 379
musculoskeletal disorders 385
nephropathy 384–5
photoallergy 384
photosensitivity 381–4
phototoxicity 381–4
skin 381–4
bacterial resistance 168
bioavailability 347
cell-killing mechanisms 130
children 356–7, 385, 445
immunocompromised 445
neonatal 445
concentration-effect relationship 407–19
pharmacokinetic aspects 416–18
protein binding 417
tissue concentration 417–18
volume of distribution 417–18
concentration-independent killing rate 410–16
animal models 411–14
clinical data 414–15
in vitro models 411
effect on bacterial adherence to eucaryotic cells 236–7 (table)
endotoxin release 248–52
fetal danger 385
gene expression affect 238
interaction with other drugs *see* interaction of fluroquines with other drugs
in vitro activity 169–77
effect onexoenzyme production of *P. aeruginosa* 246 (table)
low concentration effect on:
eukaryotic cell function 234
immune system 234
metabolism 351 (table)
P. aeruginosa resistance 276
pharmacokinetics *see* pharmacokinetics of fluoroqinolones
pregnant women 385
Q-35 457
renal elimination 350
resistance mechanisms 259–86
community-acquired pathogens 281–2
distribution 285–6
reduced intracellular accumulation *see* reduced intracellular accumulation
reduced killing 278–81
target site modification *see* target site modification
structure 456
zwitterionic compounds 340–1
5-fluro-quinolone carboxylic acid 75
2-formamido-acetophenone 39
3-formyl-ciprofloxacin 6, 67
3-formyl-danofloxacin 66
3-formyl-grepafloxacin 66
3-formyl-naphthyridones 75
3-formyl-quinolones 67, 69 (figs.)
3-formyl-tosufloxacin 66
Friedel-Crafts catalysts 16, 43
fused-ring 4 pyridone derivatives 24
future research 469

GAB 312
GABA-revoked currents 311
Gardenerella vaginalis 171, 172 (fig.)
gastrointestinal infections 425–8
genital pathogens 171
genotoxicity detection 318
gentamicin 248, 252
Glu84 123
Glu88 264
glucuronidation 350
gonorrhoea 171, 439–40
Gould-Jacobs reaction 13, 16–21, 34, 45, 76
  catalytic variant 20
  fluoroquinolone synthesis 22–3 (table)
Gram-positive cocci 174 (table), 459–62
granulocyte-macrophage colony-stimulating factor 234
grepafloxacin 37 (table)
GroEL 128
gynecological infections 441–2
gyrase-drug complex 153
GyrA gene 131, 158
  mutations 260–7
GyrB gene 131
  mutations 267–8

HA966 312
Haemophilus spp. 171
Haemophilus ducreyi 171, 172 (table)
Haemophilus influenzae 170 (table), 172 (table), 281
Haemophilus parianfluenzae 172 (table)
Hafnia alvei 172 (table)
haloaromatics 64
3-halomethyl-cephalosporin 67
heat-shock response 128–9
Helicobacter pylori 172 (table), 263 (table)
hematological malignancy 364
hemolysin activity 245
hemolytic uremic syndrome 245
hemorrhagic colitis 245
hexahydro-pyrroloisoxazoles 86 (fig.)
^{3}H-labeled AMPA 312
^{3}H-labeled glutamate 312
^{3}H-labeled kinase 312
^{3}H-labeled *N*-methyl-D-aspartate 312
hospitals, fluroquinolone-resistant strains in 282–3
host-parasite relationship 233–53
  effects on:
    adherence 235–41
    exoenzyme production 245–8
      E. coli 245
      P. aeruginosa 245–8
    slowly growing bacteria 241–5
5-hydrazino-quinolones 74
cis-/trans-4-hydoxyproline 80–1
4-hydoxy-2-pyrones 51–2
7-hydroxy-quinolone 95
4-hydroxy-quinolone carboxylic acid ester 16
5-hydroxy-quinolone carboxylic acids 74
hypoglycemia 330
hypotension 250
hypozanthine phosphoribosyltransferse 318

ICI-56780 2, 3
7-(1-imadazolyol)phenylmethyl-quinolones 21
imidazolo(3,2-a) (1,8)naphthridine carboxylic acid este 32
imipenem 251
immune system 234
IMR-32 neuroblastoma cells 311
immunoglobulins 330
interactions of fluroquinolones with othe drugs 357–78
  alterations in metabolism 365–75
    antipyrine 370
    benzodiazepines 374–5
    caffeine 366
    ciprofloxacin 366–8
    cyclosporine 372–3
    diazepam 374–5
    enoxacin 366
    fleroxacin 368
    H_2-receptor antagonists 371–2
    K^+-ATPase inhibitor 372
    lomefloxacin 368
    Na^+-ATPase inhibitor 372
    norfloxacin 368–9
    omeparazole 372
    oral contraceptive steroids 374
    orfloxacin 369
    pefloxacin 369
    phenytoin 370–1
    rifampicin 373–4
    rufloxacin 269
    sparfloxacin 369–70
    temazepam 374–5
    theophylline 365–6
    toxufloxacin 370
    warfarin 372
  alterations in renal excretion 375–6

β-lactams 375–6
probenecid 375
during absorption process 357–64
activated charcoal 364
Al^{3+}-containing steroids 359–61
bismuth subsalicylate 363
Ca^{2+}-containing antacids 359–61
chemotherapy 363–4
dairy products 359
didanosine 362
Fe^{2+} 362–3
food 359
Mg^{2+}-containing antacids 359–61
sucralfate 362
pharmacodynamic 376–7
metromidazole 377
nonsteroidal anti-inflammatory drugs 378
intra-abdominal infections 440–62
anaerobic 440–1
intracellular vacuoles 127
irloxacin 23 (table)
isatoic anhydride procedures 35–9
1-isopropenyl 64
isothiocyanates 17

J774 macrophage cell line 199

K^+-ATPase inhibitor 372
KB5246 177
kidney
benign tumors 322
hyperplasia 322
Klebsiella oxytoca 172, 237 (table)
Klebsiella pneumoniae 172 (table), 237 (table)
bactericidal activity 242 (table)
KOH 25 (fig.)
kynurenic acid 312

β-lactam 6, 67, 196
alterations in renal excretion 375–6
antibiotics 241
clearance from lungs in pneumonia 245
concentration-independent killing rate 408–9
lactic acidosis 330
Legionella spp. 174, 176 (table)
levofloxacin 5, 18, 37 (table), 319, 340
activity against *S. aureus* 460 (table)
LexA repressor 128
lipopolysaccharide 239
Listeria monocytogenes 172 (table)
liver
cirrhosis 355–6
microsomal enzyme systems 352
lomefloxacin 3–4, 5, 23 (table), 63, 309
interactions with other drugs 368
interpretive breakpoint 169 (table)
Lyme disease 458

Maalox 360
TC 360
MAC 236 (table)
macrolides 201
magnesium 142–4, 157, 239, 312
deficiency 301
malaria 458
malic acid 81
malignant external otitis 436–7
maloric acid 17
marbofloxacin 23 (table)
mar operon 277–8
mechanism B 279
melanin 314
meropenem 276
metabolic effects 330
Meth-Cohn quinolone synthesis 41–2
5-methoxy-naphthyridone carboxylic acid 75
5-methoxy-quinolone carboxylic acid 75
3-methylaminopiperidine 89
5-methyl-2,3, 4,5,6,7-hexahydro-1H-pyrrolo[3,4-c]pyridine 86 (fig.)
3-methylmercaptomethylpyrrolidines 80
3-methylmercaptopypyrrolidines 80
C4-methyl-5-methylene-2-oxo-1,3-dioxolan-4yloxy norflacin 93
5-methyl-naphthyridone caboxylic acid 75
1-methyl-piperazine 19, 27 (fig.)
1-methyl-4-quinolone-3-carboxylic acid 41, 42 (fig.)
2-methyl-quinolone-3-carboxylic acid esters 38
5-methyl-quinolone carboxylic acid 75
metoclopramide 358
metronidazole 377
mexA-mexB-oprK 276
micronuclei formation 319
minimal inhibitory concentrations 168, 235, 240 (table), 242, 407, 408
MK-801 312
molecular structure/mechanism of action 455–8
Moraxella catarhalis 171, 172 (table)
Moraxella morgani 172 (table)
Morganella morgani 237 (table)
morphine 311

mouse lymphoma test 319
mouse micronucleus test 320
moxalactam 248
muscimol 310
mutagenic effects 317–22
Mycobacterium avium 235–8, 263 (table)
Mycobacterium avium-intracellulare 464
Mycobacterium smegmatis 263 (table)
Mycobacterium tuberculosis 174, 176 (table), 263 (table), 464–5
Mycoplasma hominis 176 (table)
Mycoplasma pneumoniae 176 (table)

Na^{+}ATPase inhibitor 372
nadifloxacin 23 (table)
nalA gene 260
nalidixic acid 1, 2, 5, 13, 120 (fig.), 167, 278
  bacterial conjugation inhibition 127
  cells bearing *htp*R mutation affected by 128
  DNA degradation 127
  DNA synthesis inhibition 127
  E. coli
    adhesion increase 238
    resistance 275
    treatment 125–7
  interpretive breakpoints 169 (table)
  mutagenic effects 128
naloxone 311
naphthalenone carboxylic acid 44 (fig.), 45
naphthyridine 1, 75
1,6-naphthyridine 24
1.5-naphthyridone 21
1,6-naphthyridone 21, 78
1,&-naphthyridone 21
1,8-naphythridone 13, 21
naphthyridone carboxlic acid 21
National Committee for Clinical Laboratory Standards 119
Neisseria spp. 171
Neisseria gonorrhoae 171, 172 (table), 263 (table)
  ciprofloxacin susceptible 281
  fluroquinolone resistant 123, 281
  interpretive categories 170 (table)
Neisseria meningitis 172 (table)
nephropathy 304–6
netilmicin 251
neuroquinoron 37 (table)
neutropic patients 444–5
  febrile 444
  prophylaxis in cancer patients 444–5

new attitudes 468–9
4-nitro-fluorobenzene 64
nitromethane 68
nitro-quinolone(s) 74, 75 (fig.)
6-nitro-quinolonecarboxylic acid 100
NM394 177
nocardiosis 458
nongonococcal urethritis 440
nonsteroidal anti-inflammatory drugs 376–7
norfloxacin 13, 15, 127, 137, 167
  bactericidal potency against:
    E. coli 243 (table)
    S. aureus 244 (table)
  DNA binding 6, 138
  DNA synthesis inhibition 12
  E. coli treatment 125
  ester 66
  N-hydroxylated 94–5
  interactions with other drugs 368–9
  interpretive breakpoints 169 (table)
  intravitreal injection (rabbit) 315
  mutagenic effect 128
  pharmacokinetic parameters in;
    elderly subjects 344 (table)
    healthy subjects 344 (table)
  structure 120 (fig.), 181 (fig.)
4-nitrofluorobenzene 20
nosocomial lower respiratory tract infection 414
nosocomial pathogens 282–4
novobiocin 154
NSAIDS quinolones 311, 378
nucleophiles 74
nutritional effects 330
NY-198 309

ocular toxicity 313–15
ofloxacin 5, 6, 13, 16, 17, 26, 36 (table)
  activity against:
    E. faecalis 462 (table)
    M. tuberculosis 64 (table)
    M. avium-intracellulare 464 (table)
    S. aureus 460 (table)
    S. pneumoniae 461 (table)
  bactericidal potency against:
    E. coli 243 (table)
    S. aureus 244 (table)
  central nervous system penetration 309
  endotoxin binding absence 251
  interactions with other drugs 369
  interpretive breakpoints 169 (table)
  pharmacokinetic parameters in:
    elderly subjects 344 (table)
    healthy subjects 344 (table)

retinal toxicity 315
synthesis 17 (fig.), 27 (fig.)
OH($^-$) 24
OPC-17116 461 (table)
OprJ 276
OprK 276
OprM 276
optimal bactericidal concentration 279
oral contraceptive steroids 374
orbifloxacin 37 (table)
osteomyelitis 437
opthoymethylene derivative 26
outer membrane protein 274
oxisoxazolo[5,4-b]pyridones 21
1-(oxoalkyl)-quinolone 27, 29 (fig.)
oxoalkylquinolone carboxylic acid ester 28, 29 (fig.)
oxazolo(3,2-a)naphthyridine ester 32
1-oxides benzothiazine-2-carboxylic acids 42
oximes 80
1-(5-oxohexyl)theobromin (pentoxifylline) 368, 378
oxolinic acid 5, 13, 39, 120 (fig.), 121, 131
DNA cleavage 136
DNA-protein co-valent bond formation 137
DNA synthesis inhibition 127
isomers 38
oxolinic acid-gyrase complexes 153
3-oxopyrrolidine derivatives 80
4-oxopyrrolo(1,2-a)pyrimidine carboxylic acids 45
oxygen 6

P450 enzyme system 365
Palumbo model 143, 149
panbronchiolitis, diffuse 245
Pasteurella multiocida 172 (table)
pathogens
community acquired 281–2
nosocomial 282–4
PD117558 177
PD117596 177
PD131628 5, 177, 459, 460 (table), 461 (table), 462 (table)
pediatric patients 301–2, 356–7
ciprofloxacin for 356–7
fluroquinolones for 356–7, 385, 445
quinolones for 301–2, 466–7
pefloxacin 5, 13, 239, 350
antiadherence properties 239
interactions with other drugs 369
pelvic inflammatory disease 441–2
penems 66–7
penicillins 6
pentazcine 311
pentoxifylline (1-(5-oxohexyl)theobromin) 368, 378
peritonitis in chronic ambulatory dialysis patients 441
pharmacokinetics of fluoroquinolones 340–52
elderly subjects 350–2
ciprofloxacin 342 (table)
enoxacin 342 (table)
fleroxacin 343 (table)
lomefloxycin 352 (table)
nufloxicin 352 (table)
ofloxacin 344 (table)
healthy subjects 341–50
absorption 343–6
bioavailability 347
ciprofloxacin 341 (table)
disposition 349–50
distribution 347–9
enoxacin 342 (table)
fleoxacin 343 (table)
intravenous adminstration 346–7
lomefloxacin 343 (table)
metabolism 350
norfloxacin 344 (table)
ofloxacin 344 (table)
perfloxacin 345 (table)
renal elimination 350
sparfloxacin 345 (table)
tosufloxacin enantimeres 346 (table)
patients with hepatic failure 355–6
patients with renal failure 353–5
ciprofloxacin 353
enoxacin 354
fleroxacin 354
lomefloxacin 355
norfloxacin 354
ofloxacin 354
pefloxacin 354–5
phenylcylidine 312
L-N^6-phenylisopropyladenosine 308
5-phenyl-quinolone carboxylic acid 75
phenytoin 370–1
phospholipase C 245, 246 (table)
photocarcinogenicity 327–9
photomutagenicity 327–9
phototoxicity 322–7
reactive oxygen role 326
reduction in animals 456–7
pipemidic acid 13, 95
piperazine
derivatives 78
bicyclic 79

position-7 2
synthesis 25 (fig.)
6-piperazinyl compound 21
piromidic acid 13
plasmids 129
F CcdB protein 136
R 129
pneumocystosis 458
pneumonia 245, 432–5
sparfloxacin treatment 461
poison hypothesis 153
polycyclic quinolone carboxylic acid 26–7
polymerase blocking 153–6
polymyxin 250, 251
porins 239, 273
OmpC 273
OmpF 273–4
probenecid 375
prodrugs 66
prostatic surgery, transurethral 443
prostatitis 422–5
propteases 245, 246 (table)
alkaline 245, 246 (table)
protein binding 417
protein denaturant (SDS) 130
protein-drug interactions 121
proteinase K 130
Proteus mirabilis 172 (table), 274 (table)
Proteus rettgeri 172 (table)
Proteus vulgaris 172 (table), 274 (table)
Providencia stuartii 172 (table)
Pseudomonas aeruginosa 2, 98, 167, 235, 236 (table), 245–8
bactericidal activities 242 (table)
children with cystic fibrosis 356
ciprfloxacin susceptible 170
fluoroquinolone resistance 276
infections due to 245
outer membrane permeability 274
outer protein pattern 244
quinolone bactericidal potencies 243 (table)
quinolone resistance 263 (table)
psychoses, acute organic 306
pyrazo(5,4-H)-1,6-naphthyridines 24
pyrazol(3,4-f)quinolones 21
pyrazol(4,3-c)quinolone 71 (fig.)
pyridine 24 (fig.), 26 (fig.)
2-(4-pyridinyl)-thieno(2,3)pyridones 21
pyrido(2,3-e)asym triazines 24
pyridobenzoxazine carboxylic acid 27, 28
pyrido(3,2,1-i,j)cinnoline nucleus 101 (fig.)
pyrido(2,3-c)pyridazines 24
pyrido(2,3-d)pyrimidines 24
pyrido(2,3-b)quinoxalines 24
pyridol(1,2,3-de)(1,4)benzothiazines 21
pyrido(3,2,1-gh)(1,7)phenanthrolones 21
2-pyridone 63–4
4-pyridone-3carboxylic acid 21
4-pyridone-3-carboxylic acids synthesis 49–52
pyridophenanthrolones 21
pyridoquinolones 21
7-pyridyl-quinolone carboxylic acids 96 (fig.)
3-pyrrolidinone ketals 50
7-pyrrolidinyl-substituted quinolones 79
2-pyrrolidone-4-carboxylic acid esters 81
pyrrolo(2,3-b)pyridines 24
pyrrolo(3A-b)pyridines 24
pyrrolo(3,2-b)pyridones 21
pyrrolo(2,4-b)pyridones 21
pyrrolo(3,2,1-ij)quinolones 26

QA241 177
quaternary ammonium iodides 64
quinobenzoxazines 144
DNA binding 144–5
quinoline 1
quinoline carboxylic acid ester 16
quinolone(s) 2
absorption 308
activity against *S. aureus* 460 (table)
activity loss 66
acylation 93–4
alkylation 92–3
analogous synthesis 42–5
antibacterial activity *in vitro* 167–77
ATPase reaction 136
7-azolyl-substituted 92
bacterial DNA relaxation 129
bacterial filamentation due to 125–7
bactericidal potencies against *S aureus* 244 (table)
binding to plasma proteins 348 (table)
binding to quinolone-resistant mutants of DNA gyrase 145–6
bioavailability 308
boronate complexes 69
Camps synthesis 39–41
chemistry 13–52
cleavage induced by DNA gyrase 133 (fig.)

clinical use 421–5
decreased permeability, mutations/ phenotypes associated 271 (table)
distribution coefficients 34 (table)
DNA
  binding 138–9
  breakage-reunion interference 130–1
  cleavage induced 133–5
  damage due to 127–8
  gyrase interaction 137
  synthesis inhibition 127
  targeted 138
effects on bacteria 124–30
electron-deficient 100
first-generation 14–15
flux routes across gram-negative cell envelope 270, 272 (fig.)
N-functionalizations 94–5
future modifications 458
gyrase-DNA interaction 137
gyrase inhibitors 321
6-H- 77
half-lives 349 (table)
heat-shock response 128–9
hydrophobicity increase 239
increased efflux, mutatins/phenotypes associated 271 (table)
interference with:
  immune system 233
  phagocytic efficacy 233
intracellular vacuoles due to 127
isoelectric points 341 (table)
killing mechanism 278–81
mechanism A 130
mechanism of action 167–8
metabolites 342
Meth-Cohn synthesis 41–2
minimal inhibitory concentrations 168
mode of action 119–58
mutagenicity 128
N-nucleoside 65
paradoxical effects 150–3
pediatric patients 445
penetration into skin blister fluid 348 (table)
pH-dependent solubility 305
phototoxicity 381
plasmid elimination 129
pleiotropic interactions with host-parasite relationship 252 (fig.)
positions 1–8, 13
  1 position 63–5
  2-position 65
  3-position 66–71
  4-position 71–2
  5-position 72–6
  6-position 77–8
  7-position 78–99, 310, 456
  8-position 99–101, 456
pyrrolidine derivatives 78
resistance mutations 121
R_7 group 456, 457
ring synthesis synthesis 16–21
ryfloxacin type 92
second generation 14 (fig.), 15
self-promoted uptake 239
tautomerism capable 68
4-quinolone 2, 3
  antibacterial activity 4–5
    mechanism A–C 5
  basic structure 120 (fig.)
  "cleavage complex" assay 261
  distribution in humans 6–8
  kinetics 6–8
  minimum inhibitory concentrations 4
  outlook 8–9
  oxygen requirement 6
  peripheral chemistry 63–101
  pyrrolidine derivatives 78
  RNA snthesis inhibition 4
  therapy failure 6
  transmucosal transfer 8
quinolone carboxylic acids 34–5, 66, 68
  1-amino (aza)- 34
  1-aryl 34
  bioactive product 19, 20 (fig.)
  1,2-bridged 35
  1,8-bridged 35
  2,3-bridged 35
  1-cycloalkyl- 34
  1-cyclopropyl- 34
  derivatives synthesis 40
  1-(2,4-diflorophenyl) 34
  esters 19, 28 (fig.), 33
    synthesis 38 (table)
  1-sec alkyl- 34
  synthesis 31
  tetracyclic esters 27
  1-tert alkyl 34
  tricyclic esters 27
4-quinolone-3-carboxylic acid ester 39, 40 (fig.)
quinolone resistance determining area 262, 265, 267, 268

ranitidine 371
rats, ciprofloxacin-treated 247–8
H_2-receptor antagonists 371–2
reduced intracellular accumulation 270–8
  decreased uptake 273–5

increased efflux 275–7
max operon 277–8
renal tolerance 305
respiratory tract infections 429–38
ciproflopxacin-imipenem compared 433 (table)
polymorphism 265
rifampicin 277, 279
rifampin 127, 373–4
RNA 130
T7 polymerase 154
RO 24–81, 383, 468
R023-5068 67 (fig.)
R023-9424 67
R24-9423 468
rosaxacin 13, 96
rufloxacin 23 (table), 27, 37 (table), 340
interactions with other drugs 369

Salmonella spp. 172 (table)
salmonella gastrointeritis outbreaks 427
Salmonella typhi 263 (table), 425–6, 427 (table)
carriers 426
Salmonella typhymurium 263 (table),
R-factor R1 loss 129
S. cervisae, D7 cells 318
seizures 307, 457
sepsis 442–3
septicemia 245
Ser80 123
Ser83 121, 149, 265
Ser84 264
Ser85 264
Serratia liquiefaciens 237 (table)
Serratia marcescens 172 (table), 237 (table)
sexually transmitted diseases 439–40
Shen model 141–2, 149
shiga-like toxins 245
Shigella spp. 172 (table)
Shigella dysenterae 263 (table)
shigellosis 427–8
single-stranded conformational polymorphism 265
sinusitis 435–6
sisomicin 247
skin
rashes
structure, infection 438–9
sleep disorders 306
sodium hydride 17, 64
SOS system 125, 128, 279
DNA repair 128
gene mutations 128
proteins 128
sparfloxacin 5, 36 (table), 72, 340, 460–61
activity against:
anaerobes 463 (table)
E. faecalis 462 (table)
M. tuberculosis 464 (table)
*M*1.2 *avium-intracellulare* 464 (table)
activity against:
S. aureus 460 (table)
S. pneumniae 461
interactions with other drugs 369–70
spermatogenesis impairment 315–16
spirocyclic quinolone carboxylic acid ester 33
staphyolococcal infection 5
staphyolococci 459–60
coagulation negative 172 (table)
methicillin-resistant coagulase-negative 283
staphylococci. methicillin-resistant 459
Staphylococcus aureus 96, 167, 172, 235, 236 (table)
fluoroquinolone activity 259
fluroquinolone target 123
glutamic acid residue 265
methicillin-resistant 170, 171, 259, 459
methicillin-susceptible 170
newer quinolones against 460 (table)
optimum bactericital concentration 459
quinolone resistance 263 (table)
wild-type *gyr*A gene 261
Staphylococcus epidermidis 240, 241, 263 (table)
Staphylococcus saphrophyticus 235, 236 (table)
Stenotrophomonas maltophilia 171, 172 (table)
streptococci 460–1
Streptococcus, other 172 (table)
Streptococcus agalactiae n 172 (table)
Streptococcus bovi 172 (table)
Streptococcus faecalis 5, 237 (table)
Streptococcus pneumoniae 5, 172 (table)
interpretative categories 170 (table)
newer quinolones against 461
Streptococcus pyogenes 172 (table)
streptomycin 238
structure-side effect relationship 456
sucralfate 362

surgical prophylaxis 443–4
abdominal surgery 443–4
endoscopic retrograde cholangiopancreatography 443
transurethral prostatic surgery 443

tandem cyclizations 101
target site modification 260–9
mutations in *gyr*A 260–7
mutations in *gyr*B 267–8
mutations in other topoisomerase genes 268–9
temafloxacin 3, 457, 460
temazepam 374–5
teniposide 135
tertbutyl 63
tetracycline 277
5,6,7,8-tetrafluoro-quinolone carboxylic acids 73–4
tetrahydrofuran 76 (fig.)
tetrahydroquinoline carboxylic acid 21
theophylline 308, 365–6, 371–2
thizolo (3,2-a)naphthyridine carboxylic acid ester 32
thiazo(4,5,6)pyridine carboxylic acid ester 40 (fig.), 41
thiazolo(4,5-g)quinolone 21
7-2-thiazolyl-quinolones 21
7-thiaxolyl-quinolone carboxylic acids 97 (fig.)
thieno(2,3-f)quinolones 21
thieno(3,2-g)quinolones 21
thionyl chloride 24
tilidine 311
tobramycin 247
toluene 25 (fig.), 26 (fig.)
topoisomerase(s) 154
topoisomerase-drug complex 156
topoisomerase II
DNA relaxation 131
targeting drugs 146–8
topoisomerase IV 122, 151, 155, 158, 268, 268
mutation 269
tosufloxacin 36 (table), 41, 80, 340, 457
dipeptide derivative 94 (fig.)
interaction with other drugs 370
N-tosyl-L-prolinyl chloride 17
toxic chemicals, bacterial resistance 259–60
toxicology 297–331
toxoplasmosis 458
tramadol 311
transient electric dichromism 132
transurethral prostatic surgery 443
traveeller's diarrohea 429, 430 (table)
triethylamine 25
triethyl phosphate 64
2,4,5-trifluorobenzoylacetate ester 33
6,7,8-trifluoro-5-niitro-quinolone cargoxylic acid 72
1,2,4-trihalopbutanes 80
1,2,4-trihydroxybutane 80
tuberculosis 174

Ureaplasmaurealyticum 174, 176 (table)
urethritis, nongonococcal 440
urinary tract infections 167, 431–2
acute uncomplicated 422
complicated 422
treatment 423 (table), 424 (table)
uroma cells 239
"utopiafloxacin" 457

vancomycin 276, 277
Vibrio cholerae 172 (table)
1-vinyl 64
5-vinyl-quinolones 75, 76 (fig.)
7-vinyl-quinolone carboxylic acid 97 (fig.)

warfarin 372
WIN-57273 177, 460–1, 461 (table), 462 (table), 463 (table)

X-ray crystallography 156, 157

Y-26611 37 (table)
Yersinia enterocolica 172 (table)

Springer and the environment

At Springer we firmly believe that an international science publisher has a special obligation to the environment, and our corporate policies consistently reflect this conviction.

We also expect our business partners – paper mills, printers, packaging manufacturers, etc. – to commit themselves to using materials and production processes that do not harm the environment. The paper in this book is made from low- or no-chlorine pulp and is acid free, in conformance with international standards for paper permanency.

Printing: Saladruck, Berlin
Binding: Buchbinderei Lüderitz & Bauer, Berlin